Continuum Mechanics and Theory of Materials

Springer-Verlag Berlin Heidelberg GmbH

Physics and Astronomy ONLINE LIBRARY

http://www.springer.de/phys/

Advanced Texts in Physics

This program of advanced texts covers a broad spectrum of topics which are of current and emerging interest in physics. Each book provides a comprehensive and yet accessible introduction to a field at the forefront of modern research. As such, these texts are intended for senior undergraduate and graduate students at the MS and PhD level; however, research scientists seeking an introduction to particular areas of physics will also benefit from the titles in this collection.

Peter Haupt

Continuum Mechanics and Theory of Materials

Translated from German by Joan A. Kurth

Second Edition

With 91 Figures

Springer

Professor Dr. Peter Haupt
Institute of Mechanics
University of Kassel
Mönchebergstrasse 7
34109 Kassel
Germany

Translator
Joan A. Kurth
Silberbornweg 5
34346 Hannoversch Münden
Germany

ISSN 1439-2674

Library of Congress Cataloging-in-Publication Data applied for.

Die Deutsche Bibliothek – CIP-Einheitsaufnahme
Haupt, Peter: Continuum mechanics and theory of materials / Peter Haupt. Transl. from German by Joan A. Kurth. – 2. ed..(Advanced texts in physics) (Physics and astronomy online library)

DOI 10.1007/978-3-662-04775-0

http://www.springer.de

Originally published by Springer-Verlag Berlin Heidelberg New York in 2002
MyCopy version of the original edition 2002

Typesetting: Camera-ready copy by the author.
Cover design: *design & production* GmbH, Heidelberg

Printed on acid-free paper SPIN: 10856932 55/3141/mf 5 4 3 2 1 0
www.springer.com/mycopy

To Vivien, Jonas and Viola

Preface to the Second Edition

While the main structure of the volume is maintained, two extensions have been included in the 2nd edition: In Chap. 5 additional analytical methods concerning the continuous relaxation spectrum are included in the classical theory of linear viscoelasticity. These methods may be applied in the context of a rather new theory of finite linear viscoelasticity of incompressible isotropic materials, presented in Chap. 13 in a thermomechanical context. The consideration of anisotropic viscoplasticity is completely reformulated and extended to a general constitutive theory covering classical crystal plasticity as a special case.

The extensions would not have been possible without the basic work of *Alexander Lion* on linear and nonlinear viscoelasticity. *Thomas Kersten* supplied essential contributions to the investigation of anisotropic viscoplasticity. Valuable comments, given by *Babis Tsakmakis* on just this topic, are gratefully acknowledged.

I thank *Stefan Hartmann* and *Christian Kardelky*, who undertook the task of reading and criticizing the new parts of the manuscript.

Final thanks are addressed to Springer-Verlag and in particular to *Prof. Wolf Beiglböck* for his ongoing support.

Kassel, November 2001 Peter Haupt

Preface to the First Edition

This exposition of the theory of materials has its origins in the lectures I gave at the universities of Darmstadt and Kassel from 1978 onwards. Research projects carried out during the same period have been the source of extensive refinements to the subject-matter. The reason for adding yet another book to the existing wealth of volumes dealing with continuum mechanics was my desire to describe the phenomenological theory of material properties from my own point of view. As a result, it is without doubt a subjectively inspired and incomplete work. This particularly applies to the selection of quotations from the literature.

The text has been influenced and enhanced by the numerous discussions I had the privilege of holding with students and experts alike. I should like to thank them all sincerely for their contributions and encouragement.

My special thanks go to my academic teachers *Rudolf Trostel* [1] and *Hubertus J. Weinitschke*, [2] whose stimulating lectures convinced me at the time that continuum mechanics is a field of science worth pursuing.

I greatly appreciate the long and amicable collaboration with *Babis Tsakmakis* and *Manfred Korzeń*, during which a number of indispensable fundamental aspects emerged.

Valuable inspiration regarding the development of the thermomechanical theory of materials was given by *Roman Bonn*, *Markus Horz*, *Marc Kamlah* and *Alexander Lion*. It was *Lion*'s skill that provided the link between the theoretical modelling and experimental investigation of material behaviour. The fact that he received active support from the Institute of Mechanics'

---

[1] Trostel [1966, 1990, 1993].
[2] Weinitschke [1968].

own laboratory is mainly due to *Lothar Schreiber*, who supplied efficient measurement techniques and the optimisation procedures required for the experimental identification of material parameters. *Stefan Hartmann* and *Georg Lührs* investigated the properties of constitutive models and related field equations in terms of their numerical solutions.

I am grateful to *Kolumban Hutter* for his critical perusal of an earlier version of the manuscript. This led to many improvements and encouraged me to submit this work for publication.

My young colleagues, who checked the final version, proposed a lot of beneficial rectifications. I duly thank *Stefan Hartmann*, *Dirk Helm*, *Thomas Kersten*, *Alexander Lion*, *Anton Matzenmiller*, *Lothar Schreiber* and *Konstantin Sedlan*.

In particular, I would like to thank *Mrs Joan Kurth* for undertaking the laborious task of translating the work into the "language of modern science", which she did with considerable sensitivity and forbearance.

Finally, I wish to express my thanks to Springer-Verlag for agreeing to publish the book and for their kind cooperation.

Kassel, August 1999 Peter Haupt

Contents

Introduction 1

1 Kinematics 7

1.1 Material Bodies 7
1.2 Material and Spatial Representation 19
1.3 Deformation Gradient 23
1.4 Strain Tensors 32
1.5 Convective Coordinates 38
1.6 Velocity Gradient 42
1.7 Strain Rate Tensors 46
1.8 Strain Rates in Convective Coordinates 50
1.9 Geometric Linearisation 53
1.10 Incompatible Configurations 57
1.10.1 Euclidean Space 58
1.10.2 Non-Euclidean Spaces 63
1.10.3 Conditions of Compatibility 64

2 Balance Relations of Mechanics 75

2.1 Preliminary Remarks 75
2.2 Mass 78
2.2.1 Balance of Mass: Global Form 78
2.2.2 Balance of Mass: Local Form 79
2.3 Linear Momentum and Rotational Momentum 84
2.3.1 Balance of Linear Momentum and Rotational Momentum: Global Formulation 84
2.3.2 Stress Tensors 90
2.3.3 Stress Tensors in Convective Coordinates 95
2.3.4 Local Formulation of the Balance of Linear Momentum and Rotational Momentum 95
2.3.5 Initial and Boundary Conditions 101

2. 4 Conclusions from the Balance Equations of Mechanics 104
2. 4. 1 Balance of Mechanical Energy 105
2. 4. 2 The Principle of d'Alembert 109
2. 4. 3 Principle of Virtual Work 114
2. 4. 4 Incremental Form of the Principle of d'Alembert 115

3 Balance Relations of Thermodynamics 119
3. 1 Preliminary Remarks 119
3. 2 Energy 120
3. 3 Temperature and Entropy 125
3. 4 Initial and Boundary Conditions 130
3. 5 Balance Relations for Open Systems 132
3. 5. 1 Transport Theorem 132
3. 5. 2 Balance of Linear Momentum for Systems with Time-Dependent Mass 135
3. 5. 3 Balance Relations: Conservation Laws 137
3. 5. 4 Discontinuity Surfaces and Jump Conditions 140
3. 5. 5 Multi-Component Systems (Mixtures) 144
3. 6 Summary: Basic Relations of Thermomechanics 153

4 Objectivity 155
4. 1 Frames of Reference 155
4. 2 Affine Spaces 156
4. 3 Change of Frame: Passive Interpretation 159
4. 4 Change of Frame: Active Interpretation 162
4. 5 Objective Quantities 164
4. 6 Observer-Invariant Relations 171

5 Classical Theories of Continuum Mechanics 177
5. 1 Introduction 177
5. 2 Elastic Fluid 178
5. 3 Linear-Viscous Fluid 182
5. 4 Linear-Elastic Solid 185
5. 5 Linear-Viscoelastic Solid 188
5. 6 Perfectly Plastic Solid 227
5. 7 Plasticity with Hardening 231
5. 8 Viscoplasticity with Elastic Range 243
5. 9 Remarks on the Classical Theories 249

6 Experimental Observation and Mathematical Modelling 251
6. 1 General Aspects 251
6. 2 Information from Experiments 255
6. 2. 1 Material Properties of Steel XCrNi 18.9 255
6. 2. 2 Material Properties of Carbon-Black-Filled Elastomers 263
6. 3 Four Categories of Material Behaviour 269
6. 4 Four Theories of Material Behaviour 271
6. 5 Contribution of the Classical Theories 273

7 General Theory of Mechanical Material Behaviour 275
- 7.1 General Principles 275
- 7.2 Constitutive Equations 279
  - 7.2.1 Simple Materials 279
  - 7.2.2 Reduced Forms of the General Constitutive Equation 283
  - 7.2.3 Simple Examples of Material Objectivity 288
  - 7.2.4 Frame-Indifference and Observer-Invariance 289
- 7.3 Properties of Material Symmetry 293
  - 7.3.1 The Concept of the Symmetry Group 293
  - 7.3.2 Classification of Simple Materials into Fluids and Solids 298
- 7.4 Kinematic Conditions of Internal Constraint 305
  - 7.4.1 General Theory 305
  - 7.4.2 Special Conditions of Internal Constraint 308
- 7.5 Formulation of Material Models 311
  - 7.5.1 General Aspects 311
  - 7.5.2 Representation by Means of Functionals 312
  - 7.5.3 Representation by Means of Internal Variables 313
  - 7.5.4 Comparison 315

8 Dual Variables 317
- 8.1 Tensor-Valued Evolution Equations 317
  - 8.1.1 Introduction 317
  - 8.1.2 Objective Time Derivatives of Objective Tensors 319
  - 8.1.3 Example: Maxwell Fluid 322
  - 8.1.4 Example: Rigid-Plastic Solid with Hardening 325
- 8.2 The Concept of Dual Variables 329
  - 8.2.1 Motivation 329
  - 8.2.2 Strain and Stress Tensors (Summary) 331
  - 8.2.3 Dual Variables and Derivatives 334

9 Elasticity 345
- 9.1 Elasticity and Hyperelasticity 345
- 9.2 Isotropic Elastic Bodies 352
  - 9.2.1 General Constitutive Equation for Elastic Fluids and Solids 352
  - 9.2.2 Isotropic Hyperelastic Bodies 358
  - 9.2.3 Incompressible Isotropic Elastic Materials 363
  - 9.2.4 Constitutive Equations of Isotropic Elasticity (Examples) 365
- 9.3 Anisotropic Hyperelastic Solids 376
  - 9.3.1 Approximation of the General Constitutive Equation 376
  - 9.3.2 General Representation of the Strain Energy Function 379
  - 9.3.2 Physical Linearisation 388

10 Viscoelasticity 397
10. 1 Representation by Means of Functionals 397
10. 1. 1 Rate-Dependent Functionals with Fading Memory Properties 398
10. 1. 2 Continuity Properties and Approximations 410
10. 2 Representation by Means of Internal Variables 419
10. 2. 1 General Concept 419
10. 2. 2 Internal Variables of the Strain Type 426
10. 2. 3 A General Model of Finite Viscoelasticity 433

11 Plasticity 435
11. 1 Rate-Independent Functionals 435
11. 2 Representation by Means of Internal Variables 444
11. 3 Elastoplasticity 450
11. 3. 1 Preliminary Remarks 450
11. 3. 2 Stress-Free Intermediate Configuration 454
11. 3. 3 Isotropic Elasticity 459
11. 3. 4 Yield Function and Evolution Equations 460
11. 3. 5 Consistency Condition 463

12 Viscoplasticity 475
12. 1 Preliminary Remarks 475
12. 2 Viscoplasticity with Elastic Domain 477
12. 2. 1 A General Constitutive Model 477
12. 2. 2 Application of the Intermediate Configuration 480
12. 3 Plasticity as a Limit Case of Viscoplasticity 484
12. 3. 1 The Differential Equation of the Yield Function 484
12. 3. 2 Relaxation Property 489
12. 3. 3 Slow Deformation Processes 491
12. 3. 4 Elastoplasticity and Arclength Representation 497
12. 4 A Concept for General Viscoplasticity 499
12. 4. 1 Motivation 499
12. 4. 2 Equilibrium Stress and Overstress 500
12. 4. 3 An Example of General Viscoplasticity 501
12. 4. 4 Conclusions Regarding the Modelling of Mechanical Material Behaviour 507

13 Constitutive Models in Thermomechanics 509
13. 1 Thermomechanical Consistency 509
13. 2 Thermoelasticity 514
13. 2. 1 General Theory 514
13. 2. 2 Thermoelastic Fluid 520
13. 2. 3 Linear-Thermoelastic Solids 527
13. 3 Thermoviscoelasticity 530
13. 3. 1 General Concept 530
13. 3. 2 Thermoelasticity as a Limit Case of Thermoviscoelasticity 537
13. 3. 3 Internal Variables of Strain Type 541

13. 3. 4 Incorporation of Anisotropic Elasticity Properties 545
13. 3. 5 Incompressible Materials: An Extension of the Mooney-Rivlin Model to Thermoviscoelasticity 545
13. 4 Thermoviscoplasticity with Elastic Domain 554
13. 4. 1 Uniaxial Viscoplasticity 554
13. 4. 2 General Concept 560
13. 4. 3 Application of the Intermediate Configuration 564
13. 4. 4 Thermoplasticity as a Limit Case of Thermoviscoplasticity 568
13. 5 General Thermoviscoplasticity 577
13. 5. 1 Small Deformations 578
13. 5. 2 Finite Deformations 581
13. 5. 3 Conclusion 585
13. 6 Anisotropic Material Properties 586
13. 6. 1 Motivation 586
13. 6. 2 Axes of Elastic Anisotropy 587
13. 7 Anisotropic Viscoplasticity 591
13. 7. 1 General Considerations 591
13. 7. 2 Free Energy Function 593
13. 7. 3 Evolution Equations 596
13. 7. 4 Lattice Spin 603
13. 7. 5 Summary: A Constitutive Model of Anisotropic Viscoplasticity 606
13. 7. 6 Numerical Simulations 608
13. 7. 7 Closing Remark 618

References 619

Index 635

Notation

General Notation

$a, A, \alpha, \mathrm{a}, \mathrm{A}, \dots$	Indices, scalars, constants
$\boldsymbol{a}, \boldsymbol{A}, \dots$	Vectors
$\mathbf{a}, \mathbf{A}, \dots$	Tensors
$a_k, a^k, b_{kl}, b^{kl}, b_k{}^l, b^k{}_l, C_{kl}, \dots$	Components, matrices
$\boldsymbol{a} + \boldsymbol{b}$	Addition of vectors
$\alpha\boldsymbol{a}$	Product of scalar and vector
$\boldsymbol{0}$	Zero vector
$\boldsymbol{a} \cdot \boldsymbol{b}, a_k b^k, a_k b_k, \dots$	Scalar product of vectors
$\boldsymbol{a} \times \boldsymbol{b}$	Vector product of vectors
$\boldsymbol{a} \otimes \boldsymbol{b}, a_k b^l, a^k b_l, a_k b_l$	Tensor product of vectors
$\mathbf{A} + \mathbf{B}$	Addition of tensors
$\alpha\mathbf{A}$	Product of scalar and tensor
$\mathbf{0}$	Zero tensor
$\mathbf{A}^{\mathrm{T}}$	Transpose of $\mathbf{A}$
$\mathrm{tr}\mathbf{A}, A_{kk}, A_k{}^k, \dots$	Trace of $\mathbf{A}$
$\det\mathbf{A}$	Determinant of $\mathbf{A}$
$\mathbf{A}^{\mathrm{D}}$	Deviator of $\mathbf{A}$
$\mathbf{AB}, A_{kl}B^l{}_m, A_{kl}B_{lm}, \dots$	Product (composition) of tensors

$\mathbf{A}v$, $A_{kl}v^l$, $A_k{}^l v_l$, $A_{kl}v_l$, ...	Product of vector and tensor
$\mathbf{A}\cdot\mathbf{B}$, $A_{kl}B^{kl}$, $A_k{}^l B_k{}^l$, $A_{kl}B_{kl}$, ...	Scalar product of tensors
$\mathbf{1}$, δ^k_l , δ_{kl} , g_{kl} , g^{kl}	Identity tensor
$\mathbb{N}$	Set of natural numbers
$\mathbb{R}$	Set of real numbers
$\mathbb{R}^3$	Set of real triplets
$\mathbb{E}$	Euclidean space
$\mathbb{E}^3$	Three-dimensional Euclidean space
$\mathbb{V}$	Vector space
$\mathbb{V}^3$	Three-dimensional vector space
$\mathbb{T}$	Tangent space
$\mathscr{Lin}$	Set of second-order tensors
$\mathscr{Sym}$	Set of symmetric tensors
$\mathscr{Skw}$	Set of antisymmetric tensors
$\mathscr{Unim}$	Set of unimodular tensors
$\mathscr{Orth}$	Set of orthogonal tensors
$\mathscr{Orth}^+$	Set of orthogonal tensors with positive determinant
◊	End of a definition
♦	End of a natural law, axiom or principle
□	End of a theorem
■	End of a proof
$\mathrm{f}(x) = \mathrm{O}(x^{\mathrm{n}})$	$\lim\limits_{x \to 0} \dfrac{\mathrm{f}(x)}{x^{\mathrm{n}}} = \mathrm{A} \neq 0$ [1]
$\mathrm{f}(x) = \mathrm{o}(x^{\mathrm{n}})$	$\lim\limits_{x \to 0} \dfrac{\mathrm{f}(x)}{x^{\mathrm{n}}} = 0$
$\mathrm{f}: \mathrm{A} \longrightarrow \mathrm{B}$	Map from set A into set B
$\quad \mathrm{x} \longmapsto \mathrm{y} = \mathrm{f}(\mathrm{x})$	Transformation from $\mathrm{x} \in \mathrm{A}$ to $\mathrm{y} \in \mathrm{B}$

[1] This is a special case of the more general notation in BRONSTEIN et al. [1997], pp. 50, 51.

List of Symbols

Symbol	Name	Place of Definition
$(\,)^{\cdot}$	material time derivative	(1.39)
$\langle\cdot\,,\cdot\rangle_h$	inner product of functions	(10.11)
$\langle\cdot\rangle$	*MacCauley* bracket	(5.224)
A	amplitude	(13.218)
A_{D}	dissipated work	(5.151)
$a_{\mathrm{K}}(t)$	moving surface	(3.64)
α	coefficient of thermal expansion	(13.57)
$\boldsymbol{a}$	acceleration vector	(1.42)
$\boldsymbol{a}_{\mathrm{abs}}$	absolute acceleration	(4.78)
$\mathbf{a}$	*Finger* strain tensor	(1.88)
$\mathbf{A}$	*Almansi* strain tensor	(1.86)
$\mathbf{A}_{\mathrm{e}}$	elastic *Almansi* strain	(11.152)
$\mathbf{A}_{\mathrm{p}}$	plastic *Almansi* strain	(11.157)
$\mathbf{A}$	rate of rotation tensor	(4.39)
$\mathbf{A}_{\mathrm{k}}$	*Rivlin-Ericksen* tensors	(10.39)
B	loading function	(5.175)
$\mathcal{B}$	material body	(1.1)
β	material constant	(11.13)
β	second-order elasticity parameter	(9.83)
$\mathbf{B}$	left *Cauchy-Green* tensor	(1.69)
$\hat{\mathbf{B}}_{\mathrm{e}}$	elastic left *Cauchy-Green* tensor	(11.155)
c	specific heat capacity	(13.40)
c, b	material parameters of kinematic hardening	(5.184)
c_α	concentration	(3.97)
c_α	material parameters	(13.382)
c_{d}	specific heat at constant deformation	(13.41)
c_k	elasticity constants	(9.143)
c_p	specific heat at constant pressure	(13.66)
c_{s}	specific heat at constant stress	(13.42)
c_v	specific heat at constant volume	(13.65)
$\boldsymbol{c}(t)$	translation of a frame	(4.28)
χ	configuration	(1.1)

$\chi_R(\boldsymbol{X}, t)$	motion of a material body	(1.22)
$\chi_R{}^k(X^L, t)$	motion of a material body	(1.5)
χ_t	motion, current configuration	(1.3)
$\mathbf{C}$	right *Cauchy-Green* tensor	(1.68)
$\mathbf{C}_p$	plastic right *Cauchy-Green* tensor	(11.64)
$\mathbf{C}_{vk}$	viscous *Cauchy-Green* tensor	(10.156)
$\mathbf{C}_t(\tau)$	relative *Cauchy-Green* tensor	(13.170)
$\mathbf{C}_d^t(s)$	difference strain history	(7.29)
$\mathbb{C}$	elasticity tensor	(9.111)
C_{ijkl}	elasticity tensor	(9.112)
$\mathrm{d}\boldsymbol{A}$, $\mathrm{d}\boldsymbol{a}$	material surface element	(1.59)
$\mathrm{d}V$, $\mathrm{d}v$	material volume element	(1.60)
$\mathrm{d}\boldsymbol{X}$, $\mathrm{d}\boldsymbol{x}$	material line element	(1.51)
$\partial\mathscr{B}$	boundary of a material body	(2.86)
$\partial\chi_t$	boundary of the current configuration	(2.86)
$\partial_s\chi_t$	boundary, where the stress is prescribed	(2.86)
$\partial_u\chi_t$	boundary, where the displacement is prescribed	(2.86)
δA_e	virtual work of the external forces	(2.115)
δA_i	virtual work of the internal forces	(2.116)
δA_T	virtual work of the inertial forces	(2.117)
$\delta\mathbf{E}$	virtual *Green* strain	(2.114)
$\delta e(\mathbf{E})$	virtual change of the strain energy	(9.15)
∇	gradient operator	(1.179)
Δ	*Laplace* operator	(5.29)
$\varDelta$	specific internal dissipation	(13.5)
$\boldsymbol{D}_c$	rotational momentum	(2.23)
$\mathbf{D}$	strain rate tensor	(1.114)
e	specific internal energy	(3.1)
$e(\mathbf{E})$	specific strain energy	(9.6)
e_α	partial internal energy	(3.102)
E	elasticity modulus	(5.38)
$\hat{E}_k$	relaxation coefficient	(5.115)
$\hat{E}(z)$	continuous relaxation spectrum	(5.116)
$\mathrm{E}_\alpha(x)$	*Mittag-Leffler* function	(5.135)
ε	normal strain	(1.74)
$\varepsilon^*(t)$	complex-valued strain process	(5.147)

$\dot{\varepsilon}$	stretching	(1.119)
$\boldsymbol{e}_k$	unit vectors, current configuration	(1.7)
$\boldsymbol{E}_K$	unit vectors, reference configuration	(1.15)
e	*Piola* strain tensor	(1.78)
$\mathbf{E}$	*Green* strain tensor	(1.72)
$\mathbf{E}_\mathrm{L}$	linearised strain tensor	(1.157)
$\mathbf{E}$	linearised strain tensor	(5.37)
$\mathbf{E}^\delta$	static continuation of a strain process	(10.18)
$\mathbf{E}^t$	static continuation of a strain process	(12.48)
$\mathbf{E}_\mathrm{d}^{\mathrm{t}+\delta}$	static continuation of a strain history	(10.19)
$\mathbf{E}_\mathrm{d}^\mathrm{t}(s)$	difference strain history	(7.30)
$\mathbf{E}_\mathrm{d}^\mathrm{t}(s)$	relative strain history	(5.65)
$\mathbf{E}_\alpha(t)$	retarded strain process	(13.136)
$\mathbf{E}_{\mathrm{d}\alpha}^\mathrm{t}(s)$	retarded strain history	(10.22)
$\mathbf{E}_\mathrm{e}$	linearised elastic strain	(5.178)
$\mathbf{E}_\mathrm{i}$	linearised inelastic strain	(5.219)
$\mathbf{E}_\mathrm{i}$	inelastic *Green* strain tensor	(12.1)
$\dot{\mathbf{E}}$	*Green* strain rate tensor	(1.123)
$\mathbf{E}_\mathrm{p}$	linearised plastic strain	(5.178)
$\mathbf{E}_\mathrm{p}$	plastic *Green* strain tensor	(11.65)
$\mathbf{E}_\mathrm{t}(\tau)$	relative *Green* strain tensor	(13.171)
$\mathrm{e}_\mathrm{t}(\tau)$	relative *Piola* strain tensor	(13.172)
$\mathrm{f}(\mathscr{P}, t)$	physical quantity	(1.30)
$\dot{\mathrm{f}}(\mathscr{P}, t)$	material time derivative of a physical quantity f	(1.39)
$\hat{\mathrm{f}}(\boldsymbol{X}, t)$	material representation of a physical quantity f	(1.32)
$\bar{\mathrm{f}}(\boldsymbol{x}, t)$	spatial representation of a physical quantity f	(1.31)
$[\![\mathrm{f}]\!]$	jump of a physical quantity f	(3.85)
$F(\cdot)$	yield function	(5.174)
$f(\cdot)$	yield function	(5.180)
$f_\alpha(t)$	retarded yield function	(12.58)
$f^t(\tau)$	static continuation of the yield function	(12.49)
$\mathrm{f}^{(\alpha)}(t)$	derivative of fractional order α	(5.131)
$\varphi_\mathrm{k}(\cdot)$	material functions	(9.40)
φ_R	strain energy function with respect to R	(13.342)
$\boldsymbol{f}$	volume force density	(2.29)
$\boldsymbol{F}$	resultant force	(2.29)

$\boldsymbol{\Phi}_{t_0}(\boldsymbol{X}, t)$	motion of a material body	(1.27)
$\mathbf{f}(\cdot)$	material function	(7.121)
$\mathfrak{F}$	reduced form of the stress functional	(7.21)
$\mathbf{F}$	deformation gradient	(1.45)
$\hat{\mathbf{F}}$	transformed deformation gradient	(7.53)
$\mathbf{F}_t(\tau)$	relative deformation gradient	(13.167)
$\mathbf{F} = \hat{\mathbf{F}}_e\mathbf{F}_i$	decomposition of the deformation gradient in the theory of viscoplasticity	(12.8)
$\mathbf{F} = \hat{\mathbf{F}}_e\mathbf{F}_p$	decomposition of the deformation gradient in the theory of plasticity	(11.61)
$\mathbf{F} = \hat{\mathbf{F}}_e\mathbf{F}_v$	decomposition of the deformation gradient in the theory of viscoelasticity	(10.117)
$\mathbf{F} = \hat{\mathbf{F}}_{ek}\mathbf{F}_{vk}$	decomposition of the deformation gradient in the theory of viscoelasticity	(10.148)
$\mathbf{F}_p(\boldsymbol{X}, t)$	plastic part of the deformation gradient	(11.61)
$\mathbf{F}_i(\boldsymbol{X}, t)$	inelastic part of the deformation gradient	(12.28)
$\mathbf{F}_{vk}(\boldsymbol{X}, t)$	viscous part of the deformation gradient	(10.148)
$\widetilde{\mathbf{F}}_i$	isoclinic intermediate configuration	(13.383)
$\boldsymbol{\Phi}$	internal variable (orientation of the crystal)	(13.373)
g	specific free enthalpy	(13.8)
G	shear modulus	(13.219)
$g(t)$	memory function	(5.77)
$G(t)$	relaxation function	(5.76)
$G^*(\mathrm{i}\Omega)$	complex modulus	(5.149)
$G_1(\Omega)$	stiffness (storage) modulus	(5.145)
$G_2(\Omega)$	dissipation (loss) modulus	(5.146)
g_{ik}, g^{ik}	metric coefficients	(1.12)
$\hat{g}_{kl}$	incompatible metric	(1.225)
γ	shear strain	(5.42)
γ	specific entropy production	(3.26)
γ_{12}	shear strain	(1.76)
$\dot{\gamma}_{12}$	shearing	(1.120)
Γ	resultant entropy production	(3.26)
$\Gamma(x)$	*Eulerian* Gamma function	(5.124)
Γ^k_{ij}, Γ_{ijk}	*Christoffel* symbols	(1.170)
$\mathscr{G}_\mathrm{R}$	symmetry group with respect to R	(7.61)

$\boldsymbol{g}$	spatial temperature gradient	(3.38)
$\boldsymbol{g}_{\mathrm{R}}$	material temperature gradient	(3.41)
$\boldsymbol{G}^K$, $\boldsymbol{g}^K$	gradient vectors related to convected coordinates	(1.93)
$\boldsymbol{G}_K$, $\boldsymbol{g}_K$	tangent vectors related to convected coordinates	(1.92)
$\boldsymbol{g}_k$, $\boldsymbol{g}^k$	local base vectors	(1.10)
$\boldsymbol{G}_K$, $\boldsymbol{G}^K$	local base vectors	(1.17)
$\mathrm{g}(\cdot)$	material function	(7.119)
$\mathfrak{G}$	general stress functional for simple materials	(7.14)
$\mathfrak{G}_{\mathrm{R}}$	stress functional with respect to R	(7.60)
$\mathrm{G}^{\mathrm{t}}(s)$	relative strain history	(7.60)
G_{α}	retarded strain history	(10.21)
G^{δ}	static continuation of a strain history	(10.17)
$\mathbf{\Gamma}(\cdot)$	change of the reference configuration	(7.51)
$\hat{\mathbf{\Gamma}}$	total strain tensor	(11.66)
$\hat{\mathbf{\Gamma}}_{\mathrm{e}}$	elastic *Green* strain	(10.120)
$\hat{\mathbf{\Gamma}}_{\mathrm{p}}$	plastic *Almansi* strain	(11.68)
$\hat{\mathbf{\Gamma}}_{\mathrm{v}}$	viscous *Almansi* strain	(10.121)
h	specific enthalpy	(13.7)
$h(s)$	influence function	(10.12)
H	resultant entropy supply	(3.24)
$H(\cdot)$	cumulative spectrum	(5.119)
η	shear viscosity	(5.18)
η, η_k	viscosity of a *Maxwell* element	(5.109)
$\boldsymbol{\eta}$	virtual displacement	(2.115)
$\mathbf{H}$	displacement gradient	(1.149)
$\mathbf{H}$	unimodular transformation	(7.60)
$\mathfrak{H}$	reduced form of a stress functional	(7.27)
$\mathbb{H}(\mathbf{E})$	viscosity tensor	(10.71)
$\mathscr{H}_h$	*Hilbert* space	(10.10)
$\boldsymbol{I}$	linear momentum	(2.22)
$\mathbb{I}$	time interval	(1.22)
$\mathrm{I}_{\mathbf{B}}$, $\mathrm{II}_{\mathbf{B}}$, $\mathrm{III}_{\mathbf{B}}$	basic invariants of a second-order tensor $\mathbf{B}$	(9.41)
J_{k}	strain invariant	(9.57)
$\hat{\mathrm{J}}_{\mathrm{k}}$	strain invariant	(13.345)
k	wave number	(13.218)
k	yield stress	(5.174)

k_α	resolved yield stress	(13.401)
$k(\cdot)$	bulk creep function	(5.52)
K	bulk modulus	(5.39)
$K(\cdot)$	bulk relaxation function	(5.51)
K_{eq}	equilibrium bulk modulus	(5.62)
K	kinetic energy	(2.95)
κ	coefficient of isothermal compressibility	(13.58)
κ	internal variable	(13.212)
κ_α	resolved shear strain	(13.382)
$\mathbb{K}(\mathbf{E}(t), s)$	kernel function of finite linear viscoelasticity	(10.59)
$\boldsymbol{k}_\alpha$	partial volume force	(3.100)
ℓ_i	specific stress power per unit mass	(2.98)
L_e	power of the external forces	(2.96)
L_i	stress power	(2.97)
L	loading function	(11.60)
λ	coefficient of heat conduction	(13.37)
λ	proportionality factor (plastic multiplier)	(5.183)
$\lambda(\cdot)$	kinematic constraint	(7.86)
λ_k	eigenvalue (principal stretch)	(9.33)
$\mathscr{L}[f(t)]$	*Laplace* transformation of a function f(t)	(5.68)
$\mathbb{L}(\mathbf{E}(t), s)$	kernel function of finite linear viscoelasticity	(10.49)
$\mathfrak{L}$	linear functional	(10.52)
$\mathbf{L}$	spatial velocity gradient	(1.108)
$\mathbf{L}_n$	higher-order acceleration gradient	(10.43)
$\hat{\mathbf{L}}_i$	inelastic deformation rate	(12.14)
$\hat{\mathbf{L}}_p$	plastic deformation rate	(11.71)
$\hat{\mathbf{L}}_v$	viscous deformation rate	(10.123)
$\boldsymbol{\Lambda}(t)$	strain and temperature process	(13.100)
$\boldsymbol{\Lambda}^t(\tau)$	isothermal static continuation	(13.122)
$\mathbf{\Lambda}$	rate of change of the intermediate configuration	(8.83)
$\mathbf{\Lambda}$	tensor of heat conduction	(13.93)
m	mass	(2.3)
$m_K(t)$	time-dependent mass	(3.67)
$m(\cdot)$	shear creep function	(5.52)
μ	shear modulus	(5.38)
μ	mass flow	(3.69)

$\mu(\cdot)$	shear relaxation function	(5.51)
$\mu_{\mathrm{A,B}}(\cdot)$	relaxation functions	(13.179)
μ_{eq}	equilibrium shear modulus	(5.61)
$\boldsymbol{M}_c$	resultant torque	(2.30)
$\mathbf{M}$	structural tensor	(9.136)
ν	*Poisson* number	(5.38)
$\boldsymbol{N}, \boldsymbol{n}$	normal to a material surface	(1.84)
$\boldsymbol{N}_\alpha$	normal to the slip plane, reference configuration	(13.381)
$\boldsymbol{n}_\alpha$	normal to the slip plane, current configuration	(13.388)
$\mathbf{N}$	*Cauchy* reaction stress	(7.87)
$\tilde{\mathbf{N}}$	2nd *Piola-Kirchhoff* reaction stress	(7.88)
p	hydrostatic pressure	(5.2)
p	internal variable	(11.44)
$\mathscr{P}$	material point, material element	(1.1)
p_{i}	inelastic stress power	(13.245)
$[(\mathrm{P}, \mathrm{Q})]$	equivalence class of ordered pairs of points	(4.6)
$\mathbf{P}$	change of the reference configuration	(7.54)
$\mathbf{P}_{\mathrm{o}}$	rotation of the reference configuration	(13.343)
$\hat{\mathbf{P}}$	*Mandel* stress tensor	(13.255)
$\hat{\mathbf{P}}_{\hat{\mathbf{X}}}$	back stress of *Mandel* type	(13.257)
π	general strain measure, family 2	(8.88)
$\mathbf{\Pi}$	general strain measure, family 1	(8.84)
q, q_k	internal variable	(5.91,111)
$q_1, \dots, q_N$	array of internal variables	(7.121)
$\bar{q}_k$	equilibrium value of the internal variable q_k	(10.84)
q	heat supply per unit area of the reference configuration	(3.5)
q_{R}	heat supply per unit area of the current configuration	(3.7)
Q	resultant heat supply	(3.2)
ψ	specific free energy	(13.6)
ψ_{e}	isothermal strain energy	(13.345)
ψ_{eq}	equilibrium free energy	(13.124)
$\psi_{\mathrm{k}}(\cdot)$	material functions	(9.56)
$\boldsymbol{q}$	*Cauchy* heat flux vector	(3.10)
$\boldsymbol{q}_{\mathrm{R}}$	*Piola-Kirchhoff* heat flux vector	(3.11)

q_α	partial heat flux	(3.103)
$\mathbf{\Omega}$	general stress functional for simple fluids	(7.70)
$\mathbf{\Omega}_k$	functional	(7.125)
$\mathbf{Q}$	orthogonal tensor	(1.70)
$\bar{\mathbf{Q}}$	rotation of the intermediate configuration	(11.87)
$\mathbf{Q}(t)$	rotation of a frame of reference	(4.28)
$\boldsymbol{\Psi}$	intermediate configuration	(1.232)
r	heat supply per unit mass	(3.2)
r_α	partial heat supply	(3.102)
r	order of an influence function	(10.13)
R	reference configuration	(1.4)
$\hat{\mathrm{R}}$	transformed reference configuration	(7.52)
$R(s)$	kernel function	(11.13)
R_{IJKL}	*Riemann-Christoffel* tensor	(1.105)
ρ	mass density in the current configuration	(2.3)
ρ_α	partial density	(3.99)
ρ_{R}	mass density in the reference configuration	(2.4)
$\mathfrak{R}$	rate-independent functional	(11.5)
$\mathbf{R}$	rotational part of the deformation gradient	(1.63)
$\mathbf{R}_{\mathrm{L}}$	linearised rotation tensor	(1.158)
sgn	signum function	(5.225)
$s(t)$	arclength, accumulated strain	(11.1)
s_{p}	plastic arclength	(5.186)
s	specific entropy	(3.23)
S	entropy	(3.23)
s_{eq}	equilibrium entropy	(13.126)
s_α	partial entropy	(3.103)
$\mathscr{S}(t)$	discontinuity surface	(3.87)
σ	entropy supply per unit mass	(3.24)
σ	normal stress	(5.41)
σ_{R}	*Piola-Kirchhoff* stress	(9.94)
Σ	entropy supply per unit area	(3.24)
Σ_{R}	entropy supply per unit area	(3.25)
$\boldsymbol{s}_\alpha$	slip direction	(13.389)
$\boldsymbol{S}_\alpha$	slip direction	(13.381)
$\mathfrak{S}$	most general stress functional	(7.1)

$\mathbf{S}$	weighted *Cauchy* tensor (*Kirchhoff* stress)	(2.48)
$\hat{\mathbf{S}}$	weighted *Cauchy* stress (*Kirchhoff* stress)	(11.70)
$\mathbf{S}_{\mathrm{E}}$	extra stress	(13.175)
$\hat{\mathbf{S}}_{\mathrm{ov}k}$	partial overstress	(10.149)
σ	general stress measure, family 2	(8.90)
Σ	general stress measure, family 1	(8.86)
$\mathbb{T}_{X}, \mathbb{T}_{x}$	tangent spaces	(1.57)
Θ	absolute temperature	(3.22)
$\Theta_{\alpha}(t)$	retarded temperature process	(13.136)
τ	shear stress	(5.42)
τ_{α}	resolved shear stress	(13.396)
$\tilde{\mathbf{t}}$	convected stress tensor	(2.50)
$\tilde{\mathbf{t}}_{\mathrm{E}}$	extra stress	(13.178)
t	*Cauchy* stress vector	(2.29)
$\mathbf{T}$	*Cauchy* stress tensor	(2.43)
$\mathbf{T}_{\mathrm{eq}}$	*Cauchy* equilibrium stress	(5.63)
$\mathbf{T}_{\mathrm{ov}}$	*Cauchy* overstress	(5.67)
$\mathbf{T}_{\alpha}$	partial *Cauchy* stress	(3.100)
$\mathbf{T}_{\mathrm{R}}$	1st *Piola-Kirchhoff* tensor	(2.45)
$\tilde{\mathbf{T}}$	2nd *Piola-Kirchhoff* tensor	(2.49)
$\tilde{\mathbf{T}}_{\mathrm{E}}$	extra stress	(13.177)
$\tilde{\mathbf{T}}_{\mathrm{eq}}$	2nd *Piola-Kirchhoff* equilibrium stress	(10.87)
$\tilde{\mathbf{T}}_{\mathrm{ov}}$	2nd *Piola-Kirchhoff* overstress	(10.88)
$\overset{\circ}{\mathbf{T}}$	*Zaremba-Jaumann* derivative of $\mathbf{T}$	(8.5)
$\overset{\triangledown}{\mathbf{T}}$	contravariant *Oldroyd* rate of $\mathbf{T}$	(1.129)
$\overset{\vartriangle}{\mathbf{T}}$	covariant *Oldroyd* rate of $\mathbf{T}$	(1.128)
$\overset{\diamond}{\mathbf{T}}$	*Green-McInnis* rate of $\mathbf{T}$	(8.9)
$U(\mathscr{B}, t)$	potential energy	(9.12)
$\boldsymbol{u}$	displacement vector	(1.29)
u	efflux velocity	(3.70)
$\boldsymbol{u}_{\alpha}$	diffusion velocity	(3.110)
$\mathbf{U}$	right stretch tensor	(1.66)
v	specific volume	(9.29)
$v_{\mathrm{K}}(t)$	moving volume	(3.64)
$\boldsymbol{v}$	velocity vector	(1.36)
$v^{i}\big\|_{j}$	covariant derivative of v^{i}	(1.186)

$\boldsymbol{v}_\alpha$	partial velocity	(3.100)
$\boldsymbol{v}$	barycentric velocity	(3.109)
$\boldsymbol{v}_{\mathrm{abs}}$	absolute velocity	(4.77)
$\boldsymbol{v}_{\mathrm{k}}$	eigenvector	(9.33)
$\mathbf{V}$	left stretch tensor	(1.67)
$\boldsymbol{w}(\boldsymbol{x}, t)$	non-material velocity field	(3.63)
$\mathbf{W}$	spin tensor	(1.115)
$\boldsymbol{\Omega}$	rate of rotation tensor	(4.74)
x^k	spatial coordinates	(1.5)
X^k	material coordinates	(1.5)
ξ	backstress	(5.203)
$\boldsymbol{x}$	position vector	(1.7)
$\boldsymbol{X}$	position vector	(1.15)
$\boldsymbol{x}^*$	position vector	(4.28)
$\boldsymbol{x}_m$	centre of mass	(2.25)
$\mathbf{X}$	backstress tensor	(5.184)
$\hat{\mathbf{X}}$	backstress tensor	(11.99)
$\hat{\mathbf{X}}_k$	partial backstress tensor	(11.107)
$\widetilde{\mathbf{X}}$	backstress tensor	(11.101)
y_k	cartesian coordinates	(1.6)
Y_K	cartesian coordinates	(1.14)
$\mathbf{y}$	internal variable	(11.47)
$z(t)$	generalised arclength	(11.29)
z_{A}	relaxation time	(13.198)
z_{B}	relaxation time	(13.199)
z_k	relaxation times	(5.115)

Introduction

The object of **mechanics** is to investigate the *motions of material bodies* under the influence of *forces.* Mechanics is a branch of science generally aiming at understanding the processes of motion and presenting them in accordance with the general laws of nature governing their origins. From a mechanical point of view, all motions result from the effects of forces. This general concept also includes the special state of immobility, i.e. statics. Scientific comprehension thrives on the accumulated wealth of experience, the abstractions derived from it, and on mathematical argumentation.

A multitude of mechanical experiences can be encountered directly in our day-to-day lives. Nature presents us with the motion of the Sun, the Moon and the planets; we can observe falling motions, flowing motions, motions of waves, vibrations, deformations and fracture processes. In today's age of technology we employ a variety of tools, machines and vehicles that carry out more or less complicated motions and make such motions possible.

Abstracting the extremely diverse experience of motions calls for a language that allows for a systematic order and aids its presentation. The mechanical terms, methods and results of experiments are communicated by means of a language revolving around *mathematics.*[1] Some distinct advantages of the mathematical argumentation are the logical consistency which lends a widely acknowledged reliability to its statements, the clarity which permits a clear distinction between objective criteria and subjective

---

[1] We employ vector and tensor calculus throughout this book. An introduction to this basic mathematical tool is to be found in De Boer [1982]; Betten [1987]; Klingbeil [1989]; Lippmann [1993]; Bowen & Wang [1976]; Marsden & Hoffman [1993]; Marsden & Tromba [1995]. See also Leigh [1968] and Giesekus [1994].

assumptions and, last but not least, the precision which makes it possible to present conclusions quantitatively in the form of results of calculations.

Continuum mechanics[2] is based on the assumption that matter is continuously distributed in space. This assumption leads to the development of *field theories*. Physical terms and processes are described by means of fields. A *field* is a scalar-, vector- or tensor-valued function of space and time. For example, a material body is always represented by means of a mass density depending on space and time as well as a distribution of velocity. The influence of the outside world on a material body is also described by fields, namely by means of the volume force density and the surface force density.

In its systematic structure continuum mechanics distinguishes between general and individual statements. *General statements* refer to kinematics and balance relations, whereas *individual statements* refer to material properties.[3]

Kinematics describes the geometry of motion and deformation. Configuration is a basic term, out of which the deformation gradient, the velocity gradient and the definition of strains and strain rates emerge.

The concept of *balance equations* is founded on the *free-body principle*. The clear demarcation of a material body splits the whole of the material world into two disjoint sections: the *material body* itself and the *outside world*. In mechanics all interactions between the material body and its surroundings are represented by means of volume and surface forces. The influence of the

---

[2] There are numerous textbooks dealing with continuum mechanics; since they differ considerably in layout and order of priorities, it is impossible to make an objective choice from the wide selection available. For those confronting this field for the very first time, the introductory works by MALVERN [1969]; FUNG [1977]; BOWEN [1989]; BECKER & BÜRGER [1975]; ZIEGLER [1995] might be recommended.
Further interpretations of modern nonlinear continuum mechanics are to be found in ALTENBACH & ALTENBACH [1994]; JAUNZEMIS [1967]; ERINGEN [1962]; ERINGEN [1967]; LEIGH [1968]; WANG & TRUESDELL [1973]; CHADWICK [1976]; GURTIN [1981]; OGDEN [1984]; MARSDEN & HUGHES [1983]; BERTRAM [1989]; GIESEKUS [1994] and HOLZAPFEL [2000]. For beginners and experts alike, the monography by TRUESDELL & NOLL [1965] still retains its unrivalled position as absolutely essential reading matter, which, in conjunction with TRUESDELL & TOUPIN [1960] provides a comprehensive representation of continuum mechanics with an abundance of quotations from literature, upon which the authors comment with refined diligence. The stimulating essay by TRUESDELL [1985], written in his own inimitable style, which pertinently captures the spirit of rational continuum mechanics, also makes excellent reading. Initial-boundary-value problems are studied from the mathematical point of view in ALBER [1998]. Profound information about continuum mechanics as well as the underlying mathematical background is found in ŠILHAVÝ [1997].

[3] TRUESDELL & NOLL [1965], Sect. 1.

external world on the motion of a material body is expressed in general terms of *balance relations* for *mass*, *linear momentum* and *rotational momentum*. The principles of kinematics and balance equations are, without exception, considered to be universal laws of nature for all material systems.[4]

The particular behaviour patterns of any given material body, its so-called *material properties*, are described by statements that have to be formulated individually for each and every material, as opposed to the generally acknowledged balance equations. Three concepts are employed for representing material behaviour: *constitutive equations*, *properties of material symmetry* and *kinematic constraints*.

Constitutive equations are relations between processes of deformation and stress.[5] They do not come under the laws of nature but consist of mathematical models intended to reproduce typical behaviour patterns of real materials in a physically plausible and mathematically conclusive fashion.

The presentation of the stress-strain behaviour in terms of constitutive equations is supported by distinguishing different properties of *material symmetry*.[6] The idea behind this concept is based on the general fact that material behaviour can depend to a greater or lesser degree on the material orientation, regardless of its own special character. This dependence on the direction can be attributed in part to the microstructure of the material.

The concept of *kinematic constraints* is based on the fact that there are materials that are capable of extreme resistance to certain deformations. *Kinematic constraints* are restrictions imposed on the scope of motions assumed a priori to be possible for a material system.[7] An important example of a kinematic constraint is the assumption of incompressibility.

The mathematical modelling of material properties from the point of view of general principles and systematic methods is the object of the **theory of materials.** Generally speaking there are four distinct theories that emerge from the host of possibilities for modelling material properties, namely those of *elasticity*, *viscoelasticity*, *plasticity* and *viscoplasticity*.

Continuum mechanics covers the classical theories of mechanics under general and unifying aspects. The *classical theories* are all based on the same general principles and differ only in their constitutive equations.

[4] TRUESDELL & NOLL [1965], Sect. 16.
[5] TRUESDELL & NOLL [1965], Sect. 26.
[6] TRUESDELL & NOLL [1965], Sect. 31.
[7] TRUESDELL & NOLL [1965], Sect. 30.

Classical *hydrodynamics* is partly based on the defining equation of a perfect fluid; another section of hydrodynamics is founded on the constitutive equation for linear-viscous fluids. The definition of an isotropic elastic solid leads to the *theory of linear elasticity*. The *theory of linear viscoelasticity* is defined by a constitutive equation, according to which the present state of stress is a linear functional of the strain history to date. The classical *theory of plasticity* expounds the idea that a linear-elastic solid passes into a flowing state as soon as a critical state of stress is reached.

Every theory of continuum mechanics evolves from a combination of universal *balance relations* with individually valid *constitutive equations*. The aim of each special theory is the calculation of time-dependent fields. Some examples are the distribution of velocity and pressure in a flowing liquid or the state of displacement and stress in a solid body. The defining equations for these fields are *initial-boundary-value problems*, i.e. systems of partial differential equations to be completed by means of initial conditions and boundary conditions. In many important cases the required fields do not depend on time, as for example in connection with a stationary flow or in the statics of deformable solids. In these cases there are no initial conditions, i.e. the fields are completely defined by *boundary-value problems*.

The classical theories of continuum mechanics differ in terms of their underlying constitutive equations. The general theory of material behaviour shows that these constitutive equations are not set up disjointedly alongside each other but spring from a common root. They can be interpreted as asymptotic approximations of a much more general material description which, under restrictive assumptions, can be gained from a very fundamental constitutive equation.[8] Such assumptions may refer to continuity properties of the stress-strain relations, to symmetry properties or even to kinematic constraints. A further possibility for suitable approximations is the restriction of permitted processes to slow motions or small deformations. The results of the constitutive theory lead to a comprehensive understanding of the classical theories of continuum mechanics, while opening up possibilities for their systematic expansion.

There are numerous cases where a realistic modelling of material behaviour within the limits of mechanics is not possible. For example, experience shows that mechanical material behaviour depends on *temperature*. In order to obtain a physically consistent description of the temperature dependence, it is in no way adequate to introduce the temperature into the constitutive equations as an additional parameter. The basic terms and balance relations of thermodynamics have to be considered in addition to the natural laws of mechanics.

---

[8] TRUESDELL & NOLL [1965], Sect. 41.

It is the object of **thermodynamics**[9] to examine energy conversion processes in material bodies. In thermodynamics balance relations for *energy* and *entropy* are formulated in addition to the balance equations for mass, momentum and moment of momentum. The *energy balance (first law of thermodynamics)* states that the energy content of a material body varies according to the exchange of energy with its surroundings (work and heat). The *entropy balance* defines the *entropy production* as being the difference between the total entropy variation and the transport of entropy into the body. In this context the *dissipation principle* claims that entropy production can never be negative. In its physical sense the dissipation principle (*second law of thermodynamics*) restricts the possibilities of converting thermal energy into mechanical energy as a matter of principle. Certain physical processes, that could feasibly take place according to the balance equations of mechanics, are ruled out by the second law of thermodynamics. This general viewpoint is conclusive by virtue of a natural law and is accordingly of fundamental importance in the theory of materials as well. The second law of thermodynamics, formulated by means of the *Clausius-Duhem* inequality, implies restrictive conditions, which every special theory of material behaviour has to satisfy.[10]

The aim of the theory of materials is not so much the development of a single constitutive equation to embrace all imaginable applications. The general theory's chief goal is to present a well-ordered range of mathematical models, from which anybody using it can select one that appears to be optimal for the purpose under application, bearing the relativity of the means in mind.

This treatise is an attempt to portray the ideas, the general principles and also some concrete procedures of the *theory of materials.* This exposition is adapted from phenomenological continuum mechanics; the fundamental terms of continuum mechanics required here are therefore quoted in the appropriate form as well.

The entire presentation refers exclusively to the *classical continuum* of material points; non-classical continua such as *Cosserat* continua, oriented media, higher-gradient materials or non-local theories are not taken into

[9] In literature dealing with continuum mechanics, it is not uncommon to find the principles of thermodynamics treated as well, although usually very briefly. Introductions to thermodynamics in its own right are given by BAEHR [1973]; MÜLLER [1973]; MÜLLER [1985] and LAVENDA [1978]. OWEN [1984]; DAY [1972] and, in particular, TRUESDELL [1984a, 1984b] allow a certain insight into the school of Rational Thermodynamics. The comprehensive article by HUTTER [1977] is very readable and informative.

[10] The concept of this idea was clearly formulated by COLEMAN & NOLL [1963] and often applied thereafter with great success. It is also intended to form the basis of the thermomechanical theory of materials in this treatise.

consideration. The presentation of classical continuum mechanics in combination with more recent developments in the formulation and application of constitutive models is intended to encourage the reader to develop a comprehensive understanding of the terminological content of the constitutive theory. This, in turn, will lead to a realistic assessment of its possible achievements and present limitations. Anyone practising continuum mechanics, given a set of experimental data, should be suitably equipped for devising a constitutive model to represent the phenomena ascertained in a consistent manner and thus render such a model serviceable for practical applications in connection with a reliable prediction of the thermomechanical behaviour of material structures and systems.

1 Kinematics

1.1 Material Bodies

All statements in continuum mechanics relate to material bodies which continuously fill parts of space with matter. Matter is the actual scene of physical occurrences: a material body can be identified with different parts of space at different moments in time and is simultaneously the carrier of the physical processes. The concept that matter is continuously distributed in space is presented by the term configuration:

Definition 1.1
A *material body* $\mathscr{B} = \{ \mathscr{P} \}$ is a set of elements $\mathscr{P}$ with the following properties:

1) There is a set $\mathcal{K} = \{\chi\}$ of one-to-one mappings

$$\begin{aligned} \chi : \mathscr{B} &\longrightarrow \chi[\mathscr{B}] \subset \mathbb{R}^3 \\ \mathscr{P} &\longmapsto \chi(\mathscr{P}) = (x^1, x^2, x^3) \Leftrightarrow \mathscr{P} = \chi^{-1}(x^1, x^2, x^3) . \end{aligned} \tag{1.1}$$

Each of these mappings $\chi \in \mathcal{K}$ is called a *configuration.*

2) If $\chi_1 \in \mathcal{K}$ and $\chi_2 \in \mathcal{K}$ are two configurations, i.e.

$$\begin{aligned} \chi_1 : \mathscr{B} &\longrightarrow \chi_1[\mathscr{B}] \subset \mathbb{R}^3 \\ \mathscr{P} &\longmapsto \chi_1(\mathscr{P}) = (x^1, x^2, x^3) \end{aligned}$$

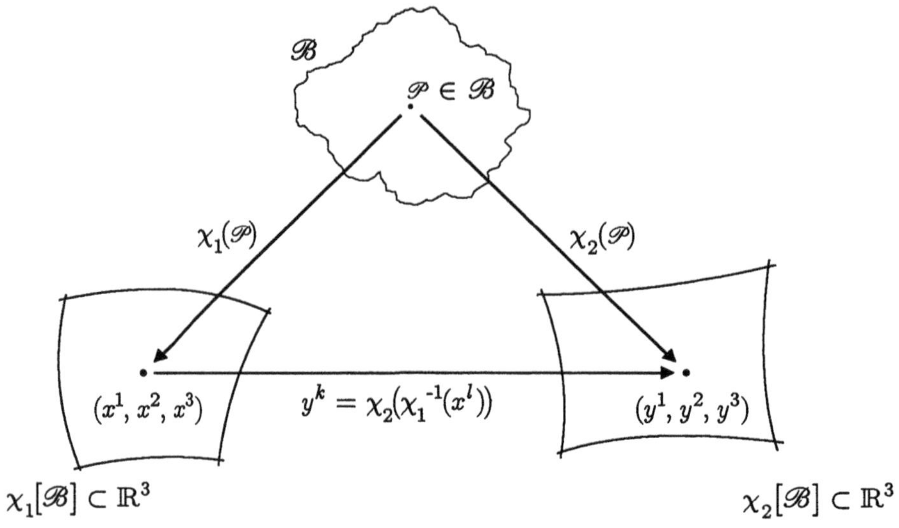

Figure 1. 1: Configurations

and

$$\begin{aligned} \chi_2 : \mathscr{B} &\longrightarrow \chi_2[\mathscr{B}] \subset \mathrm{IR}^3 \\ \mathscr{P} &\longmapsto \chi_2(\mathscr{P}) = (y^1, y^2, y^3) \ , \end{aligned}$$

then their composition

$$\begin{aligned} \chi_2 \circ \chi_1^{-1} : \quad \chi_1[\mathscr{B}] &\longrightarrow \chi_2[\mathscr{B}] \\ (x^1, x^2, x^3) &\longmapsto (y^1, y^2, y^3) = \chi_2(\chi_1^{-1}(x^1, x^2, x^3)) \end{aligned} \tag{1.2}$$

is continuously differentiable.[1] The elements $\mathscr{P} \in \mathscr{B}$ are called *particles*, *material points* or *material elements* (see Fig. 1. 1). ◇

By means of a configuration a triple of real numbers is assigned to each body particle on a one-to-one basis. Due to the definition's demand for the existence of a set $\mathcal{K}$ of configurations with the property that each composition should be continuously differentiable, the material body is

---

[1] The order of differentiability is assumed to be sufficiently high, depending on the particular context.

defined as a *differentiable manifold*.[2] The following definition ascertains that the configuration of a body during its motion depends on time.

Definition 1. 2

The *motion* of a material body is a smooth family of configurations with the time t as family parameter:

$$\begin{aligned} t \longmapsto \chi_t : \mathcal{B} &\longrightarrow \chi_t[\mathcal{B}] \subset \mathbb{R}^3 \\ \mathcal{P} &\longmapsto \chi_t(\mathcal{P}) = \left(x^1(t), x^2(t), x^3(t)\right) \Leftrightarrow \mathcal{P} = \chi_t^{-1}(x^1, x^2, x^3)\,. \end{aligned} \tag{1.3}$$

The time dependent configuration χ_t is called the *current configuration.* ◇

In order to depict the motion of a material body clearly, the abstractly defined material points have to be eliminated. This is done by emphasising one configuration from the total number of configurations imaginable.

Definition 1. 3

A *reference configuration* is a fixed configuration $\mathrm{R} \in \mathcal{K}$, selected at will,

$$\begin{aligned} \mathrm{R}: \mathcal{B} &\longrightarrow \mathrm{R}[\mathcal{B}] \subset \mathbb{R}^3 \\ \mathcal{P} &\longmapsto \mathrm{R}(\mathcal{P}) = (X^1, X^2, X^3) \Leftrightarrow \mathcal{P} = \mathrm{R}^{-1}(X^1, X^2, X^3)\,, \end{aligned} \tag{1.4}$$

which uniquely designates the material points $\mathcal{P} \in \mathcal{B}$. ◇

By selecting a reference configuration we are in a position to denote each individual material element by its *name*. The name of the particle $\mathcal{P}$ is the number triplet (X^1, X^2, X^3). At this stage the quantitative representation of the motion is a direct consequence of the above definitions: the family of configurations

---

[2] The mathematical term *differentiable manifold* is most appropriate for the precise formulation of the physical idea of a material continuum. The definition has been slightly abridged in this case to restrict the use of terminology and concentrate on the essential physical aspects. Attention is drawn to the mathematical literature for the representation of the theory of differentiable manifolds, for instance, the textbooks by BRÖCKER & JÄNICH [1973]; BISHOP & GOLDBERG [1968]. The use of concepts taken from differential geometry has more recently been employed with favourable results to achieve an exact definition of the basic terms of continuum mechanics. Refer, among others, to NOLL [1972]; MARSDEN & HUGHES [1983]; BERTRAM [1989].

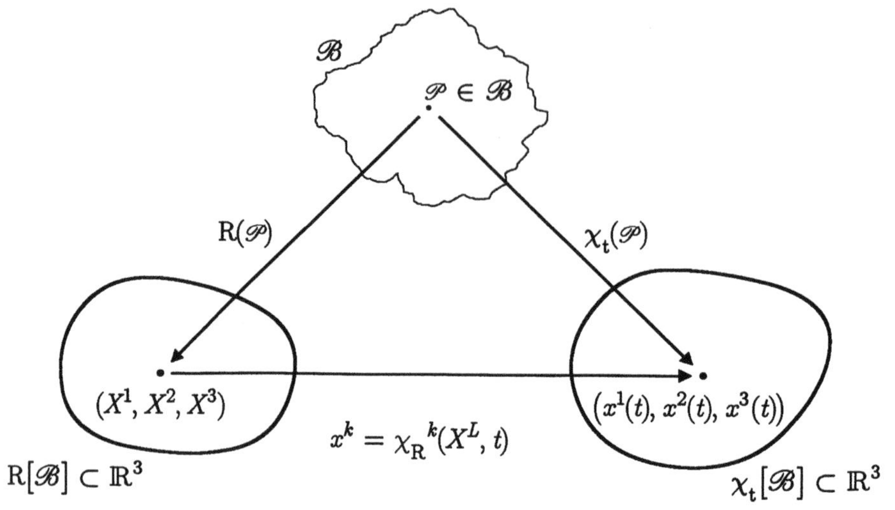

Figure 1. 2: Reference configuration and current configuration, material and spatial coordinates

$$\chi_t(\mathscr{P}) = \chi_t(R^{-1}(X^1, X^2, X^3))$$

results in three real-valued functions, each of which depends on four variables,

$$x^1 = \chi_R{}^1(X^1, X^2, X^3, t)\ , x^2 = \chi_R{}^2(X^1, X^2, X^3, t)\ , x^3 = \chi_R{}^3(X^1, X^2, X^3, t)\ . \quad (1.5)$$

For shortness we write $x^k = \chi_R{}^k(X^L, t)$ or, even shorter, $x^k = x^k(X^L, t)$, k and L taking the values 1, 2 and 3 respectively.

Definition 1. 4
In the representation

$$\begin{aligned} \chi_R{}^k: \quad & R[\mathscr{B}] \longrightarrow \chi_t[\mathscr{B}] \\ & (X^1, X^2, X^3) \longmapsto x^k = \chi_R{}^k(X^L, t) \end{aligned}$$

the number triplets $X^L = (X^1, X^2, X^3)$ and $x^k = (x^1, x^2, x^3)$ are called *material* and *spatial coordinates* of the particle $\mathscr{P}$ respectively (Fig. 1. 2). ◊

According to the definitions given so far, the *motion* is portrayed by means of a time-dependent *coordinate transformation* (1.5), i.e. a transformation of material coordinates into spatial coordinates. The definitions lead to the conclusion that the functions ${\chi_R}^k(X^L, t)$ are continuously differentiable with respect to the material coordinates. In addition, double differentiability is usually assumed with respect to t.

For further substantiation, it is necessary to give the coordinates a physical meaning. This is achieved by identifying them with points in a space with a suitable geometrical structure. In classical mechanics the location of a material body is always a domain of the three-dimensional Euclidean space $\mathbb{E}^3$ of physical observation. For this reason, the Euclidean structure is, as a rule, imprinted on the material body.

The identification of the current configuration with the Euclidean space $\mathbb{E}^3$ of physical observation allows for the introduction of Cartesian coordinates (y_1, y_2, y_3).

$$y_k = \hat{y}_k(x^1, x^2, x^3) \Leftrightarrow x^k = \hat{x}^k(y_1, y_2, y_3) \tag{1.6}$$

are components of a position vector $\boldsymbol{x}$:

$$\boldsymbol{x} = y_k \boldsymbol{e}_k \,. \tag{1.7}$$

Here, $\{\boldsymbol{e}_k\}$ is used to represent a base system consisting of 3 orthogonal unit vectors with the property $\boldsymbol{e}_k \cdot \boldsymbol{e}_l = \delta_{kl}$. ($\delta_{kl} = 1$ for $l = k$ and 0 for $l \neq k$.) The Cartesian coordinates are projections of the position vector $\boldsymbol{x}$ on the coordinate axes: $y_k = \boldsymbol{x} \cdot \boldsymbol{e}_k$. That proves the identification

$$\boldsymbol{x} = \hat{y}_k(x^1, x^2, x^3)\boldsymbol{e}_k \,, \tag{1.8}$$

i.e.

$$\boldsymbol{x} = \hat{\boldsymbol{x}}(x^1, x^2, x^3) \Longleftrightarrow x^k = \hat{x}^k(\boldsymbol{x}) \,. \tag{1.9}$$

Summarising, we have to record the following: configurations identify material points with number triplets. This serves the *designation* (reference configuration) on the one hand and the *localisation* (current configuration) on the other.

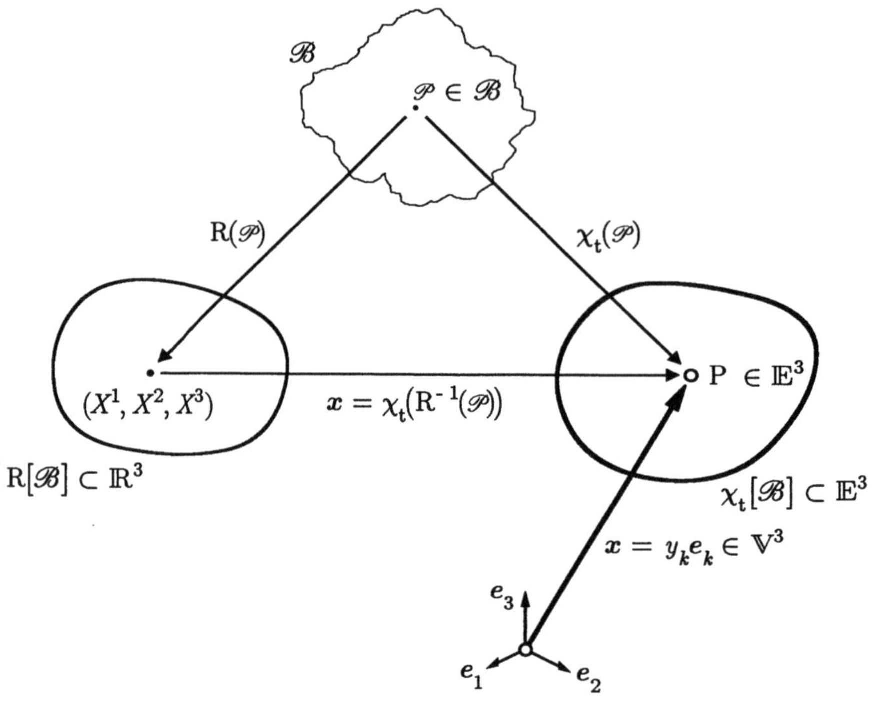

Figure 1. 3: Motion of a material body in a Euclidean space, represented in relation to a frame of reference.

The elements $\mathscr{P} \in \mathscr{B}$ of a material body should not be confused with the points of the Euclidean space where they are currently situated: during the course of its motion, every particle $\mathscr{P}$ describes a curve. This consists of time-dependent coordinates $x^k(t) \in \mathbb{R}^3$ or position vectors $\boldsymbol{x}(t) \in \mathbb{V}^3$, which can in turn be identified with points of space, $\mathrm{P} \in \mathbb{E}^3$:

$$
\begin{aligned}
(\mathscr{P}, t) \longmapsto x^k &= \chi_t{}^k(\mathscr{P}) \quad \in \mathbb{R}^3, \\
&\Updownarrow \\
\boldsymbol{x} &= \chi_t(\mathscr{P}) \quad \in \mathbb{V}^3 \;, \\
&\Updownarrow \\
\mathrm{P} &= \hat{\mathrm{P}}(\mathscr{P}, t) \in \mathbb{E}^3
\end{aligned}
$$

(Fig. 1. 3). $\mathbb{V}^3$ denotes the three-dimensional Euclidean *vector space* and $\mathbb{E}^3$ the Euclidean *point space*. A *physical observer* takes note of the current image of the body $\mathscr{B}$ in the space $\mathbb{E}^3$, the current location P of a material point $\mathscr{P}$

being illustrated by means of the position vector $\boldsymbol{x}(t) = y_k(t)\boldsymbol{e}_k$. For the sake of identifying the points $\mathrm{P} \in \mathbb{E}^3$ with vectors $\boldsymbol{x} \in \mathbb{V}^3$ or Cartesian coordinates $y_k \in \mathbb{R}^3$, a *frame of reference* has to be introduced (Fig. 1. 3).[3]

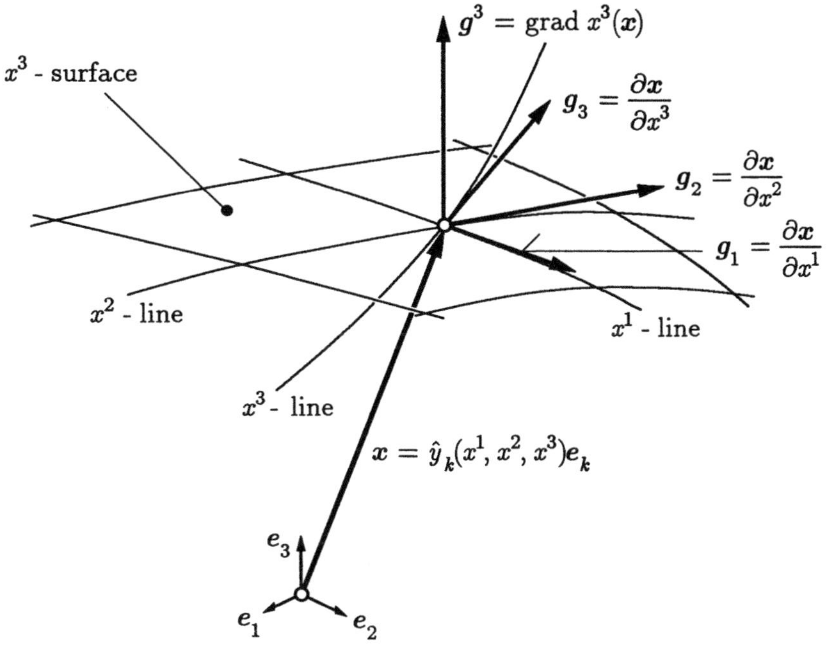

Figure 1. 4: Curvilinear coordinates

In connection with the identifications $\mathscr{P} \leftrightarrow \boldsymbol{x} \leftrightarrow \mathrm{P}$, the spatial coordinates (x^1, x^2, x^3) can be interpreted as curvilinear coordinates in the Euclidean space $\mathbb{E}^3$. The coordinates prompt us to define two reciprocal base systems, consisting of the *tangent vectors* $\{\boldsymbol{g}_k\}$,

$$\boldsymbol{g}_k = \frac{\partial \boldsymbol{x}}{\partial x^k} = \frac{\partial}{\partial x^k}\, \hat{y}_j(x^1, x^2, x^3)\boldsymbol{e}_j\,, \tag{1.10}$$

and the *gradient vectors* $\{\boldsymbol{g}^i\}$,

$$\boldsymbol{g}^i = \text{grad}\, x^i(\boldsymbol{x}) = \frac{\partial}{\partial y_l}\, \hat{x}^i(y_1, y_2, y_3)\boldsymbol{e}_l\,, \tag{1.11}$$

(Fig. 1. 4) with the property $\boldsymbol{g}^i \cdot \boldsymbol{g}_k = \delta^i_{\cdot k}$. ($\delta^i_k = 1$ for $i = k$ and 0 for $i \neq k$.)

[3] Questions arising in the context of a reference frame are to be investigated in Chap. 4.

The space-dependent base vectors give rise to the definition of covariant and contravariant metric coefficients

$$g_{ik} = \boldsymbol{g}_i \cdot \boldsymbol{g}_k = \frac{\partial \hat{y}_j}{\partial x^i}\frac{\partial \hat{y}_l}{\partial x^k}\,\delta_{jl} = \frac{\partial \hat{y}_j}{\partial x^i}\frac{\partial \hat{y}_j}{\partial x^k} \tag{1.12}$$

and

$$g^{ik} = \boldsymbol{g}^i \cdot \boldsymbol{g}^k = \frac{\partial \hat{x}^i}{\partial y_j}\frac{\partial \hat{x}^k}{\partial y_l}\,\delta_{jl} = \frac{\partial \hat{x}^i}{\partial y_j}\frac{\partial \hat{x}^k}{\partial y_j}\,, \tag{1.13}$$

which are reciprocal to each other in the sense of $g^{il}\,g_{lk} = \delta^i_k$.

It is common practice to identify the reference configuration with a Euclidean space, although this is by no means compulsory (Fig. 1. 5). During this identification the Euclidean space $\mathbb{E}^3_{\mathrm{R}}$ does not generally have to be the space $\mathbb{E}^3$ of physical observation, since the reference configuration does not indicate a localisation but rather a designation for the material particles. For the geometrical interpretation of the material coordinates X^K as curvilinear coordinates in the space $\mathbb{E}^3_{\mathrm{R}}$ we proceed in the same way as for the spatial coordinates. Based on the one-to-one mapping

$$(X^1, X^2, X^3) \leftrightarrow \mathrm{P}_{\mathrm{R}} \in \mathbb{E}^3_{\mathrm{R}}\,,$$

Cartesian coordinates

$$Y_K = \hat{Y}_K(X^1, X^2, X^3) \Leftrightarrow X^K = \hat{X}^K(Y_1, Y_2, Y_3) \tag{1.14}$$

are likewise introduced, which are components of a position vector $\boldsymbol{X}$:

$$\boldsymbol{X} = Y_K \boldsymbol{E}_K = \hat{Y}_K(X^1, X^2, X^3)\boldsymbol{E}_K\,, \tag{1.15}$$

$$\boldsymbol{X} = \hat{\boldsymbol{X}}(X^1, X^2, X^3) \Leftrightarrow X^K = \hat{X}^K(\boldsymbol{X})\,. \tag{1.16}$$

In doing so, the $\{\,\boldsymbol{E}_K\,\}$ in turn form a base system: $\boldsymbol{E}_K \cdot \boldsymbol{E}^L = \delta^L_K$. This produces the following relations:

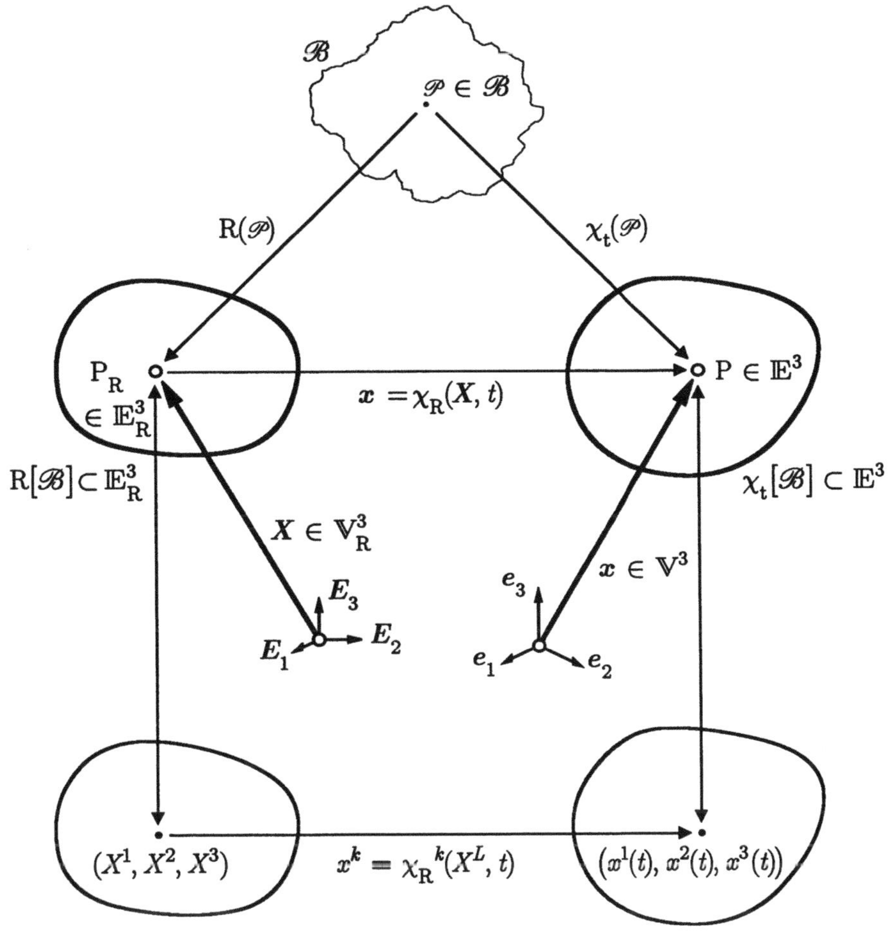

Figure 1. 5: Motion of a material body: transformation of points or transformation of coordinates

$$
\begin{aligned}
(\mathscr{P}) \longmapsto X^K &= \mathrm{R}^K(\mathscr{P}) \in \mathbb{R}^3, \\
&\Updownarrow \\
X &= \mathrm{R}(\mathscr{P}) \quad \in \mathbb{V}^3_\mathrm{R}, \\
&\Updownarrow \\
\mathrm{P}_\mathrm{R} &= \hat{\mathrm{P}}_\mathrm{R}(\mathscr{P}) \in \mathbb{E}^3_\mathrm{R}.
\end{aligned}
$$

For the curvilinear *material coordinates* corresponding definitions of tangent vectors,

$$G_K = \frac{\partial}{\partial X^K} \hat{\boldsymbol{X}}(X^1, X^2, X^3) \tag{1.17}$$

and gradient vectors,

$$\boldsymbol{G}^I = \text{Grad}\, \hat{X}^I(\boldsymbol{X}) \tag{1.18}$$

are agreed, resulting in reciprocity in the sense of $\boldsymbol{G}^I \cdot \boldsymbol{G}_K = \delta^I_K$ as well as the definition of metric coefficients $G_{IK} = \boldsymbol{G}_I \cdot \boldsymbol{G}_K$ and $G^{IK} = \boldsymbol{G}^I \cdot \boldsymbol{G}^K$ with the property $G^{IL} G_{LK} = \delta^I_K$. Using the identifications

$$(x^1, x^2, x^3) \leftrightarrow \boldsymbol{x} = \chi_t(\mathscr{P}) \leftrightarrow \mathrm{P} \in \mathbb{E}^3 \Longleftrightarrow \mathscr{P} = \chi_t^{-1}(\boldsymbol{x}) \tag{1.19}$$

or

$$(X^1, X^2, X^3) \leftrightarrow \boldsymbol{X} = \mathrm{R}(\mathscr{P}) \leftrightarrow \mathrm{P}_\mathrm{R} \in \mathbb{E}^3_\mathrm{R} \Longleftrightarrow \mathscr{P} = \mathrm{R}^{-1}(\boldsymbol{X})\,, \tag{1.20}$$

the motion can be written as follows: the relation

$$\boldsymbol{x} = \chi_t(\mathscr{P}) = \chi_t(\mathrm{R}^{-1}(\boldsymbol{X})) \tag{1.21}$$

yields the representation of the motion by means of the vector function

$$\begin{aligned} \chi_\mathrm{R} : \mathrm{R}[\mathscr{B}] \times \mathbb{I} &\longrightarrow \chi_t[\mathscr{B}] \\ (\boldsymbol{X}, t) &\longmapsto \boldsymbol{x} = \chi_\mathrm{R}(\boldsymbol{X}, t)\ , \end{aligned} \tag{1.22}$$

introducing the definition $\chi_\mathrm{R}(\boldsymbol{X}, t) = \chi_t(\mathrm{R}^{-1}(\boldsymbol{X}))$.

$\mathrm{R}[\mathscr{B}]$ now describes the image of $\mathscr{B}$ in $\mathbb{V}^3_\mathrm{R}$, $\chi_t[\mathscr{B}]$ the image in $\mathbb{V}^3$ and $\mathbb{I} = [t_0, t_1] \subset \mathbb{R}$ the interval of time during which the motion takes place. The mapping χ_R supplies the present position vector $\boldsymbol{x}$ of the material point $\mathscr{P}$, to which the vector $\boldsymbol{X}$ of the reference configuration is assigned. The special form of the function χ_R depends on the choice of reference configuration; this is suggested by means of the index R (Fig. 1. 5).

The representation $\boldsymbol{x} = \chi_\mathrm{R}(\boldsymbol{X}, t)$ of the motion as a vector function can be interpreted as a *point transformation*: different space points $\mathrm{P}_\mathrm{R} \in \mathbb{E}^3_\mathrm{R}$ and $\mathrm{P} \in \mathbb{E}^3$ correspond to the position vectors $\boldsymbol{X} \in \mathbb{V}^3_\mathrm{R}$ and $\boldsymbol{x} \in \mathbb{V}^3$. They are transformed by the representation (1.22) of the motion. The point transformation is equivalent to the *coordinate transformation* (1.5), $x^k = \chi_R{}^k(X^L, t)$ (see Fig. 1. 5).

For the purpose of their geometric interpretation the special nature of the material and the spatial coordinates has to be specified. Virtually any two curvilinear coordinate systems can be chosen to play the role of spatial or material coordinates independent of one another. The simplest way, of course, is to choose Cartesian coordinates for both configurations. In this case it is common practice to employ the symbols

$$(y_1, y_2, y_3) = (x, y, z) , \quad (Y_1, Y_2, Y_3) = (X, Y, Z) , \tag{1.23}$$

and the motion to represent as follows:

$$x = f(X, Y, Z, t) , \quad y = g(X, Y, Z, t) , \quad z = h(X, Y, Z, t) . \tag{1.24}$$

The associated base vectors are constant and identical in this particular case:

$$\{ \boldsymbol{G}_K \} \equiv \{ \boldsymbol{g}_k \} = \{ \boldsymbol{E}_K \} = \{ \boldsymbol{e}_k \} = \{ \boldsymbol{e}_x, \boldsymbol{e}_y, \boldsymbol{e}_z \} . \tag{1.25}$$

The principal independence of the spaces $\mathbb{E}^3_R$ and $\mathbb{E}^3$ or $\mathbb{V}^3_R$ and $\mathbb{V}^3$, previously ascertained, applies even for this most simple representation. The physical meaning of this independence lies in the fact that the current configuration serves to describe the position of the material point, whereas the reference configuration applies to the choice of its name: the reference configuration is not required to be occupied by the body at any time during the course of its motion.

This does not, however, rule out the possibility of choosing a special current configuration as reference configuration: since the choice of reference configuration is arbitrary, we may choose the configuration at a fixed time t_0 to be the reference configuration:

$$\boldsymbol{X} = \chi_{t_0}(\mathscr{P}) \Leftrightarrow \mathscr{P} = \chi_{t_0}{}^{-1}(\boldsymbol{X}) . \tag{1.26}$$

Under these circumstances we have $\boldsymbol{x} = \chi_t(\mathscr{P}) = \chi_t(\chi_{t_0}{}^{-1}(\boldsymbol{X}))$, i.e.

$$\boldsymbol{x} = \boldsymbol{\Phi}_{t_0}(\boldsymbol{X}, t) . \tag{1.27}$$

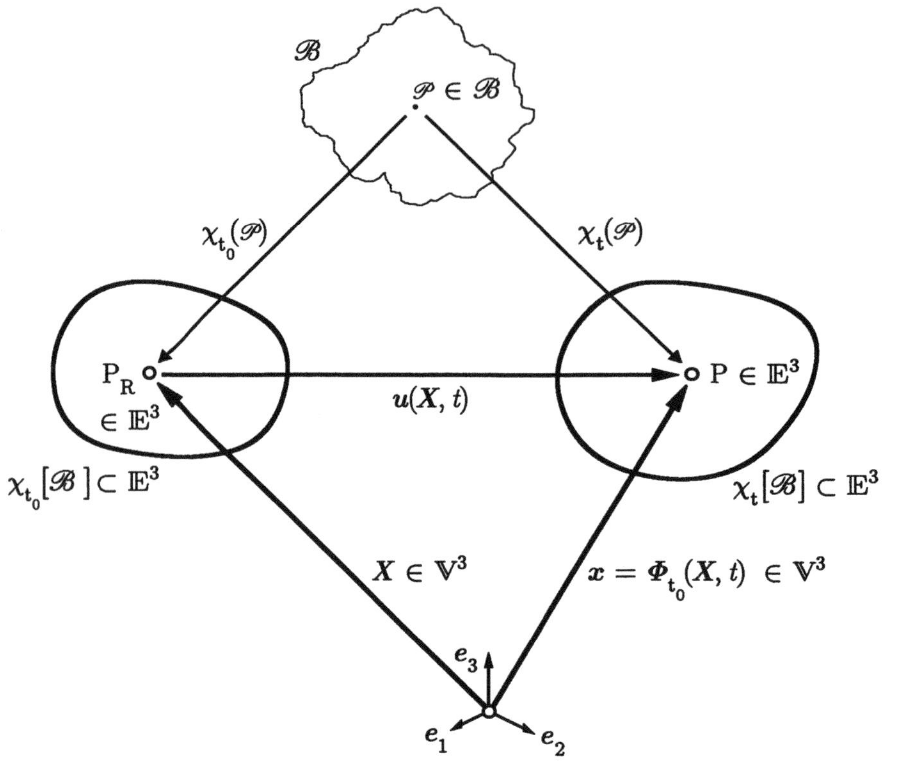

Figure 1. 6: Displacement vector

The function $\Phi_{t_0} : \chi_{t_0}[\mathscr{B}] \times \mathbb{I} \longrightarrow \chi_t[\mathscr{B}]$

$$(X, t) \longmapsto x = \Phi_{t_0}(X, t) = \chi_t(\chi_{t_0}^{-1}(X))$$

provides the present location x of the point $\mathscr{P}$, which was located at X at the time t_0 ($t_0 \leq t$). Accordingly, the identity

$$\Phi_{t_0}(X, t_0) = X \tag{1.28}$$

is valid for all $X \in \chi_{t_0}[\mathscr{B}]$.

Many representations of solid mechanics introduce the displacement vector $u = x - X$ to describe the motion. The prerequisite is the special choice of a reference configuration, occupied by the material body at some prior instant t_0: in this case the difference between the present and the past position vectors can be calculated (Fig. 1. 6):[4]

$$u(X, t) = \Phi_{t_0}(X, t) - X. \tag{1.29}$$

1. 2 Material and Spatial Representation

In continuum mechanics, matter is the intrinsic place of action with regard to physical processes: the most diverse processes take place at the individual material points $\mathscr{P} \in \mathscr{B}$. For this reason the material body $\mathscr{B}$ is the domain of definition for those physical quantities that represent the physical events. In the following definition the mapping

$$\begin{aligned} \mathrm{f}: \mathscr{B} \times \mathbb{I} &\longrightarrow \mathbb{W} \\ (\mathscr{P}, t) &\longmapsto \mathrm{w} = \mathrm{f}(\mathscr{P}, t) \end{aligned} \tag{1.30}$$

is intended to represent a function which stands for any physical quantity.

The range $\mathbb{W}$ should be a normed vector space which need not be specified any further at this stage. Ranges $\mathbb{W}$ are tensors (scalars, vectors, second order tensors, fourth order tensors etc.). In order to make the function available for a concrete representation and a mathematical analysis, the material points $\mathscr{P}$ have to be eliminated. This can be done by substituting the point $\mathscr{P}$ either by its position vector x in the current configuration according to (1.19) or by its assigned vector X in the reference configuration according to (1.20) or (1.26):

$\mathscr{P} = \chi_t^{-1}(x) = \mathrm{R}^{-1}(X)$, $\mathscr{P} = \chi_t^{-1}(x) = \chi_{t_0}^{-1}(X)$. This leads to

Definition 1. 5

The *spatial representation* of any physical quantity $\mathrm{w} = \mathrm{f}(\mathscr{P}, t)$ is expressed by means of the function

$$\bar{\mathrm{f}}: (x, t) \longmapsto \mathrm{w} = \bar{\mathrm{f}}(x, t) = \mathrm{f}(\chi_t^{-1}(x), t). \tag{1.31}$$

---

[4] In some representations of solid mechanics, the definition of the displacement field is introduced at the beginning of kinematics. This is not such a good method: not only does it lead to more complicated calculations, but, in particular, it also ignores the fact that the reference configuration is not generally a place occupied by a material body during its motion. For a variety of arguments in general continuum mechanics, it is of utmost importance to bear this fact in mind.

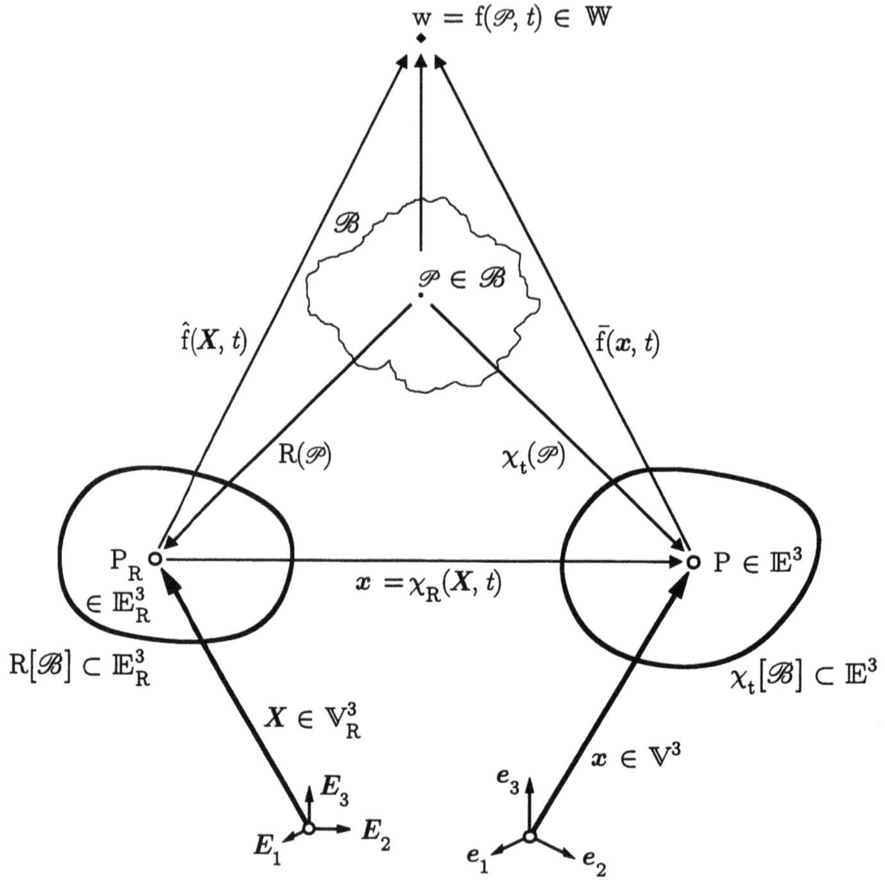

Figure 1. 7: Material and spatial representation of fields

The spatial representation is also called the *Eulerian representation.*

The *material representation* of the same quantity $\mathrm{w} = \mathrm{f}(\mathscr{P}, t)$ is expressed by means of

$$\hat{\mathrm{f}} : (\boldsymbol{X}, t) \longmapsto \mathrm{w} = \hat{\mathrm{f}}(\boldsymbol{X}, t) = \mathrm{f}(\mathrm{R}^{-1}(\boldsymbol{X}), t) \tag{1.32}$$

or

$$\hat{\mathrm{f}} : (\boldsymbol{X}, t) \longmapsto \mathrm{w} = \hat{\mathrm{f}}(\boldsymbol{X}, t) = \mathrm{f}(\chi_{\mathrm{t}_0}{}^{-1}(\boldsymbol{X}), t) \,. \tag{1.33}$$

The material representation is also called the *Lagrangian representation.* ◊

Figure 1. 7 illustrates this definition. The transition from spatial to material representation is accommodated by the following identities:

$$\hat{\mathrm{f}}(\boldsymbol{X}, t) = \bar{\mathrm{f}}(\chi_{\mathrm{R}}(\boldsymbol{X}, t), t) \,, \tag{1.34}$$

$$\bar{\mathrm{f}}(\boldsymbol{x}, t) = \hat{\mathrm{f}}(\chi_{\mathrm{R}}{}^{-1}(\boldsymbol{x}, t), t) \,. \tag{1.35}$$

The representation (1.22) of the motion, $\boldsymbol{x} = \chi_{\mathrm{R}}(\boldsymbol{X}, t)$, is a first example of a material representation. Another example is the velocity field: if

$$\boldsymbol{v} = \dot{\boldsymbol{x}}(t) = \frac{\mathrm{d}}{\mathrm{d}t}\, \chi_{\mathrm{t}}(\mathscr{P}) \tag{1.36}$$

is the velocity vector of a particle $\mathscr{P}$, then

$$\boldsymbol{v} = \hat{\boldsymbol{v}}(\boldsymbol{X}, t) = \frac{\partial}{\partial t} \chi_{\mathrm{R}}(\boldsymbol{X}, t) \tag{1.37}$$

is the material representation of the velocity field. The spatial representation is created by substituting $\boldsymbol{x}$ for $\boldsymbol{X}$:

$$\boldsymbol{v} = \bar{\boldsymbol{v}}(\boldsymbol{x}, t) = \hat{\boldsymbol{v}}(\chi_{\mathrm{R}}{}^{-1}(\boldsymbol{x}, t), t) \,. \tag{1.38}$$

When formulating balance relations in continuum mechanics, it is of utmost importance to calculate the rate of change of a physical quantity $\mathrm{w} = \mathrm{f}(\mathscr{P}, t)$ correctly. This rate of change occurring at one and the same material point $\mathscr{P} \in \mathscr{B}$ is called the material derivative:

Definition 1. 6
If $\mathrm{w}(t) = \mathrm{f}(\mathscr{P}, t)$ is a time-dependent physical quantity perceived at any given material point, then the time derivative

$$\dot{\mathrm{w}}(t) = \dot{\mathrm{f}}(\mathscr{P}, t) = \frac{\mathrm{d}}{\mathrm{d}t}\, \mathrm{f}(\mathscr{P}, t) \tag{1.39}$$

is referred to as the *material derivative* of f. ◊

It follows directly from the definition that the material derivative is the partial time derivative of the material representation:

$$\dot{\mathrm{w}}(t) = \frac{\partial}{\partial t}\,\hat{\mathrm{f}}(\boldsymbol{X}, t)\,. \tag{1.40}$$

In order to obtain the material derivative of the spatial representation, the total time derivative has to be calculated, i.e.

$$\dot{\mathrm{w}}(t) = \frac{\mathrm{d}}{\mathrm{d}t}\bar{\mathrm{f}}(\chi_{\mathrm{R}}(\boldsymbol{X}, t), t)\,.$$ This produces

$$\dot{\mathrm{w}}(t) = \frac{\partial}{\partial t}\,\bar{\mathrm{f}}(\boldsymbol{x}, t) + \left[\operatorname{grad}\bar{\mathrm{f}}(\boldsymbol{x}, t)\right]\bar{\boldsymbol{v}}(\boldsymbol{x}, t)\,. \tag{1.41}$$

According to this result, the material derivative of the spatial representation is composed of one local and one convective derivative: the *local derivative* is the partial time derivative, and the *convective derivative* is the *Gateaux* derivative of $\bar{\mathrm{f}}(\boldsymbol{x}, t)$ in the direction of $\boldsymbol{v}$:[5]

$$\left\{\operatorname{grad}\bar{\mathrm{f}}(\boldsymbol{x}, t)\right\}\,\bar{\boldsymbol{v}}(\boldsymbol{x}, t) = \left\{\frac{\mathrm{d}}{\mathrm{d}s}\bar{\mathrm{f}}(\boldsymbol{x} + s\boldsymbol{h}, t)\Big|_{s=0}\right\}\Big|_{\boldsymbol{h} = \bar{\boldsymbol{v}}(\boldsymbol{x}, t)}\,.$$

$\operatorname{grad}\bar{\mathrm{f}}(\boldsymbol{x}, t)$ is a linear mapping from $\mathbb{V}^3$ into $\mathbb{W}$.

The operator "grad" refers to the *spatial gradient*, which means differentiation with respect to the position vector $\boldsymbol{x}$ of the current configuration, i.e. differentiation with respect to the spatial coordinates.

The material derivative of the velocity vector $\boldsymbol{v}$ is the acceleration vector $\boldsymbol{a}$. The material representation of the acceleration is the partial derivative of the velocity with respect to time:

$$\boldsymbol{a} = \hat{\boldsymbol{a}}(\boldsymbol{X}, t) = \frac{\partial}{\partial t}\hat{\boldsymbol{v}}(\boldsymbol{X}, t) = \frac{\partial^2}{\partial t^2}\chi_{\mathrm{R}}(\boldsymbol{X}, t)\,. \tag{1.42}$$

For the spatial representation we have

$$\boldsymbol{a} = \bar{\boldsymbol{a}}(\boldsymbol{x}, t) = \frac{\partial}{\partial t}\bar{\boldsymbol{v}}(\boldsymbol{x}, t) + \left\{\operatorname{grad}\bar{\boldsymbol{v}}(\boldsymbol{x}, t)\right\}\,\bar{\boldsymbol{v}}(\boldsymbol{x}, t)\,. \tag{1.43}$$

[5] For the definition of the *Gateaux* derivative see LJUSTERNIK & SOBOLEV [1979], p. 308.

1. 3 Deformation Gradient

The coordinate representation $x^k = \chi_{\mathrm{R}}{}^k(X^L, t)$ of the motion of a material body is continuously differentiable in view of the general definitions 1. 1 - 1. 3. The same applies to the representation of a motion as the vector function $\boldsymbol{x} = \chi_{\mathrm{R}}(\boldsymbol{X}, t)$. There is accordingly a *Taylor* expansion for $\boldsymbol{H} \in \mathbb{V}^3_{\mathrm{R}}$,

$$\chi_{\mathrm{R}}(\boldsymbol{X} + \boldsymbol{H}, t) = \chi_{\mathrm{R}}(\boldsymbol{X}, t) + \{\mathrm{Grad}\, \chi_{\mathrm{R}}(\boldsymbol{X}, t)\}\boldsymbol{H} + |\boldsymbol{H}|\boldsymbol{r}(\boldsymbol{X}, t, \boldsymbol{H}) \,, \tag{1.44}$$

with the property $\lim\limits_{|\boldsymbol{H}| \to 0} |\boldsymbol{r}(\boldsymbol{X}, t, \boldsymbol{H})| = 0$.

The operator "Grad" refers to the *material gradient*, which means differentiation with respect to the vector $\boldsymbol{X}$ of the reference configuration, i.e. differentiation with respect to the material coordinates.

Definition 1. 7

The material gradient of motion,

$$\mathbf{F}(\boldsymbol{X}, t) = \mathrm{Grad}\, \chi_{\mathrm{R}}(\boldsymbol{X}, t) \,, \tag{1.45}$$

i.e. the linear transformation

$$\begin{aligned} \mathbf{F} : \mathbb{V}^3_{\mathrm{R}} &\longrightarrow \mathbb{V}^3 \\ \boldsymbol{H} &\longmapsto \mathbf{F}\boldsymbol{H} \end{aligned}$$

is referred to as the *deformation gradient*. ◊

A deformation gradient is assigned to every $(\boldsymbol{X}, t) \in \mathrm{R}[\mathscr{B}] \times \mathbb{I}$:

$$(\boldsymbol{X}, t) \longmapsto \mathbf{F}(\boldsymbol{X}, t) \,.$$

The deformation gradient $\mathbf{F} = \mathrm{Grad}\chi_{\mathrm{R}}(\boldsymbol{X}, t)$, occurring in the *Taylor* expansion (1.44), is the *Fréchet* derivative of $\chi_{\mathrm{R}}(\boldsymbol{X}, t)$ with respect to $\boldsymbol{X}$, a linear mapping from $\mathbb{V}^3_{\mathrm{R}}$ onto $\mathbb{V}^3$. $\mathbf{F}$ is calculated as the *Gateaux* derivative.[6] For all $\boldsymbol{H} \in \mathbb{V}^3_{\mathrm{R}}$ we have the identity

---

[6] According to a well-known theorem of functional analysis, the *Gateaux* derivative is equal to the *Fréchet* derivative, if sufficient conditions of continuity hold. Conversely, if a map is *Fréchet*-differentiable, it is also *Gateaux*-differentiable, and the two derivatives coincide. See, for example, LJUSTERNIK & SOBOLEV [1979], p. 310.

$$\mathbf{F}(\boldsymbol{X}, t)\boldsymbol{H} = \frac{\mathrm{d}}{\mathrm{d}s}\,\chi_{\mathrm{R}}(\boldsymbol{X} + s\boldsymbol{H}, t)\Big|_{s=0}\,. \tag{1.46}$$

For the purpose of calculating the component representation of the deformation gradient, vector $\boldsymbol{H}$ is identified with the tangent vectors $\boldsymbol{G}_L$, thereby obtaining from

$$\mathbf{F}(\boldsymbol{X}, t)\boldsymbol{G}_L = \frac{\mathrm{d}}{\mathrm{d}s}\,\chi_{\mathrm{R}}(\boldsymbol{X} + s\boldsymbol{G}_L, t)\Big|_{s=0}\,,$$

$$\mathbf{F}(\boldsymbol{X}, t)\boldsymbol{G}_L = \frac{\partial}{\partial X^L}\chi_{\mathrm{R}}(\boldsymbol{X}, t) = \left(\frac{\partial}{\partial X^L}\chi_{\mathrm{R}}{}^{i}(X^1, X^2, X^3, t)\right)\boldsymbol{g}_i$$

the components $F^k{}_L = \boldsymbol{g}^k \cdot \mathbf{F}\boldsymbol{G}_L = \dfrac{\partial}{\partial X^L}\chi_{\mathrm{R}}{}^{k}(X^1, X^2, X^3, t) = \dfrac{\partial x^k}{\partial X^L}$

or the component representation

$$\mathbf{F} = F^k{}_L\, \boldsymbol{g}_k \otimes \boldsymbol{G}^L = \frac{\partial x^k}{\partial X^L}\, \boldsymbol{g}_k \otimes \boldsymbol{G}^L\,. \tag{1.47}$$

The components of the deformation gradient with respect to the mixed base system $\boldsymbol{g}_k \otimes \boldsymbol{G}^L$ are identical with the elements of the functional matrix (*Jacobi* matrix) of the coordinate transformation $x^k = \chi_{\mathrm{R}}{}^{k}(X^L, t)$:[7]

$$\mathbf{F} \mathrel{\hat{=}} \begin{pmatrix} \dfrac{\partial x^1}{\partial X^1} & \dfrac{\partial x^1}{\partial X^2} & \dfrac{\partial x^1}{\partial X^3} \\[2mm] \dfrac{\partial x^2}{\partial X^1} & \dfrac{\partial x^2}{\partial X^2} & \dfrac{\partial x^2}{\partial X^3} \\[2mm] \dfrac{\partial x^3}{\partial X^1} & \dfrac{\partial x^3}{\partial X^2} & \dfrac{\partial x^3}{\partial X^3} \end{pmatrix}. \tag{1.48}$$

Since the transformation of coordinates (1.5) is invertible for all t, the *Jacobi* matrix is also invertible, i.e. its determinant is always different from zero:

$$\det\left(\frac{\partial x^k}{\partial X^L}\right) \neq 0\,. \tag{1.49}$$

The same invertibility applies to the vector function (1.22). Accordingly, it always holds that $\det\mathbf{F} \neq 0$. It is generally assumed that

$$\det\mathbf{F} > 0\,. \tag{1.50}$$

[7] Cf. LEIGH [1968], pp. 102; MARSDEN & HUGHES [1983], p. 47.

The components of the matrix (1.48) refer to a mixed base of tensors, namely the system $\{g_k \otimes G^L\}$ $(k, L \equiv 1, 2, 3)$, consisting of tangent and gradient vectors of the current and reference configuration respectively.

The component representation (1.47) can also be obtained from the transformation $x^k = \chi_R{}^k(X^L, t)$ by calculating the total differentials:

$$\mathrm{d}x^k = \frac{\partial x^k}{\partial X^L}\mathrm{d}X^L .$$

Since each of the coordinate differentials are contravariant vector components, we can specify the vectors $\mathrm{d}x = \mathrm{d}x^k g_k$ and $\mathrm{d}X = \mathrm{d}X^J G_J$, obtaining

$$\mathrm{d}x = \mathrm{d}x^k g_k = \frac{\partial x^k}{\partial X^L}\mathrm{d}X^L g_k = \frac{\partial x^k}{\partial X^L}(g_k \otimes G^L)(\mathrm{d}X^J G_J) \text{ , i.e.}$$

$$\mathrm{d}x = \mathbf{F}\,\mathrm{d}X . \tag{1.51}$$

The final step takes advantage of the property $G^L \cdot G_J = \delta^L_J$ of the reciprocal base systems.

In the equation $\mathrm{d}x = \mathbf{F}\mathrm{d}X$, the argument $\mathrm{d}X$ is any vector that replaces H in the *Gateaux* differential (1.46). The change in notation is intended to emphasise the fact that $\mathrm{d}X$ can be interpreted as a tangent vector to a smooth curve, consisting of a set of material points in the reference configuration and containing point X:

$$\alpha \longmapsto C(\alpha) \text{ , } C(\alpha_0) = X \text{ ,} \tag{1.52}$$

$$\mathrm{d}X = \frac{\mathrm{d}}{\mathrm{d}\alpha}\,C(\alpha)\mathrm{d}\alpha\bigg|_{\alpha = \alpha_0} = C'(\alpha_0)\mathrm{d}\alpha . \tag{1.53}$$

In this way it is also possible to define a smooth curve in $\mathbb{E}^3$ by means of

$$\alpha \longmapsto c(\alpha) = \chi_R(C(\alpha), t), \quad c(\alpha_0) = x . \tag{1.54}$$

We may calculate the tangent vector

$$\mathrm{d}x = \frac{\mathrm{d}}{\mathrm{d}\alpha}\,c(\alpha)\mathrm{d}\alpha\bigg|_{\alpha = \alpha_0} = c'(\alpha_0)\mathrm{d}\alpha \tag{1.55}$$

and, in accordance with the chain rule

$$c'(\alpha) = \Big\{\mathrm{Grad}\,\chi_R(C(\alpha), t)\Big\}C'(\alpha) . \tag{1.56}$$

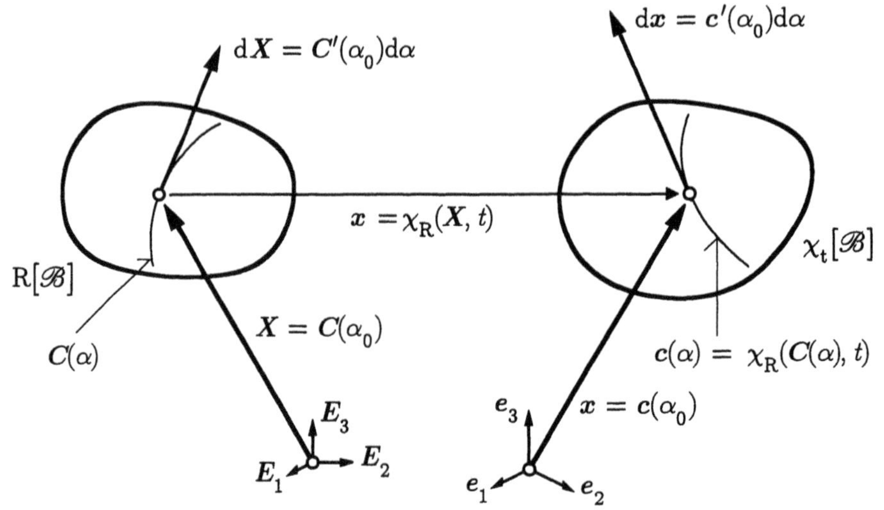

Figure 1. 8: Material line element

For $\alpha = \alpha_0$ we arrive at $\mathrm{d}x = \mathbf{F}\mathrm{d}X$ (Fig. 1. 8). The geometrical meaning of this result is described in

Definition 1. 8

Curves $\alpha \longmapsto C(\alpha) \longmapsto c(\alpha) = \chi_R(C(\alpha), t)$ are referred to as *material lines* in the reference configuration or the current configuration. The tangent vectors $\mathrm{d}X$ and $\mathrm{d}x$ at a specific material point $(C(\alpha_0) = X\,,\; c(\alpha_0) = x)$ form a *material line element* in the reference and current configuration respectively. ◇

The total set of all material line elements $\mathrm{d}X$, belonging to the point $X \in \mathbb{V}^3_R$ forms a vector space, the so-called *tangent space* $\mathbb{T}_X = \mathbb{T}_{R(\mathscr{P})}$. Along the same lines, the set of material line elements $\mathrm{d}x$ at $x = \chi_R(X, t) \in \mathbb{V}^3$ forms the *tangent space* $\mathbb{T}_x = \mathbb{T}_{\chi_t(\mathscr{P})}$.[8] As a result, the deformation gradient is a linear mapping between the two tangent spaces:

[8] The term *tangent space* specifies the loose conceptions of "infinitely (infinitesimally) close" points. A material line element of finite length is, generally speaking, not a material line connecting two material points. Instead, the tangent of a material line is its approximation in the area surrounding a point $X = C(\alpha_0)$ (Fig. 1. 8). It is only in the case of homogeneous deformations $(\mathbf{F}(X, t) = \mathbf{F}(t))$ that straight material lines remain straight, such that the material line and its tangent are identical. Cf. BRÖCKER & JÄNICH [1973], pp. 13, 19; MARSDEN & HUGHES [1983], p. 36; BERTRAM [1989], p. 45.

$$\begin{aligned} \mathbf{F}: \mathbb{T}_{\boldsymbol{X}} &\longrightarrow \mathbb{T}_{\boldsymbol{x}} \\ \mathrm{d}\boldsymbol{X} &\longmapsto \mathrm{d}\boldsymbol{x} = \mathbf{F}\mathrm{d}\boldsymbol{X} \end{aligned} \tag{1.57}$$

If a configuration happens to lie within a Euclidean space, which is usually the case, then all tangent spaces may be embedded into this all-embracing space ($\mathbb{V}^3_{\mathrm{R}}$ or $\mathbb{V}^3$) and there is no real need to distinguish between them. If a configuration is not Euclidean, however, then no embedding process exists. In this case all the tangent spaces are different, and their relationship to each other is primarily governed by the coefficients Γ^k_{ij} of the affine connection (see Sect. 1. 10). Since there are no position vectors in the non-Euclidean case, each tangent space is assigned to one triplet of material or spatial coordinates and consequently to the material point $\mathscr{P}$ itself:

$$(\mathscr{P}, t) \longmapsto \mathbf{F}(\mathscr{P}, t)\,.$$

The deformation gradient is then a linear transformation

$$\begin{aligned} \mathbf{F}: \mathbb{T}_{\mathrm{R}(\mathscr{P})} &\longrightarrow \mathbb{T}_{\chi_t(\mathscr{P})} \\ \mathrm{d}\boldsymbol{X} &\longmapsto \mathrm{d}\boldsymbol{x} = \mathbf{F}\,\mathrm{d}\boldsymbol{X}, \end{aligned}$$

existing for all $(\mathscr{P}, t)$. Attention is drawn to the fact that, whereas the notation

$$\mathbf{F} = \frac{\partial x^k}{\partial X^L}\,\boldsymbol{g}_k \otimes \boldsymbol{G}^L$$

is still appropriate, $\boldsymbol{g}_k$ and $\boldsymbol{G}^L$ can no longer be written as tangent or gradient vectors, related to curvilinear coordinate systems.

The deformation gradient contains all the local properties of a given motion; it does not merely provide the transformation of material line elements (as mentioned above) but also that of surface and volume elements:

Definition 1. 9

If $\mathrm{d}\boldsymbol{x}_1$, $\mathrm{d}\boldsymbol{x}_2$ and $\mathrm{d}\boldsymbol{X}_1$, $\mathrm{d}\boldsymbol{X}_2$ are two material line elements in the current and reference configuration respectively, the two vectors

$$\mathrm{d}\boldsymbol{a} = \mathrm{d}\boldsymbol{x}_1 \times \mathrm{d}\boldsymbol{x}_2 \quad \text{and} \quad \mathrm{d}\boldsymbol{A} = \mathrm{d}\boldsymbol{X}_1 \times \mathrm{d}\boldsymbol{X}_2$$

form a *material surface element*; for three material line elements $\mathrm{d}\boldsymbol{x}_1$, $\mathrm{d}\boldsymbol{x}_2$, $\mathrm{d}\boldsymbol{x}_3$ and $\mathrm{d}\boldsymbol{X}_1$, $\mathrm{d}\boldsymbol{X}_2$, $\mathrm{d}\boldsymbol{X}_3$ the scalars

$$\mathrm{d}v = (\mathrm{d}\boldsymbol{x}_1 \times \mathrm{d}\boldsymbol{x}_2)\cdot \mathrm{d}\boldsymbol{x}_3 \quad \text{and} \quad \mathrm{d}V = (\mathrm{d}\boldsymbol{X}_1 \times \mathrm{d}\boldsymbol{X}_2)\cdot \mathrm{d}\boldsymbol{X}_3$$

form a *material volume element* in the current and the reference configuration. ◊

The transformation of material line, surface and volume elements by means of the deformation gradient is described in the following

Theorem 1. 1

If $\mathbf{F}(\boldsymbol{X}, t) = \operatorname{Grad} \chi_R(\boldsymbol{X}, t)$ is the deformation gradient of any given motion and $(\mathrm{d}\boldsymbol{x}, \mathrm{d}\boldsymbol{X})$, $(\mathrm{d}\boldsymbol{a}, \mathrm{d}\boldsymbol{A})$, $(\mathrm{d}v, \mathrm{d}V)$ material line, surface and volume elements, the following statements apply:[9]

$$1)\ \mathrm{d}\boldsymbol{x} = \mathbf{F}\, \mathrm{d}\boldsymbol{X}\,, \tag{1.58}$$

$$2)\ \mathrm{d}\boldsymbol{a} = (\det\mathbf{F})\mathbf{F}^{\mathrm{T}\,-1}\, \mathrm{d}\boldsymbol{A}\,, \tag{1.59}$$

$$3)\ \mathrm{d}v = (\det\mathbf{F})\mathrm{d}V\,. \tag{1.60}$$

□

Proof

Statement 1) has already been established. Statement 3) is based on the algebraic identity

$$(\mathbf{A}\boldsymbol{a} \times \mathbf{A}\boldsymbol{b})\cdot \mathbf{A}\boldsymbol{c} = (\det\mathbf{A})[(\boldsymbol{a} \times \boldsymbol{b})\cdot \boldsymbol{c}]\,, \tag{1.61}$$

valid for any second order tensor $\mathbf{A}$ and vectors $\boldsymbol{a}, \boldsymbol{b}, \boldsymbol{c} \in \mathbb{V}^3$. This identity is one version of the multiplication theorem for determinants. Statement 2) results from the identity

$$\mathbf{A}^{\mathrm{T}}(\mathbf{A}\boldsymbol{a} \times \mathbf{A}\boldsymbol{b}) = (\det\mathbf{A})(\boldsymbol{a} \times \boldsymbol{b})\,,$$

which one obtains on demanding equation (1.61) for any two fixed vectors $\boldsymbol{a}$, $\boldsymbol{b}$ and all $\boldsymbol{c} \in \mathbb{V}^3$. ■

---

[9] See TRUESDELL & TOUPIN [1960], Sect. 20 for references.

Since line elements, surface elements and volume elements occur in line integrals, surface integrals and volume integrals, the theorem enables us to trace the temporal development of integrals over these material areas. From the kinematics point of view this is the basis for the formulation of balance relations.

A further fundamental assertion of the kinematics of deformable bodies is the multiplicative decomposition of the deformation gradient into orthogonal and symmetric parts.

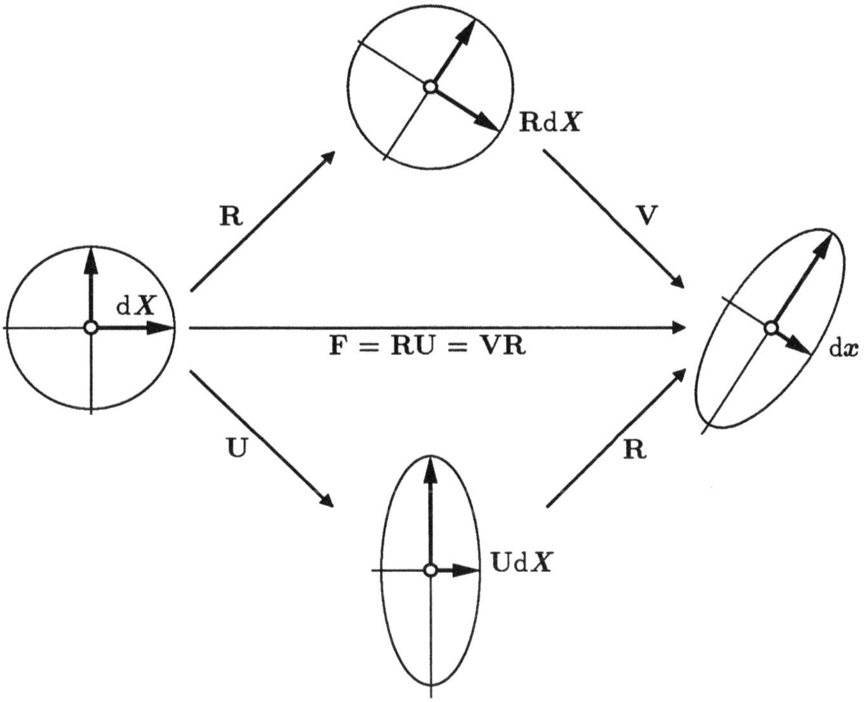

Figure 1. 9: Polar decomposition of the deformation gradient

Theorem 1. 2

The *polar decomposition* holds for the deformation gradient $\mathbf{F}$ (as in the case of every invertible second order tensor):[10]

[10] See TRUESDELL & NOLL [1965], Sect. 23.

1) **F** can be represented as the product of 2 tensors:

$$\mathbf{F} = \mathbf{RU} = \mathbf{VR}, \tag{1.63}$$

in which factors **U** and **V** are symmetric and positive definite, $\mathbf{U} = \mathbf{U}^{\mathrm{T}}$, $\boldsymbol{v} \cdot \mathbf{U}\boldsymbol{v} > 0$, $\mathbf{V} = \mathbf{V}^{\mathrm{T}}$, $\boldsymbol{v} \cdot \mathbf{V}\boldsymbol{v} > 0$, and **R** is orthogonal: $\mathbf{R}\mathbf{R}^{\mathrm{T}} = \mathbf{R}^{\mathrm{T}}\mathbf{R} = \mathbf{1}$. **U**, **V** and **R** are uniquely defined (see Fig. 1. 9).

2) **U** and **V** have the same eigenvalues: if $\boldsymbol{v}$ is an eigenvector of **U**, then $\mathbf{R}\boldsymbol{v}$ is an eigenvector of **V**. □

The proof of this theorem is based on

Theorem 1. 3
If **C** is a symmetric positive definite second order tensor, the unique positive square root exists, i.e. a tensor $\mathbf{U} = \sqrt{\mathbf{C}}$, also symmetric and positive definite, with the property $\mathbf{U}^2 = \mathbf{C}$. □

Proof

Since **C** is symmetric and positive definite, the *eigenvalue problem* $\mathbf{C}\boldsymbol{v} = \lambda \boldsymbol{v}$ yields 3 real *eigenvalues* $\lambda_i > 0$ as well as 3 orthonormal *eigenvectors* $\{\boldsymbol{v}_i\}$, $\boldsymbol{v}_i \cdot \boldsymbol{v}_k = \delta_{ik}$, and **C** has the representation

$$\mathbf{C} = \sum_{k=1}^{3} \lambda_k \boldsymbol{v}_k \otimes \boldsymbol{v}_k \,. \tag{1.64}$$

It is evident that the tensor

$$\mathbf{U} = \sum_{k=1}^{3} +\sqrt{\lambda_k}(\boldsymbol{v}_k \otimes \boldsymbol{v}_k) \tag{1.65}$$

possesses the required property: this is verified by

$$\mathbf{U}^2 = \sum_{i,k=1}^{3} \sqrt{\lambda_i}\sqrt{\lambda_k}\,(\boldsymbol{v}_i \otimes \boldsymbol{v}_i)(\boldsymbol{v}_k \otimes \boldsymbol{v}_k) = \sum_{k=1}^{3} \lambda_k \boldsymbol{v}_k \otimes \boldsymbol{v}_k = \mathbf{C} \,.$$

Uniqueness in connection with the positive definiteness is guaranteed as long as the positive sign is chosen for $\sqrt{\lambda_k}$. ■

Proof of Theorem 1. 2

1) Tensor $\mathbf{C} = \mathbf{F}^T\mathbf{F}$ is symmetric and positive definite,

$$\boldsymbol{v}\cdot(\mathbf{F}^T\mathbf{F}\boldsymbol{v}) = (\mathbf{F}\boldsymbol{v})\cdot(\mathbf{F}\boldsymbol{v}) > 0\,,$$

this verifies the existence of the unique positive square root $\mathbf{U} = \sqrt{\mathbf{C}}$. Since $\mathbf{B} = \mathbf{F}\mathbf{F}^T$ is also symmetric and positive definite, we have the unique positive root $\mathbf{V} = \sqrt{\mathbf{B}}$.

If we now define the tensor $\mathbf{R} = \mathbf{F}\mathbf{U}^{-1}$, we find that it is orthogonal, i.e. $\mathbf{R}\mathbf{R}^T = \mathbf{F}\mathbf{U}^{-1}(\mathbf{U}^{-1}\mathbf{F}^T) = \mathbf{F}(\mathbf{U}^2)^{-1}\mathbf{F}^T = \mathbf{F}(\mathbf{F}^T\mathbf{F})^{-1}\mathbf{F}^T = \mathbf{1}$; this proves $\mathbf{F} = \mathbf{R}\mathbf{U}$. Likewise, the tensor $\widetilde{\mathbf{R}} \equiv \mathbf{V}^{-1}\mathbf{F}$ is also orthogonal, i.e. $\widetilde{\mathbf{R}}\widetilde{\mathbf{R}}^T = \mathbf{V}^{-1}\mathbf{F}\mathbf{F}^T\mathbf{V}^{-1} = \mathbf{V}^{-1}\mathbf{V}^2\mathbf{V}^{-1} = \mathbf{1}$, so $\mathbf{F} = \mathbf{V}\widetilde{\mathbf{R}}$ holds too. Altogether we have shown that $\mathbf{F} = \mathbf{R}\mathbf{U} = \mathbf{V}\widetilde{\mathbf{R}}$.

The possibility of $\mathbf{R} \neq \widetilde{\mathbf{R}}$ would imply $\mathbf{F} = \mathbf{R}\mathbf{U} = \mathbf{V}\widetilde{\mathbf{R}} = \widetilde{\mathbf{R}}(\widetilde{\mathbf{R}}^T\mathbf{V}\widetilde{\mathbf{R}})$, i.e. $\mathbf{F} = \widetilde{\mathbf{R}}\widetilde{\mathbf{U}}$ or $\mathbf{F}^T\mathbf{F} = \mathbf{U}^2 = \widetilde{\mathbf{U}}^2$, meaning that the positive square root would not be unique. This is contradictory to Theorem 1. 3, leading to $\mathbf{R} = \widetilde{\mathbf{R}}$.

2) $\mathbf{U}\boldsymbol{v} = \lambda\boldsymbol{v}$ leads to $(\mathbf{R}\mathbf{U}\mathbf{R}^T)\mathbf{R}\boldsymbol{v} = \lambda\mathbf{R}\boldsymbol{v}$ and, by inserting $\mathbf{R}\mathbf{U} = \mathbf{V}\mathbf{R}$, to $\mathbf{V}(\mathbf{R}\boldsymbol{v}) = \lambda(\mathbf{R}\boldsymbol{v})$. ■

According to Theorem 1. 2, the local deformation consists of a stretch ($\mathbf{U}$) with subsequent rotation ($\mathbf{R}$) or, alternatively, a rotation ($\mathbf{R}$) with subsequent stretch ($\mathbf{V}$) (Fig. 1. 9).

Definition 1. 10

The tensor

$\mathbf{U} = \sqrt{\mathbf{C}}$ is called the right stretch tensor, (1.66)

$\mathbf{V} = \sqrt{\mathbf{B}}$ is called the left stretch tensor. (1.67)

$\mathbf{C} = \mathbf{F}^T\mathbf{F} = \mathbf{U}^2$ is termed the right *Cauchy-Green* tensor and (1.68)

$\mathbf{B} = \mathbf{F}\mathbf{F}^T = \mathbf{V}^2$ is termed the left *Cauchy-Green* tensor. (1.69)

◊

1. 4 Strain Tensors

Definition 1. 11
A *rigid body motion* is a motion of the type

$$x = \chi_R(X, t) = Q(t)(X - X_0) + x_0(t)\,, \tag{1.70}$$

$Q(t)$ being an orthogonal tensor ($QQ^T = 1$), dependent only on time. X_0 is a reference point (for example, the centre of mass) and $x_0(t)$ a vector-valued function of time, which represents the motion of the material reference point. ◊

The definition ascertains the fact that a rigid body motion consists of a rotation $Q(t)$ about a point X_0 and a translation $x_0(t)$. The rotation $Q(t)$ depends merely upon time and is also - as can be proved - independent of the choice of point X_0. The deformation gradient of a rigid body motion is accordingly spatially constant and orthogonal,

$$F(X, t) = Q(t)\,, \tag{1.71}$$

whereas the corresponding *Cauchy-Green* tensors are unit tensors,

$$C = F^TF = Q^T(t)Q(t) \equiv 1\,, B = FF^T = Q(t)Q(t)^T \equiv 1\,.$$

As far as continuum mechanics is concerned, rigid body motions are lacking in interest: attention is drawn to the deformations that deviate from rigid body motions. A motion's deviation from a rigid body motion is referred to as a *distortion* or *strain* and is represented by a properly selected second order tensor. One possibility is given by means of the following

Definition 1. 12
The tensor

$$E(X, t) = \tfrac{1}{2}(C - 1) = \tfrac{1}{2}(F^T F - 1) \tag{1.72}$$

is known as the *Green* strain tensor. ◊

The properties of the *Green* strain tensor E as regards the representation of the state of strain are explained in

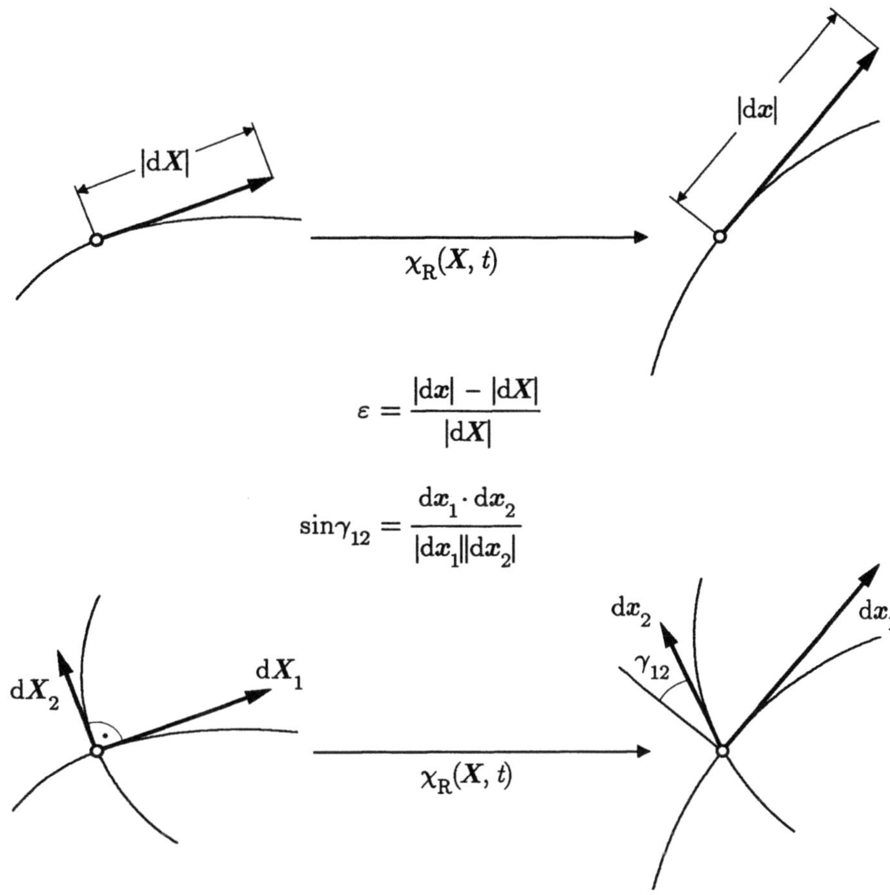

Figure 1. 10: Extension and shear

Theorem 1. 4
If $(\mathrm{d}\boldsymbol{X}, \mathrm{d}\boldsymbol{Y}) \in \mathbb{T}_{\boldsymbol{X}}$ are material line elements in the reference configuration and $(\mathrm{d}\boldsymbol{x}, \mathrm{d}\boldsymbol{y}) \in \mathbb{T}_{\boldsymbol{x}}$ their images in the current configuration, the following equation is true:

$$\mathrm{d}\boldsymbol{X}\cdot(\mathbf{E}\;\mathrm{d}\boldsymbol{Y}) = \tfrac{1}{2}\left(\mathrm{d}\boldsymbol{x}\cdot\mathrm{d}\boldsymbol{y} - \mathrm{d}\boldsymbol{X}\cdot\mathrm{d}\boldsymbol{Y}\right). \tag{1.73}$$

For the *extension* or *normal strain* of a material line element $\mathrm{d}\boldsymbol{X} = |\mathrm{d}\boldsymbol{X}|\boldsymbol{e}$, defined as the change of length divided by its length in the reference configuration,

$$\varepsilon = \frac{|\mathrm{d}\boldsymbol{x}| - |\mathrm{d}\boldsymbol{X}|}{|\mathrm{d}\boldsymbol{X}|}\,, \tag{1.74}$$

the following equation applies:

$$\varepsilon = \sqrt{1 + 2\boldsymbol{e}\cdot\mathbf{E}\boldsymbol{e}} - 1 \Leftrightarrow \boldsymbol{e}\cdot\mathbf{E}\boldsymbol{e} = \varepsilon + \frac{\varepsilon^2}{2}\,. \tag{1.75}$$

The *shear* or *shear strain* of two material line elements, $\mathrm{d}\boldsymbol{X}_1 = |\mathrm{d}\boldsymbol{X}_1|\boldsymbol{e}_1$ and $\mathrm{d}\boldsymbol{X}_2 = |\mathrm{d}\boldsymbol{X}_2|\boldsymbol{e}_2$, which are orthogonal in the reference configuration ($\boldsymbol{e}_1\cdot\boldsymbol{e}_2 = 0$), is defined by the deviation from orthogonality, i.e. by

$$\sin\gamma_{12} = \frac{\mathrm{d}\boldsymbol{x}_1\cdot\mathrm{d}\boldsymbol{x}_2}{|\mathrm{d}\boldsymbol{x}_1||\mathrm{d}\boldsymbol{x}_2|}\,, \tag{1.76}$$

and determined by

$$\sin\gamma_{12} = \frac{2\boldsymbol{e}_1\cdot\mathbf{E}\boldsymbol{e}_2}{\sqrt{1 + 2\,\boldsymbol{e}_1\cdot\mathbf{E}\boldsymbol{e}_1}\,\sqrt{1 + 2\,\boldsymbol{e}_2\cdot\mathbf{E}\boldsymbol{e}_2}} \tag{1.77}$$

(Fig. 1. 10). □

Proof

First, $\mathrm{d}\boldsymbol{x} = \mathbf{F}\mathrm{d}\boldsymbol{X}$ and $\mathrm{d}\boldsymbol{y} = \mathbf{F}\mathrm{d}\boldsymbol{Y}$ lead to $\mathrm{d}\boldsymbol{x}\cdot\mathrm{d}\boldsymbol{y} = \mathrm{d}\boldsymbol{X}\cdot(\mathbf{F}^{\mathrm{T}}\mathbf{F}\mathrm{d}\boldsymbol{Y})$ or $\mathrm{d}\boldsymbol{x}\cdot\mathrm{d}\boldsymbol{y} - \mathrm{d}\boldsymbol{X}\cdot\mathrm{d}\boldsymbol{Y} = \mathrm{d}\boldsymbol{X}\cdot[(\mathbf{F}^{\mathrm{T}}\mathbf{F} - \mathbf{1})\mathrm{d}\boldsymbol{Y}] = 2\,\mathrm{d}\boldsymbol{X}\cdot\mathbf{E}\mathrm{d}\boldsymbol{Y}$.

$\varepsilon = \dfrac{|\mathrm{d}\boldsymbol{x}| - |\mathrm{d}\boldsymbol{X}|}{|\mathrm{d}\boldsymbol{X}|}$ implies

$$(\varepsilon + 1)^2 = \frac{|\mathrm{d}\boldsymbol{x}|^2}{|\mathrm{d}\boldsymbol{X}|^2} = \frac{\mathrm{d}\boldsymbol{x}\cdot\mathrm{d}\boldsymbol{x}}{|\mathrm{d}\boldsymbol{X}|^2} = \boldsymbol{e}\cdot[(\mathbf{F}^{\mathrm{T}}\mathbf{F})\boldsymbol{e}] = \boldsymbol{e}\cdot[(\mathbf{1} + 2\,\mathbf{E})\boldsymbol{e}] = 1 + 2\,\boldsymbol{e}\cdot\mathbf{E}\boldsymbol{e}\,.$$

The definition (1.77) of the shear strain together with $\mathrm{d}\boldsymbol{X}_1\cdot\mathrm{d}\boldsymbol{X}_2 = 0$ implies

$$|\mathrm{d}\boldsymbol{x}_1||\mathrm{d}\boldsymbol{x}_2|\sin\gamma_{12} = \mathrm{d}\boldsymbol{x}_1\cdot\mathrm{d}\boldsymbol{x}_2 = \mathrm{d}\boldsymbol{X}_1\cdot[(\mathbf{1} + 2\,\mathbf{E})\mathrm{d}\boldsymbol{X}_2] = 2\,\mathrm{d}\boldsymbol{X}_1\cdot\mathbf{E}\mathrm{d}\boldsymbol{X}_2\,.$$

This, substituting $|\mathrm{d}\boldsymbol{x}_1| = (\varepsilon_1 + 1)|\mathrm{d}\boldsymbol{X}_1| = \sqrt{1 + 2\,\boldsymbol{e}_1\cdot\mathbf{E}\boldsymbol{e}_1}\,|\mathrm{d}\boldsymbol{X}_1|$ and

$|\mathrm{d}x_2| = (\varepsilon_2 + 1)|\mathrm{d}\boldsymbol{X}_2| = \sqrt{1 + 2\,\mathbf{e}_2\cdot\mathbf{E}\mathbf{e}_2}\;|\mathrm{d}\boldsymbol{X}_2|$, produces

$$2\mathbf{e}_1\cdot\mathbf{E}\mathbf{e}_2 = (\varepsilon_1 + 1)(\varepsilon_2 + 1)\sin\gamma_{12} = \sqrt{1 + 2\,\mathbf{e}_1\cdot\mathbf{E}\mathbf{e}_1}\sqrt{1 + 2\,\mathbf{e}_2\cdot\mathbf{E}\mathbf{e}_2}\;\sin\gamma_{12}.$$

■

The *Green* strain tensor $\mathbf{E}(\boldsymbol{X}, t)$, which is written as a tensor field in the material representation, operates on the tangent space $\mathbb{T}_{\boldsymbol{X}}$ or $\mathbb{T}_{\mathrm{R}(\mathscr{P})}$ of the reference configuration. Its bilinear form (1.73),

$$(\mathrm{d}\boldsymbol{X}, \mathrm{d}\boldsymbol{Y}) \longmapsto \mathrm{d}\boldsymbol{X}\cdot(\mathbf{E}\,\mathrm{d}\boldsymbol{Y}) = \tfrac{1}{2}\big(\mathrm{d}x\cdot\mathrm{d}y - \mathrm{d}\boldsymbol{X}\cdot\mathrm{d}\boldsymbol{Y}\big),$$

represents changes in length and angle of and between material line elements. By definition, $\mathbf{E}$ vanishes for rigid body motions and only for these. The norm of $\mathbf{E}$,

$$\|\mathbf{E}\| = \sqrt{\mathbf{E}\cdot\mathbf{E}} = \sqrt{\operatorname{tr}\mathbf{E}^2}\ ,$$

is therefore a measure for the magnitude of the strain, the deviation from rigid body motion.

A further possibility to represent the state of strain or distortion is proposed by

Definition 1. 13

The tensor

$$\mathbf{e}(\boldsymbol{X}, t) = \tfrac{1}{2}\big(\mathbf{C}^{-1} - \mathbf{1}\big) = \tfrac{1}{2}\big(\mathbf{F}^{-1}\mathbf{F}^{\mathrm{T}-1} - \mathbf{1}\big) \tag{1.78}$$

is called the *Piola* strain tensor.[11] ◊

The bilinear form corresponding to this tensor operates on the set of material surface elements in the reference configuration and describes the changes which they undergo. A material surface is represented in the reference configuration or the current configuration by means of

$$\Phi(\boldsymbol{X}) = C \tag{1.79}$$

or

$$\varphi(x, t) = \Phi\big(\chi_{\mathrm{R}}^{-1}(x, t)\big) = C\ . \tag{1.80}$$

[11] See HAUPT & TSAKMAKIS [1989], pp. 177. Cf. GIESEKUS [1994], p. 41.

Differentiating the identity $\Phi(\boldsymbol{X}) = \varphi(\chi_R(\boldsymbol{X}, t), t)$ with respect to $\boldsymbol{X}$ and observing the chain rule, it follows that $\mathrm{Grad}\Phi(\boldsymbol{X}) = \mathbf{F}^{\mathrm{T}}\mathrm{grad}\varphi(\boldsymbol{x}, t)$ or

$$\mathrm{grad}\varphi(\boldsymbol{x}, t) = \mathbf{F}^{\mathrm{T}\,-1}\,\mathrm{Grad}\Phi(\boldsymbol{X})\,. \tag{1.81}$$

Both gradient vectors are normal on the material surfaces $\Phi(\boldsymbol{X}) = C$ and $\varphi(\boldsymbol{x}, t) = C$. Their absolute values are the normal derivatives of the family parameter C. If

$\boldsymbol{N}_0 = \dfrac{\mathrm{Grad}\Phi(\boldsymbol{X})}{|\mathrm{Grad}\Phi(\boldsymbol{X})|}$ is the normal unit vector on the material surface

$\Phi(\boldsymbol{X}) = C$, then the derivative in the direction of $\boldsymbol{N}_0$ is given by

$$\frac{\mathrm{d}C}{\mathrm{d}N} = |\mathrm{Grad}\Phi(\boldsymbol{X})|\,.$$

Accordingly, we have $\dfrac{\mathrm{d}C}{\mathrm{d}n} = |\mathrm{grad}\varphi(\boldsymbol{x}, t)|$. By employing the abbreviations

$$\boldsymbol{n} = \mathrm{grad}\varphi(\boldsymbol{x}, t) \tag{1.82}$$

and

$$\boldsymbol{N} = \mathrm{Grad}\Phi(\boldsymbol{X})\,, \tag{1.83}$$

we arrive at

$$\boldsymbol{n} = \mathbf{F}^{\mathrm{T}\,-1}\boldsymbol{N} \tag{1.84}$$

and $\boldsymbol{n}\cdot\boldsymbol{n} = (\mathbf{F}^{\mathrm{T}\,-1}\boldsymbol{N})\cdot(\mathbf{F}^{\mathrm{T}\,-1}\boldsymbol{N}) = \boldsymbol{N}\cdot[(\mathbf{F}^{-1}\mathbf{F}^{\mathrm{T}\,-1})\boldsymbol{N}]$. This results in

$$\boldsymbol{N}\cdot\mathbf{e}\boldsymbol{N} = \tfrac{1}{2}(\boldsymbol{n}\cdot\boldsymbol{n} - \boldsymbol{N}\cdot\boldsymbol{N}) = \tfrac{1}{2}\left[\left(\frac{\mathrm{d}C}{\mathrm{d}n}\right)^2 - \left(\frac{\mathrm{d}C}{\mathrm{d}N}\right)^2\right], \tag{1.85}$$

which reflects the geometrical meaning of the *Piola* tensor $\mathbf{e}$: tensor $\mathbf{e}$ describes both the change of distances between material surfaces and the changes of their relative orientation to one another. Of course, $\mathbf{e}$ vanishes for rigid body motions as well.

Given both the strain tensors $\mathbf{E}$ and $\mathbf{e}$, one can construct two new strain tensors possessing identical properties but operating on the set of line or surface elements in the current configuration. To begin with, the transformation

$$\tfrac{1}{2}(\mathrm{d}\boldsymbol{x}\cdot\mathrm{d}\boldsymbol{y} - \mathrm{d}\boldsymbol{X}\cdot\mathrm{d}\boldsymbol{Y}) = \mathrm{d}\boldsymbol{X}\cdot(\mathbf{E}\,\mathrm{d}\boldsymbol{Y}) = \mathrm{d}\boldsymbol{x}\cdot[(\mathbf{F}^{\mathrm{T}\,-1}\mathbf{E}\mathbf{F}^{-1})\mathrm{d}\boldsymbol{y}]$$

motivates the definition of the strain tensor

$$\mathbf{A} = \mathbf{F}^{\mathrm{T}-1}\mathbf{E}\mathbf{F}^{-1} = \frac{1}{2}\mathbf{F}^{\mathrm{T}-1}(\mathbf{F}^{\mathrm{T}}\mathbf{F} - \mathbf{1})\mathbf{F}^{-1} = \frac{1}{2}\left(\mathbf{1} - \mathbf{F}^{\mathrm{T}-1}\mathbf{F}^{-1}\right).$$

Definition 1. 14
The tensor

$$\mathbf{A} = \mathbf{F}^{\mathrm{T}-1}\mathbf{E}\mathbf{F}^{-1} = \frac{1}{2}\left(\mathbf{1} - \mathbf{F}^{\mathrm{T}-1}\mathbf{F}^{-1}\right) = \frac{1}{2}\left(\mathbf{1} - \mathbf{B}^{-1}\right) \tag{1.86}$$

is termed *Almansi* strain tensor. ◊

The *Almansi* strain tensor operates on the tangent space $\mathbb{T}_{\boldsymbol{x}}$ or $\mathbb{T}_{\chi_t(\mathscr{P})}$ in the current configuration. Following its definition we conclude that

$$\mathrm{d}\boldsymbol{x} \cdot \mathbf{A}\mathrm{d}\boldsymbol{y} = \mathrm{d}\boldsymbol{X} \cdot \mathbf{E}\mathrm{d}\boldsymbol{Y}\,. \tag{1.87}$$

The transformation

$$\frac{1}{2}(\boldsymbol{n} \cdot \boldsymbol{m} - \boldsymbol{N} \cdot \boldsymbol{M}) = \boldsymbol{N} \cdot \mathbf{e}\boldsymbol{M} = \mathbf{F}^{\mathrm{T}}\boldsymbol{n} \cdot \left(\mathbf{e}\mathbf{F}^{\mathrm{T}}\boldsymbol{m}\right) = \boldsymbol{n} \cdot \left[\left(\mathbf{F}\mathbf{e}\mathbf{F}^{\mathrm{T}}\right)\boldsymbol{m}\right]$$

motivates the definition of the strain tensor

$$\mathbf{a} = \mathbf{F}\mathbf{e}\mathbf{F}^{\mathrm{T}} = \frac{1}{2}\mathbf{F}(\mathbf{F}^{-1}\mathbf{F}^{\mathrm{T}-1} - \mathbf{1})\mathbf{F}^{\mathrm{T}} = \frac{1}{2}\left(\mathbf{1} - \mathbf{F}\mathbf{F}^{\mathrm{T}}\right).$$

Definition 1. 15
The tensor

$$\mathbf{a} = \mathbf{F}\mathbf{e}\mathbf{F}^{\mathrm{T}} = \frac{1}{2}\left(\mathbf{1} - \mathbf{F}\mathbf{F}^{\mathrm{T}}\right) = \frac{1}{2}(\mathbf{1} - \mathbf{B}) \tag{1.88}$$

is called the *Finger* strain tensor.[12] ◊

The *Finger* tensor operates on the current configuration. By definition we have

$$\boldsymbol{n} \cdot \mathbf{a}\boldsymbol{m} = \boldsymbol{N} \cdot \mathbf{e}\boldsymbol{M} = \frac{1}{2}(\boldsymbol{n} \cdot \boldsymbol{m} - \boldsymbol{N} \cdot \boldsymbol{M}) \tag{1.89}$$

---

[12] Cf. GIESEKUS [1994], p. 40.

and $\boldsymbol{n}\cdot\mathbf{a}\boldsymbol{n} = \frac{1}{2}(\boldsymbol{n}\cdot\boldsymbol{n} - \boldsymbol{N}\cdot\boldsymbol{N}) = \frac{1}{2}\left[\left(\frac{\mathrm{d}C}{\mathrm{d}n}\right)^2 - \left(\frac{\mathrm{d}C}{\mathrm{d}N}\right)^2\right]$, i.e.

$$\boldsymbol{n}\cdot\mathbf{a}\boldsymbol{n} = \boldsymbol{N}\cdot\mathbf{e}\boldsymbol{N}\,. \tag{1.90}$$

1.5 Convective Coordinates

In the coordinate representation (1.5) of the motion, $x^k = \chi_R{}^k(X^L, t)$, the choice of the material and spatial coordinates is arbitrary. On principle, two independent systems of Cartesian or curvilinear coordinates may be chosen. A special choice is the introduction of convective coordinates.

Definition 1.16

A coordinate system in the space of the reference configuration, with the property that the coordinate lines in the current configurations are material lines, is called a *system of convective coordinates.*[13] ◊

A convective coordinate system is formed by conceiving the coordinate lines of the material coordinates as material lines in the reference configuration and identifying their images in the current configuration with the coordinate lines of the spatial coordinates. In this case, the tangent vectors are, of course, material line elements, whereas the gradient vectors represent material surface elements. If

$$\boldsymbol{G}_K = \frac{\partial \boldsymbol{X}}{\partial X^K}$$

are the tangent vectors of the reference configuration, according to equation (1.17), then the tangent vectors of the current configuration are given as

$$\boldsymbol{g}_K = \frac{\partial \boldsymbol{x}}{\partial X^K} = \frac{\partial}{\partial X^K}\chi_R(\boldsymbol{X}, t) = \left\{\mathrm{Grad}\chi_R\right\}\frac{\partial \boldsymbol{X}}{\partial X^K}\,, \tag{1.91}$$

$$\boldsymbol{g}_K = \mathbf{F}\boldsymbol{G}_K\,. \tag{1.92}$$

Since $\boldsymbol{G}_K\cdot\boldsymbol{G}^L = \delta_K^L$ and $\boldsymbol{g}_K\cdot\boldsymbol{g}^L = \delta_K^L$ the relation

---

[13] There are representations of continuum mechanics in which convective coordinate systems form the basis throughout. An example of this is the well-known monograph by GREEN & ZERNA [1968].

$$g^K = \mathbf{F}^{\mathrm{T}\,-1} G^K \tag{1.93}$$

holds for the gradient vectors. We accordingly write $x^K = X^K$; with respect to convective coordinates, the deformation gradient then has a very simple representation, namely

$$\mathbf{F} = \delta^K_L \, g_K \otimes G^L = g_K \otimes G^K \, . \tag{1.94}$$

We have

$$(g_K \otimes G^K)(G_L \otimes g^L) = g_K \otimes g^K \text{ and } (G_L \otimes g^L)(g_K \otimes G^K) = G_K \otimes G^K \, .$$

This implies the inverse deformation gradient

$$\mathbf{F}^{-1} = G_K \otimes g^K \tag{1.95}$$

(cf. (1.97), (1.98)). Convective coordinate systems are of lesser practical interest; they serve primarily to provide a better comprehension of the terminology employed in the kinematics of deformable bodies. The first thing one notices is that, as far as the mixed base system $\{g_K \otimes G^L\}$ is concerned, the component matrix of $\mathbf{F}$ is the identity matrix. This is not surprising, since, when comparing with the earlier representation (1.47),

$$\mathbf{F} = \frac{\partial x^i}{\partial X^K} \, g_i \otimes G^K \, ,$$

it can be seen that the convective tangent vectors g_K are connected with the tangent vectors g_i of the general spatial coordinates by means of

$$g_K = \frac{\partial x^i}{\partial X^K} \, g_i \, . \tag{1.96}$$

The basic properties of the linear map $\mathbf{F}$ are particularly easy to recognise in the simple representation of the deformation gradient, namely

$$\mathbf{F} : \mathbb{T}_X \longrightarrow \mathbb{T}_x \text{ and } \mathbf{F}^{-1} : \mathbb{T}_x \longrightarrow \mathbb{T}_X \, .$$

In particular, according to this interpretation, we have $\mathbf{F}\mathbf{F}^{-1} \neq \mathbf{F}^{-1}\mathbf{F}$:

$$\mathbf{F}\mathbf{F}^{-1} = g_K \otimes g^K = \mathbf{1} \, (= \mathbf{1}_x) \tag{1.97}$$

is the identity tensor belonging to the tangent space $\mathbb{T}_{\boldsymbol{x}}$ whereas

$$\mathbf{F}^{-1}\mathbf{F} = \boldsymbol{G}_K \otimes \boldsymbol{G}^K = \mathbf{1}\,(= \mathbf{1}_{\boldsymbol{X}}) \tag{1.98}$$

leads to the identity tensor belonging to $\mathbb{T}_{\boldsymbol{X}}$.

Cauchy-Green tensors, written in convective coordinates, are metric tensors. If $g_{KL} = \boldsymbol{g}_K \cdot \boldsymbol{g}_L$ and $g^{KL} = \boldsymbol{g}^K \cdot \boldsymbol{g}^L$ are the co- and contravariant metric coefficients of the convective coordinates, we calculate

$$\mathbf{C} = \mathbf{F}^{\mathrm{T}}\mathbf{F} = (\boldsymbol{G}^K \otimes \boldsymbol{g}_K)(\boldsymbol{g}_L \otimes \boldsymbol{G}^L) = (\boldsymbol{g}_K \cdot \boldsymbol{g}_L)(\boldsymbol{G}^K \otimes \boldsymbol{G}^L)\ , \text{ i.e.}$$

$$\mathbf{C} = \mathbf{F}^{\mathrm{T}}\mathbf{F} = g_{KL}\boldsymbol{G}^K \otimes \boldsymbol{G}^L . \tag{1.99}$$

In analogy thereto, it can be verified that

$$\mathbf{C}^{-1} = \mathbf{F}^{-1}\mathbf{F}^{\mathrm{T}-1} = g^{KL}\boldsymbol{G}_K \otimes \boldsymbol{G}_L \tag{1.100}$$

as well as

$$\mathbf{B} = \mathbf{F}\mathbf{F}^{\mathrm{T}} = G^{KL}\boldsymbol{g}_K \otimes \boldsymbol{g}_L \tag{1.101}$$

and

$$\mathbf{B}^{-1} = \mathbf{F}^{\mathrm{T}-1}\mathbf{F}^{-1} = G_{KL}\boldsymbol{g}^K \otimes \boldsymbol{g}^L . \tag{1.102}$$

The 4 strain tensors defined so far are, in this connection, nothing other than the differences between the covariant and contravariant metric tensors of the current and the reference configuration:

$$\begin{aligned}
&\textit{Green}\text{ tensor:} && \mathbf{E} = \tfrac{1}{2}(\mathbf{C} - \mathbf{1}) = \tfrac{1}{2}\left[g_{KL} - G_{KL}\right]\boldsymbol{G}^K \otimes \boldsymbol{G}^L \\
&\textit{Almansi}\text{ tensor:} && \mathbf{A} = \tfrac{1}{2}(\mathbf{1} - \mathbf{B}^{-1}) = \tfrac{1}{2}\left[g_{KL} - G_{KL}\right]\boldsymbol{g}^K \otimes \boldsymbol{g}^L \\
&\textit{Piola}\text{ tensor:} && \mathbf{e} = \tfrac{1}{2}(\mathbf{C}^{-1} - \mathbf{1}) = \tfrac{1}{2}\left[g^{KL} - G^{KL}\right]\boldsymbol{G}_K \otimes \boldsymbol{G}_L \\
&\textit{Finger}\text{ tensor:} && \mathbf{a} = \tfrac{1}{2}(\mathbf{1} - \mathbf{B}) = \tfrac{1}{2}\left[g^{KL} - G^{KL}\right]\boldsymbol{g}_K \otimes \boldsymbol{g}_L
\end{aligned} \tag{1.103}$$

The tangent (and cotangent) spaces on which these tensors operate are indicated in this representation by means of base tensors, to each of which the metric differences refer.[14]

As far as the comparison between the choice of general or convective coordinates is concerned, we can summarise as follows: the motion of a material body cannot merely be looked upon as a coordinate transformation or a point transformation but, alternatively, as a transformation of metric quantities as well. The *point transformation* defines changes of place and changes of length and angle according to (1.73). The change of place is of less interest in the *transformation of metric* $\left(x^K = X^K\right)$: the same coordinates, yet different metric tensors are assigned to one and the same material point,

$$(\mathscr{P}, t) \longmapsto \left(G_{KL}(X^J), t\right) \longmapsto g_{KL}(X^J, t) \ , \tag{1.104}$$

bearing in mind that every rigid body motion leaves the metric unaltered. In this interpretation the physical meaning of the coordinates plays a markedly subordinate role.

The interpretation of the motion as a transformation of metric includes the notion that the geometric structure of the configurations will change during the course of the motion. The metric will remain Euclidean if and only if the so-called *Riemann-Christoffel* tensor (1.217) vanishes,

$$R_{IJKL} = 0 \ , \tag{1.105}$$

which in this connection is to be formed on the basis of the metric g_{KL}, i.e. based on the components of the right *Cauchy-Green* tensor $\mathbf{C}$ or the *Green* strain tensor $\mathbf{E}$.[15] Section 1. 10 contains additional pertinent information.

The equations $R_{IJKL} = 0$ are known as *conditions of compatibility*. The validity of these conditions is guaranteed if the current configuration can be represented in a Euclidean manner, i.e. by means of position or displacement vectors. Differential geometry shows that the conditions of compatibility are also sufficient for the Euclidean structure of space: if a field of metric coefficients $g_{KL}(X^J, t)$ is given, then the conditions of compatibility $R_{IJKL} = 0$ ensure that Cartesian coordinates exist, from which the metric

[14] The 4 tensors (1.103) are exactly 4 alternatives that lend themselves to a natural definition of strain measures. It is also possible to define any number of additional strain measures by means of isotropic tensor functions. The logarithmic strain tensor or *Hencky* tensor $\mathbf{E}^{\mathrm{H}} = (1/2)\ln(\mathbf{1} + 2\mathbf{E})$ is particularly useful. See XIAO et al. [1997] and the literature cited there.

[15] See TRUESDELL & TOUPIN [1960], p. 272.

coefficients can be derived according to equations of the form (1.12). However, that means that in these cases material points are marked by means of position vectors. In the event that the conditions of compatibility are not fulfilled, a configuration cannot be identified with a Euclidean space: in this case the metric coefficients g_{KL} define a more general geometric structure.[16]

1. 6 Velocity Gradient

The definition (1.36) of the velocity of a material point $\mathscr{P} \in \mathscr{B}$,

$$\boldsymbol{v} = \dot{\boldsymbol{x}}(t) = \frac{\mathrm{d}}{\mathrm{d}t}\, \chi_t(\mathscr{P}) \,,$$

leads to the material representation (1.37) of the velocity field,

$$\boldsymbol{v} = \hat{\boldsymbol{v}}(\boldsymbol{X}, t) = \frac{\mathrm{d}}{\mathrm{d}t}\, \chi_t(\mathscr{P}) \Big|_{\mathscr{P} = \mathrm{R}^{-1}(\boldsymbol{X})}$$

and its spatial representation (1.38),

$$\boldsymbol{v} = \bar{\boldsymbol{v}}(\boldsymbol{x}, t) = \frac{\mathrm{d}}{\mathrm{d}t}\, \chi_t(\mathscr{P}) \Big|_{\mathscr{P} = \chi_t^{-1}(\boldsymbol{x})} \,.$$

The relation between these representations is the identity

$$\hat{\boldsymbol{v}}(\boldsymbol{X}, t) = \bar{\boldsymbol{v}}(\chi_{\mathrm{R}}(\boldsymbol{X}, t), t) \,. \tag{1.106}$$

Differentiation of this equation with respect to $\boldsymbol{X}$ produces

$$\mathrm{Grad}\hat{\boldsymbol{v}}(\boldsymbol{X}, t) = \mathrm{grad}\bar{\boldsymbol{v}}(\boldsymbol{x}, t)\, \mathrm{Grad}\chi_{\mathrm{R}}(\boldsymbol{X}, t) \,.$$

With $\mathrm{Grad}\chi_{\mathrm{R}} = \mathbf{F}$ and $\mathrm{Grad}\hat{\boldsymbol{v}} = \mathrm{Grad}\left(\dfrac{\partial\chi_{\mathrm{R}}}{\partial t}\right) = \dfrac{\partial}{\partial t}\mathrm{Grad}\chi_{\mathrm{R}} = \dfrac{\partial}{\partial t}\mathbf{F}(\boldsymbol{X}, t)$ we arrive at

$$\mathrm{Grad}\, \hat{\boldsymbol{v}}(\boldsymbol{X}, t) = \dot{\mathbf{F}}(\boldsymbol{X}, t) \,.$$

---

[16] See Sect. 1. 10.

Definition 1. 17
The tensor

$$\text{Grad } \hat{\boldsymbol{v}}(\boldsymbol{X}, t) = \dot{\mathbf{F}}(\boldsymbol{X}, t) \tag{1.107}$$

is called the ***material velocity gradient*** and the tensor

$$\mathbf{L}(\boldsymbol{x}, t) = \text{grad}\bar{\boldsymbol{v}}(\boldsymbol{x}, t) \tag{1.108}$$

the ***spatial velocity gradient.*** ◊

Both the material and the spatial velocity gradient are related according to $\dot{\mathbf{F}} = \mathbf{LF}$ or

$$\mathbf{L} = \dot{\mathbf{F}}\mathbf{F}^{-1} . \tag{1.109}$$

The spatial velocity gradient, in a similar way to the deformation gradient, is of utmost importance for the kinematics of deformable bodies: whereas the deformation gradient $\mathbf{F}$ represents changes of material line, surface and volume elements, the velocity gradient $\mathbf{L}$ represents the rate at which these changes take place. This is expressed by

Theorem 1. 5
The following relations hold for the time derivatives of material line elements $\mathrm{d}\boldsymbol{x}(t)$, surface elements $\mathrm{d}\boldsymbol{a}(t)$ and volume elements $\mathrm{d}v(t)$:

$$1)\ \left[\mathrm{d}\boldsymbol{x}(t)\right]^{\cdot} = \mathbf{L}\mathrm{d}\boldsymbol{x} \tag{1.110}$$

$$2)\ \left[\mathrm{d}\boldsymbol{a}(t)\right]^{\cdot} = \left[(\mathrm{div}\boldsymbol{v})\mathbf{1} - \mathbf{L}^{\mathrm{T}}\right]\mathrm{d}\boldsymbol{a} \tag{1.111}$$

$$3)\ \left[\mathrm{d}v(t)\right]^{\cdot} = (\mathrm{div}\boldsymbol{v})\ \mathrm{d}v \tag{1.112}$$

□

Proof
by means of differentiating the statements in Theorem 1. 1 with respect to timc.

Statement 1): $\mathrm{d}\boldsymbol{x} = \mathbf{F}\mathrm{d}\boldsymbol{X}$ leads to $(\mathrm{d}\boldsymbol{x})^{\cdot} = \dot{\mathbf{F}}\mathrm{d}\boldsymbol{X} = \dot{\mathbf{F}}\mathbf{F}^{-1}\mathrm{d}\boldsymbol{x} = \mathbf{L}\mathrm{d}\boldsymbol{x}$.

Statement 3): The time derivative of the volume element $\mathrm{d}v = (\det\mathbf{F})\mathrm{d}V$ is the time derivative of the determinant of the deformation gradient; this is calculated by using the chain rule and applying the relation[17]

$\frac{\mathrm{d}}{\mathrm{d}\mathbf{F}}\det\mathbf{F} = (\det\mathbf{F})\mathbf{F}^{\mathrm{T}-1}$, valid for all invertible tensors. This gives us

$$\frac{\mathrm{d}}{\mathrm{d}t}\left[\det\mathbf{F}(\boldsymbol{X}, t)\right] = \left\{\frac{\mathrm{d}}{\mathrm{d}\mathbf{F}}\det\mathbf{F}\right\}\cdot\dot{\mathbf{F}}(\boldsymbol{X}, t) = (\det\mathbf{F})\left(\mathbf{F}^{\mathrm{T}-1}\right)\cdot\dot{\mathbf{F}} = (\det\mathbf{F})\,\mathrm{tr}\left(\dot{\mathbf{F}}\mathbf{F}^{-1}\right),$$

i.e., with $\mathrm{tr}\left(\dot{\mathbf{F}}\mathbf{F}^{-1}\right) = \mathrm{tr}\mathbf{L} = \mathrm{tr}(\mathrm{grad}\boldsymbol{v}) = \mathrm{div}\boldsymbol{v}$,

$(\det\mathbf{F})^{\cdot} = (\det\mathbf{F})\mathrm{div}\boldsymbol{v}$. It follows that $(\mathrm{d}v)^{\cdot} = (\mathrm{div}\boldsymbol{v})(\det\mathbf{F})\mathrm{d}V = (\mathrm{div}\boldsymbol{v})\mathrm{d}v$.

Statement 2): Differentiating $\mathrm{d}\boldsymbol{a} = (\det\mathbf{F})\mathbf{F}^{\mathrm{T}-1}\mathrm{d}\boldsymbol{A}$ according to the product rule, we use the previous result as well as the relation

$$\left(\mathbf{F}^{\mathrm{T}-1}\right)^{\cdot} = -\,\mathbf{F}^{\mathrm{T}-1}\dot{\mathbf{F}}\mathbf{F}^{\mathrm{T}-1},$$

which results from the differentiation of $\mathbf{F}^{\mathrm{T}-1}\mathbf{F}^{\mathrm{T}} = \mathbf{1}$. We calculate

$$\begin{aligned}(\mathrm{d}\boldsymbol{a})^{\cdot} &= \left[(\det\mathbf{F})^{\cdot}\,\mathbf{F}^{\mathrm{T}-1} + (\det\mathbf{F})\left(\mathbf{F}^{\mathrm{T}-1}\right)^{\cdot}\right]\mathrm{d}\boldsymbol{A}\\ &= \left[(\det\mathbf{F})(\mathrm{div}\boldsymbol{v})\mathbf{F}^{\mathrm{T}-1} - (\det\mathbf{F})\mathbf{F}^{\mathrm{T}-1}\dot{\mathbf{F}}^{\mathrm{T}}\mathbf{F}^{\mathrm{T}-1}\right]\mathrm{d}\boldsymbol{A}\\ &= (\mathrm{div}\boldsymbol{v})\mathrm{d}\boldsymbol{a} - \left(\dot{\mathbf{F}}\mathbf{F}^{-1}\right)^{\mathrm{T}}\mathrm{d}\boldsymbol{a} = (\mathrm{div}\boldsymbol{v})\mathrm{d}\boldsymbol{a} - \mathbf{L}^{\mathrm{T}}\mathrm{d}\boldsymbol{a} = \left[(\mathrm{div}\boldsymbol{v})\mathbf{1} - \mathbf{L}^{\mathrm{T}}\right]\mathrm{d}\boldsymbol{a}\,. \quad \blacksquare\end{aligned}$$

The motion of a material body consists locally of a stretch and a rotation. This fact has already been established in Theorem 1. 2 concerning the polar decomposition of the deformation gradient. A similar statement is valid for the velocity field: the deformation rate is the sum of a stretching velocity and a rotation velocity.

Theorem 1. 6

For the spatial velocity gradient $\mathbf{L} = \mathrm{grad}\bar{\boldsymbol{v}}(\boldsymbol{x}, t)$ (as well as for every second order tensor) an additive decomposition into a symmetric and an antisymmetric part is valid:

$$\mathbf{L} = \mathbf{D} + \mathbf{W}\,, \tag{1.113}$$

[17] See TRUESDELL & NOLL [1965], eq. $(9.11)_3$.

$$\mathbf{D} = \frac{1}{2}(\mathbf{L} + \mathbf{L}^{\mathrm{T}}) \, , \mathbf{D} = \mathbf{D}^{\mathrm{T}} \, , \tag{1.114}$$

$$\mathbf{W} = \frac{1}{2}(\mathbf{L} - \mathbf{L}^{\mathrm{T}}) \, , \mathbf{W} = - \, \mathbf{W}^{\mathrm{T}} \, . \tag{1.115}$$

In these relations $\mathbf{D}$ and $\mathbf{W}$ are unique.

1) The symmetric part $\mathbf{D}$ is called the *stretching* or *strain rate tensor*. For two material line elements $\mathrm{d}\boldsymbol{x}$, $\mathrm{d}\boldsymbol{y} \in \mathbb{T}_{\boldsymbol{x}}$, $\mathbf{D}$ represents the rate of change of their scalar product:

$$\frac{1}{2}(\mathrm{d}\boldsymbol{x}\cdot\mathrm{d}\boldsymbol{y})^{\cdot} = \mathrm{d}\boldsymbol{x}\cdot\mathbf{D}\mathrm{d}\boldsymbol{y} \, . \tag{1.116}$$

2) The skew-symmetric part $\mathbf{W}$ is called the *spin* or *vorticity tensor*. The rate of change in direction $\boldsymbol{e}$ of a material line element $\mathrm{d}\boldsymbol{x} = |\mathrm{d}\boldsymbol{x}|\boldsymbol{e}$ is given by

$$\dot{\boldsymbol{e}} = \mathbf{W}\boldsymbol{e} + \big[\mathbf{D} - (\boldsymbol{e}\cdot\mathbf{D}\boldsymbol{e})\mathbf{1}\big]\boldsymbol{e} \, . \tag{1.117}$$

If $\boldsymbol{e}$ is an eigenvector of $\mathbf{D}$, then we have

$$\dot{\boldsymbol{e}} = \mathbf{W}\boldsymbol{e} \, . \tag{1.118}$$

□

Proof

The existence and uniqueness of the decomposition are self-evident. Statement 1) evolves from the property $(\mathrm{d}\boldsymbol{x})^{\cdot} = \mathbf{L}\mathrm{d}\boldsymbol{x}$: we calculate

$$\mathrm{d}\boldsymbol{x}\cdot\mathbf{D}\mathrm{d}\boldsymbol{y} = \frac{1}{2}\mathrm{d}\boldsymbol{x}\cdot\big[(\mathbf{L} + \mathbf{L}^{\mathrm{T}})\mathrm{d}\boldsymbol{y}\big] = \frac{1}{2}\big[\mathrm{d}\boldsymbol{x}\cdot(\mathrm{d}\boldsymbol{y})^{\cdot} + (\mathrm{d}\boldsymbol{x})^{\cdot}\cdot\mathrm{d}\boldsymbol{y}\big] \, .$$

In particular, for the change of the length $|\mathrm{d}\boldsymbol{x}|$ the following relation results from $\mathrm{d}\boldsymbol{x} = \mathrm{d}\boldsymbol{y}$:

$$\frac{1}{2}(\mathrm{d}\boldsymbol{x}\cdot\mathrm{d}\boldsymbol{x})^{\cdot} = |\mathrm{d}\boldsymbol{x}||\mathrm{d}\boldsymbol{x}|^{\cdot} = \mathrm{d}\boldsymbol{x}\cdot\mathbf{D}\mathrm{d}\boldsymbol{x} \, .$$

This leads to statement 2): differentiation of $\boldsymbol{e} = \dfrac{\mathrm{d}\boldsymbol{x}}{|\mathrm{d}\boldsymbol{x}|}$ produces

$$\dot{\boldsymbol{e}} = \frac{(\mathrm{d}\boldsymbol{x})^{\cdot}}{|\mathrm{d}\boldsymbol{x}|} - \frac{|\mathrm{d}\boldsymbol{x}|^{\cdot}}{|\mathrm{d}\boldsymbol{x}|^2}\mathrm{d}\boldsymbol{x} = \mathbf{L}\frac{\mathrm{d}\boldsymbol{x}}{|\mathrm{d}\boldsymbol{x}|} - \frac{\mathrm{d}\boldsymbol{x}\cdot\mathbf{D}\mathrm{d}\boldsymbol{x}}{|\mathrm{d}\boldsymbol{x}|^3}\mathrm{d}\boldsymbol{x} = \mathbf{L}\boldsymbol{e} - (\boldsymbol{e}\cdot\mathbf{D}\boldsymbol{e})\boldsymbol{e}$$

or, substituting $\mathbf{L} = \mathbf{D} + \mathbf{W}$, $\dot{e} = \mathbf{W}e + \mathbf{D}e - (e \cdot \mathbf{D}e)e$. If e is an eigenvector of $\mathbf{D}$, we have $\mathbf{D}e = \lambda e$, and the result is

$$\mathbf{D}e - (e \cdot \mathbf{D}e)e = \lambda e - \lambda e = \mathit{0}, \text{ i.e. } \dot{e} = \mathbf{W}e\,. \qquad \blacksquare$$

The spin tensor $\mathbf{W}$ represents the rate of change of those material line elements which currently coincide with the eigenvectors, the principal directions of $\mathbf{D}$. According to the theorem of polar decomposition the local deformation is multiplicatively split up into stretch and rotation. The corresponding property of the velocity field is an additive decomposition of the local deformation rate into stretching and spin.

1.7 Strain Rate Tensors

When dealing with a *rigid body motion* (1.70),

$x = \mathbf{Q}(t)(X - X_0) + x_0(t)$, the material velocity gradient is expressed by

$$\mathrm{Grad}\hat{v}(X, t) = \dot{\mathbf{F}}(X, t) = \dot{\mathbf{Q}}(t)\,,$$

and the spatial velocity gradient (1.109) by

$$\mathbf{L} = \dot{\mathbf{F}}\mathbf{F}^{-1} = \dot{\mathbf{Q}}(t)\mathbf{Q}^{\mathrm{T}}(t)\,.$$

Based on $\dot{\mathbf{Q}}\mathbf{Q}^{\mathrm{T}} = -\left(\dot{\mathbf{Q}}\mathbf{Q}^{\mathrm{T}}\right)^{\mathrm{T}}$ every rigid body motion is characterised by $\mathbf{W} = \dot{\mathbf{Q}}\mathbf{Q}^{\mathrm{T}}$ und $\mathbf{D} \equiv \mathbf{0}$. Conversely, if $\mathbf{D}$ vanishes identically, then a rigid body motion takes place.[18] This makes the symmetric part of the spatial velocity gradient a spatial strain rate tensor: the rate of change of length and angle is measured by means of $\mathbf{D}$. This is shown in the following

Theorem 1.7
1) If $\mathrm{d}x = |\mathrm{d}x|e$ is a material line element in the current configuration, its *stretching* (change of length with respect to the current length) is calculated according to

$$\dot{\varepsilon}(t) = \frac{|\mathrm{d}x|^{\dot{}}}{|\mathrm{d}x|} = e \cdot \mathbf{D}e\,. \tag{1.119}$$

[18] See TRUESDELL & TOUPIN [1960], Sect. 84 for historical references.

2) For two material line elements $\mathrm{d}\boldsymbol{x}_1 = |\mathrm{d}\boldsymbol{x}_1|\boldsymbol{e}_1$ and $\mathrm{d}\boldsymbol{x}_2 = |\mathrm{d}\boldsymbol{x}_2|\boldsymbol{e}_2$, which are orthogonal in the current configuration ($\boldsymbol{e}_1 \cdot \boldsymbol{e}_2 = 0$) and which form an angle of $\frac{\pi}{2} + \gamma_{12}(t)$ at every other instant, we have the *shearing*

$$\dot{\gamma}_{12}(t) = 2\, \boldsymbol{e}_1 \cdot \mathbf{D}\boldsymbol{e}_2 \,. \tag{1.120}$$

□

Proof

Statement 1) has already been proved in the previous theorem. In order to verify statement 2), we calculate as follows:

$$2\, \mathrm{d}\boldsymbol{x}_1 \cdot \mathbf{D}\mathrm{d}\boldsymbol{x}_2 = (\mathrm{d}\boldsymbol{x}_1 \cdot \mathrm{d}\boldsymbol{x}_2)^{\cdot} = (|\mathrm{d}\boldsymbol{x}_1||\mathrm{d}\boldsymbol{x}_2|\sin\gamma_{12})^{\cdot}$$

$$= |\mathrm{d}\boldsymbol{x}_1||\mathrm{d}\boldsymbol{x}_2|(\cos\gamma_{12})\dot{\gamma}_{12} + (|\mathrm{d}\boldsymbol{x}_1||\mathrm{d}\boldsymbol{x}_2|)^{\cdot}\, \sin\gamma_{12} \,.$$

For $\gamma_{12} = 0$ this implies $\dot{\gamma}_{12} = 2\, \boldsymbol{e}_1 \cdot \mathbf{D}\boldsymbol{e}_2$. ■

Again the theorem suggests that $\mathbf{D} = \mathbf{0}$ is necessary and sufficient for rigid body motions. The strain rate tensor $\mathbf{D} = \mathbf{D}(\boldsymbol{x}, t)$, preferably written as a tensor field in the spatial representation, operates on the tangent space $\mathbb{T}_{\boldsymbol{x}}$ of the current configuration.

It is possible to define another strain rate tensor out of the strain rate tensor $\mathbf{D}$, with identical properties but operating on the tangent space $\mathbb{T}_{\boldsymbol{X}}$ of the reference configuration. Comparing

$$\tfrac{1}{2}(\mathrm{d}\boldsymbol{x} \cdot \mathrm{d}\boldsymbol{y})^{\cdot} = \mathrm{d}\boldsymbol{x} \cdot (\mathbf{D}\, \mathrm{d}\boldsymbol{y}) = \mathrm{d}\boldsymbol{X} \cdot [(\mathbf{F}^{\mathrm{T}}\mathbf{D}\mathbf{F})\mathrm{d}\boldsymbol{Y}] \text{ and}$$

$$\tfrac{1}{2}(\mathrm{d}\boldsymbol{x} \cdot \mathrm{d}\boldsymbol{y})^{\cdot} = \mathrm{d}\boldsymbol{X} \cdot [(\mathbf{F}^{\mathrm{T}}\mathbf{D}\mathbf{F})\mathrm{d}\boldsymbol{Y}] \tag{1.121}$$

with the property (1.73) of the *Green* strain tensor, or with

$$\mathrm{d}\boldsymbol{X} \cdot (\dot{\mathbf{E}}\mathrm{d}\boldsymbol{Y}) = \tfrac{1}{2}(\mathrm{d}\boldsymbol{x} \cdot \mathrm{d}\boldsymbol{y})^{\cdot} \,, \tag{1.122}$$

produces the material strain rate tensor or the *Green* strain rate tensor

$$\dot{\mathbf{E}}(\boldsymbol{X}, t) = \tfrac{1}{2}(\dot{\mathbf{F}}^{\mathrm{T}}\mathbf{F} + \mathbf{F}^{\mathrm{T}}\dot{\mathbf{F}}) \,. \tag{1.123}$$

The material strain rate tensor $\dot{\mathbf{E}}$ is connected with the spatial strain rate $\mathbf{D}$ by means of the transformation

$$\dot{\mathbf{E}} = \mathbf{F}^{\mathrm{T}}\mathbf{D}\mathbf{F} \; ; \tag{1.124}$$

this follows from (1.121) and (1.122).
Incidentally, it is expedient at this stage to point out that the same relation exists between the strain rates $\mathbf{D}$ and $\dot{\mathbf{E}}$ as between the *Almansi* tensor $\mathbf{A}$ and the *Green* tensor $\mathbf{E}$. In analogy to

$$\mathbf{A} = \mathbf{F}^{\mathrm{T}\,-1}\mathbf{E}\mathbf{F}^{-1} \tag{1.125}$$

we also have

$$\mathbf{D} = \mathbf{F}^{\mathrm{T}\,-1}\,\dot{\mathbf{E}}\,\mathbf{F}^{-1}\,. \tag{1.126}$$

We deduce from these equations that the spatial strain rate tensor $\mathbf{D}$ can be expressed as a time rate, which is not, however, a material time derivative: by means of substitution we arrive at the identity

$$\mathbf{D} = \mathbf{F}^{\mathrm{T}\,-1}(\mathbf{F}^{\mathrm{T}}\mathbf{A}\mathbf{F})^{\cdot}\,\mathbf{F}^{-1} = \dot{\mathbf{A}} + \mathbf{F}^{\mathrm{T}\,-1}\,\dot{\mathbf{F}}^{\mathrm{T}}\mathbf{A} + \mathbf{A}\dot{\mathbf{F}}\mathbf{F}^{-1}\,,$$

or, using $\dot{\mathbf{F}}\mathbf{F}^{-1} = \mathbf{L} = \operatorname{grad}\bar{v}(\boldsymbol{x}, t)$, at

$$\mathbf{D} = \dot{\mathbf{A}} + \mathbf{L}^{\mathrm{T}}\mathbf{A} + \mathbf{A}\mathbf{L}\,. \tag{1.127}$$

The time rate on the right-hand side is called the *Oldroyd* rate. Generally speaking, we adhere to the following

Definition 1. 18
If $\mathbf{T} = \overline{\mathbf{T}}(\boldsymbol{x}, t)$ is a tensor field in the spatial representation, the time rate

$$\overset{\vartriangle}{\mathbf{T}} = \dot{\mathbf{T}} + \mathbf{L}^{\mathrm{T}}\mathbf{T} + \mathbf{T}\mathbf{L} \tag{1.128}$$

is called the *covariant (lower-convected) Oldroyd* rate; the time rate

$$\overset{\triangledown}{\mathbf{T}} = \dot{\mathbf{T}} - \mathbf{L}\mathbf{T} - \mathbf{T}\mathbf{L}^{\mathrm{T}} \tag{1.129}$$

is termed the *contravariant (upper-convected) Oldroyd* rate. ◇

The sense of these designations will become clear later on in connection with convective coordinates. A tentative use of the definition is described in

Theorem 1. 8

1) The spatial strain rate tensor $\mathbf{D}$ equals the *covariant Oldroyd* rate of the *Almansi* strain tensor

$\mathbf{A} = \frac{1}{2}\left(\mathbf{1} - \mathbf{F}^{\mathrm{T}\,-1}\mathbf{F}^{-1}\right)$: together with $\mathbf{F}^{\mathrm{T}\,-1}\mathbf{E}\mathbf{F}^{-1} = \mathbf{A}$ we have

$$\mathbf{F}^{\mathrm{T}\,-1}\,\dot{\mathbf{E}}\,\mathbf{F}^{-1} = \overset{\triangle}{\mathbf{A}} = \dot{\mathbf{A}} + \mathbf{L}^{\mathrm{T}}\mathbf{A} + \mathbf{A}\mathbf{L} = \mathbf{D}\,. \tag{1.130}$$

2) The negative strain rate tensor $-\mathbf{D}$ is the *contravariant Oldroyd* rate of the *Finger* tensor

$\mathbf{a} = \frac{1}{2}\left(\mathbf{1} - \mathbf{F}\mathbf{F}^{\mathrm{T}}\right)$: with $\mathbf{F}\mathbf{e}\mathbf{F}^{\mathrm{T}} = \mathbf{a}$ we accomplish

$$\mathbf{F}\dot{\mathbf{e}}\mathbf{F}^{\mathrm{T}} = \overset{\triangledown}{\mathbf{a}} = \dot{\mathbf{a}} - \mathbf{L}\mathbf{a} - \mathbf{a}\mathbf{L}^{\mathrm{T}} = -\,\mathbf{D}\,. \tag{1.131}$$

□

Proof

Statement 1) has already been proved. Statement 2) is the consequence of

$$\begin{aligned}\mathbf{F}\dot{\mathbf{e}}\mathbf{F}^{\mathrm{T}} &= \frac{1}{2}\mathbf{F}(\mathbf{F}^{-1}\mathbf{F}^{\mathrm{T}\,-1} - \mathbf{1})^{\cdot}\,\mathbf{F}^{\mathrm{T}} = \frac{1}{2}\mathbf{F}[(\mathbf{F}^{-1})^{\cdot}\,\mathbf{F}^{\mathrm{T}\,-1} + \mathbf{F}^{-1}(\mathbf{F}^{\mathrm{T}\,-1})^{\cdot}\,]\mathbf{F}^{\mathrm{T}}\\ &= -\,\frac{1}{2}\mathbf{F}[\mathbf{F}^{-1}\dot{\mathbf{F}}\mathbf{F}^{-1}\mathbf{F}^{\mathrm{T}\,-1} + \mathbf{F}^{-1}\mathbf{F}^{\mathrm{T}\,-1}\dot{\mathbf{F}}^{\mathrm{T}}\,\mathbf{F}^{\mathrm{T}\,-1}]\mathbf{F}^{\mathrm{T}}\\ &= -\,\frac{1}{2}[\dot{\mathbf{F}}\mathbf{F}^{-1} + \mathbf{F}^{\mathrm{T}\,-1}\dot{\mathbf{F}}^{\mathrm{T}}] = -\,\frac{1}{2}[\mathbf{L} + \mathbf{L}^{\mathrm{T}}] = -\,\mathbf{D}\,,\end{aligned}$$

and, on the other hand,

$$\begin{aligned}\mathbf{F}\dot{\mathbf{e}}\mathbf{F}^{\mathrm{T}} &= \mathbf{F}(\mathbf{F}^{-1}\,\mathbf{a}\,\mathbf{F}^{\mathrm{T}\,-1})^{\cdot}\,\mathbf{F}^{\mathrm{T}} = \dot{\mathbf{a}} + \mathbf{F}(\mathbf{F}^{-1})^{\cdot}\,\mathbf{a} + \mathbf{a}(\mathbf{F}^{\mathrm{T}\,-1})^{\cdot}\,\mathbf{F}^{\mathrm{T}}\\ &= \dot{\mathbf{a}} - \dot{\mathbf{F}}\mathbf{F}^{-1}\,\mathbf{a} - \mathbf{a}\,\mathbf{F}^{\mathrm{T}\,-1}\dot{\mathbf{F}}^{\mathrm{T}} = \dot{\mathbf{a}} - \mathbf{L}\mathbf{a} - \mathbf{a}\mathbf{L}^{\mathrm{T}} = \overset{\triangledown}{\mathbf{a}}\,.\end{aligned}$$

■

We may summarise by recapitulating that a total of four strain tensors were defined, two of which operate on the reference configuration and two on the current configuration. The attribution of a strain rate tensor to each of these strain tensors takes place naturally. The following properties result from the definitions and the geometric interpretations:

Green/ Almansi tensor

$$\mathbf{F}^{\mathrm{T}\,-1}\mathbf{E}\mathbf{F}^{-1} = \mathbf{A}$$
$$\mathbf{F}^{\mathrm{T}\,-1}\,\dot{\mathbf{E}}\,\mathbf{F}^{-1} = \overset{\vartriangle}{\mathbf{A}} = \mathbf{D} \tag{1.132}$$

$$\frac{1}{2}\left(\mathrm{d}\boldsymbol{x}\cdot\mathrm{d}\boldsymbol{y} - \mathrm{d}\boldsymbol{X}\cdot\mathrm{d}\boldsymbol{Y}\right) = \mathrm{d}\boldsymbol{X}\cdot\mathbf{E}\mathrm{d}\boldsymbol{Y} = \mathrm{d}\boldsymbol{x}\cdot\mathbf{A}\mathrm{d}\boldsymbol{y}$$

$$\frac{1}{2}\left(\mathrm{d}\boldsymbol{x}\cdot\mathrm{d}\boldsymbol{y}\right)^{\cdot} = \mathrm{d}\boldsymbol{X}\cdot\left(\dot{\mathbf{E}}\mathrm{d}\boldsymbol{Y}\right) = \mathrm{d}\boldsymbol{x}\cdot\overset{\vartriangle}{\mathbf{A}}\mathrm{d}\boldsymbol{y}$$

Piola/ Finger tensor

$$\mathbf{F}\mathbf{e}\mathbf{F}^{\mathrm{T}} = \mathbf{a}$$
$$\mathbf{F}\dot{\mathbf{e}}\mathbf{F}^{\mathrm{T}} = \overset{\triangledown}{\mathbf{a}} = -\,\mathbf{D} \tag{1.133}$$

$$\frac{1}{2}\left(\boldsymbol{n}\cdot\boldsymbol{m} - \boldsymbol{N}\cdot\boldsymbol{M}\right) = \boldsymbol{N}\cdot\mathbf{e}\boldsymbol{M} = \boldsymbol{n}\cdot\mathbf{a}\boldsymbol{m}$$

$$\frac{1}{2}\left(\boldsymbol{n}\cdot\boldsymbol{m}\right)^{\cdot} = \boldsymbol{N}\cdot\dot{\mathbf{e}}\boldsymbol{M} = \boldsymbol{n}\cdot\overset{\triangledown}{\mathbf{a}}\boldsymbol{m}$$

1. 8 Strain Rates in Convective Coordinates

The choice of convective coordinates not only affords a deeper understanding of the strain tensors but also of their strain rates. In connection with convective coordinates the temporal changes of the local base vectors of the current configuration, $\boldsymbol{g}_K = \mathbf{F}\boldsymbol{G}_K$ and $\boldsymbol{g}^K = \mathbf{F}^{\mathrm{T}\,-1}\boldsymbol{G}^K$ can be calculated:

$$\dot{\boldsymbol{g}}_K = \dot{\mathbf{F}}\boldsymbol{G}_K = \dot{\mathbf{F}}\mathbf{F}^{-1}\boldsymbol{g}_K\,,$$

$$\dot{\boldsymbol{g}}^K = \left(\mathbf{F}^{\mathrm{T}\,-1}\right)^{\cdot}\boldsymbol{G}^K = -\,\mathbf{F}^{\mathrm{T}\,-1}\,\dot{\mathbf{F}}^{\mathrm{T}}\,\mathbf{F}^{\mathrm{T}\,-1}\boldsymbol{G}^K = -\,\left(\dot{\mathbf{F}}\mathbf{F}^{-1}\right)^{\mathrm{T}}\boldsymbol{g}^K\,.$$

This leads to

$$\dot{\boldsymbol{g}}_K = \mathbf{L}\boldsymbol{g}_K \tag{1.134}$$

and

$$\dot{g}^K = -\mathbf{L}^{\mathrm{T}} g^K . \tag{1.135}$$

The velocity gradient $\mathbf{L}$ admits the representation

$$\mathbf{L} = \dot{\mathbf{F}}\mathbf{F}^{-1} = (\dot{g}_K \otimes G^K)(G_L \otimes g^L) = \dot{g}_K \otimes g^K . \tag{1.136}$$

Since strain in convective coordinates amounts to a change of metric, the rate of change for the metric of the current configuration corresponds to a strain rate. This can be confirmed straight away by means of differentiating the metric coefficients: beginning with

$\dot{g}_{KL} = (g_K \cdot g_L)^{\cdot} = \dot{g}_K \cdot g_L + g_K \cdot \dot{g}_L = \mathbf{L}g_K \cdot g_L + g_K \cdot \mathbf{L}g_L$ we proceed to

$$\dot{g}_{KL} = g_K \cdot [(\mathbf{L} + \mathbf{L}^{\mathrm{T}})g_L] = 2\, g_K \cdot \mathbf{D}g_L , \tag{1.137}$$

in other words, the strain rate tensor $\mathbf{D}$, related to convective coordinates, possesses the component representation

$$\mathbf{D} = \tfrac{1}{2}\dot{g}_{KL}\, g^K \otimes g^L . \tag{1.138}$$

Correspondingly, we calculate

$$\dot{g}^{KL} = (g^K \cdot g^L)^{\cdot} = \dot{g}^K \cdot g^L + g^K \cdot \dot{g}^L = -\, g^K \cdot [(\mathbf{L} + \mathbf{L}^{\mathrm{T}})g^L] = -\, 2\, g^K \cdot \mathbf{D}g^L .$$

As a result, $-\mathbf{D}$ gains the component representation

$$-\mathbf{D} = \tfrac{1}{2}\dot{g}^{KL}\, g_K \otimes g_L . \tag{1.139}$$

This representation could likewise have been deduced from the property $g_{KJ}g^{JL} = \delta^L_K$, whose differentiation yields the relation

$$\dot{g}_{KJ}g^{JL} = -\, g_{KJ}\dot{g}^{JL} .$$

On the other hand, the last section has shown that the strain rate tensors $\mathbf{D}$ and $-\mathbf{D}$ are *Oldroyd* rates of the strain tensors $\mathbf{A}$ and $\mathbf{a}$.

The following theorem leads to the general conclusion that the co- and contravariant *Oldroyd* rates of any given tensor field can be interpreted as relative derivatives of the component representations.

Theorem 1. 9

If $\mathbf{T} = \bar{\mathbf{T}}(x, t)$ is a tensor field in the spatial representation that is represented in relation to convective coordinates with co- or contravariant components, i.e.

$$\mathbf{T} = T_{KL}\, g^K \otimes g^L = T^{KL} g_K \otimes g_L \,,$$

the *Oldroyd* rates are *relative derivatives*: only the components are differentiated with respect to time:

$$\overset{\vartriangle}{\mathbf{T}} = \dot{\mathbf{T}} + \mathbf{L}^{\mathrm{T}}\mathbf{T} + \mathbf{T}\mathbf{L} = \dot{T}_{KL}\, g^K \otimes g^L \,, \tag{1.140}$$

$$\overset{\triangledown}{\mathbf{T}} = \dot{\mathbf{T}} - \mathbf{L}\mathbf{T} - \mathbf{T}\mathbf{L}^{\mathrm{T}} = \dot{T}^{KL} g_K \otimes g_L \,. \tag{1.141}$$

□

Proof

$$\begin{aligned}\dot{\mathbf{T}} &= (T_{KL}\, g^K \otimes g^L)^{\cdot} = \dot{T}_{KL}\, g^K \otimes g^L - T_{KL}\mathbf{L}^{\mathrm{T}} g^K \otimes g^L - T_{KL}\, g^K \otimes \mathbf{L}^{\mathrm{T}} g^L \\ &= \dot{T}_{KL}\, g^K \otimes g^L - \mathbf{L}^{\mathrm{T}}\mathbf{T} - \mathbf{T}\mathbf{L}\end{aligned}$$

$$\begin{aligned}\dot{\mathbf{T}} &= (T^{KL} g_K \otimes g_L)^{\cdot} = \dot{T}^{KL} g_K \otimes g_L + T^{KL}\mathbf{L} g_K \otimes g_L + T^{KL} g_K \otimes \mathbf{L} g_L \\ &= \dot{T}^{KL} g_K \otimes g_L + \mathbf{L}\mathbf{T} + \mathbf{T}\mathbf{L}^{\mathrm{T}}\end{aligned}$$

■

Summarising, the following component representations hold for the 4 strain tensors $\mathbf{E}$, $\mathbf{e}$ and $\mathbf{A}$, $\mathbf{a}$ and their associated rates $\dot{\mathbf{E}}$, $\dot{\mathbf{e}}$, $\overset{\vartriangle}{\mathbf{A}}$, $\overset{\triangledown}{\mathbf{a}}$ with respect to convective coordinates:

$$\begin{aligned}\mathbf{E} &= \tfrac{1}{2}[g_{KL} - G_{KL}]G^K \otimes G^L \,, \\ \dot{\mathbf{E}} &= \tfrac{1}{2}\dot{g}_{KL} G^K \otimes G^L \,,\end{aligned} \tag{1.142}$$

$$\begin{aligned}\mathbf{A} &= \tfrac{1}{2}[g_{KL} - G_{KL}]g^K \otimes g^L \,, \\ \overset{\vartriangle}{\mathbf{A}} &= \mathbf{D} = \tfrac{1}{2}\dot{g}_{KL}\, g^K \otimes g^L \,,\end{aligned} \tag{1.143}$$

$$\mathrm{e} = \frac{1}{2}\,[g^{KL} - G^{KL}]G_K \otimes G_L\,,$$
$$\dot{\mathrm{e}} = \frac{1}{2}\dot{g}^{KL}G_K \otimes G_L\,, \qquad (1.144)$$

$$\mathrm{a} = \frac{1}{2}\,[g^{KL} - G^{KL}]g_K \otimes g_L\,,$$
$$\overset{\triangledown}{\mathrm{a}} = -\,\mathrm{D} = \frac{1}{2}\dot{g}^{KL}g_K \otimes g_L\,. \qquad (1.145)$$

In these representations only the base vectors account for the differences between the tensors **E** and **A** or **e** and **a**; the matrices of the components are identical.

1. 9 Geometric Linearisation

The representation (1.29) of the motion of a material body in terms of a displacement field, starting from an initial configuration,

$$u(X, t) = x - X = \Phi_{t_0}(X, t) - X \qquad \left(\Phi_{t_0}(X, t_0) = X\right)\,, \qquad (1.146)$$

gives cause for a considerable simplification of the basic equations of continuum mechanics whenever the deformations are small. In fact, this can be assumed in many practical applications. Motions or deformations are regarded as being "small", if all the displacements and derivatives of the components of the displacement vector are small. This is explained more precisely in

Definition 1. 19
The motion of a material body consists of *small deformations*, if

$$\delta = \|\mathbf{H}\| \ll 1 \qquad (1.147)$$

applies throughout the motion as well as

$$|u(X, t)| \ll L_0\,, \qquad (1.148)$$

L_0 being a characteristic length of the body under observation. The tensor

$$\mathbf{H}(X, t) = \mathrm{Grad}\,u(X, t) \qquad (1.149)$$

is the gradient of the displacement vector and $\|\mathbf{H}\|$ is its norm, i.e.

$$\delta = \|\mathbf{H}\| = \sqrt{\mathbf{H}\cdot\mathbf{H}}\,. \tag{1.150}$$

◊

In Cartesian coordinates the norm $\|\mathbf{H}\|$ is the sum of the squares of all displacement derivatives:

$$\delta^2 = \|\mathbf{H}\|^2 = \sum_{j,k}\left(\frac{\partial u_j}{\partial y_k}\right)^2 . \tag{1.151}$$

The displacement gradient $\mathbf{H}$ is related to the deformation gradient $\mathbf{F}$: by differentiation with respect to $\boldsymbol{X}$,

$$\boldsymbol{x} = \boldsymbol{\Phi}_{t_0}(\boldsymbol{X}, t) = \boldsymbol{X} + \boldsymbol{u}(\boldsymbol{X}, t)$$

implies $\mathbf{F}(\boldsymbol{X}, t) = \mathrm{Grad}\big[\boldsymbol{X} + \boldsymbol{u}(\boldsymbol{X}, t)\big] = \mathbf{1} + \mathrm{Grad}\boldsymbol{u}(\boldsymbol{X}, t)$, i.e.

$$\mathbf{F} = \mathbf{1} + \mathbf{H}\,. \tag{1.152}$$

In the general case of a motion $\boldsymbol{x} = \chi_{\mathrm{R}}(\boldsymbol{X}, t)$ there is no evidence of the introduction of any displacement vector. To define the magnitude δ of a deformation it is expedient to define the tensor

$$\mathbf{H} = \mathbf{F} - \mathbf{1}\,,$$

which may be interpreted as the material gradient of a displacement vector. In this sense, small deformations are characterised by the fact that the deformation gradient differs only slightly from the identity tensor:[19]

$$\delta = \|\mathbf{F} - \mathbf{1}\| \ll 1\,. \tag{1.153}$$

It is now possible to express all kinematic quantities in continuum mechanics by means of the displacement gradient $\mathbf{H}$ and, wherever small deformations occur, linearise them with regard to $\mathbf{H}$. This process is called *geometric linearisation.* The practical implementation of geometric linearisation is simple: asymptotic relations are evolved, missing out all terms of the order $\mathrm{O}(\delta^2)$ (scalars), $\boldsymbol{O}(\delta^2)$ (vectors) and $\mathbf{O}(\delta^2)$ (tensors).

[19] The application of this criterion for small deformations does not require the reference configuration to be occupied at any time.

Obviously, the asymptotic relations

$$\begin{aligned}
&\mathbf{H} = \mathbf{F} - \mathbf{1} = \mathrm{O}(\delta)\,, \\
&\det\mathbf{F} = 1 + \mathrm{tr}\mathbf{H} + \mathrm{O}(\delta^2)\,, \\
&\mathbf{F}^{-1} = (\mathbf{1} + \mathbf{H})^{-1} = \mathbf{1} - \mathbf{H} + \mathrm{O}(\delta^2)\,, \\
&\mathbf{E} = \tfrac{1}{2}(\mathbf{F}^{\mathrm{T}}\mathbf{F} - \mathbf{1}) = \tfrac{1}{2}(\mathbf{H} + \mathbf{H}^{\mathrm{T}}) + \mathrm{O}(\delta^2)\,, \\
&\mathbf{A} = \tfrac{1}{2}(\mathbf{1} - \mathbf{F}^{\mathrm{T}-1}\mathbf{F}^{-1}) = \tfrac{1}{2}(\mathbf{H} + \mathbf{H}^{\mathrm{T}}) + \mathrm{O}(\delta^2)\,, \\
&\mathbf{e} = \tfrac{1}{2}\,(\mathbf{F}^{-1}\mathbf{F}^{\mathrm{T}-1} - \mathbf{1}) = -\tfrac{1}{2}(\mathbf{H} + \mathbf{H}^{\mathrm{T}}) + \mathrm{O}(\delta^2)\,, \\
&\mathbf{a} = \tfrac{1}{2}(\mathbf{1} - \mathbf{F}\mathbf{F}^{\mathrm{T}}) = -\tfrac{1}{2}(\mathbf{H} + \mathbf{H}^{\mathrm{T}}) + \mathrm{O}(\delta^2)
\end{aligned} \tag{1.154}$$

are valid. In other words, in the event of geometric linearisation, all strain tensors reduce to the symmetric part of the displacement gradient **H**.

The stretch tensors occurring in the polar decomposition $\mathbf{F} = \mathbf{R}\mathbf{U} = \mathbf{V}\mathbf{R}$ are also connected to the symmetric part of the displacement gradient: the expansions

$$\mathbf{U} = \sqrt{\mathbf{C}} = \sqrt{\mathbf{1} + \mathbf{H} + \mathbf{H}^{\mathrm{T}} + \mathrm{O}(\delta^2)} = \mathbf{1} + \tfrac{1}{2}(\mathbf{H} + \mathbf{H}^{\mathrm{T}}) + \mathrm{O}(\delta^2) \quad \text{and}$$

$$\mathbf{V} = \sqrt{\mathbf{B}} = \sqrt{\mathbf{1} + \mathbf{H} + \mathbf{H}^{\mathrm{T}} + \mathrm{O}(\delta^2)} = \mathbf{1} + \tfrac{1}{2}(\mathbf{H} + \mathbf{H}^{\mathrm{T}}) + \mathrm{O}(\delta^2)$$

imply

$$\mathbf{R} = \mathbf{F}\mathbf{U}^{-1} = (\mathbf{1} + \mathbf{H})\,[\mathbf{1} - \tfrac{1}{2}(\mathbf{H} + \mathbf{H}^{\mathrm{T}}) + \mathrm{O}(\delta^2)]\,,$$

$$\mathbf{R} = \mathbf{1} + \tfrac{1}{2}(\mathbf{H} - \mathbf{H}^{\mathrm{T}}) + \mathrm{O}(\delta^2)\,,$$

as well as

$$\mathbf{F} = \mathbf{R}\mathbf{U} = [\mathbf{1} + \tfrac{1}{2}(\mathbf{H} - \mathbf{H}^{\mathrm{T}}) + \mathrm{O}(\delta^2)][\mathbf{1} + \tfrac{1}{2}(\mathbf{H} + \mathbf{H}^{\mathrm{T}}) + \mathrm{O}(\delta^2)]\,.$$

The result

$$\mathbf{R}\mathbf{U} = \mathbf{1} + \tfrac{1}{2}(\mathbf{H} - \mathbf{H}^{\mathrm{T}}) + \tfrac{1}{2}(\mathbf{H} + \mathbf{H}^{\mathrm{T}}) + \mathrm{O}(\delta^2) \tag{1.155}$$

has the following geometric meaning. During linearisation the polar decomposition merges into the additive decomposition of the displacement gradient into symmetric and antisymmetric parts:

$$\mathbf{F} = \mathbf{1} + \mathbf{H} = \mathbf{1} + \tfrac{1}{2}(\mathbf{H} - \mathbf{H}^{\mathrm{T}}) + \tfrac{1}{2}(\mathbf{H} + \mathbf{H}^{\mathrm{T}}) = \mathbf{1} + \mathbf{R}_{\mathrm{L}} + \mathbf{E}_{\mathrm{L}} \,. \qquad (1.156)$$

This fact motivates

Definition 1. 20
The tensor

$$\mathbf{E}_{\mathrm{L}} = \tfrac{1}{2}(\mathbf{H} + \mathbf{H}^{\mathrm{T}}) \qquad (1.157)$$

is called the *linearised strain tensor* and the tensor

$$\mathbf{R}_{\mathrm{L}} = \tfrac{1}{2}(\mathbf{H} - \mathbf{H}^{\mathrm{T}}) \qquad (1.158)$$

the *linearised rotation tensor.* ◊

In order to linearise all derivatives and rates of the kinematic quantities, we have to assume *small deformation rates* in addition to *small deformations*, $\mathbf{H} = \mathrm{O}(\delta)$:

$$\dot{\mathbf{H}}(\boldsymbol{X}, t) = \mathrm{O}(\delta) \,. \qquad (1.159)$$

Then the spatial velocity gradient reduces to the material velocity gradient:

$$\mathbf{L}(\boldsymbol{x}, t) = \dot{\mathbf{F}}\mathbf{F}^{-1} = \dot{\mathbf{H}}[\mathbf{1} - \mathbf{H} + \mathrm{O}(\delta^2)] = \dot{\mathbf{H}}(\boldsymbol{X}, t) + \mathrm{O}(\delta^2) \,,$$

$$\mathbf{L}(\boldsymbol{x}, t) = \mathrm{Grad}\hat{\boldsymbol{v}}\,(\boldsymbol{X}, t) + \mathrm{O}(\delta^2) \,. \qquad (1.160)$$

The asymptotic relation $\dot{\mathbf{H}}\mathbf{H} = \mathrm{O}(\delta^2)$ has been employed for this result. The asymptotic representations for the strain rate and spin tensors,

$$\mathbf{D} = \tfrac{1}{2}(\dot{\mathbf{H}} + \dot{\mathbf{H}}^{\mathrm{T}}) + \mathrm{O}(\delta^2) \,, \qquad (1.161)$$

$$\mathbf{W} = \tfrac{1}{2}(\dot{\mathbf{H}} - \dot{\mathbf{H}}^{\mathrm{T}}) + \mathrm{O}(\delta^2) \,, \qquad (1.162)$$

evolve from the linearisation of the velocity gradient. The definition of the displacement vector, $\boldsymbol{x} = \boldsymbol{X} + \boldsymbol{u}(\boldsymbol{X}, t)$, implies for small deformations that the current configuration is very close to the reference configuration. However, this means that the difference between the material and the spatial representation is negligible. In the sense of geometric linearisation the approximation

$$\boldsymbol{u}(\boldsymbol{X}, t) \approx \boldsymbol{u}(\boldsymbol{x}, t) \tag{1.163}$$

is valid. This implies

$$\mathbf{H} = \mathrm{Grad}\boldsymbol{u}(\boldsymbol{X}, t) = \mathrm{grad}\boldsymbol{u}(\boldsymbol{x}, t) + \mathrm{O}(\delta^2)\ , \tag{1.164}$$

$$\mathbf{E}_{\mathrm{L}} = \frac{1}{2}\Big[\mathrm{grad}\boldsymbol{u}(\boldsymbol{x}, t) + \mathrm{grad}\boldsymbol{u}(\boldsymbol{x}, t)^{\mathrm{T}}\Big] + \mathrm{O}(\delta^2) \tag{1.165}$$

and

$$\dot{\mathbf{H}} = \mathrm{grad}\,\frac{\partial \boldsymbol{u}}{\partial t}(\boldsymbol{x}, t) + \mathrm{O}(\delta^2)\ . \tag{1.166}$$

Finally, for the acceleration vector, the geometric linearisation produces

$$\boldsymbol{a} = \bar{\boldsymbol{a}}(\boldsymbol{x}, t) = \frac{\partial}{\partial t}\bar{\boldsymbol{v}}(\boldsymbol{x}, t) + \big[\mathrm{grad}\ \bar{\boldsymbol{v}}(\boldsymbol{x}, t)\big]\bar{\boldsymbol{v}}(\boldsymbol{x}, t)\ ,$$

$$\boldsymbol{a} = \frac{\partial^2 \boldsymbol{u}}{\partial t^2}(\boldsymbol{x}, t) + \mathrm{O}(\delta^2)\ . \tag{1.167}$$

1. 10 Incompatible Configurations

Although it is a logical and widespread practice to identify the material body with Euclidean spaces, it cannot be ruled out that, in certain circumstances, a configuration can also identify a material body with a non-Euclidean space. That is precisely the situation in the context of intermediate configurations, introduced within the *concept of dual variables* (see Chap. 8).

1. 10. 1 Euclidean Space

In order to comprehend the generalisations required in connection with non-Euclidean spaces, it helps to take a look at those aspects of the Euclidean geometry that are essential for continuum mechanics.[20] The characteristic feature of a Euclidean space $\mathbb{E}^3$ is the *path independence of the parallel transport.* This makes it possible to determine the parallelism of straight lines, compare their lengths and consequently classify the ordered pairs of points into equivalence classes, each containing all straight lines of the same length and orientation. These equivalence classes are the geometric vectors which are identified with the Euclidean vector space $\mathbb{V}^3$. In addition, it is possible to represent the spatial points $\mathrm{P} \in \mathbb{E}^3$ one-to-one by means of *position vectors*, whose components are the Cartesian coordinates of these spatial points, $\boldsymbol{x} = y_k \boldsymbol{e}_k$, being invertible functions of the curvilinear coordinates (x^1, x^2, x^3):

$$y_k = \hat{y}_k(x^1, x^2, x^3) \Leftrightarrow x^k = \hat{x}^k(y_1, y_2, y_3) \tag{1.168}$$

(see (1.6)). Finally, every system of coordinates defines local base vectors using equation (1.10) or (1.11), i.e. by means of differentiating the position vector with respect to the coordinates (tangent vectors) or by differentiating the coordinates with respect to the position vector (gradient vectors):

$$\boldsymbol{g}_k = \frac{\partial \boldsymbol{x}}{\partial x^k} = \frac{\partial}{\partial x^k}\, \hat{y}_j(x^1, x^2, x^3)\boldsymbol{e}_j \,, \boldsymbol{g}^i = \operatorname{grad} x^i(\boldsymbol{x}) = \frac{\partial}{\partial y_l}\, \hat{x}^i(y_1, y_2, y_3)\boldsymbol{e}_l \,,$$

$$\boldsymbol{g}^i \cdot \boldsymbol{g}_k = \delta^i_k \,.$$

The local base vectors are functions of $\boldsymbol{x}$. Their partial derivatives with respect to the coordinates can be expressed as linear combinations of the vectors themselves: the proposition

$$\frac{\partial \boldsymbol{g}_i}{\partial x^j} = \Gamma^k_{ij} \boldsymbol{g}_k \tag{1.169}$$

implies

$$\Gamma^k_{ij} = \boldsymbol{g}^k \cdot \frac{\partial \boldsymbol{g}_i}{\partial x^j} = \left(\operatorname{grad} x^k\right) \cdot \frac{\partial^2 \boldsymbol{x}}{\partial x^i \partial x^j} \; ; \tag{1.170}$$

in view of $\boldsymbol{g}^k \cdot \boldsymbol{g}_i = \delta^k_i$, the relation

[20] See, for example, RASCHEWSKI [1959]; KLINGBEIL [1989].

$$\frac{\partial g^k}{\partial x^j} = - \Gamma^k_{ij} g^i \tag{1.171}$$

is also true.

The so-called *coefficients of affine connection* or *Christoffel symbols* of the second kind Γ^k_{ij} have the symmetry properties

$$\Gamma^k_{ij} = \Gamma^k_{ji} \tag{1.172}$$

in the Euclidean space and can be represented by means of the *metric coefficients* g_{ik} and g^{ik}. To this end *Christoffel symbols* of the first kind are used,

$$\Gamma_{lij} = g_l \cdot \frac{\partial g_i}{\partial x^j} = \frac{\partial x}{\partial x^l} \cdot \frac{\partial^2 x}{\partial x^i \partial x^j} , \tag{1.173}$$

which are also seen to be symmetric with respect to their indices ij, i.e. $\Gamma_{lij} = \Gamma_{lji}$. They satisfy the relations

$$\Gamma_{lij} = g_{kl} \Gamma^k_{ij} \iff \Gamma^k_{ij} = g^{kl} \Gamma_{lij} . \tag{1.174}$$

The following theorem shows that the *Christoffel symbols* are determined by the metric coefficients.

Theorem 1. 10

In the Euclidean space the *Christoffel symbols* can be calculated from the metric coefficients:[21]

$$\Gamma_{lij} = \frac{1}{2} \left(\frac{\partial g_{li}}{\partial x^j} + \frac{\partial g_{jl}}{\partial x^i} - \frac{\partial g_{ij}}{\partial x^l} \right) , \tag{1.175}$$

$$\Gamma^k_{ij} = \frac{1}{2} g^{kl} \left(\frac{\partial g_{li}}{\partial x^j} + \frac{\partial g_{jl}}{\partial x^i} - \frac{\partial g_{ij}}{\partial x^l} \right) . \tag{1.176}$$

□

Proof

To calculate Γ_{lij} one first obtains

$$\frac{\partial g_{li}}{\partial x^j} = \frac{\partial}{\partial x^j} (g_l \cdot g_i) = \Gamma_{lij} + \Gamma_{ilj} ,$$

---

[21] KLINGBEIL [1989], pp. 83.

differentiating the metric coefficients and applying (1.173). Two further equations are created by cyclic substitution of the indices *lij*, making three equations in all, namely:

$$\Gamma_{lij} + \Gamma_{ijl} = \frac{\partial g_{li}}{\partial x^j} \,, \quad \Gamma_{ijl} + \Gamma_{jli} = \frac{\partial g_{ij}}{\partial x^l} \,, \quad \Gamma_{jli} + \Gamma_{lij} = \frac{\partial g_{jl}}{\partial x^i} \,. \tag{1.177}$$

If the first and the last equation are added together and the second one subtracted, the predicted relations are confirmed. ■

This result indicates that the metric coefficients form a kind of "potential" for the coefficients of affine connection. The quantities Γ^k_{ij} or Γ_{lij} are determined by differentiation from the metric coefficients g_{ij}. The metric quantities themselves are determined by the relations (1.168) between the curvilinear and the Cartesian coordinates:

$$g_{ik} = g_i \cdot g_k = \frac{\partial \hat{y}_j}{\partial x^i} \frac{\partial \hat{y}_j}{\partial x^k} \,. \tag{1.178}$$

In the case of Cartesian coordinates we have $g_{ik} = \delta_{ik}$ and $\Gamma_{ijk} = 0$. In the case of curvilinear coordinate systems, the metric coefficients and the coefficients of affine connection define the geometric structure of a Euclidean space inasmuch as they set the scalar product and determine the differentials of vector and tensor fields.

The differentiation of vector and tensor fields can be interpreted in the Euclidean space as the application of the *gradient operator*,

$$\nabla = \text{grad} = g^k \frac{\partial}{\partial x^k} \,. \tag{1.179}$$

According to the rules of tensor analysis, the gradient operator ∇ can be coupled with tensor fields, leading to the definition of the differential operators *gradient*, *divergence* and *curl*:

$$\text{grad}\; v(x) = v \otimes \nabla = \frac{\partial v}{\partial x^k} \otimes g^k \,, \tag{1.180}$$

$$\text{div}\; v(x) = v \cdot \nabla = \nabla \cdot v = \frac{\partial v}{\partial x^k} \cdot g^k \,, \tag{1.181}$$

$$\text{div}\; \mathbf{T}(x) = \mathbf{T}\nabla = \frac{\partial \mathbf{T}}{\partial x^k} g^k \,, \tag{1.182}$$

$$\text{curl}\; v(x) = \nabla \times v = g^k \times \frac{\partial v}{\partial x^k} \,. \tag{1.183}$$

As a representative example the gradient of a vector field $v = v^i g_i = v_i g^i$, i.e.

$$v = v(x) = v^i(x^1, x^2, x^3) g_i(x^1, x^2, x^3) \quad \text{and} \quad v = v_l g^l = v_l(x^1, x^2, x^3) g^l(x^1, x^2, x^3)$$

ought to be calculated. This calls for the differentiation of the components and base vectors with respect to the coordinates. Substituting (1.169) and (1.171) and applying the product rule, one gets the results

$$\frac{\partial v}{\partial x^j} = \frac{\partial}{\partial x^j}(v^i g_i) = v^i\big|_j \, g_i \tag{1.184}$$

and

$$\frac{\partial v}{\partial x^j} = \frac{\partial}{\partial x^j}(v_l g^l) = v_i\big|_j \, g^i \, . \tag{1.185}$$

These equations contain the so-called *covariant derivatives*[22]

$$v^i\big|_j = \frac{\partial v^i}{\partial x^j} + \Gamma^i_{kj} v^k \tag{1.186}$$

and

$$v_l\big|_j = \frac{\partial v_l}{\partial x^j} - \Gamma^k_{jl} v_k \, . \tag{1.187}$$

Accordingly, we calculate the component representations for the tensor grad v:

$$\operatorname{grad} v(x) = v^i\big|_j \, g_i \otimes g^j = v_l\big|_j \, g^l \otimes g^j \, . \tag{1.188}$$

This implies

$$\operatorname{div} v(x) = v^i\big|_i \, . \tag{1.189}$$

The covariant derivatives of second and higher order tensors are formed in a like manner. To illustrate this, a tensor field in the covariant representation,

[22] KLINGBEIL [1989], p. 88.

$$\mathbf{T}(\boldsymbol{x}) = T_{ij} \boldsymbol{g}^i \otimes \boldsymbol{g}^j , \tag{1.190}$$

is differentiated: the result is

$$\frac{\partial \mathbf{T}}{\partial x^k} = \frac{\partial}{\partial x^k} \left(T_{ij} \boldsymbol{g}^i \otimes \boldsymbol{g}^j \right) = T_{ij}\big|_k \boldsymbol{g}^i \otimes \boldsymbol{g}^j . \tag{1.191}$$

In this example

$$T_{ij}\big|_k = \frac{\partial T_{ij}}{\partial x^k} - \Gamma^m_{ik} T_{mj} - \Gamma^m_{jk} T_{im} \tag{1.192}$$

is the *covariant derivative* of the covariant component T_{ij} with respect to the coordinate x^k. The tensor

$$\operatorname{grad} \mathbf{T}(\boldsymbol{x}) = \frac{\partial \mathbf{T}}{\partial x^k} \otimes \boldsymbol{g}^k = T_{ij}\big|_k \, \boldsymbol{g}^i \otimes \boldsymbol{g}^j \otimes \boldsymbol{g}^k$$

is the *gradient* of the tensor field $\mathbf{T}(\boldsymbol{x})$, a third order tensor. This produces the formula

$$\operatorname{div} \mathbf{T}(\boldsymbol{x}) = \frac{\partial \mathbf{T}}{\partial x^k} \boldsymbol{g}^k = T_{ij}\big|_k (\boldsymbol{g}^i \otimes \boldsymbol{g}^j) \boldsymbol{g}^k = g^{jk} T_{ij}\big|_k \, \boldsymbol{g}^i = T_i^{\ k}\big|_k \, \boldsymbol{g}^i \tag{1.193}$$

for the divergence of the tensor field $\mathbf{T}(\boldsymbol{x})$. In this case, the vanishing of the covariant derivatives of the metric quantities has been taken into consideration; the identity tensor is a constant:

$$\mathbf{1} = \delta^{ik} \boldsymbol{e}_i \otimes \boldsymbol{e}_k = g^{ik} \boldsymbol{g}_i \otimes \boldsymbol{g}_k \ \text{ implies } g^{ij}\big|_k = 0 .$$

An essential property of the Euclidean space is the independence of the second order covariant derivatives from the order of their calculation:[23]

$$v_i\big|_{jk} = v_i\big|_{kj} . \tag{1.194}$$

This property may be expressed using various formulations: it is in accordance with the relations

$$\frac{\partial^2}{\partial x^j \partial x^k} \left(v_l \boldsymbol{g}^l \right) = \frac{\partial^2}{\partial x^k \partial x^j} \left(v_l \boldsymbol{g}^l \right) \qquad (j, k = 1, 2, 3) , \tag{1.195}$$

[23] Klingbeil [1989], pp. 121.

the symmetry of the operator

$$\text{grad grad} = \nabla \otimes \nabla = \Big(g^k \frac{\partial}{\partial x^k}\Big) \otimes \Big(g^k \frac{\partial}{\partial x^k}\Big) \,, \tag{1.196}$$

and the fact that the operator $\nabla \times \nabla$ is the *zero operator*:

$$\nabla \times \nabla = \Big(g^k \frac{\partial}{\partial x^k}\Big) \times \Big(g^k \frac{\partial}{\partial x^k}\Big) = 0 \,. \tag{1.197}$$

These properties are only valid in connection with the Euclidean space, and in this context they are expressed by means of the symbolic notation (1.179), $\nabla = g^k \partial/\partial x^k$.

1. 10. 2 Non-Euclidean Spaces

If we fail to view the configuration of a material body as an area of a Euclidean space, there is no possibility of introducing Cartesian coordinates for representing the material points by means of position vectors. The local base vectors cannot be represented either as derivatives of the position vectors with respect to the coordinates (tangent vectors), or as derivatives of the coordinates with respect to the position vector (gradient vectors); the symbolic way of writing the gradient operator as $\nabla = g^k \partial/\partial x^k$ becomes meaningless, since a notation like this requires the validity of $g^k = \text{grad} x^k(x)$. Accordingly, it is not possible in this case to determine the metric coefficients g_{ij} and the coefficients of affine connection Γ^k_{ij} from the defining equations of the coordinates by differentiation.

In this situation the non-Euclidean structure of the space is defined by a field of covariant metric tensors,[24]

$$(x^1, x^2, x^3) \longmapsto g_{ik}(x^1, x^2, x^3) \,, \tag{1.198}$$

or, more generally, by a field of coefficients of affine connection,[25]

$$(x^1, x^2, x^3) \longmapsto \Gamma^k_{ij}(x^1, x^2, x^3) \,. \tag{1.199}$$

The fields g_{ik} and Γ^k_{ij} then form 6 or, in the most general case, 27 independent functions, which have to be transformed in a certain way when

[24] RASCHEWSKI [1959], pp. 348.
[25] RASCHEWSKI [1959], pp. 370.

changing the coordinate system. In the non-Euclidean space we generally have

$$v_i\big|_{jk} \neq v_i\big|_{kj}\,, \tag{1.200}$$

and, in the most general case, the coefficients of affine connection cannot be derived from the metric coefficients either.

The particular situation, where a configuration can only be identified with a non-Euclidean space, is to be encountered in the theory of material behaviour: various constitutive models are based on the notion of an *intermediate configuration*, evolving out of a conceived removal of the whole state of stress. Removing the total stress in this way is called a *local unloading*. For example, the thermal stresses within a heated structure may be removed by dividing (cutting) it into small pieces. In doing so, the volume elements change their shape and no longer fit together. In the language of continuum mechanics there is no unique displacement field to characterise the deformation. A configuration of this kind cannot be a Euclidean space, as more detailed considerations will show.

1. 10. 3 Conditions of Compatibility

In the Euclidean space of the current configuration the tangent vectors $\boldsymbol{g}_K$ are the derivatives of the position vector with respect to the coordinates. In the case of *convective coordinates* they are formed by applying the deformation gradient $\mathbf{F}$ to the tangent vectors $\boldsymbol{G}_K$:

$$\boldsymbol{g}_K = \frac{\partial \boldsymbol{x}}{\partial X^K} = \mathbf{F}\,\boldsymbol{G}_K. \tag{1.201}$$

The component representation of this relation with respect to an orthonormal base system $\boldsymbol{E}_I$ reads

$$\frac{\partial \boldsymbol{x}}{\partial X^K} = \frac{\partial y^I}{\partial X^K}\,\boldsymbol{E}_I \text{ und } \mathbf{F}\boldsymbol{G}_K = \mathrm{F}^I{}_K \boldsymbol{E}_I\,.$$

It leads to a system of 9 equations between the 3 functions $y^I(X^K, t)$ and the 9 functions $\mathrm{F}^I{}_K$, which are the components of the deformation gradient:

$$\frac{\partial}{\partial X^K}\, y^I(X^1, X^2, X^3, t) = F^I{}_K(X^1, X^2, X^3, t)\,, \quad I, K = 1, 2, 3\,. \tag{1.202}$$

If the 9 functions $F^I{}_K(X^1, X^2, X^3, t)$ are given, the functions $y^I(X^1, X^2, X^3, t)$ are calculable if and only if the $F^I{}_K$ fulfil the following *conditions of compatibility*:

$$\frac{\partial F^I{}_K}{\partial X^L} = \frac{\partial F^I{}_L}{\partial X^K} \,. \tag{1.203}$$

In connection with convective coordinates, the motion of a material body is represented according to (1.104) as a change of metric,

$$(\mathscr{P}, t) \longmapsto (G_{KL}(X^J), t) \longmapsto g_{KL}(X^J, t) \,,$$

with

$$g_{KL} = \boldsymbol{g}_K \cdot \boldsymbol{g}_L = (\mathbf{F}\, \boldsymbol{G}_K) \cdot (\mathbf{F}\, \boldsymbol{G}_L) = F^I{}_K F^J{}_L \delta_{IJ} \,.$$

For a complete description of the geometry of the deformation, not only the metric quantities, i. e. the strains, but also their derivatives are of importance. Whereas the base vectors $\boldsymbol{g}_K$ form the *tangent space* connected to a material point in the current configuration, the connection between neighbouring tangent spaces is represented by means of their differentiation with respect to the coordinates. Following (1.169) and (1.174) we have

$$\frac{\partial \boldsymbol{g}_K}{\partial X^L} = \frac{\partial^2 \boldsymbol{x}}{\partial X^K \partial X^L} = \Gamma_{MKL}\, \boldsymbol{g}^M \,. \tag{1.204}$$

The coefficients of affine connection Γ_{MKL} (*Christoffel symbols* of the first kind) are characterised by the derivatives of the deformation gradient with respect to the coordinates. The definition (1.173),

$$\Gamma_{MKL} = \boldsymbol{g}_M \cdot \frac{\partial \boldsymbol{g}_K}{\partial X^L} \tag{1.205}$$

implies

$$\Gamma_{MKL} = \boldsymbol{g}_M \cdot \frac{\partial}{\partial X^L} (\mathbf{F}\, \boldsymbol{G}_K) = \boldsymbol{g}_M \cdot \left[\frac{\partial \mathbf{F}}{\partial X^L}\, \boldsymbol{G}_K + \mathbf{F}\, \frac{\partial \boldsymbol{G}_K}{\partial X^L} \right] . \tag{1.206}$$

For Cartesian base vectors in the reference configuration, $\boldsymbol{G}_K = \boldsymbol{E}_K$, we deduce the relation

$$\Gamma_{MKL} = (\mathbf{F}\, \boldsymbol{E}_M) \cdot \frac{\partial \mathbf{F}}{\partial X^L}\, \boldsymbol{E}_K \tag{1.207}$$

or

$$\Gamma_{MKL} = \boldsymbol{E}_M \cdot \left[\left(\mathbf{F}^{\mathrm{T}} \frac{\partial \mathbf{F}}{\partial X^L} \right) \boldsymbol{E}_K \right] = F^I{}_M \frac{\partial F_{IK}}{\partial X^L} \,. \tag{1.208}$$

On the other hand it is possible to represent the coefficients of connection by means of the strain derivatives, i.e. by derivatives of the metric coefficients. This representation is achieved on the basis of (1.175),

$$\Gamma_{MKL} = \frac{1}{2} \left(\frac{\partial g_{KM}}{\partial X^L} + \frac{\partial g_{ML}}{\partial X^K} - \frac{\partial g_{LK}}{\partial X^M} \right) . \tag{1.209}$$

As a consequence, a system of differential equations evolves for the components $F^I{}_J$ of the deformation gradient. It depends on the derivatives of the components of the right Cauchy-Green tensor:

$$F^I{}_M \frac{\partial F_{IK}}{\partial X^L} = \frac{1}{2} \left(\frac{\partial g_{KM}}{\partial X^L} + \frac{\partial g_{ML}}{\partial X^K} - \frac{\partial g_{LK}}{\partial X^M} \right) . \tag{1.210}$$

This is a system of 27 differential equations for the 9 components $F^I{}_J$ of the deformation gradient. In general, this system is overdetermined. However, any solution that might exist would possess interesting properties:

Theorem 1. 11

If there is a solution of the system of differential equations (1.210), then it must fulfil the relations

$$F^I{}_M F_{IK} = g_{MK} , \tag{1.211}$$

as well as

$$\frac{\partial F^I{}_K}{\partial X^L} = \frac{\partial F^I{}_L}{\partial X^K} \,. \tag{1.212}$$

In other words, we have components $F^I{}_M(X^1, X^2, X^3, t)$ of a deformation gradient, which satisfy the conditions of compatibility on the one hand and belong to the given *Cauchy-Green* tensor $\mathbf{C} = \mathbf{F}^{\mathrm{T}}\mathbf{F}$ on the other. □

Proof

Interchanging the indices L and K does not alter the right-hand side of the differential equations (1.210), therefore

$$F^I_{\ M}\frac{\partial F_{IK}}{\partial X^L} = F^I_{\ M}\frac{\partial F_{IL}}{\partial X^K}$$

is valid, which includes the conditions of compatibility. By interchanging the indices M and K we obtain

$$F^I_{\ K}\frac{\partial F_{IM}}{\partial X^L} = \frac{1}{2}\left(\frac{\partial g_{MK}}{\partial X^L} + \frac{\partial g_{KL}}{\partial X^M} - \frac{\partial g_{LM}}{\partial X^K}\right); \tag{1.213}$$

adding equations (1.210) and (1.213) yields

$$F^I_{\ M}\frac{\partial F_{IK}}{\partial X^L} + F^I_{\ K}\frac{\partial F_{IM}}{\partial X^L} = \frac{\partial}{\partial X^L}\left(F^I_{\ M}F_{IK}\right) = \frac{\partial g_{MK}}{\partial X^L}\,.$$

This implies $F^I_{\ M}F_{IK} = g_{MK}$, since integration constants are not important in this case. ■

The theorem states that, in the event of a solution, the differential equations

$$\frac{\partial}{\partial X^K}\,y^I(X^1, X^2, X^3, t) = F^I_{\ K}(X^1, X^2, X^3, t) \tag{1.214}$$

for the Cartesian coordinates $y^I(X^1, X^2, X^3, t)$ can also be integrated. Accordingly, following the differential equations (1.210), we are able to calculate both the components of the deformation gradient and the components of the position vector, i.e. the complete picture of the body in the Euclidean space of the current configuration.

There remains one question to be answered, namely: what demands have to be made on the field $\mathbf{C}(\mathbf{X}, t)$ to ensure that the 9 functions $F^I_{\ K}$ are calculable using the differential equations (1.210)? The answer to this question is provided by

Theorem 1. 12

The set of equations

$$F^I_{\ M}\frac{\partial F_{IK}}{\partial X^L} = \Gamma_{MKL} \tag{1.215}$$

can be integrated if and only if the components of the right *Cauchy-Green* tensor satisfy the *conditions of compatibility*,

$$R_{IJKL} = 0\,, \tag{1.216}$$

with the definition

$$R_{IJKL} = \Gamma_{IJK,\,L} - \Gamma_{IJL,\,K} + g^{PQ}(\Gamma_{PIK}\Gamma_{QJL} - \Gamma_{PIL}\Gamma_{QJK})\,. \tag{1.217}$$

□

The quantities R_{IJKL} form the so-called *Riemann-Christoffel* tensor.

Proof

Using the abbreviation

$$\boldsymbol{\Gamma}_K = \Gamma_{IJK}\,\mathbf{G}^I \otimes \mathbf{G}^J$$

the system of equations (1.210), $F^I{}_M F_{IK,\,L} = \Gamma_{MKL}$ or $F^M{}_I F_{MJ,\,K} = \Gamma_{IJK}$, can be written symbolically in the form of

$$\mathbf{F}^{\mathrm{T}}\big(\mathbf{F}_{,\,K}\big) = \boldsymbol{\Gamma}_K\,, \tag{1.218}$$

the partial derivative with respect to a Cartesian coordinate now being defined by a comma. It is clear that for any solution $\mathbf{F}$ of these differential equations the derivatives of second order are independent of the order of calculation:

$$\mathbf{F}_{,\,KL} = \mathbf{F}_{,\,LK}\,. \tag{1.219}$$

This immediately leads to the demand made on the strain field:

$$\mathbf{F}_{,\,KL} = \big(\mathbf{F}^{\mathrm{T}\,-1}\boldsymbol{\Gamma}_K\big)_{,\,L} = -\,\mathbf{F}^{\mathrm{T}\,-1}\mathbf{F}^{\mathrm{T}}{}_{,\,L}\mathbf{F}^{\mathrm{T}\,-1}\boldsymbol{\Gamma}_K + \mathbf{F}^{\mathrm{T}\,-1}\boldsymbol{\Gamma}_{K,\,L}\,. \tag{1.220}$$

Considering the differential equation (1.218) again, we arrive at

$$\mathbf{F}_{,\,KL} = \mathbf{F}^{\mathrm{T}\,-1}\big\{\boldsymbol{\Gamma}_{K,\,L} - \mathbf{F}^{\mathrm{T}}{}_{,\,L}\mathbf{F}\,\mathbf{F}^{-1}\mathbf{F}^{\mathrm{T}\,-1}\boldsymbol{\Gamma}_K\big\}\,,$$

$$\mathbf{F}_{,\,KL} = \mathbf{F}^{\mathrm{T}\,-1}\big\{\boldsymbol{\Gamma}_{K,\,L} - \boldsymbol{\Gamma}_L{}^{\mathrm{T}}\mathbf{C}^{-1}\boldsymbol{\Gamma}_K\big\}\,.$$

This leads to

$$\mathbf{F}_{,\,KL} - \mathbf{F}_{,\,LK} = \mathbf{F}^{\mathrm{T}\,-1}\big\{\boldsymbol{\Gamma}_{K\,,\,L} - \boldsymbol{\Gamma}_{L\,,\,K} + \boldsymbol{\Gamma}_K{}^{\mathrm{T}}\mathbf{C}^{-1}\boldsymbol{\Gamma}_L - \boldsymbol{\Gamma}_L{}^{\mathrm{T}}\mathbf{C}^{-1}\boldsymbol{\Gamma}_K\big\} = \mathbf{0}\,, \tag{1.221}$$

which implies that the tensors

$$\mathbf{R}_{KL} = \mathbf{\Gamma}_{K,L} - \mathbf{\Gamma}_{L,K} + \mathbf{\Gamma}_K^{\mathrm{T}}\mathbf{C}^{-1}\mathbf{\Gamma}_L - \mathbf{\Gamma}_L^{\mathrm{T}}\mathbf{C}^{-1}\mathbf{\Gamma}_K \tag{1.222}$$

have to vanish:

$$\mathbf{R}_{KL} = \mathbf{0}. \tag{1.223}$$

Component representations provide quantities with fourfold indices R_{IJKL}, which correspond to the *Riemann-Christoffel* tensor:

$$R_{IJKL} = \Gamma_{IJK,L} - \Gamma_{IJL,K} + g^{PQ}\left(\Gamma_{PIK}\Gamma_{QJL} - \Gamma_{PIL}\Gamma_{QJK}\right).$$

These quantities have to vanish if the differential equations for $F^I{}_K$ are integrable: $R_{IJKL} = 0$.

The proof that the *conditions of compatibility* (1.216) (or (1.221)) are also sufficient for the existence of a solution of the differential equations (1.210) demands fairly close examination. This can be read up, for example, in the work of T.Y. Thomas.[26] ■

Based on these arguments, the image of a material body in a configuration is part of the Euclidean space if and only if the conditions of compatibility are valid for the strain field

$$\chi_t[\mathscr{B}] \subset \mathbb{E}^3 \Longleftrightarrow R_{IJKL} = 0. \tag{1.224}$$

The conditions of compatibility form a total of 6 independent equations.[27]

It should be borne in mind that, in the event of compatibility, the integration of the field $\mathbf{C}(\boldsymbol{X}, t)$ of *Cauchy-Green* tensors provides the whole deformation gradient, i.e. both the stretch $\mathbf{U}(\boldsymbol{X}, t)$ and the rotation $\mathbf{R}(\boldsymbol{X}, t)$.

Definition 1. 21

An *incompatible configuration* is characterised by the fact that the coordinate transformation is supplemented by a metric $\hat{g}_{kl}$ that does not fulfil the conditions of compatibility:

---

[26] Thomas [1934]. Cf. Raschewski [1959], §106.

[27] Klingbeil [1989], pp. 122.

$$(X^1, X^2, X^3, t) \quad \longmapsto \begin{cases} \hat{x}^k = \hat{\chi}_{\mathrm{R}}{}^k(X^1, X^2, X^3, t) \\ \hat{g}_{kl} = \hat{g}_{kl}(\hat{x}^1, \hat{x}^2, \hat{x}^3, t), \ \hat{R}_{ijkl} \neq 0 \end{cases} \tag{1.225}$$

◊

In the theory of material behaviour so-called *intermediate configurations* are introduced. For physical reasons these are, as a rule, *incompatible*. In this case the material body is mapped into a non-Euclidean space. Even now coordinate differentials can be calculated. These are contravariant vector components, as before, which one can consider as relating to a base in the tangent space $\mathbb{T}_{\hat{\chi}_t(\mathscr{P})}$:

$$\mathrm{d}\hat{x}^k = \frac{\partial \hat{x}^k}{\partial X^L}\,\mathrm{d}X^L\,, \tag{1.226}$$

$$\mathrm{d}\hat{x}^k \hat{\boldsymbol{g}}_k \in \mathbb{T}_{\hat{\chi}_t(\mathscr{P})}\,.$$

However, the base vectors $\hat{\boldsymbol{g}}_k$ can no longer be derived from a position vector, because it does not exist. It is that which distinguishes this situation from the state of compatibility. With regard to the base vectors $\hat{\boldsymbol{g}}_k$ one obtains the component representation of a linear transformation

$$\boldsymbol{\Psi} = \frac{\partial \hat{x}^k}{\partial X^L}\,\hat{\boldsymbol{g}}_k \otimes \boldsymbol{G}^L\,, \tag{1.227}$$

the component matrix being the functional matrix of the coordinate transformation $\hat{\chi}_{\mathrm{R}}$. For material line elements $\mathrm{d}\boldsymbol{X} = \mathrm{d}X^M \boldsymbol{G}_M$ the relation

$$\mathrm{d}\hat{\boldsymbol{x}} = \mathrm{d}\hat{x}^k \hat{\boldsymbol{g}}_k = \frac{\partial \hat{x}^k}{\partial X^L}\left(\hat{\boldsymbol{g}}_k \otimes \boldsymbol{G}^L\right)\left(\mathrm{d}X^M \boldsymbol{G}_M\right)\,,$$

is still valid, which means that the mapping

$$\begin{aligned} \boldsymbol{\Psi} : \mathbb{T}_{\mathrm{R}(\mathscr{P})} &\longrightarrow \mathbb{T}_{\hat{\chi}_t(\mathscr{P})} \\ \mathrm{d}\boldsymbol{X} &\longmapsto \mathrm{d}\hat{\boldsymbol{x}} = \boldsymbol{\Psi}\mathrm{d}\boldsymbol{X} \end{aligned} \tag{1.228}$$

transforms material line elements from the reference configuration into the incompatible (non-Euclidean) configuration $\hat{\chi}_t$.

The base vectors $\{\hat{\boldsymbol{g}}_1, \hat{\boldsymbol{g}}_2, \hat{\boldsymbol{g}}_3\}$, forming a base system in the tangent space $\mathbb{T}_{\hat{\chi}_t(\mathscr{P})}$, cannot be derived from a position vector. Nevertheless, it is possible to procure a representation of the unknown base vectors $\hat{\boldsymbol{g}}_k$ by means of the tangent vectors $\boldsymbol{G}_I$ of the reference configuration.

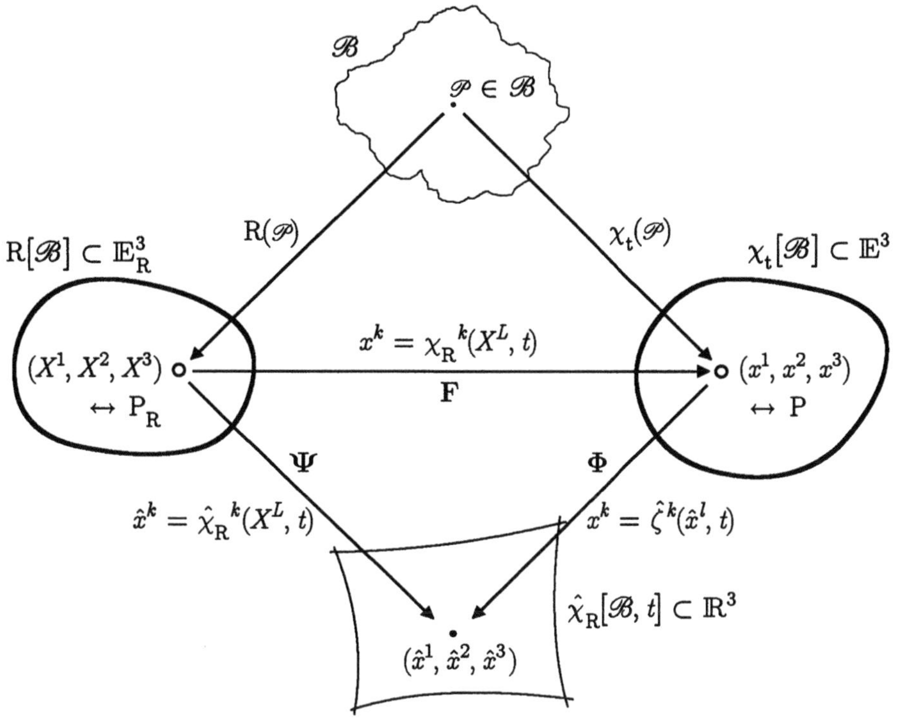

Figure 1. 11: Intermediate configuration

This representation takes the given information into consideration, i.e. it corresponds to the given metric $\hat{g}_{kl}$.[28] The preliminary assumption

$$\hat{g}_k = A^I_k G_I, \qquad \hat{g}^l = B^l_J G^J \tag{1.229}$$

implies the set of equations

$$A^I_k A^J_l G_{IJ} = \hat{g}_{kl}(\hat{x}^1, \hat{x}^2, \hat{x}^3, t), \tag{1.230}$$

from which the functions $A^I_k(\hat{x}^1, \hat{x}^2, \hat{x}^3, t)$ can be defined, since the matrix of $\hat{g}_{kl}$ is positive definite. Based on the reciprocity relation $\hat{g}_k \cdot \hat{g}^l = A^I_k B^l_I = \delta^l_k$ the matrices A^I_k und B^l_I are inverse to each other. The functions A^I_k and B^l_I

[28] The following argumentation originates from ANTHONY [1970], pp. 163.

respectively cannot, as a general rule, be integrated into a system of coordinates:

$$A^I_k \neq \frac{\partial \Xi^I}{\partial \hat{x}^k} \quad \Longleftrightarrow \quad \frac{\partial A^I_k}{\partial \hat{x}^l} \neq \frac{\partial A^I_l}{\partial \hat{x}^k} .$$

The components of the transformation (1.227),

$$\mathbf{\Psi} = \bar{\mathbf{\Psi}}(\boldsymbol{X}, t) = \frac{\partial \hat{x}^k}{\partial X^L} A^I_k \, \boldsymbol{G}_I \otimes \boldsymbol{G}^L , \tag{1.231}$$

that now refer to base vectors of a Euclidean space, no longer form a functional matrix. The results can be employed to define a so-called *intermediate configuration*.

Definition 1. 22

An *intermediate configuration* $\mathbf{\Psi}$ is a tensor field $\mathbf{\Psi}(\boldsymbol{X}, t)$ with $\det\mathbf{\Psi} \neq 0$, corresponding to a multiplicative decomposition of the deformation gradient:

$$\mathbf{F} = \left(\mathbf{F}\mathbf{\Psi}^{-1}\right)\mathbf{\Psi} = \mathbf{\Phi}\mathbf{\Psi}. \tag{1.232}$$

The field $\mathbf{\Psi}(\boldsymbol{X}, t)$ depends on the process, and we take it for granted that a set of constitutive relations is available for calculating the temporal evolution of $\mathbf{\Psi}$. The three configurations - *reference configuration*, *intermediate configuration* and *current configuration* - are related to each other by means of a composition of two coordinate transformations $\hat{\chi}_{\mathrm{R}}{}^k$ and $\hat{\zeta}^k$:

$$\hat{\chi}_{\mathrm{R}} : \quad \hat{x}^k = \hat{\chi}_{\mathrm{R}}{}^k(X^1, X^2, X^3, t) ,$$

$$\hat{\zeta} \; : \quad x^k = \hat{\zeta}^k(\hat{x}^1, \hat{x}^2, \hat{x}^3, t) ,$$

$$\chi_{\mathrm{R}} = \hat{\zeta} \circ \hat{\chi}_{\mathrm{R}} : \quad x^k = \chi_{\mathrm{R}}{}^k(X^1, X^2, X^3, t) , \tag{1.233}$$

$$x^k = \hat{\zeta}^k\Big(\hat{\chi}_{\mathrm{R}}{}^1(X^1, X^2, X^3, t), \hat{\chi}_{\mathrm{R}}{}^2(X^1, X^2, X^3, t), \hat{\chi}_{\mathrm{R}}{}^3(X^1, X^2, X^3, t), t\Big) . \tag{1.234}$$

During this process the coordinate transformation $\hat{\zeta}$ is assigned to a field of incompatible metric coefficients (Fig. 1. 11). ◊

Now the coordinate differentials $\mathrm{d}x^k$ can be calculated using the chain rule:

$$\mathrm{d}x^k = \frac{\partial x^k}{\partial \hat{x}^l}\frac{\partial \hat{x}^l}{\partial X^L}\,\mathrm{d}X^L = \frac{\partial \hat{\zeta}^k}{\partial \hat{x}^l}\frac{\partial \hat{\chi}_{\mathrm{R}}{}^l}{\partial X^L}\,\mathrm{d}X^L\,. \tag{1.235}$$

These are contravariant vector components as far as the tangent vectors $\boldsymbol{g}_k$ of the current configuration are concerned. The system of reciprocal base vectors $\hat{\boldsymbol{g}}_k$, $\hat{\boldsymbol{g}}^k$ of the tangent space $\mathbb{T}_{\hat{\chi}_t(\mathscr{P})}$ is introduced in the usual way, to produce

$$\mathrm{d}x^k\boldsymbol{g}_k = \left(\frac{\partial \hat{\zeta}^k}{\partial \hat{x}^l}\boldsymbol{g}_k \otimes \hat{\boldsymbol{g}}^l\right)\left(\frac{\partial \hat{\chi}_{\mathrm{R}}{}^m}{\partial X^L}\hat{\boldsymbol{g}}_m \otimes \boldsymbol{G}^L\right)\left(\mathrm{d}X^M\boldsymbol{G}_M\right).$$

The resulting transformations are obtained as follows:

$$\begin{aligned}\boldsymbol{\Psi} : \mathbb{T}_{\mathrm{R}(\mathscr{P})} &\longrightarrow \mathbb{T}_{\hat{\chi}_t(\mathscr{P})}\\ \mathrm{d}\boldsymbol{X} &\longmapsto \mathrm{d}\hat{\boldsymbol{x}} = \boldsymbol{\Psi}\mathrm{d}\boldsymbol{X}\,,\end{aligned} \tag{1.236}$$

$$\boldsymbol{\Psi} = \frac{\partial \hat{\chi}_{\mathrm{R}}{}^k}{\partial X^L}\hat{\boldsymbol{g}}_k \otimes \boldsymbol{G}^L = \frac{\partial \hat{\chi}_{\mathrm{R}}{}^k}{\partial X^L}A^I_k\boldsymbol{G}_I \otimes \boldsymbol{G}^L\,, \tag{1.237}$$

$$\begin{aligned}\boldsymbol{\Phi} : \mathbb{T}_{\hat{\chi}_t(\mathscr{P})} &\longrightarrow \mathbb{T}_{\chi_t(\mathscr{P})}\\ \mathrm{d}\hat{\boldsymbol{x}} &\longmapsto \mathrm{d}\boldsymbol{x} = \boldsymbol{\Phi}\,\mathrm{d}\hat{\boldsymbol{x}}\,,\end{aligned} \tag{1.238}$$

$$\boldsymbol{\Phi} = \frac{\partial \hat{\zeta}^k}{\partial \hat{x}^l}\boldsymbol{g}_k \otimes \hat{\boldsymbol{g}}^l = \frac{\partial \hat{\zeta}^k}{\partial \hat{x}^l}B^l_J\boldsymbol{g}_k \otimes \boldsymbol{G}^J\,. \tag{1.239}$$

Their composition is the deformation gradient:

$$\mathrm{d}\boldsymbol{x} = \boldsymbol{\Phi}\,\boldsymbol{\Psi}\,\mathrm{d}\boldsymbol{X} = \mathbf{F}\,\mathrm{d}\boldsymbol{X}\,. \tag{1.240}$$

The components referring to the mixed base systems $\hat{\boldsymbol{g}}_k \otimes \boldsymbol{G}^L$, $\boldsymbol{g}_k \otimes \hat{\boldsymbol{g}}^l$ are functional matrices. After the representation (1.229) of the base vectors $\hat{\boldsymbol{g}}_k$, $\hat{\boldsymbol{g}}^l$ by way of a linear combination of $\boldsymbol{G}_I$ or $\boldsymbol{G}^I$, this is, however, no longer the case.

The fact that the product of the two transformations provides the deformation gradient $\mathbf{F}$ is obvious: the matrices A^I_k, B^l_J are inverse, the bases $\boldsymbol{G}_I$, $\boldsymbol{G}^J$ reciprocal, and the chain rule refers back to the functional matrix of the motion,

$$\mathbf{F} = \boldsymbol{\Phi}\boldsymbol{\Psi} = \left(\frac{\partial\hat{\zeta}^i}{\partial\hat{x}^l}B^l_J\boldsymbol{g}_i \otimes \boldsymbol{G}^J\right)\left(\frac{\partial\hat{\chi}_{\mathrm{R}}{}^k}{\partial X^L}A^I_k\boldsymbol{G}_I \otimes \boldsymbol{G}^L\right), \tag{1.241}$$

$$\mathbf{F} = \frac{\partial\hat{\zeta}^i}{\partial\hat{x}^l}\frac{\partial\hat{\chi}_{\mathrm{R}}{}^k}{\partial X^L}B^l_JA^I_k(\boldsymbol{G}^J\cdot\boldsymbol{G}_I)\,\boldsymbol{g}_i \otimes \boldsymbol{G}^L = \frac{\partial\hat{\zeta}^i}{\partial\hat{x}^k}\frac{\partial\hat{\chi}_{\mathrm{R}}{}^k}{\partial X^L}\boldsymbol{g}_i \otimes \boldsymbol{G}^L\,,$$

i.e.

$$\mathbf{F} = \frac{\partial\chi_{\mathrm{R}}{}^i}{\partial X^L}\boldsymbol{g}_i \otimes \boldsymbol{G}^L\,.$$

On balance, it can be concluded that a complete description of a configuration demands a total of 9 functions, namely a coordinate transformation (3 functions) and a field of metric quantities (6 functions). In the case of an intermediate configuration, being on the whole incompatible, the coordinate transformation has a mathematical rather than a physical meaning: the metric coefficients are of primary physical interest, because they change during the deformation.

For this reason, as far as the modelling of inelastic material properties is concerned, the metric of the intermediate configuration has to be determined using an appropriate set of constitutive equations (flow rules). This is equivalent to determining an inelastic *Cauchy-Green* tensor or a stretch tensor (see Chaps. 10 - 12). In addition, it could be that the rotational part of the transformation $\boldsymbol{\Psi}$ also has to be determined for kinematical reasons (see Chap. 13).

2 Balance Relations of Mechanics

2. 1 Preliminary Remarks

Whereas kinematics deals with the geometry of motion and the deformation of material bodies, the purpose of kinetics is to link the sources of motion with the influence of the outside world. This interaction is described by the balance relations. The basic idea is the free-body principle, described in the following physical definition, which is complementary to Definition 1. 1 from a physical point of view.

Definition 2. 1

A *material body* $\mathcal{B}$ is defined by means of the unique demarcation of its material points $\mathcal{P} \in \mathcal{B}$ from the rest of the material world and identified with an area of the Euclidean space via its current configuration. This area is characterised by a finite volume and a piecewise smooth surface. Picture this definition as an imaginary cut, slicing the entire world into two disjoint sections, namely the *body* itself and its *environment*, the *outside world*. The environment's influence on $\mathcal{B}$ is completely substituted by appropriate physical quantities. This principal procedure is called the *principle of intersection* or *free-body principle*.[1] ◊

The use of the free-body principle in continuum physics is an abstract process and has, as such, unlimited uses: any subsets can be "sliced off", and these are material bodies as well. When this formation of subsets is reversed, a union of material bodies is formed in the sense of the set theory,

---

[1] Cf. SZABÓ [1987], p. 20; FUNG [1977], p. 11; JAUNZEMIS [1967], pp. 177, 207.

which in turn leads to material bodies too. If this abstract notion is followed up, it leads us to the concept of the entirety of material bodies as a system of sets, in which there are almost unrestricted possibilities for the set theory operations, such as the formation of unions and intersections, etc. Key words taken from the fundamental terminology, such as mass, momentum, energy, force, power, heat and entropy are functions, whose domain is the whole set of material bodies.

In order to formulate generally valid laws of nature in mathematical terms, it is necessary to have mathematical models illustrating the effect of the outside world on the present state of the material body under examination. Mechanics deals with the current state of motion represented by the terms *current configuration*, *mass*, *momentum* und *moment of momentum*; the effect of the environment is represented by the terms *force* and *torque*. If the field of thermodynamics is also incorporated under the general heading of mechanics, terms such as *temperature*, *energy* (kinetic energy and internal energy) and *entropy* will have to be added to the list of mechanical terms referring to the state of the material body. In thermomechanics the effects of the outside world extend to the mechanical and thermal energy transport, i.e. to the terms *mechanical power* and *heat*. Finally, the influence of the outside world is rounded off by the notion of *entropy supply*.

Basically, the physical quantities of mass, momentum, moment of momentum, energy and entropy are attributed to the material body $\mathscr{B}$ itself. On principle, this is based on the assumption that the physical quantities in question are additive in the sense of the measure theory. It is also assumed that these physical quantities are absolutely continuous with respect to the volume, that is to say: these quantities are zero for material bodies of zero volume (material points, lines, surfaces).[2] Additive physical quantities are represented accordingly by *volume integrals* using *density functions*. In doing so, the integration can extend over the domain of the current configuration or the domain of the reference configuration, which leads to different density functions. The effects of the outside world are also represented by integrals over density functions. They always consist of one volume and one surface share each: volume distributed shares are represented by *volume integrals* and surface distributed shares by *surface integrals*. Surface distributed quantities are absolutely continuous functions of surface, vanishing for material bodies of zero surface (points, lines).

By means of a balance relation, the temporal change of a quantity belonging to the current state of a material body is brought into a causal relationship with the effect of the outside world. There are two different versions of one and the same balance relation, namely the *global formulation* and the *local formulation*. The latter is a system of partial differential equations for

[2] See TRUESDELL & TOUPIN [1960], Sect. 155 for a more detailed analysis.

the scalar-, vector- or tensor-valued fields appertaining to the corresponding physical quantities. The local formulation of a balance relation is equivalent to the original global formulation, provided that the fields concerned possess adequate continuity properties. The principal possibility for formulating local balance statements is based on

Theorem 2. 1

Let f: $(\mathscr{P}, t) \longmapsto \mathrm{w} = \mathrm{f}(\mathscr{P}, t)$ be any physical quantity in the *spatial representation*, $\mathrm{w} = \bar{\mathrm{f}}(\boldsymbol{x}, t) = \bar{\mathrm{f}}(x^1, x^2, x^3, t)$. Assuming $\bar{\mathrm{f}}(\boldsymbol{x}, t)$ to be continuous with respect to $\boldsymbol{x}$ and

$$\iiint\limits_{v} \bar{\mathrm{f}}(\boldsymbol{x}, t)\, \mathrm{d}v = 0 \tag{2.1}$$

not only for the integration area with the volume v, but also for every other partial volume that can be selected as an integration area, then it holds that

$$\bar{\mathrm{f}}(\boldsymbol{x}, t) \equiv 0 \tag{2.2}$$

in the whole of the integration area v. □

Proof

If an inner point $\boldsymbol{x}_0$ with the property $\bar{\mathrm{f}}(\boldsymbol{x}_0, t) > 0$ (or $\bar{\mathrm{f}}(\boldsymbol{x}_0, t) < 0$) exists, then, in view of the smoothness of f, there is also a sufficiently small environment of $\boldsymbol{x}_0$, where $\bar{\mathrm{f}}(\boldsymbol{x}, t)$ does not vanish. In that case it would be possible to select this area as an integration area, resulting in an off-zero integral, which contradicts the supposition (2.1). ■

The theorem establishes that a continuous function is zero in a spatial area if the volume integral vanishes over all partial volumes. The physical meaning of this prerequisite complies with the general *postulate* requiring a *balance equation* to be valid *for all conceivable material bodies* at all times, i.e. in particular for all partial bodies that can be formed out of $\mathscr{B}$. Without this principal requirement it would not be possible to deduce local balance relations from the primary global formulations.

2. 2 Mass

2. 2. 1 Balance of Mass: Global Form

The mass of a material body is a measure for both the effect of the force of gravity (gravitational mass) and the resistance set up by a body to counteract any variation of its velocity (inertial mass).

Definition 2. 2
At any given time t, a positive scalar m, called the *mass*, is attributed to any material body $\mathscr{B}$:[3]

$$(\mathscr{B}, t) \longmapsto m(\mathscr{B}, t) = \int_{\mathscr{B}} \mathrm{d}m \,, \quad m > 0 \,. \qquad \Diamond$$

Mass is represented by a volume integral over the area of the current configuration, i.e.

$$m(\mathscr{B}, t) = \iiint_{\chi_t[\mathscr{B}]} \bar{\rho}(\boldsymbol{x}, t)\mathrm{d}v \equiv \iiint_{v} \bar{\rho}(\boldsymbol{x}, t)\mathrm{d}v \,, \tag{2.3}$$

or by means of an integral over the reference configuration:

$$m(\mathscr{B}, t) = \iiint_{\mathrm{R}[\mathscr{B}]} \hat{\rho}_{\mathrm{R}}(\boldsymbol{X}, t)\mathrm{d}V \equiv \iiint_{V} \hat{\rho}_{\mathrm{R}}(\boldsymbol{X}, t)\mathrm{d}V \,. \tag{2.4}$$

In these representations ρ_{R} and ρ are the *mass densities* in the reference or current configuration. These mass densities are defined on the material body $\mathscr{B}$:

$$\rho_{\mathrm{R}} : (\mathscr{P}, t) \longmapsto \rho_{\mathrm{R}}(\mathscr{P}, t) > 0 \,, \quad \rho : (\mathscr{P}, t) \longmapsto \rho(\mathscr{P}, t) > 0 \,.$$

The scalar field $(\boldsymbol{x}, t) \longmapsto \rho \equiv \bar{\rho}(\boldsymbol{x}, t)$ is the spatial representation of the mass density in the current configuration and, in compliance with the general definition, set up by means of

$$\bar{\rho}(\boldsymbol{x}, t) = \rho(\mathscr{P}, t)\Big|_{\mathscr{P} = \chi_t^{-1}(\boldsymbol{x})} \,. \tag{2.5}$$

The material representation of $\rho(\mathscr{P}, t)$ is constructed by means of

[3] See TRUESDELL & TOUPIN [1960], Sect. 155 for further information and references.

$$\hat{\rho}(\boldsymbol{X}, t) = \rho(\mathscr{P}, t)\Big|_{\mathscr{P} = \mathrm{R}^{-1}(\boldsymbol{X})} . \tag{2.6}$$

Between the mass densities ρ_{R} and ρ the relation

$$\rho_{\mathrm{R}} = \rho \, \det\mathbf{F} , \tag{2.7}$$

immediately following from the transformation behaviour $\mathrm{d}v = \det\mathbf{F}\mathrm{d}V$ of material volume elements, is valid. Regarded as a universal law of nature in non-relativistic mechanics, we introduce the

Balance of Mass

The mass is temporally constant for any given material body:

$$\frac{\mathrm{d}}{\mathrm{d}t}\, m(\mathscr{B}, t) = 0 \quad \Longleftrightarrow \quad m(\mathscr{B}, t) \equiv m(\mathscr{B}) . \tag{2.8}$$

♦

The balance of mass requires the material derivatives of the volume integrals to vanish,

$$\frac{\mathrm{d}}{\mathrm{d}t} \iiint\limits_{v} \bar{\rho}(\boldsymbol{x}, t)\mathrm{d}v = \frac{\partial}{\partial t} \iiint\limits_{V} \hat{\rho}_{\mathrm{R}}(\boldsymbol{X}, t)\mathrm{d}V = 0 , \tag{2.9}$$

not only for the volumes $v(\mathscr{B}, t)$ and $V(\mathscr{B})$, but also for all partial volumes.

2. 2. 2 Balance of Mass: Local Form

One implication resulting from the balance of mass is

Theorem 2. 2
The local formulation of the *mass balance* (*continuity equation*) in the material representation can be written as

$\dfrac{\partial}{\partial t}\hat{\rho}_{\mathrm{R}}(\boldsymbol{X}, t) = 0 \Longleftrightarrow \hat{\rho}_{\mathrm{R}}(\boldsymbol{X}, t) = \hat{\rho}_{\mathrm{R}}(\boldsymbol{X})$ or, equivalently, in the form

$$\hat{\rho}(\boldsymbol{X}, t)\, \det\mathbf{F}(\boldsymbol{X}, t) = \hat{\rho}_{\mathrm{R}}(\boldsymbol{X}) . \tag{2.10}$$

In the spatial representation the mass balance reads as

$$\dot{\bar{\rho}}(\boldsymbol{x}, t) + \bar{\rho}(\boldsymbol{x}, t)\,\mathrm{div}\,\bar{\boldsymbol{v}}(\boldsymbol{x}, t) = 0\,, \tag{2.11}$$

or

$$\frac{\partial}{\partial t}\bar{\rho}(\boldsymbol{x}, t) + \mathrm{div}[\bar{\rho}(\boldsymbol{x}, t)\,\bar{\boldsymbol{v}}(\boldsymbol{x}, t)] = 0\,. \tag{2.12}$$

□

Proof

First of all, the global formulation of the mass balance in the material representation leads to

$$\frac{\partial}{\partial t}\iiint\limits_{V}\hat{\rho}_{\mathrm{R}}(\boldsymbol{X}, t)\mathrm{d}V = \iiint\limits_{V}\frac{\partial}{\partial t}\hat{\rho}_{\mathrm{R}}(\boldsymbol{X}, t)\mathrm{d}V = 0$$

for all material bodies, i.e. for all partial volumes of V or $\mathrm{R}[\mathscr{B}]$. If the density ρ_{R} is a continuous function of $\boldsymbol{X}$, it follows from Theorem 2. 1 that

$$\frac{\partial}{\partial t}\hat{\rho}_{\mathrm{R}}(\boldsymbol{X}, t) = 0\,,\ \text{i.e.}\quad \rho_{\mathrm{R}} = \hat{\rho}_{\mathrm{R}}(\boldsymbol{X})\,.$$

The density of the reference configuration is constant in time. The global formulation in the spatial representation,

$$\frac{\mathrm{d}}{\mathrm{d}t}\iiint\limits_{\chi_t[\mathscr{B}]}\bar{\rho}(\boldsymbol{x}, t)\mathrm{d}v = 0\,,$$

requires the calculation of the material time derivative of an integral over an integration volume that changes in time, due to the fact that it moves with the material velocity field $\boldsymbol{v}(\boldsymbol{x}, t)$. This calculation can be carried out if the integration is performed over the volume of the reference configuration. The result is

$$\frac{\mathrm{d}}{\mathrm{d}t}\iiint\limits_{\chi_t[\mathscr{B}]}\bar{\rho}(\boldsymbol{x}, t)\mathrm{d}v = \frac{\partial}{\partial t}\iiint\limits_{\mathrm{R}[\mathscr{B}]}\hat{\rho}(\boldsymbol{X}, t)\det\mathbf{F}(\boldsymbol{X}, t)\,\mathrm{d}V$$

$$= \iiint\limits_{\mathrm{R}[\mathscr{B}]}[\dot{\hat{\rho}}(\boldsymbol{X}, t)\det\mathbf{F}(\boldsymbol{X}, t) + \hat{\rho}(\boldsymbol{X}, t)\det\mathbf{F}(\boldsymbol{X}, t)\mathrm{tr}(\dot{\mathbf{F}}\mathbf{F}^{-1})]\mathrm{d}V\,.$$

Transforming the result back to the current configuration leads to

$$\frac{\mathrm{d}}{\mathrm{d}t}\iiint\limits_{\chi_t[\mathscr{B}]}\bar{\rho}(\boldsymbol{x},t)\mathrm{d}v=\iiint\limits_{\chi_t[\mathscr{B}]}\left[\dot{\bar{\rho}}(\boldsymbol{x},t)+\bar{\rho}(\boldsymbol{x},t)\,\mathrm{div}\,\bar{\boldsymbol{v}}(\boldsymbol{x},t)\right]\mathrm{d}v\ .$$

The mass balance requires that this volume integral vanish for any choice of integration volume v:

$$\iiint\limits_{v}\left[\dot{\bar{\rho}}(\boldsymbol{x},t)+\bar{\rho}(\boldsymbol{x},t)\,\mathrm{div}\,\bar{\boldsymbol{v}}(\boldsymbol{x},t)\right]\mathrm{d}v=0\ .$$

Assuming that the integrand is continuous in space, it follows, according to the line of reasoning documented in Theorem 2. 1, that $\dot{\rho}\ +\ \rho\mathrm{div}\boldsymbol{v}=0$.

The material time derivative of the density,

$$\dot{\rho}(\boldsymbol{x},t)=\frac{\partial}{\partial t}\rho(\boldsymbol{x},t)+\big(\mathrm{grad}\rho(\boldsymbol{x},t)\big)\cdot\boldsymbol{v}(\boldsymbol{x},t)\ ,$$

and the product rule, $(\mathrm{grad}\rho)\cdot\boldsymbol{v}+\rho(\mathrm{div}\boldsymbol{v})=\mathrm{div}(\rho\boldsymbol{v})$, yield the equivalent formulation

$$\frac{\partial\rho}{\partial t}+\mathrm{div}(\rho\boldsymbol{v})=0\ .$$ ■

In the last steps the horizontal lines above ρ and $\boldsymbol{v}$ have been omitted. This practice is generally employed for simplifying the notation, provided no misunderstandings arise. Accordingly, the mathematically inprecise way of expressing the spatial representation is $\rho=\rho(\boldsymbol{x},t)$ and $\rho=\rho(\boldsymbol{X},t)$ when referring to the material representation. Similar practices are common for all other physical quantities.

The local formulation for the mass balance is called the *continuity equation*. Equation (2.10) is the continuity equation in the material representation, equations (2.11) and (2.12) the continuity equations in the spatial representation.

The local mass balance (2.12) is a partial differential equation, linking the temporal change of the mass density with the spatial derivatives of the components of the velocity vector. While the symbolically written continuity equation is valid for all coordinate systems, the formulation after the introduction of a component representation for $\boldsymbol{v}$ depends on the choice of coordinate system. To quote an example, the local mass balance when choosing Cartesian coordinates (x,y,z) reads

$$\frac{\partial \rho}{\partial t} + \frac{\partial}{\partial x}(\rho v_x) + \frac{\partial}{\partial y}(\rho v_y) + \frac{\partial}{\partial z}(\rho v_z) = 0 .$$

To abbreviate this, one can also write

$$\frac{\partial \rho}{\partial t} + \frac{\partial}{\partial x_k}(\rho v_k) = 0 \tag{2.13}$$

or even shorter $\dfrac{\partial \rho}{\partial t} + (\rho v_k)_{,k} = 0$.

When using curvilinear coordinates the partial derivatives with respect to the local coordinates have to be replaced by covariant derivatives (see (1.187)):

$$\frac{\partial \rho}{\partial t} + (\rho v_k)\big|_k = 0 . \tag{2.14}$$

Volume preserving (isochoric) motions are characterised by

$$\mathrm{d}v \equiv \mathrm{d}V \Longleftrightarrow \det\mathbf{F} \equiv 1 \tag{2.15}$$

or by

$$(\mathrm{d}v)^{\cdot} \equiv 0 \Longleftrightarrow \mathrm{div}\boldsymbol{v} = 0 . \tag{2.16}$$

In this case the continuity equation leads to

$$\dot{\rho} = \frac{\partial \rho}{\partial t} + (\mathrm{grad}\rho)\cdot \boldsymbol{v} = 0 . \tag{2.17}$$

The mass density in the case of an isochoric motion is materially constant. It is expedient to relate the densities of volume-distributed physical quantities to the mass. This simplifies the material time derivatives of the volume integrals. The local form of mass balance leads to

Theorem 2. 3

If g: $(\mathscr{P}, t) \longmapsto \mathrm{w} = \mathrm{g}(\mathscr{P}, t)$ is a physical quantity representing a density function per unit mass and $\mathrm{w} = \bar{\mathrm{g}}(\boldsymbol{x}, t) = \bar{\mathrm{g}}(x^1, x^2, x^3, t)$ the corresponding *spatial representation*, the material derivative of the volume integral over $\rho \mathrm{g}(\cdot)$ is equal to the integral over the material derivative of the field $\bar{\mathrm{g}}(\boldsymbol{x}, t)$:

$$\frac{\mathrm{d}}{\mathrm{d}t} \iiint\limits_{\chi_t[\mathcal{B}]} \bar{\mathrm{g}}(\boldsymbol{x}, t)\rho \mathrm{d}v = \iiint\limits_{\chi_t[\mathcal{B}]} \dot{\bar{\mathrm{g}}}(\boldsymbol{x}, t)\rho \mathrm{d}v \, . \tag{2.18}$$

□

Proof

Transformation to the reference configuration and application of the mass conservation yields

$$\frac{\mathrm{d}}{\mathrm{d}t} \iiint\limits_{\chi_t[\mathcal{B}]} \bar{\mathrm{g}}(\boldsymbol{x}, t)\rho(\boldsymbol{x}, t)\mathrm{d}v = \frac{\partial}{\partial t} \iiint\limits_{\mathrm{R}[\mathcal{B}]} \hat{\mathrm{g}}(\boldsymbol{X}, t)\rho_{\mathrm{R}}(\boldsymbol{X})\mathrm{d}V = \iiint\limits_{\mathrm{R}[\mathcal{B}]} \frac{\partial \hat{\mathrm{g}}}{\partial t}(\boldsymbol{X}, t)\rho_{\mathrm{R}}(\boldsymbol{X})\mathrm{d}V \, .$$

Transforming back to the current configuration gives rise to the statement

$$\frac{\mathrm{d}}{\mathrm{d}t} \iiint\limits_{\chi_t[\mathcal{B}]} \bar{\mathrm{g}}(\boldsymbol{x}, t)\rho(\boldsymbol{x}, t)\mathrm{d}v = \iiint\limits_{\mathrm{R}[\mathcal{B}]} \dot{\hat{\mathrm{g}}}(\boldsymbol{X}, t)\rho_{\mathrm{R}}(\boldsymbol{X})\mathrm{d}V = \iiint\limits_{\chi_t[\mathcal{B}]} \dot{\bar{\mathrm{g}}}(\boldsymbol{x}, t)\rho(\boldsymbol{x}, t)\mathrm{d}v \, ,$$

with the material derivative of $\bar{\mathrm{g}}$ according to equation (1.41),

$$\dot{\bar{\mathrm{g}}}(\boldsymbol{x}, t) = \frac{\partial \bar{\mathrm{g}}}{\partial t}(\boldsymbol{x}, t) + \big(\mathrm{grad}\, \bar{\mathrm{g}}(\boldsymbol{x}, t)\big)\boldsymbol{v}(\boldsymbol{x}, t) \, .$$

■

The theorem's message is more easily conveyed if the conservation of mass is written with the *mass element*

$$\mathrm{d}m = \rho \mathrm{d}v = \rho_{\mathrm{R}} \mathrm{d}V \tag{2.19}$$

in the form

$$(\mathrm{d}m)^{\cdot} = (\rho \mathrm{d}v)^{\cdot} = [\dot{\rho} + \rho \mathrm{div}\boldsymbol{v}]\mathrm{d}v = 0 \, . \tag{2.20}$$

The result is then

$$\frac{\mathrm{d}}{\mathrm{d}t} \int\limits_{\mathcal{B}} \mathrm{g}\, \mathrm{d}m = \int\limits_{\mathcal{B}} \dot{\mathrm{g}} \mathrm{d}m \, . \tag{2.21}$$

2. 3 Linear Momentum and Rotational Momentum

2. 3. 1 Balance of Linear Momentum and Rotational Momentum: Global Formulation

The linear and rotational momentum of a material body are vectors which characterise the field of velocity $\boldsymbol{v}(\boldsymbol{x}, t)$ combined with the distribution of mass $\rho(\boldsymbol{x}, t)$.

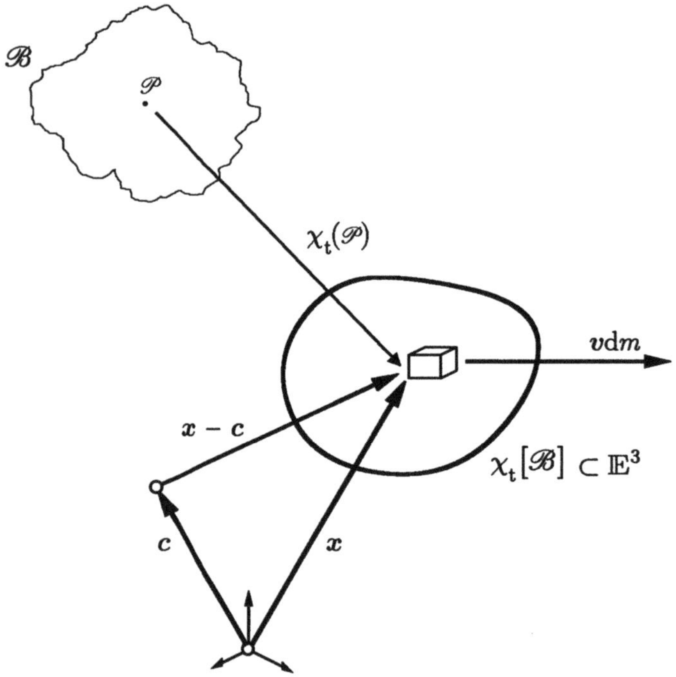

Figure 2. 1: Linear and rotational momentum

Definition 2. 3

The motion $\boldsymbol{x} = \chi_t(\mathcal{P})$ of a material body $\mathcal{B}$ determines its *momentum* or *linear momentum*

$$\boldsymbol{I} : (\mathcal{B}, t) \longmapsto \boldsymbol{I}(\mathcal{B}, t) = \int_{\mathcal{B}} \boldsymbol{v} dm$$

$$I(\mathscr{B}, t) = \iiint_{\chi_t[\mathscr{B}]} v(x, t)\rho(x, t)\mathrm{d}v = \iiint_{\mathrm{R}[\mathscr{B}]} v(X, t)\rho_{\mathrm{R}}(X)\mathrm{d}V \tag{2.22}$$

as well as its *moment of momentum* or *rotational momentum*

$$D_c : (\mathscr{B}, t) \longmapsto D_c(\mathscr{B}, t) = \int_{\mathscr{B}} (x - c) \times v\mathrm{d}m \,, \tag{2.23}$$

$$D_c(\mathscr{B}, t) = \iiint_{\chi_t[\mathscr{B}]} (x - c) \times v(x, t)\, \rho(x, t)\, \mathrm{d}v$$

$$D_c(\mathscr{B}, t) = \iiint_{\mathrm{R}[\mathscr{B}]} (\chi_{\mathrm{R}}(X, t) - c) \times v(X, t)\rho_{\mathrm{R}}(X)\mathrm{d}V \tag{2.24}$$

(see Fig. 2. 1). In the definition (2.23) of the rotational momentum c is the position vector of a fixed point in space selected at will, to which the rotational momentum refers. ◊

Definition 2. 4

The vector-valued function

$$t \longrightarrow x_m(t) = x_m(\mathscr{B}, t) = \frac{1}{m}\int_{\mathscr{B}} x\mathrm{d}m = \frac{1}{\iiint_{\chi_t[\mathscr{B}]} \rho(x, t)\mathrm{d}v} \iiint_{\chi_t[\mathscr{B}]} x\rho(x, t)\mathrm{d}v \tag{2.25}$$

is called the *centre of mass* of the material body $\mathscr{B}$. ◊

In combination with the conservation of mass, $\dot{m} = 0$, Definition 2. 3 leads to

$$I(\mathscr{B}, t) = m\dot{x}_m(\mathscr{B}, t) = mv_m(\mathscr{B}, t) \tag{2.26}$$

and

$$D_c(\mathscr{B}, t) = D_{x_m}(\mathscr{B}, t) + (x_m - c) \times I(\mathscr{B}, t)\,. \tag{2.27}$$

In this connection, the vector

$$\boldsymbol{D}_{\boldsymbol{x}_m}(\mathcal{B}, t) = \int_{\mathcal{B}} (\boldsymbol{x} - \boldsymbol{x}_m) \times \boldsymbol{v} \mathrm{d}m = \iiint_{\chi_t[\mathcal{B}]} (\boldsymbol{x} - \boldsymbol{x}_m) \times \boldsymbol{v} \rho \mathrm{d}v \tag{2.28}$$

is the rotational momentum with respect to the centre of mass $\boldsymbol{x}_m(t)$.
As shown in (2.26), the vector of linear momentum is the product of the total mass times the velocity of the centre of mass. A material body can consequently be identified with a mass-point. No such simple representation exists for the moment of momentum. According to (2.27), the moment of momentum decomposes additively, the second term being a purely translational part.

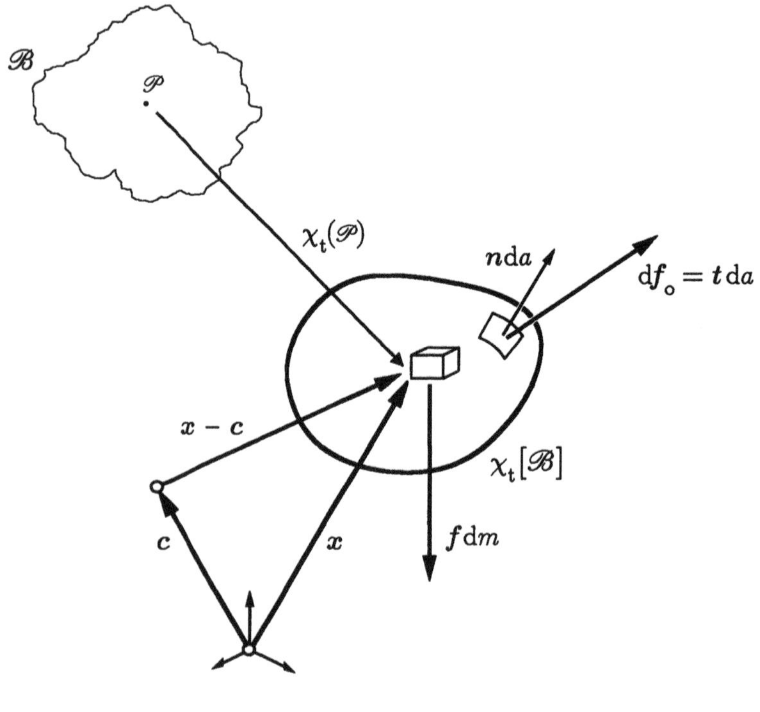

Figure 2. 2: Force and torque

The linear momentum and rotational momentum are influenced by forces acting upon a material body and thereby representing the mechanical effect of the external world in the sense of the principle of intersection. This idea is conveyed by

Definition 2. 5

According to the principle of intersection (free-body principle), a material body that is (mentally) cut out of its environment is exposed to the *resultant force*

$$\boldsymbol{F} : (\mathscr{B}, t) \longmapsto \boldsymbol{F}(\mathscr{B}, t) = \iint\limits_{\partial\chi_t[\mathscr{B}]} \boldsymbol{t}\mathrm{d}a + \iiint\limits_{\chi_t[\mathscr{B}]} \boldsymbol{f}\rho\mathrm{d}v \tag{2.29}$$

(see Fig. 2. 2). The resulting force is the sum of one part distributed over the surface and one over the volume; the volume distributed share is called the *volume force* and the integrand $\boldsymbol{f}(\mathscr{P}, t)$ *volume force density* per unit mass. The surface distributed part is called the *surface force* and the integrand $\boldsymbol{t}(\mathscr{P}, t)$ *surface force density* or *Cauchy stress vector.*

The distribution $\{\boldsymbol{t}(\cdot), \boldsymbol{f}(\cdot)\}$ of the surface and volume force densities forms a *system of forces.* Every system of forces determines a *resultant torque*

$$\boldsymbol{M}_c : (\mathscr{B}, t) \longmapsto \boldsymbol{M}_c (\mathscr{B}, t) = \iint\limits_{\partial\chi_t[\mathscr{B}]} (\boldsymbol{x} - \boldsymbol{c}) \times \boldsymbol{t}\mathrm{d}a + \iiint\limits_{\chi_t[\mathscr{B}]} (\boldsymbol{x} - \boldsymbol{c}) \times \boldsymbol{f}\rho\mathrm{d}v \, , \tag{2.30}$$

$\boldsymbol{c}$ being the position vector of a fixed point of reference. ◊

The volume force density $\boldsymbol{f} = \boldsymbol{f}(\mathscr{P}, t)$ is a vector-valued function defined on the material body $\mathscr{B}$. The stress vector $\boldsymbol{t}(\mathscr{P}, t)$, on the other hand, is merely defined on the boundary $\partial\mathscr{B}$ of $\mathscr{B}$. However, virtually every inner point of $\mathscr{B}$ can be turned into the boundary point of another material body in the sense of the principle of intersection.

Every system of forces, i.e. every conceivable distribution of volume and surface force densities, follows on from the free-body principle and represents a feasible load of the material body. There are systems of forces leading to the same resultant vectors of force and torque. For this reason the systems of forces can be divided into *equivalence classes*: two systems of forces are called *equivalent*, if they determine the same resultants. The term *equivalent force system* is independent of the choice of reference point $\boldsymbol{c}$.

A special equivalence class is formed by those systems of forces where both the resultant force and the resultant torque vanish. A force system of this kind is called a *system of equilibrium.* Both linear and rotational momentum remain temporally constant under the influence of a system of equilibrium. They vanish identically, provided that they equal zero at one moment. Accordingly, the global conditions of equilibrium, $\boldsymbol{F} = \boldsymbol{0}$ und $\boldsymbol{M}_c = \boldsymbol{0}$, are necessary (but not sufficient) conditions for the mechanical equilibrium of a material system.

The term force is not determined in full by the above definition. All the definition does is to ascertain that force is a quantity, principally represented by two vectors, namely the vector of force and the position vector of the material point where the current force is applied: any change in the force's point of application would generally alter the resultant torque.

The definition of force is not complete without linking its mathematical calculation to the characterisation of the state of motion. This is brought about by the balance equations for linear and rotational momentum. The formulation of these balance equations, which are nowadays in constant use, is attributed to Leonard Euler.[4]

Balance of Linear Momentum

The resultant force $\boldsymbol{F}$ acting on a material body causes a change of linear momentum $\boldsymbol{I}$:

$$\frac{\mathrm{d}}{\mathrm{d}t}\boldsymbol{I}(\mathscr{B}, t) = \boldsymbol{F}(\mathscr{B}, t) \tag{2.31}$$

is valid for all material bodies. ♦

Balance of Rotational Momentum

The resultant torque $\boldsymbol{M}_c$ of a system of forces acting on a material body causes a change of rotational momentum $\boldsymbol{D}_c$: for all material bodies

$$\frac{\mathrm{d}}{\mathrm{d}t}\boldsymbol{D}_c(\mathscr{B}, t) = \boldsymbol{M}_c(\mathscr{B}, t) \tag{2.32}$$

is valid. Both the vector of rotational momentum and the torque refer to one and the same spatially fixed reference point $\boldsymbol{c}$. ♦

An immediate deduction from the balance equations for mass and momentum is

Theorem 2. 4: Theorem of the Centre of Mass

The resultant force is equal to mass m times the acceleration $\ddot{\boldsymbol{x}}_m$ of the centre of mass:

$$m\ddot{\boldsymbol{x}}_m(\mathscr{B}, t) = \boldsymbol{F}\,. \tag{2.33}$$

□

[4] See TRUESDELL & TOUPIN [1960], Sect. 196.

Proof

The definition (2.25) of the centre of mass, $m\boldsymbol{x}_m(\mathcal{B}, t) = \int_{\mathcal{B}} \boldsymbol{x} dm$, in conjunction with the balance of mass (2.8),
implies $\boldsymbol{I} = m\dot{\boldsymbol{x}}_m$ and $\dot{\boldsymbol{I}} = m\ddot{\boldsymbol{x}}_m(\mathcal{B}, t)$. ■

The balance equations for mass, linear momentum and rotational momentum lead to

Theorem 2. 5: Theorem of Rotational Momentum

In the balance of rotational momentum the spatially fixed point $\boldsymbol{c}$ can be replaced by the centre of mass:

$$\dot{\boldsymbol{D}}_{\boldsymbol{x}_m}(\mathcal{B}, t) = \boldsymbol{M}_{\boldsymbol{x}_m} . \tag{2.34}$$

□

Proof

The definitions of rotational momentum and resultant torque permit a recalculation of the reference point. This yields

$$\boldsymbol{D}_c = \int_{\mathcal{B}} (\boldsymbol{x} - \boldsymbol{x}_m + \boldsymbol{x}_m - \boldsymbol{c}) \times \boldsymbol{v} dm = \boldsymbol{D}_{\boldsymbol{x}_m} + (\boldsymbol{x}_m - \boldsymbol{c}) \times \boldsymbol{I} , \tag{2.35}$$

$$\dot{\boldsymbol{D}}_c = \dot{\boldsymbol{D}}_{\boldsymbol{x}_m} + (\boldsymbol{x}_m - \boldsymbol{c}) \times \left(m\ddot{\boldsymbol{x}}_m(t)\right) \text{ and}$$

$$\boldsymbol{M}_c = \boldsymbol{M}_{\boldsymbol{x}_m} + (\boldsymbol{x}_m - \boldsymbol{c}) \times \boldsymbol{F} . \tag{2.36}$$

The balance of rotational momentum (2.31) leads to

$$\begin{aligned} \dot{\boldsymbol{D}}_c(\mathcal{B}, t) &= \dot{\boldsymbol{D}}_{\boldsymbol{x}_m}(\mathcal{B}, t) + (\boldsymbol{x}_m - \boldsymbol{c}) \times \left(m\ddot{\boldsymbol{x}}_m(\mathcal{B}, t)\right) \\ &= \boldsymbol{M}_{\boldsymbol{x}_m} + (\boldsymbol{x}_m - \boldsymbol{c}) \times \boldsymbol{F} , \end{aligned}$$

and with the theorem of the centre of mass $m\ddot{\boldsymbol{x}}_m = \boldsymbol{F}$ we arrive at

$$\dot{\boldsymbol{D}}_{\boldsymbol{x}_m} = \boldsymbol{M}_{\boldsymbol{x}_m} .$$

■

2. 3. 2 Stress Tensors

In view of the balance laws of linear momentum and mass, the resultant surface force reads

$$\iint\limits_{\partial\chi_t[\mathscr{B}]} t\mathrm{d}a = \frac{\mathrm{d}}{\mathrm{d}t}\iiint\limits_{\chi_t[\mathscr{B}]} v\rho\mathrm{d}v - \iiint\limits_{\chi_t[\mathscr{B}]} f\rho\mathrm{d}v = \iiint\limits_{\chi_t[\mathscr{B}]} (\dot{v} - f)\rho\mathrm{d}v \,.$$

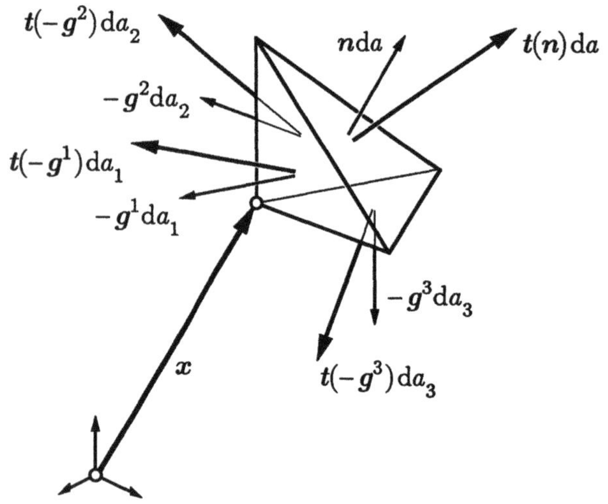

Figure 2. 3: Stress vectors

The surface force density t, which is defined on the whole surface, formed by sectioning off a body from its surroundings, is referred to as the *stress vector*. A stress vector is allotted to every point x of the boundary surface of a body in the current configuration. Elementary arguments suggest that the stress vector has to depend on quantities other than x and t, i.e. at least on the orientation of the surface element, that is on the outer unit normal n:

$$t = t(x, t, n, ...) \qquad (x \in \partial\chi_t[\mathscr{B}]) \,.$$

In continuum mechanics it is commonly assumed that the stress vector depends, apart from $x \in \partial\chi_t[\mathscr{B}]$ and t, only on n (*Cauchy* stress principle). On the basis of this assumption we set up the following

Theorem 2. 6

$\boldsymbol{t} = \boldsymbol{t}(\boldsymbol{x}, t, \boldsymbol{n})$ leads to

$$\boldsymbol{t}(\boldsymbol{x}, t, \boldsymbol{n}) = \mathbf{T}(\boldsymbol{x}, t)\boldsymbol{n} \ , \quad \boldsymbol{t} = \mathbf{T}\boldsymbol{n} \ . \tag{2.37}$$

The dependence of the stress vector on the normal vector is *linear*.[5] □

Proof

The proof of this statement gives an indication of the unlimited use of the free-body principle. It consists of applying the balance of linear momentum to a series of material bodies, whose images in the current configuration are geometrically similar tetrahedrons. Three planes of the tetrahedrons are tangent to the coordinate surfaces through $\boldsymbol{x}$, whereas the fourth plane has an arbitrary normal $\boldsymbol{n}$. These tetrahedrons all contain the common point $\boldsymbol{x}$ and their volumes form a zero sequence. For a tetrahedron out of this sequence (Fig. 2. 3), formed by the oriented surfaces $-a_1\boldsymbol{g}^1$, $-a_2\boldsymbol{g}^2$, $-a_3\boldsymbol{g}^3$ and $a\boldsymbol{n}$, the geometry produces the property $a\boldsymbol{n} - a_i\boldsymbol{g}^i = \boldsymbol{0}$, that means the given normal vector $\boldsymbol{n}$ permits the component representation

$$\boldsymbol{n} = \frac{a_i}{a}\,\boldsymbol{g}^i = n_i\boldsymbol{g}^i \ , \tag{2.38}$$

in which the covariant components n_i are the ratios of the surfaces a_i and a. A resultant surface force acts on each surface of the tetrahedron:

$$\boldsymbol{t}(-\boldsymbol{g}^1)a_1 = \iint\limits_{a_1} \boldsymbol{t}(\boldsymbol{x}, t, -\boldsymbol{g}^1)\mathrm{d}a \ , \quad \boldsymbol{t}(-\boldsymbol{g}^2)a_2 = \iint\limits_{a_2} \boldsymbol{t}(\boldsymbol{x}, t, -\boldsymbol{g}^2)\mathrm{d}a \ ,$$
$$\boldsymbol{t}(-\boldsymbol{g}^3)a_3 = \iint\limits_{a_3} \boldsymbol{t}(\boldsymbol{x}, t, -\boldsymbol{g}^3)\mathrm{d}a \ , \quad \boldsymbol{t}(\boldsymbol{n})a = \iint\limits_{a} \boldsymbol{t}(\boldsymbol{x}, t, \boldsymbol{n})\mathrm{d}a \ . \tag{2.39}$$

These equations are to be understood in the sense of the mean value theorem of the integral calculus.[6] The left-hand sides contain the values of the integrands at suitable points δ_1, δ_2, δ_3 within each integration area, i.e.

$\boldsymbol{t}(-\boldsymbol{g}^1)a_1 = \boldsymbol{t}(\delta_1, t, -\boldsymbol{g}^1)a_1$ etc.

---

[5] For further arguments and references see TRUESDELL & TOUPIN [1960], Sect. 203.

[6] See, for example, MARSDEN & TROMBA [1995], Theorem 5. 5, p. 326.

The resultant volume force is accordingly expressed by $v\boldsymbol{k} = \iiint\limits_v \rho \boldsymbol{f} \mathrm{d}v$ $\left(\boldsymbol{k} = \rho \boldsymbol{f}(\delta, t)\right)$, and the change of linear momentum by

$$\rho v \ddot{\boldsymbol{x}}_m = \frac{\mathrm{d}}{\mathrm{d}t} \iiint\limits_v \boldsymbol{v}(\boldsymbol{x}, t) \rho \mathrm{d}v = \iiint\limits_v \dot{\boldsymbol{v}}(\boldsymbol{x}, t) \rho \mathrm{d}v .$$

Under these terms the balance of linear momentum is written as

$\rho v \ddot{\boldsymbol{x}}_m = \boldsymbol{t}(\boldsymbol{n})a + \boldsymbol{t}(-\boldsymbol{g}^i)a_i + v\boldsymbol{k}$. This yields

$$\boldsymbol{t}(\boldsymbol{n})a = -\boldsymbol{t}(-\boldsymbol{g}^i)a_i + v(\rho \ddot{\boldsymbol{x}}_m - \boldsymbol{k}), \quad \text{and with } n_i = \frac{a_i}{a}$$

$$\boldsymbol{t}(\boldsymbol{n}) = -\boldsymbol{t}(-\boldsymbol{g}^i)n_i + \frac{v}{a}(\rho \ddot{\boldsymbol{x}}_m(t) - \boldsymbol{k}) . \tag{2.40}$$

The entire zero sequence of the geometrically similar tetrahedrons is now considered: it is defined by the following properties:

$$v \to 0 , \quad \frac{v}{a} \to 0 , \quad \boldsymbol{g}^i = \text{const.}, \quad \boldsymbol{n} = \text{const.}$$

Passing to the limit, we obtain the intermediate result

$$\boldsymbol{t}(\boldsymbol{n}) = -\boldsymbol{t}(-\boldsymbol{g}^i)n_i \quad \text{or} \quad \boldsymbol{t}(\boldsymbol{x}, t, \boldsymbol{n}) = -\boldsymbol{t}(\boldsymbol{x}, t, -\boldsymbol{g}^i)n_i .$$

The limit value of the stress vector, $\boldsymbol{t}(\boldsymbol{n}) = \boldsymbol{t}(\boldsymbol{x}, t, \boldsymbol{n})$, is associated with the sectional plane with the normal vector $\boldsymbol{n}$ through the point $\boldsymbol{x}$, belonging to all tetrahedons collectively. A vital implication for the conclusive proof is obtained from the intermediate result by means of successive identification of $\boldsymbol{n}$ with $\boldsymbol{g}^1, \boldsymbol{g}^2, \boldsymbol{g}^3$ ($n_i = 1$ and $n_k = 0$ for $k \neq i$):

$$\boldsymbol{t}(\boldsymbol{x}, t, \boldsymbol{g}^i) = -\boldsymbol{t}(\boldsymbol{x}, t, -\boldsymbol{g}^i) .$$

This leads to $\boldsymbol{t}(\boldsymbol{x}, t, \boldsymbol{n}) = n_i \boldsymbol{t}(\boldsymbol{x}, t, \boldsymbol{g}^i)$, which shows that the stress vector $\boldsymbol{t}(\boldsymbol{n})$ depends linearly on the normal vector $\boldsymbol{n}$. This dependence is expressed by means of the tensor

$$\mathrm{T}(\boldsymbol{x}, t) = \boldsymbol{t}(\boldsymbol{x}, t, \boldsymbol{g}^k) \otimes \boldsymbol{g}_k : \tag{2.41}$$

$$\boldsymbol{t}(\boldsymbol{x}, t, \boldsymbol{n}) = \mathrm{T}(\boldsymbol{x}, t)\boldsymbol{n} .$$

■

The stress tensor $\mathbf{T}$ represents the state of stress currently present at a material point $\mathscr{P}$,

$$\mathbf{T} : (\mathscr{P}, t) \longmapsto \mathbf{T}(\mathscr{P}, t) = \bar{\mathbf{T}}(\boldsymbol{x}, t) = \hat{\mathbf{T}}(\boldsymbol{X}, t) . \tag{2.42}$$

Definition 2. 6

The tensor $\mathbf{T}$ with the property $\boldsymbol{t} = \mathbf{T}\boldsymbol{n}$ or $\mathrm{d}\boldsymbol{f}_{\mathrm{o}} = \boldsymbol{t}(\boldsymbol{n})\mathrm{d}a = \mathbf{T}\boldsymbol{n}\mathrm{d}a = \mathbf{T}\mathrm{d}\boldsymbol{a}$ is called *Cauchy* stress tensor.[7] The components of the stress tensor in the representations

$$\mathbf{T} = T^{kl}\boldsymbol{g}_k \otimes \boldsymbol{g}_l = T_{kl}\,\boldsymbol{g}^k \otimes \boldsymbol{g}^l = T_k{}^l\boldsymbol{g}^k \otimes \boldsymbol{g}_l = T^k{}_l\,\boldsymbol{g}_k \otimes \boldsymbol{g}^l \tag{2.43}$$

are called *stresses*, either *normal stresses* ($i = k$) or *shear stresses* ($i \neq k$). ◊

The components of the *Cauchy* stress tensor possess the dimension force per area and describe the surface force acting on a material surface element in the current configuration.

Whereas the physical action of a force obviously takes place in the current configuration, its intensity can be attributed just as well to the material surface element in the reference configuration. This leads to a different stress tensor, explained by

Definition 2. 7

The tensor $\mathbf{T}_{\mathrm{R}}$ with the property

$$\boldsymbol{t}_{\mathrm{R}} = \mathbf{T}_{\mathrm{R}}\boldsymbol{n}_{\mathrm{R}} , \tag{2.44}$$

$$\mathrm{d}\boldsymbol{f}_{\mathrm{o}} = \boldsymbol{t}_{\mathrm{R}}(\boldsymbol{n}_{\mathrm{R}})\mathrm{d}A = \mathbf{T}_{\mathrm{R}}\boldsymbol{n}_{\mathrm{R}}\mathrm{d}A \tag{2.45}$$

is called the 1st *Piola-Kirchhoff* tensor.[8] ◊

Whereas the *Cauchy* stress vector $\boldsymbol{t}$ refers to the material surface element in the current configuration, the *Piola-Kirchhoff* stress vector $\boldsymbol{t}_{\mathrm{R}}$ refers to the surface element in the reference configuration. For the differential surface force we write

$$\mathrm{d}\boldsymbol{f}_{\mathrm{o}} = \boldsymbol{t}\mathrm{d}a = \boldsymbol{t}_{\mathrm{R}}\mathrm{d}A . \tag{2.46}$$

---

[7] More detailed information is given in TRUESDELL & TOUPIN [1960], Sect. 203.

[8] See TRUESDELL & TOUPIN [1960], Sect. 210.

The two stress vectors $\boldsymbol{t}$ and $\boldsymbol{t}_R$ are obviously parallel. The relationship between the two stress tensors $\mathbf{T}$ and $\mathbf{T}_R$ is a direct result of their definition, when the transformation property (1.59) of the material surface element is taken into account:

$\mathrm{d}\boldsymbol{f}_o = \mathbf{T}\boldsymbol{n}\mathrm{d}a = \mathbf{T}\mathrm{d}\boldsymbol{a} = \mathbf{T}\{(\det\mathbf{F})\mathbf{F}^{\mathrm{T}-1}\mathrm{d}\boldsymbol{A}\} = \mathbf{T}_R\mathrm{d}\boldsymbol{A}$ leads to the identity

$\{(\det\mathbf{F})\mathbf{T}\mathbf{F}^{\mathrm{T}-1}\}\mathrm{d}\boldsymbol{A} = \mathbf{T}_R\mathrm{d}\boldsymbol{A}$, valid for all vectors $\mathrm{d}\boldsymbol{A}$. This implies

$$\mathbf{T}_R = (\det\mathbf{F})\ \mathbf{T}\ \mathbf{F}^{\mathrm{T}-1}. \tag{2.47}$$

Three further stress tensors are prepared for later use:

Definition 2. 8
The stress tensor

$$\mathbf{S} = (\det\mathbf{F})\mathbf{T} \tag{2.48}$$

is called the weighted *Cauchy* tensor (or *Kirchhoff* tensor),

$$\tilde{\mathbf{T}} = \mathbf{F}^{-1}\mathbf{S}\,\mathbf{F}^{\mathrm{T}-1} \tag{2.49}$$

is called the 2nd *Piola-Kirchhoff* tensor and

$$\tilde{\mathbf{t}} = \mathbf{F}^{\mathrm{T}}\mathbf{S}\,\mathbf{F} \tag{2.50}$$

is known as the *convective stress tensor.*[9] ◊

The definition corresponds to the following relations:

$$\mathbf{S} = \frac{\rho_R}{\rho}\,\mathbf{T}\,, \tag{2.51}$$

$$\tilde{\mathbf{T}} = (\det\mathbf{F})\mathbf{F}^{-1}\mathbf{T}\mathbf{F}^{\mathrm{T}-1}\,, \tag{2.52}$$

$$\tilde{\mathbf{t}} = \mathbf{C}\,\tilde{\mathbf{T}}\,\mathbf{C}\,. \tag{2.53}$$

The physical meaning of the stress tensors $\mathbf{S}$, $\tilde{\mathbf{T}}$ and $\tilde{\mathbf{t}}$ becomes apparent in connection with the balance of the mechanical energy and also in connection with the formulation of material properties.

---

[9] Cf. TRUESDELL & NOLL [1965], Sect. 43A.

2. 3. 3 Stress Tensors in Convective Coordinates

With reference to convective coordinates, the natural component representation of the *Cauchy* stress tensor is contravariant, since the normal vector $\boldsymbol{n} = n_L \boldsymbol{g}^L$ exists as a matter of course in the covariant representation:

$$\mathbf{T} = T^{IK} \boldsymbol{g}_I \otimes \boldsymbol{g}_K, \tag{2.54}$$

$$\mathbf{S} = S^{IK} \boldsymbol{g}_I \otimes \boldsymbol{g}_K. \tag{2.55}$$

The definition of the 2nd *Piola-Kirchhoff* tensor in connection with the deformation gradient $\mathbf{F} = \boldsymbol{g}_L \otimes \boldsymbol{G}^L$ yields the component representation

$$\tilde{\mathbf{T}} = S^{IK} \boldsymbol{G}_I \otimes \boldsymbol{G}_K. \tag{2.56}$$

Based on the covariant representation of the weighted *Cauchy* tensor,

$$\mathbf{S} = S_{IK} \boldsymbol{g}^I \otimes \boldsymbol{g}^K, \tag{2.57}$$

we obtain for the convective stress tensor

$$\tilde{\mathbf{t}} = S_{IK} \boldsymbol{G}^I \otimes \boldsymbol{G}^K. \tag{2.58}$$

In the case of convective coordinates the 1st *Piola-Kirchhoff* tensor adopts a component representation in alignment with the mixed base system

$$\{\boldsymbol{g}_I \otimes \boldsymbol{G}_K\}: \mathbf{T}_{\mathrm{R}} = \mathbf{S}\mathbf{F}^{\mathrm{T}\,-1} = (S^{IK} \boldsymbol{g}_I \otimes \boldsymbol{g}_K)(\boldsymbol{g}^L \otimes \boldsymbol{G}_L),$$

$$\mathbf{T}_{\mathrm{R}} = S^{IK} \boldsymbol{g}_I \otimes \boldsymbol{G}_K. \tag{2.59}$$

2. 3. 4 Local Formulation of the Balance of Linear Momentum and Rotational Momentum

The tensor field $\mathbf{T} : (\mathscr{P}, t) \longmapsto \mathbf{T}(\mathscr{P}, t)$ describes the current distribution of stress in a material body. The boundary values of the stress tensor determine the surface force. With the stress tensor the balance of linear momentum takes on the form

$$\iiint\limits_{\chi_t[\mathscr{B}]} \dot{v}(x,t)\rho \mathrm{d}v = \iint\limits_{\partial\chi_t[\mathscr{B}]} \mathbf{T}n \mathrm{d}a + \iiint\limits_{\chi_t[\mathscr{B}]} f\rho \mathrm{d}v \tag{2.60}$$

(spatial representation) or

$$\iiint\limits_{\mathrm{R}[\mathscr{B}]} \dot{v}(X,t)\rho_{\mathrm{R}} \mathrm{d}V = \iint\limits_{\partial\mathrm{R}[\mathscr{B}]} \mathbf{T}_{\mathrm{R}}n_{\mathrm{R}} \mathrm{d}A + \iiint\limits_{\mathrm{R}[\mathscr{B}]} f\rho_{\mathrm{R}} \mathrm{d}V \tag{2.61}$$

(material representation). On the left-hand side of these equations the differentiation of linear momentum with respect to time has already been carried out. If we assume adequate continuity properties for all fields that occur, it is possible to use the global formulations to come to a conclusion concerning the local formulations, which are due to A.-L. Cauchy:[10]

Theorem 2. 7

The local form of the balance of linear momentum in the spatial representation is written

$$\rho\dot{v}(x,t) = \operatorname{div} \mathbf{T}(x,t) + \rho f\,, \tag{2.62}$$

with the material derivative of the velocity vector according to equation (1.43),

$$\dot{v}(x,t) = \frac{\mathrm{d}}{\mathrm{d}t}\, v(x,t) = \frac{\partial}{\partial t}v(x,t) + \big[\operatorname{grad} v(x,t)\big]v(x,t)\,,$$

and the spatial divergence of the *Cauchy* stress tensor,

$$\operatorname{div} \mathbf{T}(x,t) = \frac{\partial \mathbf{T}}{\partial x^k} g^k\,. \tag{2.63}$$

In the material representation the local balance of linear momentum reads

$$\rho_{\mathrm{R}}\dot{v}(X,t) = \operatorname{Div} \mathbf{T}_{\mathrm{R}}(X,t) + \rho_{\mathrm{R}} f\,, \tag{2.64}$$

with $\dot{v}(X,t) = \frac{\partial}{\partial t}v(X,t) = \frac{\partial^2}{\partial t^2}\chi_{\mathrm{R}}(X,t)$ according to equation (1.42),

[10] See TRUESDELL & TOUPIN [1960], Sect. 205.

and the material divergence of the 1st *Piola-Kirchhoff* tensor,

$$\operatorname{Div} \mathbf{T}_{\mathrm{R}}(\boldsymbol{x}, t) = \frac{\partial \mathbf{T}_{\mathrm{R}}}{\partial X^K} \boldsymbol{G}^K . \tag{2.65}$$

□

Proof

The derivation of the local balance of linear momentum is based on the possibility of expressing the resultant surface force by means of a volume integral: the divergence theorem (*Gauß* theorem) supplies

$$\iint\limits_{a} \mathbf{T}\boldsymbol{n}\mathrm{d}a = \iiint\limits_{v} \operatorname{div}\mathbf{T}\mathrm{d}v , \tag{2.66}$$

$$\iint\limits_{A} \mathbf{T}_{\mathrm{R}}\boldsymbol{n}_{\mathrm{R}}\, \mathrm{d}A = \iiint\limits_{V} \operatorname{Div}\mathbf{T}_{\mathrm{R}}\mathrm{d}V . \tag{2.67}$$

The balance of momentum (2.60) then leads to

$$\iiint\limits_{\chi_t[\mathscr{B}]} (\rho\dot{\boldsymbol{v}} - \operatorname{div} \mathbf{T}(\boldsymbol{x}, t) - \rho\boldsymbol{f})\mathrm{d}v = \boldsymbol{0} .$$

This volume integral has to vanish for any given volume of integration. If the integrand is a continuous function of space, it must be zero. That proves the spatial representation. Proving the material representation is carried out in the same way. ■

The local balance of momentum consists of three partial differential equations, which link the components of the acceleration vector with the derivatives of stresses with respect to spatial coordinates. Whereas the vector-valued differential equations formulated in symbolic notation are valid for all coordinate systems, the detailed formulation of the component equations depends on the special choice of coordinates. For example, when Cartesian coordinates are selected, the local balance of momentum in the spatial representation is written as follows:

$$\rho\left(\frac{\partial v_x}{\partial t}+\frac{\partial v_x}{\partial x}v_x+\frac{\partial v_x}{\partial y}v_y+\frac{\partial v_x}{\partial z}v_z\right)=\frac{\partial T_{xx}}{\partial x}+\frac{\partial T_{xy}}{\partial y}+\frac{\partial T_{xz}}{\partial z}+\rho f_x\,,$$

$$\rho\left(\frac{\partial v_y}{\partial t}+\frac{\partial v_y}{\partial x}v_x+\frac{\partial v_y}{\partial y}v_y+\frac{\partial v_y}{\partial z}v_z\right)=\frac{\partial T_{yx}}{\partial x}+\frac{\partial T_{yy}}{\partial y}+\frac{\partial T_{yz}}{\partial z}+\rho f_y\,, \quad (2.68)$$

$$\rho\left(\frac{\partial v_z}{\partial t}+\frac{\partial v_z}{\partial x}v_x+\frac{\partial v_z}{\partial y}v_y+\frac{\partial v_z}{\partial z}v_z\right)=\frac{\partial T_{zx}}{\partial x}+\frac{\partial T_{zy}}{\partial y}+\frac{\partial T_{zz}}{\partial z}+\rho f_z\,.$$

This can be abbreviated to

$$\rho\frac{\partial v_i}{\partial t}+\frac{\partial v_i}{\partial x_k}v_k=\frac{\partial T_{ik}}{\partial x_k}+\rho f_i \quad (2.69)$$

or even shorter to

$$\rho\frac{\partial v_i}{\partial t}+v_{i,k}v_k=T_{ik,k}+\rho f_i\,. \quad (2.70)$$

When employing curvilinear coordinates, the partial derivatives with respect to the spatial coordinates have to be replaced by covariant derivatives according to (1.187), (1.192).

In order to derive a local form of the balance of moment of momentum we proceed in a similar way. With the stress tensor the balance of rotational momentum takes on the form

$$\iiint\limits_{\chi_t[\mathcal{B}]}(\boldsymbol{x}-\boldsymbol{c})\times\dot{\boldsymbol{v}}(\boldsymbol{x},t)\rho\mathrm{d}v=\iint\limits_{\partial\chi_t[\mathcal{B}]}(\boldsymbol{x}-\boldsymbol{c})\times\mathbf{T}\boldsymbol{n}\mathrm{d}a+\iiint\limits_{\chi_t[\mathcal{B}]}(\boldsymbol{x}-\boldsymbol{c})\times\boldsymbol{f}\rho\mathrm{d}v\,, \quad (2.71)$$

bearing in mind that

$$[(\boldsymbol{x}-\boldsymbol{c})\times\boldsymbol{v}(\boldsymbol{x},t)]^{\cdot}=(\boldsymbol{x}-\boldsymbol{c})^{\cdot}\times\boldsymbol{v}(\boldsymbol{x},t)+(\boldsymbol{x}-\boldsymbol{c})\times\dot{\boldsymbol{v}}(\boldsymbol{x},t)$$

$$=(\boldsymbol{x}-\boldsymbol{c})\times\dot{\boldsymbol{v}}(\boldsymbol{x},t)\quad(\dot{\boldsymbol{c}}=\boldsymbol{0}\,,\dot{\boldsymbol{x}}=\boldsymbol{v}).$$

The local formulation of the balance of rotational momentum is the statement made by

Theorem 2. 8

The local form of the balance of rotational momentum reads

$$\mathbf{T} = \mathbf{T}^{\mathrm{T}}, \tag{2.72}$$

in other words, the *Cauchy* stress tensor is symmetric. □

Proof

To prove the symmetry of the stress tensor, we transform the resultant torque of the surface forces into a volume integral by applying the *Gauß* theorem. In order to do so, the surface torque density is written in the form

$$(\boldsymbol{x} - \boldsymbol{c}) \times \mathbf{T}\boldsymbol{n} = (\boldsymbol{x} - \boldsymbol{c}) \times \left[(T^{ij}\boldsymbol{g}_i \otimes \boldsymbol{g}_j)\boldsymbol{n}\right] = \{T^{ij}[(\boldsymbol{x} - \boldsymbol{c}) \times \boldsymbol{g}_i] \otimes \boldsymbol{g}_j\}\boldsymbol{n}\,,$$

$$(\boldsymbol{x} - \boldsymbol{c}) \times \mathbf{T}\boldsymbol{n} = \mathbf{T}_{\mathrm{M}}\boldsymbol{n}\,,$$

from which the divergence of the tensor

$$\mathbf{T}_{\mathrm{M}} = (\boldsymbol{x} - \boldsymbol{c}) \times \mathbf{T} = T^{ij}[(\boldsymbol{x} - \boldsymbol{c}) \times \boldsymbol{g}_i] \otimes \boldsymbol{g}_j \tag{2.73}$$

now has to be calculated. This is achieved by employing the definition

$$\mathrm{div}\,\mathbf{A}(\boldsymbol{x}) = \frac{\partial \mathbf{A}}{\partial x^k}\,\boldsymbol{g}^k\,, \tag{2.74}$$

valid for any tensor field $\boldsymbol{x} \longmapsto \mathbf{A}(\boldsymbol{x})$ and the product rule:

$$\mathrm{div}\mathbf{T}_{\mathrm{M}} = \left\{\frac{\partial}{\partial x^k}\left[T^{ij}((\boldsymbol{x} - \boldsymbol{c}) \times \boldsymbol{g}_i) \otimes \boldsymbol{g}_j\right]\right\}\boldsymbol{g}^k$$

$$= \left\{\frac{\partial}{\partial x^k}\left[T^{ij}(\boldsymbol{x} - \boldsymbol{c}) \times (\boldsymbol{g}_i \otimes \boldsymbol{g}_j)\right]\right\}\boldsymbol{g}^k\,,$$

$$\mathrm{div}\mathbf{T}_{\mathrm{M}} = (\boldsymbol{x} - \boldsymbol{c}) \times \left\{\left[\frac{\partial}{\partial x^k}(T^{ij}\boldsymbol{g}_i \otimes \boldsymbol{g}_j)\right]\boldsymbol{g}^k\right\} + \left[T^{ij}\left(\frac{\partial \boldsymbol{x}}{\partial x^k} \times \boldsymbol{g}_i\right) \otimes \boldsymbol{g}_j\right]\boldsymbol{g}^k\,.$$

With $\dfrac{\partial \boldsymbol{x}}{\partial x^k} = \boldsymbol{g}_k$ we obtain

$$\mathrm{div}\mathbf{T}_{\mathrm{M}} = (\boldsymbol{x} - \boldsymbol{c}) \times \mathrm{div}\mathbf{T} + T^{ik}\boldsymbol{g}_k \times \boldsymbol{g}_i\,. \tag{2.75}$$

The result is the identity

$$\mathrm{div}\left[(\boldsymbol{x} - \boldsymbol{c}) \times \mathbf{T}\right] = (\boldsymbol{x} - \boldsymbol{c}) \times \mathrm{div}\mathbf{T} - T^{ik}\boldsymbol{g}_i \times \boldsymbol{g}_k\,, \tag{2.76}$$

with which the surface integral can be transformed into a volume integral:

$$\iint_{\partial\chi_t[\mathscr{B}]} (\boldsymbol{x} - \boldsymbol{c}) \times \mathbf{T}\boldsymbol{n}\mathrm{d}a = \iiint_{\chi_t[\mathscr{B}]} \left[(\boldsymbol{x} - \boldsymbol{c}) \times \mathrm{div}\mathbf{T} - T^{ik}\boldsymbol{g}_i \times \boldsymbol{g}_k\right]\mathrm{d}v \,.$$

The balance of rotational momentum thus leads to the equation

$$\iiint_{\chi_t[\mathscr{B}]} \left\{(\boldsymbol{x} - \boldsymbol{c}) \times \left[\rho\dot{\boldsymbol{v}} - \mathrm{div}\,\mathbf{T}(\boldsymbol{x}, t) - \rho\boldsymbol{f}\right] + T^{ik}\boldsymbol{g}_i \times \boldsymbol{g}_k\right\}\mathrm{d}v = \boldsymbol{0} \tag{2.77}$$

or to $\displaystyle\iiint_{\chi_t[\mathscr{B}]} (T^{ik}\boldsymbol{g}_i \times \boldsymbol{g}_k)\mathrm{d}v = \boldsymbol{0}$,

taking the validity of the local balance of momentum (2.62) into consideration. The last equation again has to be valid for all subdomains of the current configuration. If the integrand is a continuous function of $\boldsymbol{x}$, it has to vanish identically, i.e.

$$T^{ik}\boldsymbol{g}_i \times \boldsymbol{g}_k = \boldsymbol{0} \,.$$

The corresponding component representation reads

$$(T^{23} - T^{32})\boldsymbol{g}^1 + (T^{31} - T^{13})\boldsymbol{g}^2 + (T^{12} - T^{21})\boldsymbol{g}^3 = \boldsymbol{0} \text{ or } T^{ik} = T^{ki} \,;$$

this is equivalent to $\mathbf{T} = \mathbf{T}^{\mathrm{T}}$. ■

The symmetry of the *Cauchy* stress tensor prevents the 1st *Piola-Kirchhoff* tensor $\mathbf{T}_{\mathrm{R}}$ from being symmetric: the symmetry of $\mathbf{T}$ leads to

$$\mathbf{T}_{\mathrm{R}}\mathbf{F}^{\mathrm{T}} = (\mathbf{T}_{\mathrm{R}}\mathbf{F}^{\mathrm{T}})^{\mathrm{T}} \,,$$

$$\mathbf{T}_{\mathrm{R}} = \mathbf{F}\mathbf{T}_{\mathrm{R}}^{\mathrm{T}}\mathbf{F}^{\mathrm{T}\,-1} \,. \tag{2.78}$$

Conversely, the weighted Cauchy tensor $\mathbf{S}$, the 2nd *Piola-Kirchhoff* tensor $\widetilde{\mathbf{T}}$ and the convective stress tensor $\widetilde{\mathbf{t}}$ are also symmetric:

$$\mathbf{S} = (\det\mathbf{F})\mathbf{T} \quad \Rightarrow \mathbf{S} = \mathbf{S}^{\mathrm{T}} \,, \tag{2.79}$$

$$\tilde{\mathbf{T}} = \mathbf{F}^{-1}\,\mathbf{S}\,\mathbf{F}^{\mathrm{T}-1} \quad \Rightarrow \tilde{\mathbf{T}} = \tilde{\mathbf{T}}^{\mathrm{T}}, \tag{2.80}$$

$$\tilde{\mathbf{t}} = \mathbf{F}^{\mathrm{T}}\,\mathbf{S}\,\mathbf{F} \quad \Rightarrow \tilde{\mathbf{t}} = \tilde{\mathbf{t}}^{\mathrm{T}}. \tag{2.81}$$

2. 3. 5 Initial and Boundary Conditions

The balance relations of continuum mechanics in combination with constitutive models lead to formulations of initial-boundary-value problems: the local balance of linear momentum has to be supplemented by suitably formulated initial conditions and boundary conditions.[11] With the formulation of the *initial conditions,* the *position* and *velocity distribution* of a material body are stipulated at the beginning of the motion, i.e. $\boldsymbol{x}(t_0)$ and $\boldsymbol{v}(t_0) = \dot{\boldsymbol{x}}(t_0)$ for all $\mathscr{P} \in \mathscr{B}$.

In the material representation,

$$\chi_{\mathrm{R}}(\boldsymbol{X}, t_0) = \boldsymbol{x}_0(\boldsymbol{X}) \tag{2.82}$$

and

$$\left.\frac{\partial}{\partial t}\chi_{\mathrm{R}}(\boldsymbol{X}, t)\right|_{t\,=\,t_0} = \boldsymbol{v}_0(\boldsymbol{X}) \tag{2.83}$$

are prescribed for all $\boldsymbol{X} \in \mathrm{R}[\mathscr{B}]$. In the spatial representation the initial velocity field

$$\boldsymbol{v}(\boldsymbol{x}, t_0) = \boldsymbol{v}_0(\boldsymbol{x}) \text{ for } \boldsymbol{x} \in \chi_{t_0}[\mathscr{B}] \tag{2.84}$$

is given. The initial configuration is stipulated in connection with the calculation of the path lines from the velocity field, i.e. the solution of the initial-value problem

$$\dot{\boldsymbol{x}}(t) = \boldsymbol{v}(\boldsymbol{x}(t), t)\,,\ \boldsymbol{x}(t_0) = \boldsymbol{x}_0\,. \tag{2.85}$$

---

[11] A mathematically complete formulation of boundary conditions is only possible if specific constitutive equations are adopted which close the balance relations in the sense that the number of unknown fields agrees with the number of equations. Individual examples will be discussed in Chap. 5. At this stage, the general physical notions behind the formulation of boundary conditions may be explained in the context of the general balance relations.

The general formulation of the *boundary conditions* arises out of the notion that either the motion or the stress vector can be prescribed on the boundary surface of the body. A prescription of *motion* at the boundary is called a *geometric boundary condition*, whereas the stipulation of the *stress vector* is known as a *dynamic boundary condition.*

At this stage, the motivation to formulate boundary conditions stems from a purely physical consideration. The question as to which boundary conditions have to be prescribed in a system of partial differential equations (or functional equations), in order to produce a boundary-value problem with existing solutions, has yet to be decided. In fact, this question cannot even be formulated until constitutive equations, that link the stress tensor with the motion of the material body, have been assumed. The following discussion on the subject of the boundary conditions is based exclusively on the common notion that the motion of the material body takes place under the influence of exterior forces. The surface forces have to be given as dynamic boundary conditions.

Consider the boundary of a material body divided up into two disjunctive parts, i.e. $\partial\mathcal{B} = \partial_u\mathcal{B} + \partial_s\mathcal{B}$ or

$$\partial\chi_t[\mathcal{B}] = \partial_u\chi_t[\mathcal{B}] \bigcup \partial_s\chi_t[\mathcal{B}]\,,\, \partial_u\chi_t[\mathcal{B}] \bigcap \partial_s\chi_t[\mathcal{B}] = \varnothing\,, \tag{2.86}$$

then the motion is to be given on $\partial_u\chi_t$ and the stress vector on $\partial_s\chi_t$ in the following. That means that a simultaneous prescription of motion and force is ruled out, since this would be inconsistent from a physical point of view: all that can be done is to stipulate certain velocity or displacement components at boundary points and, at the same time, such stress components as do not contradict them. An example for such a case would be the gliding of a material body along a rigid wall: in this case, the normal component of the velocity vector and, at the same time, the tangential stress on the gliding plane would be prescribed.

A *geometric boundary condition* in the material representation is the prescription of $\boldsymbol{x} = \boldsymbol{r}(\boldsymbol{X}, t)$ for all $\mathcal{P} \in \partial_u\mathcal{B}$, that means

$$\boldsymbol{x} = \chi_R(\boldsymbol{X}, t) = \boldsymbol{r}(\boldsymbol{X}, t) \text{ for } \boldsymbol{X} \in \partial_u\mathrm{R}[\mathcal{B}]\,. \tag{2.87}$$

In the spatial representation the geometric boundary condition refers to the velocity:

$$\boldsymbol{v}(\boldsymbol{x}, t) = \boldsymbol{g}(\boldsymbol{x}, t) \text{ for } \boldsymbol{x} \in \partial_u\chi_t[\mathcal{B}]\,. \tag{2.88}$$

In principle, *dynamic boundary conditions* refer to the current configuration, since the prescribed surface forces naturally act in the current configuration, i.e. on the surface $\partial_s\chi_t[\mathscr{B}]$. It must be borne in mind, however, that the current configuration is unknown prior to its calculation: in order to solve this problem, it is general practice to presume that the unknown boundary surface $\partial_s\chi_t[\mathscr{B}]$ is represented by an equation of the form $f(\boldsymbol{x}, t) = 0$. The scalar function $(\boldsymbol{x}, t) \longmapsto f(\boldsymbol{x}, t)$ is then an additional unknown. The additional equation for establishing f can be obtained from the condition that the boundary surface is and has to remain a *material surface.* A necessary and sufficient condition for a surface

$$f(\boldsymbol{x}, t) = 0 \tag{2.89}$$

representing a material surface is the *kinematic relation* $\dot{f}(\boldsymbol{x}, t) = 0$, i.e.

$$\frac{\partial}{\partial t} f(\boldsymbol{x}, t) + \boldsymbol{v}(\boldsymbol{x}, t) \cdot \operatorname{grad} f(\boldsymbol{x}, t) = 0\,,$$

which has to be valid on the unknown boundary surface, i.e. for all $(\boldsymbol{x}, t)$ with $f(\boldsymbol{x}, t) = 0$. This relation links the required function f with the material velocity field $\boldsymbol{v}(\boldsymbol{x}, t)$ that is also being sought. A dynamic boundary condition is now formulated as follows: the stress vector $\boldsymbol{t} = \mathbf{T}\boldsymbol{n}$ is prescribed for all material points $\mathscr{P} \in \partial_s\mathscr{B}$. For all $\boldsymbol{x}$ with $f(\boldsymbol{x}, t) = 0$

$$\boldsymbol{t} = \mathbf{T}\boldsymbol{n} = \bar{\boldsymbol{s}}(\boldsymbol{x}, t) \tag{2.90}$$

and

$$\frac{\partial}{\partial t} f(\boldsymbol{x}, t) + \boldsymbol{v}(\boldsymbol{x}, t) \cdot \operatorname{grad} f(\boldsymbol{x}, t) = 0 \tag{2.91}$$

should be valid.

To transfer the dynamic boundary condition into the material representation, the *Cauchy* stress vector $\boldsymbol{t}$ has to be expressed by means of the 1st *Piola-Kirchhoff* tensor. This is easily achieved by starting out from

$$\mathbf{T}\boldsymbol{n}\mathrm{d}a = \mathbf{T}_{\mathrm{R}}\boldsymbol{n}_{\mathrm{R}}\mathrm{d}A \tag{2.92}$$

and the transformation formula (1.59) for material surface elements,

$\mathrm{d}\boldsymbol{a} = (\det\mathbf{F})\mathbf{F}^{\mathrm{T}-1}\mathrm{d}\boldsymbol{A}$. Using $\mathrm{d}a = \sqrt{\mathrm{d}\boldsymbol{a}\cdot\mathrm{d}\boldsymbol{a}}$ this leads to

$$\mathrm{d}a = (\det\mathbf{F})\sqrt{\boldsymbol{n}_\mathrm{R}\cdot(\mathbf{F}^{-1}\mathbf{F}^{\mathrm{T}-1}\boldsymbol{n}_\mathrm{R})}\,\mathrm{d}A \text{ and } \boldsymbol{t} = \mathbf{T}\boldsymbol{n} = \mathbf{T}_\mathrm{R}\boldsymbol{n}_\mathrm{R}\frac{\mathrm{d}A}{\mathrm{d}a} \text{ or}$$

$$\mathbf{T}\boldsymbol{n} = \frac{1}{(\det\mathbf{F})\sqrt{\boldsymbol{n}_\mathrm{R}\cdot(\mathbf{F}^{-1}\mathbf{F}^{\mathrm{T}-1}\boldsymbol{n}_\mathrm{R})}}\mathbf{T}_\mathrm{R}\boldsymbol{n}_\mathrm{R}\,.$$

The dynamic boundary condition is thus expressed as follows:

$$\frac{1}{(\det\mathbf{F})\sqrt{\boldsymbol{n}_\mathrm{R}\cdot(\mathbf{F}^{-1}\mathbf{F}^{\mathrm{T}-1}\boldsymbol{n}_\mathrm{R})}}\mathbf{T}_\mathrm{R}\boldsymbol{n}_\mathrm{R} = \hat{\boldsymbol{s}}(\boldsymbol{X},t) \text{ for } \boldsymbol{X}\in\partial_s\mathrm{R}[\mathscr{B}]\,. \tag{2.93}$$

Here, we have $\hat{\boldsymbol{s}}(\boldsymbol{X},t) = \bar{\boldsymbol{s}}(\chi_\mathrm{R}(\boldsymbol{X},t),t)$ with $\bar{\boldsymbol{s}}(\boldsymbol{x},t)$ from (2.90).
Expressing the stress boundary condition in this exact form is not the usual procedure: when dealing with boundary-value problems in solid mechanics it is common practice to omit the *Cauchy* stress vector $\boldsymbol{t}$ and content oneself instead with prescribing the *Lagrange* stress vector $\boldsymbol{t}_\mathrm{R}$ on $\partial_s\mathscr{B}$. The simplified dynamic boundary condition is then written as

$$\mathbf{T}_\mathrm{R}\boldsymbol{n}_\mathrm{R} = \boldsymbol{s}(\boldsymbol{X},t) \text{ for } \boldsymbol{X}\in\partial_s\mathrm{R}[\mathscr{B}]\,, \tag{2.94}$$

which approximates (2.93) for small deformations of the boundary surface. There is, of course, no difference between (2.93) and (2.94) if the boundary surface is free of stress, i.e. unloaded $(\boldsymbol{s} = \boldsymbol{0})$.

2.4 Conclusions from the Balance Equations of Mechanics

Several general conclusions can be drawn from the balance equations of mechanics. These are the *balance of mechanical energy* and the *balance of virtual work*. The energy and work balances are just as universally valid as the basic balance relations themselves. They constitute a contribution towards a better physical comprehension of the balance equations: a mechanical process is not only an exchange of force and momentum but also a process of transforming mechanical power and work into energy. Moreover, the work balances form a basis for the construction of variation principles, approximate solutions and methods of numerical calculation.

2. 4. 1 Balance of Mechanical Energy

A direct conclusion from the balance equations for mass and momentum is the *balance of mechanical power and energy.* These terms are described by

Definition 2. 9

The quantity

$$K : (\mathscr{B}, t) \longmapsto K(\mathscr{B}, t) = \frac{1}{2}\int\limits_{\mathscr{B}} v^2 \mathrm{d}m = \frac{1}{2}\iiint\limits_{\chi_t[\mathscr{B}]} v^2 \rho \mathrm{d}v \tag{2.95}$$

$\left(v^2 = \boldsymbol{v}\cdot\boldsymbol{v}\right)$ is called *kinetic energy* of the material body $\mathscr{B}$ and

$$L_\mathrm{e} : (\mathscr{B}, t) \longmapsto L_\mathrm{e}(\mathscr{B}, t) = \iint\limits_{\partial\chi_t[\mathscr{B}]} \boldsymbol{t}\cdot\boldsymbol{v}\mathrm{d}a + \iiint\limits_{\chi_t[\mathscr{B}]} \boldsymbol{f}\cdot\boldsymbol{v}\rho\mathrm{d}v \tag{2.96}$$

is the *power of the external forces.* $L_\mathrm{e}(\mathscr{B}, t)$ is the power of the system $\{\boldsymbol{t}(\cdot), \boldsymbol{f}(\cdot)\}$ of surface and volume forces currently acting on the body. The quantity

$$L_\mathrm{i} : (\mathscr{B}, t) \longmapsto L_\mathrm{i}(\mathscr{B}, t) = \int\limits_{\mathscr{B}} \ell_\mathrm{i} \mathrm{d}m \ ,$$

$$L_\mathrm{i}(\mathscr{B}, t) = \int\limits_{\mathscr{B}} \frac{1}{\rho}\mathbf{T}\cdot\mathbf{D}\,\mathrm{d}m = \iiint\limits_{\chi_t[\mathscr{B}]} \Big(\frac{1}{\rho}\mathbf{T}\cdot\mathbf{D}\Big)\rho\mathrm{d}v \tag{2.97}$$

is the *power of the internal forces* or the *stress power.* The integrand

$$\ell_\mathrm{i}(\mathscr{P}, t) = \frac{1}{\rho}\mathbf{T}\cdot\mathbf{D} = \frac{1}{\rho_\mathrm{R}}\mathbf{S}\cdot\mathbf{D} \tag{2.98}$$

is called the *specific stress power.* ◊

The specific stress power possesses the dimension of power per unit mass and consists of products of stresses and strain rates:

$$\ell_\mathrm{i} = \frac{1}{\rho} T^{kl} D_{kl} \ . \tag{2.99}$$

With this definition of kinematic energy and power we come to

Theorem 2. 9

In every motion of a material body $\mathscr{B}$, where all fields are continuously differentiable, the power of the external forces is the sum of the temporal change of the kinetic energy and the stress power:

$$L_e(\mathscr{B}, t) = \dot{K}(\mathscr{B}, t) + L_i(\mathscr{B}, t) \, . \tag{2.100}$$

□

Proof

The proof of the energy balance is based on the local balance of momentum (2.62), multiplied in terms of the scalar product by the velocity vector $\boldsymbol{v}(\boldsymbol{x}, t)$:

$$\rho\boldsymbol{v}\cdot\dot{\boldsymbol{v}}(\boldsymbol{x}, t) = \boldsymbol{v}\cdot\operatorname{div}\mathbf{T}(\boldsymbol{x}, t) + \boldsymbol{v}\cdot\rho\boldsymbol{f} \, . \tag{2.101}$$

With the product rules

$$\boldsymbol{v}\cdot\dot{\boldsymbol{v}} = \frac{1}{2}(\boldsymbol{v}\cdot\boldsymbol{v})^{\cdot} \tag{2.102}$$

and

$$\operatorname{div}(\mathbf{T}^{\mathrm{T}}\boldsymbol{v}) = (\operatorname{div}\mathbf{T})\cdot\boldsymbol{v} + \mathbf{T}\cdot\operatorname{grad}\boldsymbol{v} \, , \tag{2.103}$$

(2.101) leads to the equivalent relation

$$\rho\left(\frac{v^2}{2}\right)^{\cdot} = \operatorname{div}(\mathbf{T}^{\mathrm{T}}\boldsymbol{v}) - \mathbf{T}\cdot\operatorname{grad}\boldsymbol{v} + \boldsymbol{v}\cdot\rho\boldsymbol{f} \, ; \tag{2.104}$$

this is now integrated over the entire volume of the current configuration, the initial term on the right-hand side transforming into a surface integral according to the *Gauß* theorem:

$$\iiint\limits_{\chi_t[\mathscr{B}]} \left(\frac{v^2}{2}\right)^{\cdot} \rho dv = \iint\limits_{\partial\chi_t[\mathscr{B}]} (\mathbf{T}^{\mathrm{T}}\boldsymbol{v})\cdot\boldsymbol{n}da + \iiint\limits_{\chi_t[\mathscr{B}]} \rho\boldsymbol{v}\cdot\boldsymbol{f}dv - \iiint\limits_{\chi_t[\mathscr{B}]} (\mathbf{T}\cdot\operatorname{grad}\boldsymbol{v})dv \, .$$

As a consequence of the conservation of mass, $(\rho dv)^{\cdot} = 0$, it is possible to place the material time derivative on the left-hand side in front of the integral:

$$\frac{d}{dt}\iiint\limits_{\chi_t[\mathscr{B}]}\left(\frac{v^2}{2}\right)\rho dv = \iint\limits_{\partial\chi_t[\mathscr{B}]}(\mathbf{T}\boldsymbol{n})\cdot\boldsymbol{v}da + \iiint\limits_{\chi_t[\mathscr{B}]}\rho\boldsymbol{v}\cdot\boldsymbol{f}dv - \iiint\limits_{\chi_t[\mathscr{B}]}(\mathbf{T}\cdot\mathbf{L})dv\,.$$

This corresponds to the theorem's statement, with the only difference that the specific stress power (per unit volume) is the scalar product of the stress tensor and the velocity gradient $\mathbf{L} = \text{grad}\boldsymbol{v}(\boldsymbol{x}, t) = \mathbf{D} + \mathbf{W}$. However, the stress tensor is symmetric as a result of the balance of rotational momentum, thus $\mathbf{T} = \mathbf{T}^{\mathrm{T}}$ and $\mathbf{W} = -\mathbf{W}^{\mathrm{T}}$ lead to $\mathbf{T}\cdot\mathbf{W} = 0$. This means that the spin tensor $\mathbf{W}$ makes no contribution to the stress power and following (2.51) we write: $\ell_{\mathrm{i}}(\mathscr{P}, t) = \frac{1}{\rho}\mathbf{T}\cdot\mathbf{D} = \frac{1}{\rho_{\mathrm{R}}}\mathbf{S}\cdot\mathbf{D}$. ■

Since there are various definitions for stress and strain rate tensors, there are also different ways of formulating the stress power. This is confirmed by

Theorem 2. 10

The following possibilities exist for expressing the stress power $\rho_{\mathrm{R}}\ell_{\mathrm{i}} = \mathbf{S}\cdot\mathbf{D}$, related to the volume element in the reference configuration.

1) $\rho_{\mathrm{R}}\ell_{\mathrm{i}} = \mathbf{S}\cdot\mathbf{D} = \mathbf{S}\cdot\overset{\vartriangle}{\mathbf{A}}$ (2.105)

2) $\rho_{\mathrm{R}}\ell_{\mathrm{i}} = \widetilde{\mathbf{T}}\cdot\dot{\mathbf{E}}$ (2.106)

3) $\rho_{\mathrm{R}}\ell_{\mathrm{i}} = (-\widetilde{\mathbf{t}})\cdot\dot{\mathbf{e}}$ (2.107)

4) $\rho_{\mathrm{R}}\ell_{\mathrm{i}} = (-\mathbf{S})\cdot(-\mathbf{D}) = (-\mathbf{S})\cdot\overset{\triangledown}{\mathbf{a}}$ (2.108)

□

Proof

Statement 1) has already been proved; the identification

$$\mathbf{D} = \overset{\vartriangle}{\mathbf{A}} = \dot{\mathbf{A}} + \mathbf{L}^{\mathrm{T}}\mathbf{A} + \mathbf{A}\mathbf{L}$$

of the strain rate tensor with the covariant *Oldroyd* rate of the *Almansi* tensor $\mathbf{A}$ is already familiar from (1.127) and (1.132).

Statement 2) is the result of the following conversion:

$$\mathbf{S}\cdot\mathbf{D} = \mathbf{S}\cdot\tfrac{1}{2}(\mathbf{L}+\mathbf{L}^{\mathrm{T}}) = \mathrm{tr}(\mathbf{SL}) = \mathrm{tr}(\mathbf{S}\dot{\mathbf{F}}\mathbf{F}^{-1})$$

$$= \mathrm{tr}(\mathbf{F}^{-1}\mathbf{S}\mathbf{F}^{\mathrm{T}-1}\mathbf{F}^{\mathrm{T}}\dot{\mathbf{F}}) = \widetilde{\mathbf{T}}\cdot\tfrac{1}{2}(\mathbf{F}^{\mathrm{T}}\dot{\mathbf{F}} + \dot{\mathbf{F}}^{\mathrm{T}}\,\mathbf{F}) = \widetilde{\mathbf{T}}\cdot\dot{\mathbf{E}}\,.$$

Accordingly, statement 3) is obtained as a result of

$$\mathbf{S}\cdot\mathbf{D} = \mathrm{tr}(\mathbf{S}\dot{\mathbf{F}}\mathbf{F}^{-1}) = \mathrm{tr}(\mathbf{F}^{\mathrm{T}}\mathbf{S}\mathbf{F}\mathbf{F}^{-1}\dot{\mathbf{F}}\mathbf{F}^{-1}\mathbf{F}^{\mathrm{T}-1})$$

$$= -\,\mathrm{tr}\left[\mathbf{F}^{\mathrm{T}}\mathbf{S}\mathbf{F}(\mathbf{F}^{-1})^{\cdot}\,\mathbf{F}^{\mathrm{T}-1}\right] = -\,(\mathbf{F}^{\mathrm{T}}\mathbf{S}\mathbf{F})\cdot\left[(\mathbf{F}^{-1})^{\cdot}\,\mathbf{F}^{\mathrm{T}-1}\right]$$

$$= -\,(\mathbf{F}^{\mathrm{T}}\mathbf{S}\mathbf{F})\cdot\tfrac{1}{2}\left[(\mathbf{F}^{-1})^{\cdot}\,\mathbf{F}^{\mathrm{T}-1} + \mathbf{F}^{-1}(\,\mathbf{F}^{\mathrm{T}-1})^{\cdot}\right] = (-\,\widetilde{\mathbf{t}})\cdot\dot{\mathbf{e}}\,.$$

Finally, statement 4) is based on the identity (1.131),

$$-\,\mathbf{D} = \overset{\triangledown}{\mathbf{a}} = \dot{\mathbf{a}} - \mathbf{La} - \mathbf{aL}^{\mathrm{T}}\,,$$

according to which $-\,\mathbf{D}$ is equal to the contravariant *Oldroyd* derivative of the *Finger* strain tensor $\mathbf{a}$. ■

With regard to convective coordinates the component representations (1.142) - (1.145) are valid for the strain rates,

$$\dot{\mathbf{E}} = \tfrac{1}{2}\dot{g}_{KL}\boldsymbol{G}^K\otimes\boldsymbol{G}^L\,,\quad \overset{\vartriangle}{\mathbf{A}} = \mathbf{D} = \tfrac{1}{2}\dot{g}_{KL}\boldsymbol{g}^K\otimes\boldsymbol{g}^L\,,$$

$$\dot{\mathbf{e}} = \tfrac{1}{2}\dot{g}^{KL}\boldsymbol{G}_K\otimes\boldsymbol{G}_L\,,\quad \overset{\triangledown}{\mathbf{a}} = -\,\mathbf{D} = \tfrac{1}{2}\dot{g}^{KL}\boldsymbol{g}_K\otimes\boldsymbol{g}_L\,,$$

and for the stress tensors the representations (2.55) - (2.58),

$$\widetilde{\mathbf{T}} = S^{IK}\boldsymbol{G}_I\otimes\boldsymbol{G}_K\,,\quad \mathbf{S} = S^{IK}\boldsymbol{g}_I\otimes\boldsymbol{g}_K\,,$$

$$\widetilde{\mathbf{t}} = S_{IK}\boldsymbol{G}^I\otimes\boldsymbol{G}^K\,,\quad \mathbf{S} = S_{IK}\boldsymbol{g}^I\otimes\boldsymbol{g}^K\,.$$

Consequently, we obtain just two forms of representation for the stress power:

$$\rho_{\mathrm{R}}\ell_{\mathrm{i}} = \tfrac{1}{2}S^{KL}\dot{g}_{kl} = -\,\tfrac{1}{2}S_{KL}\,\dot{g}^{KL}\,. \tag{2.109}$$

2. 4. 2 The Principle of d'Alembert

All work principles in mechanics are based on a common standard argument, described in the following lemma of variational calculus.

Theorem 2. 11

(cf. Theorem 2. 1) Let $\boldsymbol{w} : \boldsymbol{x} \longmapsto \boldsymbol{w}(\boldsymbol{x})$ be any continuous vector field,

$$\mathscr{T} = \{\boldsymbol{\eta} : \boldsymbol{x} \longrightarrow \mathbb{V}^3 \mid \boldsymbol{\eta}(\cdot) \text{ continuous for } \boldsymbol{x} \in \chi_t[\mathscr{B}]\}$$

the set of all continuous vector-valued functions, and

$$\iiint\limits_{\chi_t[\mathscr{B}]} \boldsymbol{w}(\boldsymbol{x}) \cdot \boldsymbol{\eta}(\boldsymbol{x}) \mathrm{d}v = 0 \tag{2.110}$$

be valid for all functions $\boldsymbol{\eta}(\boldsymbol{x}) \in \mathscr{T}$. Then $\boldsymbol{w}(\boldsymbol{x}) = 0$ holds throughout the integration area $\chi_t[\mathscr{B}]$. □

Proof

Let $\boldsymbol{x}_0$ be an internal point with $\boldsymbol{w}(\boldsymbol{x}_0) \neq \boldsymbol{0}$. Then, in view of the continuity of $\boldsymbol{w}$, at least a small environment of $\boldsymbol{x}_0$ exists in which $\boldsymbol{w}(\boldsymbol{x})$ does not vanish. In this case, however, it is possible to choose $\boldsymbol{\eta}(\boldsymbol{x}) = \boldsymbol{w}(\boldsymbol{x})$, and we obtain an off-zero integral, contradicting the prerequisite. ■

The principle of *d'Alembert* presents the opportunity of formulating the balance of linear and rotational momentum by means of one single scalar equation. Formulations of this kind are appropriate for developing numerical methods to solve boundary-value problems. In its physical interpretation the principle of *d'Alembert* is a balance statement concerning the so-called *virtual work* of external forces, inertia forces and stresses. Clarification of the mathematical formulation of the principle of *d'Alembert* is recommended prior to its physical application. This is the object of

Theorem 2. 12

The surface $\partial\chi_t[\mathscr{B}]$ of a material body is divided up into two disjoint parts,

$$\partial\chi_t[\mathscr{B}] = \partial_u\chi_t[\mathscr{B}] + \partial_s\chi_t[\mathscr{B}] ,$$

for which geometric and dynamic boundary conditions are prescribed:

$$x = r(X, t) \text{ for } X \in \partial_u \mathrm{R}[\mathscr{B}] \quad \left(x \in \partial_u \chi_t[\mathscr{B}]\right),$$

$$t = \mathbf{T}n = s(x, t) \text{ for } x \in \partial_s \chi_t[\mathscr{B}].$$

Moreover, let $\mathscr{T}$ be the set of all continuously differentiable vector-valued functions $x \longmapsto \eta(x) \in \mathbb{V}^3$, defined on $\chi_t[\mathscr{B}]$ and vanishing on that part of the boundary on which the motion is given,

$$\mathscr{T} = \left\{\eta : \chi_t[\mathscr{B}] \longrightarrow \mathbb{V}^3 \;\middle|\; \eta(\cdot) \text{ differentiable and } \eta(x) = 0 \text{ for } x \in \partial_u \chi_t[\mathscr{B}]\right\}.$$

Then, the *boundary-value problem of mechanics*,

$$\rho \dot{v}(x, t) = \operatorname{div} \mathbf{T}(x, t) + \rho f \,, \mathbf{T} = \mathbf{T}^{\mathrm{T}}$$

$$x = r(t) \text{ on } \partial_u \chi_t[\mathscr{B}] \quad \text{and} \quad t = \mathbf{T}n = s \text{ on } \partial_s \chi_t[\mathscr{B}] \,,$$

leads to the principle of *d'Alembert*: for all functions $\eta(x) \in \mathscr{T}$ we have

$$\iint\limits_{\partial_s \chi_t[\mathscr{B}]} s \cdot \eta \mathrm{d}a + \iiint\limits_{\chi_t[\mathscr{B}]} f \cdot \eta \rho \mathrm{d}v - \iiint\limits_{\chi_t[\mathscr{B}]} \dot{v} \cdot \eta \rho \mathrm{d}v = \iiint\limits_{\chi_t[\mathscr{B}]} \mathbf{T} \cdot \operatorname{grad} \eta \mathrm{d}v \,. \tag{2.111}$$

Conversely, if the principle of *d'Alembert* applies for all functions $\eta(x) \in \mathscr{T}$, then the outcome is the validity of both the local balance of linear and rotational momentum and a stress boundary condition. □

Proof

1) Scalar multiplication of the local balance of momentum with a function $\eta(x) \in \mathscr{T}$ yields

$$\rho(\dot{v} - f) \cdot \eta = \eta \cdot \operatorname{div} \mathbf{T} = \operatorname{div}(\mathbf{T}^{\mathrm{T}} \eta) - \mathbf{T} \cdot \operatorname{grad} \eta \,.$$

After integration over $\chi_t[\mathscr{B}]$ and use of the *Gauß* theorem we arrive at the predicted equation:

$$\iiint\limits_{\chi_t[\mathscr{B}]} [\rho(\dot{v} - f) \cdot \eta + \mathbf{T} \cdot \operatorname{grad} \eta] \mathrm{d}v = \iint\limits_{\partial \chi_t[\mathscr{B}]} (\mathbf{T}n) \cdot \eta \mathrm{d}a = \iint\limits_{\partial_s \chi_t[\mathscr{B}]} s \cdot \eta \mathrm{d}a \,.$$

With the symmetry of the stress tensor, $\mathbf{T} = \mathbf{T}^{\mathrm{T}}$, even the following equation is valid:

$$\iint\limits_{\partial_s \chi_t[\mathscr{B}]} \boldsymbol{s}\cdot\boldsymbol{\eta}\mathrm{d}a + \iiint\limits_{\chi_t[\mathscr{B}]} \rho(\boldsymbol{f} - \dot{\boldsymbol{v}})\cdot\boldsymbol{\eta}\mathrm{d}v = \iiint\limits_{\chi_t[\mathscr{B}]} \mathbf{T}\cdot\tfrac{1}{2}\left[\mathrm{grad}\boldsymbol{\eta} + (\mathrm{grad}\boldsymbol{\eta})^{\mathrm{T}}\right]\mathrm{d}v\,.$$

2) For the stress tensor $\mathbf{T}$ the decomposition

$$\mathbf{T} = \tfrac{1}{2}\left(\mathbf{T} + \mathbf{T}^{\mathrm{T}}\right) + \tfrac{1}{2}\left(\mathbf{T} - \mathbf{T}^{\mathrm{T}}\right) = \mathbf{T}^{(\mathrm{s})} + \mathbf{T}^{(\mathrm{a})}$$

is generally true, and for its scalar product with the tensor $\mathrm{grad}\boldsymbol{\eta}$, in view of $\mathbf{T}^{(\mathrm{a})}\cdot\mathrm{grad}\boldsymbol{\eta}^{(\mathrm{s})} = 0$, the conversion

$$\mathbf{T}\cdot\mathrm{grad}\boldsymbol{\eta} = \mathbf{T}^{(\mathrm{s})}\cdot\mathrm{grad}\boldsymbol{\eta} + \mathbf{T}^{(\mathrm{a})}\cdot\tfrac{1}{2}\left[\mathrm{grad}\boldsymbol{\eta} - (\mathrm{grad}\boldsymbol{\eta})^{\mathrm{T}}\right]$$

$$= \mathrm{div}\left(\mathbf{T}^{(\mathrm{s})}\boldsymbol{\eta}\right) - \boldsymbol{\eta}\cdot\mathrm{div}\mathbf{T}^{(\mathrm{s})} + \mathbf{T}^{(\mathrm{a})}\cdot\left(\mathrm{grad}\boldsymbol{\eta}\right)^{(\mathrm{a})}\,.$$

The validity of (2.111) now leads to

$$\iint\limits_{\partial_s \chi_t[\mathscr{B}]} \boldsymbol{s}\cdot\boldsymbol{\eta}\mathrm{d}a + \iiint\limits_{\chi_t[\mathscr{B}]} \rho(\boldsymbol{f} - \dot{\boldsymbol{v}})\cdot\boldsymbol{\eta}\mathrm{d}v$$

$$= \iint\limits_{\partial \chi_t[\mathscr{B}]} \left(\mathbf{T}^{(\mathrm{s})}\boldsymbol{n}\right)\cdot\boldsymbol{\eta}\mathrm{d}a - \iiint\limits_{\chi_t[\mathscr{B}]} \left[\boldsymbol{\eta}\cdot\mathrm{div}\mathbf{T}^{(\mathrm{s})} - \mathbf{T}^{(\mathrm{a})}\cdot\left(\mathrm{grad}\boldsymbol{\eta}\right)^{(\mathrm{a})}\right]\mathrm{d}v$$

or

$$\iint\limits_{\partial_s \chi_t[\mathscr{B}]} \left(\boldsymbol{s} - \mathbf{T}^{(\mathrm{s})}\boldsymbol{n}\right)\cdot\boldsymbol{\eta}\mathrm{d}a + \iiint\limits_{\chi_t[\mathscr{B}]} \left(\mathrm{div}\mathbf{T}^{(\mathrm{s})} + \rho\boldsymbol{f} - \rho\dot{\boldsymbol{v}}\right)\cdot\boldsymbol{\eta}\mathrm{d}v$$

$$= \iiint\limits_{\chi_t[\mathscr{B}]} \left[\mathbf{T}^{(\mathrm{a})}\cdot\left(\mathrm{grad}\boldsymbol{\eta}\right)^{(\mathrm{a})}\right]\mathrm{d}v\,.$$

Based on this identity, valid for all functions $\boldsymbol{\eta}(\boldsymbol{x}) \in \mathscr{T}$, it is possible to obtain the balance equations and the dynamic boundary condition by selecting appropriate subsets of $\mathscr{T}$ and applying the argumentation described in Theorem 2. 11: first we select the set of functions that vanish on the whole boundary surface and for which the gradient is symmetric, $(\mathrm{grad}\boldsymbol{\eta})^{(\mathrm{a})} = \mathbf{0}$.

For every function of this kind we have

$$\iiint\limits_{\chi_t[\mathscr{B}]} (\mathrm{div}\mathbf{T}^{(s)} + \rho f - \rho\dot{v})\cdot \eta \mathrm{d}v = 0\,,$$

which leads to the local balance of momentum, formed with the symmetric part $\mathbf{T}^{(s)}$ of $\mathbf{T}$. Furthermore, the set of functions $\eta(x)$ with symmetric gradients and different from zero on $\partial_s\chi_t[\mathscr{B}]$ leads to the stress boundary condition, formed with $\mathbf{T}^{(s)}$. Finally the set of functions $\eta(x) \in \mathscr{T}$ with skew-symmetric gradients leads to the validity of the local balance of rotational momentum $\mathbf{T}^{(a)} = \mathbf{0}$, i.e. $\mathbf{T} = \mathbf{T}^{\mathrm{T}}$. ■

A corresponding formulation of the principle of *d'Alembert* can be obtained in the material representation: multiplication of the local balance of linear momentum in the material representation (2.64) with any function $X \longmapsto \eta(X)$ from

$$\mathscr{T}_{\mathrm{R}} = \Big\{\eta : \mathrm{R}[\mathscr{B}] \longrightarrow \mathbb{V}^3 \,\Big|\, \eta(\cdot) \text{ continuously differentiable}$$
$$\text{and } \eta(X) = 0 \text{ for } x \in \partial_u\mathrm{R}[\mathscr{B}]\Big\} \; (\partial\mathrm{R}[\mathscr{B}] = \partial_u\mathrm{R}[\mathscr{B}] + \partial_s\mathrm{R}[\mathscr{B}])$$

yields $\rho_{\mathrm{R}}(\dot{v} - f)\cdot \eta = \eta\cdot \mathrm{Div}\mathbf{T}_{\mathrm{R}} = \mathrm{Div}(\mathbf{T}_{\mathrm{R}}^{\mathrm{T}}\eta) - \mathbf{T}_{\mathrm{R}}\cdot \mathrm{Grad}\eta\,.$

After integration over $\mathrm{R}[\mathscr{B}]$ and applying the *Gauß* theorem, we have

$$\iiint\limits_{\mathrm{R}[\mathscr{B}]} [\rho_{\mathrm{R}}(\dot{v} - f)\cdot \eta - \mathbf{T}_{\mathrm{R}}\cdot \mathrm{Grad}\eta]\mathrm{d}V = \iint\limits_{\partial\mathrm{R}[\mathscr{B}]} (\mathbf{T}_{\mathrm{R}}n_{\mathrm{R}})\cdot \eta \mathrm{d}A = \iint\limits_{\partial_s\mathrm{R}[\mathscr{B}]} s_{\mathrm{R}}\cdot \eta \mathrm{d}A\,.$$

Introduction of the 2nd *Piola-Kirchhoff* tensor $\tilde{\mathbf{T}} = \mathbf{F}^{-1}\mathbf{T}_{\mathrm{R}}$ leads to

$$\mathbf{T}_{\mathrm{R}}\cdot \mathrm{Grad}\eta = \mathrm{tr}(\tilde{\mathbf{T}}\mathbf{F}^{\mathrm{T}}\mathrm{Grad}\eta) = \tilde{\mathbf{T}}\cdot (\mathbf{F}^{\mathrm{T}}\mathrm{Grad}\eta)\,,$$

and we obtain the principle of *d'Alembert* in terms of the material representation:

$$\iint\limits_{\partial_s\mathrm{R}[\mathscr{B}]} s_{\mathrm{R}}\cdot \eta \mathrm{d}A + \iiint\limits_{\mathrm{R}[\mathscr{B}]} [\rho_{\mathrm{R}}(f - \dot{v})\cdot \eta]\mathrm{d}V = \iiint\limits_{\mathrm{R}[\mathscr{B}]} \tilde{\mathbf{T}}\cdot (\mathbf{F}^{\mathrm{T}}\mathrm{Grad}\eta)\mathrm{d}V\,. \tag{2.112}$$

The validity of this equation for all functions $\eta \in \mathscr{T}_{\mathrm{R}}$ follows from the field equations

$$\rho_{\mathrm{R}}\dot{v} = \mathrm{Div}(\mathbf{F}\widetilde{\mathbf{T}}) + \rho_{\mathrm{R}}f \,, \quad \widetilde{\mathbf{T}} = \widetilde{\mathbf{T}}^{\mathrm{T}} \,,$$

with the geometric and dynamic boundary conditions

$$x = r(X, t) \text{ on } \partial_u \mathrm{R}[\mathscr{B}] \quad \text{and} \quad t_{\mathrm{R}} = \mathbf{T}_{\mathrm{R}} n_{\mathrm{R}} = s_{\mathrm{R}} \text{ on } \partial_s \mathrm{R}[\mathscr{B}] \,.$$

Conversely, *d'Alembert's* principle in the form (2.112) also implies the field equations and the stress boundary conditions in the material representation. However, only the approximation (2.94) of the dynamic boundary condition arises instead of the exact formulation (2.93).
With $\widetilde{\mathbf{T}} = \widetilde{\mathbf{T}}^{\mathrm{T}}$ the material representation of *d'Alembert's* principle merges into

$$\iint\limits_{\partial_s \mathrm{R}[\mathscr{B}]} s_{\mathrm{R}} \cdot \eta \mathrm{d}A + \iiint\limits_{\mathrm{R}[\mathscr{B}]} [\rho_{\mathrm{R}}(f - \dot{v}) \cdot \eta] \mathrm{d}V = \iiint\limits_{\mathrm{R}[\mathscr{B}]} \widetilde{\mathbf{T}} \cdot \delta\mathbf{E} \mathrm{d}V \,. \tag{2.113}$$

The integrand

$$\widetilde{\mathbf{T}} \cdot \delta\mathbf{E} = \widetilde{\mathbf{T}} \cdot (\mathbf{F}^{\mathrm{T}}\mathrm{Grad}\eta) = \widetilde{\mathbf{T}} \cdot \frac{1}{2}\left[\mathbf{F}^{\mathrm{T}}\mathrm{Grad}\eta + (\mathrm{Grad}\eta)^{\mathrm{T}}\mathbf{F}\right] \,,$$

is the specific *virtual stress work*. The second factor on the right-hand side, the so-called *virtual Green strain*, is the *Gateaux* differential of the *Green* strain tensor $\mathbf{E} = \frac{1}{2}(\mathbf{F}^{\mathrm{T}}\mathbf{F} - \mathbf{1})$ in the direction of the vector field $\eta(X)$:

$$\begin{aligned} \delta\mathbf{E} &= \frac{\mathrm{d}}{\mathrm{d}s} \frac{1}{2}\left[(\mathbf{F} + s\mathrm{Grad}\eta)^{\mathrm{T}}(\mathbf{F} + s\mathrm{Grad}\eta) - \mathbf{1}\right]\Big|_{s=0} \\ &= \frac{1}{2}\left[\mathbf{F}^{\mathrm{T}}\mathrm{Grad}\eta + (\mathrm{Grad}\eta)^{\mathrm{T}}\mathbf{F}\right] \,. \end{aligned} \tag{2.114}$$

The mathematical significance of *d'Alembert's* principle is based on the fact that it represents a so-called *weak formulation* of the field equations. The statement of *d'Alembert* in particular makes more modest demands on the differentiability of the functions concerned.
The physical meaning of *d'Alembert's* principle is based on the interpretation of the functions $x \longmapsto \eta(x) \in \mathscr{T}$ or $X \longmapsto \eta(X) \in \mathscr{T}_{\mathrm{R}}$ as conceivable displacement fields starting from the current configuration. These displacement fields are called *virtual displacements*. Virtual displacements are vector fields which are compatible with the geometric constraints: they have

to be continuously differentiable to allow for the calculation of virtual strains and vanish in those portions of the boundary on which motion is prescribed. The virtual displacements $\boldsymbol{\eta}(\boldsymbol{x})$ also have differential character: derived quantities in kinematics (virtual strains) are differentials in the direction of $\boldsymbol{\eta}$. In this sense we come to

Definition 2. 10

The functions $\boldsymbol{x} \longmapsto \boldsymbol{\eta}(\boldsymbol{x}) \in \mathscr{T}$ and $\boldsymbol{X} \longmapsto \boldsymbol{\eta}(\boldsymbol{X}) \in \mathscr{T}_{\mathrm{R}}$ are called *virtual displacements.* The scalar

$$\delta A_{\mathrm{e}} = \iint\limits_{\partial_s \mathrm{R}[\mathscr{B}]} \boldsymbol{s}_{\mathrm{R}} \cdot \boldsymbol{\eta} \mathrm{d}A + \iiint\limits_{\mathrm{R}[\mathscr{B}]} \rho_{\mathrm{R}} \boldsymbol{f} \cdot \boldsymbol{\eta} \mathrm{d}V = \iint\limits_{\partial_s \chi_t[\mathscr{B}]} \boldsymbol{s} \cdot \boldsymbol{\eta} \mathrm{d}a + \iiint\limits_{\chi_t[\mathscr{B}]} \rho \boldsymbol{f} \cdot \boldsymbol{\eta} \mathrm{d}v \tag{2.115}$$

is called *virtual work of the external forces* and the scalar

$$\delta A_{\mathrm{i}} = \iiint\limits_{\mathrm{R}[\mathscr{B}]} \tilde{\mathbf{T}} \cdot \delta \mathrm{E} \mathrm{d}V = \iiint\limits_{\chi_t[\mathscr{B}]} \mathbf{T} \cdot \tfrac{1}{2} [\operatorname{grad} \boldsymbol{\eta} + (\operatorname{grad} \boldsymbol{\eta})^{\mathrm{T}}] \mathrm{d}v \tag{2.116}$$

virtual work of the internal forces or *virtual stress work.* The scalar

$$\delta A_{\mathrm{T}} = - \iiint\limits_{\mathrm{R}[\mathscr{B}]} \rho_{\mathrm{R}} \dot{\boldsymbol{v}} \cdot \boldsymbol{\eta} \mathrm{d}V = - \iiint\limits_{\chi_t[\mathscr{B}]} \rho \dot{\boldsymbol{v}} \cdot \boldsymbol{\eta} \mathrm{d}v = - \int\limits_{\mathscr{B}} \dot{\boldsymbol{v}} \cdot \boldsymbol{\eta} \mathrm{d}m \tag{2.117}$$

is known as *virtual work of the inertial forces.* ◊

These definitions lend *d'Alembert's* principle the form

$$\delta A_{\mathrm{e}} + \delta A_{\mathrm{T}} = \delta A_{\mathrm{i}} \, , \tag{2.118}$$

which states that for every given virtual displacement field the virtual work of external and inertial forces is equal to the virtual work of the stresses. To emphasise the differential character of the virtual displacements, these are frequently denoted in literature by $\delta \boldsymbol{u}$ instead of $\boldsymbol{\eta}$.

2. 4. 3 Principle of Virtual Work

For the special case of *statics,* i.e. in the case of vanishing acceleration fields, the local balance equations merge into the *conditions of equilibrium,*

$$\operatorname{div} \mathbf{T}(\boldsymbol{x}, t) + \rho \boldsymbol{f} = \boldsymbol{0} \, , \quad \mathbf{T} = \mathbf{T}^{\mathrm{T}} \, , \tag{2.119}$$

$$\mathrm{Div}(\mathbf{F}\widetilde{\mathbf{T}}) + \rho_{\mathrm{R}}\boldsymbol{f} = \boldsymbol{0}\,, \quad \widetilde{\mathbf{T}} = \widetilde{\mathbf{T}}^{\mathrm{T}}\,. \tag{2.120}$$

The boundary conditions are now time-independent:

$$\boldsymbol{x} = \boldsymbol{r}(\boldsymbol{X}) \text{ on the boundary } \partial_{\boldsymbol{u}}\chi_{\mathrm{t}}[\mathscr{B}]\,, \tag{2.121}$$

$$\boldsymbol{t} = \mathbf{T}\boldsymbol{n} = \boldsymbol{s}(\boldsymbol{x}) \text{ on the boundary } \partial_{\boldsymbol{s}}\chi_{\mathrm{t}}[\mathscr{B}] \text{ or}$$

$$\boldsymbol{t}_{\mathrm{R}} = \mathbf{F}\widetilde{\mathbf{T}}\boldsymbol{n}_{\mathrm{R}} = \boldsymbol{s}_{\mathrm{R}}(\boldsymbol{X}) \text{ on the boundary } \partial_{\boldsymbol{s}}\mathrm{R}[\mathscr{B}]\,. \tag{2.122}$$

Definition 2. 11

In the particular case of *statics* $(\dot{\boldsymbol{v}} = \boldsymbol{0}, \delta A_{\mathrm{T}} = 0)$, the principle of *d'Alembert*,

$$\delta A_{\mathrm{a}} = \delta A_{\mathrm{i}}\,, \tag{2.123}$$

is known as the *principle of virtual work.* ◊

The principle of virtual work establishes the fact that for any given virtual displacement the virtual work of the external and internal forces is equal. This statement is equivalent to the equilibrium conditions of mechanics and also supplies the stress boundary conditions in the same way as the principle of *d'Alembert.*

2. 4. 4 Incremental Form of the Principle of d'Alembert

In certain situations, for example in rate-independent plasticity (see Chap. 11), constitutive theories are set up in an incremental form, and corresponding incremental formulations of boundary-value problems may be required.

An incremental formulation of *d'Alembert's* principle can be derived by beginning with the material derivative of the local balance of momentum,

$$\rho_{\mathrm{R}}\ddot{\boldsymbol{v}} = \mathrm{Div}\dot{\mathbf{T}}_{\mathrm{R}} + \rho_{\mathrm{R}}\dot{\boldsymbol{f}}\,, \tag{2.124}$$

and multiplying it in the sense of the scalar product by any given vector function $\boldsymbol{X} \longmapsto \delta\boldsymbol{v}(\boldsymbol{X}) \in \mathscr{T}_{\mathrm{R}}$:

$$\rho_R \ddot{v} \cdot \delta v = \big(\mathrm{Div}\dot{\mathbf{T}}_R\big) \cdot \delta v + \rho_R \dot{f} \cdot \delta v \ . \tag{2.125}$$

The vector field $\delta v(\boldsymbol{X})$ now ought to possess the dimensions of a velocity and be called *virtual velocity field.* The virtual velocity $\delta v(\boldsymbol{X})$ is subjected to the same restrictions as the virtual displacement $\eta(\boldsymbol{X})$ beforehand. Partial integration and consideration of $\mathbf{T}_R = \mathbf{F}\widetilde{\mathbf{T}}$ produce

$$\mathrm{Div}\big(\dot{\mathbf{T}}_R{}^{\mathrm{T}}\delta v\big) + \rho_R(\dot{f} - \ddot{v}) \cdot \delta v = \dot{\mathbf{T}}_R \cdot \mathrm{Grad}\delta v = \big(\mathbf{F}\widetilde{\mathbf{T}}\big)^{\cdot} \cdot \mathrm{Grad}\delta v =$$

$$= \big(\mathbf{F}\dot{\widetilde{\mathbf{T}}} + \dot{\mathbf{F}}\widetilde{\mathbf{T}}\big) \cdot \mathrm{Grad}\delta v = \dot{\widetilde{\mathbf{T}}} \cdot \big(\mathbf{F}^{\mathrm{T}}\mathrm{Grad}\delta v\big) + \widetilde{\mathbf{T}} \cdot \big(\dot{\mathbf{F}}^{\mathrm{T}}\mathrm{Grad}\delta v\big) \ .$$

This can also be written in the following form

$$\mathrm{Div}\big(\dot{\mathbf{T}}_R{}^{\mathrm{T}}\delta v\big) + \rho_R(\dot{f} - \ddot{v}) \cdot \delta v = \dot{\widetilde{\mathbf{T}}} \cdot \delta\dot{\mathbf{E}} + \widetilde{\mathbf{T}} \cdot \big(\dot{\mathbf{F}}^{\mathrm{T}}\delta\dot{\mathbf{F}}\big) \ , \tag{2.126}$$

in which

$$\delta\dot{\mathbf{F}} = \frac{\mathrm{d}}{\mathrm{d}s}\Big[\mathrm{Grad}\big(v(\boldsymbol{X}, t) + s\delta v(\boldsymbol{X})\big)\Big]\Big|_{s=0} = \mathrm{Grad}\delta v(\boldsymbol{X}) \tag{2.127}$$

is the virtual velocity gradient and

$$\delta\dot{\mathbf{E}} = \delta\Big[\tfrac{1}{2}\big(\mathbf{F}^{\mathrm{T}}\dot{\mathbf{F}} + \dot{\mathbf{F}}^{\mathrm{T}}\mathbf{F}\big)\Big] = \tfrac{1}{2}\big(\mathbf{F}^{\mathrm{T}}\delta\dot{\mathbf{F}} + \delta\dot{\mathbf{F}}^{\mathrm{T}}\mathbf{F}\big) \ ,$$

$$\delta\dot{\mathbf{E}} = \tfrac{1}{2}\big(\mathbf{F}^{\mathrm{T}}\mathrm{Grad}\delta v + (\mathrm{Grad}\delta v)^{\mathrm{T}}\mathbf{F}\big) \tag{2.128}$$

the virtual strain rate.

Integration of (2.126) over the volume of the reference configuration, application of the *Gauß* theorem and observing the incremental stress boundary condition,

$$\dot{\mathbf{T}}_R n_R = \dot{t}_R = \dot{s}_R(\boldsymbol{X}, t) \text{ on the boundary } \partial_s \mathrm{R}[\mathscr{B}], \tag{2.129}$$

leads to the statement

$$\iint\limits_{\partial_s \mathrm{R}[\mathscr{B}]} \dot{s}_R \cdot \delta v \, \mathrm{d}A + \iiint\limits_{\mathrm{R}[\mathscr{B}]} \rho_R(\dot{f} - \ddot{v}) \cdot \delta v \, \mathrm{d}V = \iiint\limits_{\mathrm{R}[\mathscr{B}]} \Big(\dot{\widetilde{\mathbf{T}}} \cdot \delta\dot{\mathbf{E}} + \widetilde{\mathbf{T}} \cdot \big(\dot{\mathbf{F}}^{\mathrm{T}}\delta\dot{\mathbf{F}}\big)\Big)\mathrm{d}V \ , \tag{2.130}$$

which is valid for any given virtual velocity field $\delta v(\boldsymbol{X}) \in \mathscr{T}_R$.

This is the incremental form of *d'Alembert's* principle: the virtual power of the incremental external and inertial forces is equal to the incremental virtual stress power. This consists of the virtual change $\overset{\bullet}{\tilde{\mathbf{T}}} \cdot \delta\mathbf{E}$ in the incremental stress power

$$\ell_{\text{incr}} = \frac{1}{\rho_{\text{R}}} \overset{\bullet}{\tilde{\mathbf{T}}} \cdot \dot{\mathbf{E}} \tag{2.131}$$

and a convective part that depends on the current state of stress resulting from the geometric non-linearity.

The incremental formulation of *d'Alembert's* principle is, in the same sense as the original form, equivalent to the incremental form (2.124) of the balance of linear and rotational momentum and, consequently, equivalent to the balance equations themselves. Omitting inertia ($\ddot{\boldsymbol{v}} \approx \boldsymbol{0}$), the incremental formulation of the principle of virtual work occurs as a special case.[12]

The incremental form of the principle of *d'Alembert* suggests the definition (2.131) of the incremental stress power as the scalar product of the material derivatives of the 2nd *Piola-Kirchhoff* stress tensor and the *Green* strain. One important physical meaning of this quantity lies in the fact that it establishes a natural relationship between the stress rate $\overset{\bullet}{\tilde{\mathbf{T}}}$ and the strain rate $\dot{\mathbf{E}}$. This aspect may be viewed as a motivation for a more general concept, in which associated strain and stress rates are set up systematically. This concept is to be developed in Chap. 8; it has proved to be useful when formulating incremental stress-strain relations to represent inelastic material properties.

[12] See HAUPT & TSAKMAKIS [1988].

3 Balance Relations of Thermodynamics

3.1 Preliminary Remarks

The work of a system of forces changes a material body's kinetic energy content according to the balance of mechanical power: only a certain fraction of the power of external forces is transformed into kinetic energy; the remainder is stress power. Stress power makes two contributions to the energy budget of a material body as a general rule: part of the work done by the stresses on the strain rates is stored in the material body. Experience has shown that another part is transformed into non-mechanical energy. This part is closely related to a phenomenon which is known as dissipation. Dissipation either increases the non-mechanical energy contents or is released into the environment in the form of heat.

It is the object of *thermodynamics* to describe the energy transformations inevitably connected with mechanical processes. The incorporation of non-mechanical energy requires an extension of the basic terms of mechanics (body, motion, force) to include thermodynamical terminology: internal energy, heat, temperature and entropy. The point of adopting these terms common to thermodynamics is to maintain the kind of argumentation that has stood the test of time in continuum mechanics.

Thermodynamics contains two universally valid balance relations, also known as the first and second law. The *first law of thermodynamics* is the general *balance of energy* that postulates the equivalence of mechanical and non-mechanical work. The *second law of thermodynamics* sets fundamental limits to the possibility of transforming thermal energy into mechanical work. The second law of thermodynamics, also known as the *principle of irreversibility*, quotes evidence gleaned from general experience that natural processes cannot be reversed without leaving behind them a change in the environment of the material system under consideration.

In the formulation of the second law employed here, the local entropy production is used as a measure for the irreversibility of thermodynamical processes. The principle of irreversibility is the statement that the local entropy production must always be positive.

3. 2 Energy

The whole energy content of a body is the sum of its kinetic energy $K(\mathcal{B}, t)$ and *internal energy* $E(\mathcal{B}, t)$. This is described in

Definition 3. 1

The internal energy of a material body is an additive physical quantity which can be represented by way of a volume integral,

$$E : (\mathcal{B}, t) \longmapsto E(\mathcal{B}, t) = \int_{\mathcal{B}} e\mathrm{d}m = \iiint_{\chi_t[\mathcal{B}]} e\rho\mathrm{d}v = \iiint_{\mathrm{R}[\mathcal{B}]} e\rho_{\mathrm{R}}\mathrm{d}V\,. \tag{3.1}$$

The integrand $e : (\mathcal{P}, t) \longmapsto e(\mathcal{P}, t) = \bar{e}(\boldsymbol{x}, t) = \hat{e}(\boldsymbol{X}, t)$ is called *specific internal energy* per unit mass. ◇

The change of the energy content of a material body responds to the thermomechanical impact of the external world: following the *free-body principle* this impact is the energy supply or energy transport. The mechanical energy transport is the *power of the external forces*; the non-mechanical (thermal) energy supply is called *heat.*

Definition 3. 2

A material body is subjected to the resultant *heat supply*

$$Q : (\mathcal{B}, t) \longmapsto Q(\mathcal{B}, t) = \iint_{\partial\chi_t[\mathcal{B}]} q\mathrm{d}a + \iiint_{\chi_t[\mathcal{B}]} r\rho\mathrm{d}v$$

$$Q(\mathcal{B}, t) = \iint_{\partial\mathrm{R}[\mathcal{B}]} q_{\mathrm{R}}\mathrm{d}A + \iiint_{\mathrm{R}[\mathcal{B}]} r\rho_{\mathrm{R}}\mathrm{d}V. \tag{3.2}$$

◇

The heat supply is the sum of one surface- and one volume-distributed share; in a physical interpretation the surface-distributed share represents the heat conduction and the volume distributed share the heat radiation.

The term internal energy is complemented by linking its temporal change with the environmental impact. This relation is brought about by means of the energy balance.

Energy Balance - First Law of Thermodynamics

The mechanical power and heat suppy to a material body effect a change of its kinetic and internal energy: the relation

$$\dot{K}(\mathscr{B}, t) + \dot{E}(\mathscr{B}, t) = L_e(\mathscr{B}, t) + Q(\mathscr{B}, t) \tag{3.3}$$

holds true for all conceivable material bodies.[1] ♦

Using the definitions (2.95), (2.96), (3.1) and (3.2) for the individual physical quantities the energy balance is written as

$$\iiint\limits_{\chi_t[\mathscr{B}]} (\tfrac{1}{2}v^2 + e)^{\cdot} \rho \mathrm{d}v = \iint\limits_{\partial\chi_t[\mathscr{B}]} [(\mathbf{T}^{\mathrm{T}}\boldsymbol{v})\cdot\boldsymbol{n} + q]\mathrm{d}a + \iiint\limits_{\chi_t[\mathscr{B}]} (\boldsymbol{f}\cdot\boldsymbol{v} + r)\rho \mathrm{d}v\,. \tag{3.4}$$

The property (2.37) of the stress tensor has been inserted and, on the left-hand side, the *balance of mass* taken into consideration in the form of the differentiation rule (2.18).

According to (3.4) the surface-distributed heat supply can be represented by means of a volume integral:

$$\iint\limits_{\partial\chi_t[\mathscr{B}]} q \mathrm{d}a = \iiint\limits_{\chi_t[\mathscr{B}]} \Big[\rho(\tfrac{1}{2}v^2 + e)^{\cdot} - \mathrm{div}(\mathbf{T}^{\mathrm{T}}\boldsymbol{v}) - \rho(\boldsymbol{f}\cdot\boldsymbol{v} + r)\Big]\mathrm{d}v\,. \tag{3.5}$$

A heat flow density q is attributed to every point $\boldsymbol{x}$ on the boundary surface of a body in the current configuration. A common assumption in analogy to the surface force density is that, apart from $\boldsymbol{x} \in \partial\chi_t[\mathscr{B}]$ and t, q depends on the orientation of the surface element, i.e. on the unit normal $\boldsymbol{n}$. In compliance with this prerequisite, we recognise

[1] See TRUESDELL & TOUPIN [1960], Sect. 240.

Theorem 3. 1
$q = q(\boldsymbol{x}, t, \boldsymbol{n})$ and (3.5) lead to

$$q(\boldsymbol{x}, t, \boldsymbol{n}) = -\,\boldsymbol{q}(\boldsymbol{x}, t)\cdot\boldsymbol{n}\ , \quad q = -\,\boldsymbol{q}\cdot\boldsymbol{n}\ , \tag{3.6}$$

in other words, the heat flow density's dependence on the normal vector is *linear*. □

The minus sign in the definition of the heat flux vector is a convention. It means that the heat flow positive if the material body absorbs energy.

Proof
The proof of this statement corresponds exactly to the proof of Theorem 2. 6. ■

The surface-distributed heat supply can also be expressed by means of integration over the boundary of the body in the reference configuration:

$$\iint\limits_{\partial\chi_t[\mathscr{B}]} q\mathrm{d}a = \iint\limits_{\partial\mathrm{R}[\mathscr{B}]} q_\mathrm{R}\mathrm{d}A \tag{3.7}$$

leads to

$$-\iint\limits_{\partial\chi_t[\mathscr{B}]} \boldsymbol{q}\cdot\boldsymbol{n}\mathrm{d}a = -\iint\limits_{\partial\mathrm{R}[\mathscr{B}]} \boldsymbol{q}_\mathrm{R}\cdot\boldsymbol{n}_\mathrm{R}\mathrm{d}A\ , \tag{3.8}$$

that means

$$\boldsymbol{q}_\mathrm{R}\cdot\mathrm{d}\boldsymbol{A} = \boldsymbol{q}\cdot\mathrm{d}\boldsymbol{a} = \boldsymbol{q}\cdot\left[(\det\mathbf{F})\mathbf{F}^{\mathrm{T}\,-1}\mathrm{d}\boldsymbol{A}\right] = \left[(\det\mathbf{F})\mathbf{F}^{-1}\boldsymbol{q}\right]\cdot\mathrm{d}\boldsymbol{A}\ , \tag{3.9}$$

bearing the transformation (1.59) of a material surface element in mind. This gives rise to

Definition 3. 3
The vector $\boldsymbol{q}$ with the property $q = -\,\boldsymbol{q}\cdot\boldsymbol{n}$ or

$$q\mathrm{d}a = -\,\boldsymbol{q}\cdot\boldsymbol{n}\mathrm{d}a = -\,\boldsymbol{q}\cdot\mathrm{d}\boldsymbol{a} \tag{3.10}$$

is called the *Cauchy* heat flux vector. The vector

$$\boldsymbol{q}_{\mathrm{R}} = (\det\mathbf{F})\mathbf{F}^{-1}\boldsymbol{q} \tag{3.11}$$

with the property $q_{\mathrm{R}} = -\boldsymbol{q}_{\mathrm{R}} \cdot \boldsymbol{n}_{\mathrm{R}}$ or

$$q_{\mathrm{R}}\mathrm{d}A = -\boldsymbol{q}_{\mathrm{R}} \cdot \boldsymbol{n}_{\mathrm{R}}\mathrm{d}A = -\boldsymbol{q}_{\mathrm{R}} \cdot \mathrm{d}\boldsymbol{A} \tag{3.12}$$

is known as the *Piola-Kirchhoff* heat flux vector. ◊

The *heat flow density* q corresponds to the *stress vector* $\boldsymbol{t}$ and the *heat flux vector* $\boldsymbol{q}$ corresponds in the same way to the *stress tensor* $\mathbf{T}$. Definition 3. 3 parallels the definitions of the *Cauchy* and 1st *Piola-Kirchhoff* stress tensors: the heat flow density q or $\boldsymbol{q}$ refers to the surface element in the current configuration, whereas the heat flow density q_{R} or $\boldsymbol{q}_{\mathrm{R}}$ applies to the surface element in the reference configuration. For the differential of the surface distributed heat supply we write

$$\mathrm{d}Q_{\mathrm{o}} = q\mathrm{d}a = q_{\mathrm{R}}\mathrm{d}A = -\boldsymbol{q} \cdot \boldsymbol{n}\mathrm{d}a = -\boldsymbol{q}_{\mathrm{R}} \cdot \boldsymbol{n}_{\mathrm{R}}\mathrm{d}A\,. \tag{3.13}$$

The vector fields

$$\boldsymbol{q} : (\mathscr{P}, t) \longmapsto \boldsymbol{q}(\mathscr{P}, t) \quad \text{and} \quad \boldsymbol{q}_{\mathrm{R}} : (\mathscr{P}, t) \longmapsto \boldsymbol{q}_{\mathrm{R}}(\mathscr{P}, t) \tag{3.14}$$

indicate the current distribution of the heat flow in the body, the boundary values of the heat flux vector define the heat flow through the surface. By inserting the heat flux vector into (3.4), the energy balance takes on the form

$$\iiint\limits_{\chi_t[\mathscr{B}]} \left(\tfrac{1}{2}v^2 + e\right)^{\cdot} \rho \mathrm{d}v = \iiint\limits_{\chi_t[\mathscr{B}]} \left[\mathrm{div}(\mathbf{T}^{\mathrm{T}}\boldsymbol{v}) + \rho(\boldsymbol{f} \cdot \boldsymbol{v} + r)\right]\mathrm{d}v - \iint\limits_{\partial\chi_t[\mathscr{B}]} \boldsymbol{q} \cdot \boldsymbol{n}\mathrm{d}a \tag{3.15}$$

(spatial representation), or

$$\iiint\limits_{\mathrm{R}[\mathscr{B}]} \left(\tfrac{1}{2}v^2 + e\right)^{\cdot} \rho_{\mathrm{R}} \mathrm{d}V = \iiint\limits_{\mathrm{R}[\mathscr{B}]} \left[\mathrm{Div}(\mathbf{T}_{\mathrm{R}}{}^{\mathrm{T}}\boldsymbol{v}) + \rho_{\mathrm{R}}(\boldsymbol{f} \cdot \boldsymbol{v} + r)\right]\mathrm{d}V - \iint\limits_{\partial\mathrm{R}[\mathscr{B}]} \boldsymbol{q}_{\mathrm{R}} \cdot \boldsymbol{n}_{\mathrm{R}}\,\mathrm{d}A \tag{3.16}$$

(material representation). Once again, if we assume sufficient continuity properties for all fields that might occur, it is possible to conclude the local formulations from the global formulations of the energy balance:

Theorem 3. 2

In the spatial representation, the local formulation of the energy balance is

$$\dot{e}\,(\boldsymbol{x}, t) = -\frac{1}{\rho}\,\mathrm{div}\boldsymbol{q} + r + \frac{1}{\rho}\,\mathbf{T}\cdot\mathbf{D}\,, \tag{3.17}$$

and in the material representation

$$\dot{e}\,(\boldsymbol{X}, t) = -\frac{1}{\rho_{\mathrm{R}}}\,\mathrm{Div}\boldsymbol{q}_{\mathrm{R}} + r + \frac{1}{\rho_{\mathrm{R}}}\,\widetilde{\mathbf{T}}\cdot\dot{\mathbf{E}}\ . \tag{3.18}$$

□

Proof

In order to prove the local energy balance the *Gauß* theorem,

$$\iint\limits_{a} \boldsymbol{q}\cdot\boldsymbol{n}\mathrm{d}a = \iiint\limits_{v} \mathrm{div}\boldsymbol{q}\mathrm{d}v\ , \tag{3.19}$$

has to be applied. With the product rule (2.103) and $\mathbf{T} = \mathbf{T}^{\mathrm{T}}$, $\mathrm{div}(\mathbf{T}^{\mathrm{T}}\boldsymbol{v}) = (\mathrm{div}\mathbf{T})\cdot\boldsymbol{v} + \mathbf{T}\cdot\mathrm{grad}\boldsymbol{v} = (\mathrm{div}\mathbf{T})\cdot\boldsymbol{v} + \mathbf{T}\cdot\mathbf{D}$, the global energy balance (3.16) is transferred to

$$\iiint\limits_{\chi_t[\mathscr{B}]} \Big[\big(\rho\dot{\boldsymbol{v}} - \mathrm{div}\mathbf{T} + \rho\boldsymbol{f}\big)\cdot\boldsymbol{v} + \rho\dot{e} + \mathrm{div}\boldsymbol{q} - \rho r - \mathbf{T}\cdot\mathbf{D}\Big]\mathrm{d}v = 0\,. \tag{3.20}$$

According to the balance of linear momentum (2.62) this simplifies to

$$\iiint\limits_{\chi_t[\mathscr{B}]} \Big(\dot{e} + \frac{1}{\rho}\,\mathrm{div}\boldsymbol{q} - r - \frac{1}{\rho}\,\mathbf{T}\cdot\mathbf{D}\Big)\rho\mathrm{d}v = 0\,, \tag{3.21}$$

which has to be valid for all subsets of the current configuration. Proof of the material representation is analogous. ■

3. 3 Temperature and Entropy

The concept of *absolute temperature* can be established physically within the framework of *equilibrium thermodynamics*. This justification is connected with the idea of *entropy*, as developed by Carathéodory on the basis of an assumption pertaining to the inaccessibility of certain thermodynamic states through adiabatic processes of equilibrium.[2] The criticism expressed by Truesdell[3] regarding Carathéodory's work no doubt contains several arguments worthy of contemplation. Even if details of the mathematical proofs offered in CARATHÉODORY [1909] could be improved, it is nonetheless possible to derive from his work a physical explanation for the concept of temperature, as applicable in equilibrium thermodynamics. The substantiation of temperature and entropy according to Carathéodory applies exclusively to material systems in thermodynamic equilibrium. In the case of non-equilibrium processes, which are of course of central interest in continuum mechanics, the arguments of equilibrium thermodynamics can merely supply a physical motivation.

The fundamental terms of temperature and entropy are closely connected with the second law of thermodynamics. Contrary to the balance of momentum and energy there are, however, no generally valid and universally acknowledged formulations of the second law so far for expressing a general principle of irreversibility.[4] It is for this reason that every generalisation of these fundamental terms in connection with thermomechanical non-equilibrium processes contains elements of uncertainty and gives rise to controversies. There is no intention of enlarging upon these at this stage. Instead, any remaining uncertainty is to be counteracted here by assuming a special formulation for the principle of irreversibility that has proved its worth on the authority of collective experience at least in continuum mechanics. This is a special *entropy inequality*, also known as *Clausius-Duhem* inequality. Its formulation is based on the interpretation of the *entropy production* as a measure for the degree of irreversibility of a thermomechanical process. The entropy production is the difference between the temporal change of entropy, which a body contains, and the entropy supply, which the body absorbs from its surroundings along the lines of the *free-body principle*.

The terms *temperature* and *entropy* are introduced in a formal way at first. This procedure definitely complies with the concept of continuum mechanics, which starts with a mathematical definition of the fundamental

---

2 CARATHÉODORY [1909].

3 TRUESDELL [1984a], pp. 49.

4 HUTTER [1977], p. 4; a comparison of different formulations of the second law is given by MUSCHIK [1998b].

terms and then proceeds to complement them by means of balance relations. These balance relations later become initial-boundary-value problems in connection with appropriate assumptions with regard to the material behaviour. The solution of these problems fills the entire system of terms retrospectively with physical meaning. The temperature of a material body is a scalar-valued field which is needed to characterise the thermodynamic state of a material point:

Definition 3. 4

At all times t a *temperature* Θ is attributed to every material point $\mathscr{P}$:

$$\Theta : (\mathscr{P}, t) \longmapsto \Theta(\mathscr{P}, t) = \bar{\Theta}(\boldsymbol{x}, t) = \hat{\Theta}(\boldsymbol{X}, t) > 0 \,. \tag{3.22}$$

Its definition in equilibrium thermodynamics has a physical background; Θ is called the *thermodynamic temperature* or *absolute temperature.* ◊

The introduction of entropy is just as formal as the introduction (3.1) of the internal energy:

Definition 3. 5

Every material body $\mathscr{B}$ contains *entropy* at all times; entropy is an additive physical quantity, which can be represented by means of a volume integral:

$$S : (\mathscr{B}, t) \longmapsto S(\mathscr{B}, t) = \int_{\mathscr{B}} s\mathrm{d}m = \iiint_{\chi_t[\mathscr{B}]} s\rho\mathrm{d}v = \iiint_{\mathrm{R}[\mathscr{B}]} s\rho_{\mathrm{R}}\mathrm{d}V \,. \tag{3.23}$$

The integrand $s : (\mathscr{P}, t) \longmapsto s(\mathscr{P}, t) = \bar{s}(\boldsymbol{x}, t) = \hat{s}(\boldsymbol{X}, t)$ is called *specific entropy* per unit mass. ◊

Definition 3. 6

A material body, when sliced off mentally from its surroundings, becomes the object of the resultant entropy supply

$$H : (\mathscr{B}, t) \longmapsto H(\mathscr{B}, t) = \iint_{\partial\chi_t[\mathscr{B}]} \Sigma\mathrm{d}a + \iiint_{\chi_t[\mathscr{B}]} \sigma\rho\mathrm{d}v \tag{3.24}$$

or

$$H(\mathscr{B}, t) = \iint_{\partial\mathrm{R}[\mathscr{B}]} \Sigma_{\mathrm{R}}\mathrm{d}A + \iiint_{\mathrm{R}[\mathscr{B}]} \sigma\rho_{\mathrm{R}}\mathrm{d}V \tag{3.25}$$

and develops the entropy production

$$\Gamma\colon (\mathscr{B},\, t) \longmapsto \Gamma(\mathscr{B},\, t) = \int_{\mathscr{B}} \gamma \mathrm{d}m = \iiint_{\chi_t[\mathscr{B}]} \gamma\rho \mathrm{d}v = \iiint_{\mathrm{R}[\mathscr{B}]} \gamma\rho_{\mathrm{R}} \mathrm{d}V \,. \tag{3.26}$$

◊

The entropy supply is the sum of one surface- and one volume-distributed part. This corresponds to the interpretation of the mechanical power or heat as an energy supply or the interpretation of the resultant force as a momentum transport. Along the same lines as the stress vector and the heat flow, the surface-distributed entropy supply does not depend on location and time alone, but also on the normal vector: $\Sigma = \Sigma(\boldsymbol{x}, t, \boldsymbol{n})$.

Theorem 3. 3

$\Sigma = \Sigma(\boldsymbol{x}, t, \boldsymbol{n})$ leads to

$$\Sigma(\boldsymbol{x}, t, \boldsymbol{n}) = -\, \boldsymbol{\Sigma}(\boldsymbol{x}, t) \cdot \boldsymbol{n} \,. \tag{3.27}$$

□

Proof

Analogous to Theorem 3. 1 ■

The differential of the surface-distributed entropy supply is thus provided by

$$\mathrm{d}S_{\mathrm{o}} = \Sigma \mathrm{d}a = \Sigma_{\mathrm{R}} \mathrm{d}A = -\, \boldsymbol{\Sigma} \cdot \boldsymbol{n} \mathrm{d}a = -\, \boldsymbol{\Sigma}_{\mathrm{R}} \cdot \boldsymbol{n}_{\mathrm{R}} \mathrm{d}A \,. \tag{3.28}$$

Entropy Balance

For all material bodies the *balance of entropy* is valid,

$$\dot{S}(\mathscr{B}, t) = H(\mathscr{B}, t) + \Gamma(\mathscr{B}, t) \,, \tag{3.29}$$

i.e. the sum of the entropy supply and entropy production is equal to the temporal change of entropy. ♦

The entropy balance is primarily nothing other than a formal definition of the entropy production. The physical significance of entropy production is expressed in the second law of thermodynamics, postulated below as a universal law of nature in the following form:

Principle of Irreversibility - Second Law of Thermodynamics

Entropy production is never negative. The *entropy inequality*

$$\Gamma(\mathcal{B}, t) = \dot{S}(\mathcal{B}, t) - H(\mathcal{B}, t) \geq 0 \tag{3.30}$$

holds true for all conceivable material bodies and thermomechanical processes. ♦

Provided there are sufficient continuity properties, it is likewise possible to obtain a local formulation for the entropy inequality by applying the *Gauß* theorem to the surface-distributed entropy supply. Reiterating the standard argumentation, we obtain the following result:

Theorem 3. 4

The local formulation of the entropy balance reads as

$$\dot{s}(\boldsymbol{x}, t) = -\frac{1}{\rho}\,\mathrm{div}\boldsymbol{\Sigma} + \sigma + \gamma \tag{3.31}$$

(spatial representation) or

$$\dot{s}(\boldsymbol{X}, t) = -\frac{1}{\rho_{\mathrm{R}}}\,\mathrm{Div}\boldsymbol{\Sigma}_{\mathrm{R}} + \sigma + \gamma \tag{3.32}$$

(material representation). □

Proof

Corresponding to Theorem 3. 2. ■

In the formulation of entropy balance the *entropy supply* has to be substantiated. In *equilibrium thermodynamics of homogeneous systems* it is a proven fact that the entropy supply is the quotient of the heat transport and the absolute temperature. This identification is assumed to be valid; for processes close to equilibrium it is a good approximation. In the sense of this approximation the surface-distributed entropy supply is

$$\boldsymbol{\Sigma} = \frac{1}{\Theta}\boldsymbol{q}\,, \tag{3.33}$$

and the density of the volume-distributed entropy supply

$$\sigma = \frac{1}{\Theta}r\,. \tag{3.34}$$

The global formulation of the second law thus merges into the *Clausius-Duhem* inequality:[5]

$$\frac{\mathrm{d}}{\mathrm{dt}}\iiint\limits_{\chi_t[\mathscr{B}]} s\rho \mathrm{d}v + \iint\limits_{\partial\chi_t[\mathscr{B}]} \frac{1}{\Theta}\boldsymbol{q}\cdot\boldsymbol{n}\mathrm{d}a - \iiint\limits_{\chi_t[\mathscr{B}]} \frac{1}{\Theta} r\rho \mathrm{d}v \geq 0\,. \tag{3.35}$$

This inequality can be expressed by the local formulation

$$\gamma = \dot{s}(\boldsymbol{x}, t) + \frac{1}{\rho}\mathrm{div}\Big(\frac{\boldsymbol{q}}{\Theta}\Big) - \frac{r}{\Theta} \geq 0\,, \tag{3.36}$$

or

$$\gamma = \dot{s} + \frac{1}{\rho\Theta}\Big(\mathrm{div}\boldsymbol{q} - \rho r\Big) - \frac{1}{\rho\Theta^2}\boldsymbol{q}\cdot\boldsymbol{g} \geq 0\,, \tag{3.37}$$

where

$$\boldsymbol{g}(\boldsymbol{x}, t) = \mathrm{grad}\Theta(\boldsymbol{x}, t) \tag{3.38}$$

is the *spatial temperature gradient.* The corresponding material representation is

$$\gamma = \dot{s}(\boldsymbol{X}, t) + \frac{1}{\rho_{\mathrm{R}}}\mathrm{Div}\Big(\frac{\boldsymbol{q}_{\mathrm{R}}}{\Theta}\Big) - \frac{r}{\Theta} \geq 0\,, \tag{3.39}$$

or

$$\gamma = \dot{s} + \frac{1}{\rho_{\mathrm{R}}\Theta}\Big(\mathrm{Div}\boldsymbol{q}_{\mathrm{R}} - \rho_{\mathrm{R}} r\Big) - \frac{1}{\rho_{\mathrm{R}}\Theta^2}\,\boldsymbol{q}_{\mathrm{R}}\cdot\boldsymbol{g}_{\mathrm{R}} \geq 0\,, \tag{3.40}$$

with the *material temperature gradient*

$$\boldsymbol{g}_{\mathrm{R}}(\boldsymbol{X}, t) = \mathrm{Grad}\Theta(\boldsymbol{X}, t)\,. \tag{3.41}$$

The identity $\bar{\Theta}(\chi_{\mathrm{R}}(\boldsymbol{X}, t)) = \hat{\Theta}_{\mathrm{R}}(\boldsymbol{X}, t)$ leads to the relation

$$\mathbf{F}^{\mathrm{T}}\boldsymbol{g} = \boldsymbol{g}_{\mathrm{R}} \tag{3.42}$$

between $\boldsymbol{g}$ and $\boldsymbol{g}_{\mathrm{R}}$ according to the chain rule. Equivalent representations of the local entropy inequality (3.37) can be obtained by means of eliminating the local heat supply with the help of the energy balance (3.17) and (3.18),

---

5 Truesdell & Toupin [1960], Sect. 258.

$$-\frac{1}{\rho}\operatorname{div}\boldsymbol{q} + r = \dot{e} - \frac{1}{\rho}\mathbf{T}\cdot\mathbf{D}\,, \tag{3.43}$$

$$-\frac{1}{\rho_R}\operatorname{Div}\boldsymbol{q}_R + r = \dot{e} - \frac{1}{\rho_R}\widetilde{\mathbf{T}}\cdot\dot{\mathbf{E}}\,. \tag{3.44}$$

By means of substitution we obtain from (3.37) and (3.40) the entropy inequality in the formulations

$$\Theta\gamma = -\dot{e} + \Theta\dot{s} + \frac{1}{\rho}\mathbf{T}\cdot\mathbf{D} - \frac{1}{\rho\Theta}\boldsymbol{q}\cdot\boldsymbol{g} \geq 0 \tag{3.45}$$

and

$$\Theta\gamma = -\dot{e} + \Theta\dot{s} + \frac{1}{\rho_R}\widetilde{\mathbf{T}}\cdot\dot{\mathbf{E}} - \frac{1}{\rho_R\Theta}\boldsymbol{q}_R\cdot\boldsymbol{g}_R \geq 0\,. \tag{3.46}$$

The product

$$\delta = \Theta\gamma \tag{3.47}$$

is called *internal dissipation*; the inequality

$$\delta \geq 0\,, \tag{3.48}$$

equivalent to $\gamma \geq 0$ is called the *dissipation inequality*. The local forms (3.45) or (3.46) of the *Clausius-Duhem* inequality can be interpreted as a condition of constraint, to be fulfilled by all thermomechanical processes: the inequality presents a simple and unique criterion for checking to see whether the principle of irreversibility has been satisfied in the proposed form.

3. 4 Initial and Boundary Conditions

In the same way as the local balance of momentum the local energy balance also has to be supplemented by initial and boundary conditions. The *initial condition* usually refers to the *temperature distribution* given on $\mathrm{R}[\mathcal{B}]$ or $\chi_{t_0}[\mathcal{B}]$:

$$\Theta(\boldsymbol{X}, t_0) = \Theta_0(\boldsymbol{X}) \text{ for } \boldsymbol{X} \in \mathrm{R}[\mathcal{B}] \tag{3.49}$$

or

$$\Theta(\boldsymbol{x}, t_0) = \Theta_0(\boldsymbol{x}) \text{ for } \boldsymbol{x} \in \chi_{t_0}[\mathcal{B}]\,. \tag{3.50}$$

There are two different categories of boundary conditions: it is possible to prescribe the *temperature* ("geometric boundary conditions") or the *heat flow* ("dynamic boundary condition") at boundary points. The boundary surface of the material body is again split into two disjoint subsets,

$$\partial\chi_t[\mathscr{B}] = \partial_\Theta\chi_t[\mathscr{B}] \cup \partial_q\chi_t[\mathscr{B}] \, , \partial_\Theta\chi_t[\mathscr{B}] \cap \partial_q\chi_t[\mathscr{B}] = \varnothing \, ,$$

where the *temperature* or the *heat flow* is prescribed. The boundary condition for temperature in the material representation is

$$\Theta(\boldsymbol{X}, t) = \Theta_r(\boldsymbol{X}, t) \text{ for } \boldsymbol{X} \in \partial_\Theta \mathrm{R}[\mathscr{B}] \, . \tag{3.51}$$

In the spatial representation we have

$$\Theta(\boldsymbol{x}, t) = \Theta_r(\boldsymbol{x}, t) \quad \text{and} \quad \frac{\partial}{\partial t}\varphi(\boldsymbol{x}, t) + \boldsymbol{v}(\boldsymbol{x}, t)\cdot \mathrm{grad}\varphi(\boldsymbol{x}, t) = 0 \tag{3.52}$$

for all $\boldsymbol{x} \in \partial_\Theta\chi_t[\mathscr{B}]$ with $\varphi(\boldsymbol{x}, t) = 0$.

Here, the material boundary surface is represented by $\varphi(\boldsymbol{x}, t) = 0$, consisting of the points $\boldsymbol{x} \in \partial_\Theta\chi_t[\mathscr{B}]$.
The boundary condition for the heat flow in the spatial representation is written as

$$-\,\boldsymbol{q}\cdot\boldsymbol{n} = w(\boldsymbol{x}, t) \quad \text{and} \quad \frac{\partial}{\partial t}\psi(\boldsymbol{x}, t) + \boldsymbol{v}(\boldsymbol{x}, t)\cdot \mathrm{grad}\psi(\boldsymbol{x}, t) = 0 \tag{3.53}$$

for all $\boldsymbol{x} \in \partial_q\chi_t[\mathscr{B}]$ with $\psi(\boldsymbol{x}, t) = 0$,

the material surface being represented by $\psi(\boldsymbol{x}, t) = 0$, consisting of the points $\boldsymbol{x} \in \partial_q\chi_t[\mathscr{B}]$.

The corresponding material representation reads

$$-\frac{1}{(\det\mathbf{F})\sqrt{\boldsymbol{n}_R\cdot(\mathbf{F}^{-1}\mathbf{F}^{\mathrm{T}-1}\boldsymbol{n}_R)}}\,\boldsymbol{q}_R\cdot\boldsymbol{n}_R = w(\boldsymbol{X}, t) \text{ for } \boldsymbol{X} \in \partial_q \mathrm{R}[\mathscr{B}] \, . \tag{3.54}$$

The material representation of the boundary condition for the heat flux vector is normally applied in the linearised form only, corresponding to the stress boundary condition (2.94):

$$- \boldsymbol{q}_{\mathrm{R}} \cdot \boldsymbol{n}_{\mathrm{R}} = w(\boldsymbol{X}, t) \text{ for } \boldsymbol{X} \in \partial_q \mathrm{R}[\mathscr{B}] \, . \tag{3.55}$$

This approximation corresponds to small deformations of the boundary surface.

3. 5 Balance Relations for Open Systems

The fundamental laws of thermomechanics, the balance relations for mass (2.8), linear momentum (2.31), rotational momentum (2.32), energy (3.3) and entropy (3.29), (3.30), describe the interaction between material systems and their outside world. The idea behind this description is that a material system under observation always comprises the same material elements throughout the whole of its thermomechanical history. In this sense, the balance statements of thermomechanics relate to *closed systems*: no exchange of matter with the environment is planned in the hitherto presented formulations of the balance relations. On no account should this be regarded as a restriction in their general validity, since it turns out that a possible exchange of matter with the environment of a given *spatial domain* is already contained in the formulation of the theory so far, provided that the spatial representation is interpreted correctly. An interpretation of this kind does not merely lead to a better comprehension of the balance relations, but to more extensive possibilities for their application as well.

3. 5. 1 Transport Theorem

In the following the mapping

$$\begin{aligned} \mathrm{f} : \mathscr{B} \times \mathbb{I} \; &\longrightarrow \mathbb{W} \\ (\mathscr{P}, t) \; &\longmapsto \mathrm{w} = \mathrm{f}(\mathscr{P}, t) \end{aligned} \tag{3.56}$$

denotes a density function per unit volume of the current configuration, $\mathrm{w} = \bar{\mathrm{f}}(\boldsymbol{x}, t)$ its spatial representation and

$$\mathrm{F}(\mathscr{B}, t) = \iiint\limits_{\chi_t[\mathscr{B}]} \bar{\mathrm{f}}(\boldsymbol{x}, t) \mathrm{d}v \tag{3.57}$$

the corresponding additive physical quantity associated with a material body $\mathscr{B}$ at a time t. The material derivative of the volume integral (3.57),

$$\dot{\mathrm{F}}(\mathscr{B}, t) = \frac{\mathrm{d}}{\mathrm{d}t} \iiint\limits_{\chi_t[\mathscr{B}]} \bar{\mathrm{f}}(\boldsymbol{x}, t)\mathrm{d}v \, ,$$

is calculated using the transport theorem:

Theorem 3. 5

The following calculation rule applies for the material derivative of a volume integral:[6]

$$\frac{\mathrm{d}}{\mathrm{d}t} \iiint\limits_{\chi_t[\mathscr{B}]} \bar{\mathrm{f}}(\boldsymbol{x}, t)\mathrm{d}v = \iiint\limits_{\chi_t[\mathscr{B}]} \frac{\partial}{\partial t} \bar{\mathrm{f}}(\boldsymbol{x}, t)\mathrm{d}v + \iint\limits_{\partial\chi_t[\mathscr{B}]} (\boldsymbol{v}\cdot\boldsymbol{n})\bar{\mathrm{f}}(\boldsymbol{x}, t)\mathrm{d}a \, . \tag{3.58}$$

□

Proof

Material differentiation under the integral sign using (1.112) in combination with (1.41) yields

$$\dot{\mathrm{F}}(\mathscr{B}, t) = \iiint\limits_{\chi_t[\mathscr{B}]} \Big\{\dot{\bar{\mathrm{f}}}(\boldsymbol{x}, t) + \bar{\mathrm{f}}(\boldsymbol{x}, t)\mathrm{div}\boldsymbol{v}\Big\}\mathrm{d}v \, , \tag{3.59}$$

and finally,

$$\dot{\mathrm{F}}(\mathscr{B}, t) = \iiint\limits_{\chi_t[\mathscr{B}]} \Big\{\frac{\partial}{\partial t} \bar{\mathrm{f}}(\boldsymbol{x}, t) + \mathrm{div}\big[\bar{\mathrm{f}}(\boldsymbol{x}, t)\bar{\boldsymbol{v}}(\boldsymbol{x}, t)\big]\Big\}\mathrm{d}v \, . \tag{3.60}$$

The latter step is based on the product rule

$$\big[\mathrm{grad}\bar{\mathrm{f}}(\boldsymbol{x}, t)\big]\bar{\boldsymbol{v}}(\boldsymbol{x}, t) + \bar{\mathrm{f}}(\boldsymbol{x}, t)\mathrm{div}\bar{\boldsymbol{v}}(\boldsymbol{x}, t) = \mathrm{div}\Big\{\bar{\mathrm{f}}(\boldsymbol{x}, t)\bar{\boldsymbol{v}}(\boldsymbol{x}, t)\Big\} \, . \tag{3.61}$$

The type of the product $\mathrm{f}\boldsymbol{v}$ depends on the nature of the range $\mathbb{W}$ of the density function f:

$$\mathrm{f}\boldsymbol{v} = \begin{cases} f\boldsymbol{v} & \text{for} \quad \mathbb{W} \subset \mathbb{R} \\ \boldsymbol{f} \otimes \boldsymbol{v} \text{ for} & \mathbb{W} \subset \mathbb{V}^3 \end{cases} \, . \tag{3.62}$$

In this way the *Gauß* theorem leads directly to (3.58); this relation holds, as it stands, independent of the range of function $\mathrm{f}(\mathscr{P}, t)$. ■

[6] Truesdell & Toupin [1960], Sect. 81.

The relation (3.58) is known as the *Reynolds transport theorem* and can be interpreted as follows: the temporal change of an additive quantity consists of one local and one convective part. The local change is the integral over the partial time derivative of $\bar{\mathrm{f}}(\boldsymbol{x}, t)$; the convective change corresponds to the transport of the specific quantity f across the surface of the current configuration: the integrand $(\boldsymbol{v} \cdot \boldsymbol{n})\bar{\mathrm{f}}(\boldsymbol{x}, t)$ is the flux density per unit area.

Definition 3. 7

An *open system* is a given spatial domain, which may be fixed or temporally variable. It serves as a *volume under observation* $v_{\mathrm{K}}(t)$, bounded by a surface $a_{\mathrm{K}}(t)$. ◊

The motion of a time-dependent volume under observation can be represented by means of a non-material velocity field $(\boldsymbol{x}, t) \longmapsto \boldsymbol{w}(\boldsymbol{x}, t)$.

In this way it is no problem to calculate the temporal change of the volume integral $\mathrm{F}(\mathscr{B}, t)$, taking place in relation to the volume $v_{\mathrm{K}}(t)$, which moves with the velocity $\boldsymbol{w}$. Generalising the transport theorem (3.58), the following formula simply applies:[7]

$$\frac{\mathrm{d}_{w}}{\mathrm{d}t} \iiint\limits_{v_{\mathrm{K}}(t)} \bar{\mathrm{f}}(\boldsymbol{x}, t)\mathrm{d}v_{\mathrm{K}} = \iiint\limits_{\chi_t[\mathscr{B}]} \frac{\partial}{\partial t}\bar{\mathrm{f}}(\boldsymbol{x}, t)\mathrm{d}v + \iint\limits_{\partial\chi_t[\mathscr{B}]} (\boldsymbol{w} \cdot \boldsymbol{n})\bar{\mathrm{f}}(\boldsymbol{x}, t)\mathrm{d}a\,. \tag{3.63}$$

This derivative can obviously not be utilised in balance relations, since it is always the material derivative that has to be inserted there. Instead, still generalising (3.58), the following *generalised transport theorem* is valid:

Theorem 3. 6

Assume that $\boldsymbol{w}(\boldsymbol{x}, t)$ is a velocity field representing the motion of a time-dependent volume $v_{\mathrm{K}}(t)$ with boundary surface $a_{\mathrm{K}}(t)$, just coinciding with the current configuration $\chi_t[\mathscr{B}]$ at time t. Then the calculation rule given below holds for the material derivative of the volume integral $\mathrm{F}(\mathscr{B}, t)$:

$$\frac{\mathrm{d}}{\mathrm{d}t} \iiint\limits_{\chi_t[\mathscr{B}]} \bar{\mathrm{f}}(\boldsymbol{x}, t)\mathrm{d}v = \frac{\mathrm{d}_{w}}{\mathrm{d}t} \iiint\limits_{v_{\mathrm{K}}(t)} \bar{\mathrm{f}}(\boldsymbol{x}, t)\mathrm{d}v_{\mathrm{K}} + \iint\limits_{a_{\mathrm{K}}(t)} [(\boldsymbol{v} - \boldsymbol{w}) \cdot \boldsymbol{n}]\bar{\mathrm{f}}(\boldsymbol{x}, t)\mathrm{d}a_{\mathrm{K}}\,. \tag{3.64}$$

□

Proof

by eliminating the local derivative from (3.58) and (3.63). The volumes $v_{\mathrm{K}}(t)$ and $\chi_t[\mathscr{B}]$ as well as the surfaces $a_{\mathrm{K}}(t)$ and $\partial\chi_t[\mathscr{B}]$ are identical. ■

---

[7] Truesdell & Toupin [1960], eq. (84.4).

According to the generalised transport theorem (3.64), the material derivative of a volume integral consists of the temporal change with respect to the time-dependent volume $v_K(t)$ and the transport through the moving boundary surface $a_K(t)$. The density of the flux is determined by the normal component of the velocity difference $\boldsymbol{v} - \boldsymbol{w}$, i.e. by the very same material elements that extend beyond the boundary of the volume under observation.

3. 5. 2 Balance of Linear Momentum for Systems with Time-Dependent Mass

We seek to illuminate the physical significance of the generalised transport theorem below, using the balance of linear momentum (2.31),

$$\frac{\mathrm{d}}{\mathrm{d}t}\boldsymbol{I}(\mathcal{B}, t) = \boldsymbol{F}(\mathcal{B}, t) .$$

In the case of an open system, whose boundary surface moves with the velocity field $\boldsymbol{w}(\boldsymbol{x}, t)$, the change of linear momentum is given by

$$\frac{\mathrm{d}}{\mathrm{d}t}\boldsymbol{I}(\mathcal{B}, t) = \frac{\mathrm{d}_w}{\mathrm{d}t}\iiint\limits_{v_K(t)}\rho\boldsymbol{v}\mathrm{d}v_K + \iint\limits_{a_K(t)}[(\boldsymbol{v} - \boldsymbol{w})\cdot\boldsymbol{n}](\rho\boldsymbol{v})\mathrm{d}a_K , \tag{3.65}$$

according to the generalised transport theorem (3.64). The first term on the right-hand side represents the temporal change of linear momentum in the volume $v_K(t)$. This change can be expressed in terms of the velocity $\boldsymbol{v}_m$ of the centre of mass:

$$\frac{\mathrm{d}_w}{\mathrm{d}t}\iiint\limits_{v_K(t)}\rho\boldsymbol{v}\mathrm{d}v_K = \frac{\mathrm{d}}{\mathrm{d}t}\left\{m_K(t)\boldsymbol{v}_m(t)\right\} . \tag{3.66}$$

In this equation

$$m_K(t) = \iiint\limits_{v_K(t)}\rho\mathrm{d}v_K \tag{3.67}$$

is the time-dependent mass contained within the volume $v_K(t)$. In view of the balance of mass, we obtain

$$\frac{\mathrm{d}}{\mathrm{d}t}\iiint\limits_{\chi_t[\mathscr{B}]} \rho(\boldsymbol{x}, t)\mathrm{d}v = \frac{\mathrm{d}_w}{\mathrm{d}t}\iiint\limits_{v_\mathrm{K}(t)} \rho \mathrm{d}v_\mathrm{K} + \iint\limits_{a_\mathrm{K}(t)} [(\boldsymbol{v} - \boldsymbol{w})\cdot\boldsymbol{n}]\rho \mathrm{d}a_\mathrm{K} = 0$$

or

$$\dot{m}_\mathrm{K}(t) + \mu(t) = 0\,, \tag{3.68}$$

in other words, the mass $m_\mathrm{K}(t)$ within the volume $v_\mathrm{K}(t)$ changes in proportion to the mass flow

$$\mu(t) = \iint\limits_{a_\mathrm{K}(t)} [(\boldsymbol{v} - \boldsymbol{w})\cdot\boldsymbol{n}]\rho \mathrm{d}a_\mathrm{K} \tag{3.69}$$

through the boundary $a_\mathrm{K}(t)$.

The second term on the right-hand side of (3.65) represents the flow of linear momentum. It is advisable to express the density of linear momentum $\rho\boldsymbol{v}$ by the *efflux velocity* $\boldsymbol{u} = \boldsymbol{v} - \boldsymbol{v}_m$,

$$\rho\boldsymbol{v} = \rho(\boldsymbol{v}_m - \boldsymbol{u})\,, \tag{3.70}$$

the outcome being $\iint\limits_{a_\mathrm{K}(t)} [(\boldsymbol{v} - \boldsymbol{w})\cdot\boldsymbol{n}](\rho\boldsymbol{v})\mathrm{d}a_\mathrm{K} = \iint\limits_{a_\mathrm{K}(t)} [(\boldsymbol{v} - \boldsymbol{w})\cdot\boldsymbol{n}]\rho(\boldsymbol{v}_m - \boldsymbol{u})\mathrm{d}a_\mathrm{K}$,

$$\iint\limits_{a_\mathrm{K}(t)} [(\boldsymbol{v} - \boldsymbol{w})\cdot\boldsymbol{n}](\rho\boldsymbol{v})\mathrm{d}a_\mathrm{K} = \mu(t)\boldsymbol{v}_m(t) - \iint\limits_{a_\mathrm{K}(t)} [(\boldsymbol{v} - \boldsymbol{w})\cdot\boldsymbol{n}](\rho\boldsymbol{u})\mathrm{d}a_\mathrm{K}\,. \tag{3.71}$$

The balance of linear momentum for a system with time-dependent mass arises from the combination of equations (2.31), (3.65) and (3.66) in the form of

$$\frac{\mathrm{d}}{\mathrm{d}t}\left\{m_\mathrm{K}(t)\boldsymbol{v}_m(t)\right\} + \iint\limits_{a_\mathrm{K}(t)} [(\boldsymbol{v} - \boldsymbol{w})\cdot\boldsymbol{n}](\rho\boldsymbol{v})\mathrm{d}a_\mathrm{K} = \boldsymbol{F}(t)\,, \tag{3.72}$$

which, by introducing the *efflux velocity* $\boldsymbol{u}$ according to (3.70) and considering the mass balance (3.68), simplifies to

$$m_\mathrm{K}(t)\dot{\boldsymbol{v}}_m(t) = \boldsymbol{F}(t) + \iint\limits_{a_\mathrm{K}(t)} [(\boldsymbol{v} - \boldsymbol{w})\cdot\boldsymbol{n}](\rho\boldsymbol{u})\mathrm{d}a_\mathrm{K}\,. \tag{3.73}$$

Accordingly, the acceleration of the centre of mass of an open system is caused by the resultant force $\boldsymbol{F}(t)$ in addition to the flow of linear momentum across the boundary surface. The efflux density of linear momentum is proportional to the normal component of the difference velocity $\boldsymbol{v} - \boldsymbol{w}$ ($\boldsymbol{v}$ = material velocity on the boundary surface of the system, $\boldsymbol{w}$ = velocity of the boundary surface). The efflux of linear momentum is, of course, only different from zero in the permeable parts of the boundary surface. By contrast, impermeable walls are indicated by means of a difference velocity tangential to the boundary, i.e. by $(\boldsymbol{v} - \boldsymbol{w})\cdot\boldsymbol{n} = 0$. If we replace the efflux velocity $\boldsymbol{u} = \boldsymbol{v}_m - \boldsymbol{v}$ by its mean value $\boldsymbol{u}^*$,

$$\iint\limits_{a_K(t)} [(\boldsymbol{v} - \boldsymbol{w})\cdot\boldsymbol{n}](\rho\boldsymbol{u})\mathrm{d}a_K = \Big\{\iint\limits_{a_K(t)} [(\boldsymbol{v} - \boldsymbol{w})\cdot\boldsymbol{n}]\rho\mathrm{d}a_K\Big\}\boldsymbol{u}^*(t) = \mu(t)\boldsymbol{u}^*(t) ,$$

$$\iint\limits_{a_K(t)} [(\boldsymbol{v} - \boldsymbol{w})\cdot\boldsymbol{n}](\rho\boldsymbol{u})\mathrm{d}a_K = -\,\dot{m}_K(t)\boldsymbol{u}^*(t)\ , \text{ we obtain}$$

$$m_K(t)\dot{\boldsymbol{v}}_m(t) = \boldsymbol{F}(t) - \dot{m}_K(t)\boldsymbol{u}^*(t)\ . \tag{3.74}$$

This formulation of the balance of linear momentum applies to the motion of a rocket; the term $-\dot{m}_K\boldsymbol{u}^*$ is called the *rocket thrust*.

3. 5. 3 Balance Relations: Conservation Laws

For a general density function $(\mathscr{P}, t) \longmapsto \mathrm{w} = \mathrm{f}(\mathscr{P}, t) = \bar{\mathrm{f}}(\boldsymbol{x}, t)$, a balance relation in thermomechanics has the following general structure:[8]

$$\frac{\mathrm{d}}{\mathrm{d}t}\iiint\limits_{\chi_t[\mathscr{B}]} \bar{\mathrm{f}}(\boldsymbol{x}, t)\mathrm{d}v = \iint\limits_{\partial\chi_t[\mathscr{B}]} \Phi\boldsymbol{n}\mathrm{d}a + \iiint\limits_{\chi_t[\mathscr{B}]} (\varphi + \mathrm{p})\mathrm{d}v\ . \tag{3.75}$$

The interaction with the environment is represented by means of the general fields Φ and φ. The integrand $\Phi\boldsymbol{n}$ characterises the surface-distributed interaction as a linear function of the normal unit vector; we write

[8] Cf. TRUESDELL & TOUPIN [1960], Sect. 157.

$$\Phi n = \begin{cases} \boldsymbol{\Phi} \cdot \boldsymbol{n} & \text{for} \quad \mathbb{W} = \mathbb{R} \\ \boldsymbol{\Phi n} & \text{for} \quad \mathbb{W} = \mathbb{V}^3 \end{cases},$$

where $\mathbb{W}$ is the range of f. The term p denotes a volume-distributed *production density.* Balance statements, postulating the vanishing production term, are called *conservation laws.* In this sense, except for the balance of entropy, all balance equations of thermomechanics are *conservation laws.*

Following the standard line of reasoning, the general balance relation (3.75) leads to a local balance statement according to

Theorem 3. 7

For continuously differentiable fields f and Φ as well as continuous fields φ and p the local formulation of the general balance relation reads

$$\dot{\mathrm{f}} + \mathrm{f}\,\mathrm{div}\boldsymbol{v} - \mathrm{div}\,\Phi - \varphi = \mathrm{p}\,, \tag{3.76}$$

or

$$\frac{\partial \mathrm{f}}{\partial t} + \mathrm{div}\big[\mathrm{f}\,\boldsymbol{v} - \Phi\big] - \varphi = \mathrm{p}\,. \tag{3.77}$$

□

Proof

By employing the *Gauß* theorem and equation (3.59) we accomplish the statement

$$\iiint\limits_{\chi_t[\mathscr{B}]} \Big\{\dot{\mathrm{f}} + \mathrm{f}\,\mathrm{div}\boldsymbol{v} - \mathrm{div}\,\Phi - \varphi - \mathrm{p}\Big\} dv = 0, \tag{3.78}$$

which has to be valid for all sub-volumes of $\chi_t[\mathscr{B}]$. Applying Theorem 2. 1 this leads to (3.76). Equation (3.77) follows by employing the product rule (3.61). ■

The general relations (3.76) and (3.77) lead to a summary of all thermomechanical balance relations, substantiating each field in turn:

Conservation of Mass

$$\mathrm{f} = \rho\,, \Phi = 0\,, \varphi = 0\,, \mathrm{p} = 0$$

$$\dot{\rho} + \rho \operatorname{div} \boldsymbol{v} = 0$$

$$\frac{\partial \rho}{\partial t} + \operatorname{div}(\rho \boldsymbol{v}) = 0 \tag{3.79}$$

Conservation of Linear Momentum

$$\mathrm{f} = \rho \boldsymbol{v}\,, \Phi = \mathbf{T}\,, \varphi = \rho \boldsymbol{f}\,, \mathrm{p} = \boldsymbol{0}$$

$$\rho \dot{\boldsymbol{v}} - \operatorname{div} \mathbf{T} - \rho \boldsymbol{f} = \boldsymbol{0}$$

$$\frac{\partial}{\partial t}(\rho \boldsymbol{v}) + \operatorname{div}\Big\{\rho(\boldsymbol{v} \otimes \boldsymbol{v}) - \mathbf{T}\Big\} - \rho \boldsymbol{f} = \boldsymbol{0} \tag{3.80}$$

Conservation of Rotational Momentum

$$\mathrm{f} = (\boldsymbol{x} - \boldsymbol{c}) \times (\rho \boldsymbol{v})\,, \Phi = (\boldsymbol{x} - \boldsymbol{c}) \times \mathbf{T}\,, \varphi = (\boldsymbol{x} - \boldsymbol{c}) \times (\rho \boldsymbol{f})\,, \mathrm{p} = \boldsymbol{0}$$

$$\mathbf{T} - \mathbf{T}^{\mathrm{T}} = \mathbf{0} \tag{3.81}$$

Conservation of Energy

$$\mathrm{f} = \rho\big(e + \tfrac{1}{2}\boldsymbol{v} \cdot \boldsymbol{v}\big)\,, \Phi = \mathbf{T}^{\mathrm{T}}\boldsymbol{v} - \boldsymbol{q}\,, \varphi = \rho\big(\boldsymbol{f} \cdot \boldsymbol{v} + r\big)\,, \mathrm{p} = 0$$

$$\rho \dot{e} + \operatorname{div} \boldsymbol{q} - \rho r - \mathbf{T} \cdot \mathbf{D} = 0$$

$$\frac{\partial}{\partial t}\big(\rho e + \tfrac{1}{2}\rho v^2\big) + \operatorname{div}\Big\{\rho\big(e + \tfrac{1}{2}v^2\big)\boldsymbol{v} - \mathbf{T}^{\mathrm{T}}\boldsymbol{v} + \boldsymbol{q}\Big\} - \rho\big(\boldsymbol{f} \cdot \boldsymbol{v} + r\big) = 0 \tag{3.82}$$

Entropy Balance

$$\mathrm{f} = \rho s\,, \Phi = -\frac{1}{\Theta}\,\boldsymbol{q}\,, \varphi = \frac{1}{\Theta}\,\rho r\,, \mathrm{p} = \rho \gamma$$

$$\rho \dot{s} + \operatorname{div}\Big(\frac{\boldsymbol{q}}{\Theta}\Big) - \rho \frac{r}{\Theta} = \rho \gamma$$

$$\frac{\partial}{\partial t}(\rho s) + \operatorname{div}\Big\{\rho s \boldsymbol{v} + \frac{\boldsymbol{q}}{\Theta}\Big\} - \rho \frac{r}{\Theta} = \rho \gamma \tag{3.83}$$

The divergence terms in the local balance relations (3.80), (3.82) and (3.83) also contain, besides the contribution Φ from the interaction with the environment, such shares as stem from the material transport across the spatial boundary of the system. This becomes apparent if we integrate the balance equation (3.77) over v_K and again apply the *Gauß* theorem:

$$\iiint\limits_{v_K(t)} \frac{\partial \mathrm{f}}{\partial t}\, dv_K = \iint\limits_{a_K(t)} [\Phi - \mathrm{f}\boldsymbol{v}]\boldsymbol{n} da_K + \iiint\limits_{v_K(t)} (\varphi + \mathrm{p}) dv_K \,. \tag{3.84}$$

Here the size of the volume v_K is subject to choice. Dividing (3.84) by v_K and subsequently passing to the limit $v_K \longrightarrow 0$ leads back to (3.77). In this sense the balance relations (3.79) - (3.83) are local balance equations for open systems. With the exception of the entropy balance they have the character of conservation laws.

3. 5. 4 Discontinuity Surfaces and Jump Conditions

The assumptions regarding the smoothness of the fields needed for deriving the local balance relation (3.76) can be relaxed: it is possible to take certain discontinuities of the fields into account.

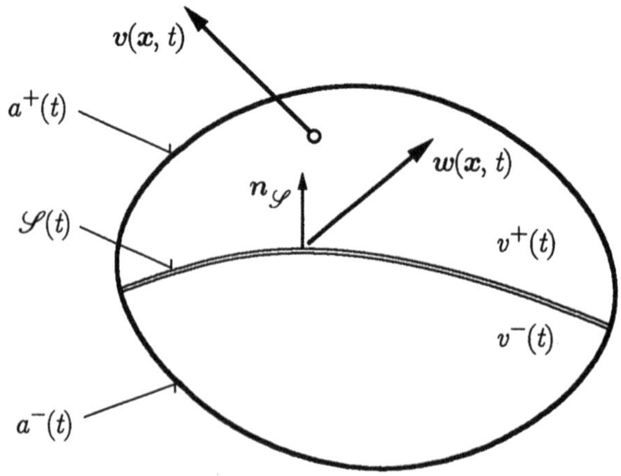

Figure 3. 1: Discontinuity Surface

The considerations set out below are based on the assumption that the discontinuities of any given field $\mathrm{w} = \mathrm{f}(\mathscr{P}, t) = \mathrm{f}(\boldsymbol{x}, t)$ are distributed over a surface area $\mathscr{S}(t)$ that travels through the material body. It is assumed that the discontinuity surface $\mathscr{S}(t)$ is smooth and possesses a unit normal $\boldsymbol{n}_{\mathscr{S}}$ at every point. We also assume that every point of the surface $\mathscr{S}(t)$ moves with the velocity $\boldsymbol{w}(\boldsymbol{x}, t)$. In general the velocity $\boldsymbol{w}(\boldsymbol{x}, t)$ differs from the material velocity $\boldsymbol{v}(\boldsymbol{x}, t)$. The discontinuity surface is sketched in Fig. 3. 1. It divides the domain of the current configuration into two parts v^+ and v^-. The part v^+ is on the side of the normal vector $\boldsymbol{n}_{\mathscr{S}}$, towards which the surface area moves and v^- on the other side. The limit values of the function $\mathrm{f}(\boldsymbol{x}, t)$ at one point of the surface are $\mathrm{f}^+(\boldsymbol{x}, t)$ and $\mathrm{f}^-(\boldsymbol{x}, t)$, depending on whether the point $\boldsymbol{x} \in \mathscr{S}(t)$ is reached from the (+) side or from the (–) side. The difference between these two limit values is the *jump* of the field $\mathrm{f}(\boldsymbol{x}, t)$, defined by

$$[\![\mathrm{f}]\!] = \mathrm{f}^+ - \mathrm{f}^- . \tag{3.85}$$

In the customary way, we assume that the jump $[\![\mathrm{f}]\!](\boldsymbol{x}, t)$ is a smooth function of $\boldsymbol{x}$ and t on the surface $\mathscr{S}(t)$. Finally, we assume that the function $\mathrm{f}(\boldsymbol{x}, t)$ is discontinuous only across the surface $\mathscr{S}(t)$. Outside $\mathscr{S}(t)$, as assumed in previous investigations, $\mathrm{f}(\boldsymbol{x}, t)$ is continuously differentiable.

Based on these assumptions, the volume integral $\mathrm{F}(\mathscr{B}, t)$ is the sum of two volume integrals,

$$\mathrm{F}(\mathscr{B}, t) = \iiint\limits_{\chi_t[\mathscr{B}]} \bar{\mathrm{f}}(\boldsymbol{x}, t)\mathrm{d}v = \iiint\limits_{v^+(t)} \bar{\mathrm{f}}(\boldsymbol{x}, t)\mathrm{d}v + \iiint\limits_{v^-(t)} \bar{\mathrm{f}}(\boldsymbol{x}, t)\mathrm{d}v , \tag{3.86}$$

which extend over the time-dependent volumes $v^+(t)$ and $v^-(t)$. Material differentiation of (3.86) using the generalised transport theorem (3.64) produces

$$\frac{\mathrm{d}}{\mathrm{d}t} \iiint\limits_{v^-(t)} \bar{\mathrm{f}}(\boldsymbol{x}, t)\mathrm{d}v = \frac{\overset{w}{\mathrm{d}}}{\mathrm{d}t} \iiint\limits_{v^-(t)} \mathrm{f}\mathrm{d}v + \iint\limits_{\mathscr{S}(t)} [(\boldsymbol{v}^- - \boldsymbol{w})\cdot\boldsymbol{n}_{\mathscr{S}}]\mathrm{f}^-\mathrm{d}a \quad \text{and}$$

$$\frac{\mathrm{d}}{\mathrm{d}t} \iiint\limits_{v^+(t)} \bar{\mathrm{f}}(\boldsymbol{x}, t)\mathrm{d}v = \frac{\overset{w}{\mathrm{d}}}{\mathrm{d}t} \iiint\limits_{v^+(t)} \mathrm{f}\mathrm{d}v - \iint\limits_{\mathscr{S}(t)} [(\boldsymbol{v}^+ - \boldsymbol{w})\cdot\boldsymbol{n}_{\mathscr{S}}]\mathrm{f}^+\mathrm{d}a .$$

For the material derivative of $\mathrm{F}(\mathscr{B}, t)$ the addition of these two equations leads to

$$\dot{\mathrm{F}}(\mathscr{B}, t) = \frac{\mathrm{d}}{\mathrm{d}t}\iiint\limits_{\chi_t[\mathscr{B}]} \bar{\mathrm{f}}(\boldsymbol{x}, t)\mathrm{d}v = \frac{\mathrm{d}}{\mathrm{d}t}\iiint\limits_{v^-(t)} \bar{\mathrm{f}}(\boldsymbol{x}, t)\mathrm{d}v + \frac{\mathrm{d}}{\mathrm{d}t}\iiint\limits_{v^+(t)} \bar{\mathrm{f}}(\boldsymbol{x}, t)\mathrm{d}v$$

$$= \frac{\mathrm{d}_{\boldsymbol{w}}}{\mathrm{d}t}\left[\iiint\limits_{v^-(t)} \mathrm{f}\mathrm{d}v + \iiint\limits_{v^+(t)} \mathrm{f}\mathrm{d}v\right]$$

$$+ \iint\limits_{\mathscr{S}(t)} \left\{[(\boldsymbol{v}^- - \boldsymbol{w})\cdot\boldsymbol{n}_{\mathscr{S}}]\mathrm{f}^- - [(\boldsymbol{v}^+ - \boldsymbol{w})\cdot\boldsymbol{n}_{\mathscr{S}}]\mathrm{f}^+\right\}\mathrm{d}a \quad \text{or to}$$

$$\dot{\mathrm{F}}(\mathscr{B}, t) = \iiint\limits_{\chi_t[\mathscr{B}]} \left\{\dot{\mathrm{f}} + \mathrm{f}\,\mathrm{div}\boldsymbol{v}\right\}\mathrm{d}v - \iint\limits_{\mathscr{S}(t)} [\![((\boldsymbol{v} - \boldsymbol{w})\cdot\boldsymbol{n}_{\mathscr{S}})\,\mathrm{f}\,]\!]\,\mathrm{d}a\,. \tag{3.87}$$

The volume integral on the right-hand side follows from the fact that the sum

$$\frac{\mathrm{d}_{\boldsymbol{w}}}{\mathrm{d}t}\iiint\limits_{v^-(t)} \mathrm{f}\mathrm{d}v + \frac{\mathrm{d}_{\boldsymbol{w}}}{\mathrm{d}t}\iiint\limits_{v^+(t)} \mathrm{f}\mathrm{d}v = \frac{\mathrm{d}_{\boldsymbol{w}}}{\mathrm{d}t}\iiint\limits_{\chi_t[\mathscr{B}]} \mathrm{f}\mathrm{d}v = \frac{\mathrm{d}}{\mathrm{d}t}\iiint\limits_{\chi_t[\mathscr{B}]} \mathrm{f}\mathrm{d}v$$

is the material derivative of the volume integral, since the total volume corresponds to the current configuration, whose boundary surface moves with the material velocity $\boldsymbol{v}(\boldsymbol{x}, t)$.

The general balance relation (3.75), addressed to the partial volumes $v^-(t)$ and $v^+(t)$ respectively, reads

$$\frac{\mathrm{d}}{\mathrm{d}t}\iiint\limits_{v^-(t)} \mathrm{f}\mathrm{d}v = \iint\limits_{a^-(t)} \boldsymbol{\Phi}\boldsymbol{n}\mathrm{d}a + \iint\limits_{\mathscr{S}(t)} \boldsymbol{\Phi}^-\boldsymbol{n}_{\mathscr{S}}\,\mathrm{d}a + \iiint\limits_{v^-(t)} (\varphi + \mathrm{p})\mathrm{d}v$$

and
$$\frac{\mathrm{d}}{\mathrm{d}t}\iiint\limits_{v^+(t)} \mathrm{f}\mathrm{d}v = \iint\limits_{a^+(t)} \boldsymbol{\Phi}\boldsymbol{n}\mathrm{d}a - \iint\limits_{\mathscr{S}(t)} \boldsymbol{\Phi}^+\boldsymbol{n}_{\mathscr{S}}\,\mathrm{d}a + \iiint\limits_{v^+(t)} (\varphi + \mathrm{p})\mathrm{d}v\,.$$

Adding these two equations together and observing the material derivative (3.87), we obtain the result

$$\iiint\limits_{\chi_t[\mathscr{B}]} \left\{\dot{\mathrm{f}} + \mathrm{f}\,\mathrm{div}\boldsymbol{v} - \mathrm{div}\,\boldsymbol{\Phi} - \varphi - \mathrm{p}\right\}\mathrm{d}v = \iint\limits_{\mathscr{S}(t)} [\![\mathrm{f}\,(\boldsymbol{v} - \boldsymbol{w})\cdot\boldsymbol{n}_{\mathscr{S}} - \boldsymbol{\Phi}\boldsymbol{n}_{\mathscr{S}}]\!]\,\mathrm{d}a\,. \tag{3.88}$$

All the while we have made use of the fact that the two areas $a^-(t)$ and $a^+(t)$, when added together, form the entire boundary surface of the current configuration:

$$a^-(t) \cup a^+(t) = \partial\chi_t[\mathscr{B}] \, .$$

Again, equation (3.88) has to be valid for all sub-domains of the current configuration. This leads to

Theorem 3. 8

At all points in the space outside the discontinuity surface, the local balance relation (3.76) is valid, whereas on the surface $\mathscr{S}(t)$ the jump condition

$$[\![\mathrm{f}\,(\boldsymbol{v} - \boldsymbol{w})\cdot \boldsymbol{n}_{\mathscr{S}} - \boldsymbol{\Phi} \boldsymbol{n}_{\mathscr{S}}]\!] = 0 \tag{3.89}$$

applies.[9] □

Proof

The integrand on the left-hand side of (3.88) is continuous outside the surface $\mathscr{S}(t)$. The result of this is the local balance relation (3.76) according to Theorem 2. 1. In view of this the relation

$$\iint\limits_{\mathscr{S}(t)} [\![\mathrm{f}\,(\boldsymbol{v} - \boldsymbol{w})\cdot \boldsymbol{n}_{\mathscr{S}} - \boldsymbol{\Phi} \boldsymbol{n}_{\mathscr{S}}]\!] \, \mathrm{d}a = 0$$

has to hold true for all sub-surfaces of $\mathscr{S}(t)$. Since the integrand is assumed to depend continuously on $\boldsymbol{x}$, the jump condition (3.89) follows by means of the analogous application of Theorem 2. 1. ■

The local balance statements in thermomechanics have to be extended to include the following jump conditions in the case of discontinuous fields in the sense of the assumptions made:

Conservation of Mass

$$[\![\rho\,(\boldsymbol{v} - \boldsymbol{w})]\!] \cdot \boldsymbol{n}_{\mathscr{S}} = 0 \tag{3.90}$$

[9] TRUESDELL & TOUPIN [1960], eq. (193.3).

Conservation of Linear Momentum

$$[\![\rho \boldsymbol{v}\,(\boldsymbol{v}-\boldsymbol{w})\cdot \boldsymbol{n}_{\mathscr{S}} - \mathbf{T}\boldsymbol{n}_{\mathscr{S}}]\!] = \boldsymbol{0}\,, \tag{3.91}$$

$$[\![\rho \boldsymbol{v}\otimes(\boldsymbol{v}-\boldsymbol{w}) - \mathbf{T}]\!]\,\boldsymbol{n}_{\mathscr{S}} = \boldsymbol{0} \tag{3.92}$$

Conservation of Rotational Momentum

$$[\![(\boldsymbol{x}-\boldsymbol{c})\times(\rho\boldsymbol{v})\otimes(\boldsymbol{v}-\boldsymbol{w}) - (\boldsymbol{x}-\boldsymbol{c})\times\mathbf{T}]\!]\,\boldsymbol{n}_{\mathscr{S}} = \boldsymbol{0}$$

If the motion $\boldsymbol{x} = \chi_{\mathrm{R}}(\boldsymbol{X}, t)$ is a continuous function, we have

$$(\boldsymbol{x}-\boldsymbol{c})\times\left\{[\![\rho\boldsymbol{v}\otimes(\boldsymbol{v}-\boldsymbol{w}) - \mathbf{T}]\!]\,\boldsymbol{n}_{\mathscr{S}}\right\} = \boldsymbol{0}\,. \tag{3.93}$$

This jump condition is fulfilled identically according to (3.92).

Conservation of Energy

$$[\![\rho(e+\tfrac{1}{2}v^2)(\boldsymbol{v}-\boldsymbol{w}) - \mathbf{T}\boldsymbol{v} + \boldsymbol{q}]\!]\cdot\boldsymbol{n}_{\mathscr{S}} = 0 \tag{3.94}$$

Entropy Balance

$$[\![\rho s(\boldsymbol{v}-\boldsymbol{w}) + \frac{\boldsymbol{q}}{\Theta}]\!]\cdot\boldsymbol{n}_{\mathscr{S}} + \gamma_{\mathscr{S}} = 0 \tag{3.95}$$

In compliance with the principle of irreversibility an entropy production per unit surface ought to be inserted into this jump condition with the property

$$\gamma_{\mathscr{S}} \geq 0\,. \tag{3.96}$$

3. 5. 5 Multi-Component Systems (Mixtures)

A careful formulation of the balance relations of thermomechanics makes it clear that different generalisations are possible for the basic terms of the continuum theory. One possible generalisation is the concept of a multi-component system. This concept is one step towards thermomechanics of a continuum with microstructure. The basic idea is that various material elements with completely different properties can be present at one and the same point in space. This notion is expounded in the following

Definition 3. 8

A *multi-component system* or so-called *mixture* consists of a finite number n of *components* or *constituents*, which are simultaneously present at every point $x \in \chi_t[\mathscr{B}]$. Every component is characterised by a *partial density* $\rho_\alpha(x, t)$ $(\alpha = 1, \cdots, n)$. If $\rho(x, t)$ is the density of the entire mixture, the ratios

$$c_\alpha(x, t) = \frac{\rho_\alpha}{\rho} \qquad (\alpha = 1, \cdots, n) \tag{3.97}$$

are called *concentrations* with respect to mass. ◊

The partial density refers to the total volume, i.e. ρ_α is not the material density of the component α: ρ_α corresponds to the mass of the component α per volume of the mixture. This definition implies that the density of the mixture is the sum of the partial densities:

$$\rho = \sum_\alpha \rho_\alpha \quad \Longleftrightarrow \quad \sum_\alpha c_\alpha = 1\,. \tag{3.98}$$

The partial densities ρ_α or the concentrations c_α may vary as a result of interactions taking place between the constituents: chemical reactions, transition from one phase to another or diffusion processes as a result of molecular migration. Other forms of interaction between the components may occur in the form of an action of force (exchange of linear momentum), mechanical work or transition of heat (energy exchange). Individual energy or entropy productions are also conceivable.

The representation of these ideas by means of an appropriate field theory demands a certain modification of the basic terminology of classical continuum mechanics.[10] A material point of any mixture does not merely contain all n constituents simultaneously: the kinematics is represented by n velocity fields $(x, t) \longmapsto v_\alpha(x, t)$ and the dynamics through n partial stress tensors $\mathbf{T}_\alpha(x, t)$ $(\alpha = 1, \cdots, n)$. For consistent modelling in terms of thermomechanics one also requires an n-quantity each of partial energies $e_\alpha(x, t)$, entropies $s_\alpha(x, t)$, heat flux vectors $q_\alpha(x, t)$, volume forces $k_\alpha(x, t)$ and volume-distributed heat supplies $r_\alpha(x, t)$. Every single component of the mixture can be regarded locally as an open system undergoing a thermomechanical exchange with the remaining n–1 systems. With regard to the general structure (3.77) of a local balance relation for open systems the

[10] The presentation follows Hutter [1977]; various contributions on the subject are to be found in Truesdell [1984a].

additional fields ought to be extended to include related production terms to represent the material, dynamic and energetic interaction between the individual constituents.

Since each constituent of a mixture is locally regarded as an open system, the balance relations (3.79) to (3.83) have to be postulated individually for each component, yet not in the form of conservation theorems: the conservation of mass, linear momentum and energy should only continue to be valid for the mixture as an entirety. In the same way the entropy production (3.83) of the whole mixture has to be positive, though individual negative contributions are acceptable on principle. By way of generalisation of the balance relations discussed so far we demand the following

Balance Relations for an n-Component System

a) The following balance relations hold for $\alpha = 1, \cdots, n$ and $\boldsymbol{x} \in \chi_t[\mathscr{B}]$:

$$\frac{\partial \rho_\alpha}{\partial t} + \operatorname{div}(\rho_\alpha \boldsymbol{v}_\alpha) = C_\alpha , \tag{3.99}$$

$$\frac{\partial}{\partial t}(\rho_\alpha \boldsymbol{v}_\alpha) + \operatorname{div}\{\rho_\alpha(\boldsymbol{v}_\alpha \otimes \boldsymbol{v}_\alpha) - \mathbf{T}_\alpha\} - \rho_\alpha \boldsymbol{f}_\alpha = \boldsymbol{m}_\alpha , \tag{3.100}$$

$$\mathbf{T}_\alpha - \mathbf{T}_\alpha^{\mathrm{T}} = \mathbf{M}_\alpha , \tag{3.101}$$

$$\frac{\partial}{\partial t}(\rho_\alpha e_\alpha + \tfrac{1}{2}\rho_\alpha v_\alpha^{\,2}) + $$

$$+ \operatorname{div}\{\rho_\alpha(e_\alpha + \tfrac{1}{2}v_\alpha^{\,2})\boldsymbol{v}_\alpha - \mathbf{T}_\alpha^{\mathrm{T}}\boldsymbol{v}_\alpha + \boldsymbol{q}_\alpha\} - \rho_\alpha(\boldsymbol{f}_\alpha \cdot \boldsymbol{v}_\alpha + r_\alpha) = p_\alpha , \tag{3.102}$$

$$\frac{\partial}{\partial t}(\rho_\alpha s_\alpha) + \operatorname{div}\left\{\rho_\alpha s_\alpha \boldsymbol{v}_\alpha + \frac{\boldsymbol{q}_\alpha}{\Theta}\right\} - \frac{\rho_\alpha r_\alpha}{\Theta} = \gamma_\alpha , \tag{3.103}$$

b) For the production terms restrictive conditions are postulated:

Conservation of Mass

$$\sum_\alpha C_\alpha = 0 \tag{3.104}$$

Conservation of Linear Momentum

$$\sum_\alpha \boldsymbol{m}_\alpha = \boldsymbol{0} \tag{3.105}$$

Conservation of Rotational Momentum

$$\sum_{\alpha} \mathbf{M}_{\alpha} = \mathbf{0} \tag{3.106}$$

Conservation of Energy

$$\sum_{\alpha} p_{\alpha} = 0 \tag{3.107}$$

Principle of Irreversibility

$$\gamma = \sum_{\alpha} \gamma_{\alpha} \geq 0 \tag{3.108}$$

♦

It is obvious that the partial fields ρ_{α}, $\boldsymbol{v}_{\alpha}$, $\mathbf{T}_{\alpha}$, e_{α}, s_{α}, $\boldsymbol{q}_{\alpha}$, $\boldsymbol{f}_{\alpha}$ and r_{α} have to be connected to the "macroscopic" fields ρ, $\boldsymbol{v}$, $\mathbf{T}$, e, s, $\boldsymbol{q}$, $\boldsymbol{f}$ and r by means of suitable definitions, to avoid the risk of statements contradicting the foundations of thermomechanics discussed so far. In particular, it must be possible to identify a multi-component system as a whole with a one-component system, to which the local balance relations (3.79) to (3.83) apply. These local balance statements or their equivalent formulations (2.11), (2.62), (3.17) and (3.31) should retain their universal validity for all material systems. The additional definitions required are to be introduced below. The first definition of this kind, equation (3.98), accounts for the conservation of mass. A further definition concerns the kinematics of partial velocity fields.

Definition 3.9
The velocity field

$$\boldsymbol{v}(\boldsymbol{x}, t) = \frac{1}{\rho}\sum_{\alpha} \rho_{\alpha}\boldsymbol{v}_{\alpha} = \sum_{\alpha} c_{\alpha}\boldsymbol{v}_{\alpha} \tag{3.109}$$

is called *barycentric velocity*; the differences

$$\boldsymbol{u}_{\alpha}(\boldsymbol{x}, t) = \boldsymbol{v}_{\alpha} - \boldsymbol{v} \;(\alpha = 1, \cdots, n) \tag{3.110}$$

are called *diffusion velocities.*

If $\mathrm{w} = \mathrm{f}(\mathscr{P}, t) = \bar{\mathrm{f}}(\boldsymbol{x}, t) \left(\equiv \mathrm{f}(\boldsymbol{x}, t)\right)$ is any given field quantity, the derivative

$$\dot{\mathrm{f}}(x, t) = \frac{\partial}{\partial t}\mathrm{f}(x, t) + [\mathrm{grad}\mathrm{f}(x, t)]v(x, t) \tag{3.111}$$

is called the *barycentric derivative.* The vector

$$\jmath_\alpha = \rho c_\alpha(v_\alpha - v) = \rho_\alpha u_\alpha \tag{3.112}$$

is called the *diffusion mass flux* of the component α. ◊

This definition leads to

Theorem 3. 9

a) The total mass density defined by (3.98) together with the barycentric velocity field according to (3.109) fulfils the conservation of mass (3.79).

b) The sum of the diffusion mass fluxes vanishes:

$$\sum_\alpha \rho_\alpha u_\alpha = 0\,. \tag{3.113}$$

c) For the concentrations c_α the balance relations

$$\rho\dot{c}_\alpha + \mathrm{div}\,\jmath_\alpha = C_\alpha \tag{3.114}$$

are valid. □

Proof

a) Adding the equations (3.99) together while taking (3.104) into consideration yields (3.98):

$$\frac{\partial}{\partial t}\sum_\alpha \rho_\alpha + \mathrm{div}\Big\{\sum_\alpha \rho_\alpha v_\alpha\Big\} = \frac{\partial\rho}{\partial t} + \mathrm{div}(\rho v) = 0\,.$$

b) (3.110) with the help of (3.109) and (3.98) leads to

$$\sum_\alpha c_\alpha u_\alpha = \sum_\alpha c_\alpha v_\alpha - \Big(\sum_\alpha c_\alpha\Big)v = v - v = 0\,.$$

c) Using the barycentric derivative of the concentration,

$$\dot{c}_\alpha = \frac{\partial c_\alpha}{\partial t} + (\operatorname{grad} c_\alpha) \cdot v \tag{3.115}$$

we can calculate

$$\frac{\partial \rho_\alpha}{\partial t} = \frac{\partial}{\partial t}(\rho c_\alpha) = c_\alpha \frac{\partial \rho}{\partial t} + \rho \frac{\partial c_\alpha}{\partial t} = - c_\alpha \operatorname{div}(\rho v) + \rho\left[\dot{c}_\alpha - (\operatorname{grad} c_\alpha) \cdot v\right] \quad \text{or}$$

$$\frac{\partial \rho_\alpha}{\partial t} = \rho \dot{c}_\alpha - \operatorname{div}(\rho_\alpha v) .$$

By inserting this into (3.99) we arrive at the formula (3.114). ■

The following definition guarantees the conservation of linear momentum:

Definition 3. 10

The total stress tensor $\mathbf{T}$ of the mixture is the sum of the partial stresses minus the diffusion stresses,

$$\mathbf{T} = \sum_\alpha \left(\mathbf{T}_\alpha - \rho_\alpha u_\alpha \otimes u_\alpha\right) . \tag{3.116}$$

The resultant volume force is the sum of the partial volume forces:

$$\rho f = \sum_\alpha \rho_\alpha f_\alpha \, . \tag{3.117}$$

◇

The definition leads to

Theorem 3. 10

With the definitions (3.116), (3.117) the conservation of linear momentum (3.80) applies. □

Proof

Adding the balances of linear momentum (3.100) together with (3.105), (3.109), (3.117) and (3.116) results in

$$\frac{\partial}{\partial t}(\rho v) - \rho f = \operatorname{div}\left\{\sum_\alpha \left(\mathbf{T}_\alpha - \rho_\alpha v_\alpha \otimes v_\alpha\right)\right\}$$

$$= \operatorname{div}\left\{\mathbf{T} + \sum_\alpha \rho_\alpha \left(u_\alpha \otimes u_\alpha - v_\alpha \otimes v_\alpha\right)\right\} = \operatorname{div}\left\{\mathbf{T} - \rho(v \otimes v)\right\} ,$$

the definition $u = v_\alpha - v$ and (3.109) once again being used in the final step. ■

The partial stress tensors $\mathbf{T}_\alpha$ do not necessarily have to be symmetric, since the equations (3.101) allow for an exchange of rotational momentum between the constituents of the mixture. However, the total stress tensor is symmetric in view of the conservation of rotational momentum (3.106).

Definition 3. 11

For the mixture as a whole the specific internal energy is expressed as

$$\rho e = \sum_\alpha \rho_\alpha \left(e_\alpha + \frac{1}{2} u_\alpha{}^2 \right), \tag{3.118}$$

the heat flow vector by means of

$$q = \sum_\alpha \left\{ q_\alpha + \rho_\alpha \left(e_\alpha + \frac{1}{2} u_\alpha{}^2 \right) u_\alpha - \mathbf{T}_\alpha^{\mathrm{T}} u_\alpha \right\}, \tag{3.119}$$

and the volume distributed heat supply as

$$\rho r = \sum_\alpha \rho_\alpha \left(r_\alpha + f_\alpha \cdot u_\alpha \right). \tag{3.120}$$

◊

This leads to

Theorem 3. 11

The definitions (3.118) to (3.120) guarantee the validity of the local conservation of energy (3.82). □

Proof

Definition (3.118) for internal energy, supported by (3.110) and (3.109), leads first to

$$\sum_\alpha \rho_\alpha e_\alpha = \rho e - \sum_\alpha \frac{1}{2} \rho_\alpha u_\alpha{}^2 = \rho e - \sum_\alpha \frac{1}{2} \rho_\alpha \left(v_\alpha{}^2 - 2(v_\alpha \cdot v) + v^2 \right),$$

$$\sum_\alpha \rho_\alpha e_\alpha = \rho e + \frac{1}{2} \rho v^2 - \sum_\alpha \frac{1}{2} \rho_\alpha v_\alpha{}^2, \quad \text{i.e.}$$

$$\sum_\alpha \rho_\alpha \left(e_\alpha + \frac{1}{2} v_\alpha{}^2 \right) = \rho \left(e + \frac{1}{2} v^2 \right). \tag{3.121}$$

Moreover, using (3.110), (3.119), (3.121) and (3.116), we calculate

$$\sum_\alpha \Big\{ q_\alpha + \rho_\alpha \Big(e_\alpha + \tfrac{1}{2} v_\alpha^{\,2} \Big) v_\alpha - \mathbf{T}_\alpha^{\mathrm{T}} v_\alpha \Big\}$$

$$= \sum_\alpha \Big\{ q_\alpha + \rho_\alpha \Big(e_\alpha + \tfrac{1}{2} v_\alpha^{\,2} \Big)(v + u_\alpha) - \mathbf{T}_\alpha^{\mathrm{T}} (v + u_\alpha) \Big\}$$

$$= \sum_\alpha \Big\{ q_\alpha - \mathbf{T}_\alpha^{\mathrm{T}} u_\alpha + \rho_\alpha \Big(e_\alpha + \tfrac{1}{2} v_\alpha^{\,2} \Big) v - \mathbf{T}_\alpha^{\mathrm{T}} v + \rho_\alpha \Big(e_\alpha + \tfrac{1}{2} v_\alpha^{\,2} \Big) u_\alpha \Big\}$$

$$= q - \sum_\alpha \rho_\alpha \Big(e_\alpha + \tfrac{1}{2} u_\alpha^{\,2} \Big) u_\alpha + \rho \Big(e + \tfrac{1}{2} v^2 \Big) v$$

$$- \Big\{ \mathbf{T}^{\mathrm{T}} + \sum_\alpha \rho_\alpha u_\alpha \otimes u_\alpha \Big) \Big\} v + \sum_\alpha \rho_\alpha \Big(e_\alpha + \tfrac{1}{2} v_\alpha^{\,2} \Big) u_\alpha$$

$$= q + \rho \Big(e + \tfrac{1}{2} v^2 \Big) v - \mathbf{T}^{\mathrm{T}} v - \tfrac{1}{2} \sum_\alpha \underbrace{\Big[u_\alpha^{\,2} + 2(u_\alpha \cdot v) - v_\alpha^{\,2} \Big]}_{= v^2} u_\alpha ,$$

that is

$$\sum_\alpha \Big\{ q_\alpha + \rho_\alpha (e_\alpha + \tfrac{1}{2} v_\alpha^{\,2}) v_\alpha - \mathbf{T}_\alpha^{\mathrm{T}} v_\alpha \Big\} = q + \rho (e + \tfrac{1}{2} v^2) v - \mathbf{T}^{\mathrm{T}} v . \tag{3.122}$$

Adding together the energy balances (3.102) and observing definition (3.120), the energy conservation (3.107) and the result (3.122), we arrive at the energy conservation for the entire mixture,

$$\frac{\partial}{\partial t} (\rho e + \tfrac{1}{2} \rho v^2) + \operatorname{div} \Big\{ q + \rho (e + \tfrac{1}{2} v^2) v - \mathbf{T}^{\mathrm{T}} v \Big\} - \rho (f \cdot v + r) = 0 \text{, i.e. (3.82).}$$

■

Definition 3. 12

The specific entropy of the entire mixture is equal to the sum of the partial entropies:

$$\rho s = \sum_\alpha \rho_\alpha s_\alpha . \tag{3.123}$$

◇

This leads to

Theorem 3. 12

The entropy production of a multi-component system is expressed as

$$\gamma = \frac{\partial}{\partial t}(\rho s) + \operatorname{div}(\rho s \boldsymbol{v}) + \operatorname{div}\boldsymbol{\Sigma} - \rho\sigma\,, \tag{3.124}$$

with the entropy flux

$$\boldsymbol{\Sigma} = \frac{1}{\Theta}\Big\{\boldsymbol{q} + \sum_\alpha \Big[\rho_\alpha\Big(\Theta s_\alpha - e_\alpha - \tfrac{1}{2}\boldsymbol{u}_\alpha{}^2\Big)\mathbf{1} + \mathbf{T}_\alpha^{\mathrm{T}}\Big]\boldsymbol{u}_\alpha\Big\} \tag{3.125}$$

and the volume-distributed entropy supply

$$\rho\sigma = \frac{1}{\Theta}\Big\{\rho r - \sum_\alpha \rho_\alpha \boldsymbol{f}_\alpha \cdot \boldsymbol{u}_\alpha\Big\}\,. \tag{3.126}$$

□

Proof

If we add the entropy balances for the individual components together with the whole entropy (3.123) and $\boldsymbol{v}_\alpha = \boldsymbol{v} + \boldsymbol{u}_\alpha$, our first result is

$$\frac{\partial}{\partial t}(\rho s) + \operatorname{div}(\rho s \boldsymbol{v}) + \sum_\alpha \Big\{\operatorname{div}\Big[\rho_\alpha s_\alpha \boldsymbol{u}_\alpha + \frac{\boldsymbol{q}_\alpha}{\Theta}\Big] - \frac{\rho_\alpha r_\alpha}{\Theta}\Big\} = \sum_\alpha \gamma_\alpha\,.$$

Together with the heat flow vector (3.119) and the volume-distributed heat supply (3.120) we get the total entropy production

$$\begin{aligned}\gamma = {} & \frac{\partial}{\partial t}(\rho s) + \operatorname{div}(\rho s \boldsymbol{v}) + \\ & + \operatorname{div}\Big\{\frac{\boldsymbol{q}}{\Theta} + \frac{1}{\Theta}\sum_\alpha \Big[\rho_\alpha\Big(\Theta s_\alpha - e_\alpha - \tfrac{1}{2}\boldsymbol{u}_\alpha{}^2\Big)\boldsymbol{u}_\alpha + \mathbf{T}_\alpha^{\mathrm{T}}\boldsymbol{u}_\alpha\Big]\Big\} \\ & - \frac{1}{\Theta}\Big\{\rho r - \sum_\alpha \rho_\alpha \boldsymbol{f}_\alpha \cdot \boldsymbol{u}_\alpha\Big\}\,.\end{aligned}$$

■

The theorem claims that, in the case of a mixture, the entropy transport can no longer be expressed in terms of the equations (3.33) and (3.34) as a quotient of heat supply and absolute temperature. These classic relations, motivated by equilibrium thermodynamics, cannot be valid in a multi-component system, since the diffusion processes taking place with $\boldsymbol{u}_\alpha \neq \boldsymbol{0}$ are typical phenomena of non-equilibrium.

3. 6 Summary: Basic Relations of Thermomechanics

Thermomechanics is created by expanding upon the basic terminology of mechanics using terms common to thermodynamics. The basic relations of this field theory are the local balance relations. For material bodies containing only one single component, they are summarised once again:

Spatial Representation

Mass $$\dot{\rho} + \rho \operatorname{div} \boldsymbol{v} = 0 \tag{3.127}$$

Linear momentum $$\rho \dot{\boldsymbol{v}}(\boldsymbol{x}, t) = \operatorname{div} \mathbf{T}(\boldsymbol{x}, t) + \rho \boldsymbol{f} \tag{3.128}$$

Rotational momentum $$\mathbf{T} = \mathbf{T}^{\mathrm{T}} \tag{3.129}$$

Energy $$\dot{e}\,(\boldsymbol{x}, t) = -\frac{1}{\rho} \operatorname{div} \boldsymbol{q} + r + \frac{1}{\rho} \mathbf{T} \cdot \mathbf{D} \tag{3.130}$$

Entropy $$-\dot{e} + \Theta \dot{s} + \frac{1}{\rho} \mathbf{T} \cdot \mathbf{D} - \frac{1}{\rho \Theta} \boldsymbol{q} \cdot \boldsymbol{g} \geq 0 \tag{3.131}$$

Material Representation

Mass $$\rho_{\mathrm{R}} = \rho_{\mathrm{R}}(\boldsymbol{X}) \tag{3.132}$$

Linear momentum $$\rho_{\mathrm{R}} \dot{\boldsymbol{v}}(\boldsymbol{X}, t) = \operatorname{Div} \mathbf{F} \widetilde{\mathbf{T}}(\boldsymbol{X}, t) + \rho_{\mathrm{R}} \boldsymbol{f} \tag{3.133}$$

Rotational momentum $$\widetilde{\mathbf{T}} = \widetilde{\mathbf{T}}^{\mathrm{T}} \tag{3.134}$$

Energy $$\dot{e}\,(\boldsymbol{X}, t) = -\frac{1}{\rho_{\mathrm{R}}} \operatorname{Div} \boldsymbol{q}_{\mathrm{R}} + r + \frac{1}{\rho_{\mathrm{R}}} \widetilde{\mathbf{T}} \cdot \dot{\mathbf{E}} \tag{3.135}$$

Entropy $$-\dot{e} + \Theta \dot{s} + \frac{1}{\rho_{\mathrm{R}}} \widetilde{\mathbf{T}} \cdot \dot{\mathbf{E}} - \frac{1}{\rho_{\mathrm{R}} \Theta^2} \boldsymbol{q}_{\mathrm{R}} \cdot \boldsymbol{g}_{\mathrm{R}} \geq 0 \tag{3.136}$$

The *spatial representation,* together with the continuity equation (3.127) and the balance of linear momentum (3.128) and energy (3.130), contains a total of 5 scalar equations to determine the fields $\boldsymbol{v}$, Θ, ρ, $\mathbf{T}$, e, $\boldsymbol{q}$, and s, covering no less than 16 unknown functions in all. The entropy inequality is of no service when determining these fields: it acts as a constraint condition to be fulfilled by all solutions of the basic equations.

The *material representation* presents the same situation except for the one difference that one equation and one unknown less occur, since the distribution of mass $\rho_R = \rho_R(\boldsymbol{X})$ is considered to be a known fact: the balance of momentum and energy contain a total of 4 equations for the unknown fields χ_R, Θ, $\tilde{\mathbf{T}}$, e, $\boldsymbol{q}_R$ and s, embracing 15 unknown functions all told.

Counting the equations and unknowns we see that the basic relations in thermomechanics are considerably underdetermined. The equations that have to be included are the so-called *constitutive equations.* Constitutive equations express the fact that, apart from the universal balance relations, there are also individual characteristics in a material body that influence its behaviour to no mean extent. In order to adjust the number of equations to the number of unknowns (without giving any indication as to solving them), 11 constitutive equations would have to be set up for a one-component system. These 11 equations can be formulated for $\mathbf{T}$, e, $\boldsymbol{q}$ and s or for the fields $\tilde{\mathbf{T}}$, e, $\boldsymbol{q}_R$ and s as well. In the case of the spatial representation 5 unknowns would still be left over, namely the density ρ, the velocity field $\boldsymbol{v}$ and the temperature distribution Θ. For the material representation 4 fields have to be determined, namely the motion χ_R and the temperature Θ. 4 or 5 scalar equations would be available for establishing these fields, namely the local *balance of mass* (spatial representation), the *balance of momentum* and the *energy balance.*

With a view to fulfilling the *Clausius-Duhem* inequality, a most essential condition for all solutions of the field equations, there are two possibilities: one way is to limit the admissible *thermodynamic processes.* The other way involves the restriction of the *constitutive equations*: from the outset one should try to formulate the constitutive equations in such a way that the validity of the *entropy inequality* is guaranteed *a priori* for every conceivable solution of the field equations. This method was devised by Coleman and Noll [11] and is to be dealt with in this volume as well (Chap. 13).[12]

In a multi-component system there would be 5n balance equations to be fulfilled. A correspondingly larger number of constitutive equations would have to be formulated to close the system of equations, i.e. to provide a number of equations corresponding to the number of unknown fields.

The general principles and concrete methods of representing material behaviour are to be discussed hereafter for one-component systems only.

---

[11] COLEMAN & NOLL [1963].

[12] See COLEMAN [1964]; MUSCHIK & EHRENTRAUT [1996].

4 Objectivity

4.1 Frames of Reference

In kinematics the current configuration of a material body is identified with an area of the three-dimensional Euclidean space of physical observation. In this way the material body is embedded in the Euclidean structure of the observation space. In particular, the points $\mathrm{P} \in \mathbb{E}^3$, where the material points $\mathscr{P} \in \mathscr{B}$ are currently located, are characterised by position vectors $\boldsymbol{x} \in \mathbb{V}^3$. In order to establish a one-to-one relation between point P and vector $\boldsymbol{x}$, it is necessary to introduce a *reference system* or *frame of reference.*[1] There are any number of possibilities for achieving this, i.e. the choice of a special reference system is always linked to a certain arbitrariness. However, we have to take into consideration what we have learned from experience, that balance relations of mechanics do not apply in the same way for all reference frames: the balance relations for linear momentum and rotational momentum refer to *inertial frames.*[2] An inertial frame is a reference system with the property that conclusions drawn from the balance relations of linear and rotational momentum correspond to results obtained in practical experiments. An example is the law of inertia, from the validity of which the term *inertial frame* is derived. Experience shows that there are reference frames that can be regarded in good approximation as inertial frames within the measuring accuracy available. Once one has found an inertial frame, then all frames that move at a constant velocity and without rotation with respect to this reference system are also inertial frames: the basic equations of classical mechanics are invariant with

---

[1] TRUESDELL & TOUPIN [1960], Sect. 13.

[2] TRUESDELL & TOUPIN [1960], Sect. 196, 196A.

respect to *Galilei* transformations. A *Galilei* transformation is a transformation between two frames of reference, related to each other by a translational motion with constant velocity. Once an inertial frame is known, the balance equations can be applied to any reference system which is not an inertial frame. In order to do so, however, it is imperative to know how the fields that appear in the balance relations transform when the frame of reference is changed. In this connection a certain transformation behaviour is denoted by the term objectivity.

The notion of *objectivity* provides a division of the physical quantities into two categories: a particular quantity is either objective or non-objective. The term objectivity in the sense of the transformation behaviour is not to be confused with the principle of material objectivity. Whereas the term *objective* denotes certain transformation properties of physical quantities,[3] the *principle of material objectivity* or *principle of material frame-indifference* refers to the structure of functional relations between field quantities (constitutive equations), which are formulated to describe material properties.[4] The principle of material frame-indifference, as stated by C. A. Truesdell and W. Noll, postulates the complete independence of the material description from the frame of reference. Nevertheless, the question may arise as to the extent to which constitutive equations may depend on the reference system.[5]

The following sections are designed to clarify the terms frame of reference, change of frame and objectivity. The principle of material frame-indifference (principle of material objectivity) is to be explained and evaluated later on in connection with the representation of material properties (Chap. 7).

4. 2 Affine Spaces

The relationship between points of space and position vectors can be represented utilising the notion of an *affine space*.

Definition 4. 1

Let $\mathbb{A} = \{\mathrm{A}, \mathrm{B}, \ldots\}$ be a given set and $\mathbb{V} = \{\boldsymbol{v}, \boldsymbol{w}, \ldots\}$ a vector space.
In addition, let

---

[3] TRUESDELL & NOLL [1965], Sect. 17.

[4] TRUESDELL & NOLL [1965], Sect. 19.

[5] This has been analysed in MÜLLER [1972, 1976]. More recent discussions on *material objectivity* are found in MUSCHIK [1998a]; BERTRAM & SVENDSEN [1997]; SVENDSEN & BERTRAM [1999].

$$\begin{aligned} \boldsymbol{A} : \ \mathbb{A} \times \mathbb{A} &\longrightarrow \mathbb{V} \\ (\mathrm{A}, \mathrm{B}) &\longmapsto \boldsymbol{v} = \boldsymbol{A}(\mathrm{A}, \mathrm{B}) \end{aligned} \tag{4.1}$$

be a mapping of $\mathbb{A} \times \mathbb{A}$ into $\mathbb{V}$. Then, in connection with the mapping $\boldsymbol{A}$, $\mathbb{A}$ is the *affine space* associated with $\mathbb{V}$ or *point space* to $\mathbb{V}$, if $\boldsymbol{A}$ has the following properties:

1) For every selected element $\mathrm{A}_0 \in \mathbb{A}$ the mapping

$$\begin{aligned} \boldsymbol{A}_{\mathrm{A}_0} : \ \mathbb{A} &\longrightarrow \mathbb{V} \\ \mathrm{A} &\longmapsto \boldsymbol{v} = \boldsymbol{A}_{\mathrm{A}_0}(\mathrm{A}) = \boldsymbol{A}(\mathrm{A}_0, \mathrm{A}) \end{aligned} \tag{4.2}$$

is bijective.

2) $\boldsymbol{A}(\mathrm{A}, \mathrm{A}) = \boldsymbol{0}$ (4.3)

holds for all $\mathrm{A} \in \mathbb{A}$.

3) $\boldsymbol{A}(\mathrm{A}, \mathrm{B}) + \boldsymbol{A}(\mathrm{B}, \mathrm{C}) = \boldsymbol{A}(\mathrm{A}, \mathrm{C})$. (4.4)

holds for all $\mathrm{A}, \mathrm{B}, \mathrm{C} \in \mathbb{A}$. ◊

In classical mechanics the affine space $\mathbb{A}$ is the three-dimensional Euclidean space $\mathbb{E}^3$ of physical observation. The most important conclusion to be drawn from the axioms of Euclidean geometry is the possibility of defining the following *equivalence relation* on the set of ordered pairs of points:

Definition 4. 2

Two pairs of points $(\mathrm{P}_1, \mathrm{P}_2)$ and $(\mathrm{Q}_1, \mathrm{Q}_2)$ $\left(\mathrm{P}_i, \mathrm{Q}_i \in \mathbb{E}^3\right)$ are called *equivalent* if the corresponding straight lines are equal in length, direction and orientation. ◊

This equivalence relation defines the division of $\mathbb{E} \times \mathbb{E}$ into equivalence classes. The whole set of equivalence classes forms a three-dimensional Euclidean vector space if the operations of addition, multiplication with a scalar and scalar product are defined geometrically in the usual way.

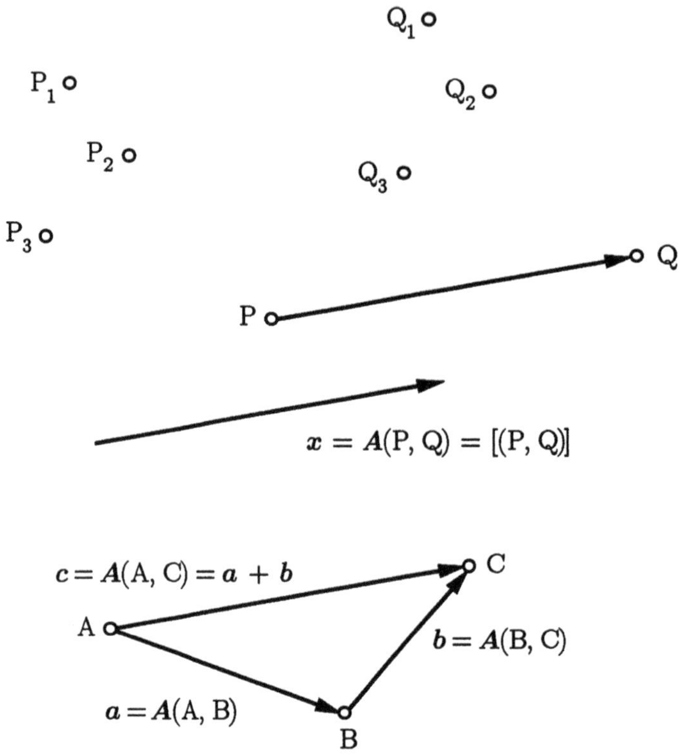

Figure 4. 1: Vectors as equivalence classes of ordered pairs of points (oriented straight lines)

This vector space is nothing other than the space of geometric vectors: a *geometric vector* is an oriented straight line or an equivalence class of ordered pairs of points. In other words, a geometric vector is the set of all oriented straight lines which can be transported into each other by means of parallel translation. In accordance with this definition a vector is commonly depicted by an arrow. This is sketched in Fig. 4. 1.

The three-dimensional Euclidean space $\mathbb{E}^3$ is the point space associated with the vector space $\mathbb{V}^3$ if the mapping $\boldsymbol{A}$ is selected as follows:

$$\begin{aligned} \boldsymbol{A} : \mathbb{E}^3 \times \mathbb{E}^3 &\longrightarrow \mathbb{V}^3 \\ (\mathrm{P}, \mathrm{Q}) &\longmapsto \boldsymbol{x} = \boldsymbol{A}(\mathrm{P}, \mathrm{Q}) = [(\mathrm{P}, \mathrm{Q})] \, . \end{aligned} \tag{4.5}$$

Here,

$$[(\mathrm{P}, \mathrm{Q})] = \{(\mathrm{P}_\alpha, \mathrm{Q}_\beta) \mid \overrightarrow{\mathrm{P}_\alpha \mathrm{Q}_\beta} = \overrightarrow{\mathrm{PQ}}\} \in \mathbb{V}^3 \tag{4.6}$$

defines an equivalence class of ordered pairs of points or an oriented straight line (see Fig. 4. 1).

It is evident that the chosen map $\boldsymbol{A}$ possesses all three properties that are stipulated in the definition of the affine space. In particular, the property (4.4) corresponds to the geometric addition of two vectors.

The interpretation of $\mathbb{E}^3$ as the point space associated with $\mathbb{V}^3$ provides two distinct possibilities for explaining the term *frame of reference*, the so-called passive and the active interpretation.

4. 3 Change of Frame: Passive Interpretation

In the following, the Euclidean space $\mathbb{E}^3$ is interpreted as the point space corresponding to $\mathbb{V}^3$ in connection with the map

$$\boldsymbol{A} : (\mathrm{P}, \mathrm{Q}) \longmapsto \boldsymbol{x} = \boldsymbol{A}(\mathrm{P}, \mathrm{Q}) = [(\mathrm{P}, \mathrm{Q})] \in \mathbb{V}^3 . \tag{4.7}$$

Definition 4. 3

A *frame of reference* $\{\mathrm{O}, \mathrm{P}_1, \mathrm{P}_2, \mathrm{P}_3\}$ according to the *passive interpretation* is the selection of 4 points $\{\mathrm{O}, \mathrm{P}_1, \mathrm{P}_2, \mathrm{P}_3\} \subset \mathbb{E}^3$, that do not lie in a plane. One of the points is called the *origin* (marked O). ◊

According to the definition the vectors

$$\boldsymbol{e}_k = \boldsymbol{A}(\mathrm{O}, \mathrm{P}_k) = [(\mathrm{O}, \mathrm{P}_k)] \tag{4.8}$$

form a base system in $\mathbb{V}^3$. Without loss of generality, we choose the points P_k such that these base vectors are orthogonal unit vectors, $\boldsymbol{e}_i \cdot \boldsymbol{e}_k = \delta_{ik}$. Definition 4. 3 serves to define the bijective map

$$\begin{aligned} \boldsymbol{A}_\mathrm{O} : \ & \mathbb{E}^3 \longrightarrow \mathbb{V}^3 \\ & \mathrm{P} \longmapsto \boldsymbol{x} = \boldsymbol{A}_\mathrm{O}(\mathrm{P}) = \boldsymbol{A}(\mathrm{O}, \mathrm{P}) , \end{aligned} \tag{4.9}$$

and we have $x = y_k e_k$ with $y_k = x \cdot e_k$. The components y_k of the space vector x are the *Cartesian coordinates* of the point $\mathrm{P} \in \mathbb{E}^3$.

Definition 4. 4

A *change of frame* according to the *passive interpretation* means substituting four other points for the ones selected:

$$\{\mathrm{O}, \mathrm{P}_1, \mathrm{P}_2, \mathrm{P}_3\} \longmapsto \{\hat{\mathrm{O}}, \hat{\mathrm{P}}_1, \hat{\mathrm{P}}_2, \hat{\mathrm{P}}_3\} . \tag{4.10}$$

◊

It is important to bear in mind that, when changing the frame of reference in the passive interpretation, the mapping $\boldsymbol{A}$ remains unaltered. (This is different in the case of the active interpretation in the next section.) A new base system is created according to Definition 4. 4,

$$\hat{e}_k = \boldsymbol{A}(\hat{\mathrm{O}}, \hat{\mathrm{P}}_k) = [(\hat{\mathrm{O}}, \hat{\mathrm{P}}_k)] , \tag{4.11}$$

in which, again, orthonormality may be taken for granted: $\hat{e}_i \cdot \hat{e}_k = \delta_{ik}$.
Under the assumption of orthonormality for both systems, there is one orthogonal tensor

$$\mathbf{Q} = e_i \otimes \hat{e}_i , \tag{4.12}$$

which conveys the base vectors towards each other:

$$\mathbf{Q}\hat{e}_k = e_k . \tag{4.13}$$

Moreover, the bijective mapping

$$\begin{aligned} \boldsymbol{A}_{\hat{\mathrm{O}}} : \; & \mathbb{E}^3 \longrightarrow \mathbb{V}^3 \\ & \mathrm{P} \longmapsto \hat{x} = \boldsymbol{A}_{\hat{\mathrm{O}}}(\mathrm{P}) = \boldsymbol{A}(\hat{\mathrm{O}}, \mathrm{P}) = [(\hat{\mathrm{O}}, \mathrm{P})] \end{aligned} \tag{4.14}$$

leads to the representation of the position vector $\hat{x} = \boldsymbol{A}(\hat{\mathrm{O}}, \mathrm{P}) = \hat{y}_k \hat{e}_k$ with the components (Cartesian coordinates of P) $\hat{y}_k = \hat{x} \cdot \hat{e}_k$. Introducing the abbreviation

$$\boldsymbol{A}(\hat{\mathrm{O}}, \mathrm{P}) = \boldsymbol{A}(\hat{\mathrm{O}}, \mathrm{O}) + \boldsymbol{A}(\mathrm{O}, \mathrm{P}) \tag{4.15}$$

$$c = \boldsymbol{A}(\hat{\mathrm{O}}, \mathrm{O}) , \tag{4.16}$$

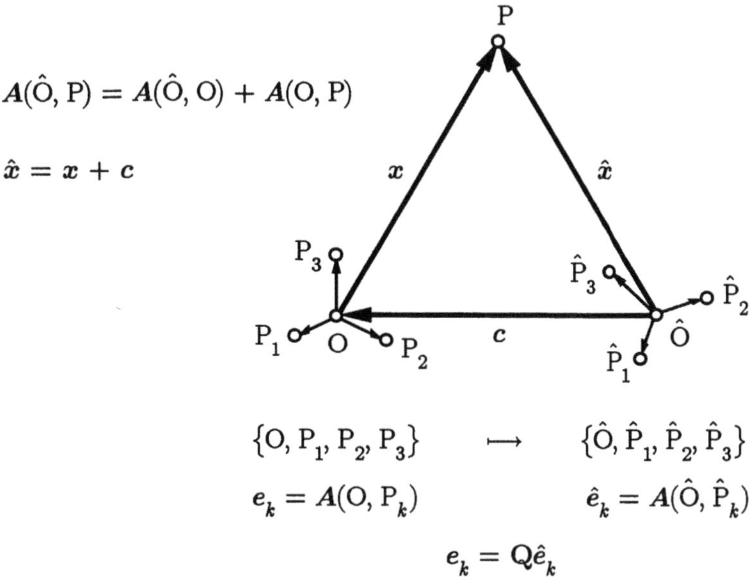

Figure 4. 2: Change of frame according to the passive interpretation

implies the relation

$$\hat{\boldsymbol{x}} = \boldsymbol{x} + \boldsymbol{c} \tag{4.17}$$

between the position vectors $\hat{\boldsymbol{x}}$ and $\boldsymbol{x}$. A transformation of the Cartesian coordinates can be derived from this: the component representation

$$\hat{y}_k\hat{\boldsymbol{e}}_k = y_l\boldsymbol{e}_l + \hat{c}_k\hat{\boldsymbol{e}}_k = y_l\mathbf{Q}\hat{\boldsymbol{e}}_l + \hat{c}_k\hat{\boldsymbol{e}}_k = y_l(\hat{\boldsymbol{e}}_k\cdot\mathbf{Q}\hat{\boldsymbol{e}}_l)\hat{\boldsymbol{e}}_k + \hat{c}_k\hat{\boldsymbol{e}}_k \quad \text{yields}$$

$$\hat{y}_k = \hat{Q}_{kl}y_l + \hat{c}_k \quad (k = 1, 2, 3)\,, \tag{4.18}$$

with $y_l = \boldsymbol{x}\cdot\boldsymbol{e}_l$, $\hat{c}_k = \boldsymbol{c}\cdot\hat{\boldsymbol{e}}_k$ and the orthogonal matrix

$$\hat{Q}_{kl} = \hat{\boldsymbol{e}}_k\cdot\mathbf{Q}\hat{\boldsymbol{e}}_l\,. \tag{4.19}$$

The coordinate transformation (4.18) is to be interpreted as a matrix equation. This equation of matrices is uniquely defined by means of the vector $\boldsymbol{c} = \boldsymbol{A}(\hat{\mathrm{O}}, \mathrm{O})$ and the orthogonal tensor $\mathbf{Q} = \boldsymbol{e}_i \otimes \hat{\boldsymbol{e}}_i$. This line of argument can be summarised in the following theorem:

Theorem 4. 1

A *change of frame* is defined by an orthogonal tensor $\mathbf{Q}$ and a vector $\boldsymbol{c}$:

$$(\mathbf{Q}, \boldsymbol{c}): \quad \{\mathrm{O}, \mathrm{P}_1, \mathrm{P}_2, \mathrm{P}_3\} \longmapsto \{\hat{\mathrm{O}}, \hat{\mathrm{P}}_1, \hat{\mathrm{P}}_2, \hat{\mathrm{P}}_3\}\,. \tag{4.20}$$

□

Proof

Vector $\boldsymbol{c}$ joins the reference points $\hat{\mathrm{O}}$ and O, whereas tensor $\mathbf{Q}$ causes the orthonormal base vectors to be rotated into each other: $\mathbf{Q}\hat{\boldsymbol{e}}_k = \boldsymbol{e}_k$. ■

The change of frame according to the passive interpretation is sketched in Fig. 4. 2.

4. 4 Change of Frame: Active Interpretation

The statement that a change of the reference system is defined by the pair ($\mathbf{Q}$, $\boldsymbol{c}$) can be interpreted in a different way. As a general rule it is this second interpretation that is applied in continuum mechanics and the theory of materials. It is based on an alternative definition of the reference frame:

Definition 4. 5

A *frame of reference* (O, $\boldsymbol{A}$) in the *active interpretation* consists of a *reference point* $\mathrm{O} \in \mathbb{E}^3$ and the *mapping*

$$\begin{aligned} \boldsymbol{A} : \mathbb{E}^3 \times \mathbb{E}^3 &\longmapsto \mathbb{V}^3 \\ (\mathrm{P}, \mathrm{Q}) &\longmapsto \boldsymbol{x} = \boldsymbol{A}(\mathrm{P}, \mathrm{Q}) = [(\mathrm{P}, \mathrm{Q})]\,. \end{aligned} \tag{4.21}$$

◊

As before, it follows from the definition that the map

$$\boldsymbol{A}_{\mathrm{O}}\colon \mathrm{P} \longmapsto \boldsymbol{x} = \boldsymbol{A}_{\mathrm{O}}(\mathrm{P}) = \boldsymbol{A}(\mathrm{O}, \mathrm{P}) \tag{4.22}$$

is bijective.

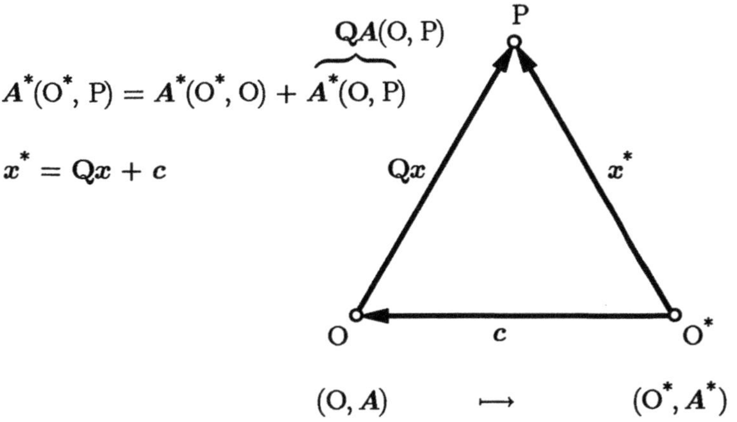

Figure 4. 3: Change of frame according to the active interpretation

Definition 4. 6

A *change of frame* according to the *active interpretation* consists of changing the reference point O and the map $\boldsymbol{A}$:

$$(\mathrm{O}, \boldsymbol{A}) \longmapsto (\mathrm{O}^*, \boldsymbol{A}^*) \,. \tag{4.23}$$

In this definition the mappings $\boldsymbol{A}$ and $\boldsymbol{A}^*$ differ by virtue of a rotation: for all $\mathrm{P}_1, \mathrm{P}_2 \in \mathbb{E}^3$ the relation

$$\boldsymbol{A}^*(\mathrm{P}_1, \mathrm{P}_2) = \mathbf{Q}\boldsymbol{A}(\mathrm{P}_1, \mathrm{P}_2) \tag{4.24}$$

is valid, with $\mathbf{Q}\mathbf{Q}^{\mathrm{T}} = \mathbf{Q}^{\mathrm{T}}\mathbf{Q} = \mathbf{1}$. ◊

Corresponding to the new frame of reference we have the bijective mapping

$$\boldsymbol{A}^*_{\mathrm{O}^*} : \mathrm{P} \longmapsto \boldsymbol{x}^* = \boldsymbol{A}^*_{\mathrm{O}^*}(\mathrm{P}) = \boldsymbol{A}^*(\mathrm{O}^*, \mathrm{P}) \,. \tag{4.25}$$

Moreover,

$$\boldsymbol{A}^*(\mathrm{O}^*, \mathrm{P}) = \boldsymbol{A}^*(\mathrm{O}^*, \mathrm{O}) + \boldsymbol{A}^*(\mathrm{O}, \mathrm{P}) = \boldsymbol{A}^*(\mathrm{O}^*, \mathrm{O}) + \mathbf{Q}\boldsymbol{A}(\mathrm{O}, \mathrm{P}) \tag{4.26}$$

holds as a self-evident property of map $\boldsymbol{A}^*$. Introducing the abbreviation $\boldsymbol{c} = \boldsymbol{A}^*(\mathrm{O}^*, \mathrm{O})$ we obtain the relations

$$\boldsymbol{x}^* = \mathbf{Q}\boldsymbol{x} + \boldsymbol{c} \tag{4.27}$$

between the position vectors $\boldsymbol{x}^* = \boldsymbol{A}^*(\mathrm{O}^*, \mathrm{P})$ and $\boldsymbol{x} = \boldsymbol{A}(\mathrm{O}, \mathrm{P})$ of one and the same point $\mathrm{P} \in \mathbb{E}^3$ with respect to the two different reference systems. The situation resulting from a change of frame according to the active interpretation is sketched in Fig. 4. 3.

Equation (4.27), as opposed to the matrix equation (4.18), is a tensor equation, independent of any choice of base vectors. Whereas changing the frame according to the *passive interpretation* leads to a *coordinate transformation*, the *active interpretation* implies a *vector transformation*:

Two different position vectors, namely $\boldsymbol{x}$ and $\boldsymbol{x}^* \in \mathbb{V}^3$, represent one and the same point $\mathrm{P} \in \mathbb{E}^3$. The vector transformation (4.27) is likewise defined by a *rotation* $\mathbf{Q}$ and a *translation* $\boldsymbol{c}$. In that respect, the statement made in Theorem 4. 1 is valid in both the passive and the active interpretation; in principle, the two interpretations are equivalent. However, as far as applications in continuum mechanics and the theory of materials are concerned, preference is given to the active interpretation.

4. 5 Objective Quantities

In its physical significance, a frame of reference represents a *physical observer*: a physical observer registers the position of a material body in the space of observation at all times and describes it by statements of distance and direction. The mathematical model for this process is precisely the mapping $\boldsymbol{A}_{\mathrm{O}}$, which identifies the points of space in a one-to-one correspondence by position vectors. Every time the frame of reference is changed, this corresponds physically to a different observer and, according to the active interpretation, to a vector transformation (4.27).

As regards physical applications in thermomechanics it is essential to take time-dependent frames of reference into consideration. In the sense of the *active interpretation* the following definition is valid for a time-dependent change of frame.

Definition 4. 7

A time-dependent *change of frame* is a transformation $(\boldsymbol{x}, t) \longmapsto (\boldsymbol{x}^*, t^*)$, defined by

$$\boldsymbol{x}^* = \mathbf{Q}(t)\boldsymbol{x} + \boldsymbol{c}(t) \tag{4.28}$$

and

$$t^* = t - a\,. \tag{4.29}$$

In this context, $t \longmapsto \big(\mathbf{Q}(t),\, \boldsymbol{c}(t)\big)$ are orthogonal tensor-valued and vector-valued functions of time $\big(\mathbf{Q}(t)\mathbf{Q}(t)^{\mathrm{T}}\big) \equiv \mathbf{1}$) and a is a real constant.[6] ◊

A transformation of the form (4.28) is called a *Euclidean* transformation since it preserves the Euclidean structure of the vector space $\mathbb{V}^3$. The scale of length in particular is invariant under Euclidean transformations. The second part of the definition expresses the fact that in classical mechanics the time scale is also independent of the reference system. Accordingly, time t may be shifted by a constant a. The functions $\mathbf{Q}(t)$ and $\boldsymbol{c}(t)$ correspond to a rigid body motion of the reference frame, consisting of a rotation and a translation (cf. (1.70)).

A subset of the *Euclidean* transformations consists of the *Galilei* transformations, defined by

$$\mathbf{Q}(t) = \mathbf{Q}_0 = \text{const. and } \boldsymbol{c}(t) = \boldsymbol{c}_0 + \boldsymbol{c}_1 t\,. \tag{4.30}$$

The *Galilei* transformation amounts to a pure translation of the reference frame with the constant velocity $\boldsymbol{c}_1$.

Definition 4. 8

Let the mapping $\mathrm{f} : \mathscr{B} \times \mathbb{I} \longrightarrow \mathbb{W}$

$$(\mathscr{P}, t) \longmapsto \mathrm{w} = \mathrm{f}(\mathscr{P}, t)$$

($\mathbb{I} \subset \mathbb{R}$) be any physical quantity or time-dependent function on a material body $\mathscr{B}$. Let the range of values $\mathbb{W}$ be either the set of scalars $\varphi \in \mathbb{R}$, vectors $\boldsymbol{v} \in \mathbb{V}^3$ or second order tensors $\mathbf{T} \in \mathrm{Lin}(\mathbb{V}^3)$. In relation to the frames of reference $(\mathrm{O}, \boldsymbol{A})$ and $(\mathrm{O}^*, \boldsymbol{A}^*)$ let the values of $\mathrm{f}(\cdot)$ be given by

$$\mathrm{w} = \mathrm{f}(\mathscr{P}, t) = \begin{cases} \varphi(\mathscr{P}, t) \\ \boldsymbol{v}(\mathscr{P}, t) \\ \mathbf{T}(\mathscr{P}, t) \end{cases} \quad \text{and} \quad \mathrm{w} = \mathrm{f}^*(\mathscr{P}, t^*) = \begin{cases} \varphi^*(\mathscr{P}, t^*) \\ \boldsymbol{v}^*(\mathscr{P}, t^*)\,, \\ \mathbf{T}^*(\mathscr{P}, t^*) \end{cases}$$

respectively. The quantity $\mathrm{f}(\cdot)$ is then called *objective* if, for every change of frame,

---

[6] Cf. TRUESDELL & NOLL [1965], eqs. (17.1,2).

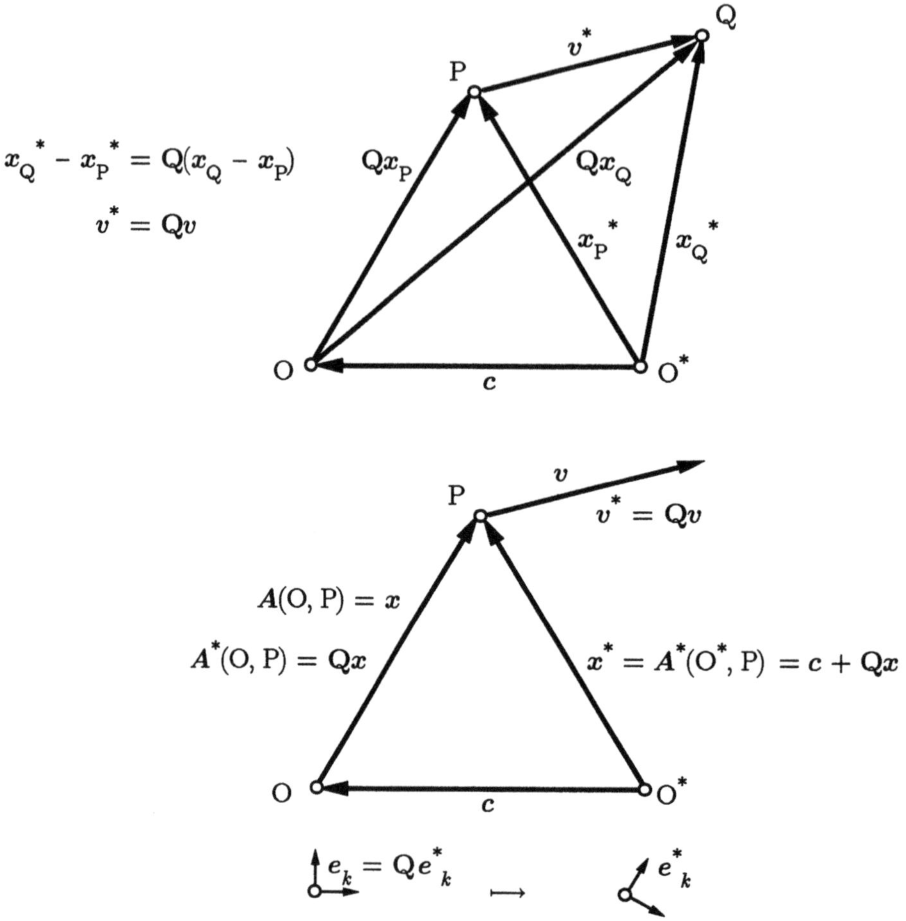

Figure 4. 4: Objectivity of a vector

$$t \longmapsto \big(\mathbf{Q}(t), \boldsymbol{c}(t)\big)\ , \qquad \boldsymbol{x}^* = \mathbf{Q}(t)\boldsymbol{x} + \boldsymbol{c}(t)\ , \qquad t^* = t - a \tag{4.31}$$

with $\det\mathbf{Q}(t) = +\,1$, the following relations are valid:

$$\varphi^*(\mathscr{P}, t^*) = \varphi(\mathscr{P}, t)\ , \text{ if } \mathbb{W} \subset \mathbb{R}\ , \tag{4.32}$$

$$\boldsymbol{v}^*(\mathscr{P}, t^*) = \mathbf{Q}(t)\boldsymbol{v}(\mathscr{P}, t)\ , \text{ if } \mathbb{W} \subset \mathbb{V}^3\ , \tag{4.33}$$

$$\mathbf{T}^*(\mathscr{P}, t^*) = \mathbf{Q}(t)\mathbf{T}(\mathscr{P}, t)\mathbf{Q}^{\mathrm{T}}(t)\ , \text{ if } \mathbb{W} \subset \mathrm{Lin}\big(\mathbb{V}^3\big)\ . \tag{4.34}$$

This definition applies not only to fields but also to functions $\mathrm{F} : (\mathscr{B}, t) \longmapsto \mathrm{F}(\mathscr{B}, t) \in \mathbb{W}$ defined on the set of all material bodies according to (3.57). ◊

The term *objectivity*, in connection with the change of frame in the sense of the active interpretation, distinguishes one particular kind of transformation behaviour. In the case of a scalar-valued function, objectivity means that its values are completely independent of the frame of reference. The motivation of this definition immediately becomes evident. It is also apparent that not all functions have this property: the absolute value of the velocity vector cannot, for example, be an objective scalar, whereas the objectivity of the mass density of a material point is readily understood.

The objectivity of a vector $\boldsymbol{v}$ does not call for its absolute invariability during a change of frame, but for the transformation property $\boldsymbol{v}^* = \mathbf{Q}\boldsymbol{v}$. The justification of this definition becomes plain if the vector $\boldsymbol{v}$ is regarded as a geometric vector, i.e. as an oriented straight line connecting two spatial points, and transformed as such, namely like the difference between two position vectors (cf. Fig. 4. 4):

$$\boldsymbol{x}_{\mathrm{Q}}^* - \boldsymbol{x}_{\mathrm{P}}^* = (\mathbf{Q}\boldsymbol{x}_{\mathrm{Q}} + \boldsymbol{c}) - (\mathbf{Q}\boldsymbol{x}_{\mathrm{P}} + \boldsymbol{c}) = \mathbf{Q}(\boldsymbol{x}_{\mathrm{Q}} - \boldsymbol{x}_{\mathrm{P}}) \,. \tag{4.35}$$

The difference between two position vectors meets the criterion for objectivity, whereas the position vector itself does not. Once one has accepted the criterion for the objectivity of a vector, the objectivity of a tensor $\mathbf{T}$ results from the requirement that an objective tensor should map objective vectors into objective vectors. Then both $\boldsymbol{w} = \mathbf{T}\boldsymbol{v}$ and $\boldsymbol{w}^* = \mathbf{T}^*\boldsymbol{v}^*$ must hold, with $\boldsymbol{v}^* = \mathbf{Q}\boldsymbol{v}$ and $\boldsymbol{w}^* = \mathbf{Q}\boldsymbol{w}$. Both $\mathbf{Q}\boldsymbol{w} = \mathbf{Q}\mathbf{T}\boldsymbol{v}$ and $\mathbf{Q}\boldsymbol{w} = \mathbf{T}^*\mathbf{Q}\boldsymbol{v}$ comply herewith, that is $\mathbf{Q}\mathbf{T}\boldsymbol{v} = \mathbf{T}^*\mathbf{Q}\boldsymbol{v}$ is valid for all $\boldsymbol{v} \in \mathbb{V}^3$. This results in $\mathbf{T}^* = \mathbf{Q}\mathbf{T}\mathbf{Q}^{\mathrm{T}}$.

Various algebraic laws, which can be verified straight away by means of simple arithmetic, are applicable for dealings with objective quantities.

Theorem 4. 2

The set of objective scalars, vectors and tensors is closed with respect to algebraic operations.

1) $(\boldsymbol{v}_1 + \boldsymbol{v}_2)^* = \boldsymbol{v}_1^* + \boldsymbol{v}_2^* = \mathbf{Q}(\boldsymbol{v}_1 + \boldsymbol{v}_2)$

2) $(\alpha\boldsymbol{v})^* = \alpha^*\boldsymbol{v}^* = \mathbf{Q}(\alpha\boldsymbol{v})$

3) $(\boldsymbol{v}_1 \cdot \boldsymbol{v}_2)^* = \boldsymbol{v}_1^* \cdot \boldsymbol{v}_2^* = \mathbf{Q}\boldsymbol{v}_1 \cdot \mathbf{Q}\boldsymbol{v}_2 = \boldsymbol{v}_1 \cdot \boldsymbol{v}_2$

4) $(\boldsymbol{v}_1 \times \boldsymbol{v}_2)^* = \boldsymbol{v}_1^* \times \boldsymbol{v}_2^* = \mathbf{Q}\boldsymbol{v}_1 \times \mathbf{Q}\boldsymbol{v}_2 = \mathbf{Q}(\boldsymbol{v}_1 \times \boldsymbol{v}_2)$

5) $(\boldsymbol{v}_1 \otimes \boldsymbol{v}_2)^* = \boldsymbol{v}_1^* \otimes \boldsymbol{v}_2^* = \mathbf{Q}\boldsymbol{v}_1 \otimes \mathbf{Q}\boldsymbol{v}_2 = \mathbf{Q}(\boldsymbol{v}_1 \otimes \boldsymbol{v}_2)\mathbf{Q}^{\mathrm{T}}$

6) $(\mathbf{T}\boldsymbol{v})^* = \mathbf{T}^*\boldsymbol{v}^* = \mathbf{Q}\mathbf{T}\mathbf{Q}^{\mathrm{T}}(\mathbf{Q}\boldsymbol{v}) = \mathbf{Q}(\mathbf{T}\boldsymbol{v})$

7) $(\mathbf{TS})^* = \mathbf{T}^*\mathbf{S}^* = (\mathbf{QTQ}^{\mathrm{T}})(\mathbf{QSQ}^{\mathrm{T}}) = \mathbf{Q}(\mathbf{TS})\mathbf{Q}^{\mathrm{T}}$

8) $(\mathbf{T} \cdot \mathbf{S})^* = \mathbf{T}^* \cdot \mathbf{S}^* = (\mathbf{QTQ}^{\mathrm{T}}) \cdot (\mathbf{QSQ}^{\mathrm{T}}) = \mathbf{T} \cdot \mathbf{S}$

■

The objectivity of a particular physical quantity is either established individually a priori by virtue of its definition or it is confirmed by calculation as a result of earlier definitions.

Definition 4. 9

The following physical quantities are objective: *mass, force, internal energy, heat, entropy, temperature, entropy flux.* ◊

The definition implies that the corresponding density functions are also objective quantities: *mass density, Cauchy stress vector, Cauchy stress tensor, volume force density, specific internal energy* and *entropy, heat flux vector.*

Theorem 4. 3

The velocity and acceleration vectors are not objective: from (4.28) we obtain

$$\boldsymbol{v}^* = \mathbf{Q}\boldsymbol{v} + \dot{\boldsymbol{c}} + \mathbf{A}(\boldsymbol{x}^* - \boldsymbol{c}) \tag{4.36}$$

and

$$\boldsymbol{a}^* = \mathbf{Q}\boldsymbol{a} + \ddot{\boldsymbol{c}} + (\dot{\mathbf{A}} + \mathbf{A}^2)(\boldsymbol{x}^* - \boldsymbol{c}) + 2\mathbf{A}\mathbf{Q}\boldsymbol{v}\,, \tag{4.37}$$

or

$$\boldsymbol{a}^* = \mathbf{Q}\boldsymbol{a} + \ddot{\boldsymbol{c}} + (\dot{\mathbf{A}} - \mathbf{A}^2)(\boldsymbol{x}^* - \boldsymbol{c}) + 2\mathbf{A}(\boldsymbol{v}^* - \dot{\boldsymbol{c}})\,. \tag{4.38}$$

Here, $\mathbf{A}$ is the rate of rotation (spin tensor):

$$\mathbf{A} = \dot{\mathbf{Q}}\mathbf{Q}^{\mathrm{T}} = -\,\mathbf{A}^{\mathrm{T}}\,. \tag{4.39}$$

□

Proof

By differentiation with respect to time: $\boldsymbol{x}^* = \mathbf{Q}(t)\boldsymbol{x} + \boldsymbol{c}(t)$ leads to

$$\boldsymbol{v}^* = (\boldsymbol{x}^*)^{\cdot} = \dot{\mathbf{Q}}\boldsymbol{x} + \mathbf{Q}\dot{\boldsymbol{x}} + \dot{\boldsymbol{c}} = \dot{\mathbf{Q}}\mathbf{Q}^{\mathrm{T}}(\boldsymbol{x}^* - \boldsymbol{c}) + \mathbf{Q}\boldsymbol{v} + \dot{\boldsymbol{c}}\ \text{, i.e.}$$
$$\boldsymbol{v}^* = \mathbf{Q}\boldsymbol{v} + \dot{\boldsymbol{c}} + \mathbf{A}(\boldsymbol{x}^* - \boldsymbol{c})\,.$$

Further differentiation yields

$$\boldsymbol{a}^* = (\boldsymbol{v}^*)^{\cdot} = \mathbf{Q}\dot{\boldsymbol{v}} + \dot{\mathbf{Q}}\boldsymbol{v} + \ddot{\boldsymbol{c}} + \dot{\mathbf{A}}(\boldsymbol{x}^* - \boldsymbol{c}) + \mathbf{A}(\dot{\boldsymbol{x}}^* - \dot{\boldsymbol{c}})\,,$$
$$\boldsymbol{a}^* = \mathbf{Q}\boldsymbol{a} + \ddot{\boldsymbol{c}} + \dot{\mathbf{A}}(\boldsymbol{x}^* - \boldsymbol{c}) + \mathbf{A}(\mathbf{Q}\boldsymbol{v} + \boldsymbol{v}^* - \dot{\boldsymbol{c}})\,.$$

With $\boldsymbol{v}^* - \dot{\boldsymbol{c}} = \mathbf{Q}\boldsymbol{v} + \mathbf{A}(\boldsymbol{x}^* - \boldsymbol{c})$ this leads to

$$\boldsymbol{a}^* = \mathbf{Q}\boldsymbol{a} + \ddot{\boldsymbol{c}} + \dot{\mathbf{A}}(\boldsymbol{x}^* - \boldsymbol{c}) + \mathbf{A}\big[2\mathbf{Q}\boldsymbol{v} + \mathbf{A}(\boldsymbol{x}^* - \boldsymbol{c})\big]\ \text{, and with}$$
$$\mathbf{Q}\boldsymbol{v} = \boldsymbol{v}^* - \dot{\boldsymbol{c}} - \mathbf{A}(\boldsymbol{x}^* - \boldsymbol{c})\ \ \text{to}$$

$$\boldsymbol{a}^* = \mathbf{Q}\boldsymbol{a} + \ddot{\boldsymbol{c}} + \dot{\mathbf{A}}(\boldsymbol{x}^* - \boldsymbol{c}) + \mathbf{A}\big[2\,(\boldsymbol{v}^* - \dot{\boldsymbol{c}}) - \mathbf{A}(\boldsymbol{x}^* - \boldsymbol{c})\big]\,.$$

■

The transformation formula (4.38) for the acceleration vector shows that $\boldsymbol{a}$ is objective for a subset of the Euclidean transformations, namely for

$$\mathbf{A} = \mathbf{0}\ \big(\Leftrightarrow \mathbf{Q}(t) = \mathbf{Q}_0 = \text{const.}\big)\ \text{and}\ \ddot{\boldsymbol{c}}(t) = \boldsymbol{0}\ \big(\Leftrightarrow \boldsymbol{c}(t) = \boldsymbol{c}_0 + \boldsymbol{c}_1 t\big)\,.$$

These are precisely the *Galilei transformations* (4.30), $\boldsymbol{x}^* = \mathbf{Q}_0\boldsymbol{x} + \boldsymbol{c}_0 + \boldsymbol{c}_1 t$.

The balance of linear momentum $m\boldsymbol{a} = \boldsymbol{F}$, with the acceleration of the centre of mass, $\boldsymbol{a} = \ddot{\boldsymbol{x}}_m(t)$, joins two vectors, which transform in different ways during a change of frame. This is to be discussed in more detail in the next section.

The transformation behaviour and consequently the objectivity of all physical quantities defined so far can easily be verified by means of direct calculation. The result is

Theorem 4. 4

A change of frame in the sense of the active interpretation induces the following transformation properties:

Deformation gradient $$\mathbf{F}^{*} = \mathbf{Q}\mathbf{F} \tag{4.40}$$

Line element $$\mathrm{d}\boldsymbol{x}^{*} = \mathbf{Q}\mathrm{d}\boldsymbol{x} \tag{4.41}$$

Surface element $$\mathrm{d}\boldsymbol{a}^{*} = \mathbf{Q}\mathrm{d}\boldsymbol{a} \tag{4.42}$$

Volume element $$\mathrm{d}v^{*} = \mathrm{d}v \tag{4.43}$$

Polar decomposition $$\mathbf{F}^{*} = \mathbf{R}^{*}\mathbf{U}^{*} = \mathbf{V}^{*}\mathbf{R}^{*} \tag{4.44}$$

$$\mathbf{R}^{*} = \mathbf{Q}\mathbf{R}\,, \qquad \mathbf{U}^{*} = \mathbf{U}\,, \quad \mathbf{V}^{*} = \mathbf{Q}\mathbf{V}\mathbf{Q}^{\mathrm{T}} \tag{4.45}$$

$$\mathbf{C}^{*} = \mathbf{C}\,, \qquad \mathbf{B}^{*} = \mathbf{Q}\mathbf{B}\mathbf{Q}^{\mathrm{T}} \tag{4.46}$$

Strain tensors $$\mathbf{E}^{*} = \mathbf{E}\,, \qquad \mathbf{A}^{*} = \mathbf{Q}\mathbf{A}\mathbf{Q}^{\mathrm{T}} \tag{4.47}$$

$$\mathbf{e}^{*} = \mathbf{e}\,, \qquad \mathbf{a}^{*} = \mathbf{Q}\mathbf{a}\mathbf{Q}^{\mathrm{T}} \tag{4.48}$$

Velocity gradient $$\mathbf{L}^{*} = \mathbf{Q}\mathbf{L}\mathbf{Q}^{\mathrm{T}} + \dot{\mathbf{Q}}\mathbf{Q}^{\mathrm{T}} \tag{4.49}$$

$$\mathbf{D}^{*} = \mathbf{Q}\mathbf{D}\mathbf{Q}^{\mathrm{T}}\,, \quad \mathbf{W}^{*} = \mathbf{Q}\mathbf{W}\mathbf{Q}^{\mathrm{T}} + \dot{\mathbf{Q}}\mathbf{Q}^{\mathrm{T}} \tag{4.50}$$

Strain rates $$(\mathbf{E}^{*})^{\cdot} = \dot{\mathbf{E}}\,, \qquad (\mathbf{e}^{*})^{\cdot} = \dot{\mathbf{e}}\,, \quad \mathbf{D}^{*} = \mathbf{Q}\mathbf{D}\mathbf{Q}^{\mathrm{T}} \tag{4.51}$$

Mass density $$\rho^{*} = \rho \tag{4.52}$$

Cauchy stress vector $$\boldsymbol{t}^{*} = \mathbf{Q}\boldsymbol{t} \tag{4.53}$$

Stress tensors $$\mathbf{T}^{*} = \mathbf{Q}\mathbf{T}\mathbf{Q}^{\mathrm{T}}\,, \quad \mathbf{S}^{*} = \mathbf{Q}\mathbf{S}\mathbf{Q}^{\mathrm{T}} \tag{4.54}$$

$$\mathbf{T}_{\mathrm{R}}{}^{*} = \mathbf{Q}\mathbf{T}_{\mathrm{R}} \tag{4.55}$$

$$\widetilde{\mathbf{T}}^{*} = \widetilde{\mathbf{T}}\,, \qquad \widetilde{\mathbf{t}}^{*} = \widetilde{\mathbf{t}} \tag{4.56}$$

Stress power $$w^{*} = w \tag{4.57}$$

$$\frac{1}{\rho^{*}}\mathbf{T}^{*}\cdot\mathbf{D}^{*} = \frac{1}{\rho}\mathbf{T}\cdot\mathbf{D} \tag{4.58}$$

$$\frac{1}{\rho_R{}^*}\tilde{\mathbf{T}}^* \cdot \dot{\mathbf{E}}^* = \frac{1}{\rho_R}\tilde{\mathbf{T}} \cdot \dot{\mathbf{E}} \tag{4.59}$$

$$\frac{1}{\rho_R{}^*}\mathbf{T}_R{}^* \cdot \dot{\mathbf{F}}^* = \frac{1}{\rho_R}\mathbf{T}_R \cdot \dot{\mathbf{F}} \tag{4.60}$$

Specific internal energy $\overset{*}{e} = e$ (4.61)

Specific entropy $\overset{*}{s} = s$ (4.62)

Heat flux vectors $\overset{*}{\boldsymbol{q}} = \mathbf{Q}\boldsymbol{q}\,, \qquad \boldsymbol{q}_R{}^* = \boldsymbol{q}_R$ (4.63)

Temperature gradients $\overset{*}{\boldsymbol{g}} = \mathbf{Q}\boldsymbol{g}\,, \qquad \boldsymbol{g}_R{}^* = \boldsymbol{g}_R$ (4.64)

□

Proof

The transformation formula (4.40) for the deformation gradient arises from the differentiation of $\boldsymbol{x}^* = \mathbf{Q}(t)\chi_R(\boldsymbol{X}, t) + \boldsymbol{c}(t)$ with respect to $\boldsymbol{X}$. For the material line elements we have $\mathrm{d}\boldsymbol{x}^* = \mathbf{F}^*\mathrm{d}\boldsymbol{X} = \mathbf{Q}\mathbf{F}\mathrm{d}\boldsymbol{X} = \mathbf{Q}\mathrm{d}\boldsymbol{x}$, i.e. (4.41). The transformation formula (4.42) of the oriented surface element emanates from the relation (1.62) for $\mathbf{A} = \mathbf{Q}$, bearing in mind that $\det\mathbf{Q} = 1$ is required for the objectivity of $\mathrm{d}\boldsymbol{a}$. A positive determinant of $\mathbf{Q}$ is also necessary for the transformation rule (4.43) of the volume element arising out of (1.62). The other relations can easily be obtained by calculation or from their definition. ■

4. 6 Observer-Invariant Relations

This section rounds off the concept of objectivity and is intended to help comprehend the invariance properties of the balance relations and the principle of material objectivity, which is to be formulated further on.

In the following, we take a look at three frames of reference $(\mathrm{O}_0, \boldsymbol{A}_0)$, $(\mathrm{O}_1, \boldsymbol{A}_1)$ and $(\mathrm{O}_2, \boldsymbol{A}_2)$. Let the frame (0)

$$\boldsymbol{A}_{\mathrm{O}_0} : \mathrm{P} \longmapsto \boldsymbol{x}_0 = \boldsymbol{A}_0(\mathrm{P}) = \boldsymbol{A}(\mathrm{O}_0, \mathrm{P})$$

be an *inertial frame*. The other two frames (1) and (2),

$$\boldsymbol{A}_{\mathrm{O}_1} : \mathrm{P} \longmapsto \boldsymbol{x}_1 = \boldsymbol{A}_1(\mathrm{O}_1, \mathrm{P}) = \mathbf{Q}_{10}\boldsymbol{A}(\mathrm{O}_1, \mathrm{P})\,,$$

$$\boldsymbol{A}_{\mathrm{O}_2} : \mathrm{P} \longmapsto \boldsymbol{x}_2 = \boldsymbol{A}_2(\mathrm{O}_2, \mathrm{P}) = \mathbf{Q}_{20}\boldsymbol{A}(\mathrm{O}_2, \mathrm{P})$$

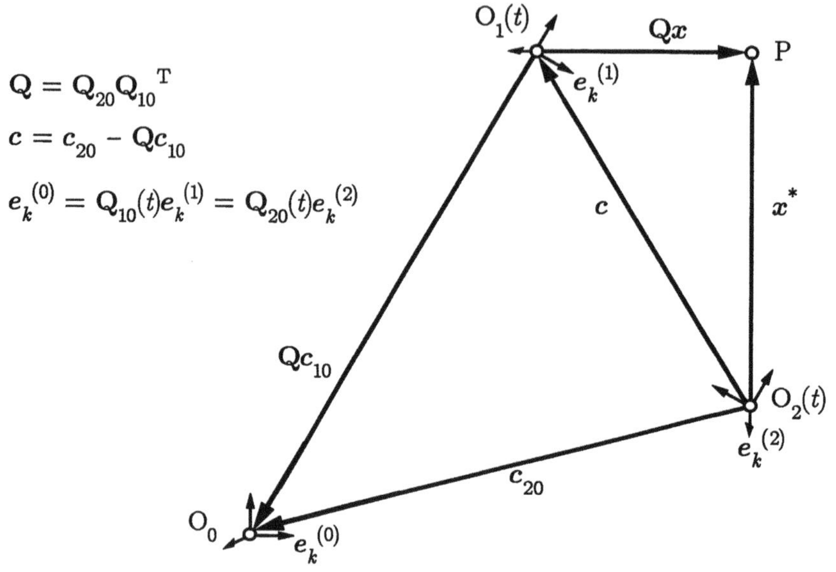

Figure 4. 5: Relative frames of reference

may move in relation to the inertial frame (0); let this motion be represented by means of functions

$$t \longmapsto \left(\mathbf{Q}_{10}(t), \boldsymbol{c}_{10}(t)\right) \text{ and } t \longmapsto \left(\mathbf{Q}_{20}(t), \boldsymbol{c}_{20}(t)\right) .$$

The frames (1) and (2) are called *relative frames of reference*, whereas the term *absolute frame of reference* is usual for the inertial frame. The choice of the three frames of reference is illustrated in Fig. 4. 5.

Both functions $\boldsymbol{c}_{10}(t)$, $\boldsymbol{c}_{20}(t)$ describe the motions of the reference points, as illustrated in Fig. 4. 5. If orthogonal base vectors are joined to the points of reference, then the base vectors of the frames (1) and (2) are rotated into the base vectors of the frame (0) by means of the tensors $\mathbf{Q}_{10}(t)$ and $\mathbf{Q}_{20}(t)$. In the sense of the active interpretation, the transformation formulae

$$\boldsymbol{x}_1 = \mathbf{Q}_{10}\boldsymbol{x}_0 + \boldsymbol{c}_{10} , \quad \boldsymbol{x}_2 = \mathbf{Q}_{20}\boldsymbol{x}_0 + \boldsymbol{c}_{20} \tag{4.65}$$

between the position vectors $\boldsymbol{x}_0$, $\boldsymbol{x}_1$, $\boldsymbol{x}_2$ are valid, as well as

$$\boldsymbol{x}_2 = \mathbf{Q}_{21}\boldsymbol{x}_1 + \boldsymbol{c}_{21} , \tag{4.66}$$

with the abbreviations

$$\mathbf{Q}_{21} = \mathbf{Q}_{20}\mathbf{Q}_{10}{}^{\mathrm{T}} \tag{4.67}$$

and

$$\boldsymbol{c}_{21} = \boldsymbol{c}_{20} - \mathbf{Q}_{20}\mathbf{Q}_{10}{}^{\mathrm{T}}\boldsymbol{c}_{10} = \boldsymbol{c}_{20} - \mathbf{Q}_{21}\boldsymbol{c}_{10} \,. \tag{4.68}$$

$\mathbf{Q}_{21}$ represents the rotation of frame (2) into frame (1) and $\boldsymbol{c}_{21} = \boldsymbol{A}_2(\mathrm{O}_2, \mathrm{O}_1)$ connects the points of reference O_2 and O_1.
Corresponding to the rotations $\mathbf{Q}_{10}$ and $\mathbf{Q}_{20}$ we introduce the skew spin tensors

$$\boldsymbol{\Omega}_{10} = \dot{\mathbf{Q}}_{10}\mathbf{Q}_{10}{}^{\mathrm{T}} \,, \quad \boldsymbol{\Omega}_{20} = \dot{\mathbf{Q}}_{20}\mathbf{Q}_{20}{}^{\mathrm{T}} \,. \tag{4.69}$$

The spin tensor corresponding to the rotation tensor $\mathbf{Q}_{21}$ reads

$$\begin{aligned}\boldsymbol{\Omega}_{21} &= \dot{\mathbf{Q}}_{21}\mathbf{Q}^{\mathrm{T}}{}_{21} = (\dot{\mathbf{Q}}_{20}\mathbf{Q}_{10}{}^{\mathrm{T}} + \mathbf{Q}_{20}\dot{\mathbf{Q}}_{10}{}^{\mathrm{T}})\mathbf{Q}_{10}\mathbf{Q}_{20}{}^{\mathrm{T}} \\ &= \dot{\mathbf{Q}}_{20}\mathbf{Q}_{20}{}^{\mathrm{T}} + \mathbf{Q}_{20}\mathbf{Q}_{10}{}^{\mathrm{T}}(\dot{\mathbf{Q}}_{10}\mathbf{Q}_{10}{}^{\mathrm{T}})^{\mathrm{T}}\mathbf{Q}_{10}\mathbf{Q}_{20}{}^{\mathrm{T}},\end{aligned}$$

that is $\boldsymbol{\Omega}_{21} = \boldsymbol{\Omega}_{20} - \mathbf{Q}_{21}\boldsymbol{\Omega}_{10}\mathbf{Q}_{21}{}^{\mathrm{T}}$.

Thus we obtain the following transformation formula for the spin tensors,

$$\boldsymbol{\Omega}_{20} = \mathbf{Q}_{21}\boldsymbol{\Omega}_{10}\mathbf{Q}_{21}{}^{\mathrm{T}} + \boldsymbol{\Omega}_{21} \,. \tag{4.70}$$

From this preparatory work we proceed to examine the transformation behaviour of the velocity and acceleration vectors. The relations between the two relative frames of reference are of particular interest. The velocity vector $\boldsymbol{v}_0$ (the velocity with respect to the inertial frame) transforms into $\boldsymbol{v}_1$ and $\boldsymbol{v}_2$; application of the transformation formula (4.36) leads to

$$\boldsymbol{v}_1 = \mathbf{Q}_{10}\boldsymbol{v}_0 + \dot{\boldsymbol{c}}_{10} + \boldsymbol{\Omega}_{10}(\boldsymbol{x}_1 - \boldsymbol{c}_{10}) \,, \quad \boldsymbol{v}_2 = \mathbf{Q}_{20}\boldsymbol{v}_0 + \dot{\boldsymbol{c}}_{20} + \boldsymbol{\Omega}_{20}(\boldsymbol{x}_2 - \boldsymbol{c}_{20}) \,.$$

$\boldsymbol{v}_0$ can be eliminated from both these equations:

$$\boldsymbol{v}_2 - \dot{\boldsymbol{c}}_{20} - \boldsymbol{\Omega}_{20}(\boldsymbol{x}_2 - \boldsymbol{c}_{20}) = \mathbf{Q}_{20}\boldsymbol{v}_0 = \mathbf{Q}_{20}\mathbf{Q}_{10}{}^{\mathrm{T}}\mathbf{Q}_{10}\boldsymbol{v}_0 \,,$$

$$\boldsymbol{v}_2 - \dot{\boldsymbol{c}}_{20} - \boldsymbol{\Omega}_{20}(\boldsymbol{x}_2 - \boldsymbol{c}_{20}) = \mathbf{Q}_{21}\{\boldsymbol{v}_1 - \dot{\boldsymbol{c}}_{10} - \boldsymbol{\Omega}_{10}(\boldsymbol{x}_1 - \boldsymbol{c}_{10})\} \,. \tag{4.71}$$

A similar result is obtained for the acceleration vector $\boldsymbol{a}_0$. Applying (4.37) we get

$$\boldsymbol{a}_1 = \mathbf{Q}_{10}\boldsymbol{a}_0 + \ddot{\boldsymbol{c}}_{10} + (\dot{\boldsymbol{\Omega}}_{10} - \boldsymbol{\Omega}_{10}{}^2)(\boldsymbol{x}_1 - \boldsymbol{c}_{10}) + 2\boldsymbol{\Omega}_{10}(\boldsymbol{v}_1 - \dot{\boldsymbol{c}}_{10}) ,$$

$$\boldsymbol{a}_2 = \mathbf{Q}_{20}\boldsymbol{a}_0 + \ddot{\boldsymbol{c}}_{20} + (\dot{\boldsymbol{\Omega}}_{20} - \boldsymbol{\Omega}_{20}{}^2)(\boldsymbol{x}_2 - \boldsymbol{c}_{20}) + 2\boldsymbol{\Omega}_{20}(\boldsymbol{v}_2 - \dot{\boldsymbol{c}}_{20}) ,$$

which, eliminating $\boldsymbol{a}_0$, leads to

$$\begin{aligned}&\boldsymbol{a}_2 - \ddot{\boldsymbol{c}}_{20} - (\dot{\boldsymbol{\Omega}}_{20} - \boldsymbol{\Omega}_{20}{}^2)(\boldsymbol{x}_2 - \boldsymbol{c}_{20}) - 2\boldsymbol{\Omega}_{20}(\boldsymbol{v}_2 - \dot{\boldsymbol{c}}_{20})\\ &\quad = \mathbf{Q}_{21}\{\boldsymbol{a}_1 - \ddot{\boldsymbol{c}}_{10} - (\dot{\boldsymbol{\Omega}}_{10} - \boldsymbol{\Omega}_{10}{}^2)(\boldsymbol{x}_1 - \boldsymbol{c}_{10}) - 2\boldsymbol{\Omega}_{10}(\boldsymbol{v}_1 - \dot{\boldsymbol{c}}_{10})\} . \end{aligned} \tag{4.72}$$

The following designations are now chosen to summarise the relations between the two relative frames of reference:

$$\begin{array}{lllll} \boldsymbol{x}_1 = \boldsymbol{x} & \boldsymbol{c}_{10} = \boldsymbol{d} & \boldsymbol{\Omega}_{10} = \boldsymbol{\Omega} & \boldsymbol{v}_1 = \boldsymbol{v} & \boldsymbol{a}_1 = \boldsymbol{a} \\ \boldsymbol{x}_2 = \boldsymbol{x}^* & \boldsymbol{c}_{20} = \boldsymbol{d}^* & \boldsymbol{\Omega}_{20} = \boldsymbol{\Omega}^* & \boldsymbol{v}_2 = \boldsymbol{v}^* & \boldsymbol{a}_2 = \boldsymbol{a}^* \end{array}$$

$$\mathbf{Q}_{21} = \mathbf{Q}_{20}\mathbf{Q}_{10}{}^{\mathrm{T}} = \mathbf{Q}$$

$$\boldsymbol{c}_{21} = \boldsymbol{c}_{20} - \mathbf{Q}_{21}\boldsymbol{c}_{10} = \boldsymbol{c}_{20} - \mathbf{Q}\boldsymbol{c}_{10} = \boldsymbol{c}$$

$$\boldsymbol{\Omega}_{21} = \boldsymbol{\Omega}_{20} - \mathbf{Q}_{21}\boldsymbol{\Omega}_{10}\mathbf{Q}_{21}{}^{\mathrm{T}} = \dot{\mathbf{Q}}\mathbf{Q}^{\mathrm{T}} = \mathbf{A}$$

Using these notations, the following relations between the relative frames of reference are valid:

$$\boldsymbol{x}^* = \mathbf{Q}\boldsymbol{x} + \boldsymbol{c} , \tag{4.73}$$

$$\boldsymbol{\Omega}^* = \mathbf{Q}\boldsymbol{\Omega}\mathbf{Q}^{\mathrm{T}} + \mathbf{A} , \tag{4.74}$$

$$\boldsymbol{v}^* - \dot{\boldsymbol{d}}^* - \boldsymbol{\Omega}^*(\boldsymbol{x}^* - \boldsymbol{d}^*) = \mathbf{Q}\{\boldsymbol{v} - \dot{\boldsymbol{d}} - \boldsymbol{\Omega}(\boldsymbol{x} - \boldsymbol{d})\} , \tag{4.75}$$

$$\begin{aligned}&\boldsymbol{a}^* - \ddot{\boldsymbol{d}}^* - (\dot{\boldsymbol{\Omega}}^* - \boldsymbol{\Omega}^{*2})(\boldsymbol{x}^* - \boldsymbol{d}^*) - 2\boldsymbol{\Omega}^*(\boldsymbol{v}^* - \dot{\boldsymbol{d}}^*)\\ &\quad = \mathbf{Q}\{\boldsymbol{a} - \ddot{\boldsymbol{d}} - (\dot{\boldsymbol{\Omega}} - \boldsymbol{\Omega}^2)(\boldsymbol{x} - \boldsymbol{d}) - 2\boldsymbol{\Omega}(\boldsymbol{v} - \dot{\boldsymbol{d}})\} . \end{aligned} \tag{4.76}$$

On the right-hand side of equations (4.75) and (4.76) we recognise the so-called *absolute velocity* vector and the *absolute acceleration* vector. The result proves

Theorem 4. 5

In the case of a *change between two relative frames of reference*, represented by

$$t \longmapsto \big(\mathbf{Q}(t), \boldsymbol{c}(t)\big)\,,$$

the *absolute velocity*

$$\boldsymbol{v}_{\text{abs}} = \boldsymbol{v} - \dot{\boldsymbol{d}} - \boldsymbol{\Omega}(\boldsymbol{x} - \boldsymbol{d}) \tag{4.77}$$

is an *objective vector*. The same applies for the *absolute acceleration*

$$\boldsymbol{a}_{\text{abs}} = \boldsymbol{a} - \ddot{\boldsymbol{d}} - (\dot{\boldsymbol{\Omega}} - \boldsymbol{\Omega}^2)(\boldsymbol{x} - \boldsymbol{d}) - 2\boldsymbol{\Omega}(\boldsymbol{v} - \dot{\boldsymbol{d}})\,. \tag{4.78}$$

Here, $-\,\boldsymbol{d}(t) = -\,\boldsymbol{c}_{10}(t)$ describes the *translation* and $-\,\boldsymbol{\Omega}(t) = -\,\dot{\mathbf{Q}}_{10}\mathbf{Q}_{10}{}^{\mathrm{T}}$ the *rate of rotation* in relation to the inertial system. ■

The minus sign in front of $\boldsymbol{d}$ and $\boldsymbol{\Omega}$ is due to the fact that $\boldsymbol{d}(t)$ represents the translation and $\boldsymbol{\Omega}(t)$ the rate of rotation of the inertial frame with respect to the relative frame of reference. Theorem 4. 5 is not a contradiction of Theorem 4. 3: the latter establishes the non-objectivity of *relative velocity* and *relative acceleration.*

The objectivity of the absolute velocity and acceleration physically means that these quantities are invariant with respect to a change of the physical observer. In this sense the balance of linear momentum, written in the form of

$$m\boldsymbol{a}_{\text{abs}} = \boldsymbol{F} \tag{4.79}$$

or

$$m\big[\boldsymbol{a} - \ddot{\boldsymbol{d}} - (\dot{\boldsymbol{\Omega}} - \boldsymbol{\Omega}^2)(\boldsymbol{x} - \boldsymbol{d}) - 2\boldsymbol{\Omega}(\boldsymbol{v} - \dot{\boldsymbol{d}})\big] = \boldsymbol{F}\,, \tag{4.80}$$

is a relation, which, in view of the objectivity of the right and the left-hand side,

$$\boldsymbol{F}^* = \mathbf{Q}\boldsymbol{F}\,, \quad m^*\boldsymbol{a}^*_{\text{abs}} = m\mathbf{Q}\boldsymbol{a}_{\text{abs}}\,, \tag{4.81}$$

is valid for all frames of reference permitted in classical mechanics. The balance of linear momentum in this interpretation is a statement that is invariant in the light of any change of physical observer. However, the physical observer has to be aware of the motion of the reference frame in relation to an inertial system.

Definition 4. 10

One relation that applies to all frames of reference in the same form is called an *observer-invariant relation.* ◊

The balance of linear momentum is an example of an observer-invariant relation. The same applies to the balance of rotational momentum, and the first law of thermodynamics; the corresponding discussion is omitted. It is possible to create other examples in connection with the representation of material properties (see Sect. 7. 2. 4).

5 Classical Theories of Continuum Mechanics

5.1 Introduction

The balance relations of thermomechanics are general laws of nature: all substances and systems of material bodies are governed by them alike. However, the balance relations for mass, momentum, energy and entropy are not sufficient for determining the fields contained in them, a fact that was established in the final remarks on the subject of balance relations (Sect. 3.6). Additional equations are therefore required for completing the balance relations and setting up initial-boundary-value problems, to permit the development of solutions. The supplementary equations needed describe the properties constituting the individual make-up of the materials and are hence termed *constitutive equations*. The physical significance of the constitutive equations is the circumstance that physical processes are not determined by the balance relations alone, but also by the individual *material properties*, which cause different substances to behave in an entirely different way, given the same external conditions. Accordingly, constitutive equations have to be distinguished from general laws of nature referring to kinematics and balance relations. They are *mathematical models*, intended to reflect the characteristic features of the actual material behaviour in an idealised form. Each time we match a set of specially selected constitutive equations to the balance relations, the outcome is a specific theory of continuum mechanics.

This chapter describes in fundamental terms how specific theories of continuum mechanics are set up. Attention is drawn to three specific constitutive equations to begin with, namely the equations defining the perfect fluid, the linear-viscous fluid and the linear-elastic isotropic solid.

These constitutive equations form the basis of the so-called *classical theories* of continuum mechanics, viz. hydrodynamics and elasticity, two fields encompassing broad areas of fluid and solid mechanics.

Two extensions to the model of linear elasticity are also included in the classical theories of continuum mechanics, namely the theories of linear viscoelasticity and plasticity.

The mathematical theories for the solution of basic equations in classical hydrodynamics, elasticity, viscoelasticity and plasticity are already highly developed but are nevertheless still undergoing research. It is of great importance for practical applications that there is a variety of known solutions and approximations within the context of these theories. This knowledge is the foundation of engineering solid and fluid mechanics. The contribution afforded by *engineering mechanics*, which is based on the classical continuum theories, remains a reliable and indispensable prop when solving countless problems from the fields of science and engineering.

5. 2 Elastic Fluid

Classical hydrodynamics is based on the defining equation of an elastic fluid. The term *fluid* is intended to signify that the following constitutive equation is suitable for expressing the mechanical properties of liquids and gases.[1]

Definition 5. 1

The constitutive equation

$$\mathbf{T} = -\,p\mathbf{1} \tag{5.1}$$

defines an *elastic fluid*, otherwise known as a *perfect fluid* or *Euler fluid.* In the case of a *barotropic fluid* the pressure p is a function of the mass density:[2]

$$p = \hat{p}(\rho)\;. \tag{5.2}$$

This *material function* represents the material properties of an elastic fluid. Under the additional assumption of *incompressibility* ($\mathrm{div}\boldsymbol{v}(\boldsymbol{x}, t) = 0$) the

[1] A detailed introduction to hydrodynamics with numerous examples of application is to be found in SPURK [1997]. SERRIN [1959] provides an extensive description of classical hydrodynamics.

[2] Cf. TRUESDELL & TOUPIN [1960], pp. 388, 712.

density ρ is substantially constant; the pressure p is no longer defined by a constitutive equation in this case, but has to be calculated from the balance of linear momentum. For an *incompressible perfect fluid* the defining equation is labelled accordingly,

$$\mathbf{T} = -\, p(x, t)\mathbf{1}\ . \tag{5.3}$$

◊

Constitutive equation (5.1) stipulates that only hydrostatic stresses can occur in the fluid, regardless of the motion or flow pattern, i.e. all phenomena of internal friction are disregarded. The shape of the material function $p = \hat{p}(\rho)$ expresses the intrinsic material properties of the fluid. The simplest specific case is a linear function: $p = c\rho$. This equation specifically defines the so-called *perfect gas*, assuming that the proportionality factor is a linear function of the absolute temperature:

$$p = R\Theta\rho\ . \tag{5.4}$$

The constant R is a *material parameter*. Applying definition (5.1), we obtain

Theorem 5. 1

The theory of perfect fluid is defined by the field equations

$$\frac{\partial \boldsymbol{v}}{\partial t} + (\mathrm{grad}\boldsymbol{v})\boldsymbol{v} = -\,\frac{1}{\rho}\,\mathrm{grad}\hat{p}(\rho) + \boldsymbol{f}\ , \tag{5.5}$$

$$\frac{\partial \rho}{\partial t} + \mathrm{div}(\rho\boldsymbol{v}) = 0\ , \tag{5.6}$$

with the *initial conditions*

$$\boldsymbol{v}(\boldsymbol{x}, t_0) = \boldsymbol{v}_0(\boldsymbol{x}) \text{ for } \boldsymbol{x} \in \chi_{t_0}[\mathcal{B}]\ , \tag{5.7}$$

$$\rho(\boldsymbol{x}, t_0) = \rho_0(\boldsymbol{x}) \text{ for } \boldsymbol{x} \in \chi_{t_0}[\mathcal{B}]\ , \tag{5.8}$$

and the following boundary conditions:

Geometric Boundary Conditions

An elastic fluid can perform any tangential motions on given boundary surfaces. It therefore holds that

$$\boldsymbol{v}\cdot\boldsymbol{n} = 0 \tag{5.9}$$

on a fixed boundary surface $\partial_u\chi_t[\mathscr{B}]$ (independent of t). Its direction is represented by the normal vector $\boldsymbol{n} = \boldsymbol{n}(\boldsymbol{x})$. More generally, a time-dependent boundary surface, described by means of a given function in the form $g(\boldsymbol{x}, t) = 0$, corresponds to the boundary condition

$$\frac{\partial}{\partial t} g(\boldsymbol{x}, t) + \boldsymbol{v}(\boldsymbol{x}, t)\cdot \operatorname{grad} g(\boldsymbol{x}, t) = 0 \tag{5.10}$$

for $\boldsymbol{x} \in \partial_u\chi_t[\mathscr{B}]$.

Dynamic Boundary Conditions
On the free surface $\partial_s\chi_t[\mathscr{B}]$, whose current shape is generally unknown, it is possible to prescribe a surface force $\boldsymbol{t}_0 = p_0(\boldsymbol{x}, t)\boldsymbol{n}(\boldsymbol{x}, t)$. For this reason the dynamic boundary conditions are written as

$$p(\boldsymbol{x}, t) = p_0(\boldsymbol{x}, t) \tag{5.11}$$

and

$$\frac{\partial}{\partial t} f(\boldsymbol{x}, t) + \boldsymbol{v}(\boldsymbol{x}, t)\cdot \operatorname{grad} f(\boldsymbol{x}, t) = 0 \tag{5.12}$$

for all $\boldsymbol{x}$ with $f(\boldsymbol{x}, t) = 0$ or $\boldsymbol{x} \in \partial_s\chi_t[\mathscr{B}]$ respectively.

The boundary surface $\partial_s\chi_t$ is represented here by means of the equation $f(\boldsymbol{x}, t) = 0$. Finding out the function f is part of the boundary value problem.

□

Proof

Inserting the constitutive equation (5.1) into the balance of linear momentum (3.128) we obtain (5.5), applying the identity

$$\operatorname{div}\big(\alpha(\boldsymbol{x}, t)\mathbf{1}\big) = \operatorname{grad}\alpha(\boldsymbol{x}, t)\ , \tag{5.13}$$

valid for all scalar fields $(\boldsymbol{x}, t) \longmapsto \alpha(\boldsymbol{x}, t)$. Besides the balance of linear momentum (3.128) the balance of mass, i.e. the continuity equation (3.127) also has to be satisfied. The geometric boundary condition corresponds to the fact that the fluid cannot penetrate a given wall. In view of the absence of friction, however, gliding along the boundary area cannot be ruled out, i.e. only the normal component of the velocity vector can be prescribed. The dynamic boundary conditions (5.11), (5.12) arise from the postulate

$$\boldsymbol{t} = \mathbf{T}\boldsymbol{n} = \boldsymbol{s}(\boldsymbol{x}, t) \text{ for } \boldsymbol{x} \in \partial_s\chi_t[\mathscr{B}] \text{ , with } \boldsymbol{t} = -p\boldsymbol{n} = -p_0\boldsymbol{n} \text{ ,}$$

and the unknown boundary surface $\partial_s\chi_t[\mathscr{B}]$, represented by the equation $f(\boldsymbol{x}, t) = 0$ has to be a material surface: one material point, located on a free surface, is obliged to remain in this boundary area. In other words its position vector $\boldsymbol{x}(t)$ has to satisfy the equation $f(\boldsymbol{x}(t), t) = 0$ for all t. The consequence is that the function $f(\boldsymbol{x}, t)$ is linked to the velocity field $\boldsymbol{v}(\boldsymbol{x}, t)$ by means of the relation

$$\frac{\partial}{\partial t} f(\boldsymbol{x}, t) + \boldsymbol{v}(\boldsymbol{x}, t) \cdot \operatorname{grad} f(\boldsymbol{x}, t) = 0 \text{ .}$$ ■

The differential equation (5.5) is called the *Euler* equation.[3] Together with the continuity equation (5.6) the *Euler* equation forms a coupled system of four partial differential equations for determining the fields $\rho(\boldsymbol{x}, t)$ and $\boldsymbol{v}(\boldsymbol{x}, t)$. In Cartesian coordinates these equations are given by

$$v_{i,t} + v_{i,k}v_k = -\frac{1}{\rho}\, p_{,i} + f_i \text{ ,} \tag{5.14}$$

$$\rho_{,t} + (\rho v_k)_{,k} = 0 \text{ .} \tag{5.15}$$

In the case of curvilinear coordinates the partial derivatives with respect to the spatial coordinates are replaced by covariant derivatives.

It should be added that the field equations (5.5), (5.6) can be simplified considerably by assuming *incompressibility* of the fluid. Then the motion turns into a volume-preserving one; the density ρ is materially constant, and the hydrostatic pressure p is a *reaction stress*, that can no longer be dependent on the density, but has to be determined from the balance of linear momentum. The balance equations for momentum and mass related to an incompressible fluid,

$$\frac{\partial \boldsymbol{v}}{\partial t} + (\operatorname{grad}\boldsymbol{v})\boldsymbol{v} = -\frac{1}{\rho} \operatorname{grad} p(\boldsymbol{x}, t) + \boldsymbol{f} \text{ ,} \tag{5.16}$$

$$\operatorname{div}\boldsymbol{v} = 0 \text{ ,} \tag{5.17}$$

[3] Cf. TRUESDELL & TOUPIN [1960], Sect. 297.

create a system of differential equations for the fields $\boldsymbol{v}(\boldsymbol{x}, t)$ and $p(\boldsymbol{x}, t)$. This system is decoupled: for instance, in the case of a potential, $\boldsymbol{v} = \text{grad}\Phi(\boldsymbol{x}, t)$, the continuity equation (5.17) (a linear equation) can be solved first and the result inserted into the balance of linear momentum (5.16) in order to calculate the distribution of pressure $p(\boldsymbol{x}, t)$. For conservative volume forces, $\boldsymbol{f} = -\,\text{grad}\varphi(\boldsymbol{x}, t)$, this calculation is carried out by means of simple integration (*Bernoulli* equation).

5. 3 Linear-Viscous Fluid

The model of the perfect fluid is an idealisation, which omits an important phenomenon per definitionem, namely the *internal friction* or *dissipation*, that is present to a greater or smaller degree in every real material. This is taken into account by the model of the linear-viscous fluid, which approaches more closely to reality in that respect.

Definition 5. 2

The constitutive equation

$$\mathbf{T} = -\,p(\rho)\mathbf{1} + 2\eta\mathbf{D} + \lambda(\text{tr}\mathbf{D})\mathbf{1} \tag{5.18}$$

defines the *linear-viscous fluid*, also known as *Newton fluid* or *Navier-Poisson fluid*.[4] In this equation, the pressure p is a function of the density, $p = \hat{p}(\rho)$, similar to the case of the perfect fluid. The tensor

$$\mathbf{D} = \frac{1}{2}\left(\text{grad}\boldsymbol{v} + (\text{grad}\boldsymbol{v})^{\text{T}}\right) \tag{5.19}$$

is the spatial strain rate tensor, defined by (1.114) and η and λ are material parameters. η is known as *shear viscosity*, and the sum $\frac{2}{3}\eta + \lambda$ is the so-called *bulk viscosity*.

If we also assume *incompressibility*, we have $\text{tr}\mathbf{D} = \text{div}\boldsymbol{v} = 0$. The constitutive equation of the incompressible *Newton* fluid reads as

$$\mathbf{T} = -\,p(\boldsymbol{x}, t)\mathbf{1} + 2\eta\mathbf{D}\,, \tag{5.20}$$

where p is no longer linked to the density ρ by means of a constitutive equation. ◊

[4] See TRUESDELL & TOUPIN [1960], p. 716.

The basic equations of the *Newton* fluid are described in

Theorem 5. 2

The theory of a linear-viscous fluid is defined by the field equations

$$\frac{\partial \boldsymbol{v}}{\partial t} + (\text{grad}\boldsymbol{v})\boldsymbol{v} = -\frac{1}{\rho}\,\text{grad}\hat{p}(\rho) + \frac{\eta}{\rho}\Delta \boldsymbol{v} + \frac{1}{\rho}(\lambda + \eta)\text{grad}(\text{div}\boldsymbol{v}) + \boldsymbol{f} \tag{5.21}$$

$$\frac{\partial \rho}{\partial t} + \text{div}(\rho \boldsymbol{v}) = 0\,, \tag{5.22}$$

with the initial conditions

$$\boldsymbol{v}(\boldsymbol{x}, t_0) = \boldsymbol{v}_0(\boldsymbol{x}) \text{ for } \boldsymbol{x} \in \chi_{t_0}[\mathcal{B}]\,, \tag{5.23}$$

$$\rho(\boldsymbol{x}, t_0) = \rho_0(\boldsymbol{x}) \text{ for } \boldsymbol{x} \in \chi_{t_0}[\mathcal{B}]\,, \tag{5.24}$$

and the following boundary conditions:

Geometric Boundary Conditions

A viscous fluid can adhere to a given boundary surface. If the boundary surface is fixed, i.e. time-independent, we write

$$\boldsymbol{v} = \boldsymbol{0} \text{ for } \boldsymbol{x} \in \partial_u\chi_t[\mathcal{B}]\,. \tag{5.25}$$

More generally,

$$\boldsymbol{v} = \boldsymbol{g}(\boldsymbol{x}, t) \tag{5.26}$$

is written for $\boldsymbol{x} \in \partial_u\chi_t[\mathcal{B}]$ in the case of a moving boundary surface, the function $\boldsymbol{g}(\boldsymbol{x}, t)$ representing the velocity of the boundary points.

Dynamic Boundary Conditions

On the free surface $\partial_s\chi_t[\mathcal{B}]$, the shape of which is normally unknown, a surface force $\boldsymbol{t}_0(\boldsymbol{x}, t) = \mathbf{T}\boldsymbol{n}$ can be prescribed: for all $\boldsymbol{x} \in \partial_s\chi_t[\mathcal{B}]$ or for all $\boldsymbol{x}$ with $f(\boldsymbol{x}, t) = 0$ we write

$$\mathbf{T}\frac{\text{grad}f}{|\text{grad}f|} = \boldsymbol{t}_0(\boldsymbol{x}, t) \tag{5.27}$$

and

$$\frac{\partial}{\partial t} f(\boldsymbol{x}, t) + \boldsymbol{v}(\boldsymbol{x}, t) \cdot \operatorname{grad} f(\boldsymbol{x}, t) = 0 \,. \tag{5.28}$$

□

Proof

By inserting the constitutive equation (5.18) into the balance of linear momentum (3.128) we obtain (5.21), the definition of the Laplacian operator,

$$\Delta \boldsymbol{v} = \operatorname{div}(\operatorname{grad}\boldsymbol{v}) = \Big[\frac{\partial}{\partial x^l}\Big(\frac{\partial \boldsymbol{v}}{\partial x^k} \otimes \boldsymbol{g}^k\Big)\Big]\boldsymbol{g}^l \,, \tag{5.29}$$

and the identity

$$\operatorname{div}\big[(\operatorname{grad}\boldsymbol{v})^{\mathrm{T}}\big] = \operatorname{grad}(\operatorname{div}\boldsymbol{v}) \,, \tag{5.30}$$

valid for all vector fields $\boldsymbol{v}(\boldsymbol{x}, t)$, having been taken into consideration.

The geometric boundary condition demands complete adhesion of the fluid to fixed or mobile walls, which is physically possible in this case, as opposed to the perfect fluid: appropriate friction forces exist to prevent it from sliding along the boundary surface.

The dynamic boundary conditions (5.27), (5.28) are the outcome of the prescription $\mathbf{T}\boldsymbol{n} = \boldsymbol{t}_0(\boldsymbol{x}, t)$ with

$$\mathbf{T}\boldsymbol{n} = -\, p(\rho)\boldsymbol{n} + 2\eta\mathbf{D}\boldsymbol{n} + \lambda(\operatorname{tr}\mathbf{D})\boldsymbol{n} \,. \tag{5.31}$$

The unknown boundary surface $\partial_{\mathscr{S}}\chi_{\mathrm{t}}[\mathscr{B}]$, represented by $f(\boldsymbol{x}, t) = 0$, is a material surface, the determination of which is part of the boundary value problem. ■

The differential equations (5.20), (5.21) are called *Navier-Stokes equations.*[5] They likewise form a coupled system of 4 partial differential equations for determining the fields $\rho(\boldsymbol{x}, t)$ and $\boldsymbol{v}(\boldsymbol{x}, t)$. In Cartesian coordinates these field equations are given by

$$\begin{aligned} &v_{i,t} + v_{i,k}v_k = -\frac{1}{\rho}\, p_{,i} + \frac{\eta}{\rho} v_{i,kk} + \frac{1}{\rho}(\eta + \lambda)v_{k,ki} + f_i \\ &\rho_{,t} + (\rho v_k)_{,k} = 0 \,. \end{aligned} \tag{5.32}$$

[5] Cf. TRUESDELL & NOLL [1965], p. 476.

Under the assumption of incompressibility the field equations (5.21) and (5.22) can again be simplified in view of tr**D** = div$\boldsymbol{v}$ = 0. The *Navier-Stokes* equations for an incompressible fluid are

$$\frac{\partial \boldsymbol{v}}{\partial t} + (\text{grad}\boldsymbol{v})\boldsymbol{v} = -\frac{1}{\rho}\,\text{grad}\;p(\boldsymbol{x},t) + \frac{\eta}{\rho}\Delta\boldsymbol{v} + \boldsymbol{f}\,, \tag{5.33}$$

$$\text{div}\boldsymbol{v} = 0 \tag{5.34}$$

and form a system of four partial differential equations for the fields $\boldsymbol{v}(\boldsymbol{x}, t)$ and $p(\boldsymbol{x}, t)$. Similar to the elastic fluid the continuity equation (5.34) is linear in $\boldsymbol{v}$ and can be solved separately. However, this fact does not facilitate the search for solutions in this case: the balance of linear momentum still contains the decisive term $\Delta\boldsymbol{v}$ and is nonlinear in the velocity field in view of the convective derivative (grad$\boldsymbol{v}$)$\boldsymbol{v}$. This fact seriously complicates the task of finding solutions.

5. 4 Linear-Elastic Solid

The classical theory of solid mechanics is the *theory of linear elasticity*. This theory is defined within the limits of geometric linearisation and is accordingly only valid for small displacements and strains.[6]

Definition 5. 3

A solid whose mechanical properties are defined by the constitutive equation

$$\mathbf{T} = 2\mu\Big[\mathbf{E} + \frac{\nu}{1-2\nu}(\text{tr}\mathbf{E})\mathbf{1}\Big] \tag{5.35}$$

is called a *linear-elastic isotropic solid*, also known as a *Hooke solid.* An equivalent definition is the constitutive equation

$$\mathbf{E} = \frac{1}{2\mu}\Big[\mathbf{T} - \frac{\nu}{1+\nu}(\text{tr}\mathbf{T})\mathbf{1}\Big]\,. \tag{5.36}$$

In these equations

---

6 An introduction is provided by TIMOSHENKO & GOODIER [1982]. Details are to be found in the monograph GURTIN [1972].

$$\mathbf{E} = \tfrac{1}{2}\left[\operatorname{grad}\boldsymbol{u} + (\operatorname{grad}\boldsymbol{u})^{\mathrm{T}}\right] \tag{5.37}$$

is the linearised strain tensor and $\boldsymbol{u} = \boldsymbol{u}(\boldsymbol{x}, t)$ the displacement field. The material and spatial representations of $\boldsymbol{u}$ are approximately identical within the limits of geometric linearisation, so that $\boldsymbol{X}$ can be replaced by $\boldsymbol{x}$. The constants μ and ν are material parameters: μ stands for the *shear modulus* and ν for the *Poisson number*.[7] The quantities obtained from them

$$E = 2\mu(1 + \nu) \tag{5.38}$$

and

$$K = \frac{2\mu(1+\nu)}{3(1-2\nu)} \tag{5.39}$$

are called *Young's elasticity modulus* and *bulk modulus* respectively. ◊

The constitutive equation (5.35) is also known under the name of *Hooke's law*.[8] In this connection the term "law" is undoubtedly misleading, yet it is quite common in engineering mechanics. Obviously, **T** and **E** always have the same eigenvectors in respect of equation (5.35). The constitutive equation (5.35) thus describes elasticity properties which are independent of direction. Solids with characteristics such as these are called *isotropic solids*. Applying the bulk modulus (5.39) *Hooke's law* can also be written in the form of

$$\mathbf{T}^{\mathrm{D}} = 2\mu\mathbf{E}^{\mathrm{D}}, \qquad \tfrac{1}{3}\operatorname{tr}\mathbf{T} = K\operatorname{tr}\mathbf{E}, \tag{5.40}$$

$\mathbf{E}^{\mathrm{D}} = \mathbf{E} - \frac{1}{3}(\operatorname{tr}\mathbf{E})\mathbf{1}$ being the deviator of **E** and $\mathbf{T}^{\mathrm{D}}$ the deviator of **T**.
The elasticity constants E, μ, ν and K, linked together by (5.38) and (5.39), have the following physical meaning in terms of homogeneous deformations: for uniaxial tension-compression loading by means of stress σ, we have a longitudinal strain ε, a transverse strain $\varepsilon_Q = -\nu\varepsilon$ and the stress-strain relation

$$\sigma = E\varepsilon\,. \tag{5.41}$$

In the case of simple shear the shear stress τ is proportional to the shear angle γ,

---

[7] See GURTIN [1972], Sect. 22.

[8] Cf. TRUESDELL & TOUPIN [1960], p. 724, footnote 1.

$$\tau = \mu\gamma \,. \tag{5.42}$$

An isotropic volume strain ε_{v} is attributed to a hydrostatic pressure,

$$-\,p = K\varepsilon_{\mathrm{v}} \,. \tag{5.43}$$

In combination with the balance of linear momentum the defining equation (5.35) leads to the basic equations of the theory of linear elasticity:

Theorem 5. 3

The theory of linear-elastic solids is defined by the field equations

$$\rho\,\frac{\partial^2 \boldsymbol{u}}{\partial t^2} = \mu\Big[\Delta\boldsymbol{u} + \frac{1}{1-2\nu}\ \mathrm{grad\ div}\boldsymbol{u}\Big] + \rho\boldsymbol{f} \tag{5.44}$$

with the initial conditions

$$\boldsymbol{u}(\boldsymbol{x}, t_0) = \boldsymbol{u}_0(\boldsymbol{x}) \text{ and } \frac{\partial \boldsymbol{u}}{\partial t}(\boldsymbol{x}, t_0) = \boldsymbol{v}_0(\boldsymbol{x}) \tag{5.45}$$

for all $\boldsymbol{x} \in \chi_{\mathrm{t}}[\mathscr{B}]$ $\Big(\mathrm{R}(\mathscr{P}) \approx \chi_{\mathrm{t}}[\mathscr{B}]\Big)$ and the following boundary conditions:

Geometric Boundary Conditions

$$\boldsymbol{u}(\boldsymbol{x}, t) = \boldsymbol{u}_{\mathrm{r}}(\boldsymbol{x}, t) \text{ for } \boldsymbol{x} \in \partial_{\boldsymbol{u}}\chi_{\mathrm{t}}[\mathscr{B}] \tag{5.46}$$

Dynamic Boundary Conditions

$$\boldsymbol{t} = \mathbf{T}\boldsymbol{n} = \boldsymbol{t}_{\mathrm{r}}(\boldsymbol{x}, t) \text{ for } \boldsymbol{x} \in \partial_{s}\chi_{\mathrm{t}}[\mathscr{B}] \tag{5.47}$$

□

Proof

Hooke's law applies for small deformations, since it contains the linearised strain tensor, which describes the state of strain merely within the limits of the geometric linearisation. Since spatial and material representations coincide in the linear approximation, the local balance of linear momentum (3.128) can be applied (instead of (3.133)). The acceleration vector has to be calculated according to the linear approximation (1.167). The *Laplace* operator $\Delta\boldsymbol{u} = \mathrm{div}(\mathrm{grad}\boldsymbol{u})$ according to (5.29) and the identities

$$\mathrm{grad}(\mathrm{div}\boldsymbol{u}) = \mathrm{div}\big[(\mathrm{grad}\boldsymbol{u})^{\mathrm{T}}\big] = \mathrm{div}\big[(\mathrm{div}\boldsymbol{u})\mathbf{1}\big] \tag{5.48}$$

lead to the field equation (5.44).

It is not necessary to know the current configuration of the boundary area $\partial_s\chi_t[\mathscr{B}]$ when formulating the dynamic boundary condition, since the difference between the current configuration and the reference configuration is negligible within the scope of the geometric linearisation. ■

For the derivation of the field equations (5.44) it is also possible to start from the material representation (3.133) of the local balance of linear momentum, bearing in mind that the stress tensors $\mathbf{T}_R = \mathbf{F}\tilde{\mathbf{T}}$ and $\mathbf{T}$ coincide in the geometric linearisation if the reference configuration is stress-free.

The differential equations (5.44) are known as *Lamé* equations or *Navier* equations.[9] They form a coupled system of 3 linear partial differential equations for determining the three components of the displacement vector $\boldsymbol{u}(\boldsymbol{x}, t)$. In Cartesian coordinates these field equations are

$$\rho u_{i,tt} = \mu\big[u_{i,kk} + \frac{1}{1-2\nu}u_{k,ki}\big] + \rho f_i \, . \tag{5.49}$$

The balance of mass contains no further unknowns: in the linear approximation, as a result of (2.7),

$\rho_R = \rho \det\mathbf{F} \approx \rho(1 + \mathrm{tr}\mathbf{H}) = \rho(1 + \mathrm{div}\boldsymbol{u})$ is valid, i.e.

$$\rho(\boldsymbol{x}, t) = \rho_R(\boldsymbol{x})(1 - \mathrm{div}\boldsymbol{u}) \, . \tag{5.50}$$

5. 5 Linear-Viscoelastic Solid

The *theory of linear viscoelasticity*[10] expands upon the theory of linear elasticity and paves the way for a more general modelling of dissipative material behaviour in comparison to a pure viscosity.

Definition 5. 4

An isotropic solid defined by the constitutive equation

$$\mathbf{T}(t) = \int_{-\infty}^{t} 2\mu(t-\tau)\mathbf{E}^{D\prime}(\tau)\mathrm{d}\tau + \left\{\int_{-\infty}^{t} K(t-\tau)[\mathrm{tr}\mathbf{E}'(\tau)]\mathrm{d}\tau\right\}\mathbf{1} \, , \tag{5.51}$$

[9] See GURTIN [1972], p. 213.

[10] NOWACKI [1965]; CHRISTENSEN [1982]; TSCHOEGL [1989]; FINDLEY, LAI & ONARAN [1976]; TOBOLSKY [1967]; LEITMAN & FISHER [1973]; GROSS [1968].

$(\cdot)' = \frac{\mathrm{d}}{\mathrm{d}\tau}(\cdot)$, is called a *linear-viscoelastic solid.* The material functions $\mu(\cdot)$ and $K(\cdot)$ are known as *shear* and *bulk relaxation functions.* An alternative definition is the constitutive equation

$$\mathbf{E}(t) = \int_{-\infty}^{t} 2m(t-\tau)\mathbf{T}^{\mathrm{D}\prime}(\tau)\mathrm{d}\tau + \left\{ \int_{-\infty}^{t} k(t-\tau)[\mathrm{tr}\mathbf{T}'(\tau)]\mathrm{d}\tau \right\}\mathbf{1} \,. \tag{5.52}$$

The material functions $m(\cdot)$ and $k(\cdot)$ are termed *shear* and *bulk creep functions.* ◇

The definitions (5.51) and (5.52) are equivalent: under appropriate mathematical assumptions the creep functions $m(\cdot)$, $k(\cdot)$ are uniquely determined if the relaxation functions $\mu(\cdot)$, $K(\cdot)$ are given and vice versa (see eqs. (5.71), (5.72) below).

The constitutive equations (5.51) and (5.52) define the *theory of linear viscoelasticity.* The theory of linear elasticity is included in the constitutive equation (5.51) as a special case, namely for constant relaxation functions, $\mu(t) = \mu_0$ and $K(t) = K_0$, and the "initial condition" $\mathbf{E}(\boldsymbol{x}, -\infty) = \mathbf{0}$.

In the case of non-constant relaxation functions the previous strain history $\tau \longmapsto \mathbf{E}(\tau)$ $(\tau \leq t)$ produces a current state of stress consisting of an additive superposition of incremental stresses,

$$\Delta_\tau \mathbf{T}(t) = 2\mu(t-\tau)\Delta\mathbf{E}^{\mathrm{D}}(\tau) + K(t-\tau)[\mathrm{tr}\Delta\mathbf{E}(\tau)]\mathbf{1} \,. \tag{5.53}$$

The stress increments $\Delta_\tau \mathbf{T}(t)$ evolve from the strain increments $\Delta\mathbf{E}(\tau)$, which took place at moments τ $(0 < \tau \leq t)$ in the past. The strain increments are multiplied by time-dependent weighting factors $\mu(t-\tau)$ or $K(t-\tau)$. Every one of these factors can be interpreted as a time-dependent shear or bulk modulus: the strain $\Delta\mathbf{E}(\tau)$, that took place at a past instant τ, leads to the time-dependent stress increment $\Delta_\tau \mathbf{T}(t)$. The constitutive equation (5.51) now expresses the supposition that all increments of stress $\Delta_\tau \mathbf{T}(t)$ can be superimposed additively to form the whole of the current state of stress $\mathbf{T}(t)$. This assumption postulates a material memory, in which the stress response is additively composed out of the individual parts of the past process of deformation. It is called *Boltzmann's principle of superposition.*[11] The physical meaning of the material functions in the constitutive equations (5.51) and (5.52) is described in

[11] Leitman & Fisher [1973], p. 12.

Theorem 5. 4
For a strain process in the form of

$$\mathbf{E}(t) = \begin{cases} \mathbf{0} & \text{for} \quad t \leq 0 \\ \mathbf{E}_0 & \text{for} \quad 0 < t < \infty \,. \end{cases} \tag{5.54}$$

equation (5.51) supplies the stress response

$$\mathbf{T}(t) = \begin{cases} \mathbf{0} & \text{for} \quad t \leq 0 \\ 2\mu(t)\mathbf{E}_0{}^{\mathrm{D}} + K(t)(\mathrm{tr}\mathbf{E}_0)\mathbf{1} & \text{for} \quad 0 < t < \infty \,. \end{cases} \tag{5.55}$$

For a stress process in the form of

$$\mathbf{T}(t) = \begin{cases} \mathbf{0} & \text{for} \quad t \leq 0 \\ \mathbf{T}_0 & \text{for} \quad 0 < t < \infty \end{cases} \tag{5.56}$$

equation (5.52) leads to the strain response

$$\mathbf{E}(t) = \begin{cases} \mathbf{0} & \text{for} \quad t \leq 0 \\ 2m(t)\mathbf{T}_0{}^{\mathrm{D}} + k(t)(\mathrm{tr}\mathbf{T}_0)\mathbf{1} & \text{for} \quad 0 < t < \infty \,. \end{cases} \tag{5.57}$$

□

Proof
The strain process (5.54) has the form of a step function. It evolves from the function

$$\mathbf{E}(t) = \begin{cases} \mathbf{0} & \text{for} \quad t \leq 0 \\ (t/T)\mathbf{E}_0 & \text{for} \quad 0 < t \leq T \\ \mathbf{E}_0 & \text{for} \quad T < t < \infty \end{cases} \tag{5.58}$$

in the limit $T \to 0$. For $t > T$ the insertion of (5.58) in (5.51) produces

$$\mathbf{T}(t) = \frac{1}{T}\int_0^T \Big\{2\mu(t - \tau)\mathbf{E}_0{}^{\mathrm{D}} + K(t - \tau)(\mathrm{tr}\mathbf{E}_0)\mathbf{1}\Big\}\mathrm{d}\tau \,. \tag{5.59}$$

If $\mu(\cdot)$ and $K(\cdot)$ are continuous functions, the mean value theorem of integration leads to

$$\mathbf{T}(t) = 2\mu(t - \vartheta_1 T)\mathbf{E}_0{}^{\mathrm{D}} + K(t - \vartheta_2 T)(\mathrm{tr}\mathbf{E}_0)\mathbf{1} \,, \tag{5.60}$$

with $0 \leq \vartheta_1 \leq 1$ and $0 \leq \vartheta_2 \leq 1$. In the limit $T \to 0$ we thereby obtain the result (5.55). Statement (5.57) is proved by the same method. ■

The theorem illustrates the physical meaning of the material functions as relaxation functions: the stress response to a strain step corresponds to a "*Hookean* law with time-dependent shear and bulk moduli". The stress relaxes from its initial state, determined by $\mu(0)$ and $K(0)$, for $t \to \infty$ to a final value, determined by the limits $\mu(\infty)$ and $K(\infty)$. In particular, $\mu(t)$ describes the temporal behaviour of the *shear relaxation* and $K(t)$ that of the *bulk relaxation*. Conversely, the material response to a stress step is a strain process, also known as *creep* and represented by the *creep functions* $m(t)$ and $k(t)$: the state of strain creeps from an initial value, which is proportional to $m(0)$ and $k(0)$, to a limit state determined by $m(\infty)$ and $k(\infty)$. The functions $m(t)$ and $k(t)$ describe the temporal behaviour of the *shear* and *bulk creep* respectively.

In this sense viscoelastic material behaviour is sometimes referred to as being *time-dependent*. This term is however somewhat misleading, since the constitutive equations (5.51) and (5.52) by no means depend on time in their mathematical structure: according to the constitutive equation (5.51) a time-dependent stress process is allotted to every time-dependent strain process in one and the same manner. However the current stress $\mathbf{T}(t)$ depends strongly on the velocity at which a given strain trajectory has been passed; it is consequently more appropriate to call the viscoelastic material behaviour *rate-dependent* rather than time-dependent.

Definition 5. 5

In the constitutive equation (5.51) the limit value of the relaxation function,

$$\mu_{\mathrm{eq}} = \lim_{t \to \infty} \mu(t) \,, \tag{5.61}$$

is called the *equilibrium shear modulus* and the limit value

$$K_{\mathrm{eq}} = \lim_{t \to \infty} K(t) \tag{5.62}$$

the *equilibrium bulk modulus.* The stress

$$\mathbf{T}_{eq} = 2\mu_{eq}\mathbf{E}^{D} + K_{eq}(\mathrm{tr}\mathbf{E})\mathbf{1} \tag{5.63}$$

is called the *state of equilibrium stress* in connection with the model of linear viscoelasticity. ◊

The constitutive equation (5.51) of linear viscoelasticity can be written in a different way, which is essential for the physical interpretation in the framework of the general theory of material behaviour:[12]

Theorem 5. 5
Assuming the initial condition $\mathbf{E}(-\infty) = \mathbf{0}$, (5.51) leads to the representation

$$\mathbf{T}(t) = 2\mu_{eq}\mathbf{E}^{D} + K_{eq}(\mathrm{tr}\mathbf{E})\mathbf{1} +$$

$$+ \int_0^\infty 2\gamma(s)\mathbf{E}_d^{tD}(s)\,\mathrm{d}s + \left\{\int_0^\infty \kappa(s)\,\mathrm{tr}\mathbf{E}_d^{t}(s)\,\mathrm{d}s\right\}\mathbf{1}\,, \tag{5.64}$$

with the *relative strain history*

$$\mathbf{E}_d^t(s) = \mathbf{E}(t - s) - \mathbf{E}(t)\,. \tag{5.65}$$

Therein the following notations are introduced:
$\gamma(s) = \mu'(s)$, $\mu_{eq} = \mu(\infty)$, $\kappa(s) = K'(s)$ and $K_{eq} = K(\infty)$. □

Proof
Integration by parts: the representation (5.51) leads to

$$\mathbf{T}(t) = 2\mu(t-\tau)\mathbf{E}^{D}(\tau)\Big|_{-\infty}^{t} - \int_{-\infty}^{t} 2\frac{\partial}{\partial\tau}\mu(t-\tau)\mathbf{E}^{D}(\tau)\mathrm{d}\tau$$

$$+ \left\{K(t-\tau)\mathrm{tr}\mathbf{E}(\tau)\Big|_{-\infty}^{t} - \int_{-\infty}^{t}\frac{\partial}{\partial\tau}K(t-\tau)\mathrm{tr}\mathbf{E}(\tau)\mathrm{d}\tau\right\}\mathbf{1}\,,$$

[12] Cf. Sect. 7. 2. 2 and Chap. 10.

$$\mathbf{T}(t) = 2\mu(0)\mathbf{E}^{\mathrm{D}}(t) + \int_{-\infty}^{t} 2\frac{\mathrm{d}}{\mathrm{d}(t-\tau)}\mu(t-\tau)\mathbf{E}^{\mathrm{D}}(\tau)\mathrm{d}\tau$$

$$+ \left\{K(0)\mathrm{tr}\mathbf{E}(t) + \int_{-\infty}^{t} \frac{\mathrm{d}}{\mathrm{d}(t-\tau)}K(t-\tau)\mathrm{tr}\mathbf{E}(\tau)\mathrm{d}\tau\right\}\mathbf{1}\,.$$

The substitution $t - \tau = s$ implies

$$\mathbf{T}(t) = 2\mu(0)\mathbf{E}^{\mathrm{D}}(t) + K(0)\big(\mathrm{tr}\mathbf{E}(t)\big)\mathbf{1}$$

$$+ \int_{0}^{\infty} 2\mu'(s)\mathbf{E}^{\mathrm{D}}(t-s)\mathrm{d}s + \left\{\int_{0}^{\infty} K'(s)\big(\mathrm{tr}\mathbf{E}(t-s)\big)\mathrm{d}s\right\}\mathbf{1}\,.$$

With the insertion of

$$\int_{0}^{\infty} \mu'(s)\mathrm{d}s = \mu(\infty) - \mu(0) \iff \mu(0)\mathbf{E}^{\mathrm{D}}(t) = \mu(\infty)\mathbf{E}^{\mathrm{D}}(t) - \mathbf{E}^{\mathrm{D}}(t)\int_{0}^{\infty} \mu'(s)\mathrm{d}s$$

and

$$\int_{0}^{\infty} K'(s)\mathrm{d}s = K(\infty) - K(0) \iff K(0)\mathrm{tr}\mathbf{E}(t) = K(\infty)\mathrm{tr}\mathbf{E}(t) - \mathrm{tr}\mathbf{E}(t)\int_{0}^{\infty} K'(s)\mathrm{d}s$$

the outcome is

$$\mathbf{T}(t) = 2\mu(\infty)\mathbf{E}^{\mathrm{D}}(t) + \int_{0}^{\infty} 2\mu'(s)\big[\mathbf{E}^{\mathrm{D}}(t-s) - \mathbf{E}^{\mathrm{D}}(t)\big]\mathrm{d}s$$

$$+ \left\{K(\infty)\mathbf{E}(t) + \int_{0}^{\infty} K'(s)\big[\mathbf{E}(t-s) - \mathbf{E}(t)\big]\mathrm{d}s\right\}\mathbf{1}\,. \qquad \blacksquare$$

The representation (5.64) can be interpreted as follows: the current stress consists of an equilibrium part and a memory part. The *equilibrium stress*

$$\mathbf{T}_{\mathrm{eq}} = 2\mu_{\mathrm{eq}}\mathbf{E}^{\mathrm{D}} + K_{\mathrm{eq}}(\mathrm{tr}\mathbf{E})\mathbf{1} \qquad (5.66)$$

depends exclusively on the present strain. The memory part, the so-called *overstress*,

$$\mathbf{T}_{\mathrm{ov}} = \int_0^\infty 2\gamma(s)\mathbf{E}_{\mathrm{d}}^{\mathrm{tD}}(s)\,\mathrm{d}s + \left\{ \int_0^\infty \kappa(s)\,\mathrm{tr}\mathbf{E}_{\mathrm{d}}^{\mathrm{t}}(s)\,\mathrm{d}s \right\}\mathbf{1}\,, \tag{5.67}$$

is a linear functional of the relative strain history $\mathbf{E}_{\mathrm{d}}^{\mathrm{t}}(s) = \mathbf{E}(t - s) - \mathbf{E}(t)$, which vanishes for temporally constant (*static*) strain processes.

The past history of the strain process is summed up by the integrals in (5.67), whereby the memory functions $\gamma(s)$ and $\kappa(s)$ are weighting functions, characterising material properties. The fact that the absolute values of the weighting functions are monotonically decreasing is a characteristic of viscoelastic behaviour: strain events are multiplied by smaller weight factors if they belong to the more distant past. In this way a *fading memory* of the material is expressed.

The general constitutive equation of linear viscoelasticity is invertible: the creep functions $m(t)$, $k(t)$ in equation (5.52) are uniquely defined by the relaxation functions $\mu(t)$ and $K(t)$ in (5.51), if the stress and strain processes vanish for $t \leq 0$. The concrete calculation of the creep functions based on the relaxation functions and vice versa can be accomplished using the method of *Laplace transformation.*[13]

The Laplace transformation of a function $t \longmapsto \mathrm{f}(t)$ with $\mathrm{f}(t) = 0$ for $t \leq 0$ is defined as follows:

$$\mathrm{f}(t) \longmapsto \hat{\mathrm{f}}(\omega) = \mathscr{L}\left[\mathrm{f}(t)\right] = \int_0^\infty \mathrm{f}(t)\mathrm{e}^{-\omega t}\mathrm{d}t\,. \tag{5.68}$$

Application to (5.51) or (5.52) together with the convolution theorem and the assumption $\mathbf{E}(\boldsymbol{x}, t) = \mathbf{0}$ or $\mathbf{T}(\boldsymbol{x}, t) = \mathbf{0}$ for $t \leq 0$ leads to

$$\hat{\mathbf{T}}(\omega) = 2\hat{\mu}(\omega)\,\omega\hat{\mathbf{E}}^{\mathrm{D}}(\omega) + \hat{K}(\omega)\,\omega\big[\mathrm{tr}\,\hat{\mathbf{E}}(\omega)\big]\mathbf{1} \tag{5.69}$$

or

$$\hat{\mathbf{E}}(\omega) = 2\hat{m}(\omega)\,\omega\hat{\mathbf{T}}^{\mathrm{D}}(\omega) + \hat{k}(\omega)\,\omega\big[\mathrm{tr}\,\hat{\mathbf{T}}(\omega)\big]\mathbf{1}\,. \tag{5.70}$$

From (5.69) and (5.70) we infer the reciprocity relations

[13] PIPKIN [1972], pp. 22; TSCHOEGL [1989], pp. 39, 560.

$$4\hat{\mu}(\omega)\,\hat{m}(\omega)\omega^2 = 1\,, \tag{5.71}$$

$$\hat{K}(\omega)\,\hat{k}(\omega)\omega^2 = 1\,. \tag{5.72}$$

As opposed to the defining equation (5.1) of the perfect fluid, the linear-viscous fluid (5.18) and the elastic solid (5.35), the general constitutive equation of linear viscoelasticity is a functional expressing a memory of the past process history. Accordingly, the insertion of the constitutive equation (5.51) into the local balance of linear momentum (3.128) no longer leads merely to a partial differential equation, but to an integro-differential equation of the form

$$\rho\,\frac{\partial^2 \boldsymbol{u}}{\partial t^2} = \operatorname{div}\left\{\int_{-\infty}^{t} 2\mu(t-\tau)\mathbf{E}^{\mathrm{D}\prime}(\tau)\mathrm{d}\tau + \left[\int_{-\infty}^{t} K(t-\tau)[\operatorname{tr}\mathbf{E}'(\tau)]\mathrm{d}\tau\right]\mathbf{1}\right\} + \rho\boldsymbol{f}\,, \tag{5.73}$$

which has to be supplemented by appropriate boundary and initial conditions. On principle, this functional equation can be transferred to a system of partial differential equations with the help of the *Laplace transformation*: application of (5.68) together with $\mathbf{E}(\boldsymbol{x},\,t) = \mathbf{0}$ for $t \leq 0$ and the initial conditions

$\boldsymbol{u}(\boldsymbol{x},\,0) = \boldsymbol{u}_0(\boldsymbol{x})$ and $\dfrac{\partial}{\partial t}\boldsymbol{u}(\boldsymbol{x},\,0) = \boldsymbol{v}_0(\boldsymbol{x})$ implies[14]

$$\rho\left[\omega^2\hat{\boldsymbol{u}}(\omega) - \omega\boldsymbol{v}_0 - \omega^2\boldsymbol{u}_0\right] = \operatorname{div}\left\{2\hat{\mu}(\omega)\,\omega\hat{\mathbf{E}}^{\mathrm{D}}(\omega) + \hat{K}(\omega)\,\omega\left[\operatorname{tr}\,\hat{\mathbf{E}}(\omega)\right]\mathbf{1}\right\} + \rho\boldsymbol{f}\,. \tag{5.74}$$

This differential equation is similar to a problem of linear elasticity; however, it depends on a complex-valued parameter and the solution of the boundary value problem has to be transformed back into the time domain.

In order to discuss the physical meaning of the constitutive equation (5.51) in more detail, we confine to the one-dimensional case and assume proportionality between the functions of shear and bulk relaxation:

$$K(t) = 2\mu(t)\,\frac{(1+\nu)}{3\,(1-2\nu)}\,. \tag{5.75}$$

The uniaxial tension process

$$\mathbf{T} = \sigma\boldsymbol{e}_x \otimes \boldsymbol{e}_x\,, \qquad \mathbf{E} = \varepsilon(\boldsymbol{e}_x \otimes \boldsymbol{e}_x) + \varepsilon_q(\boldsymbol{e}_y \otimes \boldsymbol{e}_y + \boldsymbol{e}_z \otimes \boldsymbol{e}_z)\,,$$

---

14 BRONSTEIN et al. [1997], p. 663.

then leads to $\varepsilon_q(\tau) = -\nu\varepsilon(\tau)$ for $\tau \leq t$ and to the one-dimensional stress-strain relation

$$\sigma(t) = \int_{-\infty}^{t} G(t-\tau)\varepsilon'(\tau)\mathrm{d}\tau\,, \tag{5.76}$$

with $G(s) = 2(1+\nu)\mu(s)$. If, in contrast to the proportionality assumption (5.75), an independent bulk relaxation modulus is assumed, the present lateral strain $\varepsilon_q(t)$ comes out as a functional of the strain history.
As a one-dimensional formulation of Theorem 5. 5 we have

Theorem 5. 6
For the initial condition $\varepsilon(-\infty) = 0$, (5.76) leads to

$$\sigma(t) = E\varepsilon(t) + \int_0^{\infty} g(s)\, \varepsilon_\mathrm{d}^\mathrm{t}(s)\, \mathrm{d}s\,, \tag{5.77}$$

with the one-dimensional *relative strain history*

$$\varepsilon_\mathrm{d}^\mathrm{t}(s) = \varepsilon(t-s) - \varepsilon(t)\,, \tag{5.78}$$

the memory function

$$g(s) = G'(s) \tag{5.79}$$

and the limit value of the relaxation function

$$E = G(\infty)\,. \tag{5.80}$$

□

Proof
Integration by parts supplies

$$\sigma(t) = G(t-\tau)\varepsilon(\tau)\Big|_{-\infty}^{t} - \int_{-\infty}^{t} \frac{\partial}{\partial\tau} G(t-\tau)\varepsilon(\tau)\mathrm{d}\tau$$

$$= G(0)\varepsilon(t) + \int_{-\infty}^{t} \frac{\mathrm{d}}{\mathrm{d}(t-\tau)} G(t-\tau)\varepsilon(\tau)\mathrm{d}\tau\,,$$

and the substitution $s = t - \tau$ leads to $\sigma(t) = G(0)\varepsilon(t) + \int\limits_0^\infty G'(s)\varepsilon(t-s)\mathrm{d}s.$

Inserting $G(0)\varepsilon(t) = G(\infty)\varepsilon(t) - \varepsilon(t)\int\limits_0^\infty G'(s)\mathrm{d}s$ the outcome is

$$\sigma(t) = G(\infty)\varepsilon(t) + \int\limits_0^\infty G'(s)[\varepsilon(t-s) - \varepsilon(t)]\mathrm{d}s\,. \qquad \blacksquare$$

The representation (5.77) can be interpreted in the same way as (5.64): the current stress consists of an equilibrium part and a memory part. The *equilibrium stress*

$$\sigma_{\mathrm{eq}}(t) = E\varepsilon(t) \tag{5.81}$$

depends exclusively on the present strain. The memory part, the so-called *overstress,*

$$\sigma_{\mathrm{ov}}(t) = \int\limits_0^\infty g(s)[\varepsilon(t-s) - \varepsilon(t)]\mathrm{d}s = \int\limits_0^\infty g(s)\varepsilon_{\mathrm{d}}^{\mathrm{t}}(s)\mathrm{d}s \tag{5.82}$$

is a linear functional of the relative strain history $\varepsilon_{\mathrm{d}}^{\mathrm{t}}(s) = \varepsilon(t-s) - \varepsilon(t)$, which vanishes for temporally constant (*static*) strain processes. The memory function $g(\cdot)$ in (5.82) is the derivative of the relaxation function in (5.76). These functions are material properties; they express a *fading memory* behaviour of the viscoelastic material.

In the defining equations (5.51) or (5.52) of a linear-viscoelastic isotropic solid or in the one-dimensional formulation (5.76) it is necessary to identify the material functions $\mu(\cdot)$ and $K(\cdot)$, $m(\cdot)$ and $k(\cdot)$ or $G(\cdot)$ in order to express the individual material characteristics quantitatively. This requires a more specific representation of those functions. A physically-based selection of concrete relaxation or creep functions can be motivated by recourse to *rheological models*: in the context of linear viscoelasticity rheological models are *networks* created out of (massless) linear-elastic *springs* and linear-viscous *dashpots.*[15] They serve as auxiliary tools when demonstrating the physical significance of differential equations deemed to represent the stress-strain behaviour of viscoelastic bodies.

[15] Cf. Christensen [1982], pp. 16.

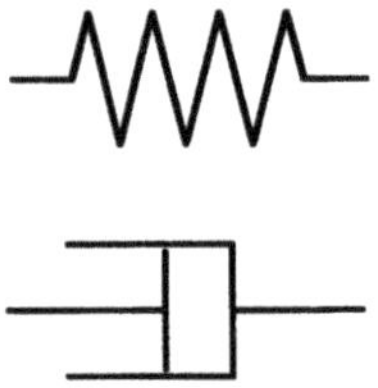

$$\iff \sigma = E\varepsilon$$

$$\iff \sigma = \eta\dot{\varepsilon}$$

Figure 5. 1: Rheological models of linear viscoelasticity: basic elements

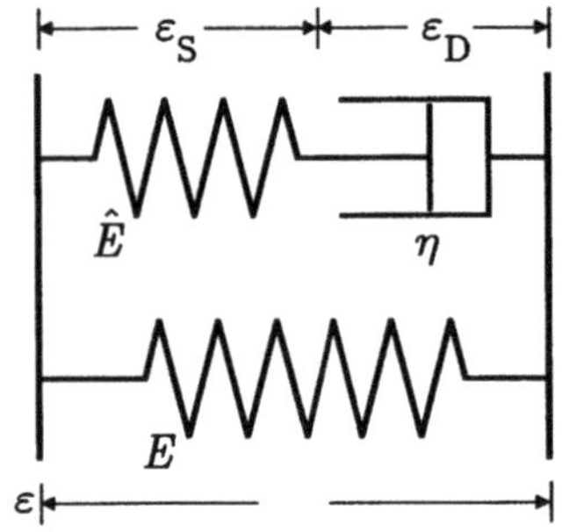

$$\iff \dot{\sigma} + \frac{\hat{E}}{\eta}\sigma = \frac{E\hat{E}}{\eta}\varepsilon + (E + \hat{E})\dot{\varepsilon}$$

Figure 5. 2: 3-parameter model of linear viscoelasticity (standard linear solid)

Every element contained in a network of this kind is a symbol denoting a mathematical relation (see Fig. 5. 1): thus the spring element stands for linear elasticity, i.e.

$$\sigma = E\varepsilon\ , \tag{5.83}$$

and the dashpot for linear viscosity, i.e.

$$\sigma = \eta\dot{\varepsilon}\ . \tag{5.84}$$

Translating a special rheological model into a one-dimensional stress-strain relation is achieved by means of elementary calculations based on the geometric *condition of compatibility* in the case of elements in series,

$$\sum \varepsilon_i = \varepsilon\ , \tag{5.85}$$

and the mechanical *condition of equilibrium* in the case of elements in parallel,

$$\sum \sigma_i = \sigma \,. \tag{5.86}$$

This procedure leads to differential equations between processes of stress $\sigma(t)$ and processes of strain $\varepsilon(t)$.

Transferring a rheological model into a differential equation ought to be explained briefly for the special case of a 3-parameter model.
The model of the 3-parameter or *standard linear solid* contains a spring and a dashpot arranged in a series with a parallel spring element. It describes the material characteristics by means of 3 parameters, namely 2 elasticity moduli E, $\hat{E}$ and one viscosity constant η (see Fig. 5. 2). In this case we write $\sigma = \sigma_1 + \sigma_2$ and $\varepsilon = \varepsilon_D + \varepsilon_S$, with $\sigma_1 = E\varepsilon$ and $\sigma_2 = \hat{E}\varepsilon_S = \eta\dot{\varepsilon}_D$. This leads to $\dot{\varepsilon} = \dot{\varepsilon}_D + \dot{\varepsilon}_S$ or

$$\dot{\varepsilon} = \frac{\sigma_2}{\eta} + \frac{\dot{\sigma}_2}{\hat{E}} \,. \tag{5.87}$$

On the other hand, the following equation is also valid:

$$\frac{E}{\eta}\varepsilon = \frac{\sigma_1}{\eta} \quad \Longleftrightarrow \quad \frac{E}{\hat{E}}\dot{\varepsilon} = \frac{\dot{\sigma}_1}{\hat{E}} \,. \tag{5.88}$$

Adding these equations together while bearing the equilibrium condition $\sigma_1 + \sigma_2 = \sigma$ in mind yields the differential equation

$$\Big(1 + \frac{E}{\hat{E}}\Big)\dot{\varepsilon} + \frac{E}{\eta}\varepsilon = \frac{\sigma}{\eta} + \frac{\dot{\sigma}}{\hat{E}} \;, \text{ i.e.}$$

$$\dot{\sigma} + \frac{\hat{E}}{\eta}\sigma = \frac{E\hat{E}}{\eta}\varepsilon + (E + \hat{E})\dot{\varepsilon} \;. \tag{5.89}$$

Alternatively, the same procedure may be carried out without eliminating the displacement in the dashpot from the constitutive description. By inserting $\varepsilon_D = q$ and $\varepsilon_S = \varepsilon - q$ into $\sigma = E\varepsilon + \hat{E}\varepsilon_S$ and $\eta\dot{\varepsilon}_D = \hat{E}\varepsilon_S$ one obtains two constitutive equations:

$$\sigma \;\; = E\varepsilon + \hat{E}(\varepsilon - q) \;, \tag{5.90}$$

$$\dot{q}(t) = \frac{\hat{E}}{\eta}\big(\varepsilon - q(t)\big) \,. \tag{5.91}$$

The first equation is a stress-strain relation depending on a so-called *internal variable* q; the second equation is a so-called *evolution equation* for

the temporal behaviour of $q(t)$. In this case the internal variable q corresponds to the displacement in the dashpot.
It is possible to integrate the differential equation (5.89) or, equivalently, to integrate (5.91) and insert the solution into (5.90). For the initial condition $\sigma(0) = 0$ the result

$$\sigma(t) = \int_0^t \left[E + \hat{E}\,\mathrm{e}^{-\frac{\hat{E}}{\eta}(t - \tau)}\right]\varepsilon'(\tau)\mathrm{d}\tau \tag{5.92}$$

determines the current stress $\sigma(t)$ in dependence on the past strain history. This equation is obviously a special case of linear viscoelasticity according to Definition 5. 4. The special relaxation function

$$G(t) = E + \hat{E}\,\mathrm{e}^{-\frac{\hat{E}}{\eta}t} \tag{5.93}$$

thus corresponds to the 3-parameter model. The relaxation function describes the temporal evolution of the stress response to a unit step in the strain, according to Theorem 5. 4. The limit value of the relaxation function,

$$E = G(\infty) = \lim_{t \to \infty} G(t) \tag{5.94}$$

determines the stress which is asymptotically reached after a sufficiently long relaxation process. This stress corresponds to a state of equilibrium in the rheological model and is accordingly termed *equilibrium stress*:

$$\sigma_{\mathrm{eq}} = E\varepsilon\,, \quad E = G(\infty)\,. \tag{5.95}$$

In the one-dimensional model (5.76) the linear stress-strain relation $\sigma_{\mathrm{eq}} = G(\infty)\varepsilon$ represents the set of all equilibrium states and is therefore called an *equilibrium relation* (see Chap. 6).

Monotonic strain processes with constant rate $\dot{\varepsilon}_0$ lead to a deviation from the equilibrium curve depending on $\dot{\varepsilon}_0$: a strain process

$$\varepsilon(t) = \begin{cases} 0 & \text{for } t < 0 \\ \dot{\varepsilon}_0 t & \text{for } t \geq 0 \end{cases} \tag{5.96}$$

leads to the stress response

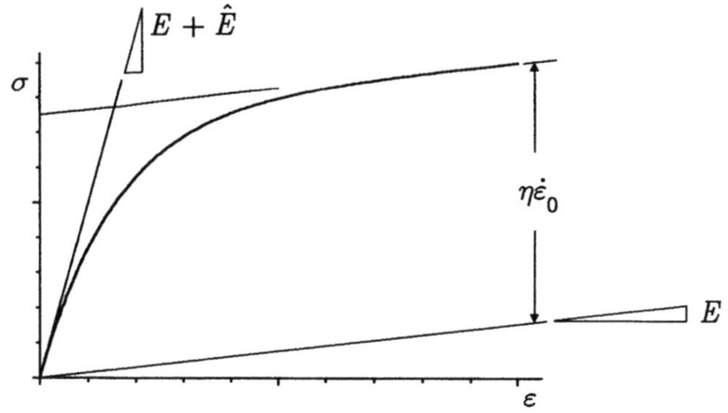

Figure 5. 3: Standard linear solid: stress response to a monotonic strain process with constant rate according to (5.99)

$$\sigma(t) = \dot{\varepsilon}_0 \int_0^t G(t-\tau)\mathrm{d}\tau = \dot{\varepsilon}_0 \int_0^t G(s)\mathrm{d}s \tag{5.97}$$

and the stress-strain curve

$$\sigma(\varepsilon) = \dot{\varepsilon}_0 \int_0^{\varepsilon/\dot{\varepsilon}_0} G(s)\mathrm{d}s \, . \tag{5.98}$$

For the standard solid we obtain the special form of

$$\sigma(\varepsilon) = E\varepsilon + \eta\dot{\varepsilon}_0\left[1 - \mathrm{e}^{-\frac{\hat{E}}{\eta\dot{\varepsilon}_0}\varepsilon}\right] . \tag{5.99}$$

The rate-dependence of the stress response is linear in $\dot{\varepsilon}_0$, as far as the asymptotic behaviour is concerned: for sufficiently high values of $\hat{E}\varepsilon/(\eta\dot{\varepsilon}_0)$ the deviation of the "dynamic" stress-strain characteristic from the equilibrium curve is proportional to $\eta\dot{\varepsilon}_0$.[16]

[16] If the viscosity η is assumed to be a function of stress, $\eta = \eta(\sigma)$, then a nonlinear rate-dependence is represented in the sense that the asymptotic value of the stress in monotonic tension depends nonlinearly on $\dot{\varepsilon}_0$ (see Sects. 10. 2. 3 and 12. 4. 3).

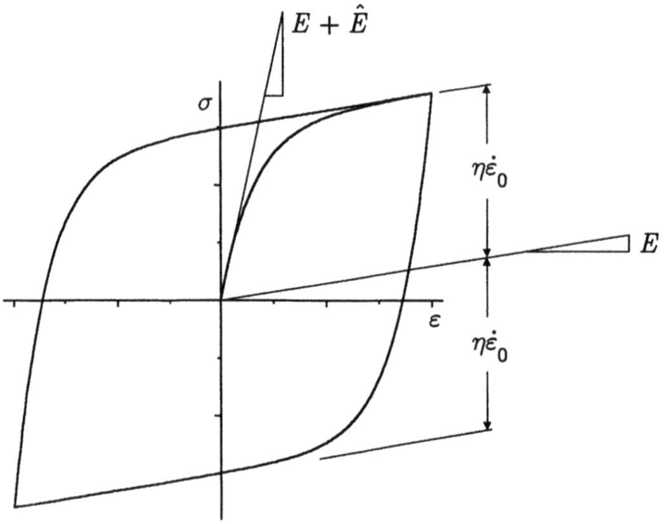

Figure 5. 4: Standard linear solid: stress response to a cyclic process with constant absolute value of the strain rate; integration of the differential equation (5.89)

In the case of sufficiently low strain rates $\dot{\varepsilon}_0$ the stress-strain curve approaches arbitrarily close to the equilibrium curve. However, its initial slope for every finite velocity $\dot{\varepsilon}_0 \neq 0$ is equal to $E + \hat{E}$ or generally speaking to $G(0)$: (5.98) and (5.99) lead to

$$\sigma'(0) = G(0) \tag{5.100}$$

and

$$\sigma'(0) = E + \hat{E}\,. \tag{5.101}$$

Figure 5. 3 illustrates this result. The initial value $G(0) = E + \hat{E}$ of the relaxation function is the so-called *spontaneous elasticity modulus.* This modulus determines the slope of the stress-strain curve in the case of velocity jumps or processes occurring at very large velocities: this is demonstrated by the differential equation (5.89), in which, for high values of $\dot{\varepsilon}$, the parts $\dot{\sigma} = (E + \hat{E})\dot{\varepsilon}$ are dominant, resulting in

$$\sigma = (E + \hat{E})\varepsilon\,. \tag{5.102}$$

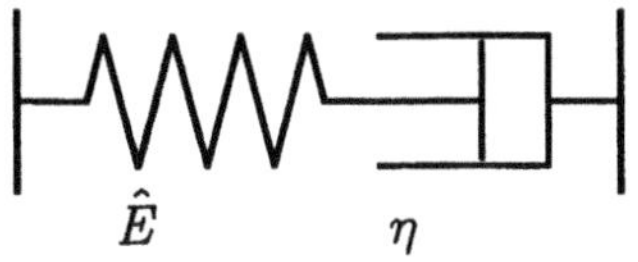

$$\Longleftrightarrow \quad \sigma + \frac{\eta}{\hat{E}}\,\dot{\sigma} = \eta\dot{\varepsilon}$$

Figure 5. 5: Maxwell model

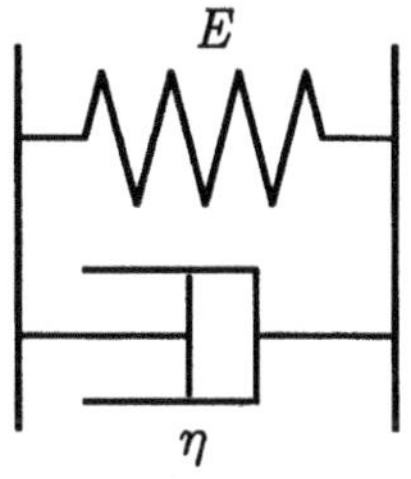

$$\Longleftrightarrow \quad \sigma = E\varepsilon + \eta\dot{\varepsilon}$$

Figure 5. 6: Kelvin model

In passing to the limit, (5.98) implies

$$\sigma(\varepsilon) \rightarrow \lim_{\dot{\varepsilon}_0 \rightarrow \infty} \dot{\varepsilon}_0 \int_0^{\varepsilon/\dot{\varepsilon}_0} G(s)\mathrm{d}s = G(0)\varepsilon\,. \tag{5.103}$$

The stress response to a cyclic strain process with constant absolute value of the strain rate is depicted in Fig. 5. 4. It shows a limit cycle that establishes itself very shortly after the first period of the cyclic strain process if the duration of the period is large in comparison to the *relaxation time* $\eta/\hat{E}$. The outcome is a hysteresis loop whose area is determined by the strain rate, and which contracts to the equilibrium curve if $\dot{\varepsilon}_0$ goes to zero.

The standard linear solid contains two more special rheological models that ought to be given special mention: firstly, the so-called *Maxwell* model is obtained for $E \rightarrow 0$ (see Fig. 5. 5):

$$\dot{\sigma} + \frac{\hat{E}}{\eta}\sigma = \hat{E}\,\dot{\varepsilon}\ . \tag{5.104}$$

The typical feature of the *Maxwell* model is the fact that the equilibrium stress is identically zero: for a constant strain, $\varepsilon(t) = \varepsilon_0$ ($\dot{\varepsilon} = 0$) following an arbitrary strain process, (5.104) leads to

$$\lim_{t\to\infty} \sigma(t) = 0 \,. \tag{5.105}$$

Conversely, constant stress $\sigma = \sigma_0$ ($\dot{\sigma} = 0$) causes a constant strain rate, i.e. a process of unlimited creep:

$$\varepsilon(t) = (\sigma_0/\hat{E}) + (\sigma_0/\eta)\, t \,. \tag{5.106}$$

The second special case contained in the 3-parameter model and arising for the limit case $\hat{E} \to \infty$ is the so-called *Kelvin* model (see Fig. 5. 6):

$$\sigma = E\varepsilon + \eta\dot{\varepsilon} \,. \tag{5.107}$$

In the case of this model the total stress σ is the sum of an equilibrium stress ($E\varepsilon$) and a viscous stress ($\eta\dot{\varepsilon}$), which is proportional to the strain rate.

In order to generalise the 3-parameter model, something that is always unavoidable when processes of relaxation and creep are to be described in greater detail, one may begin with more complicated networks of spring and dashpot elements. One simple, systematic way to proceed is to create a (2N + 1)-parameter model, conceivably comprising a single spring element and N *Maxwell* elements in parallel arrangement (Fig. 5. 7):

$$\sigma = E\varepsilon + \sum_{k=1}^{N} \sigma_k \,, \tag{5.108}$$

$$\dot{\sigma}_k(t) + \frac{\hat{E}_k}{\eta_k}\, \sigma_k(t) = \hat{E}_k\, \dot{\varepsilon}(t) \,, \qquad k = 1, ..., N \,. \tag{5.109}$$

Theorem 5. 7

A representation of the (2N + 1)-parameter model equivalent to (5.108), (5.109) is provided by

$$\sigma = E\varepsilon + \sum_{k=1}^{N} \hat{E}_k(\varepsilon - q_k) \,, \tag{5.110}$$

$$\dot{q}_k(t) = \frac{\hat{E}_k}{\eta_k}(\varepsilon(t) - q_k(t)) \,, \qquad k = 1, ..., N \,. \tag{5.111}$$

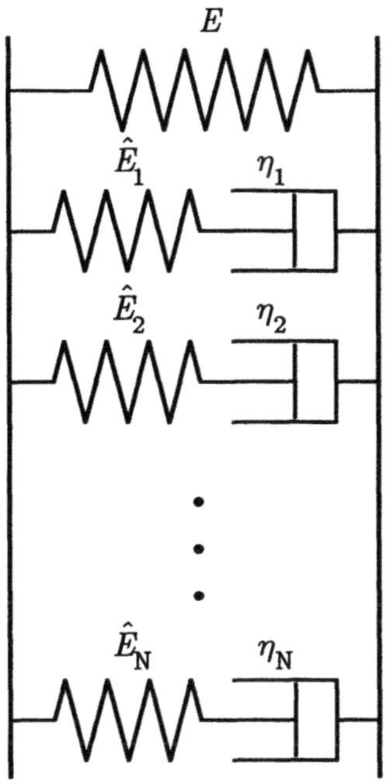

$$\Longleftrightarrow \quad \sigma = E\varepsilon + \sum_{k=1}^{N} \sigma_k$$

Figure 5. 7: (2N + 1)-parameter model

The relaxation function of the (2N + 1)-parameter model reads

$$G(t) = E + \sum_{k=1}^{N} \hat{E}_k \, e^{-\frac{\hat{E}_k}{\eta_k} t} \, . \tag{5.112}$$

□

Proof

The equivalence of equations (5.108), (5.109) and (5.110), (5.111) becomes evident when setting $\sigma_k = \hat{E}_k(\varepsilon - q_k)$. The *internal variable* q_k contained in it is the displacement (strain) in the kth dashpot. If ε_{S_k} is the elongation of the kth spring, an additive decomposition of the total strain is obtained for every *Maxwell* element,

$$\varepsilon = \varepsilon_{\text{S}_k} + q_k = \frac{\sigma_k}{\hat{E}_k} + q_k \qquad (k = 1, ..., \text{N}) \, .$$

Thus the evolution equations (5.111) for the internal variables q_k are transferred to the differential equations (5.109) for the stresses σ_k:

$$\dot{\varepsilon} - \frac{\dot{\sigma}_k}{\hat{E}_k} = \frac{\sigma_k}{\eta_k} .$$

The differential equations (5.109) can be expressed in the form of

$$\frac{\mathrm{d}}{\mathrm{d}t}\left(\mathrm{e}^{\frac{\hat{E}_k}{\eta_k}t}\,\sigma_k(t)\right) = \hat{E}_k\,\mathrm{e}^{\frac{\hat{E}_k}{\eta_k}t}\,\dot{\varepsilon}(t)\,. \tag{5.113}$$

Integration together with the initial condition $\sigma_k(0) = 0$ produces

$$\sigma_k(t) = \int_0^t \hat{E}_k \mathrm{e}^{-\frac{\hat{E}_k}{\eta_k}(t-\tau)}\,\varepsilon'(\tau)\mathrm{d}\tau\,. \tag{5.114}$$

Insertion into (5.108) and consideration of the condition $\varepsilon(0) = 0$ leads to the representation (5.76) with $G(t)$ according to equation (5.112). ∎

Definition 5. 6

In the representation (5.112) of the relaxation function,

$$G(t) = E + \sum_{k=1}^{\mathrm{N}} \hat{E}_k\,\mathrm{e}^{-\frac{t}{\mathrm{z}_k}}\,,$$

the coefficients $\hat{E}_k$ are called *relaxation coefficients* and the quotients

$$\mathrm{z}_k = \frac{\eta_k}{\hat{E}_k} \tag{5.115}$$

relaxation times. The set $\{\hat{E}_k\,,\,\mathrm{z}_k\}_{k\,=\,1,\ldots\mathrm{N}}$ of relaxation coefficients and times forms the *relaxation spectrum.*[17] ◇

[17] See CHRISTENSEN [1982], pp. 28, FINDLEY et al. 1976], pp. 64.

The material properties of a linear-viscoelastic solid are quantitatively determined by experimental identification of the relaxation spectrum.[18]

The order N of a relaxation spectrum (or creep spectrum) does not necessarily have to be a finite number; even a *continuous relaxation spectrum* in the sense of a representation

$$G(t) = E + \int_0^\infty \hat{E}(z)\, e^{-\frac{t}{z}}\, dz \tag{5.116}$$

is of physical interest. In (5.116) the relaxation time z as the variable of integration may be replaced by the relaxation frequency $\nu = \frac{1}{z}$ to give

$$G(t) = E + \int_0^\infty h(\nu)\, e^{-\nu t}\, d\nu\,, \tag{5.117}$$

with the new distribution function

$$h(\nu) = \frac{1}{\nu^2}\, \hat{E}\Big(\frac{1}{\nu}\Big)\,. \tag{5.118}$$

The representation (5.117) motivates the definition of the *cumulative spectrum*,[19]

$$H(\nu) = \int_0^\nu h(x) dx\,. \tag{5.119}$$

From (5.117) and (5.119) we obtain

$$G(t) = E + H(\nu) e^{-\nu t}\Big|_0^\infty + \int_0^\infty t H(\nu)\, e^{-\nu t}\, d\nu\,,$$

[18] An equivalent formulation of a (2N + 1)-parameter model can be derived considering a simple spring element and N *Kelvin* elements in series. In this case we have the equations

$$\varepsilon = \frac{\sigma}{E} + \sum_{k=1}^{N} \varepsilon_k \quad \text{and} \quad z_k \dot{\varepsilon}_k + \varepsilon_k = \frac{\sigma}{\hat{E}_k}\,,$$

determining the present strain as a functional of the stress history. The set of *creep coefficients (compliances)* $1/\hat{E}_k$ and *creep times* z_k then forms the *creep spectrum*.

[19] Backhaus [1997].

which leads to an alternative representation of the relaxation function, namely

$$G(t) = E + \int_0^\infty H\left(\frac{s}{t}\right) e^{-s} ds . \tag{5.120}$$

In cases of practical application, where relaxation functions have to be evaluated numerically, the representation (5.120) may be more convenient than (5.117), because the cumulative spectrum $H(\cdot)$ is a monotonically increasing function, tending asymptotically to a constant value for large values of s.

Experimental investigations into the relaxation behaviour of metals and polymers indicate that during the initial seconds of a hold time a very rapid decrease in stress is registered; thereafter, it can be observed that the stress merely decreases at a very slow pace. It can take a long time (maybe years), before a constant asymptotic value is reached. On principle, relaxation behaviour of this kind cannot be represented by means of a single exponential function, but representation by means of power functions of the type $t^{-\alpha}$ is certainly possible, $\alpha > 0$ being a real number. One example is the relaxation function[20]

$$G(t) = E + \frac{\hat{E}}{\left(1 + \frac{t}{\alpha T_0}\right)^\alpha} . \tag{5.121}$$

The shape of this function compared with the exponential function

$$G(t) = E + \hat{E} e^{-\frac{t}{T_0}} \tag{5.122}$$

is shown in Fig. 5. 8. The contrast in the decay of these functions is illustrated particularly clearly in the logarithmic scale.

A representation of the power function (5.121) can be produced by means of superimposing exponential functions; however, a continuous relaxation spectrum is required to do so. This is confirmed by

[20] See TOBOLSKY [1967], p. 219.

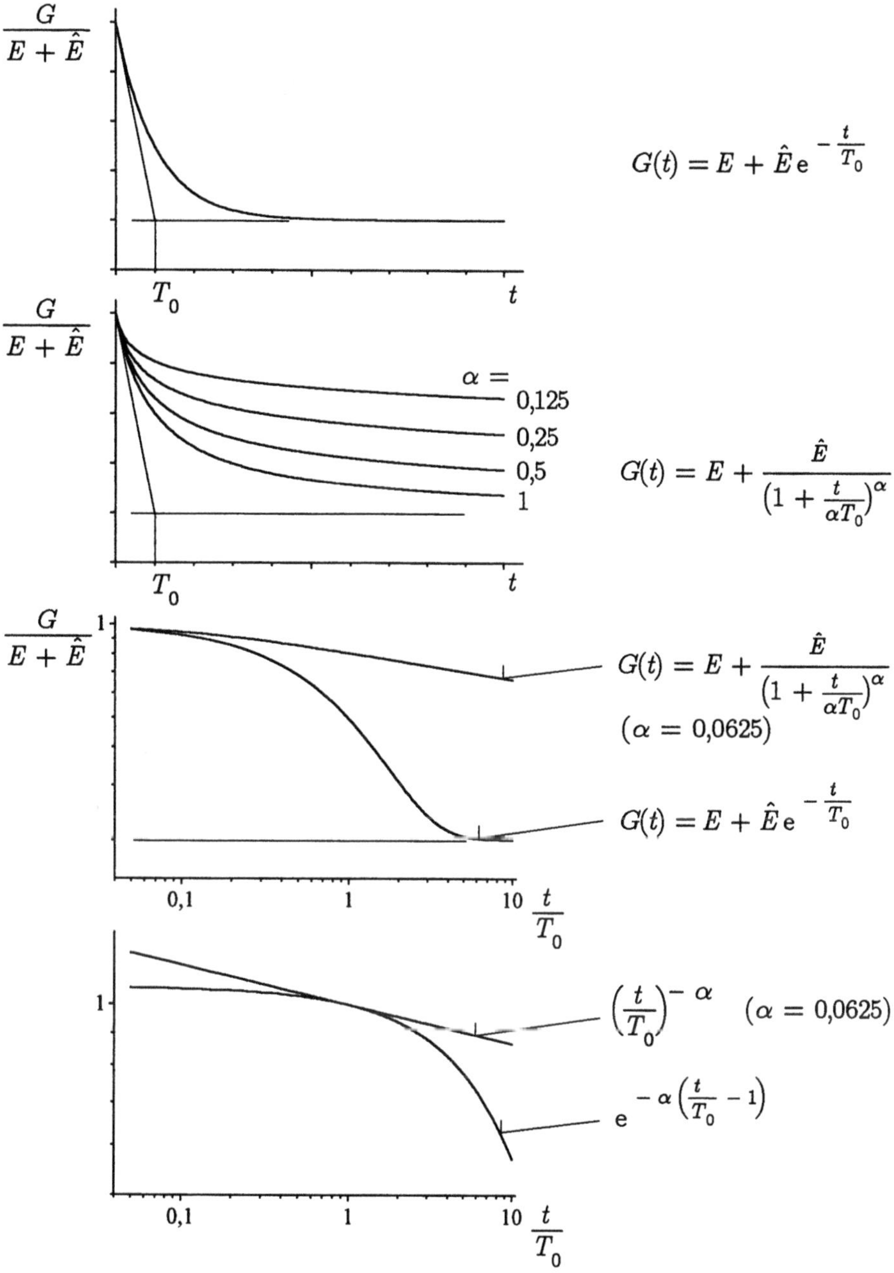

Figure 5. 8: Comparison of different relaxation functions in linear and log-log scale

Theorem 5. 8
For the relaxation function (5.121) the following representation is valid,

$$G(t) = E + \frac{\hat{E}}{\left(1 + \frac{t}{\alpha T_0}\right)^{\alpha}} = E + \hat{E}\,\frac{(\alpha T_0)^{\alpha}}{\Gamma(\alpha)} \int_0^{\infty} \frac{e^{-\frac{\alpha T_0}{z}}}{z^{\alpha+1}}\, e^{-\frac{t}{z}}\, dz\,. \tag{5.123}$$

Here, $\alpha > 0$ is a dimensionless material parameter and $\Gamma(\cdot)$ the *Eulerian* Gamma function, i.e.

$$\Gamma(x) = \int_0^{\infty} \nu^{x-1} e^{-y} dy\,. \tag{5.124}$$

□

Proof
Verification of the representation (5.123) is based on the identity

$$\frac{1}{s^{\alpha}} = \frac{1}{\Gamma(\alpha)} \int_0^{\infty} \frac{e^{-s\nu}}{\nu^{1-\alpha}}\, d\nu\,, \tag{5.125}$$

valid for all $s > 0$ and $\alpha > 0$.[21] By substituting $\nu = \frac{1}{z}$ this is modified to

$$\frac{1}{s^{\alpha}} = \frac{1}{\Gamma(\alpha)} \int_0^{\infty} \frac{e^{-\frac{s}{z}}}{z^{1+\alpha}}\, dz\,. \tag{5.126}$$

By also substituting $s = \alpha T_0 + t$ we arrive at (5.123). ■

If the relaxation time z is replaced by the relaxation frequency $\nu = \frac{1}{z}$, we have the alternative representation

$$\frac{\hat{E}}{\left(1 + \frac{t}{\alpha T_0}\right)^{\alpha}} = \hat{E}\,\frac{(\alpha T_0)^{\alpha}}{\Gamma(\alpha)} \int_0^{\infty} \frac{e^{-\alpha T_0 \nu}}{\nu^{1-\alpha}}\, e^{-\nu t}\, d\nu \tag{5.127}$$

or

$$\frac{\hat{E}}{\left(1 + \frac{t}{\alpha T_0}\right)^{\alpha}} = \int_0^{\infty} H\!\left(\frac{s}{t}\right) e^{-s}\, ds\,. \tag{5.128}$$

---

[21] BRONSTEIN et al. [1997], p. 992.

In this case the cumulative spectrum

$$H(\nu) = \hat{E}\,\frac{(\alpha T_0)^\alpha}{\Gamma(\alpha)}\int\limits_0^{\nu}\frac{\mathrm{e}^{-\alpha T_0 x}}{x^{1-\alpha}}\,\mathrm{d}x = \frac{\hat{E}}{\Gamma(\alpha)}\int\limits_0^{\alpha T_0\nu} y^{\alpha-1}\mathrm{e}^{-y}\mathrm{d}y \tag{5.129}$$

cannot be calculated analytically. However, its limit value is given as

$$H(\infty) = 1\,, \tag{5.130}$$

because of $\lim\limits_{\nu\to\infty}\int\limits_0^{\alpha T_0\nu} y^{\alpha-1}\mathrm{e}^{-y}\mathrm{d}y = \Gamma(\alpha)$.

The physical meaning of the representation (5.123) or (5.128) is the fact that a decreasing power function of t can be represented by a superposition of exponential functions. If one proceeds to replace the integral in (5.123) or (5.128) by means of an approximating sum, representation formulae can be derived, which are approximately valid for processes which are restricted to just a limited time interval: approximations of the relaxation functions (5.123), (5.128) and (5.116) lead to discrete relaxation spectra in the form of the representation (5.112).[22] In this context the continuous relaxation spectrum, depicted by the formulae (5.116), (5.117) or (5.120), corresponds to the most general formulation of one-dimensional linear viscoelasticity.

The representation of relaxation functions in the context of continuous relaxation spectra can be further promoted, if we follow a concept of generalised rheological models, which applies the idea of the derivative of fractional order. This approach has been introduced into the theory of viscoelasticity by Caputo and Mainardi;[23] the mathematical theory can be found in the textbook of Oldham and Spanier.[24] The following special definition suffices for the present situation.

Definition 5. 7

A derivative of fractional order α $(0 < \alpha < 1)$, applied to a function $t \longmapsto \mathrm{f}(t)$ with $\mathrm{f}(t) = 0$ for $t \leq 0$, is defined as

[22] See Haupt, Lion & Backhaus [2000].

[23] Caputo & Mainardi [1971].

[24] Oldham & Spanier [1974]. For further discussion of fractional derivatives and applications in viscoelasticity the reader is referred to Lion [1997c]; Lion [2000], pp.45; Drozdov [1997] or Haupt, Lion & Backhaus [2000].

$$\mathrm{f}^{(\alpha)}(t) = \frac{\mathrm{d}^\alpha}{\mathrm{d}t^\alpha}\,\mathrm{f}(t) = \frac{1}{\Gamma(1-\alpha)} \int\limits_0^t \frac{1}{(t-\tau)^\alpha}\,\mathrm{f}'(\tau)\mathrm{d}\tau\,. \tag{5.131}$$

◊

It can be shown that the derivative of fractional order $0 < \alpha < 1$ interpolates between the function itself and its first derivative: we have

$\mathrm{f}^{(0)}(t) = \mathrm{f}(t)$ and $\mathrm{f}^{(1)}(t) = \frac{\mathrm{d}}{\mathrm{d}t}\mathrm{f}(t)$.[25]

Thus, a constitutive model

$$\sigma(t) = \hat{E}\,{\mathrm{z}_0}^\alpha \frac{\mathrm{d}^\alpha}{\mathrm{d}t^\alpha}\varepsilon(t) = \frac{\hat{E}\,{\mathrm{z}_0}^\alpha}{\Gamma(1-\alpha)} \int\limits_0^t \frac{1}{(t-\tau)^\alpha}\,\varepsilon'(\tau)\mathrm{d}\tau \tag{5.132}$$

represents the idea of a fractional viscosity, interpolating between *Hooke's law* ($\alpha = 0$), $\sigma(t) = \hat{E}\varepsilon(t)$, and the classical linear viscosity ($\alpha = 1$), $\sigma(t) = \hat{E}\,\mathrm{z}_0\dot{\varepsilon}(t) = \eta\dot{\varepsilon}(t)$.

As a generalisation of the classical *Maxwell* model (5.104), a *Maxwell* model of fractional order can be defined through the relation

$$\frac{\mathrm{d}^\alpha}{\mathrm{d}t^\alpha}\,\sigma(t) + \frac{1}{{\mathrm{z}_0}^\alpha}\,\sigma(t) = \hat{E}\frac{\mathrm{d}^\alpha}{\mathrm{d}t^\alpha}\,\varepsilon(t)\,, \tag{5.133}$$

which is an integral equation of *Volterra* type according to the definition (5.131). The consequence of the generalised *Maxwell* model (5.133) is

Theorem 5. 9

The solution of eq. (5.133) reads

$$\sigma(t) = \hat{E} \int\limits_0^t \mathrm{E}_\alpha\!\left[-\left(\frac{(t-\tau)}{\mathrm{z}_0}\right)^{\alpha}\right]\varepsilon'(\tau)\mathrm{d}\tau\,, \tag{5.134}$$

where the relaxation function is related to the so-called *Mittag-Leffler* function[26]

$$x \longmapsto \mathrm{E}_\alpha(x) = \sum_{k=0}^{\infty} \frac{x^k}{\Gamma(1+\alpha k)}\,. \tag{5.135}$$

□

[25] Cf. HAUPT & LION [2001], Sect. 6.

[26] Cf. ERDELYI et al. [1954], pp.207.

Proof

Application of the *Laplace transformation* (5.68) to eq. (5.133) together with (5.131) and the formula

$$\mathscr{L}\left[\left(\frac{t}{z_0}\right)^{-\alpha}\right] = z_0\Gamma(1-\alpha)(z_0\omega)^{\alpha-1} \tag{5.136}$$

leads to $\mathscr{L}\left[z_0{}^{\alpha}\frac{d^{\alpha}}{dt^{\alpha}}\sigma(t)\right] = (z_0\omega)^{\alpha}\hat{\sigma}(\omega)$ and

$$\left[(z_0\omega)^{\alpha}+1\right]\hat{\sigma}(\omega) = \hat{E}(z_0\omega)^{\alpha}\,\hat{\varepsilon}(\omega)\,. \tag{5.137}$$

Accordingly, the solution in the ω-domain is

$$\hat{\sigma}(\omega) = \hat{E}\frac{(z_0\omega)^{\alpha}}{(z_0\omega)^{\alpha}+1}\hat{\varepsilon}(\omega)\,,$$

which can be written as

$$\hat{\sigma}(\omega) = \hat{E}\left\{\sum_{k=0}^{\infty}\frac{(-1)^k}{z_0{}^{\alpha k}\omega^{1+k\alpha}}\right\}\omega\,\hat{\varepsilon}(\omega)\,. \tag{5.138}$$

Transformation back to the t-domain yields according to the convolution theorem a stress functional of the form (5.76),

$$\sigma(t) = \int_0^t G(t-\tau)\varepsilon'(\tau)d\tau\,, \tag{5.139}$$

where the relaxation function $G(\cdot)$ has to be calculated from

$$G(t) = \hat{E}z_0\mathscr{L}^{-1}\left[\sum_{k=0}^{\infty}\frac{(-1)^k}{(z_0\omega)^{1+k\alpha}}\right]. \tag{5.140}$$

In view of (5.136) one term of the series transforms according to

$$z_0\mathscr{L}^{-1}\left[\frac{(-1)^k}{(z_0\omega)^{1+k\alpha}}\right] = \frac{(-1)^k}{\Gamma(1+\alpha k)}\left(\frac{t}{z_0}\right)^{k\alpha},$$

which leads to

$$G(t) = \hat{E}\sum_{k=0}^{\infty}\frac{(-1)^k}{\Gamma(1+\alpha k)}\left(\frac{t}{z_0}\right)^{k\alpha} = E_{\alpha}\left[-\left(\frac{t}{z_0}\right)^{\alpha}\right]. \tag{5.141}$$

■

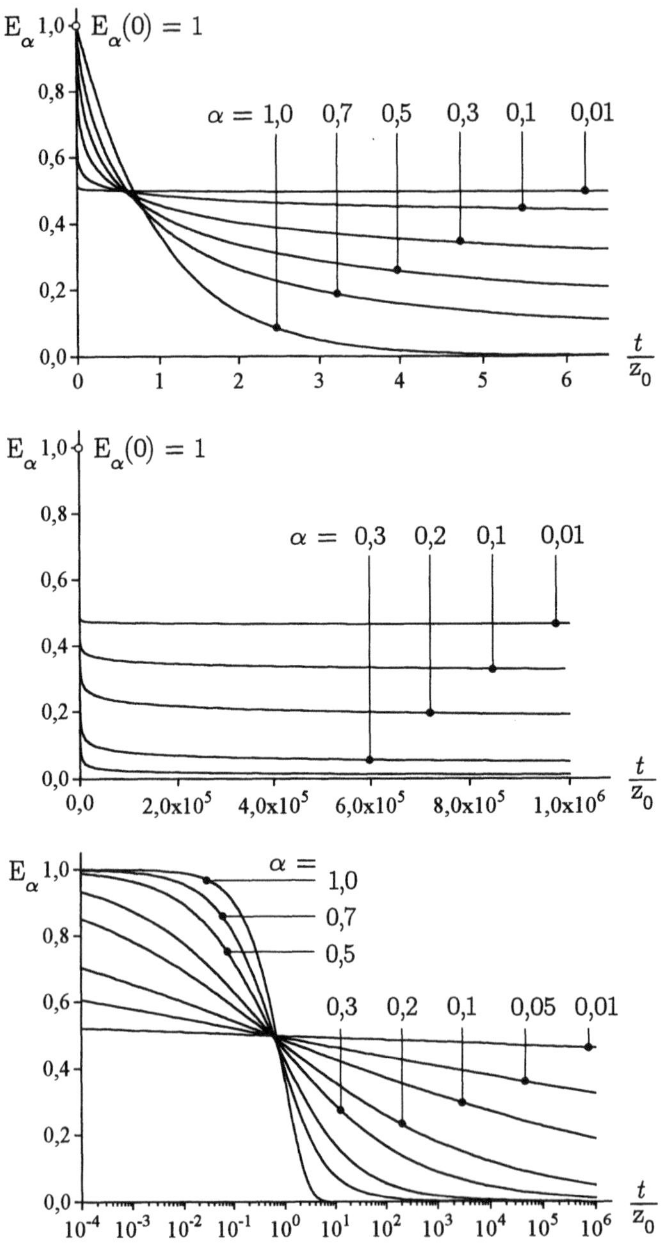

Figure 5. 9: Relaxation function (Mittag-Leffler function) defined by eq. (5.141), evaluated according to eq. (5.171). The limit value $\alpha = 1$ corresponds to the exponential function

Obviously, the relaxation function according to (5.141) coincides with the classical exponential function in the special case $\alpha = 1$. For $0 < \alpha < 1$ the function $G(t)$ has a finite value and a vertical tangent at $t = 0$. For large values of t the decrease of $G(t)$ becomes extremely slow if α is small. The shape of $G(t)$ is illustrated in Figure 5. 9 for different values of α.[27]

In addition to relaxation and creep, the theory of linear viscoelasticity is able to depict essential features of the dynamic material behaviour: the response of a material specimen to a harmonic strain process is governed by a stiffness and dissipation modulus, where both moduli depend on the frequency of the periodic strain due to the rate-dependence, which characterises viscoelasticity.[28] This is the statement of

Theorem 5. 10

The harmonic strain process

$$\varepsilon(t) = \hat{\varepsilon}\sin\Omega t\,, \tag{5.142}$$

characterised by a strain amplitude $\hat{\varepsilon}$ and a frequency Ω, assumed as an input to the uniaxial model (5.76) of viscoelasticty,

$$\sigma(t) = \int_{-\infty}^{t} G(t-\tau)\varepsilon'(\tau)\mathrm{d}\tau = G(\infty)\varepsilon(t) + \int_{-\infty}^{t}\big[G(t-\tau) - G(\infty)\big]\varepsilon'(\tau)\mathrm{d}\tau\,, \tag{5.143}$$

$\left(\varepsilon(-\infty) = 0\right)$ implies the stress response

$$\sigma(t) = \hat{\varepsilon}\big\{G_1(\Omega)\sin\Omega t + G_2(\Omega)\cos\Omega t\big\}\,. \tag{5.144}$$

Therein the frequency-dependent function

$$G_1(\Omega) = G(\infty) + \int_0^{\infty}\Omega\big[G(s) - G(\infty)\big]\sin\Omega s\,\mathrm{d}s \tag{5.145}$$

is the *stiffness* or *storage modulus* and

---

[27] It is hard to calculate $G(t)$ from the power series (5.141). Instead, a spectral representation of the form (5.120) should be applied; in this case the cumulative spectrum $H(\nu)$ is given in closed form, see eq. (5.171).

[28] See LION [2000], pp. 86.

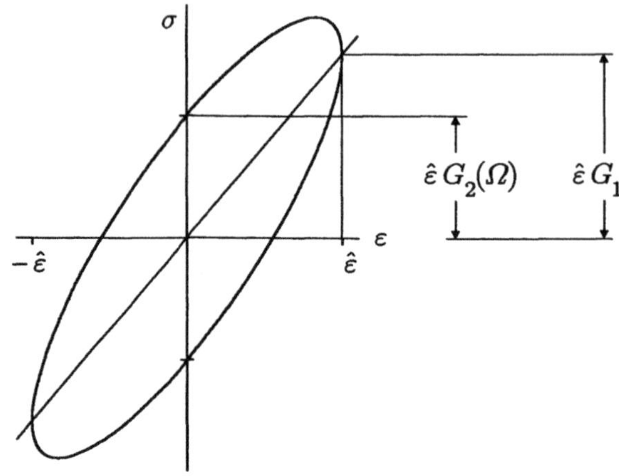

Figure 5. 10: Stress response to a harmonic strain process

$$G_2(\Omega) = \int_0^\infty \Omega\big[G(s) - G(\infty)\big]\cos\Omega s\, \mathrm{d}s \tag{5.146}$$

the *dissipation modulus* or *loss modulus*. Equivalently, a complex-valued strain process

$$\varepsilon^*(t) = \hat{\varepsilon}\mathrm{e}^{\mathrm{i}\Omega t} \tag{5.147}$$

yields the complex stress process

$$\sigma^*(t) = \hat{\varepsilon}\, G^*(\mathrm{i}\Omega)\mathrm{e}^{\mathrm{i}\Omega t} = G^*(\mathrm{i}\Omega)\varepsilon^*(t)\,, \tag{5.148}$$

with the *complex modulus* $G^*(\mathrm{i}\Omega)$, defined by

$$G^*(\mathrm{i}\Omega) = G(\infty) + \mathrm{i}\Omega\int_0^\infty \big[G(s) - G(\infty)\big]\mathrm{e}^{-\mathrm{i}\Omega s}\mathrm{d}s\,. \tag{5.149}$$

□

Proof

Inserting (5.147) into $(5.143)_2$ we obtain

$$\sigma^*(t) = \hat{\varepsilon}\left\{G(\infty) + \int_{-\infty}^{t} \mathrm{i}\Omega\big[G(t-\tau) - G(\infty)\big]\mathrm{e}^{-\mathrm{i}\Omega(t-\tau)}\mathrm{d}s\right\}\mathrm{e}^{\mathrm{i}\Omega t}\,;$$

this verifies (5.149) if $t - \tau = s$ is substituted in the integrand. The imaginary part of this equation corresponds to (5.144) (the real part is the stress response to the input $\varepsilon(t) = \hat{\varepsilon}\cos\Omega t$). ■

In order to illustrate the dynamic stress response (5.144), the parameter t may be eliminated from (5.142) and (5.144) to get a stress-strain curve. (5.144) and (5.142), i.e.

$$\frac{\sigma - G_1(\Omega)\varepsilon}{\hat{\varepsilon}\, G_2(\Omega)} = \cos\Omega t \quad \text{and} \quad \frac{\varepsilon}{\hat{\varepsilon}} = \sin\Omega t \quad \text{imply}$$

$$\left(\frac{\sigma - G_1(\Omega)\varepsilon}{\hat{\varepsilon}\, G_2(\Omega)}\right)^2 + \left(\frac{\varepsilon}{\hat{\varepsilon}}\right)^2 = 1 \quad \text{or}$$

$$\sigma = \sigma(\varepsilon) = G_1(\Omega)\,\varepsilon \pm \hat{\varepsilon}\, G_2(\Omega)\sqrt{1 - \left(\frac{\varepsilon}{\hat{\varepsilon}}\right)^2}\,. \tag{5.150}$$

This equation describes an ellipse, sketched in Figure 5. 10. During the cyclic process the stress-strain trajectory of Figure 5. 10 is passed through in clockwise direction. The mechanical work A_{D}, dissipated over one cycle of deformation, is the area of the ellipse: we have

$$A_{\mathrm{D}} = \int_0^{\frac{2\pi}{\Omega}} \sigma(t)\dot{\varepsilon}(t)\mathrm{d}t = \hat{\varepsilon}^2\Omega \int_0^{\frac{2\pi}{\Omega}} \Big[G_1(\Omega)\sin\Omega t + G_2(\Omega)\cos\Omega t\Big]\cos\Omega t\,\mathrm{d}t$$

with the result

$$A_{\mathrm{D}} = \pi\hat{\varepsilon}^2 G_2(\Omega)\,. \tag{5.151}$$

In view of a non-negative dissipated work the dissipation modulus $G_2(\Omega)$ must be non-negative for all frequencies Ω.

A simple calculation based on eq. (5.149) shows that the classical 3-parameter model, characterised by the relaxation function (5.93),

$$G(t) = E + \hat{E}\,\mathrm{e}^{-\frac{t}{z_0}}\,, \tag{5.152}$$

leads to the complex modulus

$$G^*(\mathrm{i}\Omega) = E + \hat{E}\frac{\mathrm{i}\Omega z_0}{1 + \mathrm{i}\Omega z_0}\,; \tag{5.153}$$

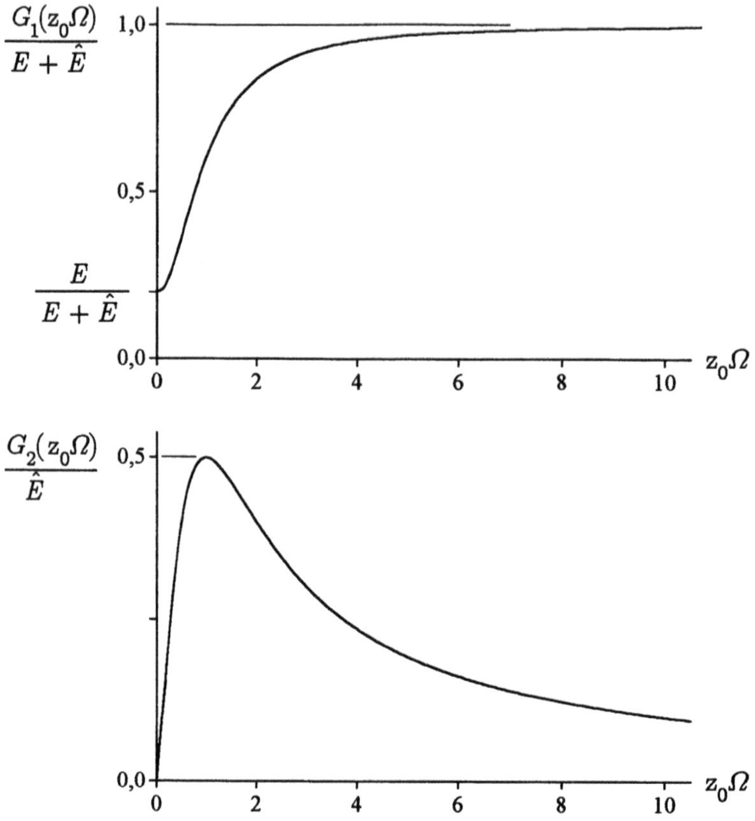

Figure 5. 11: Storage and loss modulus according to the 3-parameter model; linear scale

this implies the storage modulus

$$G_1(\Omega) = E + \hat{E}\frac{(\Omega z_0)^2}{1+(\Omega z_0)^2} \tag{5.154}$$

and the loss modulus

$$G_2(\Omega) = \hat{E}\frac{\Omega z_0}{1+(\Omega z_0)^2} \; . \tag{5.155}$$

$z_0 = \frac{\eta}{\hat{E}}$ is the relaxation time.

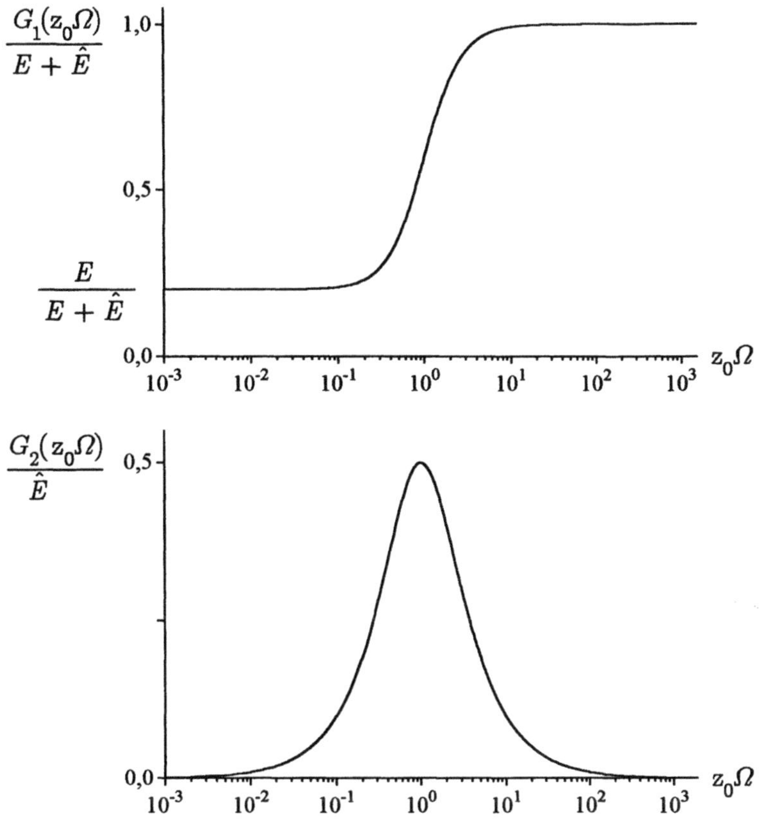

Figure 5. 12: Storage and loss modulus according to the 3-parameter model; logarithmic scale

The dynamic moduli $G_1(\Omega)$ and $G_2(\Omega)$ of the standard linear solid are depicted in Figure 5. 11. It is recognised that the storage modulus $G_1(\Omega)$ has a horizontal tangent at $\Omega = 0$. It starts with the equilibrium modulus and tends asymptotically to the spontaneous modulus $E + \hat{E}$ for large values of Ω. For small values of Ω the loss modulus behaves like $G_2(\Omega) = \eta\Omega + O(\Omega^2)$; G_2 has a maximum at the relaxation frequency $\Omega = 1/z_0$ and tends asymptotically to zero for large Ω. This behaviour of the dynamic moduli, a characteristic feature of the 3-parameter model, is even more apparent if a logarithmic scale is considered, which is shown in Figure 5. 12.

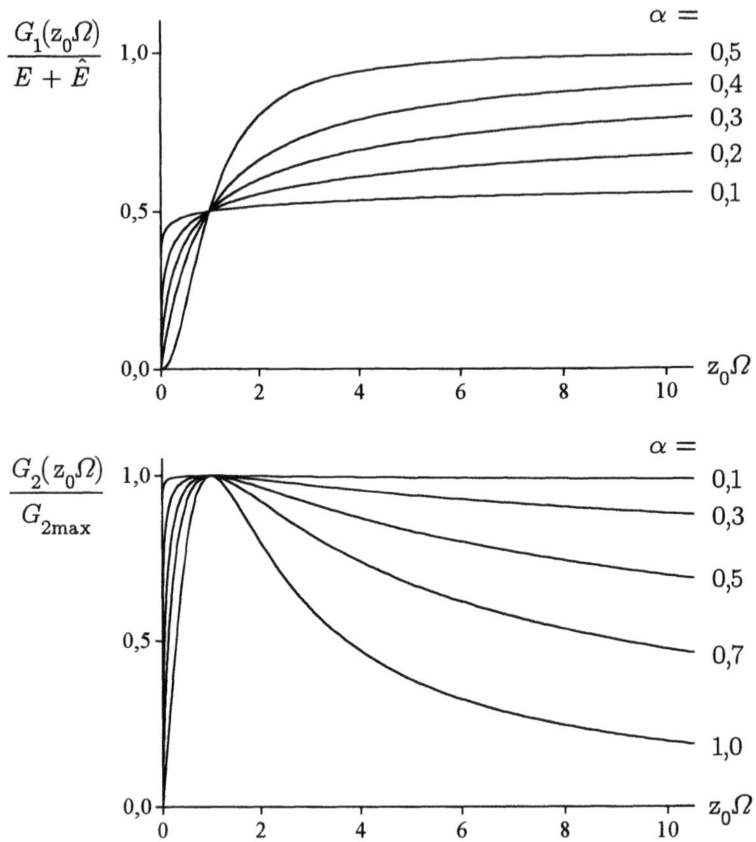

Figure 5. 13: Storage and loss modulus according to the Maxwell model of fractional order. The limit value $\alpha = 1$ implies $G_1 = 0$ and corresponds to the classical Maxwell model

The figure suggests, that a superposition of several exponentials, i.e. a broader spectrum of relaxation is required, if the stiffness and dissipation moduli are to be represented according to experimental data going over a large range of frequency.[29]

A continuous spectrum of relaxation exists in the more general case of the *Maxwell* model of fractional order. Calculating the fractional derivative of the harmonic strain process $\varepsilon^*(t) = \hat{\varepsilon}\,\mathrm{e}^{\mathrm{i}\Omega t}$ according to its definition (5.131) yields

[29] Cf. LION [1998]; HAUPT, LION & BACKHAUS [2000].

$$\frac{\mathrm{d}^\alpha}{\mathrm{d}t^\alpha}\varepsilon^*(t) = \frac{1}{\Gamma(1-\alpha)}\int_0^t \frac{1}{(t-\tau)^\alpha}\varepsilon^{*\prime}(\tau)\mathrm{d}\tau = \hat{\varepsilon}\frac{\mathrm{i}\Omega}{\Gamma(1-\alpha)}\int_0^t \frac{\mathrm{e}^{\mathrm{i}\Omega(t-s)}}{s^\alpha}\,\mathrm{d}s\,;$$

in the limit $t \to \infty$ this derivative becomes a periodic function:

$$\lim_{t\to\infty}\frac{\mathrm{d}^\alpha}{\mathrm{d}t^\alpha}\varepsilon^*(t) = \hat{\varepsilon}\frac{\mathrm{i}\Omega}{\Gamma(1-\alpha)}\left\{\int_0^\infty \frac{\mathrm{e}^{-\mathrm{i}\Omega s}}{s^\alpha}\,\mathrm{d}s\right\}\mathrm{e}^{\mathrm{i}\Omega t}\,.$$

As a result we have the asymptotic behaviour

$$\frac{\mathrm{d}^\alpha}{\mathrm{d}t^\alpha}\varepsilon^*(t) = \hat{\varepsilon}\,(\mathrm{i}\Omega)^\alpha\,\mathrm{e}^{\mathrm{i}\Omega t} = (\mathrm{i}\Omega)^\alpha\varepsilon^*(t)\,. \tag{5.156}$$

Accordingly, we conclude from $\varepsilon^*(t) = \hat{\varepsilon}\,\mathrm{e}^{\mathrm{i}\Omega t}$ and (5.133) the complex stress response $\sigma^*(t) = \hat{\varepsilon}\,G^*(\mathrm{i}\Omega)\mathrm{e}^{\mathrm{i}\Omega t}$. The complex modulus

$$G^*(\mathrm{i}\Omega) = \hat{E}\frac{(\mathrm{i}\Omega z_0)^\alpha}{1 + (\mathrm{i}\Omega z_0)^\alpha} \tag{5.157}$$

decomposes into real and imaginary parts,

$$G^*(\mathrm{i}\Omega) = G_1(\Omega) + \mathrm{i}G_2(\Omega).$$

Application of $(\mathrm{i}\Omega)^\alpha = \Omega^\alpha\left[\cos\left(\alpha\frac{\pi}{2}\right) + \mathrm{i}\,\sin\left(\alpha\frac{\pi}{2}\right)\right]$ produces the storage modulus

$$G_1(\Omega) = \hat{E}\frac{(\Omega z_0)^\alpha\cos\left(\alpha\frac{\pi}{2}\right) + (\Omega z_0)^{2\alpha}}{1 + 2(\Omega z_0)^\alpha\cos\left(\alpha\frac{\pi}{2}\right) + (\Omega z_0)^{2\alpha}} \tag{5.158}$$

and the loss modulus

$$G_2(\Omega) = \hat{E}\frac{(\Omega z_0)^\alpha\sin\left(\alpha\frac{\pi}{2}\right)}{1 + 2(\Omega z_0)^\alpha\cos\left(\alpha\frac{\pi}{2}\right) + (\Omega z_0)^{2\alpha}}\,. \tag{5.159}$$

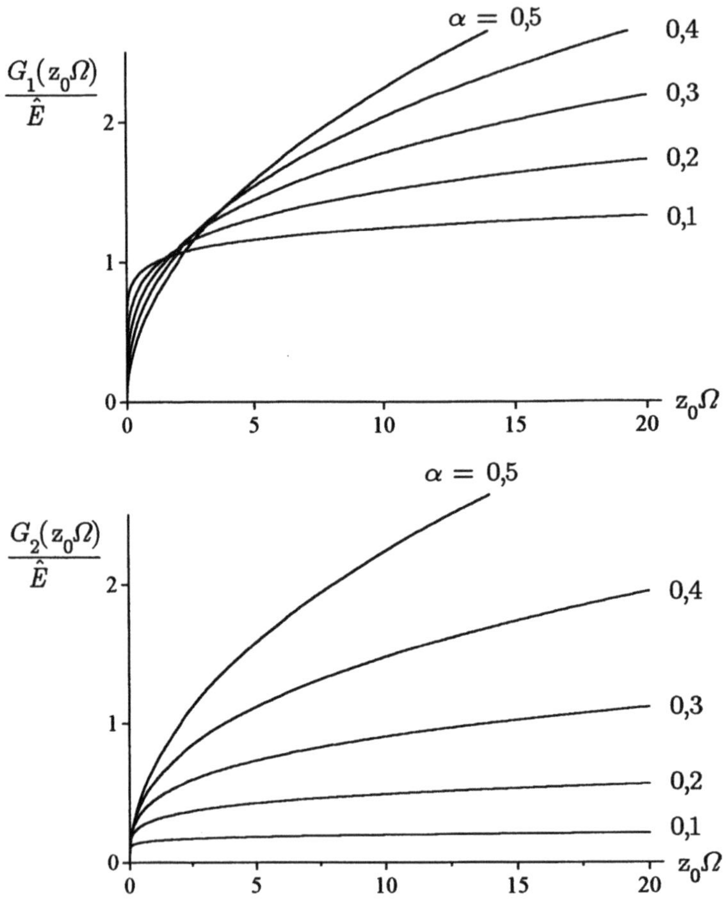

Figure 5. 14: Storage and loss modulus corresponding to a viscosity of fractional order

These moduli are illustrated in Figure 5. 13. Again, the loss modulus has a

maximum value at $\Omega = \frac{1}{z_0}$, namely $G_{2\max} = \hat{E} \frac{\sin(\alpha\frac{\pi}{2})}{2(1 + \cos(\alpha\frac{\pi}{2}))}$.

However, the decrease of G_2 for frequencies $\Omega > 1/z_0$ is extremely slow if α is sufficiently small. In the limit $\alpha \to 1$ the dynamic moduli of the fractional *Maxwell* model are equal to the moduli of the classical *Maxwell* model according to (5.154), (5.155) with $E = 0$.

In the more special case of a pure viscosity of fractional order the dynamic moduli are power functions of the frequency: the constitutive model

$$\sigma(t) = \hat{E}\, z_0{}^{\alpha} \frac{d^{\alpha}}{dt^{\alpha}} \varepsilon(t)$$

together with $\varepsilon(t) = \hat{\varepsilon}\sin\Omega t$ and $\sigma(t) = \hat{\varepsilon}\{G_1(\Omega)\sin\Omega t + G_2(\Omega)\cos\Omega t\}$ implies

$$G_1(\Omega) = \hat{E}(\Omega z_0)^{\alpha}\cos(\alpha\tfrac{\pi}{2}) \,, \tag{5.160}$$

$$G_2(\Omega) = \hat{E}(\Omega z_0)^{\alpha}\sin(\alpha\tfrac{\pi}{2}) \tag{5.161}$$

(see Fig. 5. 14). We see from (5.160) and (5.161) that in the limit $\alpha = 0$ the storage modulus G_1 tends to $\hat{E}$, whereas the loss modulus G_2 vanishes. Conversely, the limit $\alpha = 1$ implies $G_1 = 0$ and $G_2 = \hat{E}\Omega z_0$. This illustrates once more the fact that the fractional derivative interpolates between the function itself and its first derivative, i.e. the model of a fractional viscosity interpolates between *Hooke's law* and linear viscosity.

The general definition (5.149) shows that complex modulus and relaxation function are related to each other (comparable with the relation between relaxation and creep function): if a relaxation function is given, the complex modulus is determined according to (5.149). The converse statement is also valid: it is possible to determine the continuous relaxation spectrum if the complex modulus is known. This is the content of

Theorem 5. 11

The complex modulus according to eq. (5.149),

$$G^*(\mathrm{i}\Omega) = \mathrm{i}\Omega \int_0^{\infty} G(s) e^{-\mathrm{i}\Omega s} ds \,, \tag{5.162}$$

is related to the relaxation spectrum $h(\nu)$, occurring in

$$G(t) = \int_0^{\infty} h(\nu)\, e^{-\nu t} d\nu \tag{5.163}$$

(eq. (5.117) with $E = 0$) through the so-called *Stieltjes transformation*

$$\frac{G^*(\mathrm{i}\Omega)}{\mathrm{i}\Omega} = \int_0^{\infty} \frac{h(\nu)}{\nu + \mathrm{i}\Omega} d\nu \,. \tag{5.164}$$

Moreover, the inverse of this transformation is determined as the limit[30]

$$h(\omega) = \frac{1}{2\pi \mathrm{i}} \lim_{\varepsilon \to 0} \left[\frac{G^*(-\omega - \mathrm{i}\varepsilon)}{-\omega - \mathrm{i}\varepsilon} - \frac{G^*(-\omega + \mathrm{i}\varepsilon)}{-\omega + \mathrm{i}\varepsilon} \right] . \tag{5.165}$$

□

Proof

Inserting (5.163) into (5.162) yields

$$\frac{G^*(\mathrm{i}\Omega)}{\mathrm{i}\Omega} = \int_0^\infty \left\{ \int_0^\infty h(\nu)\, \mathrm{e}^{-\nu s} \mathrm{d}\nu \right\} \mathrm{e}^{-\mathrm{i}\Omega s} \mathrm{d}s$$

$$= \int_0^\infty h(\nu) \left\{ \int_0^\infty \mathrm{e}^{-(\nu + \mathrm{i}\Omega)s} \mathrm{d}s \right\} \mathrm{d}\nu = \int_0^\infty \frac{h(\nu)}{\nu + \mathrm{i}\Omega} \mathrm{d}\nu ,$$

i.e. (5.164). In order to verify (5.165), the Stieltjes transformation (5.164) is applied to calculate the difference

$$\frac{G^*(-\omega - \mathrm{i}\varepsilon)}{-\omega - \mathrm{i}\varepsilon} - \frac{G^*(-\omega + \mathrm{i}\varepsilon)}{-\omega + \mathrm{i}\varepsilon} = \int_0^\infty \left[\frac{1}{\nu - \omega - \mathrm{i}\varepsilon} - \frac{1}{\nu - \omega + \mathrm{i}\varepsilon} \right] h(\nu) \mathrm{d}\nu$$

$$= \int_0^\infty \frac{2\mathrm{i}\varepsilon}{(\nu - \omega)^2 + \varepsilon^2} h(\nu) \mathrm{d}\nu = 2\mathrm{i} \int_{-\frac{\omega}{\varepsilon}}^\infty \frac{h(\omega + \varepsilon x)}{x^2 + 1} \mathrm{d}x .$$

Transition to the limit provides

$$\lim_{\varepsilon \to 0} \int_{-\frac{\omega}{\varepsilon}}^\infty \frac{h(\omega + \varepsilon x)}{x^2 + 1} \mathrm{d}x = \int_{-\infty}^\infty \frac{\mathrm{d}x}{x^2 + 1} h(\omega) = \pi h(\omega) . \tag{5.166}$$

This proves (5.165). ■

Relation (5.165) in combination with (5.157) can be applied to calculate the relaxation spectrum of the *Mittag-Leffler* function, i.e. the relaxation spectrum corresponding to the fractional *Maxwell* model (5.133). The result is

[30] GROSS [1968]; TSCHOEGL [1989].

Theorem 5. 12

The relaxation spectrum $h(\cdot)$ of the *Mittag-Leffler* function (5.141) ,

$$G(t) = E_\alpha\left[-\left(\frac{t}{z_0}\right)^\alpha\right] = \hat{E}\sum_{k=0}^{\infty}\frac{(-1)^k}{\Gamma(1+\alpha k)}\left(\frac{t}{z_0}\right)^{k\alpha},$$

to be inserted into the representation (5.117), $G(t) = \int_0^\infty h(\nu)\, e^{-\nu t}\, d\nu$, is

$$h(\nu) = \frac{\hat{E}}{\pi\nu}\frac{(z_0\nu)^\alpha \sin(\alpha\pi)}{1 + 2(z_0\nu)^\alpha\cos(\alpha\pi) + (z_0\nu)^{2\alpha}}. \tag{5.167}$$

Moreover, the *associated* cumulative spectrum $H(\nu) = \int_0^\nu h(x)dx$ is calculated to

$$H(\nu) = \frac{\hat{E}}{\alpha\pi}\left\{\arctan\left[\frac{(z_0\nu)^\alpha + \cos(\alpha\pi)}{\sin(\alpha\pi)}\right] - \pi\left(\frac{1}{2} - \alpha\right)\right\}. \tag{5.168}$$

□

Proof

To derive the continuous spectrum (5.167) from the complex modulus (5.157),

$$G^*(i\Omega) = \hat{E}\frac{(i\Omega z_0)^\alpha}{1 + (i\Omega z_0)^\alpha},$$

the polar representation

$$\begin{aligned} -\omega - i\varepsilon &= \sqrt{\omega^2 + \varepsilon^2}\, e^{-i(\pi-\varphi)} \\ -\omega + i\varepsilon &= \sqrt{\omega^2 + \varepsilon^2}\, e^{i(\pi-\varphi)} \end{aligned} \tag{5.169}$$

with $\varphi = \arctan\frac{\varepsilon}{\omega}$ is preferable. Then, we have to calculate the difference

$$\frac{G^*(-\omega - i\varepsilon)}{-\omega - i\varepsilon} - \frac{G^*(-\omega + i\varepsilon)}{-\omega + i\varepsilon} = \hat{E}\left\{\frac{z_0^\alpha\left(\sqrt{\omega^2+\varepsilon^2}\right)^{(\alpha-1)} e^{-i(\alpha-1)(\pi-\varphi)}}{1 + z_0^\alpha\left(\sqrt{\omega^2+\varepsilon^2}\right)^\alpha e^{-i\alpha(\pi-\varphi)}}\right.$$

$$\left. - \frac{z_0^\alpha\left(\sqrt{\omega^2+\varepsilon^2}\right)^{(\alpha-1)} e^{i(\alpha-1)(\pi-\varphi)}}{1 + z_0^\alpha\left(\sqrt{\omega^2+\varepsilon^2}\right)^\alpha e^{i\alpha(\pi-\varphi)}}\right\}$$

$$= 2\mathrm{i}\hat{E}\,\mathrm{z}_0^{\alpha} \times$$
$$\times \frac{(\sqrt{\omega^2+\varepsilon^2})^{(\alpha-1)}[-\sin((\alpha-1)(\pi-\varphi)) + \mathrm{z}_0^{\alpha}(\sqrt{\omega^2+\varepsilon^2})^{\alpha}\sin(\pi-\varphi)]}{1 + 2\mathrm{z}_0^{\alpha}(\sqrt{\omega^2+\varepsilon^2})^{\alpha}\cos((\alpha(\pi-\varphi)) + \mathrm{z}_0^{2\alpha}(\sqrt{\omega^2+\varepsilon^2})^{2\alpha}} .$$

Transition to the limit, $\varepsilon \to 0 \iff \varphi \to 0$, and application of (5.165) leads to

$$2\pi \mathrm{i} h(\omega) = 2\mathrm{i}\,\hat{E}\,\mathrm{z}_0^{\alpha}\frac{\omega^{\alpha-1}\sin\alpha\pi}{1 + 2\mathrm{z}_0^{\alpha}\,\omega^{\alpha}\cos\alpha\pi + \mathrm{z}_0^{2\alpha}\omega^{2\alpha}} ,$$

i.e. to

$$h(\omega) = \frac{\hat{E}}{\pi\omega}\,\frac{(\mathrm{z}_0\omega)^{\alpha}\sin\alpha\pi}{1 + 2(\mathrm{z}_0\omega)^{\alpha}\cos\alpha\pi + (\mathrm{z}_0\omega)^{2\alpha}} .$$

The spectrum $h(\cdot)$ can be integrated in closed form. This is evident from the equivalent formulation

$$h(\omega) = \frac{\hat{E}\,\mathrm{z}_0}{\pi\sin\alpha\pi}\,\frac{(\mathrm{z}_0\omega)^{\alpha-1}}{1 + \left[\dfrac{(\mathrm{z}_0\omega)^{\alpha} + \cos\alpha\pi}{\sin\alpha\pi}\right]^2} , \tag{5.170}$$

which is immediately integrated to the result (5.168). ■

The theorem has an interesting application in so far as it provides a basis for an efficient calculation of the *Mittag Leffler* function, which occurs as the relaxation function of the *Maxwell* model of fractional order. Indeed, combination of (5.168) and (5.120) gives the representation formula[31]

$$\mathrm{E}_{\alpha}\Big[-\Big(\frac{t}{\mathrm{z}_0}\Big)^{\alpha}\Big] = \frac{\hat{E}}{\alpha\pi}\int_0^{\infty}\Big\{\arctan\Big[\frac{\big(\mathrm{z}_0\frac{s}{t}\big)^{\alpha} + \cos(\alpha\pi)}{\sin(\alpha\pi)}\Big] - \pi\Big(\frac{1}{2} - \alpha\Big)\Big\}\,\mathrm{e}^{-s}\,ds . \tag{5.171}$$

The numerical evaluation of this formula is quite easy in contrast to (5.141). It is the basis of Fig. 5. 9.

[31] Cf. HAUPT & LION [2001].

5. 6 Perfectly Plastic Solid

The concept of viscoelasticity takes several effects of "inelasticity", such as the rate-dependence, i.e. viscosity, relaxation and creep, into consideration. A completely different phenomenon is the tendency of certain solids to convert under a specific critical load from the state of elasticity to a certain state of flow, where any deformations are possible when subjected to the same load. This type of material behaviour, familiar from metals, is represented in the classical *theory of plasticity.*[32]

Definition 5. 8

An *elastic-perfectly-plastic solid* is defined by means of the following constitutive equations:

$$\dot{\mathbf{T}}^{\mathrm{D}}(t) = \begin{cases} 2\mu\Big[\dot{\mathbf{E}}^{\mathrm{D}} - \dfrac{1}{(2/3)\,k^2}\,(\mathbf{T}^{\mathrm{D}}\cdot\dot{\mathbf{E}})\mathbf{T}^{\mathrm{D}}\Big] & \text{if} \quad F = 0 \quad \text{and} \quad B > 0 \\ 2\mu\dot{\mathbf{E}}^{\mathrm{D}} & \text{in all other cases} \end{cases} , \tag{5.172}$$

$$\mathrm{tr}\mathbf{T} = 2\mu\frac{1+\nu}{1-2\nu}\,\mathrm{tr}\mathbf{E}\ . \tag{5.173}$$

Here, μ and ν denote the *shear modulus* and *Poisson number.* The quantities

$$F(\mathbf{T}, k) = \mathbf{T}^{\mathrm{D}}\cdot\mathbf{T}^{\mathrm{D}} - \tfrac{2}{3}\,k^2 \tag{5.174}$$

and

$$B = \mathbf{T}^{\mathrm{D}}\cdot\dot{\mathbf{E}} \tag{5.175}$$

are called *yield function* and *loading function.* The material constant k is called *yield stress.* ◊

The constitutive equations (5.172) are known as *Prandtl-Reuß* equations.[33] They form a system of ordinary differential equations, from which the current state of stress $\mathbf{T}(t)$ can be calculated by means of integration, provided that a time-dependent strain process $\mathbf{E}(\tau)$ $(0 \leq \tau \leq t)$ is given. The formulation of the *Prandtl-Reuß* equations gives an immediate indication that the present state of stress does not depend on the velocity at which one and the same deformation process will take place: time, being the

---

[32] HILL [1950]; PRAGER & HODGE [1954]; LIPPMANN & MAHRENHOLTZ [1967]; RECKLING [1967]; KACHANOV [1971]; LIPPMANN [1981]; BETTEN [1985]; KREISSIG [1992]; BURTH & BROCKS [1992].

[33] GEIRINGER [1973], Sect. 6.

independent variable, could be replaced by any other parameter that is a monotonic function of time. This substitution would have no influence on the solutions of the *Prandtl-Reuß* equations. The constitutive equations (5.172) and (5.173) therefore describe a *rate-independent material behaviour.*

As opposed to the other classical constitutive models in continuum mechanics, the *Prandtl-Reuß* equations are not represented by means of one single equation, but by a pair of different equations, one of which is selected for each respective case that is identified. The criterion for the distinction between different cases, the *yield condition* $F(\mathbf{T}, k) = 0$, checks to see whether a certain intensity has been reached in the state of stress. If this is not the case, then it is a question of linear-elastic behaviour. If, in addition to the yield condition

$$\mathbf{T}^{\mathrm{D}} \cdot \mathbf{T}^{\mathrm{D}} = \frac{2}{3} k^2 , \tag{5.176}$$

the *loading condition* $B = \mathbf{T}^{\mathrm{D}} \cdot \dot{\mathbf{E}} > 0$ is also fulfilled, plastic flow occurs, and the *Prandtl-Reuß* equations apply. Linear elasticity applies again upon *unloading* $(\mathbf{T}^{\mathrm{D}} \cdot \dot{\mathbf{E}} < 0)$ or for minor loads $\left(\mathbf{T}^{\mathrm{D}} \cdot \mathbf{T}^{\mathrm{D}} < (2/3)k^2\right)$.

The constitutive equations of perfect plasticity apply exclusively to the deviatoric part of the stress or strain. As far as the hydrostatic part is concerned, elastic behaviour is normally assumed, in other words, plastic changes in volume are not taken into account.[34]

It is common practice to represent the *yield condition* (5.176) by means of a surface in the space of stress tensors. This surface is known as the *yield surface.* The yield surface encloses a region within the stress space, known as the *elastic region*: for states of stress within this area the material behaviour is *linear elastic.* For states of stress on the yield surface the direction of the strain increment is the decisive factor: upon loading the state of stress remains on the yield surface and *plastic deformations* occur, that are subtracted from the total strain when calculating the stress. Upon unloading ($B < 0$) the state of stress returns to the elastic domain, and linear elasticity again becomes relevant. States of stress beyond the yield surface, that might fulfil the inequality

$$\mathbf{T}^{\mathrm{D}} \cdot \mathbf{T}^{\mathrm{D}} > \frac{2}{3} k^2 ,$$

are ruled out a priori.

In the constitutive model of perfect plasticity the two material parameters μ, ν of linear isotropic elasticity are supplemented by a further parameter k.

---

[34] BELL [1973], pp. 83, 504, 710.

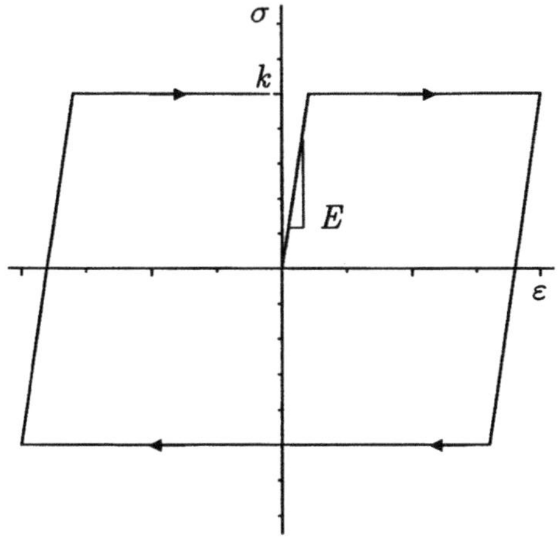

Figure 5. 15: Perfect plasticity

The physical meaning of this material constant lies in the fact that it corresponds to the *yield stress* in a uniaxial tensile test.

Theorem 5. 13

For a uniaxial tension/compression loading, defined by the stress $\sigma(t)$ and the strain $\varepsilon(t)$, the following stress-strain relation is the result of model (5.172), (5.173) of perfect plasticity:

$$\dot{\sigma} = \begin{cases} 0 \quad \text{and} \quad \dot{\varepsilon}_Q = -\frac{1}{2}\dot{\varepsilon} & \text{if} \quad \sigma^2 = k^2 \text{ and } \sigma\dot{\varepsilon} > 0 \\ E\dot{\varepsilon} \quad \text{and} \quad \dot{\varepsilon}_Q = -\nu\dot{\varepsilon} & \text{in all other cases} \end{cases} \tag{5.177}$$

□

Proof

Uniaxial loading in the x direction corresponds to

the stress tensor $\mathbf{T}(t) = \sigma(t)\mathbf{e}_x \otimes \mathbf{e}_x$,

the linearised strain tensor $\mathbf{E}(t) = \varepsilon(t)\mathbf{e}_x \otimes \mathbf{e}_x + \varepsilon_Q(t)[(\mathbf{e}_y \otimes \mathbf{e}_y + \mathbf{e}_z \otimes \mathbf{e}_z)]$ and their deviatoric parts

$$\mathbf{T}^D = \frac{2}{3}\sigma\left[\mathbf{e}_x \otimes \mathbf{e}_x - \frac{1}{2}(\mathbf{e}_y \otimes \mathbf{e}_y + \mathbf{e}_z \otimes \mathbf{e}_z)\right],$$

$$\mathbf{E}^D = \frac{2}{3}(\varepsilon - \varepsilon_Q)\left[\mathbf{e}_x \otimes \mathbf{e}_x - \frac{1}{2}(\mathbf{e}_y \otimes \mathbf{e}_y + \mathbf{e}_z \otimes \mathbf{e}_z)\right].$$

The yield and loading functions specialise to

$$F = \mathbf{T}^{\mathrm{D}} \cdot \mathbf{T}^{\mathrm{D}} - \tfrac{2}{3}k^2 = \tfrac{2}{3}\left(\sigma^2 - k^2\right) \text{ and } B = \mathbf{T}^{\mathrm{D}} \cdot \dot{\mathbf{E}} = \tfrac{2}{3}\sigma\left(\dot{\varepsilon} - \dot{\varepsilon}_{\mathrm{Q}}\right).$$

Furthermore, we obtain

$$\dot{\mathbf{E}}^{\mathrm{D}} - \frac{1}{(2/3)k^2}\left(\mathbf{T}^{\mathrm{D}} \cdot \dot{\mathbf{E}}\right)\mathbf{T}^{\mathrm{D}}$$
$$= \tfrac{2}{3}\left(\dot{\varepsilon} - \dot{\varepsilon}_{\mathrm{Q}}\right)\left[1 - \frac{\sigma^2}{k^2}\right]\left[e_x \otimes e_x - \tfrac{1}{2}\left(e_y \otimes e_y + e_z \otimes e_z\right)\right].$$

In the case of $F = 0$ we have $\sigma^2 = k^2$ and $\dot{\sigma} = 0$ according to $(5.172)_1$.

Together with (5.173) this implies $\varepsilon_{\mathrm{Q}} = -\frac{1}{2}\varepsilon$ and $B = \sigma\dot{\varepsilon}$.

All other cases lead to the elastic response $(5.172)_2$ and (5.173), i.e. to

$$\dot{\sigma} = 2\mu\left(\dot{\varepsilon} - \dot{\varepsilon}_{\mathrm{Q}}\right) \text{ and } \dot{\sigma} = 2\mu\frac{1+\nu}{1-2\nu}\left(\dot{\varepsilon} + 2\dot{\varepsilon}_{\mathrm{Q}}\right).$$

The solution of these two equations is given by $\dot{\varepsilon}_{\mathrm{Q}} = -\nu\dot{\varepsilon}$
and $\dot{\sigma} = 2\mu(1+\nu)\dot{\varepsilon} = E\dot{\varepsilon}$. ■

The stress-strain behaviour described in (5.177) is illustrated in Fig. 5. 15. In contrast to the material models of classical fluid mechanics and elasticity, the defining equations of viscoelasticity and plasticity are not algebraic equations, but *functional relations.* Thus the material model (5.51) of linear viscoelasticity consists of an integral equation, which attributes the current stress to the past strain history. The *field equations of viscoelasticity* (5.73), obtained by inserting the constitutive equation (5.51) into the local balance of linear momentum, are accordingly not partial differential equations, but *integro-differential equations.* The constitutive equations of *plasticity* are differential equations that depend on the distinction between different cases. This leads to a remarkable consequence: the demarcation of domains, in which these individual cases lead to field equations of one kind or another, is an essential part of the boundary value problem. For this reason it is not possible to write the general initial-boundary-value problem of plasticity in one explicit formulation, as with the basic equations of hydrodynamics, elasticity or viscoelasticity.

5. 7 Plasticity with Hardening

The model of perfect plasticity can be expanded by allowing the yield surface delimiting the elastic domain to alter in dependence of the history of the strain process. A change of the yield surface is known as *hardening*. Usually two methods are applied when modelling hardening properties: the so-called *kinematic hardening* comprises a pure translation of the yield surface in the stress space, whereas the *isotropic hardening* amounts to an expansion of the yield surface. A variation in shape could likewise be taken into consideration.

A more general model of rate-independent plasticity, including ideal plasticity as a special case, is explained by means of the following definition:

Definition 5. 9

An *elastic-plastic solid with hardening* is defined by the following constitutive assumptions:

Decomposition of the Strain

$$\mathbf{E} = \mathbf{E}_e + \mathbf{E}_p \tag{5.178}$$

Elasticity Relation

$$\mathbf{T} = 2\mu[\mathbf{E}_e + \frac{\nu}{1-2\nu}(\mathrm{tr}\mathbf{E}_e)\mathbf{1}] \tag{5.179}$$

Yield Function

$$f(\mathbf{T}, \mathbf{X}, k) = \frac{1}{2}\mathrm{tr}(\mathbf{T}^{\mathrm{D}} - \mathbf{X}^{\mathrm{D}})^2 - \frac{1}{3}k^2 \tag{5.180}$$

Loading Function

$$B = \frac{\partial f}{\partial \mathbf{T}} \cdot \dot{\mathbf{T}}(t) = (\mathbf{T}^{\mathrm{D}} - \mathbf{X}^{\mathrm{D}}) \cdot \dot{\mathbf{T}}(t) \tag{5.181}$$

Loading Condition

$$\frac{\partial f}{\partial \mathbf{T}} \cdot \dot{\mathbf{T}}(t) \begin{cases} > 0 \Leftrightarrow \text{loading} \\ = 0 \Leftrightarrow \text{neutral loading} \\ < 0 \Leftrightarrow \text{unloading} \end{cases} \tag{5.182}$$

Flow Rule

$$\dot{\mathbf{E}}_{\mathrm{p}}(t) = \begin{cases} \lambda \dfrac{\partial f}{\partial \mathbf{T}} = \lambda(\mathbf{T}^{\mathrm{D}} - \mathbf{X}^{\mathrm{D}}) \text{ for } f = 0 \text{ and } B > 0) \\ \mathbf{0} \qquad\qquad\qquad\qquad \text{in all other cases} \end{cases} \tag{5.183}$$

Kinematic Hardening

$$\dot{\mathbf{X}}(t) = c\dot{\mathbf{E}}_{\mathrm{p}}(t) - b\dot{s}_{\mathrm{p}}(t)\mathbf{X} \tag{5.184}$$

Isotropic Hardening

$$k = k(s_{\mathrm{p}}) \qquad (k'(s_{\mathrm{p}})) \geq 0 \tag{5.185}$$

Plastic Arclength (Accumulated Plastic Strain) $t \longmapsto s_{\mathrm{p}}(t)$

$$\dot{s}_{\mathrm{p}}(t) = \sqrt{\tfrac{2}{3}}\,\|\dot{\mathbf{E}}_{\mathrm{p}}(t)\| = \sqrt{\tfrac{2}{3}\mathrm{tr}(\dot{\mathbf{E}}_{\mathrm{p}})^2} \tag{5.186}$$

◊

The kinematic basis for the material model of classical plasticity with hardening is the additive decomposition (5.178) of the total strain into an elastic part $\mathbf{E}_{\mathrm{e}}$ and a plastic part $\mathbf{E}_{\mathrm{p}}$. The elastic part determines the state of stress by means of the linear relation (5.179). The plastic part of the strain is defined with the help of a yield function f. The so-called *yield condition*,

$$f(\mathbf{T}, \mathbf{X}, k) = 0\,, \tag{5.187}$$

defines a surface in the stress space that encloses an *elastic region*: regarding the plastic part of the strain, one initially assumes that it remains constant in time within the elastic region, i.e. for $f < 0$. Plastic strains are not produced, according to the flow rule (5.183), until both the yield condition (5.187) and the loading condition (5.182) are fulfilled. The possibility $f > 0$ is ruled out here as well. In connection with the criteria, the plastic strain rate is proportional to the normal $\partial f/\partial \mathbf{T}$ on the yield surface according to $(5.183)_1$, which is known as the *normality rule*. Due to the property that this normal is a deviator, no plastic volume changes take place. The proportionality factor λ in the normality rule cannot be a material constant, since this would be contradictory: if λ were a constant, then the yield condition $f = 0$ could not be fulfilled identically during the plastic deformation. Consequently, λ has to be determined in such a way that the

state of stress does not depart from the yield surface during plastic deformation processes. This is the so-called *consistency condition* (see below). The tensor-valued variable $\mathbf{X}$, as encountered in the yield function, is a stress tensor, the so-called *backstress* tensor, representing the centre of the yield surface in the stress space. The motion of this central point during the course of the elastoplastic process describes the translation of the yield surface, i.e. *kinematic hardening*. The temporal evolution of the yield surface in dependence of the process history is represented by differential equations, which are known as *evolution equations*. Equation (5.184) is an example of an evolution equation. The hardening model (5.184) is physically well-founded. It belongs to the fundamental assumptions in the theory of plasticity and is termed *Armstrong-Frederick* equation.[35] The evolution equation (5.184) is based on the supposition that hardening and softening processes are controlled by two mechanisms, namely a production term $c\dot{\mathbf{E}}_{\mathrm{p}}$, corresponding to the propagation and generation of dislocations, and a limiting term $b\dot{s}_{\mathrm{p}}(t)\mathbf{X}$, which indicates obstacles for the dislocation motions.
A typical feature of hardening models in plasticity is the fact that hardening effects only take place in conjunction with plastic deformations. If the state of stress lies in the elastic domain, $\mathbf{X}(t)$ is constant because $\mathbf{E}_{\mathrm{p}}(t)$ is constant. This characteristic is achieved in the evolution equation (5.184), since the first term on the right-hand side is proportional to $\dot{\mathbf{E}}_{\mathrm{p}}(t)$, and the second term contains the factor $\dot{s}_{\mathrm{p}}(t)$. The function

$$t \longmapsto s_{\mathrm{p}}(t) = \int_{t_0}^{t} \sqrt{\frac{2}{3}\mathrm{tr}(\dot{\mathbf{E}}_{\mathrm{p}}(\tau))^2}\ \mathrm{d}\tau \tag{5.188}$$

is the *accumulated plastic strain*. The geometric interpretation of the integral over the norm of the plastic strain rate motivates the term *plastic arclength*. The scalar quantity k in the yield function is the radius of the yield surface and, in its physical interpretation, the yield stress in a one-dimensional tensile test. This yield stress can also depend on the process history; the existence of such a dependence is called *isotropic hardening*. A simple model of isotropic hardening is the assumption (5.185), where the yield stress is a function of the accumulated plastic strain. This is referred to as *strain hardening*. A very similar model of isotropic hardening would result from replacing the plastic arclength by the quantity

---

[35] An evolution equation of this type was proposed in ARMSTRONG & FREDERICK [1966].

$$z(t) = \int_{t_0}^{t} (\mathbf{T}^{\mathrm{D}} - \mathbf{X}^{\mathrm{D}}) \cdot \dot{\mathbf{E}}_{\mathrm{p}}(\tau) \mathrm{d}\tau \,, \tag{5.189}$$

which can be interpreted as the work of the plastic strains on the stress difference (effective stress) $\mathbf{T}^{\mathrm{D}} - \mathbf{X}^{\mathrm{D}}$. This kind of hardening is called *work hardening*. More complicated models of isotropic hardening can, of course, also be formulated by replacing equation (5.185) by an evolution equation in the form of a differential equation for k in analogy to the model of kinematic hardening.

The proportionality factor λ, appearing in the flow rule (5.183), has to be determined in such a way that the yield function vanishes identically during the elastoplastic processes. To this aim we have recourse to

Theorem 5. 14

In connection with the constitutive equations (5.178) - (5.185) of classical plasticity, the yield condition for all plastic deformation processes is identically fulfilled if and only if

$$\lambda = \frac{3\mu}{N} (\mathbf{T}^{\mathrm{D}} - \mathbf{X}^{\mathrm{D}}) \cdot \dot{\mathbf{E}}(t) \,, \tag{5.190}$$

with

$$N = N(\mathbf{T}, \mathbf{X}, k) = k^2(2\mu + c) - kb(\mathbf{T}^{\mathrm{D}} - \mathbf{X}^{\mathrm{D}}) \cdot \mathbf{X} + \tfrac{2}{3} k^2 k'(s_{\mathrm{p}}) \,. \tag{5.191}$$

The following relation then holds true between strain and stress rates:

$$\dot{\mathbf{T}}(t) = \begin{cases} 2\mu\Big\{\dot{\mathbf{E}} - \frac{3\mu}{N}\Big[(\mathbf{T}^{\mathrm{D}} - \mathbf{X}^{\mathrm{D}}) \cdot \dot{\mathbf{E}}\Big](\mathbf{T}^{\mathrm{D}} - \mathbf{X}^{\mathrm{D}}) + \frac{\nu}{1-2\nu}(\mathrm{tr}\dot{\mathbf{E}})\mathbf{1}\Big\} \\ \qquad \text{for } \ \tfrac{1}{2}\mathrm{tr}(\mathbf{T}^{\mathrm{D}} - \mathbf{X}^{\mathrm{D}})^2 = \tfrac{1}{3}k^2 \ \text{ and } \ (\mathbf{T}^{\mathrm{D}} - \mathbf{X}^{\mathrm{D}}) \cdot \dot{\mathbf{E}} > 0 \\ 2\mu\Big[\dot{\mathbf{E}} + \frac{\nu}{1-2\nu}(\mathrm{tr}\dot{\mathbf{E}})\mathbf{1}\Big] \qquad \text{in all other cases} \end{cases} \tag{5.192}$$

□

Proof

For an elastoplastic deformation process the yield function is a function of time:

$$f(t) = f\big(\mathbf{T}(t), \mathbf{X}(t), k(s_{\mathrm{p}}(t))\big) \,.$$

This function has to vanish identically. The identical vanishing of the total derivative, the so-called *consistency condition*, is the equivalent thereof:

$$0 = \frac{\mathrm{d}}{\mathrm{d}t} f(t) = \frac{\partial f}{\partial \mathbf{T}} \cdot \dot{\mathbf{T}}(t) + \frac{\partial f}{\partial \mathbf{X}} \cdot \dot{\mathbf{X}}(t) + \frac{\partial f}{\partial k} k'(s_p)\dot{s}_p(t) .$$

Based on the specific definition (5.180) of the yield function one calculates

$$\begin{aligned}(\mathbf{T}^D - \mathbf{X}^D) \cdot \dot{\mathbf{T}}(t) &= (\mathbf{T}^D - \mathbf{X}^D) \cdot \dot{\mathbf{X}}(t) + \tfrac{2}{3} k k'(s_p)\dot{s}_p(t) \\ &= (\mathbf{T}^D - \mathbf{X}^D) \cdot \Big[c\dot{\mathbf{E}}_p(t) - b\dot{s}_p(t)\mathbf{X}\Big] + \tfrac{2}{3} k k'(s_p)\dot{s}_p(t) .\end{aligned}$$

By inserting flow rule and yield condition at this point and substituting $\dot{s}_p(t) = \frac{2}{3}\lambda k$, the outcome is

$$\begin{aligned}&(\mathbf{T}^D - \mathbf{X}^D) \cdot \dot{\mathbf{T}}(t) \\ &\quad = \lambda\Big\{c(\mathbf{T}^D - \mathbf{X}^D) \cdot (\mathbf{T}^D - \mathbf{X}^D) - \tfrac{2}{3} bk(\mathbf{T}^D - \mathbf{X}^D) \cdot \mathbf{X} + \tfrac{4}{9} k^2 k'(s_p)\Big\} ,\end{aligned}$$

$$(\mathbf{T}^D - \mathbf{X}^D) \cdot \dot{\mathbf{T}}(t) = \tfrac{2}{3}\lambda\Big\{ck^2 - bk(\mathbf{T}^D - \mathbf{X}^D) \cdot \mathbf{X} + \tfrac{2}{3} k^2 k'(s_p)\Big\} . \tag{5.193}$$

This leads to

$$\lambda = \frac{3}{2} \frac{(\mathbf{T}^D - \mathbf{X}^D) \cdot \dot{\mathbf{T}}(t)}{ck^2 - bk(\mathbf{T}^D - \mathbf{X}^D) \cdot \mathbf{X} + \frac{2}{3} k^2 k'(s_p)} . \tag{5.194}$$

By taking into consideration the elasticity relation (5.179) and the decomposition (5.178) of deformation as well, we arrive at

$$\begin{aligned}2\mu(\mathbf{T}^D - \mathbf{X}^D) \cdot \dot{\mathbf{E}}(t) &= (\mathbf{T}^D - \mathbf{X}^D) \cdot (2\mu\dot{\mathbf{E}}_p(t) + \dot{\mathbf{X}}(t)) + \tfrac{2}{3} k k'(s_p)\dot{s}_p(t) \\ &= (\mathbf{T}^D - \mathbf{X}^D) \cdot \Big\{(2\mu + c)\dot{\mathbf{E}}_p(t) - b\dot{s}_p(t)\mathbf{X}\Big\} + \tfrac{2}{3} k k'(s_p)\dot{s}_p(t) \\ &= \lambda\Big\{\tfrac{2}{3} k^2(2\mu + c) - \tfrac{2}{3} bk(\mathbf{T}^D - \mathbf{X}^D) \cdot \mathbf{X} + \tfrac{4}{9} k^2 k'(s_p)\Big\}\end{aligned}$$

or

$$\lambda = 3\mu \frac{(\mathbf{T}^D - \mathbf{X}^D) \cdot \dot{\mathbf{E}}(t)}{k^2(2\mu + c) - bk(\mathbf{T}^D - \mathbf{X}^D) \cdot \mathbf{X} + \frac{2}{3} k^2 k'(s_p)} . \tag{5.195}$$

This corroborates (5.190). It remains to be shown that the loading condition $(\mathbf{T}^D - \mathbf{X}^D) \cdot \dot{\mathbf{T}} > 0$ can be replaced by the condition $(\mathbf{T}^D - \mathbf{X}^D) \cdot \dot{\mathbf{E}} > 0$. To this end the elasticity relation (5.179) and the flow rule (5.183) are employed together with (5.190). The result is

$$(\mathbf{T}^{\mathrm{D}} - \mathbf{X}^{\mathrm{D}}) \cdot \dot{\mathbf{T}}(t) = 2\mu(\mathbf{T}^{\mathrm{D}} - \mathbf{X}^{\mathrm{D}}) \cdot (\dot{\mathbf{E}} - \dot{\mathbf{E}}_{\mathrm{p}})$$
$$= 2\mu(\mathbf{T}^{\mathrm{D}} - \mathbf{X}^{\mathrm{D}}) \cdot \dot{\mathbf{E}} - 2\mu(\mathbf{T}^{\mathrm{D}} - \mathbf{X}^{\mathrm{D}}) \cdot \left\{\frac{3\mu}{N}\left[(\mathbf{T}^{\mathrm{D}} - \mathbf{X}^{\mathrm{D}}) \cdot \dot{\mathbf{E}}\right](\mathbf{T}^{\mathrm{D}} - \mathbf{X}^{\mathrm{D}})\right\},$$

that means

$$(\mathbf{T}^{\mathrm{D}} - \mathbf{X}^{\mathrm{D}}) \cdot \dot{\mathbf{T}}(t) = 2\mu \, \frac{N - 2\mu k^2}{N} \, (\mathbf{T}^{\mathrm{D}} - \mathbf{X}^{\mathrm{D}}) \cdot \dot{\mathbf{E}} \, . \tag{5.196}$$

The sign for the prefactor depends on the numerator

$$N - 2\mu k^2 = k^2 c - kb(\mathbf{T}^{\mathrm{D}} - \mathbf{X}^{\mathrm{D}}) \cdot \mathbf{X} + \tfrac{2}{3} k^2 k'(s_{\mathrm{p}}) \, . \tag{5.197}$$

This quantity vanishes for the special case of perfect plasticity, i.e. for $c = b = k' = 0$.

In the general case $c \geq 0$, $b \geq 0$, and $k' \geq 0$ the second term on the right-hand side of (5.197) can be estimated by means of

$$|(\mathbf{T}^{\mathrm{D}} - \mathbf{X}^{\mathrm{D}}) \cdot \mathbf{X}| \leq \|(\mathbf{T}^{\mathrm{D}} - \mathbf{X}^{\mathrm{D}})\| \, \|\mathbf{X}\| \leq \|(\mathbf{T}^{\mathrm{D}} - \mathbf{X}^{\mathrm{D}})\| \sqrt{\frac{3}{2}\frac{c}{b}} = k\frac{c}{b} \, ,$$

where the inequality $\|\mathbf{X}\| \leq \sqrt{\frac{3}{2}\frac{c}{b}}$ is a conclusion from the evolution equation (5.184). It therefore holds that

$$N - 2\mu k^2 > 0,$$

if c and k' are not zero simultaneously. The two loading criteria

$(\mathbf{T}^{\mathrm{D}} - \mathbf{X}^{\mathrm{D}}) \cdot \dot{\mathbf{T}}$ and $(\mathbf{T}^{\mathrm{D}} - \mathbf{X}^{\mathrm{D}}) \cdot \dot{\mathbf{E}}$ are accordingly equivalent. ■

With the proportionality factor according to (5.195) the flow rule (5.183) reads

$$\dot{\mathbf{E}}_{\mathrm{p}}(t) = \frac{3\mu}{N}\left[(\mathbf{T}^{\mathrm{D}} - \mathbf{X}^{\mathrm{D}}) \cdot \dot{\mathbf{E}}\right](\mathbf{T}^{\mathrm{D}} - \mathbf{X}^{\mathrm{D}}) \, . \tag{5.198}$$

In connection with the remaining equations of the material model this evolution equation uniquely defines the current plastic strain $\mathbf{E}_{\mathrm{p}}(t)$, if a total strain process $t \longmapsto \mathbf{E}(t)$ is given (strain control). The same is true of stress control: (5.198) and (5.196) imply the evolution equation

$$\dot{\mathbf{E}}_{\mathrm{p}}(t) = \frac{3}{2}\,\frac{1}{N - 2\mu k^2}\left[(\mathbf{T}^{\mathrm{D}} - \mathbf{X}^{\mathrm{D}})\cdot\dot{\mathbf{T}}\right](\mathbf{T}^{\mathrm{D}} - \mathbf{X}^{\mathrm{D}})\,. \tag{5.199}$$

Obviously the last equation is not applicable in the case of perfect plasticity. In order to illustrate the physical meaning of the more general plasticity model (5.178) - (5.186), a uniaxial tension/compression process is examined. Uniaxial homogenous processes are characterised by the stress tensor

$$\mathbf{T}(t) = \sigma(t)\boldsymbol{e}_x \otimes \boldsymbol{e}_x\,, \tag{5.200}$$

whereas the strain tensor adopts the form

$$\mathbf{E}(t) = \varepsilon(t)\boldsymbol{e}_x \otimes \boldsymbol{e}_x + \varepsilon_{\mathrm{Q}}(t)(\boldsymbol{e}_y \otimes \boldsymbol{e}_y + \boldsymbol{e}_z \otimes \boldsymbol{e}_z)\,. \tag{5.201}$$

The plastic strain tensor $\mathbf{E}_{\mathrm{p}}$ is a deviator in view of the flow rule,

$$\mathbf{E}_{\mathrm{p}}(t) = \mathbf{E}_{\mathrm{p}}{}^{\mathrm{D}}(t) = \varepsilon_{\mathrm{p}}(t)\left[\boldsymbol{e}_x \otimes \boldsymbol{e}_x - \frac{1}{2}(\boldsymbol{e}_y \otimes \boldsymbol{e}_y + \boldsymbol{e}_z \otimes \boldsymbol{e}_z)\right], \tag{5.202}$$

and the same applies for the hardening tensor,

$$\mathbf{X}(t) = \mathbf{X}^{\mathrm{D}}(t) = \xi(t)\left[\boldsymbol{e}_x \otimes \boldsymbol{e}_x - \frac{1}{2}(\boldsymbol{e}_y \otimes \boldsymbol{e}_y + \boldsymbol{e}_z \otimes \boldsymbol{e}_z)\right]. \tag{5.203}$$

Corresponding to the uniaxial state of stress is the deviator

$$\mathbf{T}^{\mathrm{D}}(t) = \frac{2}{3}\sigma(t)\left[\boldsymbol{e}_x \otimes \boldsymbol{e}_x - \frac{1}{2}(\boldsymbol{e}_y \otimes \boldsymbol{e}_y + \boldsymbol{e}_z \otimes \boldsymbol{e}_z)\right]. \tag{5.204}$$

Theorem 5. 15

For uniaxial stress-strain processes the constitutive model (5.178) - (5.186) of plasticity with hardening implies the following differential equations:

$$\dot{\sigma}(t) = \begin{cases} \dfrac{EE_{\mathrm{p}}}{E + E_{\mathrm{p}}}\,\dot{\varepsilon} & \text{for } |\sigma - \frac{2}{3}\xi| = k \text{ and } (\sigma - \frac{2}{3}\xi)\dot{\varepsilon} > 0 \\ E\dot{\varepsilon} & \text{otherwise} \end{cases} \tag{5.205}$$

Therein $E = 2\mu(1 + \nu)$ is the elasticity modulus and

$$E_{\mathrm{p}} = \tfrac{3}{2}c - \tfrac{3}{2}b\,\xi\,\mathrm{sgn}(\sigma - \tfrac{3}{2}\xi) + k'(s_{\mathrm{p}}) \tag{5.206}$$

the so-called *plastic tangent modulus.* For the plastic arclength s_{p} the relation

$$\dot{s}_{\mathrm{p}}(t) = |\dot{\varepsilon}_{\mathrm{p}}(t)| \tag{5.207}$$

is valid, and for the variable ξ of kinematic hardening we have the evolution equation

$$\dot{\xi}(t) = c\dot{\varepsilon}_{\mathrm{p}}(t) - b|\dot{\varepsilon}_{\mathrm{p}}(t)|\xi(t)\ . \tag{5.208}$$

The evolution equation for the plastic strain is

$$\dot{\varepsilon}_{\mathrm{p}}(t) = \begin{cases} \dfrac{E}{E + E_{\mathrm{p}}}\,\dot{\varepsilon} & \text{for } |\sigma - \tfrac{2}{3}\xi| = k \text{ and } (\sigma - \tfrac{2}{3}\xi)\dot{\varepsilon} > 0 \\[2ex] 0 & \text{otherwise} \end{cases} \tag{5.209}$$

□

Proof

In order to insert the uniaxial process (5.200) - (5.204) into the general constitutive equations, the following quantities are calculated:

$$(\mathbf{T}^{\mathrm{D}} - \mathbf{X}^{\mathrm{D}})\cdot\dot{\mathbf{E}}(t) = (\tfrac{2}{3}\sigma - \xi)(\dot{\varepsilon} - \dot{\varepsilon}_Q)\ ,$$

$$(\mathbf{T}^{\mathrm{D}} - \mathbf{X}^{\mathrm{D}})\cdot\mathbf{X} = \tfrac{3}{2}\xi(\tfrac{2}{3}\sigma - \xi)\ ,$$

$$f(\mathbf{T}, \mathbf{X}, k) = \tfrac{3}{4}(\tfrac{2}{3}\sigma - \xi)^2 - \tfrac{1}{3}k^2\ ,$$

$$f = 0 \Leftrightarrow |\sigma - \tfrac{3}{2}\xi| = k\ .$$

Accordingly, the xx- and yy-components of the tensor-valued differential equation $(5.192)_1$ are

$$\dot{\sigma}(t) = 2\mu[\dot{\varepsilon} - \tfrac{4}{3}k^2\frac{\mu}{N}(\dot{\varepsilon} - \dot{\varepsilon}_Q) + \frac{\nu}{1 - 2\nu}(\dot{\varepsilon} + 2\dot{\varepsilon}_Q)] \tag{5.210}$$

and

$$0 = \dot{\varepsilon}_Q + \tfrac{2}{3}k^2\frac{\mu}{N}(\dot{\varepsilon} - \dot{\varepsilon}_Q) + \frac{\nu}{1 - 2\nu}(\dot{\varepsilon} + 2\dot{\varepsilon}_Q)\ . \tag{5.211}$$

The outcome of the last equation is the relation

$$\dot{\varepsilon}_Q = -\frac{\nu\frac{3N}{k^2} + 2\mu(1-2\nu)}{\frac{3N}{k^2} - 2\mu(1-2\nu)}\,\dot{\varepsilon} \tag{5.212}$$

between the axial strain $\dot{\varepsilon}(t)$ and the lateral strain $\dot{\varepsilon}_Q(t)$. By inserting this into (5.210) we obtain a differential equation to determine the stress:

$$\dot{\sigma}(t) = 2\mu(1+\nu)\,\frac{\frac{3N}{k^2} - 6\mu}{\frac{3N}{k^2} - 2\mu(1-2\nu)}\,\dot{\varepsilon}\ .$$

Equation (5.191) supplies

$$\frac{1}{k^2}\,N(\mathbf{T},\mathbf{X},k) = 2\mu + c - b\xi\,\frac{\sigma - \frac{3}{2}\xi}{|\sigma - \frac{3}{2}\xi|} + \tfrac{2}{3}k'(s_p)$$

or

$$\frac{3N}{k^2} = 6\mu + 3c - 3b\xi\,\mathrm{sgn}(\sigma - \tfrac{3}{2}\xi) + 2k'(s_p)\ . \tag{5.213}$$

The final result is

$$\dot{\sigma}(t) = \frac{2\mu(1+\nu)\{3c - 3b\xi\,\mathrm{sgn}(\sigma - \frac{3}{2}\xi) + 2k'(s_p)\}}{4\mu(1+\nu) + \{3c - 3b\xi\,\mathrm{sgn}(\sigma - \frac{3}{2}\xi) + 2k'(s_p)\}}\,\dot{\varepsilon} = \frac{EE_p}{E + E_p}\,\dot{\varepsilon}\ .$$

The loading function

$$(\mathbf{T}^D - \mathbf{X}^D)\cdot\dot{\mathbf{E}}(t) = (\tfrac{2}{3}\sigma - \xi)(\dot{\varepsilon} - \dot{\varepsilon}_Q) = (\tfrac{2}{3}\sigma - \xi)\,\frac{\frac{3N}{k^2}(1+\nu)}{\frac{3N}{k^2} - 2\mu(1-2\nu)}\,\dot{\varepsilon}$$

leads to the uniaxial loading criterion $(\sigma - \frac{3}{2}\xi)\dot{\varepsilon} > 0$.

The norm of plastic strain rate is calculated to $\|\dot{\mathbf{E}}_p(t)\| = \sqrt{\frac{3}{2}}\,|\dot{\varepsilon}_p(t)|$,

and we obtain from (5.186) the rate of the accumulated plastic strain,

$$\dot{s}_p(t) = |\dot{\varepsilon}_p(t)|\ .$$

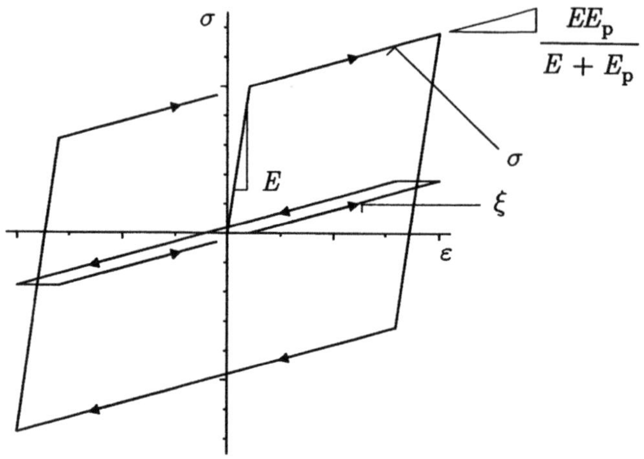

Figure 5. 16: Linear kinematic hardening

With these results the tensor-valued evolution equation (5.184) reduces to the xx-component

$$\dot{\xi}(t) = c\dot{\varepsilon}_{\mathrm{p}}(t) - b|\dot{\varepsilon}_{\mathrm{p}}(t)|\xi(t) \, .$$

For the xx-component of the flow rule (5.183) or (5.198) we have

$$\dot{\varepsilon}_{\mathrm{p}}(t) = 3\frac{\mu}{N}\left(\tfrac{2}{3}\sigma - \xi\right)^2(\dot{\varepsilon} - \dot{\varepsilon}_Q) = \tfrac{4}{3}\,k^2\frac{\mu}{N}\,\frac{\frac{3N}{k^2}(1+\nu)}{\frac{3N}{k^2} - 2\mu(1-2\nu)}\,\dot{\varepsilon}$$

$$= \frac{4\mu(1+\nu)}{\frac{3N}{k^2} - 2\mu(1-2\nu)}\,\dot{\varepsilon}$$

$$= \frac{4\mu(1+\nu)}{4\mu(1+\nu) + 3c - 3b\xi\,\mathrm{sgn}(\sigma - \frac{3}{2}\xi) + 2k'(s_{\mathrm{p}})}\,\dot{\varepsilon} = \frac{E}{E+E_{\mathrm{p}}}\,\dot{\varepsilon} \, . \qquad \blacksquare$$

Just two special cases should be discussed for the purpose of illustrating the physical significance of this model.

Linear Kinematic Hardening: $c \neq 0, b = 0, k'(s_p) = 0$
In this particular case the plastic tangent modulus according to (5.206) is a constant:

$$E_p = \frac{3}{2}c \,. \tag{5.214}$$

As a result, one obtains a field of stress-strain characteristics defined by two straight lines. This is illustrated in Fig. 5. 16. The representation of the elastoplastic stress-strain curve by means of two straight lines is called *bilinear approximation.*
The model of linear kinematic hardening describes an increase in yield stress through plastic deformation. After tensile loading into the plastic range and unloading, followed by compression, the absolute value of the yield stress is smaller than the initial yield stress. In this sense the model of linear kinematic hardening is able to reproduce a very special dependence of the yield stress on the loading history, which is called a *Bauschinger* effect. With that, however, the capacity of the bilinear approximation is exhausted: during cyclic deformation processes the form of these hysteresis curves is always reproduced, regardless of any previous process history (see Fig. 5. 16).

Nonlinear Kinematic Hardening: $c \neq 0, b \neq 0, k'(s_p) = 0$.
Using the more general evolution equation (5.208) with $b \neq 0$ we can represent a nonlinear hardening, that means a curvilinear stress-strain characteristic. It is very easy to visualise the path of this curve by reverting to the yield condition

$$|\sigma - \frac{3}{2}\xi| = k \,. \tag{5.215}$$

In the case of monotonic tension we have $\sigma > \frac{3}{2}\xi$, and (5.215) reduces to

$$\sigma = \frac{3}{2}\xi + k \,, \tag{5.216}$$

i.e. a constant yield stress is added to the current value of $(3/2)\xi$. By integrating the evolution equation (5.208) it is possible to calculate the kinematic hardening variable ξ as a function of the plastic strain: (5.208) combined with $\dot{\varepsilon}_p(t) > 0$ leads to the differential equation

$$\xi'(\varepsilon_p) + b\xi(\varepsilon_p) = c \tag{5.217}$$

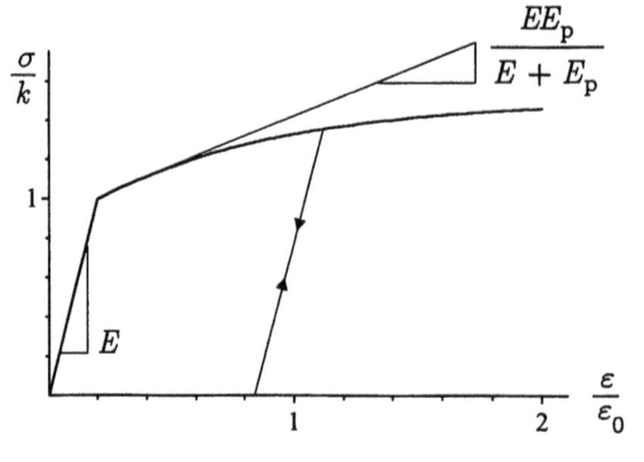

Figure 5. 17: Nonlinear kinematic hardening; integration of the differential equations (5.205) - (5.209) for a monotonic strain process

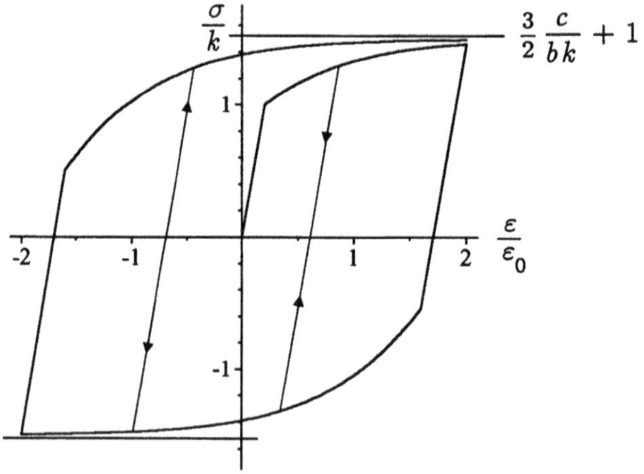

Figure 5. 18: Nonlinear kinematic hardening for cyclic deformations; integration of differential equations (5.205) - (5.209) for one strain cycle

with the solution

$$\xi(\varepsilon_p) = \frac{c}{b}\left[1 - e^{-b\varepsilon_p}\right] . \tag{5.218}$$

The shape of the σ-ε characteristic as a solution of equations (5.205) - (5.209) is illustrated in Fig. 5. 17. In the case of linear kinematic hardening the yield stress is a linear function of ε_{p} (Fig. 5. 16). According to the nonlinear kinematic hardening model the yield stress is limited (Fig. 5. 17): its upper bound is $\frac{3}{2}\frac{c}{b} + k$ (Fig. 5. 18). In order to calculate the stress-strain curve for arbitrary deformation processes the differential equations (5.205) - (5.208) have to be integrated in steps, evaluating the yield and loading conditions.

One result that can be obtained for cyclic strain processes is sketched in Fig. 5. 18. We observe a stress-strain behaviour with typical features of the hardening model:

1) The hardening variable ξ cannot exceed a maximum of $\frac{3c}{2b}$.

2) The response to a cyclic strain process is a hysteresis loop.

3) Cyclic hardening and softening effects are qualitatively represented by the model; however, the degree of hardening during the hysteresis loops is much too small in comparison with experimental data.

These features indicate fundamental limits of the nonlinear hardening model (5.184).[36]

The constitutive equations of the classical plasticity theory represent *rate-independent* material behaviour: the current stress response depends on the path of the past deformation process, but it is independent of the velocity of the process. The property of rate-independence is obvious from the general equations (5.172) and (5.192), where first order time derivatives appear on both sides, so that any transformation of the time scale, which can also be interpreted as a change in the rate of the process, cancels out of the equations.

5. 8 Viscoplasticity with Elastic Range

The classical theories of viscoelasticity and plasticity represent inelastic material behaviour from a completely different angle. The theory of linear viscoelasticity models rate-dependent material behaviour: all unlimited relaxation processes lead to equilibrium stresses, which are linear functions of the strains. Cyclic loadings lead to rate-dependent hysteresis loops. By contrast, the theory of rate-independent plasticity is physically nonlinear. It reproduces hysteresis behaviour as well, but has no potential for describing

[36] It is possible to achieve a more realistic modelling of cyclic hardening and softening within the concept of the generalised arclength (see Sect. 11. 2).

the influence of the deformation rate and relaxation properties. To compensate, however, the theory of plasticity is in a position to represent hysteresis effects, which do not depend on the velocity of the deformation processes. The *viscoplasticity model* to be discussed in the following was proposed at an early stage and therefore counts as one of the classical theories of continuum mechanics. It is an attempt to combine the particular advantages of viscoelasticity and plasticity.

The model of viscoplasticity grows out of the theory of plasticity by changing the flow rule (5.183): one waives the requirement of the yield condition $f = 0$ in the case of inelastic processes, i.e. positive values of the yield function are permitted. As a consequence, the proportionality factor λ in equation (5.183) becomes a quantity which one is at liberty to use for describing material properties: instead of λ, which previously had to be determined from the consistency condition, a material constant or even a material function is inserted. Because it relates the inelastic deformation rate to the stress, it can be interpreted as a viscosity. A variation of this kind brings about the desired effect of removing the rate-independence of the material response according to the model: the flow rule determines an inelastic strain rate, which can be understood as a creep rate. In this context it seems reasonable to make the following assumption: the further the state of stress lies from the elastic domain, the larger the inelastic deformation rate is. This motivates

Definition 5. 10

A *viscoplastic solid* is characterised by means of the following set of constitutive equations:[37]

Decomposition of the Deformation

$$\mathbf{E} = \mathbf{E}_e + \mathbf{E}_i \tag{5.219}$$

Elasticity Relation

$$\mathbf{T} = 2\mu[\mathbf{E}_e + \frac{\nu}{1-2\nu}(\mathrm{tr}\mathbf{E}_e)\mathbf{1}] \tag{5.220}$$

Yield Function

$$f(\mathbf{T} - \mathbf{X}, k) = \sqrt{\tfrac{3}{2}}\,\|\mathbf{T}^D - \mathbf{X}^D\| - k \tag{5.221}$$

[37] This constitutive model is attributed to PERZYNA [1963]. The principal idea goes back to HOHENEMSER & PRAGER [1932]. See PRAGER [1961], p. 126.

Associated Flow Rule

$$\dot{\mathbf{E}}_{\mathrm{i}}(t) = \frac{1}{\eta}\left\langle\frac{1}{k_0} f(\mathbf{T} - \mathbf{X}, k)\right\rangle^{\mathrm{m}} \frac{\partial f}{\partial \mathbf{T}} = \frac{1}{\eta}\left\langle\frac{1}{k_0} f\right\rangle^{\mathrm{m}} \sqrt{\frac{3}{2}}\,\frac{\mathbf{T}^{\mathrm{D}} - \mathbf{X}^{\mathrm{D}}}{\left\|\mathbf{T}^{\mathrm{D}} - \mathbf{X}^{\mathrm{D}}\right\|} \tag{5.222}$$

Kinematic Hardening

$$\dot{\mathbf{X}}(t) = c\dot{\mathbf{E}}_{\mathrm{i}}(t) - b\sqrt{\frac{2}{3}}\,\left\|\dot{\mathbf{E}}_{\mathrm{i}}(t)\right\|\mathbf{X}$$

$$\dot{\mathbf{X}}(t) = \frac{1}{\eta}\left\langle\frac{1}{k_0} f\right\rangle^{\mathrm{m}}\left\{c\sqrt{\frac{3}{2}}\,\frac{\mathbf{T}^{\mathrm{D}} - \mathbf{X}^{\mathrm{D}}}{\left\|\mathbf{T}^{\mathrm{D}} - \mathbf{X}^{\mathrm{D}}\right\|} - b\mathbf{X}\right\} \tag{5.223}$$

MacCauley Bracket

$$\langle x\rangle = \begin{cases} x \text{ for } x \geq 0 \\ 0 \text{ for } x < 0 \end{cases} \tag{5.224}$$

◊

In analogy to plasticity the total strain tensor $\mathbf{E}$ is the sum of one elastic and one inelastic part. The elastic part determines the stress by means of the elasticity relation (5.220). The yield function (5.221) determines the elastic domain in the space of stress tensors by means of $f \leq 0$. The equation $f = 0$ defines the so-called *static yield surface*. States of stress lying beyond this yield surface create inelastic deformations according to the flow rule (5.222). The flow rule determines an inelastic strain rate, the absolute value of which is assumed to be proportional to the power f^{m}, if the value of the yield function is positive. For negative or vanishing values of the yield function the inelastic strain remains constant, a fact that is guaranteed by introducing the *MacCauley* bracket (5.224) into the flow rule. k_0 has the dimension of stress and may be a constant or a function of internal variables. The quantity η is a material parameter with the physical meaning of a viscosity. The stress tensor $\mathbf{X}^{\mathrm{D}}$, which occurs in the yield function (5.221), is the central point of the static yield surface. The evolution equation (5.223) corresponds exactly to the earlier assumption (5.184) for kinematic hardening in rate-independent plasticity. A further evolution equation could be inserted as well to describe isotropic hardening.

In order to study the physical significance of the material model (5.219) - (5.223) we examine a uniaxial stress-strain process again,

$$\mathbf{T}(t) = \sigma(t)\mathbf{e}_x \otimes \mathbf{e}_x\,, \quad \mathbf{X}(t) = \xi(t)\Big[\mathbf{e}_x \otimes \mathbf{e}_x - \tfrac{1}{2}(\mathbf{e}_y \otimes \mathbf{e}_y + \mathbf{e}_z \otimes \mathbf{e}_z)\Big]\,,$$

$$\mathbf{E}(t) = \varepsilon(t)\mathbf{e}_x \otimes \mathbf{e}_x + \varepsilon_{\mathrm{Q}}(t)(\mathbf{e}_y \otimes \mathbf{e}_y + \mathbf{e}_z \otimes \mathbf{e}_z)\,,$$

$$\mathbf{E}_{\mathrm{i}}(t) = \varepsilon_{\mathrm{i}}(t)\Big[\mathbf{e}_x \otimes \mathbf{e}_x - \tfrac{1}{2}(\mathbf{e}_y \otimes \mathbf{e}_y + \mathbf{e}_z \otimes \mathbf{e}_z)\Big]\,.$$

This is accounted for in

Theorem 5. 16

In connection with the constitutive equations (5.219) - (5.223) of viscoplasticity uniaxial processes are represented by the following differential equations:

$$\dot\sigma(t) = 2\mu(1+\nu)\Big[\dot\varepsilon(t) - \tfrac{1}{\eta}\Big\langle\tfrac{1}{k_0}\,(|\sigma - \tfrac{3}{2}\xi| - k)\Big\rangle^{\mathrm{m}}\mathrm{sgn}(\sigma - \tfrac{3}{2}\xi)\Big]\,, \tag{5.225}$$

$$\dot\xi(t) = \tfrac{1}{\eta}\Big\langle\tfrac{1}{k_0}\,(|\sigma - \tfrac{3}{2}\xi| - k)\Big\rangle^{\mathrm{m}}\Big[c\,\mathrm{sgn}(\sigma - \tfrac{3}{2}\xi) - b\xi\Big]\,. \tag{5.226}$$

□

Proof

The stress difference

$$\mathbf{T}^{\mathrm{D}} - \mathbf{X}^{\mathrm{D}} = \Big(\tfrac{2}{3}\sigma - \xi\Big)\Big[\mathbf{e}_x \otimes \mathbf{e}_x - \tfrac{1}{2}(\mathbf{e}_y \otimes \mathbf{e}_y + \mathbf{e}_z \otimes \mathbf{e}_z)\Big]$$

leads to the yield function $f = \sqrt{\tfrac{3}{2}}\,\|(\mathbf{T}^{\mathrm{D}} - \mathbf{X}^{\mathrm{D}})\| - k = |\sigma - \tfrac{3}{2}\xi| - k$.

Moreover, (5.220) with (5.219) gives us the component equations

$$\dot\sigma(t) = 2\mu[\dot\varepsilon(t) - \dot\varepsilon_{\mathrm{i}}(t) + \frac{\nu}{1-2\nu}(\dot\varepsilon(t) + 2\dot\varepsilon_{\mathrm{Q}}(t))]\,, \tag{5.227}$$

$$0 = \dot\varepsilon_{\mathrm{Q}}(t) + \tfrac{1}{2}\dot\varepsilon_{\mathrm{i}}(t) + \frac{\nu}{1-2\nu}(\dot\varepsilon(t) + 2\dot\varepsilon_{\mathrm{Q}}(t))\,. \tag{5.228}$$

Equations (5.222) and (5.223) supply the evolution equations for the inelastic strain rate $\varepsilon_{\mathrm{i}}(t)$,

$$\dot\varepsilon_{\mathrm{i}}(t) = \tfrac{1}{\eta}\Big\langle\tfrac{1}{k_0}\,(|\sigma - \tfrac{3}{2}\xi| - k)\Big\rangle^{\mathrm{m}}\,\mathrm{sgn}(\sigma - \tfrac{3}{2}\xi)\,, \tag{5.229}$$

and the hardening variable $\xi(t)$,

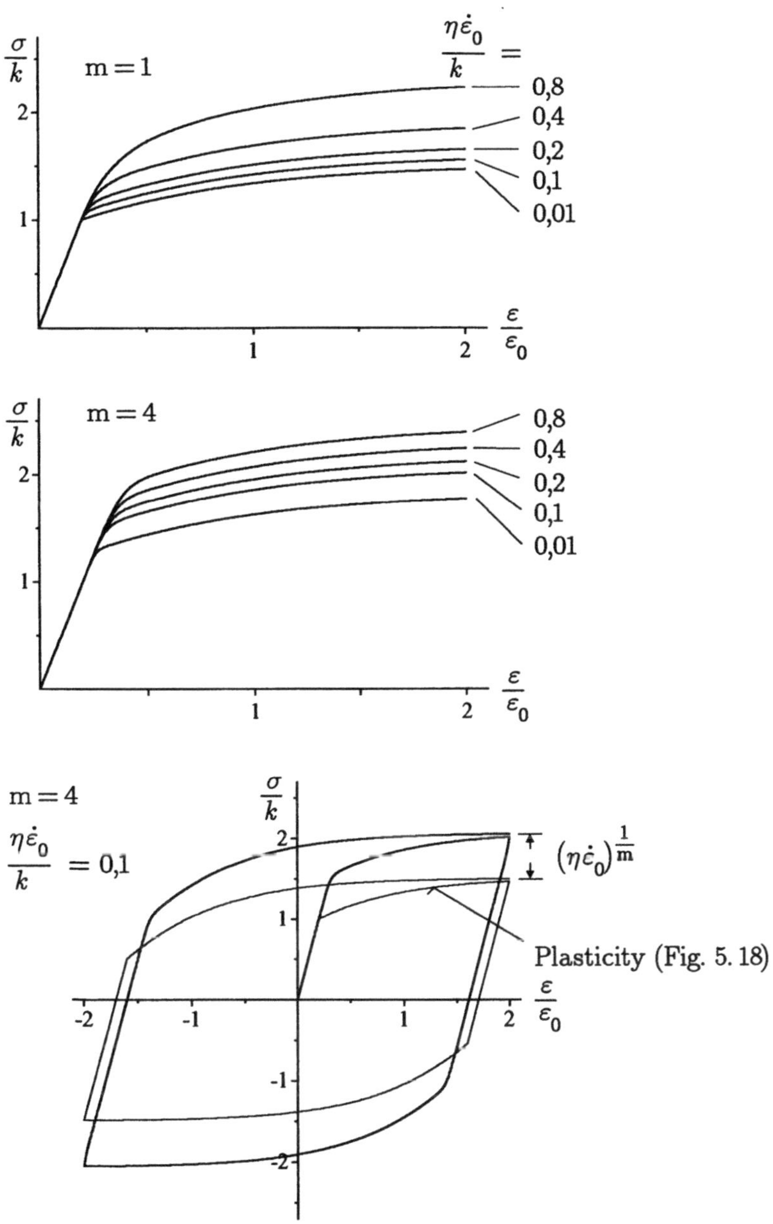

Figure 5. 19: Representation of rate-dependence for the material model of viscoplasticity; integration of differental equations (5.225) and (5.226) for monotonic and cyclic processes with constant strain rate

$$\dot{\xi}(t) = c\dot{\varepsilon}_i(t) - b|\dot{\varepsilon}_i(t)|\xi \,. \tag{5.230}$$

Based on the yy-component (5.228) one calculates for the transverse strain

$$\dot{\varepsilon}_Q(t) = -\nu\dot{\varepsilon}(t) - \frac{1}{2}(1 - 2\nu)\dot{\varepsilon}_i(t) \,. \tag{5.231}$$

It therefore holds that $\dot{\varepsilon} + 2\dot{\varepsilon}_Q = (1 - 2\nu)(\dot{\varepsilon} - \dot{\varepsilon}_i)$ and the xx-component (5.227) reads $\dot{\sigma}(t) = 2\mu(1 + \nu)(\dot{\varepsilon} - \dot{\varepsilon}_i)$. ■

The one-dimensional equations (5.225) and (5.226) provide information on the representation of nonlinear rate-dependence: for this purpose, based on the initial conditions $\sigma(0) = 0$ and $\varepsilon(0) = 0$, a monotonic strain process with constant strain rate $\dot{\varepsilon}_0 \neq 0$ is investigated. In this context, the functions $\dot{\sigma}(t)$ and $\dot{\xi}(t)$ are positive as well (but not constant). In particular, σ and ξ tend towards stationary values σ_{st} and ξ_{st}, for which the differential equations (5.225) and (5.226) provide the relations

$$0 = \dot{\varepsilon}_0 - \frac{1}{\eta}\left[\frac{1}{k_0}\left(\sigma_{st} - \frac{3}{2}\xi_{st} - k\right)\right]^{m} , \tag{5.232}$$

$$0 = c - b\xi_{st} \,. \tag{5.233}$$

For the asymptotic limit value of the stress these lead to

$$\sigma_{st} = k + \frac{3}{2}\frac{c}{b} + (\eta\dot{\varepsilon}_0)^{\frac{1}{m}} \,. \tag{5.234}$$

Figure 5. 19 illustrates the result: the model describes for $m \neq 1$ a nonlinear rate-dependence, inasmuch as the asymptotic limit value of stress increases by the value of $(\eta\dot{\varepsilon}_0)^{(1/m)}$.

Regardless of the numerical value of the strain rate $\dot{\varepsilon}_0$, the stress-strain curve is always continuously differentiable when leaving the elastic domain: for every finite strain rate a continuous tangent with a slope $E = 2\mu(1 + \nu)$ belongs to the strain $\varepsilon = k/E$. This is suggested by (5.225). In rate-independent plasticity, the corresponding tangent is on principle discontinuous (cf. Fig. 5. 18).[38]

[38] It is a general result, that the evolution equations of plasticity can be understood as an asymptotic limit of viscoplasticity if deformation processes evolve sufficiently slow or, equivalently, for sufficiently small values of the viscosity η. This will be discussed in Sect. 12. 3 for the purely mechanical theory and in Sect. 13. 4. 2 in a thermodynamical context.

5. 9 Remarks on the Classical Theories

As opposed to fluid mechanics, the classical theories of solid mechanics are defined within the scope of geometric linearisation: the theory of linear elasticity employs the linearised strain tensor, and the distinction between the stress tensors $\mathbf{T}$ and $\mathbf{T}_R$ in the balance of linear momentum is disregarded. The consequence of geometric linearisation is the equating of the current configuration and reference configuration. This process of equating is implemented in the formulation of both the field equations and the dynamic boundary conditions; for linear elasticity and viscoelasticity it leads to a linear initial-boundary-value problem. The advantage of the simpler solvability in the case of these boundary value problems (one result of linearity is the superposition property) is offset by the disadvantage that all solutions bear physical significance for small deformations only. This argument also applies for the classical theories of plasticity and viscoplasticity. From a pragmatic point of view this is not always a drawback, since in many technical applications of solid mechanics only small deformations are significant.

Contrary to solid mechanics, no such assumptions of approximation are made in classical fluid mechanics, apart from the specialised constitutive equations. Accordingly, the boundary value problems in fluid mechanics are nonlinear. In the field equations the nonlinearity is expressed in the convective derivative of the velocity vector as well as in the coupling of the velocity and density fields by the mass balance. The dynamic boundary conditions in particular, with their exact formulation, harbour their own high-ranking nonlinearity. It is true that they do not have any effect in many practical problems of fluid mechanics, where there are just geometric boundary conditions. However, the nonlinearity of the stress boundary condition presents considerable mathematical problems when *free surfaces* have to be taken into account. These problems would not occur in the case of geometric linearisation; on the other hand, the variety of phenomena described by means of boundary value problems of this kind would be seriously reduced. It should be noted that the classical theories of fluid mechanics are exact theories in the context of continuum mechanics, whereas the theories of linear elasticity and viscoelasticity as well as the classical theories of plasticity and viscoplasticity can only have the character of an approximation.

The recent development of continuum mechanics has led to the setting-up of a general *theory of material behaviour*. One aspect that has proved to be most beneficial is the idea that the current state of stress in a material body is determined by the past history of its deformation process. This aspect is also evident in the classical theories of continuum mechanics: the constitutive equations of viscoelasticity, plasticity and viscoplasticity are functional relations, i.e. differential or integral equations: given a strain process (possibly in connection with initial conditions), the current stress is determined by the integration of constitutive equations. The constitutive equations likewise determine a present state of strain as a material response to every given stress process.

The constitutive equations leading to the classical theories of continuum mechanics are not positioned haphazardly alongside each other, despite their apparently arbitrary definitions and the marked difference in their appearance. A general theory of material behaviour shows that they originate from a common source: the classical constitutive equations can be regarded as *approximations* of one general constitutive model, all of which are valid subject to *restrictive assumptions*. These restrictive assumptions refer on the one hand to the structure of constitutive equations, for example *continuity* properties of *constitutive functionals* or properties of *material symmetry*. On the other hand the restrictive assumptions can also refer to the confinement of permissible processes, such as *slow motions* or *small deformations*. This general outcome of the *theory of materials*, which is yet to be accounted for, is not merely of theoretical interest, but also of practical importance, since it produces systematic possibilities for generalising the classical theories and discovering new fields of application for continuum mechanics.

6 Experimental Observation and Mathematical Modelling

6.1 General Aspects

Constitutive equations are mathematical models intended to reflect the principal features of real material behaviour in an idealised form. The basic question that immediately springs to mind is which aspects of a material body's behaviour can be regarded as essential. Depending on the individual case, the answer to this is bound to be influenced by subjective evaluations as well and is consequently not devoid of a certain arbitrariness. One operation that is subject to a high level of intent is the process of *idealisation*, which filters out those aspects of reality that are to be disregarded. After all, the mathematical model is intended to simplify the exeedingly complicated reality and produce a conclusive theory to render the behaviour of material bodies and systems predictable.

One should not suppose that it is the aim of the theory of material behaviour to develop a single, universal constitutive equation for describing every conceivable phenomenon. Rather the aim of the theory is to provide a classified range of possibilities from which a user can select the constitutive model that appears to be the most convenient in relation to an intended field of application, bearing in mind the relativity of the means.

It is of fundamental importance for the application of the theory of material behaviour to identify the *material parameters* by means of experimental tests.

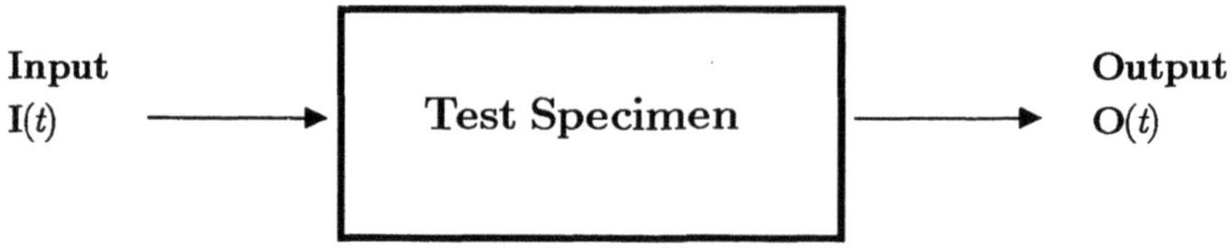

$\mathbf{I}(t) = \varepsilon(t)$ ***Strain Control*** $\mathbf{O}(t) = \sigma(t)$

$\mathbf{I}(t) = \sigma(t)$ ***Stress Control*** $\mathbf{O}(t) = \varepsilon(t)$

Figure 6. 1: Test specimen as operator

The characteristic material parameters, which a specific constitutive model possesses for the quantitative representation of the material properties, have to be determinable by practicable experiments.[1]

A specimen, undergoing laboratory research to identify its thermo-mechanical properties, proves to be an extremely complicated system in the light of close examination - under a sufficiently strong microscope, for instance. Every experimental test in pursuit of mathematical description and application of the results obtained would be a lost cause without radical simplification. The *experimental identification of material properties* in the sense of phenomenological continuum mechanics takes a number of idealisations for granted. As a general rule, the first one is the assumption that it is possible to produce spatially homogeneous states of strain and stress in macroscopically homogeneous test specimens: controlling and measuring strains and stresses in the case of homogeneous deformations has the same significance as conducting and measuring displacements and resultant forces.[2] If one accepts the implementability of these assumptions, a *test specimen* can be seen as an *operator* from the point of view of continuum mechanics, assigning an output process to every input process (see Fig. 6. 1). The input and output processes are both functions of time. In the case of strain control, the test specimen is subjected to time-dependent displacements (strains), causing resultant forces and torques (stresses) by way of material response. It is the other way round with stress control: the

[1] A comprehensive presentation of experimental work in continuum mechanics is to be found in BELL [1973].

[2] If these assumptions do not apply, inhomogeneous displacement fields have to be evaluated, see KREISSIG [1998].

test specimen reacts to the loading of time-dependent forces and torques (stresses) with displacements corresponding to time-dependent states of strain. Every *experiment* provides information on the *test specimen*, and it is the aim of the experiments to obtain as comprehensive a picture of the operator "test specimen" as possible. Then, it is the task of the constitutive theory to construct another operator, namely a *mathematical relation* to connect input and output processes; this relation - restricted to the experimental situation - ought to approach closely to the essential characteristics of the operator "test specimen".

A material model is naturally not restricted to reproducing data obtained in special experimental tests. Over and above the experimental data, which served to identify the material parameters, a material model has to be in a position to embrace general, three-dimensional deformation processes extending far beyond the selected phenomena simulated in the laboratory. It is only under these circumstances that a theory of continuum mechanics can evolve out of the combination of constitutive equations and balance relations to allow the calculation of the thermomechanical behaviour of material systems.

In *solid mechanics* one-dimensional tension and torsion tests are commonly carried out to determine material behaviour. Under ideal conditions a homogeneous uniaxial state of stress prevails in the *tension/compression test* in the middle domain of a thin rod with constant cross-section. A *torsion test* carried out on a cylindrical tube produces a homogeneous shear stress distribution if the wall is thin enough. In combined tension and torsion tests on thin-walled tubes it is possible, up to a point, to drive and measure two-dimensional states of stress and strain. Servo-controlled testing equipment makes light work of producing virtually any input process; the material responses can be registered simultaneously using electronic media. This provides a multitude of possibilities for making the macroscopic material behaviour accessible to experimental observation. The principal possibilities are, of course, considerably confined due to the technical limitations of the available testing devices and specimens: narrow restrictions are imposed on the attainable process velocities, to avoid the disturbing effects of inertia or influences resulting from temperature changes. In particular, the magnitude of the deformations that can be obtained is limited for reasons of stability: tension/compression rods may, for example, show necking or buckling phenomena when critical loading levels are reached, and tubes under torsion can become dented if the active torques exceed critical values.

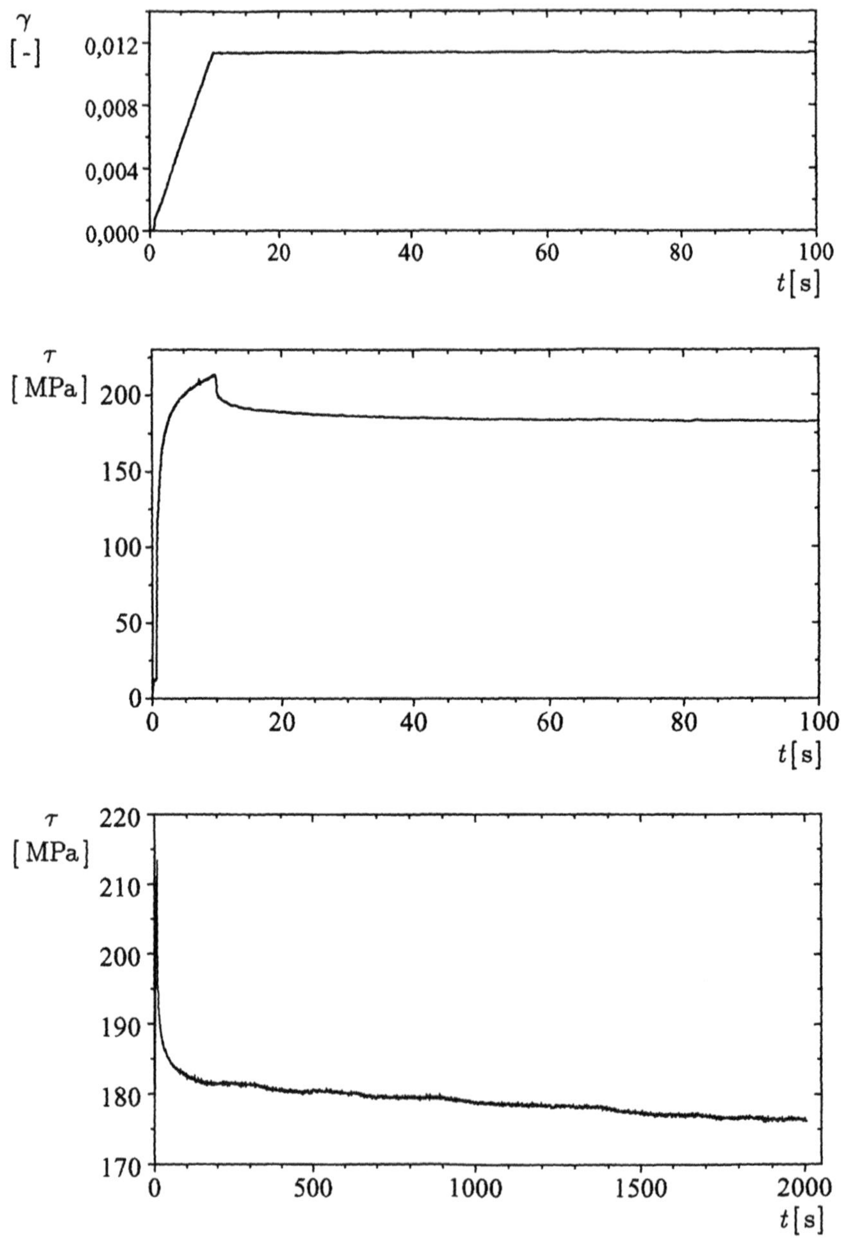

Figure 6. 2: Monotonic loading with hold time, strain control: relaxation

6. 2 Information from Experiments

Even the question of up to what point it might be adequate to investigate the material properties of solids, based exclusively on one-dimensional (and perhaps a few special two-dimensional) tests, is a fundamental problem.

As far as the *classical theories* of continuum mechanics are concerned, this problem is overcome in most instances. There is no generally applicable solution, however. Instead, we have to be content with checking from case to case whether an experimentally identifiable material description is possible, which would lead to a qualitative and quantitative representation of those phenomena deemed to be essential.

This section is intended to illustrate a selection of experimental data obtained in the laboratory of the Institute of Mechanics at the University of Kassel.[3] The results of the tests documented are of exemplary nature and are intended only to provide hints as to what ought to be expected of material models, if material behaviour is to be reproduced accurately in at least a number of essential points.

6. 2. 1 Material Properties of Steel XCrNi 18.9

The following figures illustrate the behaviour of thin-walled tubes made of XCrNi 18.9 steel, as measured. This material is suitable for demonstrating a range of phenomena, the modelling of which can be an integral part of the theory of material behaviour.

Figure 6. 2 documents the results of a strain-controlled *torsion test* with a final shear of $\gamma_0 = 0{,}0114$ [-], approached at a rate of $\dot{\gamma}_0 = 0{,}00114$ $[\mathrm{s}^{-1}]$. The shear was then kept constant. The resulting shear stress is a function of time. In particular, it is time-dependent in that section of the test in which the prescribed shear strain is constant: the recorded shear stress $\tau(t)$ decreases monotonically. This process is known as *relaxation*. Whether the relaxation process has reached its limit after 2000 seconds remains open, as the lower diagram of Fig. 6. 2 shows. One can assume that the stress would eventually tend towards a time-independent equilibrium value, were the experiment to be continued indefinitely. The shape of the relaxation curve indicates that the limit value is different from zero.

[3] The experiments were carried out by Dr. A. Lion and Dr. L. Schreiber, with the support of Dipl.-Ing. G. Linek and W. Zugreif.

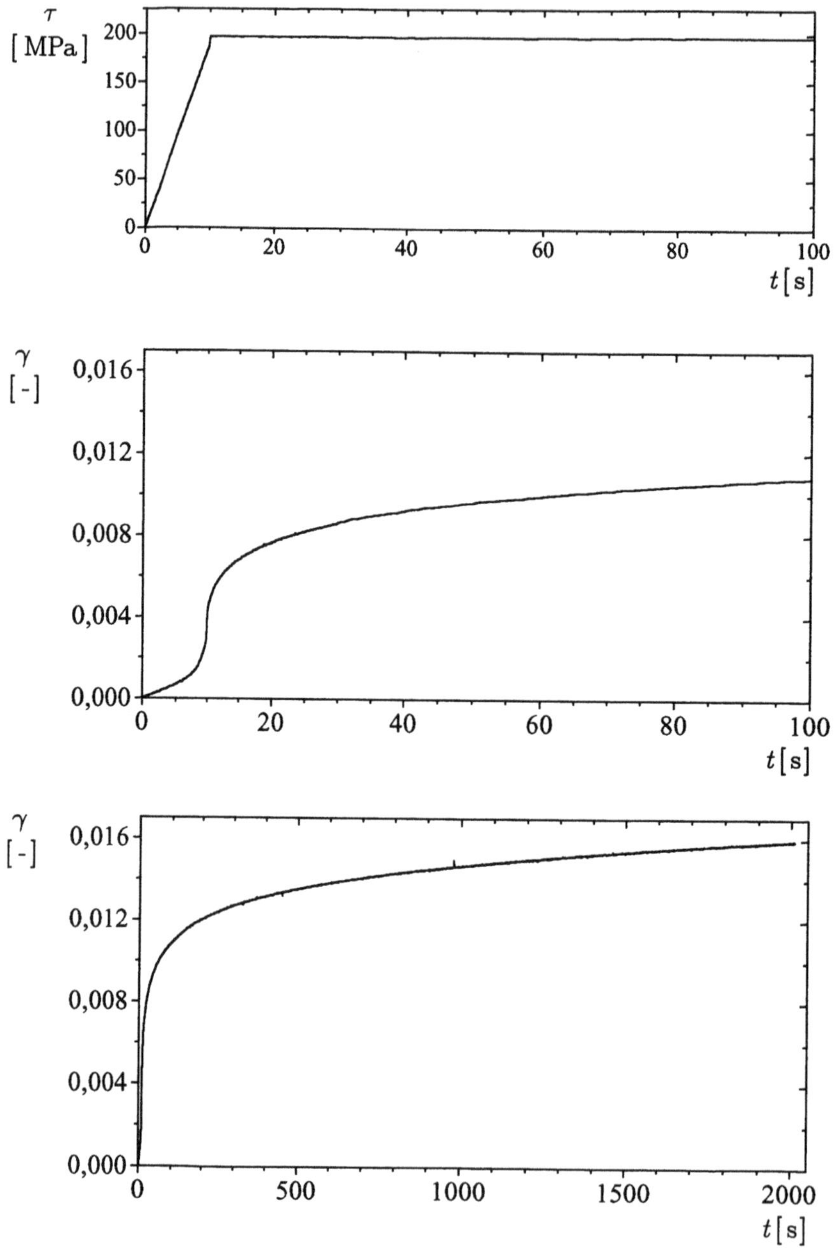

Figure 6. 3: Monotonic loading with hold time, stress control: creep

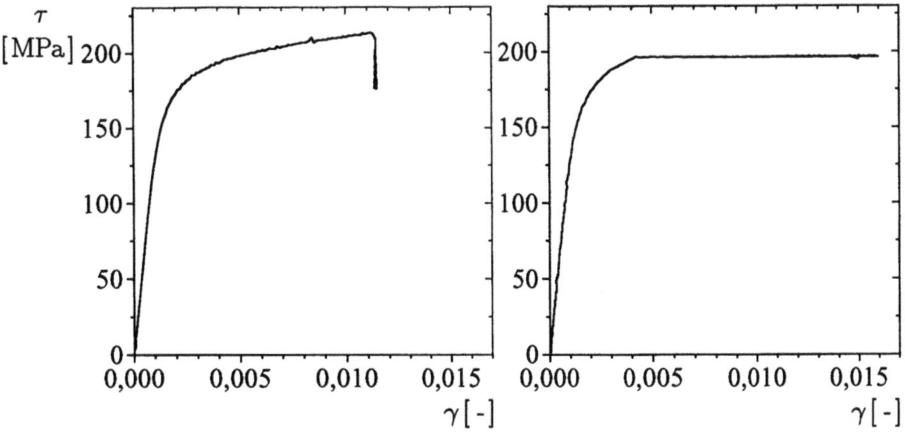

Figure 6. 4: Relaxation and creep test

Figure 6. 3 illustrates a similar experiment under stress control. A shear stress of $\tau_0 = 196{,}5$ [Mpa] is applied at a constant rate of $\dot{\tau}_0 = 19{,}65$ [MPa s^{-1}] and subsequently kept constant. It is obvious that at the point where the given stress $\sigma(t)$ is constant, the response $\gamma(t)$ of the test specimen also depends on time.

This process is called *creep*. The bottom section of Fig. 6. 3 clearly shows that the creep process is by no means completed after a hold time of 2000[s]. One can only guess in this case that the shear $\gamma(t)$ might tend towards a finite limit value if the experiment were prolonged indefinitely, but this is uncertain.

Eliminating the time from the representation of the test data, which is usual in solid mechanics, leads to a summary of input and output processes in stress-strain diagrams. These summaries are set out in Fig. 6. 4 for the relaxation and creep tests illustrated in Fig. 6. 2 and Fig. 6. 3.

Figure 6. 5 illustrates a series of six strain-controlled torsion tests with constant shear rate, which is varied over five powers of ten.[4] A nonlinear dependence of the stress-strain curves on the process rate $\dot{\gamma}_0$ becomes evident. In the sixth experiment 20 hold times lasting 2000 [s] each are introduced, during which relaxation processes take place. The termination points of relaxation still lie below the curve appertaining to the smallest shear rate, as shown in the diagram below.

[4] All stress-strain curves related to XCrNi 18.9 are quoted from the thesis of LION [1994], pp. 16. See also HAUPT & LION [1995].

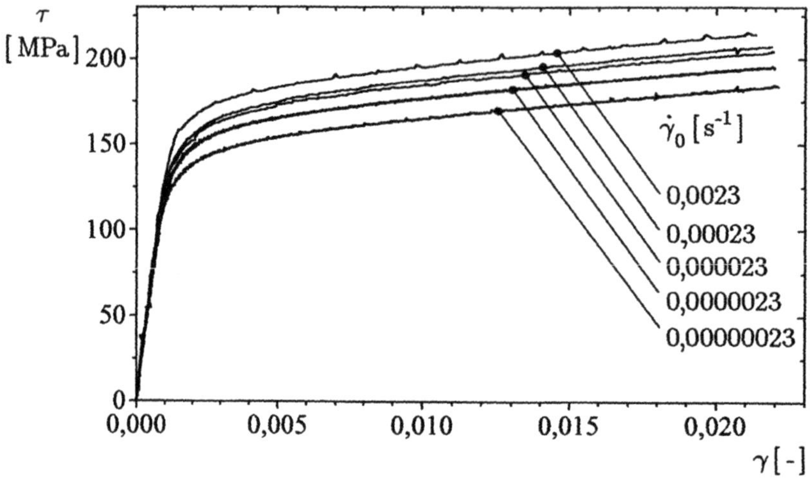

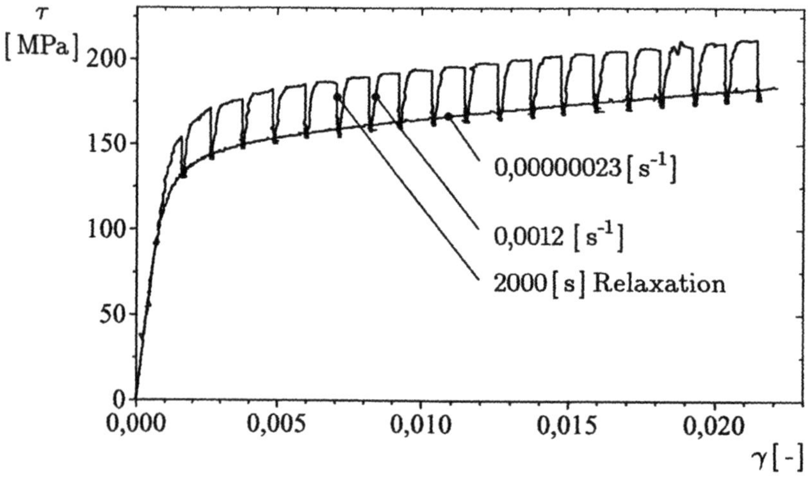

Figure 6. 5: Strain control: rate-dependence and relaxation

A series of six torsion tests under stress control is illustrated in Fig. 6. 6. In the sixth experiment several hold times lasting 2000 [s] are introduced once more, during which creep processes take place. Here too, the phenomenon of nonlinear dependence on the stress rate becomes evident as well as the fact that the points where the creep processes were discontinued lie underneath the curve belonging to the smallest stress rate.

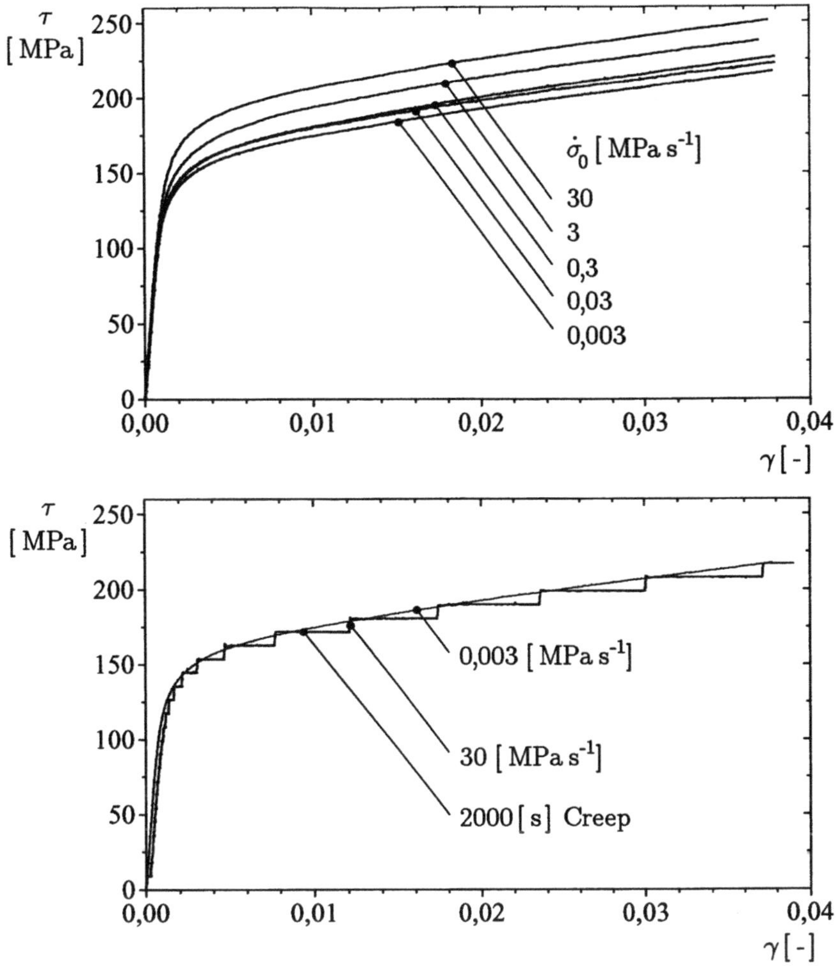

Figure 6. 6: Stress control: rate-dependence and creep

The relaxation and creep curves shown in Figs. 6. 5 and 6. 6 are the outcome of dynamic processes taking place on the microscopic level within the material. By its very nature, a mathematical description is always bound to be highly deficient if it has to be done on a macroscopic level. In spite of this fundamental difficulty, there are nevertheless quite dependable aspects that can be pursued by a phenomenological modelling of the local material behaviour.

Among these is the basic assumption that a dynamic process always comes to a standstill in a *passive system* if the external conditions are kept constant. That is to say the process tends towards a state of equilibrium. If the relative distortion of the end cross-sections of the specimen is regarded as an external condition, then one can appreciate the hypothesis that an infinitely prolonged relaxation process has to end in a time-independent *equilibrium state* of stress and deformation. It can accordingly be assumed that, provided the hold times are long enough, the termination points of relaxation in Fig. 6. 5 represent *states of equilibrium* in an asymptotic approximation.

Definition 6. 1

During the course of a strain-controlled process of relaxation a temporally constant state of stress is reached asymptotically, representing a *state of equilibrium* in the material. This state of stress is called *equilibrium stress.* ◊

In the relaxation experiment illustrated in Figs. 6. 2 and 6. 4 the relaxation process would terminate in a state of equilibrium, characterised by the given shear $\gamma_0 = 0{,}0114$ [-] and the limit value of the shear stress. In addition, it may also depend on the previous history of the shear process and not only on γ_0. The stress reached by the termination of the relaxation process, $\tau(2000[\text{s}]) = 176{,}2$ [MPa], is an approximation for this limit value, the so-called *equilibrium stress.*

Whereas, on principle, a process of relaxation always tends to a state of equilibrium, similar behaviour does not always occur during creep processes. At least the test data from Fig. 6. 3, as opposed to Fig. 6. 2, does not lead to any such supposition.

The whole set of equilibrium states of stress, attributed to a controlled deformation process, can be summarised utilising the term equilibrium relation.

Definition 6. 2

The set of strain processes and equilibrium stresses, attainable by means of relaxation processes, forms the *equilibrium relation.* ◊

The concept of associating a material with its equilibrium relation is quite a general idea. What is meant, basically, is a relation that contains all strain and stress states, connected to each other by means of relaxation processes.

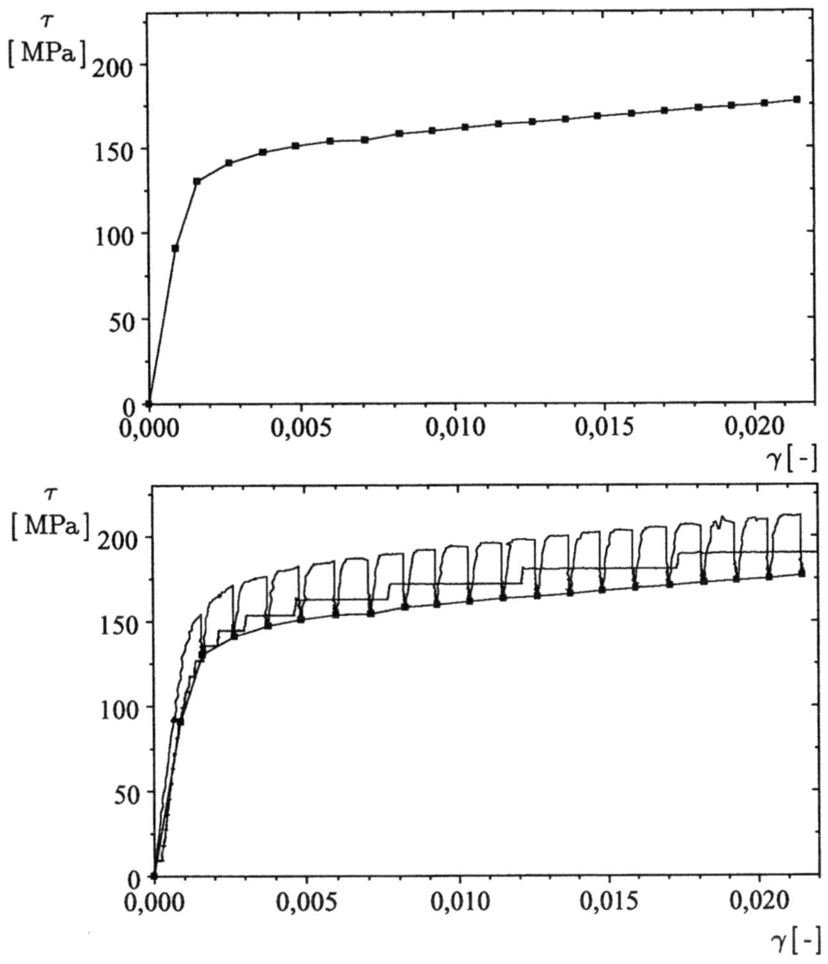

Figure 6. 7: Tension test: equilibrium curve

It is obvious that this set is merely an abstract conception, since it can never be completely substantiated by means of experimental investigations. By restricting oneself to the very special situation of a torsion test, the idea of an equilibrium relation is best illustrated by means of the *equilibrium curve*, in this case using the line joining the termination points of relaxation in Fig. 6. 5.

This is illustrated in Fig. 6. 7. As opposed to the relaxation processes which inevitably lead to a state of equilibrium, this can only be reached by a creep process if the equilibrium curve contains the corresponding state of stress.

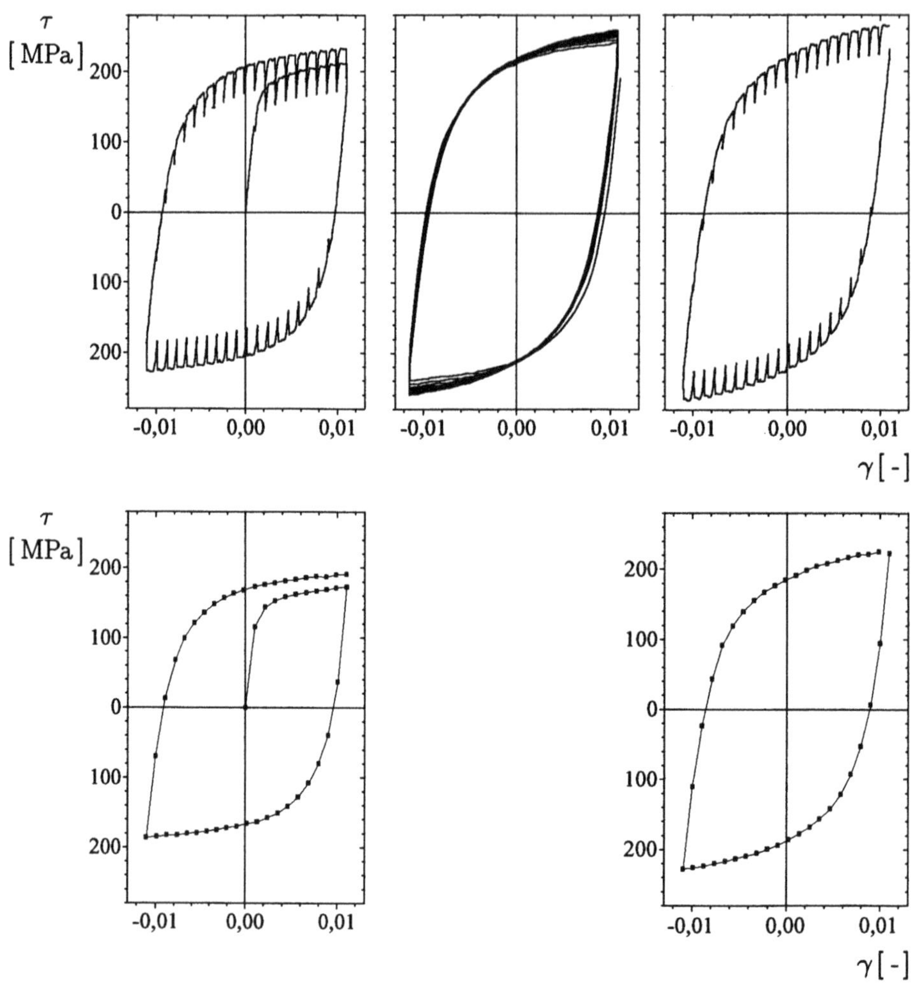

Figure 6. 8: Cyclic tests with hold times: history-dependent equilibrium hysteresis

If this is not the case, one can expect unlimited creep; this can, however, be terminated quite naturally after a finite period of time due to material failure. In the lower section of Fig. 6. 7 the equilibrium curve, resulting from relaxation processes (Fig. 6. 5), is compared with the creep processes in Fig. 6. 6. The comparison shows that the individual creep processes would likewise drift into states of equilibrium, since the applied stresses meet the equilibrium curve. (In the illustration this only remains open for the final creep process with $\tau_0 = 190$ [MPa]).

Experimental tests show that it usually takes a lot longer to arrive at a state of equilibrium in a creep process than it does in a relaxation process. Referring back to Fig. 6. 5 it can be seen that it would probably be unrealistic to attempt any direct measurement of the equilibrium curve by stipulating an adequately low strain rate, since this would demand extremely slow process rates. Instead, it is rational to carry out strain-controlled relaxation tests to determine equilibrium curves by means of experiments.

Observations based on experiments indicate that, in general, the equilibrium stress cannot be a function of the current strain alone, but that it depends on the process history. This is confirmed by Fig. 6. 8: the top section shows a strain-controlled torsion test with a constant absolute value of the shear rate and inserted hold times. By driving a series of shear increments, each kept constant for 2000 [s], a complete cycle of relaxation processes, consisting of loading, unloading and reloading, is passed through. This is succeeded by about 50 cycles without hold times, but also possessing a constant absolute value of the strain rate. In the last cycle the same hold times have been reinstated. Figure 6. 8 shows that the equilibrium curve, which can be attributed to the first cyclic relaxation test, is a hysteresis loop, and that this hysteresis loop has altered its shape during the further course of the process, in other words it depends on the process history.[5]

These observations show typical phenomena of macroscopic material behaviour: *rate-dependence*, *relaxation*, *creep* and *equilibrium hysteresis*.

6. 2. 2 Material Properties of Carbon-Black-Filled Elastomers

Rubber-like materials are able to undergo very large deformations without the integrity of the material suffering too badly. In order to determine the material properties, uniaxial tensile tests on rods with a constant cross-section are appropriate. The strains can amount to several 100%. Compression tests can also be carried out; however, the degree of compressive strain is limited in view of instability problems (buckling).

The following figures illustrate the behaviour measured on cylindrical tensile rods made of filled rubber.[6]

---

[5] Experimental evidence for the existence of rate-independent equilibrium hysteresis is demonstrated in TSOU & QUESNEL [1982].

[6] The results are taken from LION [1996]. Further experimental investigations concerning tension and tension-torsion tests on cylindrical specimens are reported in HAUPT & SEDLAN [2000].

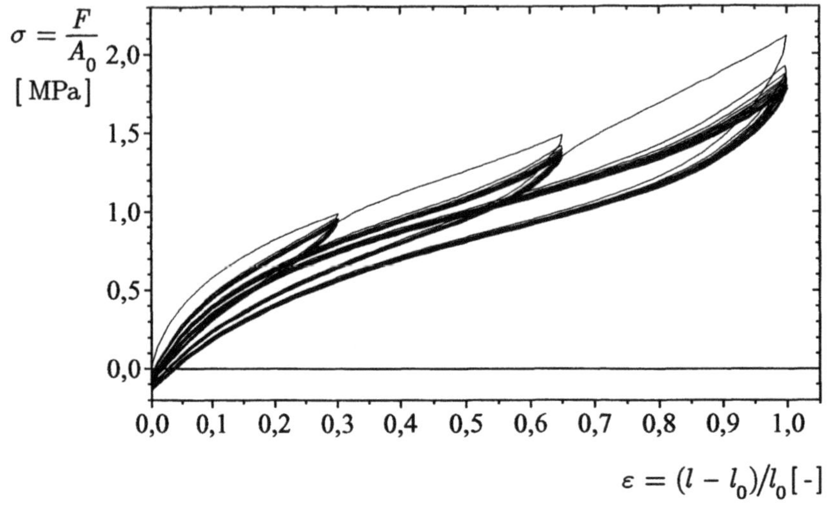

Figure 6. 9: Strain-controlled cyclic tensile test on a new specimen

Figure 6. 9 illustrates a cyclic tensile test with three different amplitudes under strain control with strain rate $\dot{\varepsilon}_0 = \pm\, 0{,}02\,[\mathrm{s}^{-1}]$. 12 cycles in each of 3 different strain amplitudes were run in succession. The tensile stress $\sigma = F/A_0$ referring to the initial cross-section is shown in dependence of the strain $\varepsilon = (l - l_0)/l_0$.

What is conspicuous is the difference between the first and second cycles following every change in amplitude: after the first cycle a considerable softening effect can be recognised, which takes place at every increase in strain amplitude. This softening effect is due to damaging processes within the material and is referred to in specialist literature as the *Mullins* effect.[7] In order to separate the *Mullins* effect from the other phenomena of interest, a cyclic preprocess or *training process*, destined to lead the material to a stable stationary state, is usually conducted prior to the stress-strain experiment. A training process of 12 cycles between the limits $\varepsilon = -\,0{,}3$ in compression and $\varepsilon = 1{,}0$ in tension was carried out before each of the other tests, which explains why the *Mullins* effect is no longer present in the subsequent figures.[8]

Figure 6. 10 shows a series of strain-controlled tension tests with a constant strain rate, which is varied over 4 orders of magnitude. A rate-dependence of the stress response is observed.

---

[7] See the literature cited by Lion [1996], p. 154.

[8] Lion [1996], pp. 155.

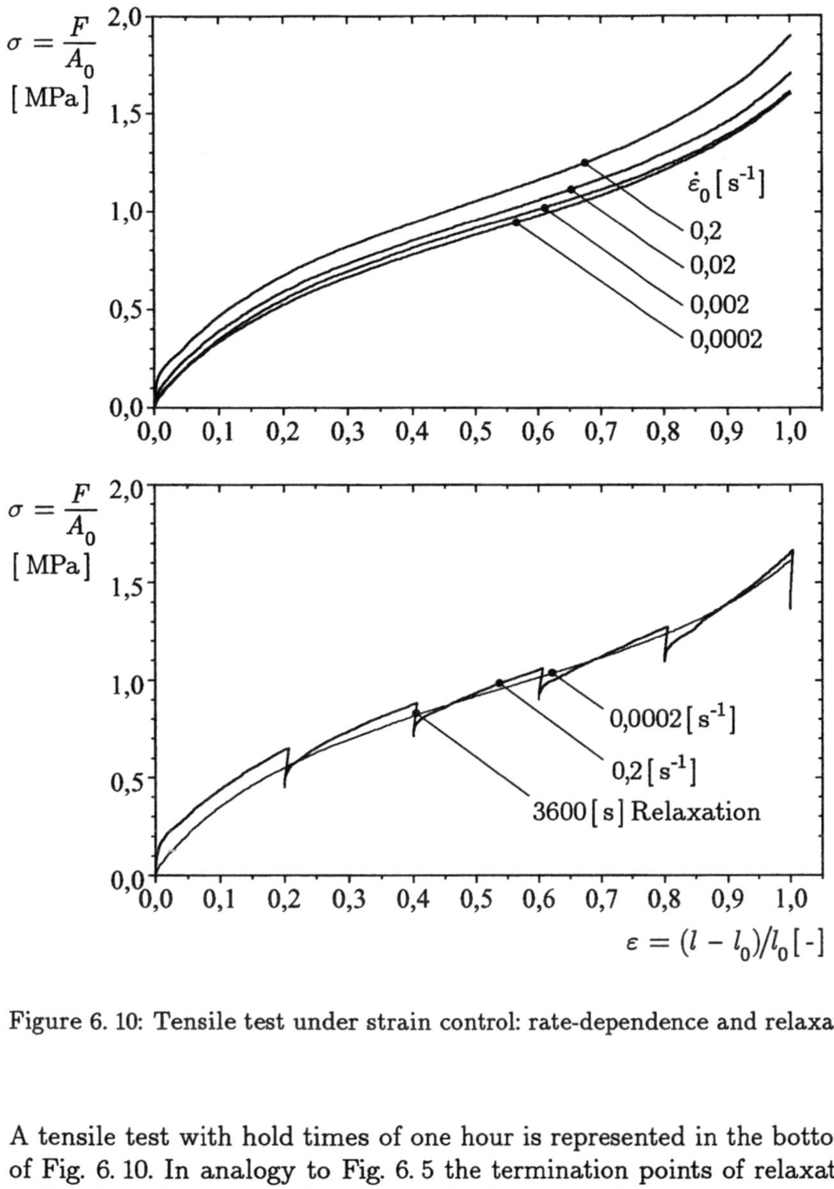

Figure 6. 10: Tensile test under strain control: rate-dependence and relaxation

A tensile test with hold times of one hour is represented in the bottom half of Fig. 6. 10. In analogy to Fig. 6. 5 the termination points of relaxation lie well below the curve belonging to the lowest strain rate. A corresponding series of compression tests is shown in Fig. 6. 11.

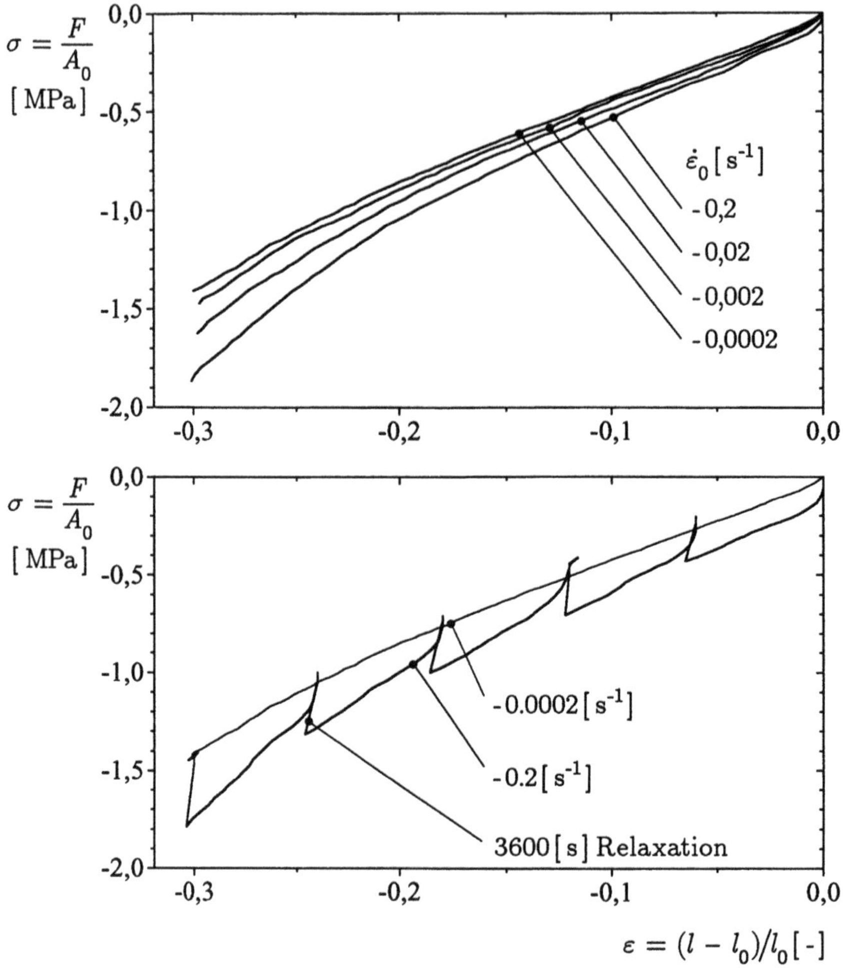

Figure 6. 11: Compression tests under strain control: rate-dependence and relaxation

The termination points of relaxation are to be regarded in asymptotic approximation as equilibrium points. This is illustrated in Fig. 6. 12. Figure 6. 13 shows strain-controlled cyclic tensile tests with a strain amplitude of 0,3 [-] and three different mean strains. We see that the cyclic stress-strain curves are largely dependent on the mean strain. In the last cycle hold times of one hour are inserted, during which relaxation takes place.

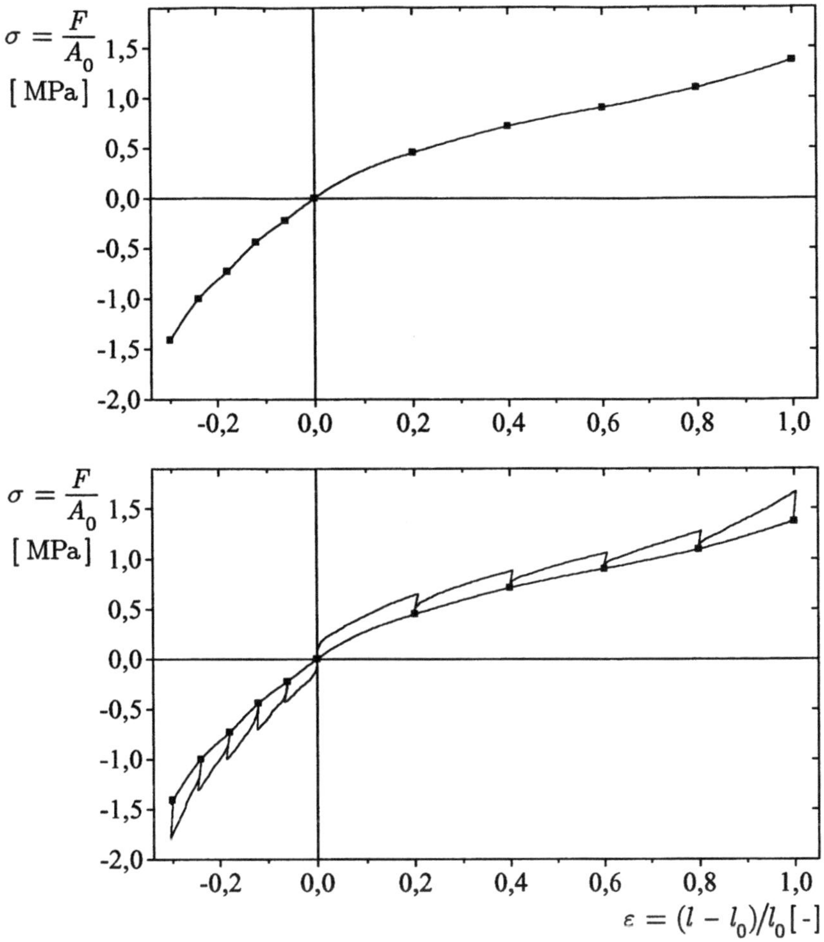

Figure 6. 12: Tension and compression test: equilibrium curve

The curves indicate that there is an equilibrium hysteresis in this case as well. However, this effect is weakly pronounced in comparison with Fig. 6. 8. In Fig. 6. 14 the experiments from Fig. 6.13 are summarised, making it clear that all effects of hysteresis are relatively small in this material.

We conclude that the typical phenomena of the material behaviour, namely *rate-dependence*, *relaxation* and *equilibrium hysteresis*, are also visible, albeit in different magnitudes, in the case of polymeric materials.[9]

[9] Experimental work and constitutive modelling related to the inelastic behaviour of filled rubbers are presented in MIEHE & KECK [1998].

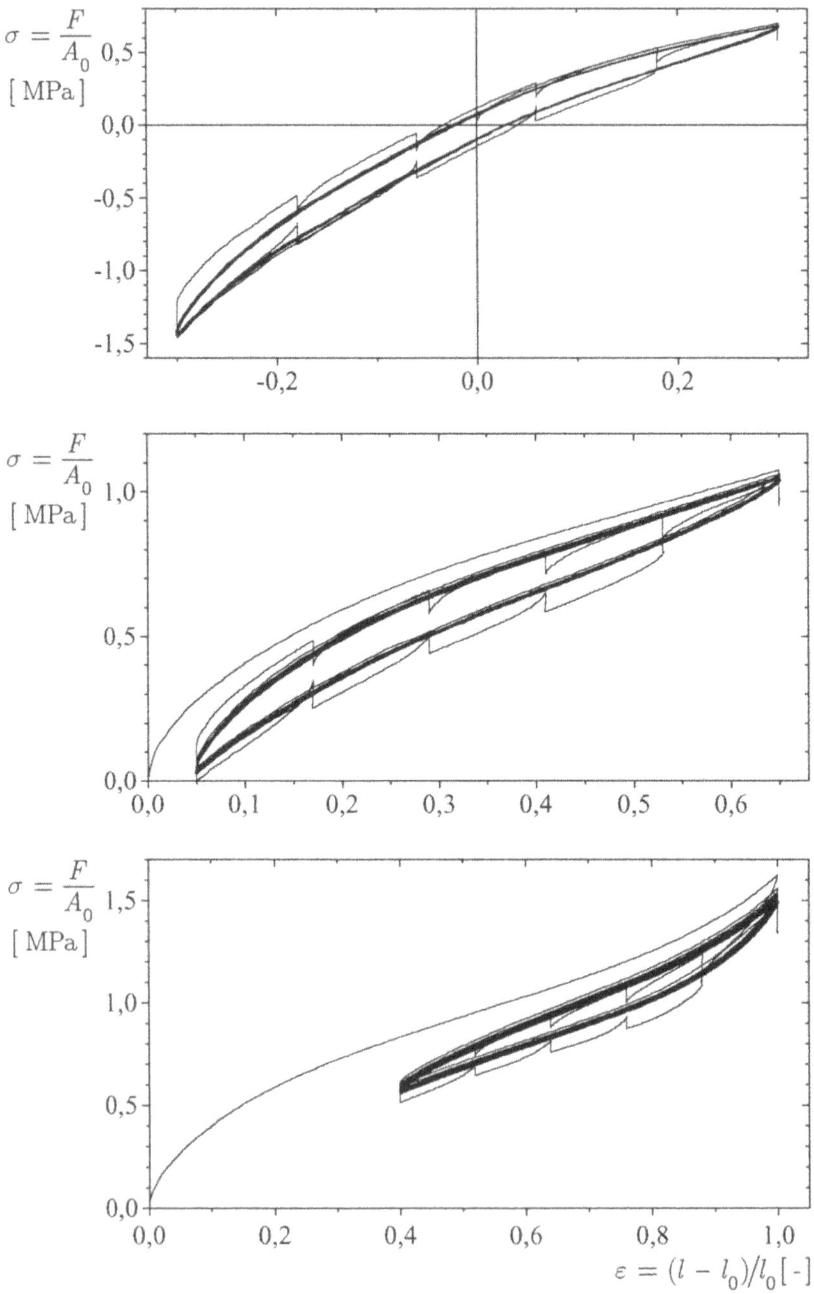

Figure 6. 13: Cyclic tests with different mean strains

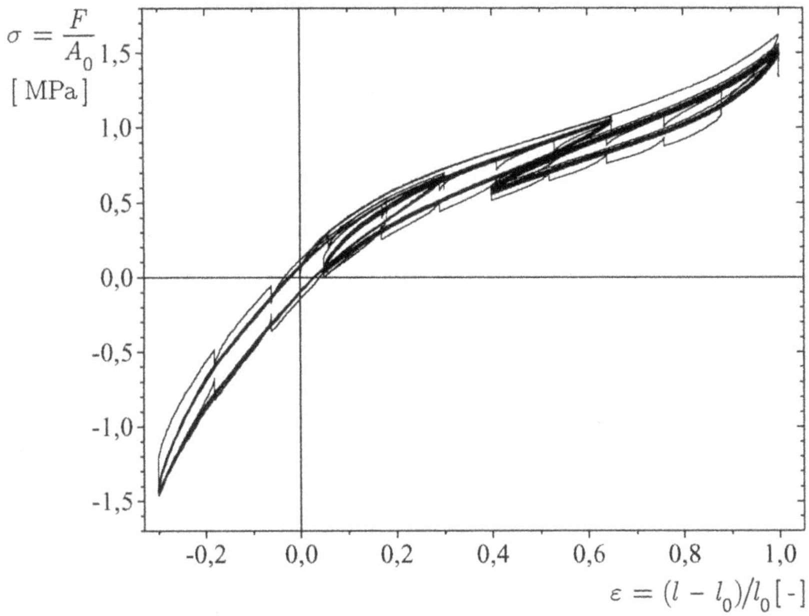

Figure 6. 14: Cyclic tests using filled rubber: summary

6. 3 Four Categories of Material Behaviour

Experimental observations of material behaviour should not be influenced by any hasty interpretations or judgements, which might promote rushing into the choice of specific material models. It would be a precipitate conclusion, for instance, to interpret as plasticity any deviation of the stress-strain curve from linearity; terms used in the theory of material behaviour, such as *elastic strain*, *plastic strain*, *creep strain*, *yield limit*, *elastic domain* etc. almost amount to prejudices in this connection and ought to be avoided in favour of an unbiased interpretation of the experimental data: not until after the selection and complete formulation of a concrete material model does it make sense to use this kind of terminology. In the evaluation and interpretation of experimental data, only those aspects should be taken into consideration that do not depend on the modelling in mind, but are the direct outcome of the actual results of experiments.

When faced with an unbiased observation of mechanical material behaviour based on an adequate collection of experimental data, the first impression is one of a seemingly unfathomable multitude of diverse behaviours.

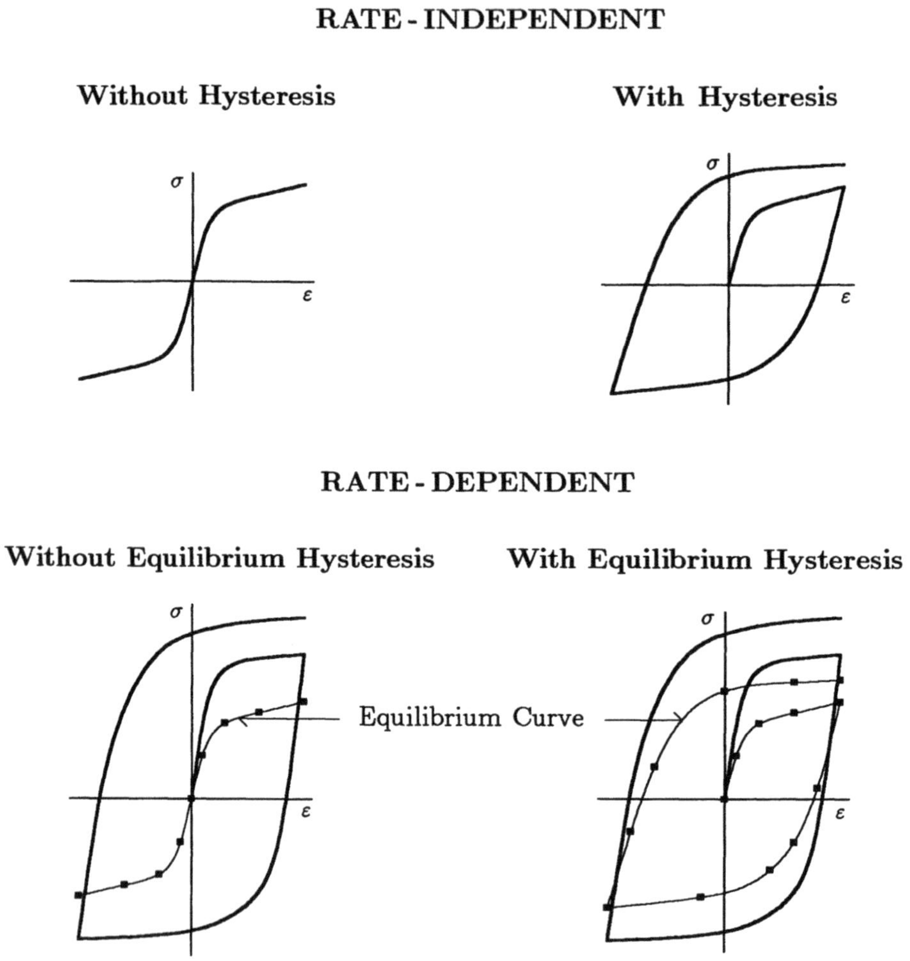

Figure 6. 15: Experimental observation: 4 categories of material behaviour

At this stage it is possible and helpful to introduce a classification for the sake of clarity.

To begin with, the data on hand can give evidence as to whether or not the stress-strain curves depend on the velocity at which the experiments were carried out. If the material responses are not dependent on the speed of the input processes, then it is possible to check whether there are hysteresis effects, or whether these can be ignored.

Should the experimental data indicate a significant dependence of the material response on the rate of the input process, then attention should be turned to the equilibrium curve that has become accessible as a result of relaxation experiments. The equilibrium curve may indicate hysteresis - or it may not. This allows us to divide the observable behaviour tendencies of materials into 4 different categories:[10]

- *rate-independent without hysteresis*
- *rate-independent with hysteresis*
- *rate-dependent without equilibrium hysteresis*
- *rate-dependent with equilibrium hysteresis*

This is a simple pattern for systematically going through the wealth of available material responses. Figure 6. 15 provides an illustration.

Once a sufficient number of experiments has been carried out, the test data provides an immediate indication as to which of the four categories the behaviour of the test specimen fits into. On the one hand this classification is based on *objective criteria*. On the other hand it is also influenced by *subjective points of view*, namely the decision as to which details of the material behaviour observed should be disregarded during the mathematical modelling of the material properties.

6. 4 Four Theories of Material Behaviour

There are precisely four categories of mathematical models, namely the theories of *elasticity*, *plasticity*, *viscoelasticity* and *viscoplasticity*, to match the four categories of material behaviour observed in experiments.

Definition 6. 3

One distinguishes between 4 different theories of material behaviour. These are characterised as follows:

- The theory of *elasticity* describes *rate-independent material behaviour without hysteresis.*
- The theory of *plasticity* describes *rate-independent material behaviour with hysteresis.*
- The theory of *viscoelasticity* describes *rate-dependent material behaviour without equilibrium hysteresis.*
- The theory of *viscoplasticity* describes *rate-dependent material behaviour with equilibrium hysteresis.* ◊

[10] Cf. HAUPT [1993].

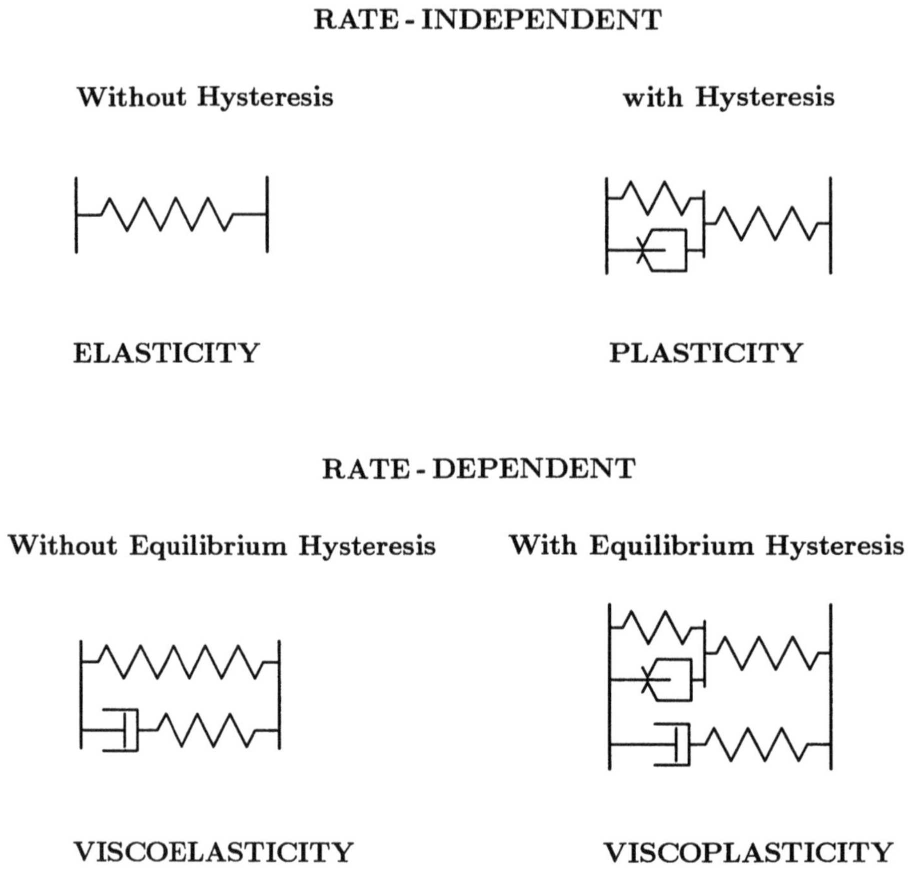

Figure 6. 16: Mathematical modelling: 4 theories of material behaviour

The general definition for the 4 classes of material theories is illustrated in Fig. 6. 16. The physical definitions of the individual theories are symbolised by *rheological models.* These models in Fig. 6. 16 consist of simple *networks* of (massless) *elastic springs*, *viscous dashpots* and *Coulomb friction elements.* The rheological models are not intended to denote concrete differential equations here (as opposed to linear viscoelasticity): in this table they merely serve as *symbols* for marking the physical significance and message of the four different classes of constitutive theory.

Elasticity is clearly the simplest of all constitutive models. It is restricted to the basic phenomenon of the restoring force, which all material bodies put up to counteract a deformation. Another basic phenomenon, internal friction or dissipation, is covered by *viscoelasticity.* Internal friction is also represented by *plasticity*, but from a completely different point of view. *Viscoplasticity* is the most general theory, being equipped, on principle, to handle all macroscopically observable phenomena of material behaviour.

The test data on steel XCrNi 18.9, shown in this chapter, provides a clear case for modelling the behaviour of this substance within the theory of *viscoplasticity.* A similar decision could theoretically be reached for rubber too. However, given the relatively narrow hystereses in this field, one might also regard *nonlinear elasticity* as a suitable theory for reproducing a considerable part of the material behaviour. If there is a reason for including viscous friction in the model, however, one could omit the small portion of the equilibrium hysteresis, for its part, and would then find *viscoelasticity* to be the appropriate theory. On the other hand, the material behaviour of carbon-black-filled rubber has to be modelled according to the theory of *viscoplasticity*, if the equilibrium hysteresis is to be taken into account. This example demonstrates that both objective and subjective aspects can play a role in the description of a material, employing constitutive theories.

6. 5 Contribution of the Classical Theories

The 4 basic concepts of material modelling - elasticity, plasticity, viscoelasticity, viscoplasticity - are realised by the classical theories of continuum mechanics, albeit in one particular respect.

The *perfect fluid*, defined by (5.1), is a material model belonging to the category of elasticity, that purely and simply describes the volume changes caused by pressure.

The *linear-viscous fluid* according to (5.18) and (5.20) should be classed under viscoelasticity. However, it is a very special representative of this category, being nothing but a three-dimensional formulation of the *Kelvin* model (5.90) suitable for application in fluid mechanics.

The *isotropic linear-elastic solid*, according to its definition (5.35), leads to a linear stress-strain curve for one-dimensional processes.

The *elastic-perfectly plastic solid*, defined by (5.114) limits the validity domain of linear elasticity: if a critical yield stress is reached, plastic flow occurs under loading, whereas *Hooke's law* applies again upon unloading. Perfect plasticity describes a rate-independent hysteresis for one-dimensional loading according to equation (5.119). That means, all points of the stress-strain curve are equilibrium points. Introducing isotropic and kinematic hardening, which incorporates a dependence of the elastic domain on the process history, brings about a considerable improvement in the model of perfect plasticity.

The *linear-viscoelastic solid* in its definition through (5.51) offers a host of possibilities for representing rate-dependent material behaviour. This theory contains the model of linear elasticity as an asymptotic limit, namely as an equilibrium relation. In conformity with the constitutive equations (5.51), (5.52), (5.72) and (5.105) the material response in linear viscoelasticity depends on the aforegoing history of the input process. This conception proves to be most rewarding, since its contribution to the development of an all-embracing theory of material properties is considerable.[11]

The *viscoplastic solid*, defined by equations (5.161) - (5.166), can be interpreted as a composition of two rheological elements: one is a linear-elastic spring and the other a nonlinear damping element, representing a combination of *Coulomb* friction and nonlinear viscosity.

If one compares the conclusions drawn from the classical theories of continuum mechanics with experimental test results, only a few of which have been alluded to in this chapter, it will become apparent that the classical constitutive models do reflect significant aspects of the material behaviour observed. On the other hand, considerable discrepancies come to light, which for the most part cannot be resolved within the context of the classical theories. These discrepancies immediately become apparent upon comparing the corresponding stress-strain curves. Moreover, the restriction to geometric linearisation has to be recognised as a basic flaw of the classical theories of solid mechanics, since it rules out their application to thermomechanic processes with large deformations. All these aspects speak in favour of a comprehensive theory of phenomenological material behaviour based on the general principles of thermomechanics. This general theory is to be expounded in the chapters which follow.

---

[11] See COLEMAN & NOLL [1961]

7 General Theory of Mechanical Material Behaviour

7.1 General Principles

The *theory of material behaviour* (*material theory*) is a division of continuum mechanics. The aim of the material theory is to provide general principles and systematic methods for constructing mathematical models suitably representing the individual properties of material bodies. The modelling of material behaviour can be realised on three levels: *constitutive equations*, *material symmetry properties* and *conditions of kinematic constraint*.

A *constitutive equation* is a relation representing the individual response of any material element to a given input process. In continuum mechanics a constitutive equation correlates processes of deformation and stress. Either the stress history or the deformation history can be regarded as the input process, or even a combination of the two, as far as this makes physical sense. In advanced theories the input processes and material responses are more complex: for example, if the temperature-dependence of the material properties is to be taken into account, the basic aspects of *thermodynamics* have to be observed. Together with mechanics, thermodynamics then forms the basis for a material description, in connection with which further variables such as temperature, temperature gradient, internal energy, entropy, heat flux and entropy flux arise. The material behaviour can be modelled with the help of these quantities within the context of *thermo-mechanics*.

In the case of *elasticity* all constitutive equations are represented by *material functions*. However, material functions are not sufficient for modelling inelastic material behaviour: for substantiating the general categories of viscoelasticity, plasticity and viscoplasticity, functional relations have to be depicted. *Functionals* can either be represented explicitly, for example by means of integrals over the temporal history of the input process or implicitly using differential equations. The classical theories of continuum mechanics give hints as to what explicit or implicit representations of functional stress-strain relations can look like. In the general theory of material behaviour the specific constitutive equations of the classical theories are incorporated into a general framework.

Constitutive equations are not the only means of representing material behaviour. The term *material symmetry* is intended to express the fact that material behaviour can be dependent on the material direction, regardless of any concrete stress-strain relation. A general concept to permit a representation of the dependence on direction is to analyse the constitutive equation's dependence on the reference configuration: if a stress-strain relation remains invariant under certain changes of reference configuration, then there is a property of material symmetry. A systematic investigation of material symmetries allows us to divide the so-called *simple materials* into fluids and solids. In the sense of this classification a *fluid* is characterised by different properties of symmetry than a *solid*. In the whole set of solids one has to distinguish between different levels of symmetry. When representing properties of material symmetry, the question of the specific structure of the constitutive equations remains open. The classification of material bodies according to their symmetry is of a general validity. On principle it embraces all four categories of material behaviour, i.e. elasticity, viscoelasticity, plasticity and viscoplasticity.

A *kinematic constraint* is a restriction of a material body's motions, definable as a material property *a priori*. A kinematic constraint is an idealisation that can be introduced regardless of any specific form of stress-strain relation. It is also independent of any special property of material symmetry. The best known example of a kinematic constraint of is the assumption of *incompressibility*, which occurs in hydromechanics and rubber elasticity: an incompressible material body can only effect volume preserving motions, whereas the material behaviour can be completely arbitrary. The most restrictive form of internal constraint leads to the definition of a *rigid body*. Rigid bodies possess no other mechanical properties than the constraint condition of rigidity, which makes them somewhat trivial for the material theory of continuum mechanics.

With regard to a systematic construction of material models by means of constitutive equations, symmetry properties and constraint conditions, it makes sense to take note of general ideas and principles that have proved to be useful, logical or even absolutely necessary throughout the development of the constitutive theory.[1] Three general principles deserve to be emphasised.

- *Determinism*
- *Local Action*
- *Frame-Indifference*

1) The *principle of determinism* states that the current state of stress $\mathbf{T}(\mathscr{P}, t)$ in one material point $\mathscr{P} \in \mathscr{B}$. is uniquely determined by the past history of a body's motion.

2) The *principle of local action* demands that the state of stress in the material point $\mathscr{P} \in \mathscr{B}$ only be influenced by the history of motion of this point's environment and not by the motion of all body particles.

The extremely general principle of determinism is limited by the principle of local action. A restriction of this kind is doubtlessly of utmost necessity, since it would hardly be possible otherwise to formulate and apply constitutive equations definitely.
A scientifically based description of material behaviour has to be independent of arbitrarily selected means of representation such as coordinate systems, base vectors or frames of reference. Accordingly, constitutive equations have to possess suitable *invariance properties*. One indispensable requirement would be that every constitutive equation should qualify as an *observer-invariant relation* (see Sect. 4. 6). This requisite proves to be too weak, however, in terms of practical applications. For this reason it is common for thermomechanics to make more stringent demands. These are best known under the heading of *material frame-indifference*.

3) The *principle of material frame-indifference* or *material objectivity* postulates that every representation of material properties should be independent of the frame of reference, that is to say, a constitutive equation is invariant with respect to any change of frame. Nor does it include any information about the motion of the applied reference frame in relation to the inertial frame upon which the representation of a thermomechanical process is based. The principle of material frame-indifference is an assumption restricting the mathematical form of the constitutive equations, as will be

---

[1] TRUESDELL & NOLL [1965], Sect. 14; LEIGH [1968], pp. 139. Cf. also TRUESDELL & TOUPIN [1960], Sect. 293.

demonstrated in a later section. The assumption of frame-indifference can be regarded as generally binding to the extent that the formal invariance of the constitutive equations with respect to the reference frame is without a doubt absolutely essential. On the other hand, the postulation that no information concerning the frame of reference (e.g. its rate of translation or rotation) should occur in a constitutive equation does not necessarily have to be viewed as a compelling law of nature.[2]

The three principles - *determinism*, *local action* and *frame-indifference* - are not the only demands that have to be made on a physically significant constitutive model.[3] Even the general structure of continuum mechanics enforces the *compatibility with the balance relations*: no constitutive equation is allowed to contradict the balance relations of continuum mechanics and thermodynamics. In this connection compatibility with the balance of rotational momentum is doubtlessly a trivial matter, since this merely requires symmetry of the stress tensor in the light of classical continua. By contrast the question of compatibility with the balance of linear momentum can only be answered for very specific cases: it is hardly possible to make general statements regarding the mathematical character and the solutions of the initial-boundary-value problem which arises from the combination of a material model with the balance of linear momentum. The same applies for compatibility with the energy balance of thermodynamics. Conversely, compatibility with the entropy balance is not only indispensable but often an easier prospect: for certain categories of material models the formulation of constitutive equations can be arranged *a priori* in such a way that compliance with the *principle of irreversibility* is guaranteed.

Further assumptions that support the construction of material models could be named, such as the *principle of equipresence*,[4] the *principle of fading memory* and others. As a whole, the question as to which generally applicable demands would be necessary and sufficient to secure the physical consistency and plausibility of a material model in every aspect, has remained unanswered so far. This question has been described by C. A. Truesdell and W. Noll as the *main open problem of the theory of material behaviour*.[5] With regard to this question the theory of material behaviour cannot be viewed as a closed subject of research: all there is so far is a handful of systematic aspects which can prove to be useful, if observed.

The general theory of material behaviour, as it is outlined in the following sections, is mainly due to W. Noll.[6]

---

[2] See MÜLLER [1976], eq. (4.7).

[3] TRUESDELL & TOUPIN [1960], pp. 700.

[4] TRUESDELL & TOUPIN [1960], pp. 704; TRUESDELL & NOLL [1965], p. 359.

[5] TRUESDELL & NOLL [1965], Sect. 20.

[6] NOLL [1958]. An improved formulation of the general theory is found in NOLL [1972].

7. 2 Constitutive Equations

7. 2. 1 Simple Materials

According to the *principle of determinism*, the first statement is that the current state of stress in a material point $\mathscr{P}$ depends on the past motion of the entire material body:[7]

$$(\mathscr{P}, t) \longmapsto \mathbf{T}(\mathscr{P}, t) = \underset{\substack{\tau \leqq t \\ \mathscr{Q} \in \mathscr{B}}}{\mathfrak{S}} \left[\chi_\tau(\mathscr{Q}); \mathscr{P}, t\right] . \tag{7.1}$$

The functional $\mathfrak{S}$ characterises the mechanical properties of the body $\mathscr{B}$ in the broadest possible sense. In the most general formulation (7.1), $\mathfrak{S}$ incorporates the entire past motion of all material particles $\mathscr{Q} \in \mathscr{B}$. The stress functional $\mathfrak{S}$ may vary from point to point, corresponding to an inhomogeneous distribution of the material properties. An explicit dependence of the functional $\mathfrak{S}$ on time t is also conceivable.

After a change of the reference system, represented by the transformations (4.28), (4.29),

$$\boldsymbol{x}^* = \mathbf{Q}(t)\boldsymbol{x} + \boldsymbol{c}(t) , \quad t^* = t - a , \tag{7.2}$$

it might be possible for a different constitutive equation to assume validity, namely

$$\mathbf{T}^*(\mathscr{P}, t) = \underset{\substack{\tau \leqq t^* \\ \mathscr{Q} \in \mathscr{B}}}{\mathfrak{S}^*} \left[\chi_\tau^*(\mathscr{Q}); \mathscr{P}, t^*\right] , \tag{7.3}$$

Here, $\mathfrak{S}^*$ is a different constitutive functional, that could in addition depend on $\mathbf{Q}(\tau)$ and $\boldsymbol{c}(\tau)$ in the form of an observer-invariant relation.

The *principle of material frame-indifference* demands that the constitutive equation (7.1) be completely independent of the choice of reference frame: in connection with the freedom to select any reference system, and using relation (7.1), the constitutive equation

$$\mathbf{T}^*(\mathscr{P}, t) = \underset{\substack{\tau \leqq t \\ \mathscr{Q} \in \mathscr{B}}}{\mathfrak{S}} \left[\chi_\tau^*(\mathscr{Q}); \mathscr{P}, t^*\right] \tag{7.4}$$

[7] Truesdell & Noll [1965], Sect. 26.

should hold, or in the abbreviated form, the identity $\mathfrak{S}^* = \mathfrak{S}$. It is possible to derive a restrictive condition for stress functional $\mathfrak{S}$ from this.

Theorem 7.1

The *principle of material frame-indifference* implies the following restrictive condition: for any given orthogonal tensor-valued functions

$$\tau \longmapsto \mathbf{Q}(\tau) \qquad \left(\mathbf{Q}(\tau)\mathbf{Q}^{\mathrm{T}}(\tau) = \mathbf{Q}^{\mathrm{T}}(\tau)\mathbf{Q}(\tau) \equiv \mathbf{1}\,,\ \det\mathbf{Q}(\tau) \equiv +1\right),$$

vector-valued functions $\tau \longmapsto \mathbf{c}(\tau)$ and scalars a, the general stress functional $\mathfrak{S}$ meets the following identity:

$$\mathbf{Q}(t) \underset{\substack{\tau \leqq t \\ \mathcal{Q} \in \mathcal{B}}}{\mathfrak{S}} \left[\chi_\tau(\mathcal{Q}); \mathcal{P}, t\right]\mathbf{Q}^{\mathrm{T}}(t) = \underset{\substack{\tau \leqq t \\ \mathcal{Q} \in \mathcal{B}}}{\mathfrak{S}} \left[\mathbf{Q}(\tau)\chi_\tau(\mathcal{Q}) + \mathbf{c}(\tau); \mathcal{P}, t - a\right]. \tag{7.5}$$

This implies

$$\mathbf{T}(\mathcal{P}, t) = \underset{\substack{\tau \leqq t \\ \mathcal{Q} \in \mathcal{B}}}{\mathfrak{S}} \left[\chi_\tau(\mathcal{Q}) - \chi_\tau(\mathcal{P}); \mathcal{P}\right]. \tag{7.6}$$

□

Proof

A change of frame according to the active interpretation is represented by the transformation formulae (7.2). The corresponding transformation of the *Cauchy* stress tensor is (4.54), thus:

$$\mathbf{T}^*(\mathcal{P}, t) = \mathbf{Q}(t)\mathbf{T}(\mathcal{P}, t)\mathbf{Q}^{\mathrm{T}}(t)\,.$$

For a given material point $\mathcal{P} \in \mathcal{B}$ one defines a particular change by means of

$$\mathbf{Q}(\tau) \equiv \mathbf{1}\,, \qquad \mathbf{c}(\tau) = -\,\chi_\tau(\mathcal{P})\,. \tag{7.7}$$

In this situation, the restrictive condition (7.5) leads to the statement

$$\underset{\substack{\tau \leqq t \\ \mathcal{Q} \in \mathcal{B}}}{\mathfrak{S}} \left[\chi_\tau(\mathcal{Q}); \mathcal{P}, t\right] = \underset{\substack{\tau \leqq t \\ \mathcal{Q} \in \mathcal{B}}}{\mathfrak{S}} \left[\chi_\tau(\mathcal{Q}) - \chi_\tau(\mathcal{P}); \mathcal{P}, t - a\right].$$

Since this identity holds for all a, the choice of $a = t$ leads to (7.6). ■

The current state of stress in $\mathcal{P} \in \mathcal{B}$ is accordingly a functional of the history of the *relative motion* of all particles $\mathcal{Q} \in \mathcal{B}$ with respect to $\mathcal{P}$. Relation (7.6) is only a necessary conclusion and not at all sufficient for material frame-indifference (see below).

As a consequence of material frame-indifference, the stress functional (7.6) cannot depend explicitly on time t. This basic time-independence does not mean that experimental evidence, that seems to document time-dependent material behaviour, is disregarded: all that has been established is that the material behaviour associated with mechanical processes is independent of time. Temporal variations in material properties can certainly be triggered by non-mechanical processes, such as ageing, phase transitions, chemical reactions or changes in temperature. If the representation of these phenomena were required, additional variables would have to be included in the constitutive functional $\mathfrak{S}$. The whole theory would have to be expanded to maintain time-independent constitutive equations in place of the apparent time-dependence. An example of a modification of this kind is the generalisation of mechanics to thermomechanics by means of the introduction of temperature, entropy and additional balance relations.

The assumption of *local action* offers the possibility of restricting the constitutive equation (7.6), which is much too general for any practical application. The state of stress in a given point $\mathcal{P}$ is only influenced by the local history of motion, that is to say by the relative motion of the material points contained in the close neighbourhood of $\mathcal{P}$. In this sense (7.6) is replaced by

$$\mathbf{T}(\mathcal{P}, t) = \underset{\substack{\tau \leqq t \\ \mathcal{Q} \in \mathcal{U}(\mathcal{P})}}{\mathfrak{S}} \left[\chi_\tau(\mathcal{Q}) - \chi_\tau(\mathcal{P}); \mathcal{P}\right] . \tag{7.8}$$

The term *close neighbourhood* $\mathcal{U}(\mathcal{P})$ can be defined more precisely by developing the relative motion of the particles surrounding $\mathcal{P}$ into a *Taylor* series. Beginning with the more concrete formulation

$$\mathbf{T}(\boldsymbol{X}, t) = \underset{\substack{\tau \leqq t \\ \boldsymbol{Y} \in \mathcal{U}(\boldsymbol{X})}}{\mathfrak{S}} \left[\chi_{\mathrm{R}}(\boldsymbol{Y}, \tau) - \chi_{\mathrm{R}}(\boldsymbol{X}, \tau); \boldsymbol{X}\right] \tag{7.9}$$

related to a reference configuration R(·), the *Taylor* expansion of the relative motion reads

$$\chi_{\mathrm{R}}(\boldsymbol{Y}, \tau) - \chi_{\mathrm{R}}(\boldsymbol{X}, \tau) = \overset{\langle 1\rangle}{\mathbf{F}}(\boldsymbol{X}, \tau)[\boldsymbol{Y} - \boldsymbol{X}] + \frac{1}{2!}\overset{\langle 2\rangle}{\mathbf{F}}(\boldsymbol{X}, \tau)[\boldsymbol{Y} - \boldsymbol{X}, \boldsymbol{Y} - \boldsymbol{X}] + \cdots$$
$$+ \frac{1}{\mathrm{n}!}\overset{\langle \mathrm{n}\rangle}{\mathbf{F}}(\boldsymbol{X}, \tau)[\boldsymbol{Y} - \boldsymbol{X}, \ldots, \boldsymbol{Y} - \boldsymbol{X}] + \cdots . \tag{7.10}$$

Here, the tensor

$$\overset{\langle 1\rangle}{\mathbf{F}}(\boldsymbol{X}, t) = \mathbf{F}(\boldsymbol{X}, t) = \operatorname{Grad}\chi_{\mathrm{R}}(\boldsymbol{X}, t) \tag{7.11}$$

is the deformation gradient with the property

$$\overset{\langle 1\rangle}{\mathbf{F}}(\boldsymbol{X}, \tau)[\boldsymbol{Y} - \boldsymbol{X}] = \frac{\partial x^k}{\partial X^L}(Y^L - X^L)\boldsymbol{g}_k . \tag{7.12}$$

Considering the gradients of higher order,

$$\begin{aligned}
&\overset{\langle 2\rangle}{\mathbf{F}}(\boldsymbol{X}, \tau)[\boldsymbol{Y} - \boldsymbol{X}, \boldsymbol{Y} - \boldsymbol{X}] = \frac{\partial^2 x^k}{\partial X^L \partial X^M}(Y^L - X^L)(Y^M - X^M)\boldsymbol{g}_k ,\\
&\vdots \\
&\overset{\langle \mathrm{n}\rangle}{\mathbf{F}}(\boldsymbol{X}, \tau)[\boldsymbol{Y} - \boldsymbol{X}, \dots, \boldsymbol{Y} - \boldsymbol{X}] = \frac{\partial^{\mathrm{n}} x^k}{\partial X^{L_1} \dots \partial X^{L_{\mathrm{n}}}}(Y^{L_1} - X^{L_1}) \cdots (Y^{L_{\mathrm{n}}} - X^{L_{\mathrm{n}}})\boldsymbol{g}_k ,
\end{aligned} \tag{7.13}$$

the *Taylor* expansion would lead to a constitutive functional depending, in principle, on all deformation gradients $\overset{\langle \mathrm{k}\rangle}{\mathbf{F}}$. If an approximation of the local motion with n deformation gradients appears sufficient, one can define a so-called *material of grade* n.[8] Most theories in continuum mechanics refer to the special case of n = 1:

Definition 7.1
A material body complying to the general constitutive equation

$$\mathbf{T}(\mathcal{P}, t) = \underset{\tau \leqq t}{\mathfrak{G}}\left[\mathbf{F}(\mathcal{P}, \tau); \mathcal{P}\right] \tag{7.14}$$

or

$$\mathbf{T}(\boldsymbol{X}, t) = \underset{\tau \leqq t}{\mathfrak{G}}\left[\mathbf{F}(\boldsymbol{X}, \tau); \boldsymbol{X}\right] , \tag{7.15}$$

(a material of grade 1) is called a *simple material.* ◇

The definition of a simple material sharply tightens up the principle of local action: the zone of points surrounding $\mathcal{P}$ which has an influence on the current state of stress in $\mathcal{P}$ is *infinitesimal* (i.e. the *tangent space*). This

[8] TRUESDELL & NOLL [1965], p. 60.

corresponds to a *linear approximation* of the relative motion, represented by the first order deformation gradient $\mathbf{F}(\boldsymbol{X}, t)$.

In this connection it is interesting to note the obvious non-symmetry, when considering the influence of *space* and *time* on the material response: there is no reason to doubt the practicality of the infinite tightening up of the principle of local action. However, the introduction of a similar principle in a temporal sense would be unphysical. It is a known fact that the typical range of atomic forces measures about 10^{-10}m, which means that all *long-distance effects* can be disregarded without risk of error. On the other hand, we observe viscoelastic and viscoplastic material behaviour, which reflects a material's memory over *long spaces of time*.

The existence of a material's long-term memory is demonstrated in numerous experiments. The mathematical representation of a material's recollection of mechanical processes of finite duration is to be found even in classical theories of continuum mechanics. *Linear viscoelasticity*, for instance, allows the representation of relaxation and creep properties using *functionals*. According to the defining equation (5.51) and in particular equation (5.64), the current state of stress depends on the entire past history of the deformation process. Without a representation of this kind, i.e. on the mere basis of linear elasticity, one could only model relaxation and creep effects by means of time-dependent material parameters. Corresponding arguments apply to the classical theories of plasticity and viscoplasticity, where ordinary differential equations are formulated to set up functional relationships between stress and strain processes.

7. 2. 2 Reduced Forms of the General Constitutive Equation

In the following considerations the material point $\mathscr{P} \in \mathscr{B}$ is always arbitrary, however it is selected at will and is not included as an argument for that reason. In place of the detailed representation (7.14),

$$\mathbf{T}(\mathscr{P}, t) = \underset{\tau \leqq t}{\mathfrak{G}} \left[\mathbf{F}(\mathscr{P}, \tau); \mathscr{P}\right] ,$$

we use the abbreviated form

$$\mathbf{T}(t) = \underset{\tau \leqq t}{\mathfrak{G}} \left[\mathbf{F}(\tau)\right] , \tag{7.16}$$

which frequently appears written in the form of

$$\mathbf{T}(t) = \underset{s \geqq 0}{\mathfrak{G}} \left[\mathbf{F}^{\mathrm{t}}(s)\right] , \tag{7.17}$$

where the past deformation history is represented by

$$s \longmapsto \mathbf{F}^{\mathrm{t}}(s) = \mathbf{F}(t - s) . \tag{7.18}$$

The independent variable s ($s \geq 0$) symbolises the length of time over which the past deformation event occurred.

The definition of a simple material was admittedly motivated by the assumptions of determinism, material frame-indifference and local action. However, the representation (7.14) is only a necessary conclusion, which is still not sufficient for material frame-indifference. At this stage, frame-indifference has not been completely guaranteed. Instead it enforces restrictive conditions on the stress functional $\mathfrak{G}$ in (7.14), (7.16) and (7.17).

With (7.16) every change of frame makes the constitutive equation

$$\mathbf{T}^{*}(t) = \underset{\tau \leqq t}{\mathfrak{G}} \left[\mathbf{F}^{*}(\tau)\right] \tag{7.19}$$

valid, in other words, for all orthogonal tensor-valued functions $\tau \longmapsto \mathbf{Q}(\tau)$ $\left(\mathbf{Q}\mathbf{Q}^{\mathrm{T}} = \mathbf{Q}^{\mathrm{T}}\mathbf{Q} \equiv \mathbf{1} \text{ , det } \mathbf{Q} \equiv +1\right)$ and for all deformation histories $\tau \longmapsto \mathbf{F}(\tau)$ one has to insist on the identity

$$\mathbf{Q}(t) \underset{\tau \leqq t}{\mathfrak{G}} \left[\mathbf{F}(\tau)\right]\mathbf{Q}^{\mathrm{T}}(t) = \underset{\tau \leqq t}{\mathfrak{G}} \left[\mathbf{Q}(\tau)\mathbf{F}(\tau)\right] . \tag{7.20}$$

This requirement, imposed on the stress functional $\mathfrak{G}$, can be fulfilled thus:

Theorem 7. 2

For a simple material, defined by the general constitutive equation (7.16), the *principle of material frame-indifference* is fulfilled if and only if the general constitutive equation takes on the so-called *reduced form*

$$\widetilde{\mathbf{T}}(t) = \underset{\tau \leqq t}{\mathfrak{F}} \left[\mathbf{C}(\tau)\right] \tag{7.21}$$

or

$$\widetilde{\mathbf{T}}(t) = \underset{\tau \leqq t}{\mathfrak{F}} \left[\mathbf{E}(\tau)\right] . \tag{7.22}$$

In this reduced form the stress functional $\mathfrak{F}$ can be selected at will: material frame-indifference does not require any further restrictions. □

Proof

For the deformation gradient we have the polar decomposition (1.60): $\mathbf{F}(\tau) = \mathbf{R}(\tau)\mathbf{U}(\tau)$. For the specific choice of $\mathbf{Q}(\tau) = \mathbf{R}^{\mathrm{T}}(\tau)$ equation (7.20) yields

$$\mathbf{R}^{\mathrm{T}}(t) \underset{\tau \leqq t}{\mathfrak{G}} [\mathbf{F}(\tau)]\mathbf{R}(t) = \underset{\tau \leqq t}{\mathfrak{G}} [\mathbf{R}^{\mathrm{T}}(\tau)\mathbf{F}(\tau)] = \underset{\tau \leqq t}{\mathfrak{G}} [\mathbf{U}(\tau)] ,$$

or

$$\mathbf{T}(t) = \mathbf{R}(t) \underset{\tau \leqq t}{\mathfrak{G}} [\mathbf{U}(\tau)]\mathbf{R}^{\mathrm{T}}(t) . \tag{7.23}$$

This means that the present state of stress $\mathbf{T}(t)$ can only depend on the present rotation $\mathbf{R}(t)$ and, apart from that, only on the right stretch tensor's history $\mathbf{U}(\tau)$, in lieu of $\mathbf{F}(\tau)$.

The representation (7.23) is also sufficient for material frame-indifference: a change of frame implies the transformations

$$\mathbf{T}^* = \mathbf{Q}\mathbf{T}\mathbf{Q}^{\mathrm{T}} , \mathbf{R}^* = \mathbf{Q}\mathbf{R} , \mathbf{U}^* = \mathbf{U} \text{ and } \mathbf{T}^*(t) = \mathbf{R}^*(t) \underset{\tau \leqq t}{\mathfrak{G}} [\mathbf{U}^*(\tau)]\mathbf{R}^{*\mathrm{T}}(t) .$$

Representation (7.23) is less suitable for practical applications, since the polar decomposition of $\mathbf{F}$ would have to be worked out beforehand. To derive a more practical representation, one can take another look at the polar decomposition: substituting $\mathbf{R}(t) = \mathbf{F}(t)\mathbf{U}^{-1}(t)$ we obtain

$$\mathbf{T}(t) = \mathbf{F}(t)\mathbf{U}^{-1}(t) \underset{\tau \leqq t}{\mathfrak{G}} [\mathbf{U}(\tau)]\mathbf{U}^{-1}(t)\mathbf{F}^{\mathrm{T}}(t) .$$

Applying the definition $\widetilde{\mathbf{T}} = (\det\mathbf{F})\mathbf{F}^{-1}\mathbf{T}\mathbf{F}^{\mathrm{T}-1}$ of the 2nd *Piola-Kirchhoff* tensor and considering $\det\mathbf{F} = \det\mathbf{U}$ we arrive at

$$\widetilde{\mathbf{T}}(t) = (\det\mathbf{U})\, \mathbf{U}^{-1}(t) \underset{\tau \leqq t}{\mathfrak{G}} [\mathbf{U}(\tau)]\mathbf{U}^{-1}(t) . \tag{7.24}$$

Now we have $\mathbf{U}^2 = \mathbf{C} = \mathbf{F}^{\mathrm{T}}\mathbf{F} = \mathbf{1} + 2\mathbf{E}$, which means that the right-hand side of (7.24) can be expressed as a functional of the past history of the right *Cauchy-Green* tensor $\mathbf{C}(\tau)$ or the *Green* strain tensor $\mathbf{E}(\tau)$. Conversely, representation (7.22) is also sufficient for *material frame-indifference* since, under a change of reference frame, in view of $\widetilde{\mathbf{T}}^*(t) = \widetilde{\mathbf{T}}(t)$ and $\mathbf{E}^* = \mathbf{E}$, (7.22) also implies

$$\widetilde{\mathbf{T}}^*(t) = \underset{\tau \leqq t}{\mathfrak{F}} [\mathbf{E}^*(\tau)] . \qquad \blacksquare$$

Under the assumption of *material frame-indifference* the following hint is deemed to be of importance, namely that a first fundamental step is to connect the right *Cauchy-Green* tensor $\mathbf{C}$ or the *Green* strain tensor $\mathbf{E}$ to the 2nd *Piola-Kirchhoff* tensor $\tilde{\mathbf{T}}$ in constitutive equations: the two variables $(\tilde{\mathbf{T}}, \mathbf{E})$ form a natural link as far as the formulation of material models is concerned. The balance (2.100) of mechanical energy and the stress power (2.106) occurring therein, $\rho_{\mathrm{R}} \ell_{\mathrm{i}} = \tilde{\mathbf{T}} \cdot \dot{\mathbf{E}}$, confirm the natural attribution of $\tilde{\mathbf{T}}$ and $\mathbf{E}$. The variables $\tilde{\mathbf{T}}$ and $\mathbf{E}$ can be regarded as *conjugate* and may be utilised for a fresh definition of a simple material. This definition automatically complies with the requirement of *material frame-indifference*.

Definition 7. 2

A simple material is represented by a functional relation of the form

$$\underset{\substack{\tau_1 \leqq t \\ \tau_2 \leqq t}}{\mathfrak{R}} \left[\tilde{\mathbf{T}}(\tau_1), \mathbf{E}(\tau_2); \boldsymbol{X}\right] = \mathbf{0} \,. \tag{7.25}$$

In the context of the principle of determinism it is assumed that, given a strain process, this relation uniquely defines a current state of stress:

$$\mathbf{E}(\tau)\Big|_{\tau \leqq t} \longmapsto \tilde{\mathbf{T}}(t) = \underset{\tau \leqq t}{\mathfrak{F}} \left[\mathbf{E}(\tau)\right] . \tag{7.26}$$

◊

The reduced forms (7.21) and (7.22) do not offer the only prospect of considering *material frame-indifference*: there is a complementary range of equivalent reduced forms of the general constitutive equation.[9]

Theorem 7. 3

The following reduced forms are likewise necessary and sufficient for the validity of *material frame-indifference*:

$$\tilde{\mathbf{T}}(t) = \underset{s \geqq 0}{\mathfrak{H}} \left[\mathbf{C}_{\mathrm{d}}^{\mathrm{t}}(s); \mathbf{C}(t)\right] = \mathbf{g}(\mathbf{C}(t)) + \underset{s \geqq 0}{\mathfrak{G}} \left[\mathbf{C}_{\mathrm{d}}^{\mathrm{t}}(s); \mathbf{C}(t)\right] , \tag{7.27}$$

$$\tilde{\mathbf{T}}(t) = \underset{s \geqq 0}{\mathfrak{H}} \left[\mathbf{E}_{\mathrm{d}}^{\mathrm{t}}(s); \mathbf{E}(t)\right] = \mathbf{g}(\mathbf{E}(t)) + \underset{s \geqq 0}{\mathfrak{G}} \left[\mathbf{E}_{\mathrm{d}}^{\mathrm{t}}(s); \mathbf{E}(t)\right] . \tag{7.28}$$

[9] Noll [1958]; Truesdell & Noll [1965], Sect. 29.

Here, by means of

$$s \longmapsto \mathbf{C}_{\mathrm{d}}^{\mathrm{t}}(s) = \mathbf{C}^{\mathrm{t}}(s) - \mathbf{C}(t) = \mathbf{C}(t-s) - \mathbf{C}(t) \tag{7.29}$$

and

$$s \longmapsto \mathbf{E}_{\mathrm{d}}^{\mathrm{t}}(s) = \mathbf{E}^{\mathrm{t}}(s) - \mathbf{E}(t) = \mathbf{E}(t-s) - \mathbf{E}(t) \tag{7.30}$$

the strain history is represented as the difference between the past and present states of strain; the functional $\mathfrak{G}$ vanishes for static strain histories:

$$\underset{s \geqq 0}{\mathfrak{G}}\left[\mathbf{0}(s); \mathbf{C}\right] = \mathbf{0}\,. \tag{7.31}$$

□

Proof

From (7.29) we obtain

$$\mathbf{C}(t-s) = \mathbf{C}_{\mathrm{d}}^{\mathrm{t}}(s) + \mathbf{C}(t)\ , \text{ i.e.}$$

the stress functional depends on the difference history $\mathbf{C}_{\mathrm{d}}^{\mathrm{t}}(s)$ and on the present strain $\mathbf{C}(t)$:

$$\tilde{\mathbf{T}}(t) = \underset{s \geqq 0}{\mathfrak{H}}\left[\mathbf{C}_{\mathrm{d}}^{\mathrm{t}}(s); \mathbf{C}(t)\right].$$

If we define the function $\mathbf{g}(\mathbf{C}) = \underset{s \geqq 0}{\mathfrak{H}}\left[\mathbf{0}(s); \mathbf{C}(t)\right]$ and the functional

$$\underset{s \geqq 0}{\mathfrak{G}}\left[\mathbf{C}_{\mathrm{d}}^{\mathrm{t}}(s); \mathbf{C}\right] = \underset{s \geqq 0}{\mathfrak{H}}\left[\mathbf{C}_{\mathrm{d}}^{\mathrm{t}}(s); \mathbf{C}\right] - \underset{s \geqq 0}{\mathfrak{H}}\left[\mathbf{0}(s); \mathbf{C}\right],$$

we obtain (7.27). The argumentation remains unaltered if the *Green* tensor $\mathbf{E}$ is inserted in place of the right *Cauchy-Green* tensor $\mathbf{C}$:

$$\mathbf{E} = \tfrac{1}{2}(\mathbf{C} - \mathbf{1})\,, \quad \mathbf{E}_{\mathrm{d}}^{\mathrm{t}}(s) = \tfrac{1}{2}\mathbf{C}_{\mathrm{d}}^{\mathrm{t}}(s) = \tfrac{1}{2}\big(\mathbf{C}(t-s) - \mathbf{C}(t)\big)\,.$$

■

According to this theorem, the current state of stress is the sum of a so-called elastic part and a memory part. The *elastic part* is a function of the current state of strain and the *memory part* a functional of the strain history. The memory part also includes the present strain and vanishes for strain processes that are temporally constant in the sense of $\mathbf{E}(t-s) \equiv \mathbf{E}(t)$ for all $s \geq 0$ (static processes).

The strain history is expressed by (7.30) as the difference between past and present states of strain, $\mathbf{E}_{\mathrm{d}}^{\mathrm{t}}(s) = \mathbf{E}(t - s) - \mathbf{E}(t)$. The function $s \longmapsto \mathbf{E}_{\mathrm{d}}^{\mathrm{t}}(s)$ operates on the reference configuration and pursues the history of the changing lengths and angles of material line elements:

$$\frac{1}{2}\left[(\mathrm{d}\boldsymbol{x}\cdot \mathrm{d}\boldsymbol{y})(t - s) - (\mathrm{d}\boldsymbol{x}\cdot \mathrm{d}\boldsymbol{y})(t)\right] = \mathrm{d}\boldsymbol{X}\cdot \mathbf{E}_{\mathrm{d}}^{\mathrm{t}}(s)\mathrm{d}\boldsymbol{Y}\,. \tag{7.32}$$

7. 2. 3 Simple Examples of Material Objectivity

By substantiating relation (7.25) or (7.26) it is straight away possible to formulate a great number of constitutive equations, which are invariant under every change of reference frame. It can be perceived that the constitutive equations of classical solid mechanics (e.g. linear elasticity, perfect plasticity etc.) only show this invariance in a first approximation for sufficiently modest rotations of the reference frame and small deformations. This characteristic is caused by the fact that the *Cauchy* stress tensor in constitutive equations of classical solid mechanics only depends on the symmetric part of the displacement gradient and not on its antisymmetric part. The latter contains an infinitesimal rotation, as is shown in the approximation (1.155) of the polar decomposition.

A classical constitutive equation in continuum mechanics, that fulfils the principle of material frame-indifference exactly, is the defining equation (5.18) of a linear-viscous fluid:

$$\mathbf{T} = -\,p(\rho)\mathbf{1} + 2\eta\mathbf{D} + \lambda(\mathrm{tr}\mathbf{D})\mathbf{1}\,. \tag{7.33}$$

This is obvious in the light of the transformation formulae $\mathbf{T}^{*} = \mathbf{Q}\mathbf{T}\mathbf{Q}^{\mathrm{T}}$ and $\mathbf{D}^{*} = \mathbf{Q}\mathbf{D}\mathbf{Q}^{\mathrm{T}}$. The connection with the reduced form (7.27) is also apparent: the definition of the linear-viscous fluid emerges from the general constitutive equation (7.27) by the special choice of the elastic part

$$\mathrm{g}(\mathrm{C}) = -\,\sqrt{\det\mathrm{C}}\;p\Big(\frac{\rho_{\mathrm{R}}}{\sqrt{\det\mathrm{C}}}\Big)\mathrm{C}^{-1} \tag{7.34}$$

and the memory part

$$\underset{s\,\geqq\,0}{\mathfrak{G}}\left[\mathrm{C}_{\mathrm{d}}^{\mathrm{t}}(s);\,\mathrm{C}\right] = \sqrt{\det\mathrm{C}}\Big\{\eta\mathrm{C}^{-1}\dot{\mathrm{C}}\mathrm{C}^{-1} + \frac{\lambda}{2}\,\mathrm{tr}(\dot{\mathrm{C}}\mathrm{C}^{-1})\mathrm{C}^{-1}\Big\}\,. \tag{7.35}$$

For the spatial strain rate tensor $\mathbf{D}$, according to equation (1.126) with (1.72), we can write

$$\mathbf{D} = \tfrac{1}{2}\mathbf{F}^{\mathrm{T}\,-\,1}\dot{\mathbf{C}}\mathbf{F}^{-\,1}\,. \tag{7.36}$$

With differentiation of (7.29) with respect to s this yields

$$\mathbf{D}(t) = -\,\tfrac{1}{2}\mathbf{F}^{\mathrm{T}\,-\,1}(t)\mathbf{C}_{\mathrm{d}}^{\mathrm{t}\,\prime}(0)\;\mathbf{F}^{-\,1}(t)\,. \tag{7.37}$$

A general constitutive equation for a nonlinear elastic solid can be obtained by equating the memory part in the reduced form (7.28) to zero:

$$\widetilde{\mathbf{T}}(t) = \mathbf{g}(\mathbf{E})\,. \tag{7.38}$$

A nonlinear viscoelastic solid could be defined by means of the constitutive equation

$$\widetilde{\mathbf{T}}(t) = \mathbf{g}(\mathbf{E}) + \mathbf{h}(\mathbf{E}, \dot{\mathbf{E}}), \tag{7.39}$$

with $\mathbf{h}(\mathbf{E}, \mathbf{0}) = \mathbf{0}$.

A classical example of a non-mechanical constitutive equation is the *Fourier* model of heat conduction, an isotropic relation between the spatial *temperature gradient* $g = \mathrm{grad}\Theta(x, t)$ and the *Cauchy* heat flux vector:

$$q = -\,\lambda(\Theta, |g|)g\,. \tag{7.40}$$

This constitutive equation also satisfies the principle of *material frame-indifference*, as is immediately recognisable from the transformation formulae (4.63), $q^* = \mathbf{Q}\,q$, and (4.64), $g^* = \mathbf{Q}\,g$.

7. 2. 4 Frame-Indifference and Observer-Invariance

All reduced forms mentioned in this chapter are *observer-invariant relations* in terms of Definition 4. 10: they are invariant under a given change of frame. Moreover, the reduced forms do not contain any information regarding the absolute motion of the reference system in use. Data of this kind could occur in observer-invariant relations in general, as the example of the balance of linear momentum, representing an observer-invariant relation in the particular version (4.80), shows.

On principle, one might wonder whether it is possible and physically significant to formulate constitutive equations which constitute more general, observer-invariant relations and are thus beyond the range of the reduced forms.

The theoretical possibility of setting up observer-invariant constitutive equations may be analysed with the help of individual examples. To this end we call to mind the spatial velocity gradient, which is not an objective tensor according to equation (4.49):

$$\mathbf{L}^* = \mathbf{Q}\mathbf{L}\mathbf{Q}^{\mathrm{T}} + \dot{\mathbf{Q}}\mathbf{Q}^{\mathrm{T}} = \mathbf{Q}\mathbf{L}\mathbf{Q}^{\mathrm{T}} + \mathbf{A}\,. \tag{7.41}$$

If we replace the relative rate of rotation $\mathbf{A}$ according to (4.74), $\mathbf{A} = \boldsymbol{\Omega}^* - \mathbf{Q}\boldsymbol{\Omega}\mathbf{Q}^{\mathrm{T}}$, we obtain

$$\mathbf{L}^* = \mathbf{Q}\mathbf{L}\mathbf{Q}^{\mathrm{T}} + \boldsymbol{\Omega}^* - \mathbf{Q}\boldsymbol{\Omega}\mathbf{Q}^{\mathrm{T}}\,, \tag{7.42}$$

i.e. the tensor $\mathbf{L} - \boldsymbol{\Omega}$ is objective:

$$\mathbf{L}^* - \boldsymbol{\Omega}^* = (\mathbf{L} - \boldsymbol{\Omega})^* = \mathbf{Q}(\mathbf{L} - \boldsymbol{\Omega})\mathbf{Q}^{\mathrm{T}}\,. \tag{7.43}$$

Consequently, not only $\mathbf{D}^* = \mathbf{Q}\mathbf{D}\mathbf{Q}^{\mathrm{T}}$ is valid, but also

$$\mathbf{W}^* - \boldsymbol{\Omega}^* = (\mathbf{W} - \boldsymbol{\Omega})^* = \mathbf{Q}(\mathbf{W} - \boldsymbol{\Omega})\mathbf{Q}^{\mathrm{T}}\,; \tag{7.44}$$

the difference $\mathbf{W} - \boldsymbol{\Omega}$ is likewise objective. As opposed to the rotation rate tensor itself, the difference between the rotation rate $\mathbf{W}$ and the rotation rate $\boldsymbol{\Omega}$ of the reference systems relative to the inertial frame is objective. It is therefore possible to construct some examples to set up constitutive equations, which are equally valid as observer-invariant relations in all references frames.

Viscous Fluid

A constitutive equation following the pattern

$$\mathbf{T} = -\,p(\rho)\mathbf{1} + \mathbf{f}(\mathbf{D}, \mathbf{W} - \boldsymbol{\Omega})\,,\ \mathbf{f}(\mathbf{0}, \mathbf{0}) = \mathbf{0}, \tag{7.45}$$

could be regarded as a definition of a viscous fluid. $\mathbf{f}(\cdot, \cdot)$ is an *isotropic tensor function* of two variables, a tensor-valued function with the property

$$\mathbf{f}(\mathbf{Q}\mathbf{D}\mathbf{Q}^{\mathrm{T}}, \mathbf{Q}(\mathbf{W} - \boldsymbol{\Omega})\mathbf{Q}^{\mathrm{T}}) = \mathbf{Q}\mathbf{f}(\mathbf{D}, \mathbf{W} - \boldsymbol{\Omega})\mathbf{Q}^{\mathrm{T}},$$

valid for all orthogonal tensors $\mathbf{Q}$. The following constitutive equation of a modified *Navier-Stokes* fluid represents a special case within this material category:

$$\mathbf{T} = -\,p(\rho)\mathbf{1} + 2\eta(\|\mathbf{W} - \boldsymbol{\Omega}\|)\mathbf{D} + \lambda(\|\mathbf{W} - \boldsymbol{\Omega}\|)(\mathrm{tr}\mathbf{D})\mathbf{1}\,. \tag{7.46}$$

As a characteristic of this fluid, its viscosity depends on the relative rotation rate $\mathbf{W} - \boldsymbol{\Omega}$.

Viscous Solid

A constitutive equation corresponding to the form

$$\widetilde{\mathbf{T}} = \mathbf{g}(\mathbf{E}, \dot{\mathbf{E}}, \mathbf{F}^{\mathrm{T}}(\mathbf{W} - \boldsymbol{\Omega})\mathbf{F}) \tag{7.47}$$

is also an observer-invariant relation. A special case of this equation would be an isotropic elastic solid, whose shear modulus μ depends on the rate of relative rotation:

$$\widetilde{\mathbf{T}} = 2\mu(\|\mathbf{F}^{\mathrm{T}}(\mathbf{W} - \boldsymbol{\Omega})\mathbf{F}\|)\Big[\mathbf{E} + \frac{\nu}{1-2\nu}(\mathrm{tr}\mathbf{E})\mathbf{1}\Big]\,. \tag{7.48}$$

Heat Conduction

Fourier's Law of heat conduction (7.40), $\boldsymbol{q} = -\,\lambda\boldsymbol{g}$, could be generalised as follows:

$$\boldsymbol{q} = -\,\lambda\boldsymbol{g} + \mu(\mathbf{W} - \boldsymbol{\Omega})\boldsymbol{g}\,. \tag{7.49}$$

It is plain to see that this constitutive equation is an observer-invariant relation. Its physical meaning is that heat flux vector and temperature gradient are not parallel even in the case of isotropic material behaviour: one part of the heat flux vector is orthogonal to the temperature gradient $\boldsymbol{g}$ according to

$$\boldsymbol{g}\cdot(\mathbf{W} - \boldsymbol{\Omega})\boldsymbol{g} = 0 \tag{7.50}$$

and depends on the relative rotation rate $\mathbf{W} - \boldsymbol{\Omega}$.

All these examples are fairly theoretical. The dependence of the viscosity of a fluid or the elasticity constant of a solid on the rotation rate of the reference system would doubtlessly be difficult to sustain, when restricted to macroscopic systems. In the case of heat conduction it is hard to envisage coming up with any experimental evidence of the heat flux vector's orthogonal part.[10]

The observer-invariant relations discussed contain information on the motion of the reference frame with regard to its rotation rate $\Omega(t)$. As far as the balance of linear momentum (4.80) is concerned, this information is, of course, indispensable. Accordingly, a constitutive equation would have to be formulated in the form of an observer-invariant relation if it were appropriate to assign finite masses or *properties of inertia* to the points $\mathscr{P}$ of a material body. As a rule, this is not done in phenomenological continuum mechanics. Instead, one assumes that the motion of the reference system exerts no influence on the representation of the material characteristics, i.e. the material properties are invariant with respect to superimposed rigid body motions. This assumption is synonymous with the assumption of *material frame-indifference*, which normally forms the foundation of any description of material behaviour.

If one queries the necessity of an invariance postulate in connection with the formulation of constitutive equations, there is not likely to be any difficulty in reaching an agreement as to a generally binding *principle of observer-invariance*: material properties cannot depend on the choice of representation tools used to portray them; in particular, their representations have to remain *form-invariant* with respect to any change of observer or reference system. The generally accepted *principle of material frame-indifference* narrows down the *principle of observer-invariance* to the assumption that the mathematical representation of material characteristics should remain unchanged when overlaid by any given *rigid body motion*.[11] Although such an assumption cannot be regarded as being a compelling law of nature, based on what has been said so far, it is of such a high physical plausibility that it is difficult to doubt it. The great application value of the assumption of frame-indifference lies in the very simple opening for deriving *reduced forms* from the general constitutive equation (7.1), which comply with this assumption for every conceivable mechanical (and thermomechanical) process, without any further restriction.

---

[10] However, a kinetic theory is capable of suggesting observer-invariant constitutive relations. This was shown by MÜLLER [1976], p. 124. A theoretical investigation of frame-dependent constitutive relations is to be found in MUSCHIK [1998a]. A physically interesting application of observer-invariant constitutive relations is provided in SADIKI & HUTTER [1996]. A further example is mentioned in BALKE [1998].

[11] In this sense, two formulations of *material frame-indifference* are to be found in LEIGH [1968], pp. 143. See also BERTRAM & SVENDSEN [1997].

7. 3 Properties of Material Symmetry

7. 3. 1 The Concept of the Symmetry Group

A constitutive equation ought to be independent of the reference frame on the grounds of the assumption of material frame-indifference. However, a dependence of the material behaviour on the choice of reference configuration cannot be ruled out: a change of reference configuration generally influences the form of the constitutive equation. If this is not the case for special changes, then there is a *property of material symmetry*. A change of the reference configuration is established by means of a one-to-one mapping

$$\boldsymbol{X} \longmapsto \hat{\boldsymbol{X}} = \boldsymbol{\Gamma}(\boldsymbol{X}) \Leftrightarrow \boldsymbol{X} = \boldsymbol{\Gamma}^{-1}(\hat{\boldsymbol{X}}) \,, \tag{7.51}$$

assigning the vector $\hat{\boldsymbol{X}}$ to a material point $\mathscr{P} \in \mathscr{B}$ instead of $\boldsymbol{X}$. This transforms the representation $\boldsymbol{x} = \chi_{\mathrm{R}}(\boldsymbol{X}, t)$, describing the motion of a material body, to the equivalent representation $\boldsymbol{x} = \chi_{\hat{\mathrm{R}}}(\hat{\boldsymbol{X}}, t)$. Corresponding to the definition

$$\boldsymbol{x} = \chi_{\hat{\mathrm{R}}}(\hat{\boldsymbol{X}}, t) = \chi_{\mathrm{R}}(\boldsymbol{\Gamma}^{-1}(\hat{\boldsymbol{X}}), t)$$

we have the identity

$$\chi_{\mathrm{R}}(\boldsymbol{X}, t) = \chi_{\hat{\mathrm{R}}}(\boldsymbol{\Gamma}(\boldsymbol{X}), t) \,, \tag{7.52}$$

valid for all $\boldsymbol{X} \in \mathrm{R}[\mathscr{B}]$. Differentiating this identity leads to the relation

$$\mathbf{F} = \hat{\mathbf{F}}\mathbf{P} \tag{7.53}$$

between the deformation gradients $\mathbf{F}(\boldsymbol{X}, t) = \mathrm{Grad}\chi_{\mathrm{R}}(\boldsymbol{X}, t)$ and

$\hat{\mathbf{F}}(\hat{\boldsymbol{X}}, t) = \mathrm{Grad}\; \chi_{\hat{\mathrm{R}}}(\hat{\boldsymbol{X}}, t)$. In this context the mapping

$$\mathbf{P} = \mathbf{P}(\boldsymbol{X}) = \mathrm{Grad}\boldsymbol{\Gamma}(\boldsymbol{X}) \tag{7.54}$$

is the gradient of the transformation $\boldsymbol{\Gamma}$, which relates the two reference configurations R and $\hat{\mathrm{R}}$.

In order to examine the dependence of a constitutive equation on the choice of reference configuration, it is useful to revert to the definition (7.15) of a simple material:[12]

$$\mathbf{T}(t) = \underset{\tau \leqq t}{\mathfrak{G}_{\mathrm{R}}}[\mathbf{F}(\tau)]\,. \tag{7.55}$$

The subscript R indicates that the functional $\mathfrak{G}_{\mathrm{R}}$ may depend on the choice of reference configuration. This dependence arises out of the identity

$$\mathbf{T}(t) = \underset{\tau \leqq t}{\mathfrak{G}_{\mathrm{R}}}[\mathbf{F}(\tau)] = \underset{\tau \leqq t}{\mathfrak{G}_{\hat{\mathrm{R}}}}[\hat{\mathbf{F}}(\tau)]\,.$$

In connection with the transformation (7.53), this leads to

$$\underset{\tau \leqq t}{\mathfrak{G}_{\hat{\mathrm{R}}}}[\hat{\mathbf{F}}(\tau)] = \underset{\tau \leqq t}{\mathfrak{G}_{\mathrm{R}}}[\hat{\mathbf{F}}(\tau)\mathbf{P}] \tag{7.56}$$

or

$$\underset{\tau \leqq t}{\mathfrak{G}_{\mathrm{R}}}[\mathbf{F}(\tau)] = \underset{\tau \leqq t}{\mathfrak{G}_{\hat{\mathrm{R}}}}[\mathbf{F}(\tau)\mathbf{P}^{-1}]\,. \tag{7.57}$$

In the theory of materials those changes of reference configuration that leave the material functional invariant are of particular interest, i.e. those for which

$$\underset{\tau \leqq t}{\mathfrak{G}_{\mathrm{R}}}[\mathbf{F}(\tau)] = \underset{\tau \leqq t}{\mathfrak{G}_{\mathrm{R}}}[\mathbf{F}(\tau)\mathbf{P}^{-1}] \tag{7.58}$$

is valid. One should bear in mind that this equation is to be understood as an identity which holds for all deformation processes $\tau \longmapsto \mathbf{F}(\tau)$. The material point $\mathcal{P} \in \mathcal{B}$ remains fixed in all these considerations.

In connection with simple materials it is sufficient to consider only the gradient $\mathbf{P}$ of the mapping $\boldsymbol{\Gamma}$, i.e. just local changes of the reference configuration. Although a change of the reference configuration is only of terminological importance, since merely the names of the material points are altered, it is also possible to attribute a physical significance to it by interpreting the function $\boldsymbol{X} \longmapsto \hat{\boldsymbol{X}} = \boldsymbol{\Gamma}(\boldsymbol{X})$ as an imaginary static deformation superimposed on the real motion.

---

[12] The presentation throughout this section follows TRUESDELL & NOLL [1965], pp. 76.

In connection with this interpretation one cannot expect a change in volume to leave the material response unaltered: for this reason, only those changes of reference configuration are selected for the analysis of material symmetry properties, as leave the volume locally constant:

$$|\det \mathbf{P}| = |\det(\mathrm{Grad}\, \boldsymbol{\Gamma}(\boldsymbol{X}))| = 1\,. \tag{7.59}$$

This argument prompts

Definition 7. 3

A *unimodular tensor* $\mathbf{H}$ $(|\det\ \mathbf{H}| = 1)$ represents a *property of material symmetry*, if it leaves the stress functional $\mathfrak{G}_{\mathrm{R}}$ invariant, i.e. if the following identity is satisfied for all deformation processes $\tau \longmapsto \mathbf{F}(\tau)$ $(\tau \leq t)$:

$$\underset{\tau \leqq t}{\mathfrak{G}_{\mathrm{R}}}[\mathbf{F}(\tau)\mathbf{H}] = \underset{\tau \leqq t}{\mathfrak{G}_{\mathrm{R}}}[\mathbf{F}(\tau)]\,. \tag{7.60}$$

◇

This definition leads to

Theorem 7. 4

The set

$$\mathscr{g}_{\mathrm{R}} = \Big\{\mathbf{H} \;\Big|\; \det\mathbf{H} = 1 \text{ and } \underset{\tau \leqq t}{\mathfrak{G}_{\mathrm{R}}}[\mathbf{F}(\tau)] = \underset{\tau \leqq t}{\mathfrak{G}_{\mathrm{R}}}[\mathbf{F}(\tau)\mathbf{H}]\Big\} \tag{7.61}$$

of unimodular tensors representing the properties of material symmetry of the constitutive functional $\mathfrak{G}_{\mathrm{R}}$ forms a group, which is called the *symmetry group* of the constitutive functional with respect to the reference configuration R. □

Proof

For $\mathbf{H}_{1,2} \in \mathscr{g}_{\mathrm{R}}$ the identities

$$\underset{\tau \leqq t}{\mathfrak{G}_{\mathrm{R}}}[\mathbf{F}(\tau)\mathbf{H}_1] = \underset{\tau \leqq t}{\mathfrak{G}_{\mathrm{R}}}[\mathbf{F}(\tau)] \quad \text{and} \quad \underset{\tau \leqq t}{\mathfrak{G}_{\mathrm{R}}}[\mathbf{F}(\tau)\mathbf{H}_2] = \underset{\tau \leqq t}{\mathfrak{G}_{\mathrm{R}}}[\mathbf{F}(\tau)]$$

are valid. This leads to

$$\underset{\tau \leqq t}{\mathfrak{G}_{\mathrm{R}}}[\mathbf{F}(\tau)] = \underset{\tau \leqq t}{\mathfrak{G}_{\mathrm{R}}}[\mathbf{F}(\tau)\mathbf{H}_1] = \underset{\tau \leqq t}{\mathfrak{G}_{\mathrm{R}}}[(\mathbf{F}(\tau)\mathbf{H}_1)\mathbf{H}_2] = \underset{\tau \leqq t}{\mathfrak{G}_{\mathrm{R}}}[\mathbf{F}(\tau)(\mathbf{H}_1\mathbf{H}_2)]\,,$$

if $\mathbf{F}(\tau)\mathbf{H}_1$ is also interpreted as a deformation process. Therefore,

$\mathbf{H}_1 \in \mathscr{g}_R$ and $\mathbf{H}_2 \in \mathscr{g}_R$ implies $\mathbf{H}_1\mathbf{H}_2 \in \mathscr{g}_R$.

On the other hand, $\mathbf{H} \in \mathscr{g}_R$ implies $\mathbf{H}^{-1} \in \mathscr{g}_R$, since

$$\underset{\tau \leqq t}{\mathfrak{G}_R}[\mathbf{F}(\tau)\mathbf{H}] = \underset{\tau \leqq t}{\mathfrak{G}_R}[\mathbf{F}(\tau)] = \underset{\tau \leqq t}{\mathfrak{G}_R}[(\mathbf{F}(\tau)\mathbf{H})\mathbf{H}^{-1}] \,.$$

$\mathbf{1} \in \mathscr{g}_R$ is self-evident. ■

The following theorem demonstrates how the symmetry group of a simple material depends on the reference configuration:

Theorem 7. 5
If $\mathbf{\Gamma}$: $\mathrm{R} \longrightarrow \hat{\mathrm{R}} \mid X \longmapsto \hat{X} = \mathbf{\Gamma}(X)$ is a change of reference configuration, $\mathbf{P}(X) = \mathrm{Grad}\mathbf{\Gamma}(X)$ and $\mathscr{g}_R$ as well as $\mathscr{g}_{\hat{R}}$ are the symmetry groups of the functionals $\mathfrak{G}_R$ and $\mathfrak{G}_{\hat{R}}$, we have the relation

$$\mathscr{g}_{\hat{R}} = \mathbf{P}\mathscr{g}_R\mathbf{P}^{-1}. \tag{7.62}$$

□

Proof
(i) $\mathscr{g}_{\hat{R}} \subset \mathbf{P}\mathscr{g}_R\mathbf{P}^{-1}$:

For $\hat{\mathbf{H}} \in \mathscr{g}_{\hat{R}}$ according to Definition 7. 3 and equation (7.56), it follows that

$$\underset{\tau \leqq t}{\mathfrak{G}_{\hat{R}}}[\hat{\mathbf{F}}(\tau)] = \underset{\tau \leqq t}{\mathfrak{G}_{\hat{R}}}[\hat{\mathbf{F}}(\tau)\hat{\mathbf{H}}] = \underset{\tau \leqq t}{\mathfrak{G}_R}[\hat{\mathbf{F}}(\tau)\hat{\mathbf{H}}\mathbf{P}] = \underset{\tau \leqq t}{\mathfrak{G}_R}[\hat{\mathbf{F}}(\tau)\mathbf{P}\mathbf{P}^{-1}\hat{\mathbf{H}}\mathbf{P}] \,.$$

Substituting $\underset{\tau \leqq t}{\mathfrak{G}_{\hat{R}}}[\hat{\mathbf{F}}(\tau)] = \underset{\tau \leqq t}{\mathfrak{G}_R}[\mathbf{F}(\tau)]$ and $\hat{\mathbf{F}}(\tau)\mathbf{P} = \mathbf{F}(\tau)$ this leads to

$$\underset{\tau \leqq t}{\mathfrak{G}_R}[\mathbf{F}(\tau)] = \underset{\tau \leqq t}{\mathfrak{G}_R}[\mathbf{F}(\tau)(\mathbf{P}^{-1}\hat{\mathbf{H}}\mathbf{P})] \,, \text{ i.e. } \mathbf{P}^{-1}\hat{\mathbf{H}}\mathbf{P} = \mathbf{H} \in \mathscr{g}_R \,.$$

Hence $\hat{\mathbf{H}} \in \mathscr{g}_{\hat{R}}$ can be represented by $\mathbf{P}$ and a tensor $\mathbf{H} \in \mathscr{g}_R$ in the form of $\hat{\mathbf{H}} = \mathbf{P}\mathbf{H}\mathbf{P}^{-1}$.

(ii) $\mathbf{P}\mathscr{g}_{\mathrm{R}}\mathbf{P}^{-1} \subset \mathscr{g}_{\hat{\mathrm{R}}}$:

$\mathbf{H} \in \mathscr{g}_{\mathrm{R}}$ leads to $\underset{\tau \leqq t}{\mathfrak{G}_{\mathrm{R}}}[\mathbf{F}(\tau)] = \underset{\tau \leqq t}{\mathfrak{G}_{\mathrm{R}}}[\mathbf{F}(\tau)\mathbf{H}]$ as well as

$$\underset{\tau \leqq t}{\mathfrak{G}_{\mathrm{R}}}[\mathbf{F}(\tau)] = \underset{\tau \leqq t}{\mathfrak{G}_{\hat{\mathrm{R}}}}[\hat{\mathbf{F}}(\tau)] = \underset{\tau \leqq t}{\mathfrak{G}_{\hat{\mathrm{R}}}}[\mathbf{F}(\tau)\mathbf{H}\mathbf{P}^{-1}] \text{ and}$$

$$\underset{\tau \leqq t}{\mathfrak{G}_{\hat{\mathrm{R}}}}[\hat{\mathbf{F}}(\tau)] = \underset{\tau \leqq t}{\mathfrak{G}_{\hat{\mathrm{R}}}}[\mathbf{F}(\tau)\mathbf{P}^{-1}\mathbf{P}\mathbf{H}\mathbf{P}^{-1}] = \underset{\tau \leqq t}{\mathfrak{G}_{\hat{\mathrm{R}}}}[\hat{\mathbf{F}}(\tau)(\mathbf{P}\mathbf{H}\mathbf{P}^{-1})] .$$

For every tensor $\mathbf{H} \in \mathscr{g}_{\mathrm{R}}$ it also holds that $\hat{\mathbf{H}} = \mathbf{P}\mathbf{H}\mathbf{P}^{-1} \in \mathscr{g}_{\hat{\mathrm{R}}}$. ■

The symmetry group should be understood as a material property. It is attributed to a material point $\mathscr{P} \in \mathscr{B}$ itself and depends on the choice of reference configuration in such a way that it applies to all materials. The physical interpretation of the theorem is the conclusion that the properties of material symmetry alter according to equation (7.62) if any given (static) deformation is attached to the motion. In this argumentation, neither $\mathbf{P} \in \mathscr{g}_{\mathrm{R}}$ nor $|\det\mathbf{P}| = 1$ is assumed.

The examination of the material symmetry properties was based on the general definition (7.15) of a simple material. The results can be transferred to the reduced forms that allow for the principle of material frame-indifference.

Theorem 7. 6

If we are given the constitutive equation of a simple material in the reduced form (7.21),

$$\tilde{\mathbf{T}}(t) = \underset{\tau \leqq t}{\mathfrak{F}_{\mathrm{R}}}[\mathbf{C}(\tau)] ,$$

or, equivalently, in the form of (7.27),

$$\tilde{\mathbf{T}}(t) = \mathbf{g}_{\mathrm{R}}(\mathbf{C}(t)) + \underset{s \geqq 0}{\mathfrak{G}_{\mathrm{R}}}[\mathbf{C}_{\mathrm{d}}^{t}(s); \mathbf{C}(t)] \qquad \Big(\underset{s \geqq 0}{\mathfrak{G}_{\mathrm{R}}}[\mathbf{0}(s); \mathbf{C}(t)] = \mathbf{0} \Big), \tag{7.63}$$

the tensor $\mathbf{H}$ ($\det\mathbf{H} = +1$) is an element of the symmetry group $\mathscr{g}_{\mathrm{R}}$, if and only if the relation

$$\mathbf{H} \underset{\tau \leqq t}{\mathfrak{F}_{\mathrm{R}}}[\mathbf{H}^{\mathrm{T}}\mathbf{C}(\tau)\mathbf{H}]\mathbf{H}^{\mathrm{T}} = \underset{\tau \leqq t}{\mathfrak{F}_{\mathrm{R}}}[\mathbf{C}(\tau)] \tag{7.64}$$

holds for all strain histories, or

$$\mathbf{H}\mathbf{g}_{\mathrm{R}}(\mathbf{H}^{\mathrm{T}}\mathbf{C}(t)\mathbf{H})\mathbf{H}^{\mathrm{T}} + \mathbf{H} \underset{s \geqq 0}{\mathfrak{G}_{\mathrm{R}}}[\mathbf{H}^{\mathrm{T}}\mathbf{C}_{\mathrm{d}}^{t}(s)\mathbf{H};\, \mathbf{H}^{\mathrm{T}}\mathbf{C}(t)\mathbf{H}]\mathbf{H}^{\mathrm{T}}$$

$$= \mathbf{g}_{\mathrm{R}}(\mathbf{C}(t)) + \underset{s \geqq 0}{\mathfrak{G}_{\mathrm{R}}}[\mathbf{C}_{\mathrm{d}}^{t}(s);\, \mathbf{C}(t)]\,. \tag{7.65}$$

□

Proof

In view of $\tilde{\mathbf{T}}(t) = \underset{\tau \leqq t}{\mathfrak{F}_{\mathrm{R}}}[\mathbf{C}(\tau)]$ or $\mathbf{T}(t) = \frac{1}{\det\mathbf{F}}\mathbf{F}(t) \underset{\tau \leqq t}{\mathfrak{F}_{\mathrm{R}}}[\mathbf{F}^{\mathrm{T}}(\tau)\mathbf{F}(\tau)]\mathbf{F}^{\mathrm{T}}(t)$

and the identity (7.60),

$\mathbf{H} \in \mathscr{g}_{\mathrm{R}}$ is equivalent to

$$\frac{1}{\det\mathbf{F}}\mathbf{F} \underset{\tau \leqq t}{\mathfrak{F}_{\mathrm{R}}}[\mathbf{F}^{\mathrm{T}}(\tau)\mathbf{F}(\tau)]\mathbf{F}^{\mathrm{T}} = \frac{1}{\det(\mathbf{F}\mathbf{H})}\mathbf{F}\mathbf{H} \underset{\tau \leqq t}{\mathfrak{F}_{\mathrm{R}}}[\mathbf{H}^{\mathrm{T}}\mathbf{F}^{\mathrm{T}}(\tau)\mathbf{F}(\tau)\mathbf{H}]\mathbf{H}^{\mathrm{T}}\mathbf{F}^{\mathrm{T}}.$$

With $\det\mathbf{H} = +1$ this is (7.64). ■

The condition $\det\mathbf{H} = +1$ is necessary for proving the theorem, i.e. only those elements of the symmetry group with a positive determinant are included. A modified definition of the symmetry group could be based on the identity (7.64) instead of (7.60), omitting this restriction. With $\mathbf{H} \in \mathscr{g}_{\mathrm{R}}$ there would then be $-\mathbf{H} \in \mathscr{g}_{\mathrm{R}}$ as well. Following the more general criterion (7.60), $-\mathbf{H} \in \mathscr{g}_{\mathrm{R}}$ could not be concluded from $\mathbf{H} \in \mathscr{g}_{\mathrm{R}}$.

7. 3. 2 Classification of Simple Materials into Fluids and Solids

The concept of the symmetry group can be used to classify the variety of simple materials according to their symmetry properties.[13] One practical use of this classification is the fact that the material description may become easier when there are more symmetry properties. Fluids possess the highest degree of symmetry, whereas the symmetry properties of solids can vary considerably.

---

[13] TRUESDELL & NOLL [1965], Sect. 32, 33.

Definition 7. 4

A simple material whose symmetry group is equal to the total set of unimodular tensors, i.e.

$$\mathcal{g} = \mathcal{Unim} \,, \tag{7.66}$$

is called a *fluid.*

If there is a reference configuration R, such that the symmetry group $\mathcal{g}_{\mathrm{R}}$ is a subgroup of the group of orthogonal tensors,

$$\mathcal{g}_{\mathrm{R}} \subset \mathcal{Orth} \,, \tag{7.67}$$

this material is a *solid.* If the symmetry group $\mathcal{g}_{\mathrm{R}}$ is the whole group of orthogonal tensors. i.e.

$$\mathcal{g}_{\mathrm{R}} = \mathcal{Orth} \,, \tag{7.68}$$

then this material is an *isotropic solid.* ◊

The above definition of a fluid manages without distinguishing any particular reference configuration; on account of (7.62), the symmetry group remains the same for any changes of reference configuration:

$$\mathcal{g} = \mathcal{Unim} = \mathbf{P}\, \mathcal{Unim}\, \mathbf{P}^{-1} \,. \tag{7.69}$$

The subscript R has been omitted here for this reason. In order to define the solid, however, one has to distinguish a so-called *undistorted reference configuration*: the tensor $\mathbf{H} = \mathbf{PQP}^{-1}$, constructed with an orthogonal tensor $\mathbf{Q}$, is admittedly unimodular but not orthogonal, i.e. $\mathcal{g}_{\mathrm{R}} \subset \mathcal{Orth}$ implies $\mathcal{g}_{\hat{\mathrm{R}}} \not\subset \mathcal{Orth}$. If a change of reference configuration is interpreted as a static deformation which is attached to a motion, the statement $\mathcal{g}_{\hat{\mathrm{R}}} \not\subset \mathcal{Orth}$ is synonymous with the assertion that a material body appears to lose its characteristic symmetry properties through deformation.

The definition of the fluid considerably simplifies the material description, as the following theorem indicates.

Theorem 7. 7

For simple fluids the following reduced form of the general constitutive equation is valid:

$$\mathbf{T}(t) = -\,p(\rho)\mathbf{1} + \underset{s \geqq 0}{\mathfrak{Q}}\left[\mathbf{G}^{t}(s); \rho\right]. \tag{7.70}$$

In this context $\mathfrak{Q}$ is an isotropic functional of the relative strain history

$$\mathbf{G}^{t}(s) = \mathbf{F}^{\mathrm{T}\,-1}(t)\mathbf{C}_{\mathrm{d}}^{t}(s)\mathbf{F}^{-1}(t) = \mathbf{F}^{\mathrm{T}\,-1}(t)[\mathbf{C}(t-s) - \mathbf{C}(t)]\mathbf{F}^{-1}(t)\,,$$

$$\mathbf{G}^{t}(s) = \left[\mathbf{F}(t-s)\mathbf{F}^{-1}(t)\right]^{\mathrm{T}}\left[\mathbf{F}(t-s)\mathbf{F}^{-1}(t)\right] - \mathbf{1}\,. \tag{7.71}$$

For all orthogonal tensors $\mathbf{Q}$ the functional $\mathfrak{Q}$ satisfies the identity

$$\mathbf{Q}\,\underset{s \geqq 0}{\mathfrak{Q}}\left[\mathbf{G}^{t}(s); \rho\right]\mathbf{Q}^{\mathrm{T}} = \underset{s \geqq 0}{\mathfrak{Q}}\left[\mathbf{Q}\mathbf{G}^{t}(s)\mathbf{Q}^{\mathrm{T}}; \rho\right]. \tag{7.72}$$

Moreover $\mathfrak{Q}$ vanishes for static processes:

$$\underset{s \geqq 0}{\mathfrak{Q}}\left[\mathbf{0}(s); \rho\right] = \mathbf{0}\,. \tag{7.73}$$

□

Proof

The verification of representation (7.70) is based on the reduced form (7.27):

$$\tilde{\mathbf{T}}(t) = \mathbf{g}(\mathbf{C}) + \underset{s \geqq 0}{\mathfrak{G}}\left[\mathbf{C}_{\mathrm{d}}^{t}(s); \mathbf{C}\right].$$

On account of the general definition of a fluid and Theorem 7. 6 the identity

$$\begin{aligned} &\mathbf{g}(\mathbf{C}) + \underset{s \geqq 0}{\mathfrak{G}}\left[\mathbf{C}_{\mathrm{d}}^{t}(s); \mathbf{C}\right] \\ &\quad = \mathbf{H}\mathbf{g}(\mathbf{H}^{\mathrm{T}}\mathbf{C}\mathbf{H})\mathbf{H}^{\mathrm{T}} + \mathbf{H}\,\underset{s \geqq 0}{\mathfrak{G}}\left[\mathbf{H}^{\mathrm{T}}\mathbf{C}_{\mathrm{d}}^{t}(s)\mathbf{H}; \mathbf{H}^{\mathrm{T}}\mathbf{C}\mathbf{H}\right]\mathbf{H}^{\mathrm{T}} \end{aligned} \tag{7.74}$$

holds for all tensors $\mathbf{H} \in \mathcal{U}nim$. If we insert the special unimodular tensor

$$\mathbf{H} = (\det\mathbf{F})^{\frac{1}{3}}\,\mathbf{F}^{-1} = J^{\frac{1}{3}}\,\mathbf{F}^{-1} \tag{7.75}$$

into this identity, we arrive at

$$\begin{aligned} \tilde{\mathbf{T}}(t) = {} & J^{\frac{2}{3}}\,\mathbf{F}^{-1}\mathbf{g}(J^{\frac{2}{3}}\,\mathbf{F}^{\mathrm{T}\,-1}\mathbf{C}\mathbf{F}^{-1})\mathbf{F}^{\mathrm{T}\,-1} \\ & + J^{\frac{2}{3}}\,\mathbf{F}^{-1}\,\underset{s \geqq 0}{\mathfrak{G}}\left[J^{\frac{2}{3}}\,\mathbf{F}^{\mathrm{T}\,-1}\mathbf{C}_{\mathrm{d}}^{t}(s)\mathbf{F}^{-1}; J^{\frac{2}{3}}\,\mathbf{F}^{\mathrm{T}\,-1}\mathbf{C}\mathbf{F}^{-1}\right]\mathbf{F}^{\mathrm{T}\,-1}, \end{aligned}$$

and with $\widetilde{\mathbf{T}} = J\mathbf{F}^{-1}\mathbf{T}\mathbf{F}^{\mathrm{T}-1}$ and $\mathbf{F}^{\mathrm{T}-1}\mathbf{C}_{\mathrm{d}}^{\mathrm{t}}(s)\mathbf{F}^{-1} = \mathbf{G}^{\mathrm{t}}(s)$ we get

$$\mathbf{T}(t) = J^{-\frac{1}{3}}\left\{\mathbf{g}(J^{\frac{2}{3}}\mathbf{1}) + \underset{s \geqq 0}{\mathfrak{G}}\left[J^{\frac{2}{3}}\mathbf{G}^{\mathrm{t}}(s); J^{\frac{2}{3}}\mathbf{1}\right]\right\} .$$

Expressing J by means of the continuity equation $J = \rho_{\mathrm{R}}/\rho$ and introducing new designations, the final constitutive equation reads

$$\mathbf{T} = \mathbf{f}(\rho) + \underset{s \geqq 0}{\mathbf{\Omega}}\left[\mathbf{G}^{\mathrm{t}}(s); \rho\right] .$$

For temporally constant (static) deformation histories, $\mathbf{G}^{\mathrm{t}}(s) \equiv \mathbf{0}$, the state of stress (static stress) depends exclusively on the current mass density ρ:

$$\mathbf{T}_{\mathrm{st}} = \mathbf{f}(\rho) . \tag{7.76}$$

In compliance with material frame-indifference, $\mathbf{T}^{*}_{\mathrm{st}} = \mathbf{f}(\rho^{*})$ also has to hold true. This yields the identity

$$\mathbf{Q}\mathbf{f}(\rho)\mathbf{Q}^{\mathrm{T}} = \mathbf{f}(\rho) , \tag{7.77}$$

valid for all $\mathbf{Q} \in \mathcal{O}rth^{+}$. Accordingly, $\mathbf{Q}e$ would be an eigenvector of $\mathbf{f}(\rho)$ if e is an eigenvector, $\mathbf{Q} \in \mathcal{O}rth$ being arbitrary. That means, however, that every vector is an eigenvector of $\mathbf{f}(\rho)$; hence $\mathbf{f}(\rho)$ has to be a spherical tensor. The usual notation is $\mathbf{T}_{\mathrm{st}} = -p(\rho)\mathbf{1}$. The scalar p stands for *hydrostatic pressure.* If we proceed once more to apply the principle of frame-indifference to the representation

$$\mathbf{T} = -p(\rho)\mathbf{1} + \underset{s \geqq 0}{\mathbf{\Omega}}\left[\mathbf{G}^{\mathrm{t}}(s); \rho\right] ,$$

obtained by this method, observing the transformation property

$$\mathbf{G}^{\mathrm{t}*}(s) = \mathbf{Q}\mathbf{G}^{\mathrm{t}}(s)\mathbf{Q}^{\mathrm{T}} ,$$

we obtain (7.72) which establishes the isotropy of the functional $\mathbf{\Omega}$. ■

In the general constitutive equation of a simple fluid the elastic part of the stress is a hydrostatic pressure, which is a function of mass density. The memory part is an isotropic functional of the relative strain history, which also depends on the present mass density. The function $s \longmapsto \mathbf{G}^{\mathrm{t}}(s)$

measures the past strain of a material element relative to the current configuration. As a result, the general constitutive equation of a simple fluid is not related to a preferential reference configuration.

For an *isotropic solid* there is, according to Definition 7. 4, a reference configuration R, for which the symmetry group is the whole set of orthogonal tensors, $\mathcal{g}_R = \mathcal{Orth}$. In (7.65) the *Cauchy-Green* tensor **C** may be replaced by the *Green* strain tensor **E**, so that the relations

$$\mathbf{Q}\mathbf{g}(\mathbf{E})\mathbf{Q}^T = \mathbf{g}(\mathbf{Q}\mathbf{E}\mathbf{Q}^T) \tag{7.78}$$

and

$$\mathbf{Q} \underset{s \geqq 0}{\mathfrak{G}} \left[\mathbf{E}_d^t(s); \mathbf{E}(t)\right]\mathbf{Q}^T = \underset{s \geqq 0}{\mathfrak{G}} \left[\mathbf{Q}\mathbf{E}_d^t(s)\mathbf{Q}^T; \mathbf{Q}\mathbf{E}(t)\mathbf{Q}^T\right] \tag{7.79}$$

are valid for all orthogonal tensors **Q**. The elastic part of the stress is an isotropic function of **E** (or **C**), and the memory part is an isotropic functional of $\mathbf{E}_d^t(s)$ and **E** or of $\mathbf{C}_d^t(s)$ and **C**. Another variation of the general constitutive equation for isotropic solids is substantiated by the following

Theorem 7. 8

The general constitutive equation for an isotropic solid can be written as

$$\mathbf{T}(t) = \mathbf{f}(\mathbf{B}) + \underset{s \geqq 0}{\mathfrak{F}} \left[\mathbf{G}^t(s); \mathbf{B}(t)\right] . \tag{7.80}$$

In this formulation, the elastic part **f** is an isotropic function of the left *Cauchy-Green* tensor $\mathbf{B} = \mathbf{F}\mathbf{F}^T$ and $\mathfrak{F}$ an isotropic functional with

$$\underset{s \geqq 0}{\mathfrak{F}} \left[\mathbf{0}(s); \mathbf{B}\right] = \mathbf{0} . \tag{7.81}$$

□

Proof

The polar decomposition $\mathbf{F} = \mathbf{R}\mathbf{U} = \mathbf{V}\mathbf{R}$ leads to $\mathbf{C} = \mathbf{R}^T\mathbf{B}\mathbf{R}$ and

$$\mathbf{C}_d^t(s) = \mathbf{F}^T\mathbf{G}^t(s)\mathbf{F} = \mathbf{R}^T\mathbf{V}\mathbf{G}^t(s)\mathbf{V}\mathbf{R} \qquad (\mathbf{U}^2 = \mathbf{C},\ \mathbf{V}^2 = \mathbf{B}) .$$

In addition, we have $\tilde{\mathbf{T}} = (\det\mathbf{V})\mathbf{R}^T\mathbf{V}^{-1}\mathbf{T}\mathbf{V}^{-1}\mathbf{R}$, and the reduced form (7.27) can be rewritten in terms of the *Cauchy* stress tensor:

$$(\det\mathbf{V})\mathbf{T} = \mathbf{VRg}(\mathbf{R}^T\mathbf{BR})\mathbf{R}^T\mathbf{V} + \mathbf{VR}\ \underset{s \geqq 0}{\mathfrak{G}}\left[\mathbf{R}^T\mathbf{VG}^t(s)\mathbf{VR};\ \mathbf{R}^T\mathbf{BR}\right]\mathbf{R}^T\mathbf{V}\,,$$

$$(\det\mathbf{V})\mathbf{T} = \mathbf{Vg}(\mathbf{B})\mathbf{V} + \mathbf{V}\ \underset{s \geqq 0}{\mathfrak{G}}\left[\mathbf{VG}^t(s)\mathbf{V};\ \mathbf{B}\right]\mathbf{V}\,,$$

taking advantage of the isotropy in the final step. The representation (7.80) follows on from $\mathbf{V} = \sqrt{\mathbf{B}}$ with new designations. The isotropy of $\mathbf{g}$ and $\mathfrak{G}$ implies the isotropy of $\mathbf{f}$ and $\mathfrak{F}$. ■

The memory part of the stress contains the present state of strain $\mathbf{B}$ and vanishes for processes which are constant in time, $\mathbf{G}^t(s) \equiv \mathbf{0}$ (static deformations). The relative strain history

$$s \longmapsto \mathbf{G}^t(s) = \mathbf{F}^{T-1}(t)\left[\mathbf{C}(t-s) - \mathbf{C}(t)\right]\mathbf{F}^{-1}(t)\,, \tag{7.82}$$

operates on the current configuration and pursues the temporal change of material line elements,

$$\left[(\mathrm{d}\boldsymbol{x}\cdot\mathrm{d}\boldsymbol{y})(t-s) - (\mathrm{d}\boldsymbol{x}\cdot\mathrm{d}\boldsymbol{y})(t)\right] = \mathrm{d}\boldsymbol{x}(t)\cdot\mathbf{G}^t(s)\mathrm{d}\boldsymbol{y}(t) \tag{7.83}$$

(Compare this to the corresponding property (7.32) of the strain history $\mathbf{E}^t_d(s)$, operating on the reference configuration). The concept of material symmetry is completed by the following

Definition 7. 5

A *simple material* is called *isotropic* if a reference configuration exists so that $\mathcal{O}rth \subset \mathcal{g}_R$. ◊

According to this definition a simple fluid is always isotropic; the physical meaning of the definition is, however, not exploited by this obvious statement. It arises in conjunction with a theorem taken from algebra, according to which the group of orthogonal tensors represents the largest proper subgroup of unimodular tensors; in other words, if one selects any given tensor $\mathbf{H} \in \mathcal{U}nim$, $\mathbf{H} \notin \mathcal{O}rth$, one can prove that $\mathcal{O}rth$ and $\mathbf{H}$ can never create a proper subgroup of $\mathcal{U}nim$, but unfailingly $\mathcal{U}nim$ itself: $\mathcal{O}rth$ is *maximal* in $\mathcal{U}nim$.[14]

[14] BRAUER [1965], NOLL [1965].

The physical meaning of this mathematical theorem can be condensed as follows in connection with Definitions 7. 4 and 7. 5:

An isotropic simple material is either a fluid or a solid.

Solids are defined by $\mathscr{g}_R \subset \mathscr{Orth}$ in connection with an appropriate (natural) reference configuration. There are solids whose symmetry groups are proper subgroups of the orthogonal group:

Definition 7. 6

A solid with $\mathscr{g}_R \subset \mathscr{Orth}$ and $\mathscr{g}_R \neq \mathscr{Orth}$ ($\mathscr{g}_R$ proper subgroup of $\mathscr{Orth}$) is called an *anisotropic solid.* ◊

It is evident that the smallest symmetry group only contains one element, i.e.

$$\mathscr{g}_R = \{\mathbf{1}\}\,. \tag{7.84}$$

This minimal symmetry group is likewise independent of any reference configuration. A material body with this trivial symmetry group would possess no material symmetry. Theoretically, it is possible to visualise an infinite number of subgroups between $\mathscr{g}_R = \{\mathbf{1}\}$ and $\mathscr{g}_R = \mathscr{Orth}$, that is, there is an infinite spectrum of mathematically conceivable anisotropic solids. In this connection, the question naturally arises as to the existence of a physical significance in all of these possible subgroups. This question cannot be answered from the point of view of continuum mechanics, but only by looking at the micromechanical structure of the solid itself. Suffice it to say that anisotropic solids often have a crystalline structure: they consist of crystals in which the atoms are arranged in regular lattice patterns. The symmetry properties of a crystal lattice correspond to a special group of rotations, i.e. a given symmetry group defines a special type of anisotropic solid. Since only a finite number of crystal lattices exists in nature, there can basically only be a finite number of subgroups of the orthogonal group that are of physical significance for the theory of material behaviour. More detailed information, coupled with the representation of elastic material behaviour, is given in Sect. 9. 3.

7. 4 Kinematic Conditions of Internal Constraint

7. 4. 1 General Theory

A third independent possibility for modelling material properties is the *a priori* assumption of restrictions on the local motion. Material bodies may develop a disproportionally strong resistance to certain types of deformation. At the same time they remain relatively submissive when subjected to other loads. It amounts to an idealisation of this behaviour to represent a material's proven unyieldingness from the point of view of kinematics by introducing a *condition of internal constraint*, which excludes those deformations a priori. The theory of internal constraints, developed by W. Noll, follows the doctrine of analytical mechanics:[15] one assumes that reaction stresses, caused by kinematic constraints, perform no work in any motions that are compatible with the constraint conditions. This determines the direction of the reaction stress. The intensity of the reaction stress then results from the balance of linear momentum in combination with appropriate boundary conditions.

Definition 7. 7

A *condition of a simple internal constraint* is a restriction on local motions, represented by an equation of the form

$$\gamma(\mathbf{F}) = 0 \tag{7.85}$$

or

$$\lambda(\mathbf{C}) = 0\,. \tag{7.86}$$

◊

The definition (7.86) is compatible with the principle of material frame-indifference, since the condition of constraint is expressed by means of the right *Cauchy-Green* tensor $\mathbf{C}$. The further evolution of the concept is based on the two assumptions that follow, representing a so-called *modified principle of determinism.*

Assumption 1

A simple material whose local motion is restricted by the internal *constraint* $\gamma(\mathbf{F}) = 0$ or $\lambda(\mathbf{C}) = 0$ is represented by the general constitutive equation

[15] TRUESDELL & NOLL [1965], Sect. 30. A thermomechanical theory is found in GREEN et al. [1970]. See BERTRAM & HAUPT [1976].

$$\mathbf{T} = \mathbf{N} + \underset{\tau \leqq t}{\mathfrak{G}} [\mathbf{F}(\tau)] \tag{7.87}$$

or

$$\widetilde{\mathbf{T}} = \widetilde{\mathbf{N}} + \underset{\tau \leqq t}{\mathfrak{F}} [\mathbf{C}(\tau)] \,. \tag{7.88}$$

The functionals $\mathfrak{G}$ and $\mathfrak{F}$ are defined on the set of deformation processes which satisfy the constraint condition $\gamma(\mathbf{F}) = 0$ or $\lambda(\mathbf{C}) = 0$. The tensors $\mathbf{N}$ and $\widetilde{\mathbf{N}}$, the so-called *reaction stresses*, remain undetermined: the parts $\mathbf{N}$ and $\widetilde{\mathbf{N}}$ do not depend on the deformation history by means of a constitutive equation.

Assumption 2

The *reaction stress tensor*

$$\mathbf{N} = \mathbf{T} - \underset{\tau \leqq t}{\mathfrak{G}} [\mathbf{F}(\tau)] \quad \text{or} \quad \widetilde{\mathbf{N}} = \widetilde{\mathbf{T}} - \underset{\tau \leqq t}{\mathfrak{F}} [\mathbf{C}(\tau)]$$

performs no work: the power of the reaction stress vanishes for every motion that is compatible with the constraint $\gamma(\mathbf{F}) = 0$ or $\lambda(\mathbf{C}) = 0$:

$$\mathbf{N} \cdot \mathbf{D} = 0 \,, \tag{7.89}$$

$$\widetilde{\mathbf{N}} \cdot \dot{\mathbf{C}} = 0 \,. \tag{7.90}$$

Combined with these assumptions, the definition of the constraint condition leads to

Theorem 7. 9

The reaction stresses are only determined up to a proportionality factor: they allow the representation

$$\mathbf{N} = \alpha \frac{\mathrm{d}}{\mathrm{d}\mathbf{F}} \gamma(\mathbf{F}) \mathbf{F}^{\mathrm{T}} \tag{7.91}$$

or

$$\widetilde{\mathbf{N}} = \beta \frac{\mathrm{d}}{\mathrm{d}\mathbf{C}} \lambda(\mathbf{C}) \,. \tag{7.92}$$

Factors α and β therein are not related to the deformation process by means of constitutive equations. □

Proof

Differentiation of $\lambda(\mathbf{C}) = 0$ supplies the identity

$$\frac{\mathrm{d}\lambda}{\mathrm{d}\mathbf{C}} \cdot \dot{\mathbf{C}}(t) = 0 \,,$$

valid for all t. If $\mathbf{C}_0$ $\left(\lambda(\mathbf{C}_0) = 0\right)$ is a state of strain and

$$\mathbf{\Gamma}_0 = \left.\frac{\mathrm{d}\lambda}{\mathrm{d}\mathbf{C}}\right|_{\mathbf{C} = \mathbf{C}_0}$$

the gradient of $\lambda(\cdot)$, the set of all admissible strain rates,

$$\mathbb{V}_{\mathbf{C}_0} = \{\dot{\mathbf{C}} \,|\, \mathbf{\Gamma}_0 \cdot \dot{\mathbf{C}} = 0\},$$

is a vector space of dimension 5. Geometrically speaking, this vector space is the tangential plane of the surface $\lambda(\mathbf{C}) = 0$ in the space of strain tensors at point $\mathbf{C}_0$. The tensor $\mathbf{\Gamma}_0$ is the normal of the surface at this point. According to the assumption $\tilde{\mathbf{N}} \cdot \dot{\mathbf{C}} = 0$ the reaction stress $\tilde{\mathbf{N}}$ is also orthogonal to all tensors $\dot{\mathbf{C}} \in \mathbb{V}_{\mathbf{C}_0}$; therefore $\tilde{\mathbf{N}}$ is parallel to the normal $\mathbf{\Gamma}_0$, i.e. $\tilde{\mathbf{N}} = \beta\,\mathbf{\Gamma}_0$. The verification of representation (7.91) is done in an analogous manner. ■

For a simple material with one constraint condition the general constitutive equation reads

$$\mathbf{T} = \alpha \frac{\mathrm{d}}{\mathrm{d}\mathbf{F}} \gamma(\mathbf{F}) \mathbf{F}^{\mathrm{T}} + \underset{\tau \leqq t}{\mathfrak{G}} \left[\mathbf{F}(\tau)\right] \tag{7.93}$$

or

$$\tilde{\mathbf{T}} = \beta \frac{\mathrm{d}}{\mathrm{d}\mathbf{C}} \lambda(\mathbf{C}) + \underset{\tau \leqq t}{\mathfrak{F}} \left[\mathbf{C}(\tau)\right] . \tag{7.94}$$

If there are several conditions of internal constraint, the theory is extended along the same lines. We define

$$\gamma_k(\mathbf{F}) = 0 \,, \qquad k = 1, ..., n \tag{7.95}$$

or

$$\lambda_k(\mathbf{C}) = 0 \,, \qquad k = 1, ..., n \tag{7.96}$$

and obtain

$$\mathbf{T} = \sum_{k=1}^{n} \alpha_k \frac{\mathrm{d}}{\mathrm{dF}} \gamma_k(\mathbf{F})\mathbf{F}^{\mathrm{T}} + \underset{\tau \leqq t}{\mathfrak{G}} [\mathbf{F}(\tau)] \tag{7.97}$$

or

$$\tilde{\mathbf{T}} = \sum_{k=1}^{n} \beta_k \frac{\mathrm{d}}{\mathrm{dC}} \lambda_k(\mathbf{C}) + \underset{\tau \leqq t}{\mathfrak{F}} [\mathbf{C}(\tau)] \, . \tag{7.98}$$

It is obvious that the number n of constraint conditions cannot be greater than 6: for n = 6 there is no degree of freedom left to allow a strain process.

7. 4. 2 Special Conditions of Internal Constraint

Incompressibility

The condition of *incompressibility* is the *a priori* assumption that the volume remains locally constant in every motion. For the case of incompressibility we employ $\gamma(\mathbf{F}) = \det\mathbf{F} - 1$ or

$$\lambda(\mathbf{C}) = \det\mathbf{C} - 1 \, . \tag{7.99}$$

Differentiation yields $\frac{\mathrm{d}}{\mathrm{dF}}\gamma(\mathbf{F}) = (\det\mathbf{F})\mathbf{F}^{\mathrm{T}\,-1}$ or

$$\frac{\mathrm{d}}{\mathrm{dC}}\lambda(\mathbf{C}) = (\det\mathbf{C})\mathbf{C}^{-1} \, , \tag{7.100}$$

and from (7.91) or (7.92) we obtain

$$\mathbf{N} = \alpha\mathbf{1} = -\,p\mathbf{1} \tag{7.101}$$

or

$$\tilde{\mathbf{N}} = \beta\mathbf{C}^{-1} = -\,p\mathbf{C}^{-1} \, . \tag{7.102}$$

It is common practice to denote the proportionality factor by $-\,p$ in the event of incompressibility. The *hydrostatic pressure p* is the *reaction stress* resulting from this constraint condition. The general constitutive equation for an incompressible simple material accordingly reads

$$\mathbf{T} = - p\mathbf{1} + \underset{\tau \leqq t}{\mathfrak{G}} [\mathbf{F}(\tau)] \tag{7.103}$$

or

$$\tilde{\mathbf{T}} = - p\mathbf{C}^{-1} + \underset{\tau \leqq t}{\mathfrak{F}} [\mathbf{C}(\tau)] , \tag{7.104}$$

where any symmetry properties - as discussed in Sect. 7. 3 - can be attributed to the functionals $\mathfrak{G}$ and $\mathfrak{F}$. For an incompressible simple fluid in particular, the most general constitutive equation is

$$\mathbf{T} = - p\mathbf{1} + \underset{s \geqq 0}{\mathfrak{Q}} [\mathbf{G}^t(s)] . \tag{7.105}$$

Here, the hydrostatic pressure p is not determined by a constitutive equation, as opposed to (7.70): since the mass density is materially constant in the case of incompressibility, it can no longer be used as an argument; this also applies to the memory functional $\mathfrak{Q}$.

Inextensibility in a Material Direction e

The condition of internal constraint in this case reads

$$\gamma(\mathbf{F}) = |\mathbf{F}e| - 1 \text{ or}$$

$$\lambda(\mathbf{C}) = e \cdot \mathbf{C}e - 1 , \tag{7.106}$$

and we obtain for the constitutive equation the general form

$$\mathbf{T} = \alpha \, \mathbf{F}e \otimes \mathbf{F}e + \underset{\tau \leqq t}{\mathfrak{G}} [\mathbf{F}(\tau)] \tag{7.107}$$

or

$$\tilde{\mathbf{T}} = \beta \, e \otimes e + \underset{\tau \leqq t}{\mathfrak{F}} [\mathbf{C}(\tau)] . \tag{7.108}$$

In these relations the scalars α and β have the physical meaning of a tension or compression stress, occurring as a reaction stress in inextensible fibres, which are defined by means of the vector field $e = e(\boldsymbol{X})$.

Shear Rigidity Between Two Material Directions

The constraint condition that the angle between two given material directions $\mathbf{e}_1(\boldsymbol{X})$ and $\mathbf{e}_2(\boldsymbol{X})$ is maintained throughout each motion complies with the formulation

$$\gamma(\mathbf{F}) = \mathbf{F}\mathbf{e}_1 \cdot \mathbf{F}\mathbf{e}_2 - \mathbf{e}_1 \cdot \mathbf{e}_2 \text{ or}$$

$$\lambda(\mathbf{C}) = \mathbf{e}_1 \cdot \mathbf{C}\mathbf{e}_2 - \mathbf{e}_1 \cdot \mathbf{e}_2 \,. \tag{7.109}$$

Hence we obtain

$$\mathbf{T} = \alpha\left(\mathbf{F}\mathbf{e}_1 \otimes \mathbf{F}\mathbf{e}_2 + \mathbf{F}\mathbf{e}_2 \otimes \mathbf{F}\mathbf{e}_1\right) + \underset{\tau \leqq t}{\mathfrak{G}}\left[\mathbf{F}(\tau)\right] \tag{7.110}$$

or

$$\tilde{\mathbf{T}} = \beta\left(\mathbf{e}_1 \otimes \mathbf{e}_2 + \mathbf{e}_2 \otimes \mathbf{e}_1\right) + \underset{\tau \leqq t}{\mathfrak{F}}\left[\mathbf{C}(\tau)\right] . \tag{7.111}$$

Rigid Body

The above examples can be combined, but no more than 6 constraint conditions are possible at a time. In fact, the definition of 6 conditions of constraint, for example in the normed specification,

$$\lambda_{ij}(\mathbf{C}) = \mathbf{e}_i \cdot \mathbf{C}\mathbf{e}_j - \delta_{ij} \; (i, j = 1, 2, 3) \tag{7.112}$$

$\left(\mathbf{e}_i \cdot \mathbf{e}_j = \delta_{ij}\right)$ leads to the constitutive equation

$$\tilde{\mathbf{T}} = \tilde{\mathbf{N}} = \sum_{i,j=1}^{3} \alpha_{ij} \mathbf{e}_i \otimes \mathbf{e}_j \,, \tag{7.113}$$

according to which the entire state of stress is a reaction stress, i.e. it remains completely undefined. In any case 6 conditions of constraint define a *rigid body motion*, as (7.112) specifically defines the identity $\mathbf{C}(t) \equiv \mathbf{1}$. If only rigid body motions are possible for a material body per definitionem, then there is no constitutive equation left for the state of stress: it remains completely undetermined.

7. 5 Formulation of Material Models

7. 5. 1 General Aspects

Most constitutive theories in continuum mechanics refer to simple materials. Usually they apply only those constitutive equations which conform to the assumption of *material frame-indifference.* More general, observer-invariant representations of material behaviour could be contemplated, but we are not dealing with them here, as long as no physical necessity can be detected. The reduced forms (7.25), (7.26) and (7.28),

$$\underset{\substack{\tau_1 \leqq t \\ \tau_2 \leqq t}}{\mathfrak{R}} \left[\tilde{\mathbf{T}}(\tau_1), \mathbf{E}(\tau_2); X\right] = \mathbf{0} \,, \tag{7.114}$$

$$\tilde{\mathbf{T}}(t) = \underset{\tau \leqq t}{\mathfrak{F}} \left[\mathbf{E}(\tau)\right] , \tag{7.115}$$

$$\tilde{\mathbf{T}}(t) = \mathbf{g}(\mathbf{E}) + \underset{s \geqq 0}{\mathfrak{G}} \left[\mathbf{E}_{\mathrm{d}}^{\mathrm{t}}(s); \mathbf{E}\right] , \tag{7.116}$$

or the more special representations (7.70) and (7.80),

$$\mathbf{T}(t) = -\, p(\rho)\mathbf{1} + \underset{s \geqq 0}{\mathfrak{Q}} \left[\mathbf{G}^{\mathrm{t}}(s); \rho\right] , \tag{7.117}$$

$$\mathbf{T}(t) = \mathbf{f}(\mathbf{B}) + \underset{s \geqq 0}{\mathfrak{F}} \left[\mathbf{G}^{\mathrm{t}}(s); \mathbf{B}\right] , \tag{7.118}$$

contain extensive possibilities for a systematic development of constitutive models, in which all four categories of material behaviour - elasticity, viscoelasticity, plasticity, viscoplasticity - are covered.

The category of *elasticity* (rate-independent material behaviour without hysteresis) offers the easiest situation: constitutive modelling of elastic material behaviour can be done with the help of appropriate material functions

$$\mathbf{g} : \mathbf{E} \longmapsto \tilde{\mathbf{T}} = \mathbf{g}(\mathbf{E}) \,. \tag{7.119}$$

A given tensor-valued function $\mathbf{g}$ represents the entire material behaviour of an elastic body.

Inelastic material behaviour - this includes every deviation from elasticity - is fundamentally associated with memory properties: the current state of stress no longer depends on the present state of strain alone, but also on the way in which this came about. That means that *inelastic materials* are in possession of a *memory* with regard to the process history.

A material model in mechanics consists of a set of constitutive equations, relating strain processes and stresses. By means of the special formulation of these constitutive equations, properties of material symmetry are also expressed: for example, the mathematical representation of the tensor function $\mathbf{g}(\cdot)$, which determines the properties of an elastic body, depends on the symmetry characteristics of the material. Further influence can possibly be exerted on the form of the material function $\mathbf{g}(\cdot)$ by assumptions regarding conditions of internal constraint, such as incompressibility. These aspects also apply for the representation of the functional relations depicting inelastic material behaviour.

There are two approaches for describing inelastic material properties by means of constitutive equations, the *method of functionals* and the *method of internal variables.*

7. 5. 2 Representation by Means of Functionals

Using the representation method of *functionals*, constitutive equations are formulated *explicitly* by way of functional relations. The easiest constitutive equation of inelasticity is the definition (5.18) of a linear-viscous fluid. The classical constitutive equation

$$\mathbf{T} = -\,p(\rho)\mathbf{1} + 2\eta\mathbf{D} + \lambda(\mathrm{tr}\mathbf{D})\mathbf{1}\,,$$

is a constitutive functional of the type (7.27), if $\mathbf{g}$ and $\mathfrak{G}$ are written in the (somewhat unfamiliar) form (7.34) and (7.35). On the other hand, it is also a special case of the general constitutive equation (7.70), if the strain rate tensor is expressed in the form

$$\mathbf{D} = -\,\frac{1}{2}\,\frac{\mathrm{d}}{\mathrm{d}s}\,\mathbf{G}^{\mathrm{t}}(s)\bigg|_{s=0}\,, \tag{7.120}$$

which evolves out of (7.82) by differentiation (see also (7.37)). In this case the functional is of differential type. The memory of a viscous material is infinitesimal: only an infinitely small part of the process, namely the strain rate tensor $\mathbf{D}$ as the tangent to the process history, has any influence on the current stress.

To illustrate an explicitly formulated functional of integral type, the model (5.51) of linear viscoelasticity serves as an example, if the linearised strain tensor is replaced by the *Green* strain (1.72) and the *Cauchy* stress by the 2nd *Piola-Kirchhoff* tensor (2.49). Nonlinear functionals can be formulated by integrating nonlinear functions of the past strain or applying double or multiple integrals.[16] There is liberal scope for setting up functional relations; constitutive equations can belong to the differential, integral or mixed type categories.

The memory properties generally characterising inelastic material behaviour can vary considerably in their intensity. This influences the mathematical structure of the functional stress-strain relations. As an example, the material memory that is represented within the category of *viscoelasticity* takes finite periods of time into account but fades away in proportion to the time that has elapsed since deformation took place. The material memory is much more pronounced, however, in the theory of *rate-independent plasticity*. The most general class of theories, *viscoplasticity*, expresses both short and long-term memory properties.

7. 5. 3 Representation by Means of Internal Variables

When employing the method of *internal variables* for the representation of inelastic material properties, functional relations of the types (7.114), (7.115), (7.116), (7.117) and (7.118) are defined *implicitly* using ordinary differential equations for so-called *internal variables*.[17] The (2N+1) parameter model of linear viscoelasticity is a characteristic example of this approach: the defining equations (5.110) and (5.111), equivalent to equations (5.108) and (5.109), lead to the functional formulation (5.76) with the relaxation function $G(t)$ according to (5.112), obtained by integrating the differential equations. Another example taken from the classical theories is the viscoplasticity model, defined by equations (5.219) - (5.224). As opposed to linear visco-elasticity, it is not possible to integrate the evolution equations (5.222) and (5.223) in a closed form.

The starting-point for modelling material properties by means of internal variables is the assumption that the stress tensor can be represented as a function of the current strain tensor, which simultaneously depends on a finite number of parameters $q_1, \dots, q_N$:

---

[16] See, for example, LOCKETT [1972], p. 65.

[17] COLEMAN & GURTIN [1967]; LUBLINER [1973]. A comprehensive representation of the theory of internal variables is found in MAUGIN & MUSCHIK [1994]. Cf. MAUGIN [1990].

$$\tilde{\mathbf{T}} = \mathbf{f}(\mathbf{E}, q_1, \dots, q_N) \,. \tag{7.121}$$

The parameters $q_1, \dots, q_N$ are called *internal variables.* They represent a material state that depends on the process history. To determine the current values of the internal variables, a system of ordinary differential equations is postulated:

$$\dot{q}_k(t) = f_k\big(\mathbf{E}(t), q_1(t), \dots, q_N(t)\big) \,, \quad k = 1, \dots, N \,. \tag{7.122}$$

These differential equations are termed *evolution equations.* Evolution equations specify the temporal changes of the internal variables depending on their current values and an input history. This can either be the current strain $\mathbf{E}$ or even the stress $\tilde{\mathbf{T}}$, if we assume that the stress-strain relation (7.121) is invertible for fixed values of the q_k, i.e.

$$\mathbf{E} = \mathbf{f}^{-1}(\tilde{\mathbf{T}}, q_1, \dots, q_N) = \mathbf{g}(\tilde{\mathbf{T}}, q_1, \dots, q_N) \,. \tag{7.123}$$

The evolution equations can then be written as follows:

$$\dot{q}_k(t) = g_k\big(\tilde{\mathbf{T}}(t), q_1(t), \dots, q_N(t)\big) \,, \quad k = 1, \dots, N \,. \tag{7.124}$$

The internal variables provide a tool to represent the memory properties on which inelastic material behaviour is generally based. The accuracy of such a representation depends on how many internal variables are selected in a special case. This number N can be looked upon as the dimension of a *state space.*

The evolution equations are additional constitutive equations completing the stress-strain relation (7.121). The right-hand sides $f_k(\cdot)$ or $g_k(\cdot)$ are material functions, modelling memory properties of the material. Which kind of behaviour is actually modelled depends on the structure of the evolution equations and on the physical meaning of the internal variables.

For any given strain process $\tau \longmapsto \mathbf{E}(\tau)$ $(t_0 \leq \tau \leq t)$ with initial conditions $q_k(t_0) = q_{k0}$ the evolution equations (7.122) determine the present values of the internal variables as functionals of the strain history:

$$q_k(t) = \underset{t_0 \leq \tau \leq t}{\mathbf{\Omega}_k} \big[\mathbf{E}(\tau); q_{k0}, \dots, q_{N0}\big] \,. \tag{7.125}$$

The general integral (7.125) of the evolution equations (7.122) can be inserted into the stress relation (7.121) to obtain an explicit representation of the constitutive equation in the form of (7.26). However, an explicit representation of the functional (7.125) cannot be given (as opposed to linear viscoelasticity). Solutions to evolution equations, which are usually nonlinear, can only be determined numerically.

The mathematical structure of the evolution equations (7.122) or (7.124) is influenced by the decision as to which category of material behaviour the modelling should take place in.

On principle, the temporal evolution of the internal variables is the outcome of dynamical processes taking place within the material's microstructure in connection with its inelasticity. The physical meaning of the internal variables lies in the scope for representing these dynamic processes in such a way that their macroscopic effects become visible. At the same time the evolution equations draw an idealised, considerably simplified picture of reality. In the evolution equations (5.111) of linear viscoelasticity the internal variables q_k have a very clear meaning: they are piston displacements in dashpots. In the classical model of viscoplasticity, (5.219) - (5.224), the internal variables are tensors, namely the inelastic strain tensor $\mathbf{E}_{\mathrm{i}}$ and the stress tensor $\mathbf{X}$ of kinematic hardening.

7. 5. 4 Comparison

If we compare the representation by means of functionals with the representation by means of internal variables, we can detect a certain similarity. There is a constitutive functional to correspond to every set of evolution equations in a constitutive theory with internal variables. Even if it is not possible, as a rule, to write down this functional explicitly, it can always be calculated numerically since its existence has been established mathematically. Conversely, if a particular functional is prescribed, it may be possible to find evolution equations whose general integral corresponds to this given functional. In the classical theory of linear viscoelasticity, the representations with functionals and those with internal variables are equivalent.

Functionals are occasionally the better choice, when it comes to discussing the general aspects of the theory of material behaviour, since the fundamental dependencies are particularly visible. On the other hand, the internal variable method has the advantage of better prospects in view of representation and practical application. In particular, the selection of the number and structure of the internal variables and the substantiation of their evolution equations can be guided by physical aspects that are not available in the representation theory of functionals. The fact that it is normally impossible to calculate the general integral of a set of evolution equations explicitly does not constitute a disadvantage in view of modern numerical integration techniques. Therefore, the representation of inelastic material behaviour by means of internal variables is given preference in matters connected with the development of concrete material models.

8 Dual Variables

8.1 Tensor-Valued Evolution Equations

8.1.1 Introduction

The general theory of material behaviour shows that wide scope exists for formulating material models, neatly contained in a perceivably well-ordered system: the categories of material behaviour - elasticity, viscoelasticity, plasticity and viscoplasticity - demand differently structured functionals and evolution equations combined with an appropriate regard for the properties of material symmetry and kinematic conditions of constraint.

One of the most important conclusions resulting from the assumption of objectivity is the fact that the *Green* strain tensor $\mathbf{E}$ (or the right *Cauchy-Green* tensor $\mathbf{C} = \mathbf{1} + 2\mathbf{E}$) has to be joined to the 2nd *Piola-Kirchhoff* tensor $\widetilde{\mathbf{T}}$ to guarantee invariance when the reference frame changes (or when undergoing of superimposed rigid body motions). Since the material time derivatives of these tensors, in addition to $\mathbf{E}$ and $\widetilde{\mathbf{T}}$, also remain unaltered during a change of frame, it is in accordance with the assumption of objectivity to use the material time derivatives to establish evolution equations for internal variables, which are either of *Green* strain or *Piola-Kirchhoff* stress type. In view of the material description's independence of the reference system, there should not be any objections, therefore, to using exclusively type $\mathbf{E}$, $\widetilde{\mathbf{T}}$, or $\dot{\mathbf{E}}$, $\dot{\widetilde{\mathbf{T}}}$ tensors, in both material functionals and evolution equations. That would mean that it is of no consequence whether the internal variables $q_1, \ldots, q_N$ in the evolution equations (7.122) are scalars or tensor components: the formulation of these evolution equations would remain the same during any change of the reference system. Accordingly, the possibility of developing material models within the frame of terminology depicted so far would continue without restrictions. However, upon closer examination, it becomes clear that limiting the variables to type $\mathbf{E}$, $\widetilde{\mathbf{T}}$ etc. would prove to be too one-sided in terms of the physical meaning of the constitutive equations and the internal variables, since tensors of this kind are exclusively related to the reference configuration: experience in dealing with constitutive models shows that it can be rational to apply tensors as internal variables that operate on other configurations, for example on the current configuration or on appropriate intermediate configurations. The defining equation (7.33) of the linear-viscous fluid is an instance of a constitutive equation naturally formulated using tensors of the current configuration.

For isotropic solids the general constitutive equation (7.80) is applicable: the current *Cauchy* stress tensor is a functional of the relative strain history $\mathbf{G}^t(\cdot)$, which also depends on the current strain $\mathbf{A}$:

$$\mathbf{T}(t) = \mathbf{f}(\mathbf{A}(t)) + \underset{s \geqq 0}{\mathfrak{F}} \left[\mathbf{G}^t(s); \mathbf{A}(t)\right] . \tag{8.1}$$

In this reduced form the left *Cauchy-Green* tensor $\mathbf{B}$ has been replaced by the *Almansi* tensor $\mathbf{A} = (1/2)(\mathbf{1} - \mathbf{B}^{-1})$ as opposed to (7.80), to make it clear that all tensors here ($\mathbf{T}$, $\mathbf{A}$ and $\mathbf{G}^t$) operate on the current configuration. If one wishes to represent a functional of this type implicitly within the scope of the internal variable method, the question arises as to which structure the corresponding evolution equations should take on for internal variables of the *Cauchy* stress type or *Almansi* strain type.

In particular in viscoelasticity and plasticity, where evolution equations are established for tensor-valued internal variables, and where intermediate configurations may have to be introduced in connection with a decomposition of the deformation, this question is of greater importance. Guidelines as to which kinematic and dynamic variables ought to be joined to each other by means of constitutive equations are desirable here. Unfortunately, the general theory of material behaviour does not provide an exhaustive answer to this question, since the catalogue of demands made on material models is as yet incomplete. In this situation the *concept of dual variables and derivatives* gives recommendations for assigning the kinematic and dynamic variables, i.e. stresses and strains. These are inspired by the balance equations of mechanics and consequently a natural guide.[1]

[1] The concept of dual variables and derivatives was originally proposed in TSAKMAKIS [1987]. The subsequent considerations follow HAUPT & TSAKMAKIS [1989]. A further development is presented in HAUPT & TSAKMAKIS [1996] and further discussed under mathematical points of view in TSAKMAKIS [2001a]. See also SVENDSEN & TSAKMAKIS [1994] and VAN DER GIESSEN & KOLLMANN [1996a, 1996b].

8. 1. 2 Objective Time Derivatives of Objective Tensors

The time rates of the internal variables as a function of their present values and a driving quantity are determined by evolution equations. Since evolution equations are constitutive equations, they have to be invariant with respect to any change of reference frame, according to the assumption of material frame-indifference. This requirement of invariance can be met if the rate of change of an internal variable possesses the same transformation behaviour as the internal variable itself, when the reference system changes. Bearing this aspect in mind, any formulation of evolution equations calls for the definition of an appropriate time rate. In connection with vector- or tensor-valued internal variables, suitable rates can differ from the usual material time derivative. For example, this is the case when objective tensors are applied as internal variables.

If $(\boldsymbol{x}, t) \longmapsto \mathbf{T}(\boldsymbol{x}, t)$ is an objective tensor field, for instance the *Cauchy* stress or any internal variable of *Cauchy* stress or *Almansi* strain type, it is transformed under a change of reference system according to equation (4.54),

$$\mathbf{T}^*(\boldsymbol{x}^*, t^*) = \mathbf{Q}(t)\mathbf{T}(\boldsymbol{x}, t)\mathbf{Q}^{\mathrm{T}}(t)\,, \tag{8.2}$$

$t \longmapsto \mathbf{Q}(t)$ being an orthogonal tensor-valued function representing the time-dependent change of frame. Obviously, the material time derivative of an objective tensor is not objective, for one calculates

$$(\mathbf{T}^*)^{\cdot} = (\mathbf{Q}\mathbf{T}\mathbf{Q}^{\mathrm{T}})^{\cdot} = \mathbf{Q}\dot{\mathbf{T}}\mathbf{Q}^{\mathrm{T}} + \dot{\mathbf{Q}}\mathbf{T}\mathbf{Q}^{\mathrm{T}} + \mathbf{Q}\mathbf{T}\dot{\mathbf{Q}}^{\mathrm{T}}\,,$$

$$(\mathbf{T}^*)^{\cdot} = \mathbf{Q}\dot{\mathbf{T}}\mathbf{Q}^{\mathrm{T}} + \dot{\mathbf{Q}}\mathbf{Q}^{\mathrm{T}}\mathbf{Q}\mathbf{T}\mathbf{Q}^{\mathrm{T}} + \mathbf{Q}\mathbf{T}\mathbf{Q}^{\mathrm{T}}(\dot{\mathbf{Q}}\mathbf{Q}^{\mathrm{T}})^{\mathrm{T}}\,. \tag{8.3}$$

An objective tensor rate can be obtained from this relation by expressing the reference system's rate of rotation $\dot{\mathbf{Q}}\mathbf{Q}^{\mathrm{T}}$ by a skew-symmetric tensor that is directly connected to the deformation rate. To this end, the antisymmetric part of the spatial velocity gradient is well suited: $(4.50)_2$ leads to

$$\dot{\mathbf{Q}}\mathbf{Q}^{\mathrm{T}} = \mathbf{W}^* - \mathbf{Q}\mathbf{W}\mathbf{Q}^{\mathrm{T}}\,,$$

and the result obtained with $\mathbf{W}^{\mathrm{T}} = -\mathbf{W}$ and $\mathbf{W}^{*\mathrm{T}} = -\mathbf{W}^*$ is

$$(\mathbf{T}^*)^{\cdot} = \mathbf{Q}\dot{\mathbf{T}}\mathbf{Q}^{\mathrm{T}} + (\mathbf{W}^* - \mathbf{Q}\mathbf{W}\mathbf{Q}^{\mathrm{T}})\mathbf{Q}\mathbf{T}\mathbf{Q}^{\mathrm{T}} + \mathbf{Q}\mathbf{T}\mathbf{Q}^{\mathrm{T}}(\mathbf{W}^* - \mathbf{Q}\mathbf{W}\mathbf{Q}^{\mathrm{T}})^{\mathrm{T}}$$

$$= \mathbf{Q}\dot{\mathbf{T}}\mathbf{Q}^{\mathrm{T}} + \mathbf{W}^*\mathbf{T}^* - \mathbf{Q}\mathbf{W}\mathbf{T}\mathbf{Q}^{\mathrm{T}} - \mathbf{T}^*\mathbf{W}^* + \mathbf{Q}\mathbf{T}\mathbf{W}\mathbf{Q}^{\mathrm{T}} .$$

This gives us the transformation formula

$$(\mathbf{T}^*)^{\cdot} - \mathbf{W}^*\mathbf{T}^* + \mathbf{T}^*\mathbf{W}^* = \mathbf{Q}(\dot{\mathbf{T}} - \mathbf{W}\mathbf{T} + \mathbf{T}\mathbf{W})\mathbf{Q}^{\mathrm{T}}, \tag{8.4}$$

which indicates that the tensor $\dot{\mathbf{T}} - \mathbf{W}\mathbf{T} + \mathbf{T}\mathbf{W}$ is objective.

Definition 8. 1

If $(x, t) \longmapsto \mathbf{T}(x, t)$ is an objective tensor field in the spatial representation, then the likewise objective tensor field

$$\overset{\circ}{\mathbf{T}}(x, t) = \dot{\mathbf{T}} - \mathbf{W}\mathbf{T} + \mathbf{T}\mathbf{W} \tag{8.5}$$

is called the *Zaremba-Jaumann* derivative of $\mathbf{T}$.[2] ◊

If a material derivative vanishes, then the corresponding function on a fixed material point $\mathscr{P} \in \mathscr{B}$ is constant in time. In lieu of this characteristic, the following theorem holds for the *Jaumann* rate.

Theorem 8. 1

If the *Jaumann* rate (8.5) is zero for a tensor field $\mathbf{T}(x, t)$, then the eigenvalues of $\mathbf{T}$ are constant in time. □

Proof

$\overset{\circ}{\mathbf{T}}(x, t) = \mathbf{0}$ implies

$$\dot{\mathbf{T}} = \mathbf{W}\mathbf{T} - \mathbf{T}\mathbf{W} \tag{8.6}$$

for the material derivative. If one proceeds to calculate the time derivative of the invariants

$$\mathrm{tr}(\mathbf{T}^k) \quad (k = 1, 2, 3, \dots),$$

then, inserting (8.6), we obtain terms like

---

[2] See ERINGEN [1962], p. 252. In the following the term *Jaumann* rate will be used.

$$\mathrm{tr}(\dot{\mathbf{T}}\,\mathbf{T}^{k-1}) = \mathrm{tr}(\mathbf{W}\mathbf{T}^{k} - \mathbf{T}\mathbf{W}\mathbf{T}^{k-1}) = \mathrm{tr}[(\mathbf{W}\mathbf{T} - \mathbf{T}\mathbf{W})\mathbf{T}^{k-1}] = 0\,.$$

It therefore holds that

$$\frac{\mathrm{d}}{\mathrm{d}t}\,\mathrm{tr}(\mathbf{T}^{k}) = 0\,.$$

Due to this result, the basic invariants and, accordingly, the eigenvalues of $\mathbf{T}$ are materially constant. ■

Forming the *Jaumann* rate would basically solve the problem of finding an objective time rate for an objective tensor field. However the solution is not unique: it is possible to find other objective rates, for example, the so-called *Oldroyd* rates[3] (1.128) and (1.129),

$$\overset{\vartriangle}{\mathbf{T}} = \dot{\mathbf{T}} + \mathbf{L}^{\mathrm{T}}\mathbf{T} + \mathbf{T}\mathbf{L}\,, \tag{8.7}$$

$$\overset{\triangledown}{\mathbf{T}} = \dot{\mathbf{T}} - \mathbf{L}\mathbf{T} - \mathbf{T}\mathbf{L}^{\mathrm{T}}\,, \tag{8.8}$$

or the rate according to *Green* and *McInnis*,[4]

$$\overset{\diamond}{\mathbf{T}} = \dot{\mathbf{T}} - \dot{\mathbf{R}}\mathbf{R}^{\mathrm{T}}\mathbf{T} + \mathbf{T}\,\dot{\mathbf{R}}\mathbf{R}^{\mathrm{T}}\,. \tag{8.9}$$

These rates also provide objective tensors. The *Jaumann* rate is clearly included in these examples: the identities

$\mathbf{L} = \mathbf{D} + \mathbf{W}$ and $\mathbf{L} = \dot{\mathbf{F}}\mathbf{F}^{-1} = \dot{\mathbf{R}}\mathbf{R}^{\mathrm{T}} + \mathbf{R}\dot{\mathbf{U}}\mathbf{U}^{-1}\mathbf{R}^{\mathrm{T}}$ lead to

$$\overset{\triangledown}{\mathbf{T}} = \overset{\circ}{\mathbf{T}} - (\mathbf{D}\mathbf{T} + \mathbf{T}\mathbf{D})\,, \tag{8.10}$$

$$\overset{\vartriangle}{\mathbf{T}} = \overset{\circ}{\mathbf{T}} + (\mathbf{D}\mathbf{T} + \mathbf{T}\mathbf{D})\,, \tag{8.11}$$

and with $\dot{\mathbf{R}}\mathbf{R}^{\mathrm{T}} = \mathbf{W} - \frac{1}{2}\mathbf{R}(\dot{\mathbf{U}}\mathbf{U}^{-1} - \mathbf{U}^{-1}\dot{\mathbf{U}})\mathbf{R}^{\mathrm{T}}$ we have

$$\overset{\diamond}{\mathbf{T}} = \overset{\circ}{\mathbf{T}} + \tfrac{1}{2}\mathbf{R}(\dot{\mathbf{U}}\mathbf{U}^{-1} - \mathbf{U}^{-1}\dot{\mathbf{U}})\mathbf{R}^{\mathrm{T}}\mathbf{T} - \mathbf{T}\tfrac{1}{2}\mathbf{R}(\dot{\mathbf{U}}\mathbf{U}^{-1} - \mathbf{U}^{-1}\dot{\mathbf{U}})\mathbf{R}^{\mathrm{T}}. \tag{8.12}$$

Apart from these two examples, any number of other objective derivatives could be quoted. It seems that, of all the objective derivatives, the *Jaumann*

---

[3] OLDROYD [1950].
[4] GREEN & MCINNIS [1967].

rate is the simplest. Nevertheless, it is not generally recommendable to use it in evolution equations for tensor-valued internal variables, as the special cases discussed below will show.

8. 1. 3 Example: Maxwell Fluid

A simple example of a differential equation for an objective tensorial quantity is taken from the three-dimensional generalisation of the *Maxwell* model, which was defined in the one-dimensional form by means of equation (5.104),

$$\lambda\dot{\sigma} + \sigma = \eta\dot{\varepsilon}$$

($\lambda = \eta/\hat{E}$). A 3-dimensional formulation of this constitutive model using the *Cauchy* stress tensor and observing the assumption of objectivity, could look like this:[5]

$$\lambda\overset{\circ}{\mathbf{T}} + \mathbf{T} = \eta\mathbf{D}\ . \tag{8.13}$$

$\mathbf{T}$ can be interpreted as the viscous stress in a viscoelastic fluid of *Maxwell* type. In order to examine this constitutive model, we take a look at the special case of a simple shear. A simple shear in the xy-plane corresponds to the spatially constant velocity gradient

$$\mathbf{L}(t) = \dot{\gamma}(t)\mathbf{e}_x \otimes \mathbf{e}_y\ , \tag{8.14}$$

where $\gamma(t)$ is the tangent of the shear angle. The corresponding stress tensor is two-dimensional and deviatoric. It consists of a shear stress τ and a normal stress σ:

$$\mathbf{T}(t) = \sigma(t)(\mathbf{e}_x \otimes \mathbf{e}_x - \mathbf{e}_y \otimes \mathbf{e}_y) + \tau(t)(\mathbf{e}_x \otimes \mathbf{e}_y + \mathbf{e}_y \otimes \mathbf{e}_x)\ . \tag{8.15}$$

We calculate

$$\overset{\circ}{\mathbf{T}} = (\dot{\sigma} - \dot{\gamma}\tau)(\mathbf{e}_x \otimes \mathbf{e}_x - \mathbf{e}_y \otimes \mathbf{e}_y) + (\dot{\tau} + \dot{\gamma}\sigma)(\mathbf{e}_x \otimes \mathbf{e}_y + \mathbf{e}_y \otimes \mathbf{e}_x)\ ,$$

$$\mathbf{D} = \frac{1}{2}\dot{\gamma}(\mathbf{e}_x \otimes \mathbf{e}_y + \mathbf{e}_y \otimes \mathbf{e}_x)$$

[5] HAUPT & TSAKMAKIS [1989], pp. 166.

and obtain, by insertion into the tensor-valued differential equation (8.13), two equations, namely

$$\lambda(\dot{\sigma} - \dot{\gamma}\tau) + \sigma = 0 \tag{8.16}$$

and

$$\lambda(\dot{\tau} + \dot{\gamma}\sigma) + \tau = \frac{1}{2}\eta\dot{\gamma} \,. \tag{8.17}$$

These equations can be integrated if the shear rate is kept constant:

$$\dot{\gamma}(t) = \dot{\gamma}_0 = \text{const.} \tag{8.18}$$

Elimination of σ in this special case leads to a second order differential equation:

$$\ddot{\tau} + \frac{2}{\lambda}\dot{\tau} + \left(\frac{1}{\lambda^2} + \dot{\gamma}_0{}^2\right)\tau = \eta\,\frac{\dot{\gamma}_0}{2\lambda^2} \,. \tag{8.19}$$

A solution for the initial conditions $\tau(0) = 0$ and $\sigma(0) = 0$ or $\dot{\tau}(0) = \eta\,\dfrac{\dot{\gamma}_0}{2\lambda}$ is given by

$$\tau(t) = \frac{\eta\dot{\gamma}_0}{2(1 + \lambda^2\dot{\gamma}_0{}^2)}\left[1 + (\lambda\dot{\gamma}_0\,\sin\dot{\gamma}_0 t - \cos\dot{\gamma}_0 t)e^{-\frac{t}{\lambda}}\right] . \tag{8.20}$$

The temporal development of the shear stress, according to this solution, is characterised for every value of the relaxation time λ by a damped oscillation with the frequency

$$\nu = \frac{\dot{\gamma}_0}{2\pi} \,.$$

This result appears to be rather unphysical. Another way of representing the *Maxwell* model in three dimensions is by using an *Oldroyd* rate, i.e. a constitutive equation of the form $\lambda\overset{\triangle}{\mathbf{T}} + \mathbf{T} = \eta\mathbf{D}$ or

$$\lambda(\dot{\mathbf{T}} + \mathbf{L}^{\mathrm{T}}\mathbf{T} + \mathbf{T}\mathbf{L}) + \mathbf{T} = \eta\mathbf{D} \,. \tag{8.21}$$

According to this constitutive equation the shear in the xy-plane produces a stress tensor, which is again two-dimensional but not a deviator:

$$\mathbf{T}(t) = \sigma_{xx}\mathbf{e}_x \otimes \mathbf{e}_x + \sigma_{yy}\,\mathbf{e}_y \otimes \mathbf{e}_y + \tau(\mathbf{e}_x \otimes \mathbf{e}_y + \mathbf{e}_y \otimes \mathbf{e}_x) \,.$$

The *Oldroyd* rate of $\mathbf{T}$ is

$$\overset{\triangle}{\mathbf{T}} = \dot{\sigma}_{xx}\mathbf{e}_x \otimes \mathbf{e}_x + (\dot{\sigma}_{yy} + 2\tau\dot{\gamma})\mathbf{e}_y \otimes \mathbf{e}_y + (\dot{\tau} + \sigma_{xx}\dot{\gamma})(\mathbf{e}_x \otimes \mathbf{e}_y + \mathbf{e}_y \otimes \mathbf{e}_x)\,,$$

and we obtain from (8.21) three independent component equations:

$$\lambda\dot{\sigma}_{xx} + \sigma_{xx} = 0\,, \tag{8.22}$$

$$\lambda(\dot{\tau} + \dot{\gamma}\sigma_{xx}) + \tau = \frac{1}{2}\eta\dot{\gamma}\,, \tag{8.23}$$

$$\lambda(\dot{\sigma}_{yy} + 2\dot{\gamma}\tau) + \sigma_{yy} = 0\,. \tag{8.24}$$

The shear rate is again assumed to be constant,

$$\dot{\gamma}(t) = \dot{\gamma}_0 = \text{const.}$$

For the initial conditions $\sigma_{xx}(0) = \sigma_{yy}(0) = \tau(0) = 0$ the first equation yields $\sigma_{xx}(t) = 0$; accordingly, the second equation yields the shear stress

$$\tau(t) = \frac{1}{2}\eta\dot{\gamma}_0\left[1 - \mathrm{e}^{-\frac{t}{\lambda}}\right]. \tag{8.25}$$

The third equation delivers the non-vanishing normal stress $\sigma_{yy}(t)$. If we compare the results (8.20) and (8.25), we notice that the *Jaumann* rate of the *Cauchy* stress tensor in combination with the strain rate tensor $\mathbf{D}$ provides a shear stress that tends towards a stationary value in a fading oscillation. By contrast, the shear stress calculated using the *Oldroyd* rate reaches its stationary value in a monotonic course. In view of the simple physical situation that a spring-dashpot model represents, the oscillating shear stress hardly appears to be physically plausible: in contrast to this the monotonic development seems to be rather more reliable, a fact that can be regarded as an advantage of the *Oldroyd* rate.

Another advantage of the *Oldroyd* rate is the fact that a 3-dimensional evolution equation of the form (8.21) can be integrated in a closed form: the identities

$$\dot{\mathbf{T}} + (\dot{\mathbf{F}}\mathbf{F}^{-1})^{\mathrm{T}}\mathbf{T} + \mathbf{T}(\dot{\mathbf{F}}\mathbf{F}^{-1}) = \mathbf{F}^{\mathrm{T}\,-1}(\mathbf{F}^{\mathrm{T}}\mathbf{T}\mathbf{F})^{\cdot}\,\mathbf{F}^{-1} \text{ and}$$

$$\mathbf{F}^{\mathrm{T}}\mathbf{D}\mathbf{F} = \dot{\mathbf{E}} = \frac{1}{2}\dot{\mathbf{C}}$$

allow the constitutive equation (8.21) to be written as

$$\lambda(\mathbf{F}^{\mathrm{T}}\mathbf{T}\mathbf{F})^{\cdot} + \mathbf{F}^{\mathrm{T}}\mathbf{T}\mathbf{F} = \frac{1}{2}\eta\dot{\mathbf{C}}\,. \tag{8.26}$$

Integrating this tensor-valued differential equation provides

$$\mathbf{F}^{\mathrm{T}}\mathbf{T}\mathbf{F}\, \mathrm{e}^{\frac{t}{\lambda}} = \frac{\eta}{2\lambda} \int_{-\infty}^{t} \mathrm{e}^{\frac{\tau}{\lambda}} \mathbf{C}'(\tau)\mathrm{d}\tau = -\frac{\eta}{2\lambda^2} \int_{-\infty}^{t} \mathrm{e}^{\frac{\tau}{\lambda}} [\mathbf{C}(\tau) - \mathbf{C}(t)]\mathrm{d}\tau$$

and leads to the result

$$\mathbf{T}(t) = -\frac{\eta}{2\lambda^2} \int_{-\infty}^{t} \mathrm{e}^{-\frac{t-\tau}{\lambda}} [\mathbf{F}^{\mathrm{T}-1}\mathbf{C}(\tau)\mathbf{F}^{-1} - \mathbf{1}]\mathrm{d}\tau$$

or

$$\mathbf{T}(t) = -\frac{\eta}{2\lambda^2} \int_{0}^{\infty} \mathrm{e}^{-\frac{t-\tau}{\lambda}} \mathbf{G}^{\mathrm{t}}(s)\mathrm{d}s\,. \tag{8.27}$$

[6]

Accordingly, the current stress is a linear functional of the relative strain history $\mathbf{G}^{\mathrm{t}}(s) = \mathbf{F}^{\mathrm{T}-1}(t)\mathbf{C}(t-s)\mathbf{F}^{-1}(t) - \mathbf{1}$. The result is a special case of the constitutive equations (8.1) or (7.80), which apply to isotropic materials in general.

It is not possible to obtain a simple result like this using the *Jaumann* rate (8.5).[7]

8. 1. 4 Example: Rigid-Plastic Solid with Hardening

A further example illustrating the properties of objective tensor derivatives in connection with tensor-valued evolution equations arises out of a generalisation of geometrically linear plasticity with hardening to incorporate large deformations. For simplicity, the elastic strains are disregarded in the following, so that no decomposition of the strain takes place. This procedure backs up the assumption that a material body is either rigid or the yield and loading conditions are fulfilled at every point of the body to enable deformations to take place everywhere. Then the material model consists of a yield condition, e.g.

$$\tfrac{1}{2}(\mathbf{T} - \mathbf{X})^{\mathrm{D}} \cdot (\mathbf{T} - \mathbf{X})^{\mathrm{D}} = \tfrac{1}{3}k^2\,, \tag{8.28}$$

---

[6] HAUPT & TSAKMAKIS [1989], eq. (1.9).

[7] HAUPT & TSAKMAKIS [1989], eq. (1.7).

and a flow rule, assumed to take the form of

$$\mathbf{D} = \lambda(\mathbf{T} - \mathbf{X})^{\mathrm{D}} = \sqrt{\frac{3}{2}}\,\frac{\sqrt{\mathbf{D}\cdot\mathbf{D}}}{k}\,(\mathbf{T} - \mathbf{X})^{\mathrm{D}}\,. \tag{8.29}$$

The tensor $\mathbf{X}$, representing the centre of the yield surface in the stress space, is an internal variable of *Cauchy* stress type. It describes the kinematic hardening in connection with an evolution equation, in which an objective time derivative is to be employed. A possible generalisation of the *Armstrong-Frederick* model (5.184) using the *Jaumann* rate would be the evolution equation

$$\overset{\circ}{\mathbf{X}} = c\mathbf{D} - b\dot{s}\mathbf{X}\,, \tag{8.30}$$

where $s(t)$ is the accumulated strain or the so-called *arclength*, defined by

$$\dot{s}(t) = \sqrt{\frac{2}{3}\mathbf{D}\cdot\mathbf{D}}\,. \tag{8.31}$$

At this stage the material model of a rigid-plastic solid is completely defined. The flow rule (8.29) leads to an explicit equation for the stress deviator:

$$\mathbf{T}^{\mathrm{D}} = \sqrt{\frac{2}{3}}\,k\,\frac{\mathbf{D}}{\sqrt{\mathbf{D}\cdot\mathbf{D}}} + \mathbf{X}^{\mathrm{D}}\,. \tag{8.32}$$

By inserting the simple shear $\mathbf{L}(t) = \dot{\gamma}(t)\mathbf{e}_x \otimes \mathbf{e}_y \mathrel{\hat{=}} \dot{\gamma}\begin{pmatrix} 0 & 1 \\ 0 & 0 \end{pmatrix}$,

(8.32) corresponds to the matrix equation

$$\begin{pmatrix} \sigma & \tau \\ \tau & -\sigma \end{pmatrix} = \frac{k}{\sqrt{3}}\frac{\dot{\gamma}}{|\dot{\gamma}|}\begin{pmatrix} 0 & 1 \\ 1 & 0 \end{pmatrix} + \begin{pmatrix} \xi & \eta \\ \eta & -\xi \end{pmatrix}. \tag{8.33}$$

The evolution equation (8.30) for the hardening tensor $\mathbf{X}$ yields the component relations

$$\dot{\xi} - \dot{\gamma}\eta + \frac{b}{\sqrt{3}}\,|\dot{\gamma}|\,\xi = 0\,, \tag{8.34}$$

$$\dot{\eta} + \dot{\gamma}\xi + \frac{b}{\sqrt{3}}\,|\dot{\gamma}|\,\eta = \frac{c}{2}\dot{\gamma}\,. \tag{8.35}$$

For a monotonic shear process ($\dot{\gamma} > 0$) time t can be replaced by the shear strain γ, resulting in

$$\xi'(\gamma) - \eta(\gamma) + \frac{b}{\sqrt{3}}\,\xi(\gamma) = 0\,, \tag{8.36}$$

$$\eta'(\gamma) + \xi(\gamma) + \frac{b}{\sqrt{3}}\,\eta(\gamma) = \frac{c}{2}\,, \tag{8.37}$$

where the notation $(\ )' = \dfrac{\mathrm{d}}{\mathrm{d}\gamma}(\ )$ is used.

By eliminating ξ, a second order differential equation for $\eta(\gamma)$ emerges:

$$\eta''(\gamma) + 2\frac{b}{\sqrt{3}}\,\eta'(\gamma) + \Big(1 + \frac{b^2}{3}\Big)\eta(\gamma) = \frac{bc}{2\sqrt{3}}\,. \tag{8.38}$$

The solution for the initial conditions $\eta(0) = 0$ and $\eta'(0) = \frac{c}{2}$ $\big(\xi(0) = 0\big)$ is

$$\eta(\gamma) = \frac{1}{2\sqrt{3}\,(1 + \frac{b^2}{3})}\,\Big\{bc + \mathrm{e}^{-\frac{b}{\sqrt{3}}\gamma}\Big[c\sqrt{3}\sin\gamma - bc\cos\gamma\Big]\Big\}\,. \tag{8.39}$$

It leads to the shear stress

$$\tau(\gamma) = \frac{k}{\sqrt{3}} + \eta(\gamma)\,. \tag{8.40}$$

We see that an oscillating shear stress is apparent, which seems to be barely plausible in view of the monotonic deformation process and the simple physical situation. In the special case of linear kinematic hardening $(b = 0)$ the damping of the oscillation vanishes and the shear stress response

$$\tau(\gamma) = \frac{k}{\sqrt{3}} + \frac{c}{2}\sin\gamma\,, \tag{8.41}$$

characterised by an oscillation of constant amplitude, occurs.
A possible alternative to the *Jaumann* rate is the *Oldroyd* rate (8.8),

$$\overset{\triangledown}{\mathbf{X}} = \dot{\mathbf{X}} - \mathbf{L}\mathbf{X} - \mathbf{X}\mathbf{L}^{\mathrm{T}}\,.$$

Using the evolution equation

$$\overset{\triangledown}{\mathbf{X}} = c\mathbf{D} - b\dot{s}\mathbf{X} \tag{8.42}$$

in lieu of (8.30) leads to a hardening tensor

$$\mathbf{X} \mathrel{\hat{=}} \begin{pmatrix} \xi & \eta \\ \eta & \zeta \end{pmatrix}, \tag{8.43}$$

which is no longer a deviator. Consequently (8.42) provides 3 component relations:

$$\begin{aligned} &\xi'(\gamma) - 2\eta(\gamma) + \frac{b}{\sqrt{3}}\,\xi(\gamma) = 0\,, \\ &\eta'(\gamma) + \zeta(\gamma) + \frac{b}{\sqrt{3}}\,\eta(\gamma) = \frac{c}{2}\,, \\ &\zeta'(\gamma) + \Big(\frac{b}{\sqrt{3}} - 1\Big)\xi(\gamma) = 0\,. \end{aligned} \tag{8.44}$$

Integration with the initial conditions $\xi(0) = \eta(0) = \zeta(0) = 0$ yields

$$\eta(\gamma) = \frac{\sqrt{3}}{2}\,\frac{c}{b}\left[1 - \mathrm{e}^{-\frac{b}{\sqrt{3}}\gamma}\right]. \tag{8.45}$$

This is a monotonic shear stress response:

$$\tau(\gamma) = \frac{k}{\sqrt{3}} + \frac{\sqrt{3}}{2}\,\frac{c}{b}\left[1 - \mathrm{e}^{-\frac{b}{\sqrt{3}}\gamma}\right]. \tag{8.46}$$

For the special case of $b = 0$ (linear kinematic hardening) the shear stress is a linear function of γ:

$$\tau(\gamma) = \frac{k}{\sqrt{3}} + \frac{c}{2}\,\gamma\,. \tag{8.47}$$

Similar to the *Maxwell* fluid, the evolution equation for kinematic hardening is also integrable, when the *Oldroyd* rate is applied for the stress derivative. Beginning with the identities

$$\dot{\mathbf{X}} - (\dot{\mathbf{F}}\mathbf{F}^{-1})\mathbf{X} - \mathbf{X}(\dot{\mathbf{F}}\mathbf{F}^{-1})^{\mathrm{T}} = \mathbf{F}(\mathbf{F}^{-1}\mathbf{X}\mathbf{F}^{\mathrm{T}-1})^{\cdot}\,\mathbf{F}^{\mathrm{T}}$$

and

$$\mathbf{F}^{-1}\mathbf{D}\mathbf{F}^{\mathrm{T}-1} = \mathbf{F}^{-1}\mathbf{F}^{\mathrm{T}-1}\mathbf{F}^{\mathrm{T}}\mathbf{D}\mathbf{F}\mathbf{F}^{-1}\mathbf{F}^{\mathrm{T}-1} = \frac{1}{2}\,\mathbf{C}^{-1}\dot{\mathbf{C}}\mathbf{C}^{-1} = -\,\frac{1}{2}(\mathbf{C}^{-1})^{\cdot}\,,$$

the differential equation (8.42), $\overset{\triangledown}{\mathbf{X}} + b\dot{s}\mathbf{X} = c\mathbf{D}$, merges into

$$(\mathbf{F}^{-1}\mathbf{X}\mathbf{F}^{\mathrm{T}-1})^{\cdot} + b\dot{s}(\mathbf{F}^{-1}\mathbf{X}\mathbf{F}^{\mathrm{T}-1}) = -\,\frac{c}{2}(\mathbf{C}^{-1})^{\cdot}\,. \tag{8.48}$$

Dividing by $\dot{s}$ and introducing the notation $(\,)' = \frac{\mathrm{d}}{\mathrm{d}s}(\,)$, this becomes

$$\mathrm{e}^{-bs}\Big(\mathrm{e}^{bs}\mathbf{F}^{-1}\mathbf{X}\mathbf{F}^{\mathrm{T}-1}\Big)' = -\tfrac{c}{2}\big(\mathbf{C}^{-1}\big)' . \tag{8.49}$$

Integration yields the functional representation

$$\mathbf{F}^{-1}\mathbf{X}\mathbf{F}^{\mathrm{T}-1} = -\int_0^s \tfrac{c}{2}\,\mathrm{e}^{-b(s-\sigma)}\big(\mathbf{C}^{-1}(\sigma)\big)'\,\mathrm{d}\sigma . \tag{8.50}$$

For the special case of $b = 0$ we have linear kinematic hardening, and the representation is reduced to

$$\mathbf{F}^{-1}\mathbf{X}\mathbf{F}^{\mathrm{T}-1} = -\tfrac{c}{2}\big[\mathbf{C}^{-1}(s) - \mathbf{1}\big] ,$$

or with $\mathbf{F}\mathbf{C}^{-1}\mathbf{F} = \mathbf{F}(\mathbf{F}^{\mathrm{T}}\mathbf{F})^{-1}\mathbf{F}^{\mathrm{T}} = \mathbf{1}$ and $\mathbf{F}\mathbf{F}^{\mathrm{T}} = \mathbf{B}$ to

$$\mathbf{X} = \tfrac{c}{2}\big(\mathbf{B} - \mathbf{1}\big) . \tag{8.51}$$

Both examples motivate the use of *Oldroyd* rates in tensor-valued evolution equations. General recommendations can be drawn from these results, as the following section shows.

8. 2 The Concept of Dual Variables

8. 2. 1 Motivation

A non-monotonic shear stress during monotonic shear processes cannot be excluded on principle: anisotropic material properties or material instabilities, which could lead to non-monotonic stress-strain curves, are definitely conceivable. However, the modelling ought to ensure that instabilities and non-monotonies are depicted in the way real experiments have suggested. In this connection a period cycle of 2π, that follows from the use of the *Jaumann* rate combined with the strain rate $\mathbf{D}$, appears to be physically unrealistic. Elementary constitutive models like the *Maxwell* body or rigid-plasticity should be expected to provide a monotonic response to monotonic loading processes. This demand can be viewed as a restrictive condition for the choice of variables and their time derivatives. The principle of objectivity fails to provide any further indications on this account, since it allows any number of realisations for tensor-valued evolution equations. The examples under examination suggest that the *Jaumann* rate is not the best choice for a tensor rate, although it represents the simplest formula for an objective derivative.

In this situation further criteria are required for selecting appropriate kinematic and dynamic variables including their time rates: the formulation of physically meaningful constitutive models calls for a concept that successfully correlates kinematic and dynamic variables and their time rates.

Appropriate guidelines can be inferred from the balance equations of mechanics. The balance of mechanical energy (2.100), an implication of the conservation laws for mass, linear momentum and rotational momentum, includes the power of the stresses at the rates of deformation. According to (2.106) the specific stress power can be written as a scalar product of the 2nd *Piola-Kirchhoff* tensor $\widetilde{\mathbf{T}}$ and the material time derivative of the *Green* strain tensor $\mathbf{E}$:

$$\ell_{\mathrm{i}} = \frac{1}{\rho_{\mathrm{R}}} \widetilde{\mathbf{T}} \cdot \dot{\mathbf{E}} \,. \tag{8.52}$$

This indicates that the two tensors $\widetilde{\mathbf{T}}$ and $\mathbf{E}$ are naturally associated with each other. This is also expressed by the specific virtual work of the stresses,

$$\delta \mathrm{w}_{\mathrm{i}} = \frac{1}{\rho_{\mathrm{R}}} \widetilde{\mathbf{T}} \cdot \delta \mathbf{E} \,, \tag{8.53}$$

contained in the principle of *d'Alembert* (2.118) and the principle of virtual displacements (2.123). The specific virtual stress work is the scalar product of stress tensor $\widetilde{\mathbf{T}}$ and the virtual strain (2.114), which is the differential of the strain tensor $\mathbf{E}$ in the direction of virtual displacement.

The incremental formulation (2.130) of *d'Alembert's* principle leads to the incremental virtual stress power $\dot{\widetilde{\mathbf{T}}} \cdot \delta \dot{\mathbf{E}}$ that motivates the scalar product $\dot{\widetilde{\mathbf{T}}} \cdot \dot{\mathbf{E}}$ of stress and strain rates. Finally, the complementary stress power $\dot{\widetilde{\mathbf{T}}} \cdot \mathbf{E}$ is also of significance, since it complements the stress power in the sense of the relation

$$\dot{\widetilde{\mathbf{T}}} \cdot \mathbf{E} + \widetilde{\mathbf{T}} \cdot \dot{\mathbf{E}} = (\widetilde{\mathbf{T}} \cdot \mathbf{E})^{\boldsymbol{\cdot}} \,. \tag{8.54}$$

The representation (8.52) of the stress power shows that $\widetilde{\mathbf{T}}$ and $\mathbf{E}$ are work conjugate variables. The natural bond between $\widetilde{\mathbf{T}}$ and $\mathbf{E}$ is, however, only partly described by the term *conjugate variables*; there are in addition further aspects regarding the complementary and incremental stress powers. For this reason the quantities $\widetilde{\mathbf{T}}$ and $\mathbf{E}$ will be referred to as *dual variables* in the following.

The concept of dual variables is based on the assumption that, when introducing stress and strain tensors that operate on configurations other than the reference configuration, the physically significant scalar products

$\widetilde{\mathbf{T}} \cdot \dot{\mathbf{E}}$, $\dot{\widetilde{\mathbf{T}}} \cdot \mathbf{E}$, $\widetilde{\mathbf{T}} \cdot \mathbf{E}$, and $\dot{\widetilde{\mathbf{T}}} \cdot \dot{\mathbf{E}}$

remain invariant.

8. 2. 2 Strain and Stress Tensors (Summary)

In Chap. 1 a total of 4 different strain tensors were introduced:

Green tensor $\mathbf{E} = \frac{1}{2}(\mathbf{C} - \mathbf{1})\,,$ (1.72)

Almansi tensor $\mathbf{A} = \frac{1}{2}(\mathbf{1} - \mathbf{B}^{-1})\,,$ (1.86)

Piola tensor $\mathbf{e} = \frac{1}{2}(\mathbf{C}^{-1} - \mathbf{1})\,,$ (1.78)

Finger tensor $\mathbf{a} = \frac{1}{2}(\mathbf{1} - \mathbf{B})\,.$ (1.88)

These definitions are not quite arbitrary: their component representations in terms of convective coordinates show that the strain components correspond to the differences in the metric coefficients of the current and the reference configurations. The components are either contravariant metric quantities ($\mathbf{E}$, $\mathbf{A}$) or covariant metric quantities ($\mathbf{e}$, $\mathbf{a}$). Moreover, 2 of these tensors operate on the reference configuration ($\mathbf{E}$, $\mathbf{e}$) and 2 on the current configuration ($\mathbf{A}$, $\mathbf{a}$). It is therefore not surprising to find precisely 4 strain tensors, if one takes the difference between the metric coefficients to be the strain measure:

$$
\begin{aligned}
\mathbf{E} &= \frac{1}{2}\left[g_{KL} - G_{KL}\right]\boldsymbol{G}^K \otimes \boldsymbol{G}^L\,,\\
\mathbf{A} &= \frac{1}{2}\left[g_{KL} - G_{KL}\right]\boldsymbol{g}^K \otimes \boldsymbol{g}^L\,,\\
\mathbf{e} &= \frac{1}{2}\left[g^{KL} - G^{KL}\right]\boldsymbol{G}_K \otimes \boldsymbol{G}_L\,,\\
\mathbf{a} &= \frac{1}{2}\left[g^{KL} - G^{KL}\right]\boldsymbol{g}_K \otimes \boldsymbol{g}_L\,.
\end{aligned} \qquad (1.103)
$$

According to their general definitions, the following transformation formulae are valid:

$$\mathbf{A} = \mathbf{F}^{\mathrm{T}\,-1}\mathbf{E}\mathbf{F}^{-1}\,, \tag{1.86}$$

$$\mathbf{a} = \mathbf{F}\mathbf{e}\mathbf{F}^{\mathrm{T}}\,. \tag{1.88}$$

These can be obtained using the relations (1.92), $g_K = \mathbf{F}G_K$ and (1.93), $g^K = \mathbf{F}^{\mathrm{T}\,-1}\,G^K$, which hold for convective coordinates.

In addition, as a result of (1.130) and (1.131), the spatial strain rate tensor **D** emerges from the material derivative of **E** or **e** by means of the same transformations as the *Oldroyd* rate of **A** or **a**:

$$\mathbf{D} = \overset{\triangle}{\mathbf{A}} = \mathbf{F}^{\mathrm{T}\,-1}\,\dot{\mathbf{E}}\,\mathbf{F}^{-1}\,, \tag{1.130}$$

$$-\,\mathbf{D} = \overset{\triangledown}{\mathbf{a}} = \mathbf{F}\dot{\mathbf{e}}\mathbf{F}^{\mathrm{T}}\,. \tag{1.131}$$

In connection with these strain tensors and rates, 3 stress tensors, introduced in Chap. 2, are of interest. These are

the weighted *Cauchy* tensor (*Kirchhoff* stress) $\mathbf{S} = (\det\mathbf{F})\mathbf{T}\,,$ (2.48)

the 2nd *Piola-Kirchhoff* tensor $\widetilde{\mathbf{T}} = \mathbf{F}^{-1}\,\mathbf{S}\,\mathbf{F}^{\mathrm{T}\,-1}$ (2.49)

and the *convective stress tensor* $\widetilde{\mathbf{t}} = \mathbf{F}^{\mathrm{T}}\,\mathbf{S}\,\mathbf{F}\,.$ (2.50)

One of these stress tensors operates on the current configuration (**S**) and two on the reference configuration ($\widetilde{\mathbf{T}}, \widetilde{\mathbf{t}}$). This also becomes plain from the component representations with respect to convective coordinates:

$$\mathbf{S} = S^{IK} g_I \otimes g_K\,, \tag{2.55}$$

$$\mathbf{S} = S_{IK} g^I \otimes g^K\,, \tag{2.57}$$

$$\widetilde{\mathbf{T}} = S^{IK} G_I \otimes G_K\,, \tag{2.56}$$

$$\widetilde{\mathbf{t}} = S_{IK} G^I \otimes G^K\,. \tag{2.58}$$

The transformation formulae (2.49) and (2.50) for $\widetilde{\mathbf{T}}$ and $\widetilde{\mathbf{t}}$ lead to the contra- or covariant representation of **S**:

$$\mathbf{F}\widetilde{\mathbf{T}}\mathbf{F}^{\mathrm{T}} = \mathbf{S} = S^{IK}\boldsymbol{g}_I \otimes \boldsymbol{g}_K\,, \tag{8.55}$$

$$\mathbf{F}^{\mathrm{T}\,-1}\widetilde{\mathbf{t}}\,\mathbf{F}^{-1} = \mathbf{S} = S_{IK}\boldsymbol{g}^I \otimes \boldsymbol{g}^K\,. \tag{8.56}$$

The diverse strain and stress tensors are coupled by means of the specific stress power per unit volume of the reference configuration:

$$\rho_{\mathrm{R}}\ell_{\mathrm{i}} = \widetilde{\mathbf{T}}\cdot\dot{\mathbf{E}} = \mathbf{S}\cdot\mathbf{D}\,,$$

$$\rho_{\mathrm{R}}\ell_{\mathrm{i}} = \widetilde{\mathbf{T}}\cdot\dot{\mathbf{E}} = \mathbf{S}\cdot\overset{\vartriangle}{\mathbf{A}}\,, \tag{8.57}$$

$$\rho_{\mathrm{R}}\ell_{\mathrm{i}} = (-\,\widetilde{\mathbf{t}}\,)\cdot\dot{\mathbf{e}} = (-\,\mathbf{S})\cdot\overset{\triangledown}{\mathbf{a}}\,. \tag{8.58}$$

A complete list of such invariance properties is contained in the following

Theorem 8. 2

1) For the pairs of variables ($\widetilde{\mathbf{T}}$, $\mathbf{E}$) and ($\mathbf{S}$, $\mathbf{A}$) the following relations are valid:

$$\widetilde{\mathbf{T}}\cdot\mathbf{E} = \mathbf{S}\cdot\mathbf{A}\,, \tag{8.59}$$

$$\widetilde{\mathbf{T}}\cdot\dot{\mathbf{E}} = \mathbf{S}\cdot\overset{\vartriangle}{\mathbf{A}}\,, \tag{8.60}$$

$$\dot{\widetilde{\mathbf{T}}}\cdot\mathbf{E} = \overset{\triangledown}{\mathbf{S}}\cdot\mathbf{A}\,, \tag{8.61}$$

$$(\widetilde{\mathbf{T}}\cdot\mathbf{E})^{\boldsymbol{\cdot}} = \overset{\triangledown}{\mathbf{S}}\cdot\mathbf{A} + \mathbf{S}\cdot\overset{\vartriangle}{\mathbf{A}}\,, \tag{8.62}$$

$$\dot{\widetilde{\mathbf{T}}}\cdot\dot{\mathbf{E}} = \overset{\triangledown}{\mathbf{S}}\cdot\overset{\vartriangle}{\mathbf{A}}\,. \tag{8.63}$$

2) For the pairs of variables ($\widetilde{\mathbf{t}}$, $\mathbf{e}$) and ($\mathbf{S}$, $\mathbf{a}$) we have the relations

$$\widetilde{\mathbf{t}}\cdot\mathbf{e} = \mathbf{S}\cdot\mathbf{a}\,, \tag{8.64}$$

$$\widetilde{\mathbf{t}}\cdot\dot{\mathbf{e}} = \mathbf{S}\cdot\overset{\triangledown}{\mathbf{a}}\,, \tag{8.65}$$

$$\dot{\widetilde{\mathbf{t}}}\cdot\mathbf{e} = \overset{\vartriangle}{\mathbf{S}}\cdot\mathbf{a}\,, \tag{8.66}$$

$$(\widetilde{\mathbf{t}}\cdot\mathbf{e})^{\boldsymbol{\cdot}} = \overset{\vartriangle}{\mathbf{S}}\cdot\mathbf{a} + \mathbf{S}\cdot\overset{\triangledown}{\mathbf{a}}\,, \tag{8.67}$$

$$\dot{\widetilde{\mathbf{t}}}\cdot\dot{\mathbf{e}} = \overset{\vartriangle}{\mathbf{S}}\cdot\overset{\triangledown}{\mathbf{a}}\,. \tag{8.68}$$

□

Proof

All these equations are the direct outcome of the transformation formulae:

$$\mathbf{A} = \mathbf{F}^{T\,-1}\,\mathbf{E}\,\mathbf{F}^{-1}\,, \quad \overset{\triangle}{\mathbf{A}} = \mathbf{F}^{T\,-1}\,\dot{\mathbf{E}}\,\mathbf{F}^{-1}\,, \tag{8.69}$$

$$\mathbf{S} = \mathbf{F}\tilde{\mathbf{T}}\mathbf{F}^{T}\,, \quad \overset{\triangledown}{\mathbf{S}} = \mathbf{F}\dot{\tilde{\mathbf{T}}}\mathbf{F}^{T}\,, \tag{8.70}$$

$$\mathbf{a} = \mathbf{F}\mathbf{e}\mathbf{F}^{T}\,, \quad \overset{\triangledown}{\mathbf{a}} = \mathbf{F}\dot{\mathbf{e}}\mathbf{F}^{T}\,, \tag{8.71}$$

$$\mathbf{S} = \mathbf{F}^{T\,-1}\tilde{\mathbf{t}}\,\mathbf{F}^{-1}\,, \quad \overset{\triangle}{\mathbf{S}} = \mathbf{F}^{T\,-1}\dot{\tilde{\mathbf{t}}}\,\mathbf{F}^{-1}\,. \tag{8.72}$$

The transformation formulae for the stress rates arise from the identities

$$\mathbf{F}\dot{\tilde{\mathbf{T}}}\mathbf{F}^{T} = \mathbf{F}(\mathbf{F}^{-1}\mathbf{S}\,\mathbf{F}^{T\,-1})^{\cdot}\,\mathbf{F}^{T} = \dot{\mathbf{S}} - \mathbf{L}\mathbf{S} - \mathbf{S}\mathbf{L}^{T} = \overset{\triangledown}{\mathbf{S}} \tag{8.73}$$

and

$$\mathbf{F}^{T\,-1}\dot{\tilde{\mathbf{t}}}\,\mathbf{F}^{-1} = \mathbf{F}^{T\,-1}(\mathbf{F}^{T}\mathbf{S}\mathbf{F})^{\cdot}\,\mathbf{F}^{-1} = \dot{\mathbf{S}} + \mathbf{L}^{T}\mathbf{S} + \mathbf{S}\mathbf{L} = \overset{\triangle}{\mathbf{S}}\,. \tag{8.74}$$

■

The theorem states that it is not only the pairs of variables

$$(\tilde{\mathbf{T}}, \mathbf{E}), (\mathbf{S}, \mathbf{A}), (\tilde{\mathbf{t}}, \mathbf{e}), (\mathbf{S}, \mathbf{a})$$

which are naturally coupled, but also their derivatives

$$(\dot{\tilde{\mathbf{T}}}, \dot{\mathbf{E}}), (\overset{\triangledown}{\mathbf{S}}, \overset{\triangle}{\mathbf{A}}), (\dot{\tilde{\mathbf{t}}}, \dot{\mathbf{e}}), (\overset{\triangle}{\mathbf{S}}, \overset{\triangledown}{\mathbf{a}})\,.$$

8. 2. 3 Dual Variables and Derivatives

In the theory of materials it is often useful to formulate constitutive models or evolution equations for tensor-valued quantities that refer neither to the reference configuration nor the current configuration, but to other configurations that may accompany the motion of a material body. The results of the previous section can be generalised to this end. An appropriate tool for implementing a suitable generalisation is the notion of an intermediate configuration, which was proposed in Sect. 1. 10. An *intermediate configuration* according to Definition 1. 22 is a tensor field

$$(\boldsymbol{X}, t) \longmapsto \boldsymbol{\Psi}(\boldsymbol{X}, t)\,, \ \det\boldsymbol{\Psi} \neq 0 \tag{8.75}$$

(cf. (1.227)); this tensor field amounts to a multiplicative decomposition of the deformation gradient according to (1.232),

$$\mathbf{F}(X,t) = (\mathbf{F}\,\boldsymbol{\Psi}^{-1})\boldsymbol{\Psi} = \boldsymbol{\Phi}\boldsymbol{\Psi}\,, \tag{8.76}$$

simultaneously inducing a field of metric quantities which is generally incompatible, i.e. non-Euclidean. The field $\boldsymbol{\Psi}(X,t)$ usually depends on the process history, and it can be assumed that a set of constitutive equations is available for calculating $\boldsymbol{\Psi}(X,t)$. The tensor field $\boldsymbol{\Psi}(X,t)$ can more or less be selected at will, in accordance with the material properties to be described. For physical reasons it is necessary that the condition

$$(\dot{\boldsymbol{\Psi}}\boldsymbol{\Psi}^{-1}) + (\dot{\boldsymbol{\Psi}}\boldsymbol{\Psi}^{-1})^{\mathrm{T}} = 0 \tag{8.77}$$

is fulfilled in the case of a *rigid body motion*. The point of this restriction is that the metric induced by $\boldsymbol{\Psi}$ in a rigid body motion should remain constant in time.

An intermediate configuration $\boldsymbol{\Psi}$ defines a system of local base vectors,

$$\hat{\boldsymbol{g}}_K = \boldsymbol{\Psi}\boldsymbol{G}_K\,, \tag{8.78}$$

$$\hat{\boldsymbol{g}}^K = \boldsymbol{\Psi}^{\mathrm{T}-1}\boldsymbol{G}^K, \tag{8.79}$$

with the property $\hat{\boldsymbol{g}}_K \cdot \hat{\boldsymbol{g}}^L = \delta^L_K$.

It therefore holds that

$$\boldsymbol{\Psi} = \hat{\boldsymbol{g}}_K \otimes \boldsymbol{G}^K \Longleftrightarrow \boldsymbol{\Psi}^{-1} = \boldsymbol{G}_K \otimes \hat{\boldsymbol{g}}^K\,. \tag{8.80}$$

Integral curves to the 3 vector fields $\hat{\boldsymbol{g}}_K(X,t)$ exist, but these 3 families of vector lines do not form a system of coordinate lines. This is obvious, since the metric quantities

$$\hat{g}_{KL} = \hat{\boldsymbol{g}}_K \cdot \hat{\boldsymbol{g}}_L = \boldsymbol{G}_K \cdot \boldsymbol{\Psi}^{\mathrm{T}}\boldsymbol{\Psi}\boldsymbol{G}_L \tag{8.81}$$

do not generally represent a Euclidean metric. It is equally obvious that for the specific example

$$\boldsymbol{\Psi}(X, t) = \mathbf{F}(X, t) \tag{8.82}$$

(intermediate configuration = current configuration) the vector fields $\hat{g}_K = g_K$ are integrable as coordinate lines; in fact, the special case of $\boldsymbol{\Psi} = \mathbf{F}$ corresponds to the choice of convective coordinates. Of all the imaginable strain and stress tensors, two families can be emphasised using the following definition:

Definition 8. 2

Let the tensor field $(X, t) \longmapsto \boldsymbol{\Psi}(X, t)$ be any given intermediate configuration and

$$\boldsymbol{\Lambda} = \dot{\boldsymbol{\Psi}}\boldsymbol{\Psi}^{-1} \tag{8.83}$$

its relative rate of change. The two families of strain and stress tensors, $(\boldsymbol{\Pi}, \boldsymbol{\Sigma})$ and $(\boldsymbol{\pi}, \boldsymbol{\sigma})$, defined in the following, are called *dual strain and stress tensors*; the strain and stress rates $(\overset{\vartriangle}{\boldsymbol{\Pi}}, \overset{\triangledown}{\boldsymbol{\Sigma}})$ and $(\overset{\triangledown}{\boldsymbol{\pi}}, \overset{\vartriangle}{\boldsymbol{\sigma}})$ are *dual derivatives* or *dual strain* and *stress rates.*

Family 1

$$\boldsymbol{\Pi} = \boldsymbol{\Psi}^{\mathrm{T}\,-1}\,\mathbf{E}\,\boldsymbol{\Psi}^{-1} \tag{8.84}$$

$$\overset{\vartriangle}{\boldsymbol{\Pi}} = \boldsymbol{\Psi}^{\mathrm{T}\,-1}\,\dot{\mathbf{E}}\,\boldsymbol{\Psi}^{-1} = \dot{\boldsymbol{\Pi}} + \boldsymbol{\Lambda}^{\mathrm{T}}\boldsymbol{\Pi} + \boldsymbol{\Pi}\boldsymbol{\Lambda} \tag{8.85}$$

$$\boldsymbol{\Sigma} = \boldsymbol{\Psi}\,\tilde{\mathbf{T}}\,\boldsymbol{\Psi}^{\mathrm{T}} \tag{8.86}$$

$$\overset{\triangledown}{\boldsymbol{\Sigma}} = \boldsymbol{\Psi}\,\dot{\tilde{\mathbf{T}}}\,\boldsymbol{\Psi}^{\mathrm{T}} = \dot{\boldsymbol{\Sigma}} - \boldsymbol{\Lambda}\boldsymbol{\Sigma} - \boldsymbol{\Sigma}\boldsymbol{\Lambda}^{\mathrm{T}} \tag{8.87}$$

Family 2

$$\boldsymbol{\pi} = \boldsymbol{\Psi}\,\mathbf{e}\,\boldsymbol{\Psi}^{\mathrm{T}} \tag{8.88}$$

$$\overset{\triangledown}{\boldsymbol{\pi}} = \boldsymbol{\Psi}\,\dot{\mathbf{e}}\,\boldsymbol{\Psi}^{\mathrm{T}} = \dot{\boldsymbol{\pi}} - \boldsymbol{\Lambda}\boldsymbol{\pi} - \boldsymbol{\pi}\boldsymbol{\Lambda}^{\mathrm{T}} \tag{8.89}$$

$$\boldsymbol{\sigma} = \boldsymbol{\Psi}^{\mathrm{T}\,-1}\,\tilde{\mathbf{t}}\,\boldsymbol{\Psi}^{-1} \tag{8.90}$$

$$\overset{\vartriangle}{\boldsymbol{\sigma}} = \boldsymbol{\Psi}^{\mathrm{T}\,-1}\,\dot{\tilde{\mathbf{t}}}\,\boldsymbol{\Psi}^{-1} = \dot{\boldsymbol{\sigma}} + \boldsymbol{\Lambda}^{\mathrm{T}}\boldsymbol{\sigma} + \boldsymbol{\sigma}\boldsymbol{\Lambda} \tag{8.91}$$

◊

The transformed strain tensor $\boldsymbol{\Pi}$ is a *Green* tensor operating on the intermediate configuration: (1.73), (8.84) and $\mathrm{d}\hat{\boldsymbol{x}} = \boldsymbol{\Psi}\mathrm{d}\boldsymbol{X}$ lead to

$$\frac{1}{2}\left(\mathrm{d}\boldsymbol{x}\cdot\mathrm{d}\boldsymbol{x} - \mathrm{d}\boldsymbol{X}\cdot\mathrm{d}\boldsymbol{X}\right) = \mathrm{d}\boldsymbol{X}\cdot\left(\mathbf{E}\,\mathrm{d}\boldsymbol{X}\right) = \mathrm{d}\hat{\boldsymbol{x}}\cdot\boldsymbol{\Pi}\mathrm{d}\hat{\boldsymbol{x}}\ . \tag{8.92}$$

The transformed *Piola* strain tensor $\boldsymbol{\pi}$ possesses the corresponding property: (1.85), combined with the definition (8.88) and the normal vector $\hat{\boldsymbol{n}} = \boldsymbol{\Psi}^{\mathrm{T}\,-\,1}\boldsymbol{N}$, produces

$$\frac{1}{2}\left(\boldsymbol{n}\cdot\boldsymbol{n} - \boldsymbol{N}\cdot\boldsymbol{N}\right) = \boldsymbol{N}\cdot\mathbf{e}\boldsymbol{N} = \hat{\boldsymbol{n}}\cdot\boldsymbol{\pi}\hat{\boldsymbol{n}}\ . \tag{8.93}$$

For the *Green* strain tensor operating on the intermediate configuration we obtain additive decompositions: (8.84) in combination with the multiplicative decomposition (8.76) implies the decomposition

$$\boldsymbol{\Pi} = \frac{1}{2}\left(\boldsymbol{\Phi}^{\mathrm{T}}\boldsymbol{\Phi} - \boldsymbol{\Psi}^{\mathrm{T}\,-\,1}\boldsymbol{\Psi}^{-\,1}\right) = \boldsymbol{\Pi}_{\mathrm{a}} + \boldsymbol{\Pi}_{\mathrm{b}} \tag{8.94}$$

into the two parts

$$\boldsymbol{\Pi}_{\mathrm{a}} = \frac{1}{2}\left(\boldsymbol{\Phi}^{\mathrm{T}}\boldsymbol{\Phi} - \mathbf{1}\right) \tag{8.95}$$

and

$$\boldsymbol{\Pi}_{\mathrm{b}} = \frac{1}{2}\left(\mathbf{1} - \boldsymbol{\Psi}^{\mathrm{T}\,-\,1}\boldsymbol{\Psi}^{-\,1}\right). \tag{8.96}$$

For the *Piola* tensor we apply (8.88) and (8.76) and obtain

$$\boldsymbol{\pi} = \frac{1}{2}\left(\boldsymbol{\Phi}^{-\,1}\boldsymbol{\Phi}^{\mathrm{T}\,-\,1} - \boldsymbol{\Psi}\boldsymbol{\Psi}^{\mathrm{T}}\right) = \boldsymbol{\pi}_{\mathrm{a}} + \boldsymbol{\pi}_{\mathrm{b}}\,. \tag{8.97}$$

Thus $\boldsymbol{\pi}$ is the sum of the parts

$$\boldsymbol{\pi}_{\mathrm{a}} = \frac{1}{2}\left(\boldsymbol{\Phi}^{-\,1}\boldsymbol{\Phi}^{\mathrm{T}\,-\,1} - \mathbf{1}\right) \tag{8.98}$$

and

$$\boldsymbol{\pi}_{\mathrm{b}} = \frac{1}{2}\left(\mathbf{1} - \boldsymbol{\Psi}\boldsymbol{\Psi}^{\mathrm{T}}\right). \tag{8.99}$$

The dual strain and stress rates (8.85), (8.87), (8.89) and (8.91) are defined by the same transformations applied to the tensors themselves. When calculating the dual derivatives, *Oldroyd* rates occur which are formed using tensor $\boldsymbol{\Lambda} = \dot{\boldsymbol{\Psi}}\boldsymbol{\Psi}^{-1}$ instead of the velocity gradient $\mathbf{L} = \dot{\mathbf{F}}\mathbf{F}^{-\,1}$. This becomes apparent in the context of the identities

$$\overset{\triangle}{\Pi} = \Psi^{T-1}\dot{E}\,\Psi^{-1} = \Psi^{T-1}(\Psi^{T}\Pi\Psi)^{\cdot}\,\Psi^{-1}\,, \tag{8.100}$$

$$\overset{\triangle}{\Pi} = \dot{\Pi} + \Lambda^{T}\Pi + \Pi\Lambda \tag{8.101}$$

and

$$\overset{\triangledown}{\Sigma} = \Psi\,\dot{\tilde{T}}\,\Psi^{T} = \Psi(\Psi^{-1}\Pi\Psi^{T-1})^{\cdot}\,\Psi^{T}\,, \tag{8.102}$$

$$\overset{\triangledown}{\Sigma} = \dot{\Sigma} - \Lambda\Sigma - \Sigma\Lambda^{T}\,, \tag{8.103}$$

as well as

$$\overset{\triangledown}{\pi} = \Psi\,\dot{e}\,\Psi^{T} = \Psi\,(\Psi^{-1}\pi\Psi^{T-1})^{\cdot}\,\Psi^{T}\,, \tag{8.104}$$

$$\overset{\triangledown}{\pi} = \Psi\,\dot{e}\,\Psi^{T} = \dot{\pi} - \Lambda\pi - \pi\Lambda^{T} \tag{8.105}$$

and

$$\overset{\triangle}{\sigma} = \Psi^{T-1}\dot{\tilde{t}}\,\Psi^{-1} = \Psi^{T-1}(\Psi^{T}\sigma\Psi)^{\cdot}\,\Psi^{-1}\,, \tag{8.106}$$

$$\overset{\triangle}{\sigma} = \dot{\sigma} + \Lambda^{T}\sigma + \sigma\Lambda\,. \tag{8.107}$$

The relations between the strain and stress tensors and their time derivatives, which one obtains in connection with the general definition (8.76) of an intermediate configuration, $\mathbf{F} = \mathbf{\Phi\Psi}$, are summarised in the form of a tabulation in Figs. 8. 1 to 8. 4 (cf. Fig. 1.11).

By virtue of their definition, the dual variables and derivatives are characterised in the light of their invariance properties:

Theorem 8. 3

1) For the dual variables (Π, Σ) and their dual derivatives the following relations are valid:

$$\tilde{T}\cdot E = \Sigma\cdot\Pi\,, \tag{8.108}$$

$$\tilde{T}\cdot\dot{E} = \Sigma\cdot\overset{\triangle}{\Pi}\,, \tag{8.109}$$

$$\dot{\tilde{T}}\cdot E = \overset{\triangledown}{\Sigma}\cdot\Pi\,, \tag{8.110}$$

$$(\tilde{T}\cdot E)^{\cdot} = \overset{\triangledown}{\Sigma}\cdot\Pi + \Sigma\cdot\overset{\triangle}{\Pi}\,, \tag{8.111}$$

$$\dot{\tilde{T}}\cdot\dot{E} = \overset{\triangledown}{\Sigma}\cdot\overset{\triangle}{\Pi}\,. \tag{8.112}$$

$$\mathbf{E} = \tfrac{1}{2}\left(\mathbf{F}^{\mathrm{T}}\mathbf{F} - \mathbf{1}\right)$$
$$\mathbf{E} = \mathbf{E}_{\mathrm{a}} + \mathbf{E}_{\mathrm{b}}$$
$$\mathbf{E}_{\mathrm{a}} = \tfrac{1}{2}\left(\mathbf{F}^{\mathrm{T}}\mathbf{F} - \boldsymbol{\Psi}^{\mathrm{T}}\boldsymbol{\Psi}\right)$$
$$\mathbf{E}_{\mathrm{b}} = \tfrac{1}{2}\left(\boldsymbol{\Psi}^{\mathrm{T}}\boldsymbol{\Psi} - \mathbf{1}\right)$$
$$\dot{\mathbf{E}} = \frac{\mathrm{d}}{dt}\left(\tfrac{1}{2}\mathbf{F}^{\mathrm{T}}\mathbf{F}\right)$$
$$\dot{\mathbf{E}} = \dot{\mathbf{E}}_{\mathrm{a}} + \dot{\mathbf{E}}_{\mathrm{b}}$$
$$\dot{\mathbf{E}}_{\mathrm{a}} = \tfrac{1}{2}\left((\mathbf{F}^{\mathrm{T}}\mathbf{F})^{\cdot} - (\boldsymbol{\Psi}^{\mathrm{T}}\boldsymbol{\Psi})^{\cdot}\right)$$
$$\dot{\mathbf{E}}_{\mathrm{b}} = \tfrac{1}{2}(\boldsymbol{\Psi}^{\mathrm{T}}\boldsymbol{\Psi})^{\cdot}$$

$\mathbf{F}^{\mathrm{T}-1}(\cdot)\mathbf{F}^{-1}$ →

$$\mathbf{A} = \tfrac{1}{2}\left(\mathbf{1} - \mathbf{F}^{\mathrm{T}-1}\mathbf{F}^{-1}\right)$$
$$\mathbf{A} = \mathbf{A}_{\mathrm{a}} + \mathbf{A}_{\mathrm{b}}$$
$$\mathbf{A}_{\mathrm{a}} = \tfrac{1}{2}\left(\mathbf{1} - \boldsymbol{\Phi}^{\mathrm{T}-1}\boldsymbol{\Phi}^{-1}\right)$$
$$\mathbf{A}_{\mathrm{b}} = \tfrac{1}{2}\left(\boldsymbol{\Phi}^{\mathrm{T}-1}\boldsymbol{\Phi}^{-1} - \mathbf{F}^{\mathrm{T}-1}\mathbf{F}^{-1}\right)$$
$$\overset{\triangle}{\mathbf{A}} = \mathbf{F}^{\mathrm{T}-1}\dot{\mathbf{E}}\,\mathbf{F}^{-1} = \boldsymbol{\Phi}^{\mathrm{T}-1}\overset{\triangle}{\boldsymbol{\Pi}}\boldsymbol{\Phi}^{-1}$$
$$\overset{\triangle}{\mathbf{A}} = \dot{\mathbf{A}} + \mathbf{L}^{\mathrm{T}}\mathbf{A} + \mathbf{A}\mathbf{L} = \tfrac{1}{2}\left(\mathbf{L} + \mathbf{L}^{\mathrm{T}}\right)$$
$$\overset{\triangle}{\mathbf{A}} = \overset{\triangle}{\mathbf{A}}_{\mathrm{a}} + \overset{\triangle}{\mathbf{A}}_{\mathrm{b}}$$
$$\overset{\triangle}{\mathbf{A}}_{\mathrm{a}} = \dot{\mathbf{A}}_{\mathrm{a}} + \mathbf{L}^{\mathrm{T}}\mathbf{A}_{\mathrm{a}} + \mathbf{A}_{\mathrm{a}}\mathbf{L}$$
$$\overset{\triangle}{\mathbf{A}}_{\mathrm{b}} = \dot{\mathbf{A}}_{\mathrm{b}} + \mathbf{L}^{\mathrm{T}}\mathbf{A}_{\mathrm{b}} + \mathbf{A}_{\mathrm{b}}\mathbf{L}$$

↓ $\boldsymbol{\Psi}^{\mathrm{T}-1}(\cdot)\boldsymbol{\Psi}^{-1}$

↑ $\boldsymbol{\Phi}^{\mathrm{T}-1}(\cdot)\boldsymbol{\Phi}^{-1}$

$$\boldsymbol{\Pi} = \tfrac{1}{2}\left(\boldsymbol{\Phi}^{\mathrm{T}}\boldsymbol{\Phi} - \boldsymbol{\Psi}^{\mathrm{T}-1}\boldsymbol{\Psi}^{-1}\right)$$
$$\boldsymbol{\Pi} = \boldsymbol{\Pi}_{\mathrm{a}} + \boldsymbol{\Pi}_{\mathrm{b}}$$
$$\boldsymbol{\Pi}_{\mathrm{a}} = \tfrac{1}{2}\left(\boldsymbol{\Phi}^{\mathrm{T}}\boldsymbol{\Phi} - \mathbf{1}\right)$$
$$\boldsymbol{\Pi}_{\mathrm{b}} = \tfrac{1}{2}\left(\mathbf{1} - \boldsymbol{\Psi}^{\mathrm{T}-1}\boldsymbol{\Psi}^{-1}\right)$$
$$\overset{\triangle}{\boldsymbol{\Pi}} = \boldsymbol{\Psi}^{\mathrm{T}-1}\dot{\mathbf{E}}\,\boldsymbol{\Psi}^{-1} = \boldsymbol{\Phi}^{\mathrm{T}}\overset{\triangle}{\mathbf{A}}\boldsymbol{\Phi}$$
$$\overset{\triangle}{\boldsymbol{\Pi}} = \dot{\boldsymbol{\Pi}} + \boldsymbol{\Lambda}^{\mathrm{T}}\boldsymbol{\Pi} + \boldsymbol{\Pi}\boldsymbol{\Lambda}$$
$$\overset{\triangle}{\boldsymbol{\Pi}} = \overset{\triangle}{\boldsymbol{\Pi}}_{\mathrm{a}} + \overset{\triangle}{\boldsymbol{\Pi}}_{\mathrm{b}}$$
$$\overset{\triangle}{\boldsymbol{\Pi}}_{\mathrm{a}} = \dot{\boldsymbol{\Pi}}_{\mathrm{a}} + \boldsymbol{\Lambda}^{\mathrm{T}}\boldsymbol{\Pi}_{\mathrm{a}} + \boldsymbol{\Pi}_{\mathrm{a}}\boldsymbol{\Lambda}$$
$$\overset{\triangle}{\boldsymbol{\Pi}}_{\mathrm{b}} = \dot{\boldsymbol{\Pi}}_{\mathrm{b}} + \boldsymbol{\Lambda}^{\mathrm{T}}\boldsymbol{\Pi}_{\mathrm{b}} + \boldsymbol{\Pi}\boldsymbol{\Lambda}_{\mathrm{b}} = \tfrac{1}{2}\left(\boldsymbol{\Lambda} + \boldsymbol{\Lambda}^{\mathrm{T}}\right)$$

Figure 8. 1: Family 1: Strain tensors and rates

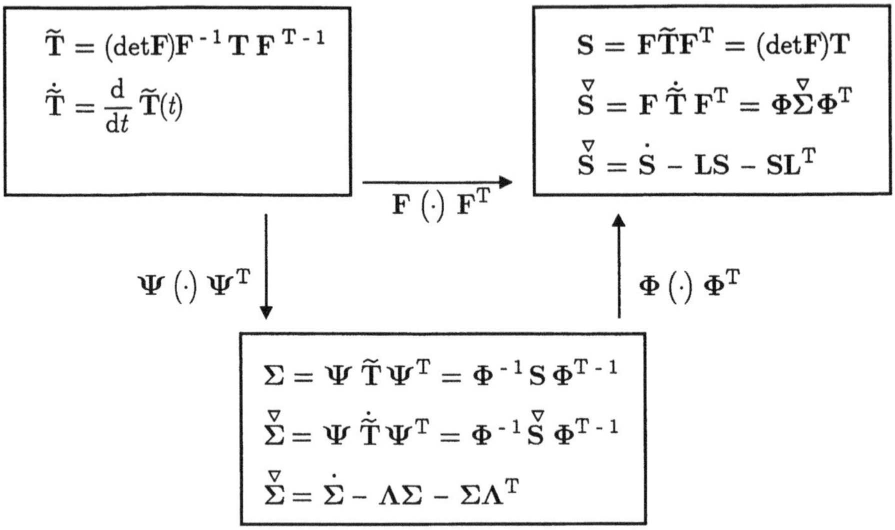

Figure 8. 2: Family 1: Stress tensors and rates

2) For the dual variables $(\boldsymbol{\pi}, \boldsymbol{\sigma})$ we have the relations

$$\tilde{\mathbf{t}} \cdot \mathbf{e} = \boldsymbol{\sigma} \cdot \boldsymbol{\pi}\,, \tag{8.113}$$

$$\tilde{\mathbf{t}} \cdot \dot{\mathbf{e}} = \boldsymbol{\sigma} \cdot \overset{\triangledown}{\boldsymbol{\pi}}\,, \tag{8.114}$$

$$\dot{\tilde{\mathbf{t}}} \cdot \mathbf{e} = \overset{\vartriangle}{\boldsymbol{\sigma}} \cdot \boldsymbol{\pi}\,, \tag{8.115}$$

$$(\tilde{\mathbf{t}} \cdot \mathbf{e})^{\cdot} = \overset{\vartriangle}{\boldsymbol{\sigma}} \cdot \boldsymbol{\pi} + \boldsymbol{\sigma} \cdot \overset{\triangledown}{\boldsymbol{\pi}}\,, \tag{8.116}$$

$$\dot{\tilde{\mathbf{t}}} \cdot \dot{\mathbf{e}} = \overset{\vartriangle}{\boldsymbol{\sigma}} \cdot \overset{\triangledown}{\boldsymbol{\pi}}\,. \tag{8.117}$$

□

Proof

Statement 1) results from the definitions (8.86), $\mathbf{\Sigma} = \mathbf{\Psi}\ \tilde{\mathbf{T}}\ \mathbf{\Psi}^{\mathrm{T}}$, and (8.84),

$$\mathbf{\Pi} = \mathbf{\Psi}^{\mathrm{T}-1}\mathbf{E}\,\mathbf{\Psi}^{-1}\,,$$

in conjunction with (8.85) and (8.87),

$$\overset{\vartriangle}{\mathbf{\Pi}} = \mathbf{\Psi}^{\mathrm{T}-1}\dot{\mathbf{E}}\,\mathbf{\Psi}^{-1}\,, \quad \overset{\triangledown}{\mathbf{\Sigma}} = \mathbf{\Psi}\,\dot{\tilde{\mathbf{T}}}\mathbf{\Psi}^{\mathrm{T}}\,.$$

$$\mathbf{e} = \frac{1}{2}\left(\mathbf{F}^{-1}\mathbf{F}^{T-1} - \mathbf{1}\right)$$
$$\mathbf{e} = \mathbf{e}_a + \mathbf{e}_b$$
$$\mathbf{e}_a = \frac{1}{2}\left(\mathbf{F}^{-1}\mathbf{F}^{T-1} - \boldsymbol{\Psi}^{-1}\boldsymbol{\Psi}^{T-1}\right)$$
$$\mathbf{e}_b = \frac{1}{2}\left(\boldsymbol{\Psi}^{-1}\boldsymbol{\Psi}^{T-1} - \mathbf{1}\right)$$
$$\dot{\mathbf{e}} = \frac{d}{dt}\left(\frac{1}{2}\mathbf{F}^{-1}\mathbf{F}^{T-1}\right)$$
$$\dot{\mathbf{e}} = \dot{\mathbf{e}}_a + \dot{\mathbf{e}}_b$$
$$\dot{\mathbf{e}}_a = \frac{1}{2}\left(\mathbf{F}^{-1}\mathbf{F}^{T-1} - (\boldsymbol{\Psi}^{-1}\boldsymbol{\Psi}^{T-1}\right)^{\cdot}$$
$$\dot{\mathbf{e}}_b = \frac{1}{2}\left(\boldsymbol{\Psi}^{-1}\boldsymbol{\Psi}^{T-1}\right)^{\cdot}$$

$\xrightarrow{\ \mathbf{F}\,(\cdot)\,\mathbf{F}^T\ }$

$$\mathbf{a} = \frac{1}{2}\left(\mathbf{1} - \mathbf{F}\mathbf{F}^T\right)$$
$$\mathbf{a} = \mathbf{a}_a + \mathbf{a}_b$$
$$\mathbf{a}_a = \frac{1}{2}\left(\mathbf{1} - \boldsymbol{\Phi}\boldsymbol{\Phi}^T\right)$$
$$\mathbf{a}_b = \frac{1}{2}\left(\boldsymbol{\Phi}\boldsymbol{\Phi}^T - \mathbf{F}\mathbf{F}^T\right)$$
$$\overset{\triangledown}{\mathbf{a}} = \mathbf{F}\dot{\mathbf{e}}\mathbf{F}^T = \boldsymbol{\Phi}\,\overset{\triangledown}{\boldsymbol{\pi}}\,\boldsymbol{\Phi}^T$$
$$\overset{\triangledown}{\mathbf{a}} = \dot{\mathbf{a}} - \mathbf{L}\mathbf{a} - \mathbf{a}\mathbf{L}^T = -\frac{1}{2}\left(\mathbf{L} + \mathbf{L}^T\right)$$
$$\overset{\triangledown}{\mathbf{a}} = \overset{\triangledown}{\mathbf{a}}_a + \overset{\triangledown}{\mathbf{a}}_b$$
$$\overset{\triangledown}{\mathbf{a}}_a = \dot{\mathbf{a}}_a - \mathbf{L}\mathbf{a}_a - \mathbf{a}_a\mathbf{L}^T$$
$$\overset{\triangledown}{\mathbf{a}}_b = \dot{\mathbf{a}}_b - \mathbf{L}\mathbf{a}_b - \mathbf{a}_b\mathbf{L}^T$$

$\boldsymbol{\Psi}\,(\cdot)\,\boldsymbol{\Psi}^T \downarrow$ $\qquad$ $\uparrow \boldsymbol{\Phi}\,(\cdot)\,\boldsymbol{\Phi}^T$

$$\boldsymbol{\pi} = \frac{1}{2}\left(\boldsymbol{\Phi}^{-1}\boldsymbol{\Phi}^{T-1} - \boldsymbol{\Psi}\boldsymbol{\Psi}^T\right)$$
$$\boldsymbol{\pi} = \boldsymbol{\pi}_a + \boldsymbol{\pi}_b$$
$$\boldsymbol{\pi}_a = \frac{1}{2}\left(\boldsymbol{\Phi}^{-1}\boldsymbol{\Phi}^{T-1} - \mathbf{1}\right)$$
$$\boldsymbol{\pi}_b = \frac{1}{2}\left(\mathbf{1} - \boldsymbol{\Psi}\boldsymbol{\Psi}^T\right)$$
$$\overset{\triangledown}{\boldsymbol{\pi}} = \boldsymbol{\Psi}\,\dot{\mathbf{e}}\,\boldsymbol{\Psi}^T = \boldsymbol{\Phi}^{-1}\overset{\triangledown}{\mathbf{a}}\,\boldsymbol{\Phi}^{T-1}$$
$$\overset{\triangledown}{\boldsymbol{\pi}} = \dot{\boldsymbol{\pi}} - \boldsymbol{\Lambda}\boldsymbol{\pi} - \boldsymbol{\pi}\boldsymbol{\Lambda}^T$$
$$\overset{\triangledown}{\boldsymbol{\pi}} = \overset{\triangledown}{\boldsymbol{\pi}}_a + \overset{\triangledown}{\boldsymbol{\pi}}_b$$
$$\overset{\triangledown}{\boldsymbol{\pi}}_a = \dot{\boldsymbol{\pi}}_a - \boldsymbol{\Lambda}\boldsymbol{\pi}_a - \boldsymbol{\pi}_a\boldsymbol{\Lambda}^T$$
$$\overset{\triangledown}{\boldsymbol{\pi}}_b = \dot{\boldsymbol{\pi}}_b - \boldsymbol{\Lambda}\boldsymbol{\pi}_b - \boldsymbol{\pi}_b\boldsymbol{\Lambda}^T = -\frac{1}{2}\left(\boldsymbol{\Lambda} + \boldsymbol{\Lambda}^T\right)$$

Figure 8. 3: Family 2: Strain tensors and rates

$$\tilde{\mathbf{t}} = (\det\mathbf{F})\mathbf{F}^{\mathrm{T}}\mathbf{T}\mathbf{F}$$

$$\dot{\tilde{\mathbf{t}}} = \frac{\mathrm{d}}{\mathrm{d}t}\tilde{\mathbf{t}}(t)$$

$$\xrightarrow{\ \mathbf{F}^{\mathrm{T}-1}(\cdot)\mathbf{F}^{-1}\ }$$

$$\mathbf{S} = \mathbf{F}^{\mathrm{T}-1}\tilde{\mathbf{t}}\,\mathbf{F}^{-1} = (\det\mathbf{F})\mathbf{T}$$

$$\overset{\triangle}{\mathbf{S}} = \mathbf{F}^{\mathrm{T}-1}\dot{\tilde{\mathbf{t}}}\,\mathbf{F}^{-1} = \boldsymbol{\Phi}^{\mathrm{T}-1}\overset{\triangle}{\sigma}\,\boldsymbol{\Phi}^{-1}$$

$$\overset{\triangle}{\mathbf{S}} = \dot{\mathbf{S}} + \mathbf{L}^{\mathrm{T}}\mathbf{S} + \mathbf{S}\mathbf{L}$$

$$\boldsymbol{\Psi}^{\mathrm{T}-1}(\cdot)\boldsymbol{\Psi}^{-1}\ \downarrow \qquad \uparrow\ \boldsymbol{\Phi}^{\mathrm{T}-1}(\cdot)\boldsymbol{\Phi}^{-1}$$

$$\sigma = \boldsymbol{\Psi}^{\mathrm{T}-1}\tilde{\mathbf{t}}\,\boldsymbol{\Psi}^{-1} = \boldsymbol{\Phi}^{\mathrm{T}}\mathbf{S}\,\boldsymbol{\Phi}$$

$$\overset{\triangle}{\sigma} = \boldsymbol{\Psi}^{\mathrm{T}-1}\dot{\tilde{\mathbf{t}}}\,\boldsymbol{\Psi}^{-1} = \boldsymbol{\Phi}^{\mathrm{T}}\overset{\triangle}{\mathbf{S}}\,\boldsymbol{\Phi}$$

$$\overset{\triangle}{\sigma} = \dot{\sigma} + \boldsymbol{\Lambda}^{\mathrm{T}}\sigma + \sigma\boldsymbol{\Lambda}$$

Figure 8. 4: Family 2: Stress tensors and rates

Statement 2) follows accordingly on the basis of (8.90),

$\sigma = \boldsymbol{\Psi}^{\mathrm{T}-1}\tilde{\mathbf{t}}\,\boldsymbol{\Psi}^{-1}$, and (8.88), $\pi = \boldsymbol{\Psi}\,\mathbf{e}\,\boldsymbol{\Psi}^{\mathrm{T}}$, in conjunction with

(8.91) and (8.89), $\overset{\triangle}{\sigma} = \boldsymbol{\Psi}^{\mathrm{T}-1}\dot{\tilde{\mathbf{t}}}\,\boldsymbol{\Psi}^{-1}$ and $\overset{\nabla}{\pi} = \boldsymbol{\Psi}\,\dot{\mathbf{e}}\,\boldsymbol{\Psi}^{\mathrm{T}}$. ■

The theorem establishes one physical meaning of the dual variables: those scalars that are significant in connection with work or energy balances ($\dot{\tilde{\mathbf{T}}}\cdot\mathbf{E}$, $\tilde{\mathbf{T}}\cdot\dot{\mathbf{E}}$, $\dot{\tilde{\mathbf{T}}}\cdot\mathbf{E}$, $\tilde{\mathbf{T}}\cdot\mathbf{E}$), stay invariant under the transformation $\boldsymbol{\Psi}$, i.e. they are the same in every intermediate configuration.
As regards a second physical meaning, i.e. the geometrical interpretation of the dual variables and their dual rates, we refer to the component representations with respect to convective coordinates:

$$\mathbf{E} = \frac{1}{2}[g_{KL} - G_{KL}]\boldsymbol{G}^K \otimes \boldsymbol{G}^L ,$$

$$\tilde{\mathbf{T}} = S^{IK}\boldsymbol{G}_I \otimes \boldsymbol{G}_K .$$

With $\boldsymbol{\Psi}\boldsymbol{G}_K = \hat{\boldsymbol{g}}_K$, $\boldsymbol{\Psi}^{\mathrm{T}-1}\boldsymbol{G}^K = \hat{\boldsymbol{g}}^K$ the definitions (8.84) and (8.86) of dual stresses and strains lead to

$$\Pi = \frac{1}{2}[g_{KL} - G_{KL}]\hat{g}^K \otimes \hat{g}^L \tag{8.118}$$

and

$$\Sigma = S^{KL}\hat{g}_K \otimes \hat{g}_L . \tag{8.119}$$

Moreover, by differentiating (8.118) and (8.119) with respect to time, according to the product, and taking

$$\dot{\hat{g}}_K = \dot{\Psi}\Psi^{-1}\hat{g}_K \tag{8.120}$$

and

$$\dot{\hat{g}}^K = -(\dot{\Psi}\Psi^{-1})^{\mathrm{T}}\hat{g}^K \tag{8.121}$$

into consideration, the *Oldroyd* rates are recognised as relative derivatives (cf. Theorem 1. 9):

$$\overset{\triangle}{\Pi} = \frac{1}{2}\dot{g}_{KL}\hat{g}^K \otimes \hat{g}^L , \tag{8.122}$$

$$\overset{\triangledown}{\Sigma} = \dot{S}^{KL}\hat{g}_K \otimes \hat{g}_L . \tag{8.123}$$

We accordingly find

$$\pi = \frac{1}{2}[g^{KL} - G^{KL}]\hat{g}_K \otimes \hat{g}_L , \tag{8.124}$$

$$\sigma = S_{KL}\hat{g}^K \otimes \hat{g}^L \tag{8.125}$$

and

$$\overset{\triangledown}{\pi} = \frac{1}{2}\dot{g}^{KL}\hat{g}_K \otimes \hat{g}_L , \tag{8.126}$$

$$\overset{\triangle}{\sigma} = \dot{S}_{KL}\hat{g}^K \otimes \hat{g}^L . \tag{8.127}$$

The special case of $\mathbf{\Psi} = \mathbf{F}$, which leads to the current configuration as a trivial example of an intermediate configuration, has been described in detail in the previous section.

Another example is provided by the identification $\mathbf{\Psi} = \mathbf{R}$ with $\mathbf{R}$ from the polar decomposition $\mathbf{F} = \mathbf{RU} = \mathbf{VR}$. As a consequence of the definition $\mathbf{\Lambda} = \dot{\mathbf{R}}\mathbf{R}^{\mathrm{T}}$, dual rates of the *Green-McInnis* type defined in (8.9) arise:

$$\overset{\diamond}{\Sigma} = \dot{\Sigma} - \dot{\mathbf{R}}\mathbf{R}^{\mathrm{T}}\Sigma + \Sigma\dot{\mathbf{R}}\mathbf{R}^{\mathrm{T}} . \tag{8.128}$$

In this case the transformed stress is $\boldsymbol{\Sigma} = \mathbf{R}\tilde{\mathbf{T}}\mathbf{R}^{\mathrm{T}} = \mathbf{V}^{-1}\mathbf{S}\mathbf{V}^{-1}$, dual to the strain $\boldsymbol{\Pi} = \frac{1}{2}(\mathbf{B} - \mathbf{1})$.[8]

The *Jaumann* rate (8.5), applied to the stress tensor $\boldsymbol{\Sigma} = \mathbf{P}\tilde{\mathbf{T}}\mathbf{P}^{\mathrm{T}}$,

$$\overset{\circ}{\boldsymbol{\Sigma}} = \dot{\boldsymbol{\Sigma}} - \mathbf{W}\boldsymbol{\Sigma} + \boldsymbol{\Sigma}\mathbf{W}$$

can also be accommodated in the pattern of dual variables, although an intermediate configuration $\boldsymbol{\Psi}(\boldsymbol{X}, t) = \mathbf{P}(\boldsymbol{X}, t)$ would have to be introduced, which would first have to be calculated from the tensor-valued differential equation

$$\dot{\mathbf{P}}(\boldsymbol{X}, t) = \mathbf{W}(\boldsymbol{X}, t)\mathbf{P}(\boldsymbol{X}, t)\ . \tag{8.129}$$

The general integral of the system (8.129) of ordinary differential equations is a functional of the process history, which determines the present value of $\mathbf{P}$ through the history of the vorticity tensor $\mathbf{W} = \frac{1}{2}(\mathbf{L} - \mathbf{L}^{\mathrm{T}})$.
Accomplishing this integration and the subsequent work involved with corresponding dual stress and strain quantities $(\boldsymbol{\Psi} = \mathbf{P})$ would, however, be so complex that probably nobody would seriously consider attempting the task. Within the concept of dual variables and derivatives, the application of the *Jaumann* rate appears to be somewhat unnatural.
In the examples of evolution equations given in Sects. 8. 1. 3 and 8. 1. 4, dual derivatives have been employed, namely out of the 2nd family in the case of the *Maxwell* fluid (8.21) and out of the 1st family in the case of the rigid-plastic solid (8.29), (8.30).

The conclusion (8.27) drawn from the definition (8.21) of the *Maxwell* fluid shows that this aspect can be realised by selecting an *Oldroyd* rate. The same applies for conclusion (8.50), which can be drawn from the definitions (8.28), (8.29) of a rigid-plastic solid with kinematic hardening according to (8.42). These results speak in favour of the application of dual variables and derivatives when formulating material models.[9]

---

[8] HAUPT & TSAKMAKIS [1989], p. 182.

[9] Dual variables and derivatives have been applied successfully in various cases: see BONN [1992]; BONN & HAUPT [1995]; HARTMANN [1993]; TSAKMAKIS [1996]; LÜHRS [1997]; LÜHRS et al. [1997]; LION [1996, 1997a, 1997b, 1998, 1999]; LION & SEDLAN [1998]; KAMLAH & TSAKMAKIS [1999].

9 Elasticity

9.1 Elasticity and Hyperelasticity

The theory of *elasticity* models rate-independent material behaviour without hysteresis. An elastic material is distinguished by the property that the current state of stress depends only on the current state of deformation, so that the memory part $\mathfrak{G}$ in the reduced form (7.27) and (7.28) vanishes identically. A memory going back to past deformation processes does not exist; the most general constitutive equation in the theory of elasticity reads

$$\tilde{\mathbf{T}} = \mathbf{g}(\mathbf{C}) \tag{9.1}$$

or

$$\tilde{\mathbf{T}} = \mathbf{g}(\mathbf{E}) \,. \tag{9.2}$$

Here, $\mathbf{g}(\cdot)$: $\mathcal{Sym}(\mathbb{V}^3) \longrightarrow \mathcal{Sym}(\mathbb{V}^3)$ is a symmetric tensor-valued function of the *Cauchy-Green* tensor $\mathbf{C}$ or *Green* strain tensor $\mathbf{E}$.

The concept of elasticity expresses the idea that a material element instantaneously reverts to its initial configuration once the external forces vanish. Moreover, an elastic structure is expected to give back the entire work done by forces and absorbed during a loading process, upon unloading. This characteristic is not automatically guaranteed by the general constitutive equation (9.2) of elasticity; it is an additional condition which has to be satisfied by the material function $\mathbf{g}(\cdot)$. The mathematical formulation of this condition is based on the balance (2.100) of mechanical energy,

$$L_{\mathrm{e}}(\mathcal{B}, t) = \dot{K}(\mathcal{B}, t) + L_{\mathrm{i}}(\mathcal{B}, t)\,, \tag{9.3}$$

in which $L_{\mathrm{i}}(\mathcal{B}, t) = \int\limits_{\mathcal{B}} \ell_{\mathrm{i}} \mathrm{d}m$ is the stress power and

$$\ell_{\mathrm{i}} = \frac{1}{\rho_{\mathrm{R}}} \widetilde{\mathbf{T}} \cdot \dot{\mathbf{E}} \tag{9.4}$$

the specific stress power per unit mass. The restrictive condition applying to the material function $\mathbf{g}(\cdot)$ originates from the demand for the stress power to be the material time derivative of a scalar-valued function.

Definition 9. 1

A material body, whose mechanical behaviour is represented by a constitutive equation of the form (9.1) or (9.2), is called *elastic material.* An elastic material characterised by the existence of a scalar-valued function $\mathbf{E} \longmapsto e(\mathbf{E})$ with the property

$$\frac{1}{\rho_{\mathrm{R}}} \widetilde{\mathbf{T}} = \frac{\mathrm{d}}{\mathrm{d}\mathbf{E}} e(\mathbf{E})\,, \tag{9.5}$$

is called *hyperelastic.*[1] The function $\mathbf{E} \longmapsto e(\mathbf{E})$ is called *specific strain energy* per unit mass and the volume integral

$$E(\mathcal{B}, t) = \int\limits_{\mathcal{B}} e(\mathbf{E}) \mathrm{d}m \tag{9.6}$$

strain energy of the body $\mathcal{B}$. ◊

The definition states that in the case of a hyperelastic material the stresses are derived from a *potential.* This potential is the *specific strain energy.* It gives rise to

Theorem 9. 1

An elastic material is hyperelastic if and only if the stress power is equal to the material time derivative of the strain energy, depending on the state of strain, $e = e(\mathbf{E})$:

[1] Hyperelasticity is also called *Green* elasticity, see e.g. OGDEN [1984], p. 206. Elasticity without the existence of a strain energy and a potential relation (9.5) is called *Cauchy* elasticity, OGDEN [1984], pp. 175.

$$\ell_{\mathrm{i}} = \frac{1}{\rho_{\mathrm{R}}} \tilde{\mathbf{T}} \cdot \dot{\mathbf{E}} = \frac{\mathrm{d}}{\mathrm{d}t}\, e(\mathbf{E}(t)) \,, \tag{9.7}$$

$$L_{\mathrm{i}} = \frac{\mathrm{d}}{\mathrm{d}t} \int_{\mathscr{B}} e \,\mathrm{d}m = \dot{E}(\mathscr{B}, t) \,. \tag{9.8}$$

For hyperelastic bodies the power of the external forces according to (2.96) is equal to the temporal change of its total energy; this is the sum of the kinetic energy K according to (2.95) and the strain energy E according to (9.6):

$$L_{\mathrm{e}} = (K + E)^{\cdot} \,. \tag{9.9}$$

□

Proof

1) If hyperelasticity holds, then the potential property $\frac{1}{\rho_{\mathrm{R}}} \tilde{\mathbf{T}} = \frac{\mathrm{d}}{\mathrm{d}\mathbf{E}} e(\mathbf{E})$ yields the specific stress power

$$\ell_{\mathrm{i}} = \frac{1}{\rho_{\mathrm{R}}} \tilde{\mathbf{T}} \cdot \dot{\mathbf{E}} = \left[\frac{\mathrm{d}}{\mathrm{d}\mathbf{E}} e(\mathbf{E})\right] \cdot \dot{\mathbf{E}}(t) = \frac{\mathrm{d}}{\mathrm{d}t} e(\mathbf{E}(t)); \quad \text{thus we have } L_{\mathrm{i}} = \dot{E} \,.$$

2) Conversely, $\ell_{\mathrm{i}} = \frac{\mathrm{d}}{\mathrm{d}t} e(\mathbf{E}(t))$ implies $\ell_{\mathrm{i}} = \frac{\mathrm{d}}{\mathrm{d}\mathbf{E}} e(\mathbf{E}) \cdot \dot{\mathbf{E}}(t)$,

and the definition (9.4) of the stress power leads to the identity

$$\left[\frac{1}{\rho_{\mathrm{R}}} \tilde{\mathbf{T}} - \frac{\mathrm{d}}{\mathrm{d}\mathbf{E}} e(\mathbf{E})\right] \cdot \dot{\mathbf{E}}(t) = 0 \,.$$

Since this identity has to hold for all strain rates $\dot{\mathbf{E}}(t)$, it implies (9.5). The generally valid energy balance (2.100), combined with (9.8), then implies the more specific formulation (9.9). ■

The power of the system of forces acting upon a hyperelastic body brings about a change of mechanical energy, which is the sum of kinetic energy K and strain energy E. The special formulation (9.9) of the energy balance shows that an elastic body returns all the mechanical work accomplished, if and only if it is hyperelastic. In this sense the conversion of energy taking place in a hyperelastic system can be regarded as being *reversible*. In addition, if the acting system of forces can be derived from a potential energy, then a more specific statement is obtained, namely the *conservation of mechanical energy*.

Definition 9. 2
A system of forces $\left\{s_{\mathrm{R}}(\cdot), f(\cdot)\right\}$, consisting of a surface force density

$$s_{\mathrm{R}} = \mathbf{T}_{\mathrm{R}} n_{\mathrm{R}} = s_{\mathrm{R}}(x) \ , \ x \in \partial_s \chi_{\mathrm{t}}[\mathcal{B}]^2$$

and a volume force density

$$f = f(x) \ , \ x \in \chi_{\mathrm{t}}[\mathcal{B}]$$

is called a *conservative system of forces*, if two potentials $\Phi(x)$ and $\varphi(x)$ exist with the property

$$s_{\mathrm{R}}(x) = -\ \mathrm{grad}\Phi(x) \tag{9.10}$$

and

$$f(x) = -\ \mathrm{grad}\varphi(x) \ . \tag{9.11}$$

The scalar-valued function

$$U(\mathcal{B}, t) = \iint\limits_{\partial_2 \mathrm{R}[\mathcal{B}]} \Phi \mathrm{d}A + \iiint\limits_{\mathrm{R}[\mathcal{B}]} \varphi \rho_{\mathrm{R}} \mathrm{d}V \tag{9.12}$$

is referred to as the *potential energy* of the system of forces $\left\{s_{\mathrm{R}}(\cdot), f(\cdot)\right\}$.

◊

Observing Theorem 9. 1 this definition leads to

Theorem 9. 2
The power of a conservative system of forces is the negative rate of change of the potential energy:

$$L_{\mathrm{a}} = -\frac{\mathrm{d}}{\mathrm{d}t} U(\mathcal{B}, t) \ . \tag{9.13}$$

In the case of a hyperelastic body under the influence of a conservative system of forces the *total energy* $K + E + U$ remains constant in time:

$$\frac{\mathrm{d}}{\mathrm{d}t}\Big(K(t) + E(t) + U(t)\Big) = 0 \ . \tag{9.14}$$

□

---

[2] Boundary conditions were explained in Sect. 2. 3. 5.

Proof

The material derivative of the potential energy (9.12) is calculated to

$$\frac{\mathrm{d}}{\mathrm{d}t}U(\mathscr{B}, t) = \iint\limits_{\partial_2 \mathrm{R}[\mathscr{B}]} \dot{\Phi}\mathrm{d}A + \iiint\limits_{\mathrm{R}[\mathscr{B}]} \dot{\varphi}\rho_{\mathrm{R}}\mathrm{d}V$$

$$= \iint\limits_{\partial_2 \mathrm{R}[\mathscr{B}]} (\mathrm{grad}\Phi)\cdot\dot{\boldsymbol{x}}\mathrm{d}A + \iiint\limits_{\mathrm{R}[\mathscr{B}]} (\mathrm{grad}\varphi)\cdot\dot{\boldsymbol{x}}\rho_{\mathrm{R}}\mathrm{d}V$$

$$= -\iint\limits_{\partial_2 \mathrm{R}[\mathscr{B}]} (\boldsymbol{s}_{\mathrm{R}}\cdot\boldsymbol{v})\mathrm{d}A - \iiint\limits_{\mathrm{R}[\mathscr{B}]} (\boldsymbol{f}\cdot\boldsymbol{v})\rho_{\mathrm{R}}\mathrm{d}V = -L_{\mathrm{a}}(\mathscr{B}, t)\,.$$

Insertion into the energy balance (9.9) leads to $-\dot{U} = (K + E)^{\cdot}$ and (9.14).

■

A hyperelastic body under conservative loadings is a *conservative system*: its mechanical behaviour is characterised by the *conservation of mechanical energy.*

In connection with *d'Alembert's* principle (2.111) and the principle of virtual work (2.123) the differential of the strain energy arises in hyperelasticity:

Theorem 9. 3

The following three statements are equivalent:

(A) An elastic body is hyperelastic.

(B) For all virtual displacements $\boldsymbol{\eta}(\boldsymbol{X})$ the virtual work of the stresses is equal to the virtual change (the total differential) of the strain energy:

$$\frac{1}{\rho_{\mathrm{R}}}\tilde{\mathbf{T}}\cdot\delta\mathbf{E} = \delta e(\mathbf{E})\,. \tag{9.15}$$

(C) The virtual work of the stresses along any closed curve in the strain space vanishes:

$$\oint \tilde{\mathbf{T}}\cdot\delta\mathbf{E} = 0\,. \tag{9.16}$$

□

Proof

1) The starting point for the proof is the fact that a differential form

$$\sum_{i=1}^{n} f_i(x_1, \dots, x_n)\mathrm{d}x_i$$

is the total differential of a function $F(x_1, \dots, x_n)$ if and only if the functions f_i are the partial derivatives of F with respect to the independent variables x_i:

$$\sum_{i=1}^{n} f_i(x_1, \dots, x_n)\mathrm{d}x_i = \mathrm{d}F(x_1, \dots, x_n) \quad \Leftrightarrow \quad f_i = \frac{\partial F}{\partial x_i} .$$

This leads to the equivalence of the statements (A) and (B):

$$\frac{1}{\rho_{\mathrm{R}}}\widetilde{T}_{ik}\,\delta E_{ik} = \delta e(E_{11}, \dots, E_{33}) \quad \Longleftrightarrow \quad \frac{1}{\rho_{\mathrm{R}}}\widetilde{T}_{ik} = \frac{\partial}{\partial E_{ik}} e(E_{11}, \dots, E_{33}) ,$$

i.e., $\frac{1}{\rho_{\mathrm{R}}}\widetilde{\mathbf{T}}\cdot\delta\mathbf{E} = \delta e(\mathbf{E}) \quad \Longleftrightarrow \quad \frac{1}{\rho_{\mathrm{R}}}\widetilde{\mathbf{T}} = \frac{\mathrm{d}}{\mathrm{d}\mathbf{E}} e(\mathbf{E})$.

2) (B) implies (C).

Since the specific strain energy e depends uniquely upon $\mathbf{E}$, the line integral over $\delta e(\mathbf{E})$ vanishes along any closed curve,

$$\oint \frac{1}{\rho_{\mathrm{R}}}\widetilde{\mathbf{T}}\cdot\delta\mathbf{E} = \oint \left\{\frac{\mathrm{d}}{\mathrm{d}\mathbf{E}} e(\mathbf{E})\right\}\cdot\delta\mathbf{E} = \oint \delta e(\mathbf{E}) = 0 .$$

3) (C) implies (A).

According to (C) the line integral $\displaystyle\int_{\mathbf{E}_0}^{\mathbf{E}} \widetilde{\mathbf{T}}\cdot\delta\mathbf{E}$

between two states of strain $\mathbf{E}_0$ and $\mathbf{E}$ is independent from the curve connecting $\mathbf{E}_0$ and $\mathbf{E}$. A function can thus be defined as

$$\mathbf{E} \longmapsto e(\mathbf{E}) = \frac{1}{\rho_{\mathrm{R}}} \int_{\mathbf{E}_0}^{\mathbf{E}} \widetilde{\mathbf{T}}(\mathbf{E})\cdot\delta\mathbf{E} ,$$

which uniquely depends on $\mathbf{E}$. Now, the *Gateaux* derivative of this function in the direction of any symmetric tensor $\mathbf{H}$ is calculated:

$$\delta e(\mathbf{E} \mid \mathbf{H}) = \rho_{\mathrm{R}}\frac{\mathrm{d}e}{\mathrm{d}\mathbf{E}}\cdot\mathbf{H} = \frac{\mathrm{d}}{\mathrm{d}s}\int_{\mathbf{E}_0}^{\mathbf{E}+s\mathbf{H}} \widetilde{\mathbf{T}}(\mathbf{E})\cdot\delta\mathbf{E}\,\Bigg|_{s=0} .$$

Rearranging the integral, we obtain

$$\int_{\mathbf{E}_0}^{\mathbf{E}+s\mathbf{H}} \widetilde{\mathbf{T}}(\mathbf{E})\cdot\delta\mathbf{E} = \int_{\mathbf{E}_0}^{\mathbf{E}} \widetilde{\mathbf{T}}\cdot\delta\mathbf{E} + \int_{0}^{s} \widetilde{\mathbf{T}}(\mathbf{E}+\sigma\mathbf{H})\cdot\mathbf{H}\mathrm{d}\sigma\,,$$

and after differentiation and setting $s = 0$, the identity

$$\rho_{\mathrm{R}}\frac{\mathrm{d}e}{\mathrm{d}\mathbf{E}}\cdot\mathbf{H} = \widetilde{\mathbf{T}}\cdot\mathbf{H}\,,$$

valid for all symmetric tensors $\mathbf{H}$. This implies hyperelasticity. ■

The principle of virtual work (2.123) merges into the statement

$$\delta A_{\mathrm{a}} = \delta E \tag{9.17}$$

for hyperelastic bodies. If we assume that the acting system of forces is derivable from a potential energy, i.e. $\delta A_{\mathrm{a}} = -\,\delta U$, then we obtain

$$\delta(U + E) = 0\,. \tag{9.18}$$

This version of the principle of virtual displacements can be interpretated as follows: the potential energy of a hyperelastic body under conservative loading assumes a stationary value in any state of equilibrium.[3] In nonlinear elasticity, this can either be a minimum or a maximum or even a saddle point.

Summarising, we ascertain that the physically based concept of hyperelasticity represents a rational restriction of the more general term of *Cauchy* elasticity. This restriction expresses the aspects of energy conservation and reversibility, which one connects with the notion of elasticity.

The specific strain energy can also be expressed as a function of the *Piola* strain tensor $\mathbf{e} = \frac{1}{2}(\mathbf{1} - \mathbf{C}^{-1})$. It is then the potential of the convective stress tensor $\widetilde{\mathbf{t}} = \mathbf{F}^{\mathrm{T}}\mathbf{S}\mathbf{F} = \mathbf{C}\widetilde{\mathbf{T}}\mathbf{C}$. In addition to Definition 9. 1 we set up

[3] OGDEN [1984], pp. 310.

Theorem 9. 4
For a hyperelastic material the potential relation

$$\frac{1}{\rho_R}\tilde{\mathrm{t}} = -\frac{\mathrm{d}}{\mathrm{d\,e}}e(\mathrm{e}) \tag{9.19}$$

between the *Piola* strain and the convective stress is equivalent to (9.5). □

Proof
Starting with the assumption $e = e(\mathrm{e})$ we calculate the material derivative

$$\dot{e}(\mathrm{e}(t)) = \frac{\mathrm{d}e}{\mathrm{de}} \cdot \dot{\mathrm{e}}(t)\,. \tag{9.20}$$

On the other hand, with (8.58), $\rho_R \ell_i = (-\tilde{\mathrm{t}}) \cdot \dot{\mathrm{e}}$, we also have

$$\dot{e}(\mathrm{e}(t)) = -\frac{1}{\rho_R}\tilde{\mathrm{t}} \cdot \dot{\mathrm{e}}(t)\,, \tag{9.21}$$

and obtain the identity

$$\left(\frac{1}{\rho_R}\tilde{\mathrm{t}} + \frac{\mathrm{d}e}{\mathrm{de}}\right) \cdot \dot{\mathrm{e}} = 0$$

for all strain rates $\dot{\mathrm{e}}$. ■

9. 2 Isotropic Elastic Bodies

9. 2. 1 General Constitutive Equation for Elastic Fluids and Solids

With respect to a natural reference configuration, the material properties of an isotropic elastic body can be represented by the general constitutive equation

$$\tilde{\mathrm{T}} = \mathrm{g}(\mathrm{E}) \tag{9.22}$$

or

$$\mathrm{T} = \mathrm{f}(\mathrm{B})\,. \tag{9.23}$$

$\mathbf{g}(\cdot)$ and $\mathbf{f}(\cdot)$ are isotropic tensor functions: for all $\mathbf{Q} \in \mathcal{O}rth$ the identities

$$\mathbf{g}(\mathbf{QEQ}^{\mathrm{T}}) = \mathbf{Qg}(\mathbf{E})\mathbf{Q}^{\mathrm{T}} \tag{9.24}$$

and

$$\mathbf{f}(\mathbf{QBQ}^{\mathrm{T}}) = \mathbf{Qf}(\mathbf{B})\mathbf{Q}^{\mathrm{T}} \tag{9.25}$$

are valid. The property of isotropy makes a more concrete representation of the general constitutive equation possible: a fluid is characterised by the greatest degree of symmetry, namely by a symmetry group that is the whole set of unimodular transformations: $g = \mathcal{U}nim$. The most general constitutive equation for an elastic fluid according to (7.70) ($\boldsymbol{\Omega} = \mathbf{0}$) is

$$\mathbf{T} = -\, p(\rho)\mathbf{1}\ , \tag{9.26}$$

where the mass density ρ depends on the state of strain,

$$\rho = \frac{\rho_{\mathrm{R}}}{\sqrt{\det \mathbf{B}}}\ . \tag{9.27}$$

According to (9.26), the most general constitutive equation of an elastic fluid is reduced to a single scalar-valued function of one variable.

An elastic fluid is also hyperelastic: using the constitutive equation $\mathbf{T} = -\, p(\rho)\mathbf{1}$ the stress power is calculated to be

$$\ell_{\mathrm{i}} = \frac{1}{\rho}\mathbf{T}\cdot\mathbf{D} = -\,\frac{1}{\rho}\,p(\rho)(\mathbf{1}\cdot\mathbf{D}) = -\,\frac{1}{\rho}\,p\ \mathrm{div}\boldsymbol{v} = p\,\frac{\dot{\rho}}{\rho^2}\ ,$$

$$\ell_{\mathrm{i}} = -\,p\Big(\frac{1}{\rho}\Big)^{\cdot} = -\,p\dot{v}\ . \tag{9.28}$$

In the last equation the *specific volume*

$$v = \frac{1}{\rho} \tag{9.29}$$

has been introduced. If we write the constitutive equation for the pressure as $p = p(v)$, then we obtain

$$\ell_{\mathrm{i}} = -\,p(v)\dot{v} = \dot{e}\big(v(t)\big)\ , \tag{9.30}$$

that is, the hyperelasticity of the elastic fluid. (9.30) implies the strain energy

$$e(v) = -\int p(v)\mathrm{d}v \tag{9.31}$$

as well as the potential property

$$p(v) = -\frac{\mathrm{d}}{\mathrm{d}v}\, e(v)\,. \tag{9.32}$$

If a solid is isotropic, a natural reference configuration R exists with the property that the symmetry group is the set of all orthogonal transformations according to (7.68), i.e. $\mathscr{g}_{\mathrm{R}} = \mathscr{Orth}$. An isotropic tensor function, defined by (9.25), is also called a *coaxial function*: every eigenvector of $\mathbf{B}$ is also an eigenvector of $\mathbf{f}(\mathbf{B})$.

Theorem 9. 5

If $\mathbf{f} : \mathbf{B} \longmapsto \mathbf{T} = \mathbf{f}(\mathbf{B})$ is an isotropic tensor function in the sense of (9.25), then the eigenvectors of $\mathbf{B}$ and $\mathbf{f}(\mathbf{B})$ are identical.[4] □

Proof

Since the left *Cauchy-Green* tensor $\mathbf{B}$ is symmetric, there are three real eigenvalues λ_{k} and an orthonormal system of eigenvectors,

$\{v_1, v_2, v_3\}$, $v_{\mathrm{i}} \cdot v_{\mathrm{k}} = \delta_{\mathrm{ik}}$, so that we have the spectral representation

$$\mathbf{B} = \sum_{\mathrm{k}=1}^{3} \lambda_{\mathrm{k}} v_{\mathrm{k}} \otimes v_{\mathrm{k}}\,. \tag{9.33}$$

Moreover, there is an orthogonal tensor that inverts one of these eigenvectors, leaving the others invariant, for instance $\mathbf{Q}_1$ with

$$\mathbf{Q}_1 v_1 = -\,v_1\,, \quad \mathbf{Q}_1 v_2 = v_2\,, \quad \mathbf{Q}_1 v_3 = v_3\,. \tag{9.34}$$

Obviously, $\mathbf{B}$ remains invariant under $\mathbf{Q}_1$ in the sense of

$$\mathbf{Q}_1 \mathbf{B} \mathbf{Q}_1^{\mathrm{T}} = \mathbf{B}\,. \tag{9.35}$$

Due to isotropy, the same property applies for $\mathbf{f}(\mathbf{B})$:

$$\mathbf{f}(\mathbf{B}) = \mathbf{f}(\mathbf{Q}_1 \mathbf{B} \mathbf{Q}_1^{\mathrm{T}}) = \mathbf{Q}_1 \mathbf{f}(\mathbf{B}) \mathbf{Q}_1^{\mathrm{T}}\,. \tag{9.36}$$

(9.36) implies $\mathbf{Q}_1\mathbf{f}(\mathbf{B}) = \mathbf{f}(\mathbf{B})\mathbf{Q}_1$. Bearing (9.34) in mind, three conclusions can now be drawn:

---

[4] Cf. Truesdell & Noll [1965], p. 32.

$$\mathbf{Q}_1[\mathbf{f}(\mathbf{B})\boldsymbol{v}_1] = -\,\mathbf{f}(\mathbf{B})\boldsymbol{v}_1\,,\quad \mathbf{Q}_1[\mathbf{f}(\mathbf{B})\boldsymbol{v}_2] = \mathbf{f}(\mathbf{B})\boldsymbol{v}_2\,,\quad \mathbf{Q}_1[\mathbf{f}(\mathbf{B})\boldsymbol{v}_3] = \mathbf{f}(\mathbf{B})\boldsymbol{v}_3\,. \tag{9.37}$$

Consequently, $\mathbf{Q}_1$ effects a reflection on the $(\boldsymbol{v}_2, \boldsymbol{v}_3)$ plane, both for $\boldsymbol{v}_1$ and for $\mathbf{f}(\mathbf{B})\boldsymbol{v}_1$; the two vectors $\boldsymbol{v}_1$ and $\mathbf{f}(\mathbf{B})\boldsymbol{v}_1$ must therefore be parallel:

$$\mathbf{f}(\mathbf{B})\boldsymbol{v}_1 = \mu_1 \boldsymbol{v}_1\,. \tag{9.38}$$

Accordingly, $\boldsymbol{v}_1$ is also an eigenvector of $\mathbf{f}(\mathbf{B})$ and μ_1 is the corresponding eigenvalue. Similar statements apply to the remaining eigenvectors. ■

Compared with an elastic fluid, the characterisation of the material properties of an isotropic solid demands considerably more information, as the following representation theorem shows.[5]

Theorem 9. 6

Every isotropic tensor function $\mathbf{f} : \mathbf{B} \longmapsto \mathbf{T} = \mathbf{f}(\mathbf{B})$ can be represented in the form of

$$\mathbf{f}(\mathbf{B}) = \varphi_0 \mathbf{1} + \varphi_1 \mathbf{B} + \varphi_2 \mathbf{B}^2\,. \tag{9.39}$$

In this representation, the coefficients φ_k are scalar-valued functions of the three basic invariants of $\mathbf{B}$:

$$\varphi_k = \varphi_k(\mathrm{I}_\mathbf{B}, \mathrm{II}_\mathbf{B}, \mathrm{III}_\mathbf{B}) \quad (k = 1, 2, 3)\,, \tag{9.40}$$

$$\mathrm{I}_\mathbf{B} = \mathrm{tr}\mathbf{B} = \lambda_1 + \lambda_2 + \lambda_3\,,$$

$$\mathrm{II}_\mathbf{B} = \tfrac{1}{2}\left[(\mathrm{tr}\mathbf{B})^2 - \mathrm{tr}(\mathbf{B}^2)\right] = \lambda_1\lambda_2 + \lambda_2\lambda_3 + \lambda_3\lambda_1\,, \tag{9.41}$$

$$\mathrm{III}_\mathbf{B} = \det\mathbf{B} = \lambda_1\lambda_2\lambda_3\,.$$ □

Proof

Due to the isotropy of the tensor function $\mathbf{f}(\cdot)$ we have in addition to (9.33)

$$\mathbf{f}(\mathbf{B}) = \sum_{k=1}^{3} \mu_k \boldsymbol{v}_k \otimes \boldsymbol{v}_k\,, \tag{9.42}$$

the eigenvalues μ_k depending on the eigenvalues λ_i of $\mathbf{B}$:

$$\mu_k = f_k(\lambda_1, \lambda_2, \lambda_3) \qquad (k = 1, 2, 3)\,. \tag{9.43}$$

[5] TRUESDELL & NOLL [1965], Sects. 12, 47, 48. Cf. OGDEN [1984], pp. 190.

The three functions f_k are not independent of each other, since they can be reduced to a single function. Isotropy substantiates the following relations:

$$\mu_1 = f_1(\lambda_1, \lambda_2, \lambda_3) = f_1(\lambda_1, \lambda_3, \lambda_2) ,$$

$$\mu_2 = f_2(\lambda_1, \lambda_2, \lambda_3) = f_1(\lambda_2, \lambda_3, \lambda_1) = f_1(\lambda_2, \lambda_1, \lambda_3) ,$$

$$\mu_3 = f_3(\lambda_1, \lambda_2, \lambda_3) = f_1(\lambda_3, \lambda_1, \lambda_2) = f_1(\lambda_3, \lambda_2, \lambda_1) ,$$

There is therefore a function $\mu = f(\xi, \eta, \zeta)$ with the property

$$f(\xi, \eta, \zeta) = f(\xi, \zeta, \eta) \tag{9.44}$$

and

$$\mu_1 = f(\lambda_1, \lambda_2, \lambda_3) , \quad \mu_2 = f(\lambda_2, \lambda_3, \lambda_1) , \quad \mu_3 = f(\lambda_3, \lambda_1, \lambda_2) . \tag{9.45}$$

The function $f(\cdot)$ represents the material properties of an isotropic elastic solid. The isotropic function $\mathbf{B} \longmapsto \mathbf{f}(\mathbf{B})$ can be represented in the form of (9.39) if the coefficients φ_k are uniquely defined by the principal stresses μ_k. This demand leads to the system of linear equations

$$\begin{aligned} \varphi_0 + \varphi_1\lambda_1 + \varphi_2\lambda_1^2 = \mu_1 &\quad = f(\lambda_1, \lambda_2, \lambda_3) , \\ \varphi_0 + \varphi_1\lambda_2 + \varphi_2\lambda_2^2 = \mu_2 &\quad = f(\lambda_2, \lambda_3, \lambda_1) , \\ \varphi_0 + \varphi_1\lambda_3 + \varphi_2\lambda_3^2 = \mu_3 &\quad = f(\lambda_3, \lambda_1, \lambda_2) . \end{aligned} \tag{9.46}$$

The solution reads[6]

$$\varphi_k = \frac{V_k}{V} \qquad (k = 0, 1, 2) , \tag{9.47}$$

with the determinants

$$V = \begin{vmatrix} 1 & \lambda_1 & \lambda_1^2 \\ 1 & \lambda_2 & \lambda_2^2 \\ 1 & \lambda_3 & \lambda_3^2 \end{vmatrix} = (\lambda_1 - \lambda_2)(\lambda_2 - \lambda_3)(\lambda_3 - \lambda_1) , \tag{9.48}$$

6 TRUESDELL & NOLL [1965], eq. (48. 5).

$$V_0 = \begin{vmatrix} \mu_1 & \lambda_1 & \lambda_1^2 \\ \mu_2 & \lambda_2 & \lambda_2^2 \\ \mu_3 & \lambda_3 & \lambda_3^2 \end{vmatrix} = \mu_1\lambda_2\lambda_3(\lambda_3 - \lambda_2) + \mu_2\lambda_3\lambda_1(\lambda_1 - \lambda_3) + \\ + \mu_3\lambda_1\lambda_2(\lambda_2 - \lambda_1) \,, \tag{9.49}$$

$$V_1 = \begin{vmatrix} 1 & \mu_1 & \lambda_1^2 \\ 1 & \mu_2 & \lambda_2^2 \\ 1 & \mu_3 & \lambda_3^2 \end{vmatrix} = \mu_1(\lambda_2^2 - \lambda_3^2) + \mu_2(\lambda_3^2 - \lambda_1^2) + \mu_3(\lambda_1^2 - \lambda_2^2) \,, \tag{9.50}$$

$$V_2 = \begin{vmatrix} 1 & \lambda_1 & \mu_1 \\ 1 & \lambda_2 & \mu_2 \\ 1 & \lambda_3 & \mu_3 \end{vmatrix} = \mu_1(\lambda_3 - \lambda_2) + \mu_2(\lambda_1 - \lambda_3) + \mu_3(\lambda_2 - \lambda_1) \,. \tag{9.51}$$

Provided that all eigenvalues λ_i differ from one another, we have $V \neq 0$, and the coefficients φ_i are unique functions of the eigenvalues $(\lambda_1, \lambda_2, \lambda_3)$, or functions of the basic invariants $(\mathrm{I}_\mathbf{B}, \mathrm{II}_\mathbf{B}, \mathrm{III}_\mathbf{B})$. This indicates that an isotropic tensor function permits the representation (9.39) provided that all eigenvalues are different. Vice-versa, it is evident that (9.39) defines an isotropic function.

If two eigenvalues of **B** coincide, the solution of the system (9.46) is no longer unique, since the equations then become linearly dependent: for instance, $\lambda_2 = \lambda_3$, leads to $\mu_2 = \mu_3$ according to (9.45). Consequently, the second and third equation of (9.46) are identical, and the value of one coefficient can be selected at will, for instance $\varphi_2 = 0$. In the case of $\lambda_1 = \lambda_2 = \lambda_3$ two coefficients are open to choice, for example $\varphi_1 = \varphi_2 = 0$. Whether such a choice is physically rational is debatable. The non-determined function values should be chosen to ensure the continuity of the φ_i coefficients' dependence on the basic invariants. In view of the structure of the solution (9.46), a continuous continuation of the functions $\varphi_i(\lambda_1, \lambda_2, \lambda_3)$ or $\varphi_i(\mathrm{I}_\mathbf{B}, \mathrm{II}_\mathbf{B}, \mathrm{III}_\mathbf{B})$ may be possible; whether this can always be achieved is left open here.[7] ■

According to this representation theorem, the material properties of an isotropic elastic solid in the sense of *Cauchy elasticity* are expressed by means of three scalar-valued functions of three independent variables.

[7] Cf. TRUESDELL & NOLL [1965], pp. 33, 34.

The general constitutive equation (9.39) for isotropic elastic solids can be written in various equivalent forms: by applying the *Cayley-Hamilton* equation,

$$\mathbf{B}^3 - \mathrm{I}_{\mathbf{B}}\mathbf{B}^2 + \mathrm{II}_{\mathbf{B}}\mathbf{B} - \mathrm{III}_{\mathbf{B}}\mathbf{1} = \mathbf{0}\,, \tag{9.52}$$

the square $\mathbf{B}^2$ can be expressed as

$$\mathbf{B}^2 = \mathrm{I}_{\mathbf{B}}\mathbf{B}^1 - \mathrm{II}_{\mathbf{B}}\mathbf{1} + \mathrm{III}_{\mathbf{B}}\mathbf{B}^{-1}\,, \tag{9.53}$$

and (with other material functions φ_i) the equivalent representation

$$\mathbf{T} = \mathbf{f}(\mathbf{B}) = \varphi_0\mathbf{1} + \varphi_1\mathbf{B} + \varphi_{-1}\mathbf{B}^{-1} \tag{9.54}$$

is obtainable. The general constitutive equation (9.22) corresponds to the representation

$$\tilde{\mathbf{T}} = \mathbf{g}(\mathbf{E}) = \psi_0\mathbf{1} + \psi_1\mathbf{E} + \psi_2\mathbf{E}^2 \tag{9.55}$$

using

$$\psi_k = \psi_k(\mathrm{J}_1, \mathrm{J}_2, \mathrm{J}_3) \quad (k = 0, 1, 2)\,. \tag{9.56}$$

In the formulation (9.55) and (9.56) the basic invariants have been replaced by an equivalent system of independent invariants, namely:

$$\begin{aligned} \mathrm{J}_1 &= \mathrm{tr}\mathbf{E} = \mathrm{I}_{\mathbf{E}}\,, \\ \mathrm{J}_2 &= \tfrac{1}{2}\mathrm{tr}\mathbf{E}^2 = \tfrac{1}{2}\mathrm{I}_{\mathbf{E}}^{\,2} - \mathrm{II}_{\mathbf{E}}\,, \\ \mathrm{J}_3 &= \tfrac{1}{3}\mathrm{tr}\mathbf{E}^3 = \tfrac{1}{3}\mathrm{I}_{\mathbf{E}}^{\,3} - \mathrm{I}_{\mathbf{E}}\mathrm{II}_{\mathbf{E}} + \mathrm{III}_{\mathbf{E}}\,. \end{aligned} \tag{9.57}$$

(9.55) and (9.56) are a useful representation in situations where asymptotic approximations of the general constitutive equations, which are valid for finite deformations, are to be derived by *Taylor* expansion.

9. 2. 2 Isotropic Hyperelastic Bodies

As opposed to fluids, hyperelasticity does not automatically apply for elastic solids. If a solid is hyperelastic, the constitutive equation for the stress tensor is derivable from a potential:

$$\frac{1}{\rho_R}\widetilde{\mathbf{T}} = \frac{\mathrm{d}}{\mathrm{d}\mathbf{E}}e(\mathbf{E}) \Longleftrightarrow \frac{1}{\rho_R}\widetilde{\mathbf{t}} = -\frac{\mathrm{d}}{\mathrm{d}\,\mathbf{e}}e(\mathbf{e})\,. \tag{9.58}$$

The following theorem shows that in the case of *hyperelasticity* the material symmetry properties are represented by the symmetry group of the strain energy.

Theorem 9. 7

For hyperelastic materials an orthogonal tensor $\mathbf{Q}$ is an element of the symmetry group, i.e. $\mathbf{Q} \in \mathscr{g}_{\mathrm{R}}$, if and only if

$$e(\mathbf{Q}\mathbf{E}\mathbf{Q}^{\mathrm{T}}) = e(\mathbf{E}) \tag{9.59}$$

is valid for all $\mathbf{E}$.[8] □

Proof

The proof is based on identity (9.5), $\widetilde{\mathbf{T}} = \mathbf{g}(\mathbf{E}) = \rho_R \dfrac{\mathrm{d}}{\mathrm{d}\mathbf{E}}e(\mathbf{E})$,

and applies the following statement:

If $f : \mathbf{S} \longmapsto f(\mathbf{S})$ is a scalar-valued tensor function and $\mathbf{A}$ a constant tensor, then we have the relation

$$\frac{\partial}{\partial \mathbf{S}} f(\mathbf{A}\mathbf{S}\mathbf{A}^{\mathrm{T}}) = \mathbf{A}^{\mathrm{T}} \frac{\mathrm{d}f(\mathbf{A}\mathbf{S}\mathbf{A}^{\mathrm{T}})}{\mathrm{d}(\mathbf{A}\mathbf{S}\mathbf{A}^{\mathrm{T}})}\mathbf{A}\,. \tag{9.60}$$

It is possible to prove this statement by choosing a component representation with regard to Cartesian coordinates. Given

$T_{lj} = A_{lr}S_{rs}A_{js}$ $\quad (\mathbf{T} = \mathbf{A}\mathbf{S}\mathbf{A}^{\mathrm{T}})$ and employing the chain rule, we arrive at

$$\frac{\partial}{\partial S_{ik}} f(T_{lj}) = \frac{\partial f}{\partial T_{lj}}\frac{\partial T_{lj}}{\partial S_{ik}} = \frac{\partial f}{\partial T_{lj}}\frac{\partial}{\partial S_{ik}}(A_{lr}A_{js}S_{rs}) = \frac{\partial f}{\partial T_{lj}} A_{lr}A_{js}\delta_{ri}\delta_{sk}$$

and conclude $\quad \dfrac{\partial f}{\partial S_{ik}} = A_{li}\dfrac{\partial f}{\partial T_{lj}}A_{jk}\,.$

This result is the component representation of (9.60).

8 TRUESDELL & NOLL [1965], Sect. 85.

Proof of the Theorem

1) If the property $e(\mathbf{QEQ}^{\mathrm{T}}) = e(\mathbf{E})$ is fulfilled for one $\mathbf{Q} \in \mathcal{O}rth$, then it holds that

$$\frac{\mathrm{d}}{\mathrm{dE}}\, e(\mathbf{E}) = \frac{\partial}{\partial \mathbf{E}}\, e(\mathbf{QEQ}^{\mathrm{T}}) = \mathbf{Q}^{\mathrm{T}} \frac{\mathrm{d}e(\mathbf{QEQ}^{\mathrm{T}})}{\mathrm{d}(\mathbf{QEQ}^{\mathrm{T}})}\, \mathbf{Q}$$

according to (9.60). However, that means

$\mathbf{g}(\mathbf{E}) = \mathbf{Q}^{\mathrm{T}}\mathbf{g}(\mathbf{QEQ}^{\mathrm{T}})\mathbf{Q}$, i.e. $\mathbf{Q} \in \mathcal{g}_{\mathrm{R}}$, in terms of the general criterion (7.65).

2) On the other hand, $\mathbf{Q} \in \mathcal{g}_{\mathrm{R}}$ or $\mathbf{Q}\mathbf{g}(\mathbf{E})\mathbf{Q}^{\mathrm{T}} = \mathbf{g}(\mathbf{QEQ}^{\mathrm{T}})$ implies

$$\mathbf{Q}\left\{\frac{\mathrm{d}}{\mathrm{dE}}\, e(\mathbf{E})\right\}\mathbf{Q}^{\mathrm{T}} = \frac{\mathrm{d}e(\mathbf{QEQ}^{\mathrm{T}})}{\mathrm{d}(\mathbf{QEQ}^{\mathrm{T}})}\,.$$

According to (9.60) this identity converts into

$$\mathbf{Q}\frac{\mathrm{d}e(\mathbf{E})}{\mathrm{dE}}\mathbf{Q}^{\mathrm{T}} = \mathbf{Q}\left\{\frac{\partial}{\partial \mathbf{E}}\, e(\mathbf{QEQ}^{\mathrm{T}})\right\}\mathbf{Q}^{\mathrm{T}}\,, \text{ i.e. } \frac{\mathrm{d}e(\mathbf{E})}{\mathrm{dE}} = \frac{\partial}{\partial \mathbf{E}}\, e(\mathbf{QEQ}^{\mathrm{T}})\,.$$

Integration yields $e(\mathbf{E}) = e(\mathbf{QEQ}^{\mathrm{T}}) + e_0$.

The constant e_0 can be set at zero. ■

The symmetry group $\mathcal{g}_{\mathrm{R}}$ of a hyperelastic body, according to the theorem, is equal to the symmetry group of the strain energy. For an isotropic hyperelastic body the strain energy is an isotropic function of the strain tensor. This can only depend on the basic invariants (9.41),

$$e(\mathbf{E}) = e(\mathrm{I}_{\mathbf{E}}, \mathrm{II}_{\mathbf{E}}, \mathrm{III}_{\mathbf{E}})\,, \tag{9.61}$$

or equivalently on the invariants J_k in (9.57):

$$e(\mathbf{E}) = e(\mathrm{J}_1, \mathrm{J}_2, \mathrm{J}_3)\,. \tag{9.62}$$

In the case of isotropic hyperelasticity one single scalar function, depending on three basic invariants, is needed for portraying the material properties. The three material functions in the representations (9.39) and (9.55) are therefore not independent of each other but are derived from the strain energy.

Theorem 9. 8

The general constitutive equation of isotropic hyperelasticity reads

$$\frac{1}{\rho_R}\tilde{\mathbf{T}} = \left(\frac{\partial e}{\partial \mathrm{I}_\mathbf{E}} + \mathrm{I}_\mathbf{E}\frac{\partial e}{\partial \mathrm{II}_\mathbf{E}} + \mathrm{II}_\mathbf{E}\frac{\partial e}{\partial \mathrm{III}_\mathbf{E}}\right)\mathbf{1} - \left(\frac{\partial e}{\partial \mathrm{II}_\mathbf{E}} + \mathrm{I}_\mathbf{E}\frac{\partial e}{\partial \mathrm{III}_\mathbf{E}}\right)\mathbf{E} + \frac{\partial e}{\partial \mathrm{III}_\mathbf{E}}\,\mathbf{E}^2 \tag{9.63}$$

or

$$\frac{1}{\rho_R}\tilde{\mathbf{T}} = \frac{\partial e}{\partial \mathrm{J}_1}\mathbf{1} + \frac{\partial e}{\partial \mathrm{J}_2}\mathbf{E} + \frac{\partial e}{\partial \mathrm{J}_3}\mathbf{E}^2\,, \tag{9.64}$$

where e is given in the form of (9.61) or (9.62). □

Proof

Using the chain rule, the outcome is

$$\frac{1}{\rho_R}\tilde{\mathbf{T}} = \frac{\partial e}{\partial \mathrm{I}_\mathbf{E}}\frac{\mathrm{d}\mathrm{I}_\mathbf{E}}{\mathrm{d}\mathbf{E}} + \frac{\partial e}{\partial \mathrm{II}_\mathbf{E}}\frac{\mathrm{d}\mathrm{II}_\mathbf{E}}{\mathrm{d}\mathbf{E}} + \frac{\partial e}{\partial \mathrm{III}_\mathbf{E}}\frac{\mathrm{d}\mathrm{III}_\mathbf{E}}{\mathrm{d}\mathbf{E}}\,. \tag{9.65}$$

Inserting the derivatives of the basic invariants,

$$\begin{aligned}
\frac{\mathrm{d}\mathrm{I}_\mathbf{E}}{\mathrm{d}\mathbf{E}} &= \frac{\mathrm{d}}{\mathrm{d}\mathbf{E}}\,\mathrm{tr}\mathbf{E} &&= \mathbf{1}\,,\\
\frac{\mathrm{d}\mathrm{II}_\mathbf{E}}{\mathrm{d}\mathbf{E}} &= \frac{\mathrm{d}}{\mathrm{d}\mathbf{E}}\,\tfrac{1}{2}[(\mathrm{tr}\mathbf{E})^2 - \mathrm{tr}(\mathbf{E}^2)] &&= \mathrm{I}_\mathbf{E}\mathbf{1} - \mathbf{E}\,,\\
\frac{\mathrm{d}\mathrm{III}_\mathbf{E}}{\mathrm{d}\mathbf{E}} &= \frac{\mathrm{d}}{\mathrm{d}\mathbf{E}}\,\mathrm{det}\mathbf{E} &&= \mathrm{II}_\mathbf{E}\mathbf{1} - \mathrm{I}_\mathbf{E}\mathbf{E} + \mathbf{E}^2\,,
\end{aligned} \tag{9.66}$$

leads to (9.63). Alternatively, (9.64) results from $e(\mathbf{E}) = e(\mathrm{J}_1, \mathrm{J}_2, \mathrm{J}_3)$,

$$\frac{1}{\rho_R}\tilde{\mathbf{T}} = \frac{\partial e}{\partial \mathrm{J}_1}\frac{\mathrm{d}\mathrm{J}_1}{\mathrm{d}\mathbf{E}} + \frac{\partial e}{\partial \mathrm{J}_2}\frac{\mathrm{d}\mathrm{J}_2}{\mathrm{d}\mathbf{E}} + \frac{\partial e}{\partial \mathrm{J}_3}\frac{\mathrm{d}\mathrm{J}_3}{\mathrm{d}\mathbf{E}}\,, \tag{9.67}$$

with

$$\begin{aligned}
\frac{\mathrm{d}\mathrm{J}_1}{\mathrm{d}\mathbf{E}} &= \frac{\mathrm{d}}{\mathrm{d}\mathbf{E}}\,\mathrm{tr}\mathbf{E} &&= \mathbf{1}\,,\\
\frac{\mathrm{d}\mathrm{J}_2}{\mathrm{d}\mathbf{E}} &= \frac{\mathrm{d}}{\mathrm{d}\mathbf{E}}\,\tfrac{1}{2}\mathrm{tr}\mathbf{E}^2 &&= \mathbf{E}\,,\\
\frac{\mathrm{d}\mathrm{J}_3}{\mathrm{d}\mathbf{E}} &= \frac{\mathrm{d}}{\mathrm{d}\mathbf{E}}\,\tfrac{1}{3}\mathrm{tr}\mathbf{E}^3 &&= \mathbf{E}^2\,.
\end{aligned} \tag{9.68}$$

■

The strain energy of an isotropic hyperelastic material can also be expressed as a function of the *Almansi* tensor or the *Finger* tensor. The corresponding potential relations are described in

Theorem 9. 9

An isotropic hyperelastic material can be represented by

$$e = e(\mathbf{A}) = e(\mathrm{I}_{\mathbf{A}}, \mathrm{II}_{\mathbf{A}}, \mathrm{III}_{\mathbf{A}}) \,, \tag{9.69}$$

$$\frac{1}{\rho_{\mathrm{R}}}\mathbf{S} = \mathbf{B}^{-1}\frac{\mathrm{d}e(\mathbf{A})}{\mathrm{d}\mathbf{A}} \,, \tag{9.70}$$

with $\mathbf{A} = \frac{1}{2}(\mathbf{1} - \mathbf{B}^{-1})$, as well as by

$$e = e(\mathbf{a}) = e(\mathrm{I}_{\mathbf{a}}, \mathrm{II}_{\mathbf{a}}, \mathrm{III}_{\mathbf{a}}) \,, \tag{9.71}$$

$$\frac{1}{\rho_{\mathrm{R}}}\mathbf{S} = -\,\mathbf{B}\frac{\mathrm{d}e(\mathbf{a})}{\mathrm{d}\mathbf{a}} \,, \tag{9.72}$$

with $\mathbf{a} = \frac{1}{2}(\mathbf{1} - \mathbf{B})$. The representation

$$e(\mathbf{B}) = e(\mathrm{I}_{\mathbf{B}}, \mathrm{II}_{\mathbf{B}}, \mathrm{III}_{\mathbf{B}}) \tag{9.73}$$

finally leads to the potential relation

$$\frac{1}{\rho_{\mathrm{R}}}\mathbf{S} = 2\mathbf{B}\frac{\mathrm{d}e(\mathbf{B})}{\mathrm{d}\mathbf{B}} \,. \tag{9.74}$$

(The *Cauchy* stress tensors **S** and **T** are related by (2.48), $\frac{1}{\rho_{\mathrm{R}}}\mathbf{S} = \frac{1}{\rho}\mathbf{T}$.) □

Proof

$e(t) = e(\mathbf{A}(t))$, with $\mathbf{A} = \frac{1}{2}(\mathbf{1} - \mathbf{F}^{\mathrm{T}-1}\mathbf{F}^{-1})$, yields

$$\frac{\mathrm{d}}{\mathrm{d}t}e(\mathbf{A}(t)) = \frac{\mathrm{d}e(\mathbf{A})}{\mathrm{d}\mathbf{A}}\cdot\dot{\mathbf{A}} = -\frac{1}{2}\frac{\mathrm{d}e(\mathbf{A})}{\mathrm{d}\mathbf{A}}\cdot\left(\mathbf{F}^{\mathrm{T}-1}\mathbf{F}^{-1}\right)^{\cdot}$$

$$= \frac{1}{2}\frac{\mathrm{d}e(\mathbf{A})}{\mathrm{d}\mathbf{A}}\cdot\left(\mathbf{F}^{\mathrm{T}-1}\dot{\mathbf{F}}^{\mathrm{T}}\,\mathbf{F}^{\mathrm{T}-1}\mathbf{F}^{-1} + \mathbf{F}^{\mathrm{T}-1}\mathbf{F}^{-1}\dot{\mathbf{F}}\,\mathbf{F}^{-1}\right)$$

$$= \frac{1}{2}\frac{\mathrm{d}e(\mathbf{A})}{\mathrm{d}\mathbf{A}}\cdot\left(\mathbf{L}^{\mathrm{T}}\mathbf{B}^{-1} + \mathbf{B}^{-1}\mathbf{L}\right) = \frac{1}{2}\,\mathrm{tr}\left[\frac{\mathrm{d}e(\mathbf{A})}{\mathrm{d}\mathbf{A}}\mathbf{B}^{-1}\left(\mathbf{L}^{\mathrm{T}} + \mathbf{L}\right)\right],$$

i.e. $\frac{d}{dt} e(\mathbf{A}(t)) = \left(\mathbf{B}^{-1} \frac{de(\mathbf{A})}{d\mathbf{A}}\right) \cdot \mathbf{D}$, the relation

$$\frac{de}{d\mathbf{A}} \mathbf{A} = \mathbf{A} \frac{de}{d\mathbf{A}} \tag{9.75}$$

being employed, which holds for every scalar-valued isotropic function $e = e(\mathbf{A})$. On the other hand, according to (2.105), we have

$\dot{e}(t) = \ell_i = \frac{1}{\rho_R} \mathbf{S} \cdot \mathbf{D}$. This proves the potential relation (9.70).

The representation $e(t) = e(\mathbf{a}(t))$, with $\mathbf{a} = \frac{1}{2}(\mathbf{1} - \mathbf{F}\mathbf{F}^T)$, leads to

$$\begin{aligned}
\dot{e}(t) &= \frac{de}{d\mathbf{a}} \cdot \dot{\mathbf{a}} = -\frac{1}{2} \frac{de}{d\mathbf{a}} \cdot (\mathbf{F}\mathbf{F}^T)^{\cdot} \\
&= -\frac{1}{2} \frac{de}{d\mathbf{a}} \cdot \left(\dot{\mathbf{F}}\mathbf{F}^{-1}\mathbf{F}\mathbf{F}^T + \mathbf{F}\mathbf{F}^T\mathbf{F}^{T-1}\dot{\mathbf{F}}^T\right) \\
&= -\frac{1}{2} \frac{de}{d\mathbf{a}} \cdot \left(\mathbf{L}\mathbf{B} + \mathbf{B}\mathbf{L}^T\right) = -\left(\mathbf{B} \frac{de}{d\mathbf{a}}\right) \cdot \mathbf{D} = \frac{1}{\rho_R} \mathbf{S} \cdot \mathbf{D} .
\end{aligned}$$

This implies (9.72). Here, too, the relation $\frac{de}{d\mathbf{a}} \mathbf{a} = \mathbf{a} \frac{de}{d\mathbf{a}}$ has been used.

Proof of equation (9.74) is provided in the same way:

$$\begin{aligned}
\dot{e}(t) &= \dot{e}(\mathbf{B}(t)) = \frac{de}{d\mathbf{B}} \cdot \dot{\mathbf{B}} = \frac{de}{d\mathbf{B}} \cdot (\mathbf{F}\mathbf{F}^T)^{\cdot} \\
&= \frac{de}{d\mathbf{B}} \cdot \left(\dot{\mathbf{F}}\mathbf{F}^{-1}\mathbf{F}\mathbf{F}^T + \mathbf{F}\mathbf{F}^T\mathbf{F}^{T-1}\dot{\mathbf{F}}^T\right) \\
&= \frac{de}{d\mathbf{B}} \cdot \left(\mathbf{L}\mathbf{B} + \mathbf{B}\mathbf{L}^T\right) = 2\left(\mathbf{B} \frac{de}{d\mathbf{B}}\right) \cdot \mathbf{D} = \frac{1}{\rho_R} \mathbf{S} \cdot \mathbf{D} .
\end{aligned}$$

■

9. 2. 3 Incompressible Isotropic Elastic Materials

The amount of information required for a material's description decreases if the constraint condition of incompressibility is assumed. This is particularly evident in the case of isotropy:

Theorem 9. 10

For an incompressible isotropic elastic material the most general constitutive equation

$$\mathbf{T} = -p(x, t)\mathbf{1} + \varphi_1 \mathbf{B} + \varphi_{-1} \mathbf{B}^{-1} \tag{9.76}$$

applies. The material functions φ_k depends solely upon the first two invariants of **B**:

$$\varphi_k = \varphi_k(I_B, II_B) \quad (k = -1, 1) \tag{9.77}$$

$$\left(I_B = \mathrm{tr}\mathbf{B}\ ,\ II_B = \tfrac{1}{2}[(\mathrm{tr}\mathbf{B})^2 - \mathrm{tr}(\mathbf{B}^2)]\ \right).$$

An incompressible elastic fluid is represented by the constitutive equation

$$\mathbf{T} = -\ p(\boldsymbol{x}, t)\mathbf{1}\ . \tag{9.78}$$

In neither of the representations (9.76) and (9.78) is the hydrostatic pressure $p(\boldsymbol{x}, t)$ connected to the state of deformation by a constitutive equation, but has to be determined from the balance of linear momentum and the boundary conditions. □

Proof

In the case of incompressibility $III_B = \det\mathbf{B} \equiv 1$ applies, so that this variable drops out of the material functions. As described above (Sect. 7. 4), the state of stress includes an undetermined reaction stress, which is a hydrostatic pressure according to (7.101). Since the mass density is materially constant and the pressure undetermined, the material function φ_0 in (9.39) becomes superfluous. The same applies for the material function $p(\rho)$ in (9.26). ■

The material properties of an incompressible, isotropic elastic solid are completely described by two material functions, depending on two variables. Both functions φ_1 and φ_{-1} are derivable from the specific strain energy in the case of hyperelasticity. Applying the potential relation (9.74) to incompressible isotropic hyperelastic materials along the same lines leads to the constitutive equation

$$\mathbf{T} = -\ p\mathbf{1} + 2\rho_0\,\mathbf{B}\frac{d}{d\mathbf{B}}\,e(I_B, II_B)\ . \tag{9.79}$$

Differentiation and application of the *Cayley-Hamilton* equation (9.52) with $III_B = 1$ produces

$$\mathbf{B}\frac{d}{d\mathbf{B}}\,e(I_B, II_B) = \frac{\partial e}{\partial I_B}\mathbf{B} + \frac{\partial e}{\partial II_B}(I_B\mathbf{B} - \mathbf{B}^2)\ ,$$

$$\mathbf{B}\frac{\mathrm{d}}{\mathrm{d}\mathbf{B}}\,e(\mathrm{I}_{\mathbf{B}},\mathrm{II}_{\mathbf{B}}) = \Big(\frac{\partial e}{\partial \mathrm{I}_{\mathbf{B}}} + \mathrm{I}_{\mathbf{B}}\frac{\partial e}{\partial \mathrm{II}_{\mathbf{B}}}\Big)\mathbf{B} - \frac{\partial e}{\partial \mathrm{II}_{\mathbf{B}}}\Big(\mathrm{I}_{\mathbf{B}}\mathbf{B} - \mathrm{II}_{\mathbf{B}}\mathbf{1} + \mathbf{B}^{-1}\Big)$$

$$= \frac{\partial e}{\partial \mathrm{I}_{\mathbf{B}}}\mathbf{B} - \frac{\partial e}{\partial \mathrm{II}_{\mathbf{B}}}\mathbf{B}^{-1} + \mathrm{II}_{\mathbf{B}}\frac{\partial e}{\partial \mathrm{II}_{\mathbf{B}}}\mathbf{1}\,.$$

Again, the last term can be added to the undetermined pressure, resulting in the general constitutive equation for an incompressible isotropic hyperelastic solid:

$$\mathbf{T} = -\,p\mathbf{1} + 2\rho_0\left(\frac{\partial e}{\partial \mathrm{I}_{\mathbf{B}}}\mathbf{B} - \frac{\partial e}{\partial \mathrm{II}_{\mathbf{B}}}\mathbf{B}^{-1}\right). \tag{9.80}$$

9. 2. 4 Constitutive Equations of Isotropic Elasticity (Examples)

Apart from the case of incompressible fluids, material properties of elastic solids are always described by material functions. Prior to applying the theory to concrete problems, these functions have to be identified by means of appropriate experimental tests. To keep this task within reasonable limits, one can assume that the material functions are often needed only in a restricted domain of the arguments involved. In this context approximations of material functions containing just the smallest possible number of independent material constants are called for. The domain of validity of such an individual theory of elasticity is, of course, restricted. Special constitutive equations of isotropic hyperelasticity can be constructed by introducing concrete approximations for the material functions.

A systematic procedure is the *Taylor* expansion of material functions in the representations (9.39), (9.55), or even in (9.63), (9.64). The *Taylor* expansion allows the derivation of asymptotic approximations of a given order. The number of material parameters naturally increases with the order of approximation.

Theorem 9. 11

A linear-elastic isotropic material in the sense of *physical linearisation* is completely described by means of two elasticity constants; the constitutive equation reads

$$\tilde{\mathbf{T}} = 2\mu\Big[\mathbf{E} + \frac{\nu}{1-2\nu}(\mathrm{tr}\mathbf{E})\mathbf{1}\Big]\,. \tag{9.81}$$

In this representation the material constants μ and ν correspond to the classical shear modulus and *Poisson* number.

A linear-elastic isotropic material is also hyperelastic. The strain energy pertaining to (9.81) is

$$\rho_R e(\mathbf{E}) = \mu[\mathbf{E}\cdot\mathbf{E} + \frac{\nu}{1-2\nu}(\mathrm{tr}\mathbf{E})^2] \,. \tag{9.82}$$

□

Proof

Physical linearisation of representation (9.55) yields

$$\widetilde{\mathbf{T}} = \frac{\partial\psi_0}{\partial \mathrm{J}_1}(0,0,0)\,\mathrm{J}_1\mathbf{1} + \psi_1(0,0,0)\mathbf{E} \,.$$

Given the notations $\psi_1(0,0,0) = 2\mu$, $\psi_0(0,0,0) = 2\mu\,\frac{\nu}{1-2\nu}$,

this produces the stress-strain relation (9.81), which is derivable from the strain energy (9.82). ■

The elasticity relation (9.81) generalises *Hooke*'s law (5.35) to finite deformations. At the same time it is a first order approximation of the general constitutive equation (9.55). However, the deformations must not be too large if approximation (9.81) is to be employed (see below).

The property of hyperelasticity is not generally valid for isotropic materials: comparing equation (9.55) with (9.63) or (9.64) shows that the three material functions ψ_k are not independent of one another, but are derived from a single function, the strain energy. This reduces the number of independent material parameters.

The additional assumption of *incompressibility* causes a further reduction in the number of material constants involved in the approximations. In this connection an approximation of the second order is of special interest, since it involves just two material parameters.

Theorem 9. 12

For incompressible isotropic materials an asymptotic approximation of the second order is given by

$$\mathbf{T} = -p\mathbf{1} + \mu\Big[\big(\beta + \tfrac{1}{2}\big)\mathbf{B} + \big(\beta - \tfrac{1}{2}\big)\mathbf{B}^{-1}\Big] + \mathrm{O}(\delta^3) \,, \tag{9.83}$$

where $\delta = \|\mathbf{H}\| = \sqrt{\mathbf{H}\cdot\mathbf{H}}$ is the norm of the displacement gradient according to (1.150). The elastic body defined by this constitutive equation is hyperelastic; the corresponding approximation of the specific strain energy is

$$\rho_0 e(\mathbf{B}) = \tfrac{1}{2}\mu\Big[\Big(\tfrac{1}{2} + \beta\Big)(\mathrm{I}_{\mathbf{B}} - 3) + \Big(\tfrac{1}{2} - \beta\Big)(\mathrm{II}_{\mathbf{B}} - 3)\Big] + 0(\delta^4)\,. \tag{9.84}$$

This strain energy is *positive definite* if and only if $\mu > 0$ and $-\frac{1}{2} \leq \beta \leq \frac{1}{2}$. The elasticity parameter μ corresponds to the classical shear modulus, whereas the dimensionless material constant β represents second-order effects.[9] □

Proof

From the condition of incompressibility,

$\mathrm{III}_{\mathbf{B}} \equiv 1$ and $1 \equiv \det\mathbf{F} = \det(\mathbf{1} + \mathbf{H})$, in connection with the expansion

$\det(\mathbf{1} + \mathbf{H}) = 1 + \mathrm{I}_{\mathbf{H}} + \mathrm{II}_{\mathbf{H}} + \mathrm{III}_{\mathbf{H}}$, we obtain the condition

$$\mathrm{I}_{\mathbf{H}} + \mathrm{II}_{\mathbf{H}} + \mathrm{III}_{\mathbf{H}} = 0\,, \tag{9.85}$$

and thus

$$\mathrm{I}_{\mathbf{H}} = -\,(\mathrm{II}_{\mathbf{H}} + \mathrm{III}_{\mathbf{H}}) = 0(\delta^2)\,. \tag{9.86}$$

Accordingly, the invariants $\mathrm{I}_{\mathbf{B}}$ and $\mathrm{II}_{\mathbf{B}}$ satisfy the following asymptotic relations:

$$\mathrm{I}_{\mathbf{B}} - 3 = 0(\delta^2)\,, \quad \mathrm{II}_{\mathbf{B}} - 3 = 0(\delta^2)\,. \tag{9.87}$$

If we consider the stress-strain relation (9.76), $\mathbf{T} = -\,p\mathbf{1} + \varphi_1\mathbf{B} + \varphi_{-1}\mathbf{B}^{-1}$ and insert the *Taylor* expansions

$$\varphi_k(\mathrm{I}_{\mathbf{B}}, \mathrm{II}_{\mathbf{B}}) = \varphi_k(3,3) + \frac{\partial\varphi_k}{\partial \mathrm{I}_{\mathbf{B}}}(3,3)\,(\mathrm{I}_{\mathbf{B}} - 3) + \frac{\partial\varphi_k}{\partial \mathrm{II}_{\mathbf{B}}}(3,3)\,(\mathrm{II}_{\mathbf{B}} - 3) + 0(\delta^3)$$

$(k = -1, 1)$ and $\mathbf{B} = \mathbf{1} + \mathbf{0}(\delta)$, we obtain the approximation

$$\mathbf{T} = -\,p\mathbf{1} + \varphi_1(3,3)\mathbf{B} + \varphi_{-1}(3,3)\mathbf{B}^{-1} + \mathbf{0}(\delta^3)\,. \tag{9.88}$$

[9] TRUESDELL & NOLL [1965], Sect. 95.

This again makes use of the fact that the hydrostatic pressure in the constitutive equation remains undetermined. The notation

$$\varphi_1(3,3) = \mu\left(\beta + \tfrac{1}{2}\right) , \quad \varphi_{-1}(3,3) = \mu\left(\beta - \tfrac{1}{2}\right) \tag{9.89}$$

leads to

$$\mathbf{T} = -p\mathbf{1} + \tfrac{1}{2}\mu\left(\mathbf{B} - \mathbf{B}^{-1}\right) + \mu\beta\left(\mathbf{B} + \mathbf{B}^{-1}\right) + 0(\delta^3) . \tag{9.90}$$

The asymptotic relations $\mathbf{B} - \mathbf{B}^{-1} = 2\left(\mathbf{H} + \mathbf{H}^{\mathrm{T}}\right) + 0(\delta^2)$ and $\mathbf{B} + \mathbf{B}^{-1} = 2\mathbf{1} + 0(\delta^2)$ indicate that the material constant μ determines the part of the stress that is of first order in $\mathbf{H}$, whereas the term multiplied by β only contains quantities of the order $0(\delta^2)$. That proves the elasticity relation (9.83).
Substituting the potential relation (9.79) and differentiation of (9.84) using the *Cayley-Hamilton* equation (9.52) leads to

$$\begin{aligned}\mathbf{T} + p\mathbf{1} &= +\,2\rho_0\,\mathbf{B}\frac{de}{d\mathbf{B}} = \mu\mathbf{B}\left[\left(\beta + \tfrac{1}{2}\right)\mathbf{1} + \left(\beta - \tfrac{1}{2}\right)\left(\mathrm{I}_{\mathbf{B}}\mathbf{1} - \mathbf{B}\right)\right] \\ &= \mu\left[\left(\beta + \tfrac{1}{2}\right)\mathbf{B} - \left(\beta - \tfrac{1}{2}\right)\left(\mathrm{II}_{\mathbf{B}}\mathbf{1} - \mathbf{B}^{-1}\right)\right] \\ &= \mu\left[-\left(\beta - \tfrac{1}{2}\right)\mathrm{II}_{\mathbf{B}}\mathbf{1} + \left(\beta + \tfrac{1}{2}\right)\mathbf{B} + \left(\beta - \tfrac{1}{2}\right)\mathbf{B}^{-1}\right] .\end{aligned}$$

In the case of incompressibility, the invariants of $\mathbf{B}$ are minimal for $\lambda_1 = \lambda_2 = 1$ and bounded from below:

$$\begin{aligned}\mathrm{I}_{\mathbf{B}} &= \lambda_1 + \lambda_2 + \lambda_3 = \lambda_1 + \lambda_2 + \frac{1}{\lambda_1\lambda_2} \geq 3 , \\ \mathrm{II}_{\mathbf{B}} &= \lambda_1\lambda_2 + \lambda_2\lambda_3 + \lambda_3\lambda_1 = \lambda_1\lambda_2 + \frac{1}{\lambda_1} + \frac{1}{\lambda_2} \geq 3 .\end{aligned} \tag{9.91}$$

Thus $e > 0$ is equivalent to $\mu > 0$ and $-\frac{1}{2} \leq \beta \leq \frac{1}{2}$. ■

Definition 9. 3
The incompressible, isotropic elastic body defined by means of (9.83) is known as *Mooney-Rivlin* material. For the special case of $\beta = \frac{1}{2}$ the constitutive equation

$$\mathbf{T} = -p\mathbf{1} + \mu\mathbf{B} \tag{9.92}$$

defines the *Neo-Hooke* material. ◊

The definition of the *Neo-Hooke* material generalises *Hooke's* Law to finite deformations, but only subject to the constraint condition of incompressibility.

The *Mooney-Rivlin* model is an asymptotic approximation of the general constitutive equation (9.78), containing all second order terms in the displacement gradient. The constitutive equation (9.83) without the error term is also of service as a defining equation for a special perfectly elastic body, since it fulfils all the general principles of the theory of materials. The corresponding statement applies to the special case (9.92) of the *Neo-Hooke* model, which represents a first order approximation on the one hand and an ideal material on the other.

Test calculations carried out on homogeneous deformations and investigations of analytical solutions, which are well-known for incompressible isotropic materials,[10] offer hints as to the physical meaning of the *Neo-Hooke* and *Mooney-Rivlin* models. The following theorem restricts itself in an exemplary way to stress-strain curves resulting from the *Mooney-Rivlin* model for spatially homogeneous uniaxial processes.

Theorem 9. 13

For uniaxial processes the *Mooney-Rivlin* model (9.83) supplies the following stress-strain curve:

$$\sigma_{\mathrm{R}} = \mu\left(\beta + \tfrac{1}{2}\right)\left(\lambda - \frac{1}{\lambda^2}\right) + \mu\left(\beta - \tfrac{1}{2}\right)\left(\frac{1}{\lambda^3} - 1\right). \tag{9.93}$$

The xx-component of the 1st *Piola-Kirchhoff* stress tensor,

$$\sigma_{\mathrm{R}} = \frac{F}{A_0} \quad \left(= T_{\mathrm{R}xx}\right), \tag{9.94}$$

is the force related to the surface A_0 in the reference configuration and $\lambda = 1 + \varepsilon$ the stretch. □

---

[10] As outlined in TRUESDELL & NOLL [1965], Sects. 56, 57, there are 5 families of *universal solutions*, i.e. analytical solutions, valid for all incompressibe isotropic elastic materials. It is worth noting that these universal solutions are also applicable for arbitrary simple materials, if they are incompressible and isotropic. This was shown by CARROLL [1967] and FOSDICK [1968]. See also W.C. MÜLLER [1969].

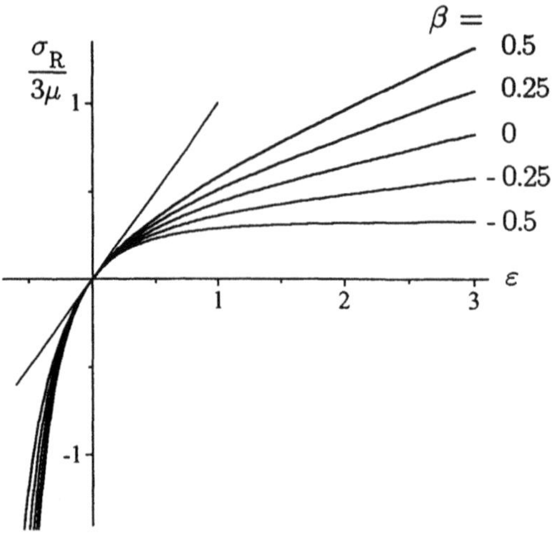

Figure 9. 1: One-dimensional stress-strain curves for the *Mooney-Rivlin* model (9.83), (9.93)

Proof

In combination with incompressibility, uniaxial tension corresponds to the deformation gradient

$$\begin{aligned} \mathbf{F} &= \lambda \mathbf{e}_x \otimes \mathbf{e}_x + \lambda_Q (\mathbf{e}_y \otimes \mathbf{e}_y + \mathbf{e}_z \otimes \mathbf{e}_z) \\ &= \lambda \mathbf{e}_x \otimes \mathbf{e}_x + \frac{1}{\sqrt{\lambda}} (\mathbf{e}_y \otimes \mathbf{e}_y + \mathbf{e}_z \otimes \mathbf{e}_z) \end{aligned}$$

and the *Cauchy-Green* tensor

$$\mathbf{B} = \lambda^2 \mathbf{e}_x \otimes \mathbf{e}_x + \frac{1}{\lambda} (\mathbf{e}_y \otimes \mathbf{e}_y + \mathbf{e}_z \otimes \mathbf{e}_z) .$$

By inserting this into the constitutive model (9.83) we get

$$\begin{aligned} T_{xx} &= -p + \mu\Big(\beta + \tfrac{1}{2}\Big)\lambda^2 + \mu\Big(\beta - \tfrac{1}{2}\Big)\frac{1}{\lambda^2} , \\ T_{yy} &= T_{zz} = -p + \mu\Big(\beta + \tfrac{1}{2}\Big)\frac{1}{\lambda} + \mu\Big(\beta - \tfrac{1}{2}\Big)\lambda . \end{aligned} \tag{9.95}$$

The boundary condition $T_{yy} = T_{zz} = 0$ implies the hydrostatic pressure p; from (9.95) we obtain

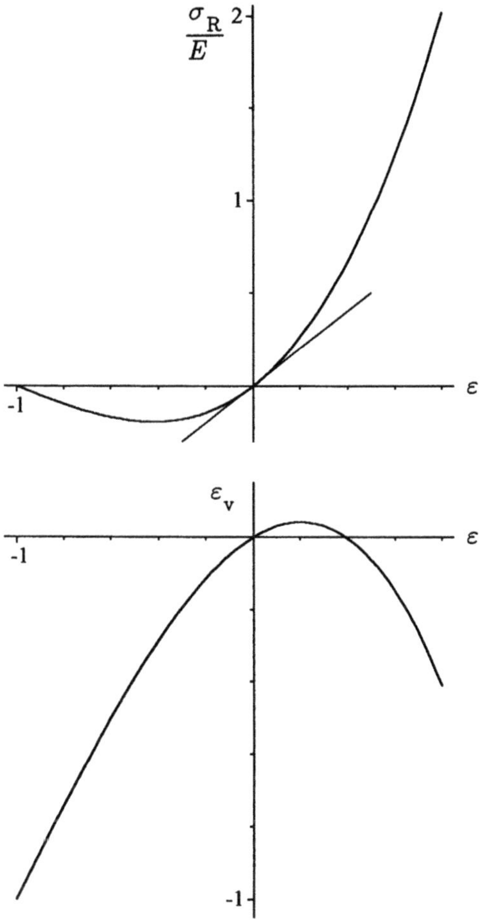

Figure 9. 2: One-dimensional stress-strain curve according to (9.81), (9.97); the stress and the volume strain show that this constitutive equation cannot be suitable for larger strains from a physical point of view.

$$T_{xx} = \mu\left(\beta + \tfrac{1}{2}\right)\left(\lambda^2 - \tfrac{1}{\lambda}\right) + \mu\left(\beta - \tfrac{1}{2}\right)\left(\tfrac{1}{\lambda^2} - \lambda\right) .$$

The *Piola-Kirchhoff* stress σ_R is calculated using $\mathbf{T}_R = \mathbf{T}\mathbf{F}^{T\,-1}$ to give

$\sigma_R = \dfrac{1}{\lambda} T_{xx}$, which leads to (9.93). ■

Figure 9. 1 illustrates the result: for the admissible values of $-\frac{1}{2} \leq \beta \leq \frac{1}{2}$ we obtain a set of monotonically increasing curves.

In an attempt to construct a most simple constitutive model of finite elasticity, which is equipped to allow for large deformations and compressibility, one could be reminded of the physically linearised constitutive equation (9.81), which is a first order approximation of the general constitutive equation (9.55). It turns out, however, that this model is only of limited validity.

Corresponding to uniaxial tension we have the *Green* strain tensor

$$\mathbf{E} = \tfrac{1}{2}(\lambda^2 - 1)\, \boldsymbol{e}_x \otimes \boldsymbol{e}_x + \tfrac{1}{2}(\lambda_Q{}^2 - 1)\, (\boldsymbol{e}_y \otimes \boldsymbol{e}_y + \boldsymbol{e}_z \otimes \boldsymbol{e}_z)\,. \tag{9.96}$$

The lateral stretch λ_Q has to be calculated from the boundary condition $T_{Ryy} = T_{Rzz} = 0$ or from $\widetilde{T}_{yy} = \widetilde{T}_{zz} = 0$. Inserting this into (9.81) we obtain the tension stress

$$\widetilde{T}_{xx} = \mu\Big\{(\lambda^2 - 1) + \frac{\nu}{1 - 2\nu}\big[(\lambda^2 - 1) + 2(\lambda_Q{}^2 - 1)\big]\Big\}\,.$$

The boundary condition

$$\widetilde{T}_{yy} = \mu\Big\{(\lambda_Q{}^2 - 1) + \frac{\nu}{1 - 2\nu}\big[(\lambda^2 - 1) + 2(\lambda_Q{}^2 - 1)\big]\Big\} = 0 \quad \text{leads to}$$

$$(\lambda_Q{}^2 - 1) = -\,\nu(\lambda^2 - 1) \text{ and } \widetilde{T}_{xx} = \mu(1 + \nu)(\lambda^2 - 1)\,.$$

From $\mathbf{T}_R = \mathbf{F}\widetilde{\mathbf{T}}$ and $\sigma_R = \lambda \widetilde{T}_{xx}$ one arrives at the stress-strain characteristic

$$\sigma_R = \mu(1 + \nu)(\lambda^2 - 1)\lambda = 2\mu(1 + \nu)\Big(\varepsilon + \tfrac{1}{2}\varepsilon^2\Big)(1 + \varepsilon)\,. \tag{9.97}$$

This material model is, as evidenced by Fig. 9. 2, physically meaningful for small amounts of strain[11] only: it hardly seems to be credible, largely due to the fact that the absolute value of the compressive stress does not increase monotonically and even approaches zero for $\varepsilon \to -1$. In addition, the predicted volume strain ε_v decreases for elongations larger than 0.2, which appears to be physically inacceptable. Although the constitutive equation (9.81) satisfies the principle of material frame-indifference, it is only suitable as an approximation for small strains and not as a defining equation for an ideal material.

---

[11] up to 20%.

A physically more plausible constitutive model for elastic material behaviour with compressibility can be obtained on the basis of (9.92), if the reaction stress $-p\mathbf{1}$ is replaced by a function of the volume strain, $\det\mathbf{F} = \sqrt{\det\mathbf{B}}$. The strain energy in this case would be

$$2\rho_{\mathrm{R}} e(\mathbf{B}) = f\big(\sqrt{\det\mathbf{B}}\,\big) + \mu\,\frac{\mathrm{tr}\mathbf{B}}{\sqrt[3]{\det\mathbf{B}}}\,. \tag{9.98}$$

This strain energy consists of two terms. The first term results from the volume change and the second term is the energy produced by the change of shape. $f(\cdot)$ is a material function describing compressibility. The background for this formulation is a multiplicative decomposition of the left *Cauchy-Green* tensor $\mathbf{B}$ into a spherical part (change of volume) and a unimodular part (change of shape):

$$\mathbf{B} = \Big\{(\det\mathbf{B})^{\frac{1}{3}}\,\mathbf{1}\Big\}\overline{\mathbf{B}}\,, \quad \overline{\mathbf{B}} = \Big\{(\det\mathbf{B})^{-\frac{1}{3}}\Big\}\mathbf{B}\,. \tag{9.99}$$

Differentiation of (9.98) according to (9.74) leads to a generalised *Neo-Hooke* model.

Theorem 9. 14

The strain energy (9.98) implies the constitutive equation

$$\mathbf{S} = \tfrac{1}{2}\sqrt{\det\mathbf{B}}\;g\big(\sqrt{\det\mathbf{B}}\,\big)\mathbf{1} + \mu(\det\mathbf{B})^{-\frac{1}{3}}\Big[\mathbf{B} - \tfrac{1}{3}(\mathrm{tr}\mathbf{B})\mathbf{1}\Big]\,, \tag{9.100}$$

with $g(x) = f'(x)$. The material constant μ is the classical shear modulus. For uniaxial processes (9.100) yields the stress-strain characteristic

$$\sigma_{\mathrm{R}} = \sigma_{\mathrm{R}}(\lambda) = \tfrac{3}{2}\lambda_{\mathrm{Q}}{}^{2}\,g\big(\lambda\lambda_{\mathrm{Q}}{}^{2}\big)\,; \tag{9.101}$$

here, the transversal stretch $\lambda_{\mathrm{Q}} = \lambda_{\mathrm{Q}}(\lambda)$ has to be calculated as a function of the longitudinal stretch from the equation

$$\lambda_{\mathrm{Q}}{}^{2} = \lambda^{2} - \frac{3}{2\mu}\big(\lambda\lambda_{\mathrm{Q}}{}^{2}\big)^{\frac{5}{3}}\,g\big(\lambda\lambda_{\mathrm{Q}}{}^{2}\big)\,. \tag{9.102}$$

□

Proof

By means of differentiation with respect to $\mathbf{B}$, (9.98) implies

$$2\rho_{\mathrm{R}}\frac{de}{d\mathbf{B}} = f'\big(\sqrt{\det\mathbf{B}}\,\big)\,\tfrac{1}{2}\sqrt{\det\mathbf{B}}\,\mathbf{B}^{-1} + \mu\Big\{(\det\mathbf{B})^{-\frac{1}{3}}\mathbf{1} - \tfrac{1}{3}(\det\mathbf{B})^{-\frac{1}{3}}(\mathrm{tr}\mathbf{B})\mathbf{B}^{-1}\Big\}\,.$$

Given (9.74) this leads to (9.100). For uniaxial tension of a compressible specimen we have the *Cauchy-Green* tensor

$$\mathbf{B} = \lambda^2 e_x \otimes e_x + \lambda_Q{}^2(e_y \otimes e_y + e_z \otimes e_z) ;$$

inserting this into (9.100) yields the axial stress

$$S_{xx} = \tfrac{1}{2}\lambda\lambda_Q{}^2 g(\lambda\lambda_Q{}^2) + \tfrac{2}{3}\mu(\lambda^2 - \lambda_Q{}^2)(\lambda^2\lambda_Q{}^4)^{-\frac{1}{3}} .$$

The boundary condition

$$S_{yy} = \tfrac{1}{2}\lambda\lambda_Q{}^2 g(\lambda\lambda_Q{}^2) - \tfrac{1}{3}\mu(\lambda^2 - \lambda_Q{}^2)(\lambda^2\lambda_Q{}^4)^{-\frac{1}{3}} \stackrel{!}{=} 0$$

is the equation (9.102) to determine the lateral stretch $\lambda_Q(\lambda)$. Given this equation and $\sigma_R = \frac{1}{\lambda}S_{xx}$ we obtain (9.101). ■

The special choice of[12]

$$f(\sqrt{\det\mathbf{B}}) = K\left(\ln\sqrt{\det\mathbf{B}}\right)^2 \tag{9.103}$$

$\Longleftrightarrow \frac{1}{2}\sqrt{\det\mathbf{B}}\; g(\sqrt{\det\mathbf{B}}) = K\ln\sqrt{\det\mathbf{B}}$ corresponds to the strain energy

$$2\rho_R e(\mathbf{B}) = K\left(\ln\sqrt{\det\mathbf{B}}\right)^2 + \mu\,\frac{\mathrm{tr}\mathbf{B}}{\sqrt[3]{\det\mathbf{B}}} \tag{9.104}$$

and the constitutive equation

$$\mathbf{S} = K\ln\sqrt{\det\mathbf{B}}\;\mathbf{1} + \mu(\det\mathbf{B})^{-\frac{1}{3}}\left[\mathbf{B} - \tfrac{1}{3}(\mathrm{tr}\mathbf{B})\mathbf{1}\right] , \tag{9.105}$$

where K corresponds to the classical bulk modulus. The uniaxial stress-strain behaviour is represented for this special case by means of

$$\sigma_R = 3K\,\frac{\ln(\lambda\lambda_Q{}^2)}{\lambda} ; \tag{9.106}$$

λ_Q is the solution of the nonlinear equation

$$\lambda_Q{}^2 = \lambda^2 - \frac{3K}{\mu}(\lambda\lambda_Q{}^2)^{\frac{2}{3}}\ln(\lambda\lambda_Q{}^2) . \tag{9.107}$$

[12] Other proposals are found in HOLZAPFEL [2000], pp. 244.

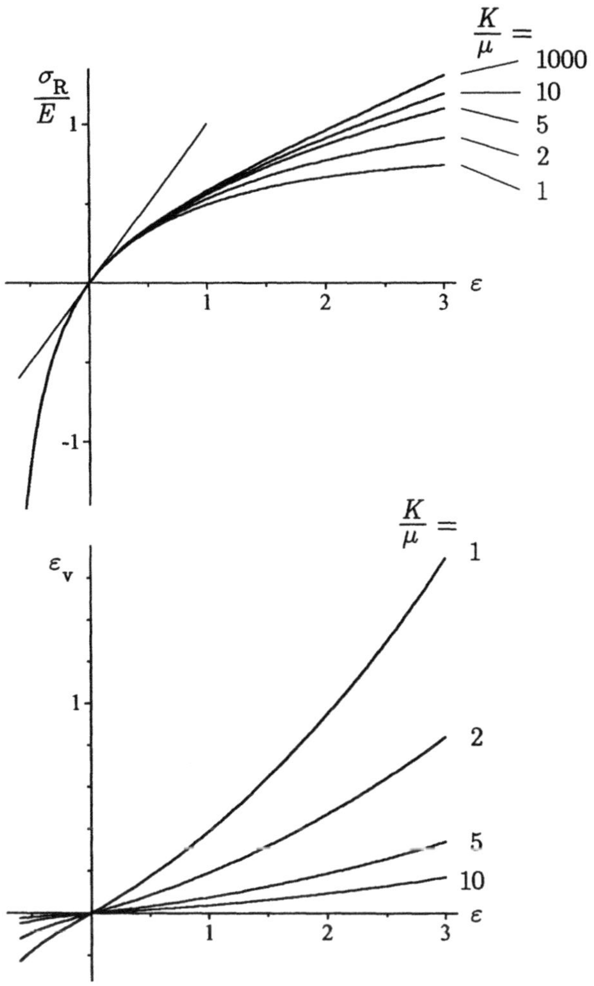

Figure 9. 3: One-dimensional stress-strain curves according to (9.105), (9.106)

The stress-strain curves for different values of K/μ are illustrated in Fig. 9. 3. Comparing these to the curves in Fig. 9. 2 confirms that the approximation (9.81) is not appropriate for describing the behaviour of the volume strain in uniaxial tension. Instead, (9.105) or the more general relation (9.100) supplies a physically meaningful extension of the *Neo-Hooke* model for representing compressibility.[13]

[13] Similar elasticity relations have been applied by SIMO [1988], SIMO & PISTER [1984] and LÜHRS et al. [1997], eq. (3. 6).

There are many arguments in favour of modelling the elasticity of compressible and incompressible isotropic materials on the basis of strain energy functions. On principle, the strain energy is assumed to be a function of the strain invariants (e.g. a power series)[14] or, equivalently, as a function of the principal strains.[15]

9. 3 Anisotropic Hyperelastic Solids

9. 3. 1 Approximation of the General Constitutive Equation

The *Taylor* expansion of the general constitutive equation (9.2) of elasticity reads as follows:

$$\tilde{\mathbf{T}} = \mathbf{g}(\mathbf{E}) = \mathbf{g}(0) + \left\{ \frac{\mathrm{d}\mathbf{g}}{\mathrm{d}\mathbf{E}}\bigg|_{\mathbf{E}=0} \right\} [\mathbf{E}] + \left\{ \frac{\mathrm{d}^2\mathbf{g}}{\mathrm{d}\mathbf{E}^2}\bigg|_{\mathbf{E}=0} \right\} [\mathbf{E}, \mathbf{E}] + \mathrm{O}(\mathbf{E}^3) \ . \quad (9.108)$$

Truncating the *Taylor* series after the linear term leads to the so-called *physical linearisation*. Of course, physical linearisation takes place with respect to a strain measure, which in itself depends nonlinearly on the state of displacement. A physical linearisation with regard to the *Green* strain tensor $\mathbf{E}$ yields an approximation of first order, i.e.

$$\tilde{\mathbf{T}} = \tilde{\mathbf{T}}_0 + \mathbb{C}[\mathbf{E}] + \mathrm{O}(\delta^2) \ . \quad (9.109)$$

$\tilde{\mathbf{T}}_0 = \mathbf{g}(\mathbf{0})$ is the state of stress of the reference configuration and

$$\mathbb{C} = \frac{\mathrm{d}}{\mathrm{d}\mathbf{E}} \mathbf{g}(\mathbf{E})\bigg|_{\mathbf{E}=0} \quad (9.110)$$

the so-called *elasticity tensor*. $\mathbb{C}$ is a tensor of fourth order, i.e. a linear mapping of the space of symmetric tensors into itself:

[14] OGDEN [1984], eq. (4. 3. 52).

[15] OGDEN [1984], eq. (4. 3. 55); TRELOAR [1975]. For further proposals see LURIE [1991], Chap. 5.

$$\begin{aligned} \mathbb{C} : \mathcal{Sym}(\mathbb{V}^3) &\longrightarrow \mathcal{Sym}(\mathbb{V}^3) \\ \mathbf{E} &\longmapsto \mathbb{C}[\mathbf{E}] \quad . \end{aligned}$$

If the reference configuration is stress-free, i.e. $\widetilde{\mathbf{T}}_0 = \mathbf{0}$, the physical linearisation of the general elasticity equation (9.2) is given by

$$\widetilde{\mathbf{T}} = \mathbb{C}[\mathbf{E}] \; . \tag{9.111}$$

Constitutive relations like (9.109) or (9.111) are confined to moderate strains; for large deformations, their physical relevance may be lost. This is suggested by Fig. 9. 2, where an implication of the special case (9.81) is depicted.

The component representation of (9.111) with respect to Cartesian coordinates reads

$$\widetilde{T}_{ij} = C_{ijkl} E_{kl} \; . \tag{9.112}$$

Since the elasticity tensor $\mathbb{C}$ maps symmetric tensors into symmetric tensors, its components have the following symmetry properties:

$$C_{ijkl} = C_{jikl} = C_{ijlk} \, . \tag{9.113}$$

The symmetric tensors $\widetilde{\mathbf{T}}$ and $\mathbf{E}$ each have 6 independent components.

A more compact version of the notation can be achieved by formulating the tensor equation (9.111) and (9.112) as a matrix equation between 6-dimensional vectors:

$$\underline{\widetilde{T}} = \underline{\underline{C}}\, \underline{E} \; . \tag{9.114}$$

To obtain concrete representations of the matrix equation (9.114) it is important to decide how the strain and stress components should be arranged as column vectors. No single convention exists for this purpose. The following possibility is one of several:

$$\begin{bmatrix} \tilde{T}_{11} \\ \tilde{T}_{22} \\ \tilde{T}_{33} \\ \tilde{T}_{12} \\ \tilde{T}_{23} \\ \tilde{T}_{31} \end{bmatrix} = \begin{bmatrix} C_{1111} & C_{1122} & C_{1133} & C_{1112} & C_{1123} & C_{1131} \\ C_{2211} & C_{2222} & C_{2233} & C_{2212} & C_{2223} & C_{2231} \\ C_{3311} & C_{3322} & C_{3333} & C_{3312} & C_{3323} & C_{3331} \\ C_{1211} & C_{1222} & C_{1233} & C_{1212} & C_{1223} & C_{1231} \\ C_{2311} & C_{2322} & C_{2333} & C_{2312} & C_{2323} & C_{2331} \\ C_{3111} & C_{3122} & C_{3133} & C_{3112} & C_{3123} & C_{3131} \end{bmatrix} \begin{bmatrix} E_{11} \\ E_{22} \\ E_{33} \\ 2E_{12} \\ 2E_{23} \\ 2E_{31} \end{bmatrix} . \quad (9.115)$$

The elasticity matrix $\underline{\underline{C}}$ in (9.115) is known as the *Voigt* matrix.[16]
In the case of isotropy, the 36 components C_{ijkl} can be traced back to just 2 elasticity constants (μ, ν). If the material behaviour is *anisotropic*, then this is obviously no longer possible, but even in the case of anisotropy not all components are independent.

Theorem 9. 15

A linear-elastic material (in terms of physical linearisation) is characterised by a maximum of 36 material parameters, and in the case of hyperelasticity by a maximum of 21 parameters. The strain energy in the case of physical linearisation is a quadratic function of the strain tensor:

$$\rho_R e(\mathbf{E}) = \tfrac{1}{2}\mathbf{E} \cdot \mathbb{C}[\mathbf{E}] = \tfrac{1}{2}\tilde{\mathbf{T}} \cdot \mathbf{E} . \quad (9.116)$$

□

Proof

In the matrix equation (9.115) the 6×6 elasticity matrix is equipped with 36 elements. In the case of *hyperelasticity* a further symmetry property exists for the elasticity parameters in addition to (9.113).

The property $\dfrac{\partial^2 e}{\partial E_{ij}\,\partial E_{kl}} = \dfrac{\partial^2 e}{\partial E_{kl}\,\partial E_{ij}}$ leads to

$$\frac{\partial \tilde{T}_{ij}}{\partial E_{kl}} = \frac{\partial \tilde{T}_{kl}}{\partial E_{ij}} \quad (9.117)$$

and finally to

[16] Cf. GURTIN [1972], p. 87. A slightly different notation is used there.

$$C_{ijkl} = C_{klij} \quad \Longleftrightarrow \quad \mathbb{C} = \mathbb{C}^{\mathrm{T}} . \tag{9.118}$$

Thus the *Voigt* elasticity matrix contains a maximum of 21 different components. Differentiation of the bilinear form

$$\rho_{\mathrm{R}} e(\mathbf{E}) = \frac{1}{2} C_{ijkl} E_{ij} E_{kl} ,$$

taking advantage of the additional symmetry property (9.118), leads directly to (9.112),

$$\widetilde{T}_{ij} = C_{ijkl} E_{kl} . \qquad \blacksquare$$

9. 3. 2 General Representation of the Strain Energy Function

In the case of a hyperelastic solid an orthogonal tensor $\mathbf{Q}$ is an element of the symmetry group $\mathscr{g}_{\mathrm{R}}$, if and only if $\mathbf{Q}$ leaves the strain energy invariant in the sense of (9.59):

$$\mathbf{Q} \in \mathscr{g}_{\mathrm{R}} \Longleftrightarrow e(\mathbf{Q}\mathbf{E}\mathbf{Q}^{\mathrm{T}}) = e(\mathbf{E}) \quad \text{for all } \mathbf{E} .$$

If a hyperelastic solid is isotropic, then the strain energy is a function of the invariants, $e(\mathbf{E}) = e(\mathrm{I}_{\mathbf{E}}, \mathrm{II}_{\mathbf{E}}, \mathrm{III}_{\mathbf{E}})$, or $e(\mathbf{E}) = e(\mathrm{J}_1, \mathrm{J}_2, \mathrm{J}_3)$. The basic invariants

$$\begin{aligned}
\mathrm{I}_{\mathbf{E}} &= \mathrm{tr}\mathbf{E} = E_{11} + E_{22} + E_{33} , \\
\mathrm{II}_{\mathbf{E}} &= \frac{1}{2}\left[(\mathrm{tr}\mathbf{E})^2 - \mathrm{tr}(\mathbf{E}^2)\right] = E_{11}E_{22} + E_{22}E_{33} + E_{33}E_{11} \\
&\quad - E_{12}{}^2 - E_{23}{}^2 - E_{31}{}^2 , \\
\mathrm{III}_{\mathbf{E}} &= \det\mathbf{E} = E_{11}E_{22}E_{33} + E_{12}E_{23}E_{31} + E_{13}E_{21}E_{32} \\
&\quad - E_{11}E_{23}E_{32} - E_{12}E_{21}E_{33} - E_{13}E_{22}E_{31} ,
\end{aligned} \tag{9.119}$$

are scalar-valued functions of $\mathbf{E}$, which remain invariant under every orthogonal transformation of $\mathbf{E}$. The three basic invariants form an independent, complete system of functions of that kind. The same applies for the equivalent set of invariants according to (9.57).

According to Definition 7. 6, an anisotropic solid is characterised by a symmetry group that does not contain all orthogonal transformations but a proper subgroup of the orthogonal group.[17] A general representation of anisotropic hyperelasticity demands that for every given symmetry group $\mathscr{g}_R \subset \mathscr{Orth}$ ($\mathscr{g}_R \neq \mathscr{Orth}$) a complete system of independent scalar-valued functions is known, which remain invariant under all transformations $\mathbf{Q} \in \mathscr{g}_R$. The specific strain energy then has to depend on these invariants.

From the mathematical point of view, an infinite number of subgroups of $\mathscr{Orth}$ is possible. Accordingly, any number of representations of anisotropic hyperelasticity could be thought up. For physical reasons, however, there is just a small number of subgroups bearing physical significance. The reason for this is that most anisotropic solids are made up of crystals. The structure of crystals is characterised by a periodic pattern of the atoms forming a 3-dimensional lattice. For purely geometrical reasons the symmetry properties of those lattices are denoted by groups of rotations that mostly contain a finite number of elements.

In all, 32 categories of crystals and 11 different types of crystal lattices are known.[18] Including transverse isotropy and full isotropy this leads to 13 different symmetry groups. Assuming a polynomial dependence of the strain energy $e(\mathbf{E})$, Smith and Rivlin indicated a complete system of independent invariants $J_1(\mathbf{E})$, $J_2(\mathbf{E})$, ... , $J_N(\mathbf{E})$ for each of these 13 symmetry groups, so that the strain energy describes the characteristic of the elasticity in each case.[19] The acquired results are set out below, following Truesdell and Noll. The reader is referred to the literature for proof of the subsequent statements.

To define a particular symmetry group it is sufficient to specify its generating elements. The group is then created by means of an unlimited performance of the group operations. As is evidenced in (9.59), not only a rotation $\mathbf{Q}$ belongs to the symmetry group of a hyperelastic material but also $-\mathbf{Q}$ (rotation plus reflection). Accordingly, the only generating elements that have to be mentioned in each case are the reflection $-\mathbf{1}$ and the proper rotations. The smallest symmetry group is evidently provided by $\{\mathbf{1}, -\mathbf{1}\}$. This subgroup is independent of the reference configuration and expresses the fact that there are no symmetry properties in the material whatsoever.

For the quantitative description of symmetry properties, the orientation of a crystal, which may vary from point to point, is to be represented by means of three orthogonal unit vectors $\boldsymbol{i}, \boldsymbol{j}, \boldsymbol{k}$.

In the following tables, a list of generating elements and the order n (number of elements) is given for every one of the 13 symmetry groups as

---

[17] Cf. Section 7. 3.

[18] See TRUESDELL & NOLL [1965], Sect. 33.

[19] SMITH & RIVLIN [1958]. Further literature is cited in TRUESDELL & NOLL [1965], Sects. 33 and 85. Cf. HORZ [1994], pp. 29. A general survey of methods to describe anisotropic material properties by means of tensor functions is given in BETTEN [1998].

well as the appropriate system of invariants, sufficient for representing the strain energy.[20]

A generating element $\mathbf{Q}_e^{\varphi}$ represents the rotation of an angle φ, whereas the unit vector e, represented in the base system $\{i, j, k\}$, gives the direction of the axis of rotation. In the invariants, the components of the *Green* strain tensor $\mathbf{E}$ also refer to these base vectors. The base $\{i, j, k\}$ corresponds to the axes of the crystal and is the basis for defining the axes of the elastic anisotropy. For the 11 different types of lattices the strain energy is a function of the invariants J_k, or (according to *Smith* and *Rivlin*) a polynomial in these variables:

$$e = e(J_1, J_2, J_3, .., J_N) . \tag{9.120}$$

$$J_k = J_k(\mathbf{E}) \,, \quad \text{invariant under } \mathscr{G}_R \quad (k = 1, \cdots, N) \tag{9.121}$$

1) Triclinic System

Generating elements	$-\mathbf{1}, \mathbf{1}$		$n = 2$
Invariants J_k			
E_{11}	E_{22}	E_{33}	
E_{12}	E_{23}	E_{31}	

(9.122)

2) Monoclinic System

Generating elements	$-\mathbf{1}, \mathbf{Q}_k$		$n = 4$
Invariants J_k			
E_{11}	E_{22}	E_{33}	E_{12}
${E_{13}}^2$	${E_{23}}^2$	$E_{13}E_{23}$	

(9.123)

[20] The information is taken from TRUESDELL & NOLL [1965], p. 312. A more recent discussion on the topic is to be found in XIAO [1996a, 1996b, 1996c, 1997].

3) Rhombic System

Generating elements	$-\mathbf{1}, \mathbf{Q}_i^{\pi}, \mathbf{Q}_j^{\pi}$	n = 8
Invariants J_k		
E_{11}	E_{22}	E_{33}
$E_{23}{}^2$	$E_{13}{}^2$	$E_{12}{}^2$
$E_{12}E_{23}E_{13}$		

(9.124)

4) Tetragonal System

Generating elements	$-\mathbf{1}, \mathbf{Q}_k^{\frac{\pi}{2}}$	n = 8
Invariants J_k		
$E_{11}+E_{22}$	E_{33}	
$E_{13}{}^2+E_{23}{}^2$	$E_{11}E_{22}$	
$E_{12}{}^2$	$E_{12}(E_{11}-E_{22})$	$E_{11}E_{23}{}^2+E_{22}E_{13}{}^2$
$E_{12}E_{23}E_{13}$	$E_{12}(E_{13}{}^2-E_{23}{}^2)$	$E_{13}E_{23}(E_{11}-E_{22})$
$E_{23}E_{13}(E_{23}{}^2-E_{13}{}^2)$	$E_{13}{}^2E_{23}{}^2$	

(9.125)

5) Tetragonal System

Generating elements	$-\mathbf{1}, \mathbf{Q}_k^{\frac{\pi}{2}}, \mathbf{Q}_i^{\pi}$	n = 16
Invariants J_k		
$E_{11}+E_{22}$	E_{33}	
$E_{13}{}^2+E_{23}{}^2$	$E_{11}E_{22}$	$E_{12}{}^2$
$E_{11}E_{23}{}^2+E_{22}E_{13}{}^2$	$E_{12}E_{23}E_{13}$	$E_{13}{}^2E_{23}{}^2$

(9.126)

6) Cubic System $p = \sqrt{\frac{1}{3}}(i + j + k)$

Generating elements	$-1, Q_i^{\pi}, Q_j^{\pi}, Q_p^{2\frac{\pi}{3}}$	n = 24

Invariants J_k	
$E_{11} + E_{22} + E_{33}$	
$E_{22}E_{33} + E_{33}E_{11} + E_{11}E_{22}$	$E_{23}^2 + E_{13}^2 + E_{12}^2$
$E_{11}E_{22}E_{33}$	$E_{23}E_{13}E_{12}$
$E_{33}E_{22}^2 + E_{11}E_{33}^2 + E_{22}E_{11}^2$	$E_{13}^2E_{33} + E_{12}^2E_{11} + E_{23}^2E_{22}$
$E_{22}E_{12}^2 + E_{33}E_{23}^2 + E_{11}E_{13}^2$	$E_{13}^2E_{12}^2 + E_{12}^2E_{23}^2 + E_{23}^2E_{13}^2$
$E_{23}^2E_{22}E_{33} + E_{13}^2E_{33}E_{11} + E_{12}^2E_{11}E_{22}$	
$E_{11}E_{22}E_{13}^2 + E_{22}E_{33}E_{12}^2 + E_{33}E_{11}E_{23}^2$	
$E_{11}E_{13}^2E_{12}^2 + E_{22}E_{12}^2E_{23}^2 + E_{33}E_{23}^2E_{13}^2$	
$E_{23}^2E_{13}^2E_{22} + E_{13}^2E_{12}^2E_{33} + E_{12}^2E_{23}^2E_{11}$	
$E_{12}^2E_{13}^4 + E_{23}^2E_{12}^4 + E_{13}^2E_{23}^4$	

(9.127)

7) Cubic System

Generating elements	$-1, Q_i^{\frac{\pi}{2}}, Q_j^{\frac{\pi}{2}}, Q_k^{\frac{\pi}{2}}$	n = 48

Invariants J_k	
$E_{11} + E_{22} + E_{33}$	
$E_{22}E_{33} + E_{33}E_{11} + E_{11}E_{22}$	$E_{23}^2 + E_{13}^2 + E_{12}^2$
$E_{11}E_{22}E_{33}$	$E_{23}E_{13}E_{12}$
$E_{13}^2E_{33} + E_{12}^2E_{11} + E_{23}^2E_{22} + E_{22}E_{12}^2 + E_{33}E_{23}^2 + E_{11}E_{13}^2$	
$E_{23}^2E_{22}E_{33} + E_{13}^2E_{33}E_{11} + E_{12}^2E_{11}E_{22}$	
$E_{13}^2E_{12}^2 + E_{12}^2E_{23}^2 + E_{23}^2E_{13}^2$	
$E_{11}E_{13}^2E_{12}^2 + E_{22}E_{12}^2E_{23}^2 + E_{33}E_{23}^2E_{13}^2$	

(9.128)

8) Hexagonal System

Generating elements	$-1, \mathbf{Q}_k^{2\frac{\pi}{3}}$	n = 6

Invariants J_k		
$E_{11} + E_{22}$	E_{33}	
$E_{11}E_{22} - E_{12}{}^2$	$E_{13}{}^2 + E_{23}{}^2$	
$(E_{22} - E_{11})E_{23} - 2E_{12}E_{13}$		$(E_{11} - E_{22})E_{13} - 2E_{12}E_{23}$
$E_{23}(E_{23}{}^2 - 3E_{13}{}^2)$		$E_{13}(E_{13}{}^2 - 3E_{23}{}^2)$
$E_{11}[(E_{11} + 3E_{22})^2] - 12E_{12}{}^2]$		
$3E_{12}(E_{11} - E_{22})^2 - 4E_{12}{}^3$		
$E_{22}E_{13}{}^2 + E_{11}E_{23}{}^2 - 2E_{23}E_{13}E_{12}$		
$E_{23}[(E_{11} + E_{22})^2 + 4(E_{12}{}^2 - E_{22}{}^2)] + 8E_{11}E_{12}E_{13}$		
$E_{13}[(E_{11} + E_{22})^2 + 4(E_{12}{}^2 - E_{22}{}^2)] - 8E_{11}E_{12}E_{23}$		
$(E_{11} - E_{22})E_{23}E_{13} + E_{12}(E_{23}{}^2 - E_{13}{}^2)$		

(9.129)

9) Hexagonal System

Generating elements	$-1, \mathbf{Q}_i^{\pi}, \mathbf{Q}_k^{2\frac{\pi}{3}}$	n = 12

Invariants J_k		
$E_{11} + E_{22}$	E_{33}	
$E_{11}E_{22} - E_{12}{}^2$	$E_{13}{}^2 + E_{23}{}^2$	
$(E_{11} - E_{22})E_{23} + 2E_{12}E_{13}$		
$E_{11}[(E_{11} + 3E_{22})^2 - 12E_{12}{}^2]$		$E_{23}(E_{23}{}^2 - 3E_{13}{}^2)$
$E_{11}E_{13}{}^2 + E_{22}E_{23}{}^2 + 2E_{23}E_{13}E_{12}$		
$E_{23}[(E_{11} + E_{22})^2 - 4(E_{22}{}^2 - E_{12}{}^2)] + 8E_{11}E_{12}E_{13}$		

(9.130)

10) Hexagonal System

Generating elements	$-\mathbf{1}\,,\ \mathbf{Q}_k^{\frac{\pi}{3}}$	n = 12

Invariants J_k

$$E_{11} + E_{22} \qquad E_{33} \qquad E_{11}E_{22} - E_{12}^{\;2} \qquad E_{13}^{\;2} + E_{23}^{\;2}$$
$$E_{11}[(E_{11} + 3E_{22})^2 - 12E_{12}^{\;2}] \qquad E_{11}E_{23}^{\;2} + E_{22}E_{13}^{\;2} - 2E_{23}E_{13}E_{12}$$
$$E_{12}(E_{13}^{\;2} - E_{23}^{\;2}) + (E_{22} - E_{11})E_{13}E_{23} \qquad 3E_{12}(E_{11} - E_{22})^2 - 4E_{12}^{\;3}$$
$$E_{13}^{\;2}[(E_{11} + E_{22})^2 - 4(E_{22}^{\;2} - E_{12}^{\;2})] - 2E_{11}[(E_{11} + 3E_{22})(E_{13}^{\;2} + E_{23}^{\;2}) - 4E_{23}E_{13}E_{12}]$$
$$E_{23}E_{13}[(E_{11} + E_{22})^2 - 4(E_{22}^{\;2} - E_{12}^{\;2})] + 4E_{11}E_{12}(E_{23}^{\;2} - E_{13}^{\;2})$$
$$E_{11}(E_{13}^{\;4} + 3E_{23}^{\;4}) + 2E_{22}E_{13}^{\;2}(E_{13}^{\;2} + 3E_{23}^{\;2}) - 8E_{12}E_{23}E_{13}^{\;3}$$
$$E_{12}[(E_{13}^{\;2} + E_{23}^{\;2})^2 + 4E_{23}^{\;2}(E_{13}^{\;2} - E_{23}^{\;2})] - 4E_{13}^{\;3}E_{23}(E_{11} - E_{22})$$
$$E_{13}E_{23}[3(E_{13}^{\;2} - E_{23}^{\;2})^2 - 4E_{13}^{\;2}E_{23}^{\;2}] \qquad E_{13}^{\;2}(E_{13}^{\;2} - 3E_{23}^{\;2})^2$$

(9.131)

11) Hexagonal System

Generating elements	$-\mathbf{1}\,,\ \mathbf{Q}_i^{\pi}\,,\ \mathbf{Q}_k^{\frac{\pi}{3}}$	n = 24

Invariants J_k

$$E_{11} + E_{22} \qquad E_{33}$$
$$E_{11}E_{22} - E_{12}^{\;2} \qquad E_{13}^{\;2} + E_{23}^{\;2}$$
$$E_{11}[(E_{11} + 3E_{22})^2] - 12E_{12}^{\;2}]$$
$$E_{11}E_{23}^{\;2} + E_{22}E_{13}^{\;2} - 2E_{23}E_{13}E_{12}$$
$$E_{13}^{\;2}(E_{13}^{\;2} - 3E_{23}^{\;2})^2$$
$$E_{13}^{\;2}[(E_{11} + E_{22})^2 - 4(E_{22}^{\;2} - E_{12}^{\;2})] - 2E_{11}[(E_{11} + 3E_{22})(E_{13}^{\;2} + E_{23}^{\;2}) - 4E_{23}E_{13}E_{12}]$$
$$E_{11}(E_{13}^{\;4} + 3E_{23}^{\;4}) + 2E_{22}E_{13}^{\;2}(E_{13}^{\;2} + 3E_{23}^{\;2}) - 8E_{12}E_{23}E_{13}^{\;3}$$

(9.132)

The following symmetry group is not motivated by a crystal structure, but by the notion that an axis of rotational symmetry exists in the material. That axis can, for instance, result from a unidirectional fibre reinforcement.

Transverse Isotropy

Generating elements	$-\mathbf{1}\,,\ \mathbf{Q}_{\boldsymbol{k}}^{\varphi}$	$n = \infty$
Invariants J_k		
$E_{11} + E_{22}$	E_{33}	
$E_{11}E_{22} - E_{12}{}^2$	$E_{13}{}^2 + E_{23}{}^2$	
$\det\mathbf{E}$		

(9.133)

It is also possible to obtain this system of invariants by assuming the strain energy to be an isotropic function of two variables, i.e.

$$e = e(\mathbf{E}, \mathbf{M})\,, \tag{9.134}$$

with the property

$$e(\mathbf{E}, \mathbf{M}) = e(\mathbf{QEQ}^{\mathrm{T}}, \mathbf{QMQ}^{\mathrm{T}}) \tag{9.135}$$

for all $\mathbf{Q} \in \mathcal{O}rth$.

$$\mathbf{M} = \boldsymbol{k} \otimes \boldsymbol{k} \tag{9.136}$$

is a tensor denoting the direction $\boldsymbol{k}$ under consideration ($\boldsymbol{k} \cdot \boldsymbol{k} = 1$).
A complete system of invariants for an isotropic function (9.135) is given by

$$\mathrm{tr}\mathbf{E},\ \mathrm{tr}\mathbf{E}^2,\ \mathrm{tr}\mathbf{E}^3,\ \mathrm{tr}\mathbf{M},\ \mathrm{tr}\mathbf{M}^2,\ \mathrm{tr}\mathbf{M}^3,\ \mathrm{tr}\mathbf{ME},\ \mathrm{tr}\mathbf{ME}^2,\ \mathrm{tr}\mathbf{M}^2\mathbf{E}\,. \tag{9.137}$$ [21]

In view of the specific choice $\mathbf{M} = \boldsymbol{k} \otimes \boldsymbol{k}$ some of these quantities cease to apply, leaving just the following set of invariants, which is equivalent to (9.133).

[21] TRUESDELL & NOLL [1965], eq. (11. 22).

Transverse Isotropy

$$
\begin{array}{|l|}
\hline
\mathrm{tr}\mathbf{E} = E_{11} + E_{22} + E_{33} \\
\mathrm{tr}\mathbf{E}^2 = E_{11}{}^2 + E_{22}{}^2 + E_{33}{}^2 + 2E_{12}{}^2 + 2E_{23}{}^2 + 2E_{13}{}^2 \\
\mathrm{tr}(\mathbf{ME}) = E_{33} \\
\mathrm{tr}(\mathbf{ME}^2) = E_{13}{}^2 + E_{32}{}^2 + E_{33}{}^2 \\
\det\mathbf{E} \\
\hline
\end{array}
\tag{9.138}
$$

The alternative representation of transverse isotropy using (9.135) indicates that anisotropic material properties can also be represented by isotropic functions dependent on additional tensor-valued variables. These extra tensors are called *structural variables* because they represent the internal structure of the material. The number and type of the structural variables depend on the type of the symmetry group. In the case of transverse isotropy the structural variable $\mathbf{M}$ is easy to find.[22]

The representation (9.134) has the advantage that it is valid for any choice of preferred direction. It is also easy to generalise: if more than one preferred direction exists within the material, for example, owing to the existence of various layers with fibres of different directions $\boldsymbol{e}_1, \boldsymbol{e}_2, \dots$, the strain energy could be represented as an isotropic function of several variables,[23]

$$e = e(\mathbf{E}, \mathbf{M}_1, \mathbf{M}_2, \cdots), \tag{9.139}$$

$$\mathbf{M}_1 = \boldsymbol{e}_1 \otimes \boldsymbol{e}_1 \, , \mathbf{M}_2 = \boldsymbol{e}_2 \otimes \boldsymbol{e}_2 \, , \cdots . \tag{9.140}$$

For the sake of completeness in this compilation we repeat the case of isotropic behaviour.

[22] It is much harder to specify the structural variables in the case of more general anisotropies; see ZHANG & RYCHLEWSKI [1990]; HORZ [1994].

[23] The representation theorems for these tensor functions can be taken from TRUESDELL & NOLL [1965], Sect. 11.

Isotropy

Generating elements	$-1, \mathbf{Q}_i^{\alpha}, \mathbf{Q}_j^{\beta}, \mathbf{Q}_k^{\gamma}$	n = ∞
Invariants J_k		
$E_{11} + E_{22} + E_{33}$		
$E_{11}E_{22} + E_{22}E_{33} + E_{33}E_{11} - E_{12}{}^2 - E_{23}{}^2 - E_{13}{}^2$		
$E_{11}E_{22}E_{33} + E_{12}E_{23}E_{13} + E_{13}E_{21}E_{32} - E_{11}E_{23}E_{32} - E_{12}E_{21}E_{33} - E_{13}E_{22}E_{13}$		

(9.141)

9. 3. 2 Physical Linearisation

Constitutive equations for anisotropic hyperelastic solids are created along the lines of (9.5) by means of differentiation of the strain energy with respect to the components of the strain tensor:

$$\widetilde{T}_{ij} = \rho_{\mathrm{R}} \frac{\partial e}{\partial E_{ij}} . \tag{9.142}$$

Since the invariant systems for the physically relevant symmetry groups upon which the strain energy depends are known, there is no difficulty in obtaining explicit representations for nonlinear elasticity. These can also be asymptotic approximations of any given order. For a physical linearisation it is necessary to develop an approximation of the strain energy including all second order terms. This leads to a characteristic number of material parameters, depending on the type of symmetry group involved. It is clear that there are at least two and at the most 21 elasticity constants in the physical linearisation. Closer details are set out in the following theorem. The statements of the theorem are the direct outcome of (9.142) in connection with the systems of invariants (9.122) to (9.141).

Theorem 9. 16

Applying physical linearisation, the 13 symmetry groups lead to the following constitutive equations for the strain energy and stresses, listed in eqs. (9.143) to (9.162):[24]

---

[24] Cf. GURTIN [1872], pp. 87.

1) Triclinic System

$$
\begin{aligned}
\rho_R e = {} & \tfrac{1}{2} c_1 E_{11}{}^2 + c_2 E_{11} E_{22} + c_3 E_{11} E_{33} + 2c_4 E_{11} E_{12} + 2c_5 E_{11} E_{23} \\
& + 2c_6 E_{11} E_{13} + \tfrac{1}{2} c_7 E_{22}{}^2 + c_8 E_{22} E_{33} + 2c_9 E_{22} E_{12} + 2c_{10} E_{22} E_{23} \\
& + 2c_{11} E_{22} E_{13} + \tfrac{1}{2} c_{12} E_{33}{}^2 + 2c_{13} E_{33} E_{12} + 2c_{14} E_{33} E_{23} + 2c_{15} E_{22} E_{13} \\
& + c_{16} E_{12}{}^2 + 2c_{17} E_{12} E_{23} + 2c_{18} E_{12} E_{13} + c_{19} E_{23}{}^2 + 2c_{20} E_{23} E_{13} + c_{21} E_{13}{}^2
\end{aligned}
\tag{9.143}
$$

$$
\begin{bmatrix} \tilde{T}_{11} \\ \tilde{T}_{22} \\ \tilde{T}_{33} \\ \tilde{T}_{12} \\ \tilde{T}_{23} \\ \tilde{T}_{31} \end{bmatrix}
=
\begin{bmatrix}
c_1 & c_2 & c_3 & c_4 & c_5 & c_6 \\
c_2 & c_7 & c_8 & c_9 & c_{10} & c_{11} \\
c_3 & c_8 & c_{12} & c_{13} & c_{14} & c_{15} \\
2c_4 & 2c_9 & 2c_{13} & c_{16} & c_{17} & c_{18} \\
2c_5 & 2c_{10} & 2c_{14} & c_{17} & c_{19} & c_{20} \\
2c_6 & 2c_{11} & 2c_{15} & c_{18} & c_{20} & c_{21}
\end{bmatrix}
\begin{bmatrix} E_{11} \\ E_{22} \\ E_{33} \\ 2E_{12} \\ 2E_{23} \\ 2E_{31} \end{bmatrix}
\tag{9.144}
$$

To avoid misunderstandings, it should be noted that the apparent non-symmetry of the elasticity matrix in the notation (9.144) in contrast to (9.115) is due to the fact that these two representations refer to a different mathematical structure. The symmetric matrix in (9.115) is the component representation of the tensor equation (9.111), whereas (9.144) relates two 6-dimensional vectors by means of a 6×6 matrix. In analogy to the components of the elasticity tensor $\mathbb{C}$, the elements of this matrix are also second order derivatives of the strain energy. For example, the [1,6] element is given by

$$
c_6 = \frac{\partial \tilde{T}_{11}}{\partial (2E_{31})} = \rho_R \frac{\partial^2 e}{\partial E_{11} \partial (2E_{31})} ,
$$

whereas the [6,1] element is

$$
2c_6 = \frac{\partial \tilde{T}_{31}}{\partial E_{11}} = \rho_R \frac{\partial^2 e}{\partial E_{31} \partial E_{11}} .
$$

2) Monoclinic System

$$\begin{aligned}\rho_{\mathrm{R}}e = {} & \tfrac{1}{2}c_1 E_{11}{}^2 + c_2 E_{11}E_{22} + c_3 E_{11}E_{33} + 2c_4 E_{11}E_{12} \\ & + \tfrac{1}{2}c_5 E_{22}{}^2 + c_6 E_{22}E_{33} + 2c_7 E_{22}E_{12} + \tfrac{1}{2}c_8 E_{33}{}^2 + 2c_9 E_{33}E_{12} \\ & + c_{10} E_{12}{}^2 + c_{11} E_{23}{}^2 + 2c_{12} E_{13}E_{23} + c_{13} E_{13}{}^2\end{aligned} \tag{9.145}$$

$$\begin{bmatrix}\tilde{T}_{11}\\ \tilde{T}_{22}\\ \tilde{T}_{33}\\ \tilde{T}_{12}\\ \tilde{T}_{23}\\ \tilde{T}_{31}\end{bmatrix} = \begin{bmatrix} c_1 & c_2 & c_3 & c_4 & 0 & 0\\ c_2 & c_5 & c_6 & c_7 & 0 & 0\\ c_3 & c_6 & c_8 & c_9 & 0 & 0\\ 2c_4 & 2c_7 & 2c_9 & c_{10} & 0 & 0\\ 0 & 0 & 0 & 0 & c_{11} & c_{12}\\ 0 & 0 & 0 & 0 & c_{12} & c_{13}\end{bmatrix}\begin{bmatrix}E_{11}\\ E_{22}\\ E_{33}\\ 2E_{12}\\ 2E_{23}\\ 2E_{31}\end{bmatrix} \tag{9.146}$$

3) Rhombic System

$$\begin{aligned}\rho_{\mathrm{R}}e = {} & \tfrac{1}{2}c_1 E_{11}{}^2 + c_2 E_{11}E_{22} + c_3 E_{11}E_{33} + \tfrac{1}{2}c_4 E_{22}{}^2 + c_5 E_{22}E_{33} \\ & + \tfrac{1}{2}c_6 E_{33}{}^2 + c_7 E_{12}{}^2 + c_8 E_{23}{}^2 + c_9 E_{13}{}^2\end{aligned} \tag{9.147}$$

$$\begin{bmatrix}\tilde{T}_{11}\\ \tilde{T}_{22}\\ \tilde{T}_{33}\\ \tilde{T}_{12}\\ \tilde{T}_{23}\\ \tilde{T}_{31}\end{bmatrix} = \begin{bmatrix} c_1 & c_2 & c_3 & 0 & 0 & 0\\ c_2 & c_4 & c_5 & 0 & 0 & 0\\ c_3 & c_5 & c_6 & 0 & 0 & 0\\ 0 & 0 & 0 & c_7 & 0 & 0\\ 0 & 0 & 0 & 0 & c_8 & 0\\ 0 & 0 & 0 & 0 & 0 & c_9\end{bmatrix}\begin{bmatrix}E_{11}\\ E_{22}\\ E_{33}\\ 2E_{12}\\ 2E_{23}\\ 2E_{31}\end{bmatrix} \tag{9.148}$$

4) Tetragonal System

$$\rho_R e = \frac{1}{2}c_1(E_{11}{}^2 + E_{22}{}^2) + c_2 E_{11}E_{22} + c_3(E_{11} + E_{22})E_{33}$$

$$+ 2c_4 E_{12}(E_{11} - E_{22}) + \frac{1}{2}c_5 E_{33}{}^2 + c_6 E_{12}{}^2 + c_7(E_{13}{}^2 + E_{23}{}^2) \tag{9.149}$$

$$\begin{bmatrix} \tilde{T}_{11} \\ \tilde{T}_{22} \\ \tilde{T}_{33} \\ \tilde{T}_{12} \\ \tilde{T}_{23} \\ \tilde{T}_{31} \end{bmatrix} = \begin{bmatrix} c_1 & c_2 & c_3 & c_4 & 0 & 0 \\ c_2 & c_1 & c_3 & -c_4 & 0 & 0 \\ c_3 & c_3 & c_5 & 0 & 0 & 0 \\ 2c_4 & -2c_4 & 0 & c_6 & 0 & 0 \\ 0 & 0 & 0 & 0 & c_7 & 0 \\ 0 & 0 & 0 & 0 & 0 & c_7 \end{bmatrix} \begin{bmatrix} E_{11} \\ E_{22} \\ E_{33} \\ 2E_{12} \\ 2E_{23} \\ 2E_{31} \end{bmatrix} \tag{9.150}$$

5) Tetragonal System

$$\rho_R e = \frac{1}{2}c_1(E_{11}{}^2 + E_{22}{}^2) + c_2 E_{11}E_{22} + c_3(E_{11} + E_{22})E_{33}$$

$$+ \frac{1}{2}c_4 E_{33}{}^2 + c_5 E_{12}{}^2 + c_6(E_{13}{}^2 + E_{23}{}^2) \tag{9.151}$$

$$\begin{bmatrix} \tilde{T}_{11} \\ \tilde{T}_{22} \\ \tilde{T}_{33} \\ \tilde{T}_{12} \\ \tilde{T}_{23} \\ \tilde{T}_{31} \end{bmatrix} = \begin{bmatrix} c_1 & c_2 & c_3 & 0 & 0 & 0 \\ c_2 & c_1 & c_3 & 0 & 0 & 0 \\ c_3 & c_3 & c_4 & 0 & 0 & 0 \\ 0 & 0 & 0 & c_5 & 0 & 0 \\ 0 & 0 & 0 & 0 & c_6 & 0 \\ 0 & 0 & 0 & 0 & 0 & c_6 \end{bmatrix} \begin{bmatrix} E_{11} \\ E_{22} \\ E_{33} \\ 2E_{12} \\ 2E_{23} \\ 2E_{31} \end{bmatrix} \tag{9.152}$$

6) Cubic System = 7) Cubic System

$$\rho_{\mathrm{R}} e = \frac{1}{2} c_1 (E_{11}{}^2 + E_{22}{}^2 + E_{33}{}^2) + c_2 (E_{22} E_{33} + E_{33} E_{11} + E_{11} E_{22}) + c_3 (E_{23}{}^2 + E_{13}{}^2 + E_{12}{}^2) \tag{9.153}$$

$$\begin{bmatrix} \tilde{T}_{11} \\ \tilde{T}_{22} \\ \tilde{T}_{33} \\ \tilde{T}_{12} \\ \tilde{T}_{23} \\ \tilde{T}_{31} \end{bmatrix} = \begin{bmatrix} c_1 & c_2 & c_2 & 0 & 0 & 0 \\ c_2 & c_1 & c_2 & 0 & 0 & 0 \\ c_2 & c_2 & c_1 & 0 & 0 & 0 \\ 0 & 0 & 0 & c_3 & 0 & 0 \\ 0 & 0 & 0 & 0 & c_3 & 0 \\ 0 & 0 & 0 & 0 & 0 & c_3 \end{bmatrix} \begin{bmatrix} E_{11} \\ E_{22} \\ E_{33} \\ 2E_{12} \\ 2E_{23} \\ 2E_{31} \end{bmatrix} \tag{9.154}$$

8) Hexagonal System

$$\rho_{\mathrm{R}} e = \frac{1}{2} c_1 (E_{11} + E_{22})^2 + c_3 (E_{11} + E_{22}) E_{33} + \frac{1}{2} c_6 E_{33}{}^2 + (c_2 - c_1)(E_{11} E_{22} - E_{12}{}^2) + c_7 (E_{13}{}^2 + E_{23}{}^2) - 2c_4 [(E_{22} - E_{11}) E_{23} - 2E_{12} E_{13}] + 2c_5 [(E_{11} - E_{22}) E_{13} - 2E_{12} E_{23}] \tag{9.155}$$

$$\begin{bmatrix} \tilde{T}_{11} \\ \tilde{T}_{22} \\ \tilde{T}_{33} \\ \tilde{T}_{12} \\ \tilde{T}_{23} \\ \tilde{T}_{31} \end{bmatrix} = \begin{bmatrix} c_1 & c_2 & c_3 & 0 & c_4 & c_5 \\ c_2 & c_1 & c_3 & 0 & -c_4 & -c_5 \\ c_3 & c_3 & c_6 & 0 & 0 & 0 \\ 0 & 0 & 0 & c_1 - c_2 & -2c_5 & 2c_4 \\ 2c_4 & -2c_4 & 0 & -2c_5 & c_7 & 0 \\ 2c_5 & -2c_5 & 0 & 2c_4 & 0 & c_7 \end{bmatrix} \begin{bmatrix} E_{11} \\ E_{22} \\ E_{33} \\ 2E_{12} \\ 2E_{23} \\ 2E_{31} \end{bmatrix} \tag{9.156}$$

9) Hexagonal System

$$\rho_{\mathrm{R}} e = \frac{1}{2} c_1 (E_{11} + E_{22})^2 + c_3 (E_{11} + E_{22}) E_{33} + \frac{1}{2} c_5 E_{33}{}^2$$
$$+ (c_2 - c_1)(E_{11} E_{22} - E_{12}{}^2)$$
$$+ c_6 (E_{13}{}^2 + E_{23}{}^2) + 2c_4 [(E_{11} - E_{22}) E_{23} + 2E_{12} E_{13}] \tag{9.157}$$

$$\begin{bmatrix} \tilde{T}_{11} \\ \tilde{T}_{22} \\ \tilde{T}_{33} \\ \tilde{T}_{12} \\ \tilde{T}_{23} \\ \tilde{T}_{31} \end{bmatrix} = \begin{bmatrix} c_1 & c_2 & c_3 & 0 & c_4 & 0 \\ c_2 & c_1 & c_3 & 0 & -c_4 & 0 \\ c_3 & c_3 & c_5 & 0 & 0 & 0 \\ 0 & 0 & 0 & c_1 - c_2 & 0 & 2c_4 \\ 2c_4 & -2c_4 & 0 & 0 & c_6 & 0 \\ 0 & 0 & 0 & 2c_4 & 0 & c_6 \end{bmatrix} \begin{bmatrix} E_{11} \\ E_{22} \\ E_{33} \\ 2E_{12} \\ 2E_{23} \\ 2E_{31} \end{bmatrix} \tag{9.158}$$

10) Hexagonal System = 11) Hexagonal System = 12) Transversal Isotropy

$$\rho_{\mathrm{R}} e = \frac{1}{2} c_1 (E_{11} + E_{22})^2 + (c_2 - c_1)(E_{11} E_{22} - E_{12}{}^2)$$
$$+ c_3 (E_{11} + E_{22}) E_{33} + \frac{1}{2} c_4 E_{33}{}^2 + c_5 (E_{13}{}^2 + E_{23}{}^2) \tag{9.159}$$

$$\begin{bmatrix} \tilde{T}_{11} \\ \tilde{T}_{22} \\ \tilde{T}_{33} \\ \tilde{T}_{12} \\ \tilde{T}_{23} \\ \tilde{T}_{31} \end{bmatrix} = \begin{bmatrix} c_1 & c_2 & c_3 & 0 & 0 & 0 \\ c_2 & c_1 & c_3 & 0 & 0 & 0 \\ c_3 & c_3 & c_4 & 0 & 0 & 0 \\ 0 & 0 & 0 & c_1 - c_2 & 0 & 0 \\ 0 & 0 & 0 & 0 & c_5 & 0 \\ 0 & 0 & 0 & 0 & 0 & c_5 \end{bmatrix} \begin{bmatrix} E_{11} \\ E_{22} \\ E_{33} \\ 2E_{12} \\ 2E_{23} \\ 2E_{31} \end{bmatrix} \tag{9.160}$$

13) Isotropy

$$\rho_R e = \frac{1}{2}(c_1 + c_2)(E_{11} + E_{22} + E_{33})^2$$

$$+ c_1(E_{12}^2 + E_{23}^2 + E_{13}^2 - E_{11}E_{22} - E_{22}E_{33} - E_{33}E_{11}) \quad (9.161)$$

$$\begin{bmatrix} \tilde{T}_{11} \\ \tilde{T}_{22} \\ \tilde{T}_{33} \\ \tilde{T}_{12} \\ \tilde{T}_{23} \\ \tilde{T}_{31} \end{bmatrix} = \begin{bmatrix} c_1 + c_2 & c_2 & c_2 & 0 & 0 & 0 \\ c_2 & c_1 + c_2 & c_2 & 0 & 0 & 0 \\ c_2 & c_2 & c_1 + c_2 & 0 & 0 & 0 \\ 0 & 0 & 0 & c_1 & 0 & 0 \\ 0 & 0 & 0 & 0 & c_1 & 0 \\ 0 & 0 & 0 & 0 & 0 & c_1 \end{bmatrix} \begin{bmatrix} E_{11} \\ E_{22} \\ E_{33} \\ 2E_{12} \\ 2E_{23} \\ 2E_{31} \end{bmatrix} \quad (9.162)$$

■

These constitutive relations demonstrate that certain differences in the symmetry properties do not show up after physical linearisation, since the invariants of third or higher order are excluded. The development of higher order approximations, which are needed for the representation of finite deformations of anisotropic materials, does not create any fundamental difficulties: all it needs is more extensive calculation. Of course, it leads to a considerably greater number of elasticity parameters that are necessary for describing the material behaviour.

According to the above results 10 different classes of elasticity relations exist, if physical linearisation is considered. This follows in a straightforward manner by differentiating the second-order approximation of the strain energy functions depending in each special case on the appropriate set of invariants.

However, in contrast to this well-known result, there are other argumentations found in the literature, which suggest that only 8 different classes of elasticity relations can be distinguished in linear elasticity.[25] Thus, in terms of the present notation it has been shown that the two elasticity matrices

[25] This is pointed out in BÖHLKE [2001] and the literature cited there; see in particular TING [1996]; COWIN [1995] and FORTE & VIANELLO [1996]. See in addition CHADWICK et al. [2001].

(9.150) and (9.152) of the tetragonal systems are equivalent,[26] as well as the matrices (9.156) and (9.158) of the hexagonal systems.[27]

It should be noted that the proof of those redundancies between the 10 elasticity matrices rests on the linear elasticity relation $\tilde{\mathbf{T}} = \mathbb{C}[\mathbf{E}]$ and not on the representation of the strain energy in the context of hyperelasticity.

If we look at the tetragonal systems 4) and 5) and their invariants of first and second order, we see that the recollection (9.125) includes one more invariant in comparison to (9.126), namely the term $E_{12}(E_{11} - E_{22})$. Obviously, this term is independent from the other invariants.

The two hexagonal systems 8) and 9) show a similar situation: comparing (9.129) and (9.130), the invariant $(E_{11} - E_{22})E_{13} - 2E_{12}E_{23}$ is recognised to be the only difference in the first and second-order terms. Again, no dependency in relation to the other invariants of (9.129) seems to be perceptible.

One may conclude, that different arguments (sets of equivalent classes of elasticity tensors on the one side[28] and sets of strain invariants on the other side[29]) lead to different results. A final clarification of this apparent contradiction remains open.

---

[26] Cowin [1995], p. 222.

[27] Cowin [1995], p. 223.

[28] Forte & Vianello [1996].

[29] Eqs. (9.125), (9.126) and (9.129), (9.130).

10 Viscoelasticity

10.1 Representation by Means of Functionals

The theory of *viscoelasticity* models rate-dependent material behaviour without equilibrium hysteresis. A systematic development of special material models of viscoelasticity can be based on the general constitutive equations for simple materials, in which the total stress response is split into an elastic part and a memory part. This is expressed by the reduced form (7.28),

$$\tilde{\mathbf{T}}(t) = \mathbf{g}(\mathbf{E}) + \underset{s \geqq 0}{\mathfrak{G}}\left[\mathbf{E}_{\mathrm{d}}^{\mathrm{t}}(s); \mathbf{E}\right], \tag{10.1}$$

or by the representations (7.80) and (7.70),

$$\mathbf{T}(t) = \mathbf{f}(\mathbf{B}) + \underset{s \geqq 0}{\mathfrak{F}}\left[\mathbf{G}^{\mathrm{t}}(s); \mathbf{B}(t)\right], \tag{10.2}$$

$$\mathbf{T}(t) = -\,p(\rho)\mathbf{1} + \underset{s \geqq 0}{\mathfrak{Q}}\left[\mathbf{G}^{\mathrm{t}}(s); \rho\right], \tag{10.3}$$

which describe the behaviour of isotropic solids and fluids. The mechanical process history is represented here in terms of the relative strain history,

$$s \longmapsto \mathbf{E}_{\mathrm{d}}^{\mathrm{t}}(s) = \mathbf{E}(t-s) - \mathbf{E}(t) = \tfrac{1}{2}\big(\mathbf{C}(t-s) - \mathbf{C}(t)\big), \tag{10.4}$$

$$s \longmapsto \mathbf{G}^{\mathrm{t}}(s) = 2\mathbf{F}^{\mathrm{T}\,-1}(t)\left[\mathbf{E}(t-s) - \mathbf{E}(t)\right]\mathbf{F}^{-1}(t). \tag{10.5}$$

According to (7.32) and (7.83) these tensor functions compare material line elements at the current time t and past time $t - s$:

$$(\mathrm{d}\boldsymbol{x}\cdot\mathrm{d}\boldsymbol{y})(t-s) - (\mathrm{d}\boldsymbol{x}\cdot\mathrm{d}\boldsymbol{y})(t) = 2\mathrm{d}\boldsymbol{X}\cdot\mathbf{E}_{\mathrm{d}}^{\mathrm{t}}(s)\mathrm{d}\boldsymbol{Y} = \mathrm{d}\boldsymbol{x}(t)\cdot\mathbf{G}^{\mathrm{t}}(s)\mathrm{d}\boldsymbol{x}(t)\,. \tag{10.6}$$

The memory parts $\mathfrak{G}$ and $\mathfrak{F}$ of the stress are normalised, so that they vanish for static deformation processes $\big(\mathbf{E}_{\mathrm{d}}^{\mathrm{t}}(s) = \mathbf{G}^{\mathrm{t}}(s) \equiv \mathbf{0}\big)$,

$$\underset{s\,\geqq\,0}{\mathfrak{G}}\big[\mathbf{0}(s);\,\mathbf{C}\big] = \underset{s\,\geqq\,0}{\mathfrak{F}}\big[\mathbf{0}(s);\,\mathbf{B}\big] = \mathbf{0}\,. \tag{10.7}$$

In the theory of viscoelasticity it can be assumed that the tensor functions $\mathbf{g}(\mathbf{E})$ or $\mathbf{f}(\mathbf{B})$ represent the equilibrium relation in full, in other words the memory functionals $\mathfrak{F}$ and $\mathfrak{G}$ do not contribute to the equilibrium stress. The normalisation (10.7) is a necessary condition for this property. In order to establish sufficient conditions, the notion of rate-dependence has to be defined more precisely.

Definition 10. 1

A material model is called *rate-dependent* if the present state of stress depends on the rate at which one and the same deformation process was performed. A rate-dependent material model represents the *fading memory property*, when reflecting the phenomenon of relaxation. ◇

Modelling viscoelasticity demands the representation of rate-dependent functionals with fading memory properties. In the face of the general constitutive equations (10.1), (10.2) and (10.3) all constitutive models of viscoelasticity are characterised by the fact that the response of the memory functional always relaxes to zero, if the state of strain is kept constant following any process.

10. 1. 1 Rate-Dependent Functionals with Fading Memory Properties

A general theory of rate-dependent functionals with fading memory properties was evolved by B. D. Coleman and W. Noll.[1] With regard to its application in modelling viscoelasticity, the physically essential aspects are reflected by the first and mathematically simplest version of this theory.[2]

[1] Coleman & Noll [1960]; see also Coleman & Noll [1961].

[2] Later versions, more sophisticated in their mathematical formulation, have been proposed in Coleman & Mizel [1966]; Mizel & Wang [1966].

The following considerations can therefore content themselves with expounding the simplest formulation of Coleman and Noll's fading memory theory. They refer to tensor-valued functionals of the type

$$\mathbf{G}(s) \longmapsto \mathbf{T} = \underset{s \geqq 0}{\mathfrak{F}} \left[\mathbf{G}(s)\right] , \tag{10.8}$$

which occur in the reduced forms (7.80) and (7.28). Due to (10.7) or (7.31) these functionals vanish for the zero function:

$$\underset{s \geqq 0}{\mathfrak{F}} \left[\mathbf{0}(s)\right] = \mathbf{0} . \tag{10.9}$$

The arguments of $\mathfrak{F}$ are symmetric tensor-valued functions $s \longmapsto \mathbf{G}(s)$ $(0 \leq s < \infty)$ with the property $\mathbf{G}(0) = \mathbf{0}$. In its application to viscoelasticity the argument $\mathbf{G}(\cdot)$ corresponds to the relative strain history $\mathbf{G}^t(\cdot)$ or to the difference history $\mathbf{E}_d^t(\cdot)$. The value $\mathbf{T}$ of the functional is the current *Cauchy* stress $\mathbf{T}(t)$ or the 2nd *Piola-Kirchhoff* stress $\tilde{\mathbf{T}}(t)$. In order to define the fading memory property, the domain of the functional $\mathfrak{F}$ has to be specified.

Definition 10. 2

The domain of the functional $\mathfrak{F}$ is the *Hilbert space* $\mathscr{H}_h$ of the square-integrable, tensor-valued functions $s \longmapsto \mathbf{G}(s)$,

$$\mathscr{H}_h = \left\{\mathbf{G}(s) \middle| \ \|\mathbf{G}(s)\|_h < \infty\right\} , \tag{10.10}$$

with the inner product

$$\left\langle \mathbf{G}(s)_1, \mathbf{G}_2(s)\right\rangle_h = \int_0^\infty \mathbf{G}_1(s) \cdot \mathbf{G}_2(s) h^2(s) \mathrm{d}s \tag{10.11}$$

and the norm

$$\|\mathbf{G}(s)\|_h = \left\{ \int_0^\infty |\mathbf{G}(s)|^2 h^2(s) \mathrm{d}s \right\}^{1/2} . \tag{10.12}$$

The scalar-valued function $s \longmapsto h(s)$ with $h(s) > 0$ and $h(0) = 1$ is called an *influence function* of the order $\mathrm{r} > \frac{1}{2}$; it is a positive real-valued function with the following characteristics:

1) $h(s)$ is continuous in $[0 \leq s < \infty)$.

2) $h(s)$ decays to zero in the sense that

$$\lim_{s \to \infty} s^{\mathrm{r}} h(s) = 0 \tag{10.13}$$

and $s^{\mathrm{r}}h(s)$ is a monotonically decreasing function for $s \geq s_1$. ◊

The definition implies that $s^{\mathrm{r}}h(s)$ is bounded for all $s \in [0, \infty)$, i.e.

$$s^{\mathrm{r}}h(s) \leq \mathrm{N} . \tag{10.14}$$

The demand $\mathrm{r} > \frac{1}{2}$ on the order of the influence function guarantees that the function space $\mathscr{H}_h$ contains at least all bounded tensor functions. The norm $\|\mathrm{G}(s)\|_h$ allocates to every relative deformation history a positive number that represents the recollection of all past deformation events. Throughout, the fading memory property is taken into consideration by means of the influence function $h(s)$, which decreases monotonically for large s and accordingly evaluates the more distant past with smaller weight factors. An influence function of the order r would be

$$h(s) = \frac{1}{(1 + s)^{\beta}} \tag{10.15}$$

for $\beta > \mathrm{r}$. In contrast to this example, a decreasing exponential function

$$h(s) = \mathrm{e}^{-\beta s} \tag{10.16}$$

defines an influence function of arbitrary order. For modelling material properties the detailed shape of the influence function $h(\cdot)$ has no relevance. It is only the order r which influences the domain of the constitutive functional $\mathfrak{F}$ and its mathematical properties.

The physical significance of the memory norm (10.12) is emphasised by additional conclusions that can be drawn from the definition. These deductions refer to static continuations as well as retarded deformation processes.

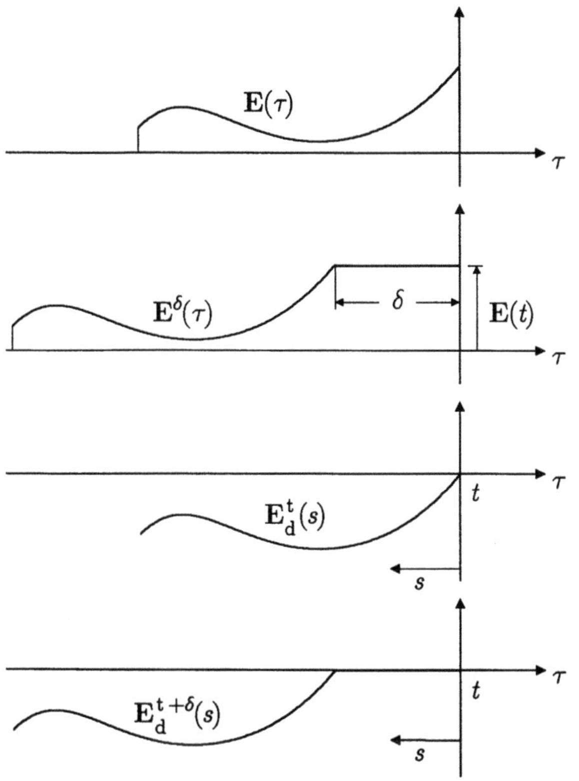

Figure 10. 1: Static continuation

Definition 10. 3

Given a deformation process $s \longmapsto \mathbf{G}(s) \in \mathscr{H}_h$ and a time interval $\delta > 0$, then the modified process

$$\mathbf{G}^{\delta}(s) = \begin{cases} \mathbf{0} & \text{for } 0 \leq s < \delta \\ \mathbf{G}(s - \delta) & \text{for } \delta \leq s < \infty \end{cases} \tag{10.17}$$

is called the *static continuation of* $\mathbf{G}(\cdot)$ by the amount δ.[3] ◇

---

[3] Coleman [1964], p. 23.

The concept of the static continuation is illustrated in Fig. 10. 1 using the example of the difference history $\mathbf{E}_{\mathrm{d}}^{\mathrm{t}}(s) = \mathbf{E}(t-s) - \mathbf{E}(t)$. A given strain process

$\tau \longmapsto \mathbf{E}(\tau)$ $(\tau \leq t)$ and $\delta > 0$ leads to the static continuation

$$\mathbf{E}^{\delta}(\tau) = \begin{cases} \mathbf{E}(t) & \text{for} \quad t-\delta \leq \tau \leq t \\ \mathbf{E}(\tau+\delta) & \text{for} \quad -\infty \leq \tau \leq t-\delta \end{cases}, \tag{10.18}$$

written as $(\tau = t - s)$

$$\mathbf{G}^{\delta}(s) = \mathbf{E}_{\mathrm{d}}^{\mathrm{t}+\delta}(s) = \begin{cases} \mathbf{0} & \text{for } 0 \leq s < \delta \\ \mathbf{E}_{\mathrm{d}}^{\mathrm{t}}(s-\delta) & \text{for } \delta \leq s < \infty \end{cases}, \tag{10.19}$$

with $s = t - \tau$. As a typical property of the memory norm (10.12), the norm of the static continuation vanishes in passing to the limit $\delta \rightarrow \infty$.

Theorem 10. 1

a) $\mathbf{G}(\cdot) \in \mathscr{H}_h$ implies $\mathbf{G}^{\delta}(\cdot) \in \mathscr{H}_h$.

b) The norm of a static continuation vanishes for $\delta \rightarrow \infty$:

$$\lim_{\delta \rightarrow \infty} \left\| \mathbf{G}^{\delta}(s) \right\|_h = 0 \,. \tag{10.20}$$

□

Proof

Statement a) arises from

$$\left\| \mathbf{G}^{\delta}(s) \right\|_h^2 = \int_{\delta}^{\infty} |\mathbf{G}(s-\delta)|^2 h^2(s) \mathrm{d}s = \int_{0}^{\infty} |\mathbf{G}(s)|^2 h^2(s+\delta) \mathrm{d}s < \infty \,.$$

For bounded process histories $\left(|\mathbf{G}(s)| \leq M\right)$ we have

$\left\| \mathbf{G}^{\delta}(s) \right\|_h^2 \leq M \int_{\delta}^{\infty} h^2(s) \mathrm{d}s$, which gives rise to statement b). ■

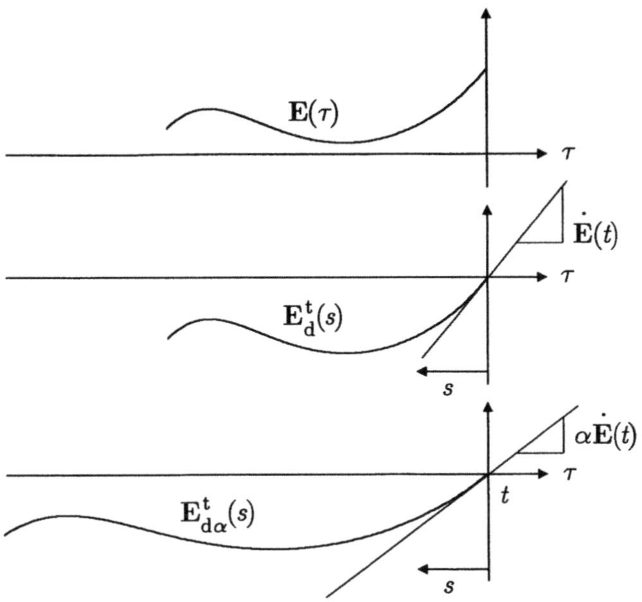

Figure 10. 2: Retarded strain history

During the static continuation a deformation process is shifted further and further into the past. The monotonically decreasing influence function $h(s)$ ensures that the corresponding norm keeps on diminishing until it vanishes in the limit $\delta \to \infty$. A similar effect occurs in connection with the slowing down of a process history.

Definition 10. 4
Given any deformation history $s \longmapsto \mathbf{G}(s) \in \mathscr{H}_h$ and a number α $(0 < \alpha < 1)$, then the new function

$$\mathbf{G}_\alpha(s) = \mathbf{G}(\alpha s) \tag{10.21}$$

is called the *retarded deformation history* with retardation factor α.[4] ◊

The physical meaning of this definition can, for example, be gleaned from the formulation

---

[4] COLEMAN & NOLL [1960], p. 358.

$$\mathbf{E}^{\mathrm{t}}_{\mathrm{d}\alpha}(s) = \mathbf{E}(t - \alpha s) - \mathbf{E}(t) \tag{10.22}$$

and is illustrated in Fig. 10. 2. Differentiation with respect to s and equating s to zero leads to the current strain rate multiplied by the retardation factor $\alpha < 1$:

$$\frac{\mathrm{d}}{\mathrm{d}s}\,\mathbf{E}^{\mathrm{t}}_{\mathrm{d}\alpha}(s)\bigg|_{s=0} = -\,\alpha\dot{\mathbf{E}}(t)\,. \tag{10.23}$$

The following theorem was established by Coleman and Noll.[5] They have shown that, in the case of a retarded process, the *Taylor* expansion of the relative strain history represents an asymptotic approximation, if the error term is evaluated by means of the memory norm (10.12).

Theorem 10. 2

If $h(s)$ is an influence function of a given order, $\mathrm{r} > \mathrm{n} + \frac{1}{2}$ ($\mathrm{n} \in \mathbb{N}$), $s \longmapsto \mathbf{G}(s) \in \mathscr{H}_h$ a process history, $\mathbf{G}_\alpha(s) = \mathbf{G}(\alpha s)$ its retardation with factor α ($0 < \alpha < 1$), then we have the *Taylor* expansion

$$\mathbf{G}_\alpha(s) = \sum_{\mathrm{k}=1}^{\mathrm{n}} \frac{s^{\mathrm{k}}}{\mathrm{k}!}\,\mathbf{G}^{(\mathrm{k})}_\alpha(0) + \mathrm{o}(\alpha^{\mathrm{n}}) \tag{10.24}$$

or

$$\mathbf{G}(\alpha s) = \sum_{\mathrm{k}=1}^{\mathrm{n}} \frac{(\alpha s)^{\mathrm{k}}}{\mathrm{k}!}\,\mathbf{G}^{(\mathrm{k})}(0) + \mathrm{o}(\alpha^{\mathrm{n}})\,. \tag{10.25}$$

In this expansion the order of magnitude of the error term, $\mathrm{o}(\alpha^{\mathrm{n}})$, refers to the norm $\|\cdot\|_h$, defined by (10.12). □

Proof

The validity of the *Taylor* approximation is based on the properties of the influence function, of which one first has to convince oneself.

If we are given an influence function $h(s)$ of order r and two real numbers σ and α ($\sigma > 0$, $0 < \alpha < 1$), then a positive number M_σ exists, which is independent of α and has the property

$$\frac{h(\frac{s}{\alpha})}{\alpha^{\mathrm{r}} h(s)} \leq M_\sigma\,, \tag{10.26}$$

valid for all $s \geq \sigma$.[6]

---

[5] COLEMAN & NOLL [1960], Theorem 1.

[6] COLEMAN & NOLL [1960], eq. (2.1).

In fact, the property that $s^r h(s)$ decreases monotonically for $s \geq s_1$, results in the inequality

$$\left(\frac{s}{\alpha}\right)^r h\left(\frac{s}{\alpha}\right) \leq s^r h(s) \quad \text{or} \quad \frac{h\left(\frac{s}{\alpha}\right)}{\alpha^r h(s)} \leq 1 ,$$

valid for $s \geq s_1$. Thus relation (10.26) is true, provided that $\sigma \geq s_1$. A bound M_σ also exists for $\sigma < s_1$, since the influence function $h(s)$ is continuous.
The magnitude of the error term can now be investigated. We have to show

$$\lim_{\alpha \to 0} \frac{1}{\alpha^n} \left\|\mathbf{R}(\alpha s)\right\|_h = 0 , \tag{10.27}$$

i.e.

$$\lim_{\alpha \to 0} \frac{1}{\alpha^{2n}} \int_0^\infty |\mathbf{R}(\alpha s)|^2 h^2(s) ds = 0 ,$$

or

$$\lim_{\alpha \to 0} \frac{1}{\alpha^{2n+1}} \int_0^\infty |\mathbf{R}(s)|^2 h^2\left(\frac{s}{\alpha}\right) ds = 0 . \tag{10.28}$$

In this equation

$$\mathbf{R}(s) = \mathbf{G}(s) - \sum_{k=1}^{n} \frac{s^k}{k!} \mathbf{G}_\alpha^{(k)}(0) \tag{10.29}$$

is the deviation of the *Taylor* expansion from the given process $\mathbf{G}(s)$. Proving the relation (10.28) can be achieved in 4 steps.

$$\text{(i) } \mathbf{G}_\alpha(s) \in \mathscr{H}_h \tag{10.30}$$

In addition to $\mathbf{G}(s)$ the retarded process history $\mathbf{G}_\alpha(s)$ is also contained in the *Hilbert* space $\mathscr{H}_h$:

$$\left\|\mathbf{G}_\alpha(s)\right\|_h^2 = \int_\delta^\infty |\mathbf{G}(\alpha s)|^2 h^2(s) ds = \frac{1}{\alpha} \int^\infty |\mathbf{G}(\tau)|^2 h^2(\tau) \frac{h^2\left(\frac{\tau}{\alpha}\right)}{h^2(\tau)} ds ,$$

$$\left\|\mathbf{G}_\alpha(s)\right\|_h^2 \leq \frac{1}{\alpha} \left\|\mathbf{G}_\alpha(s)\right\|_h^2 \sup \frac{h^2\left(\frac{\tau}{\alpha}\right)}{h^2(\tau)} < \infty .$$

(ii) $s^k \mathbf{1} \in \mathscr{H}_h$ for $k = 1, \cdots, n$ (10.31)

This statement is the outcome of the property $\int_0^1 (s^r h(s))^2 ds < \infty$ (continuity of $h(s)$) and (10.14), $s^r h(s) \leq N$:

$$\int_1^\infty (s^k h(s))^2 ds = \int_1^\infty (s^r h(s))^2 s^{2(k-r)} ds \leq N^2 \int_1^\infty s^{2(k-r)} ds < \infty$$

for $2(k - r) + 1 < 0$, i.e. $r > k + \frac{1}{2}$.

(iii) $\mathbf{R}(\alpha s) \in \mathscr{H}_h$ and $\mathbf{R}(s) \in \mathscr{H}_h$ (10.32)

The error term (10.29) of the *Taylor* expansion of $\mathbf{G}(s)$ has the order of magnitude

$$\mathbf{R}(s) = o(s^n), \tag{10.33}$$

i.e. for a given $\varepsilon > 0$ a real number $\sigma(\varepsilon)$ exists, so that for $0 \leq s \leq \sigma(\varepsilon)$

$$|\mathbf{R}(s)| < \varepsilon s^n \tag{10.34}$$

can be written. To prove the claim (10.28) it is expedient to decompose the integral into two parts. For the first part we obtain the estimation

$$\frac{1}{\alpha^{2n+1}} \int_0^{\sigma(\varepsilon)} |\mathbf{R}(s)|^2 h^2\left(\frac{s}{\alpha}\right) ds$$

$$\leq \underbrace{\frac{\varepsilon^2}{\alpha^{2n+1}} \int_0^{\sigma(\varepsilon)} \left[s^n h\left(\frac{s}{\alpha}\right)\right]^2 ds}_{\varepsilon^2 \int_0^{\sigma(\varepsilon)} \left[\left(\frac{s}{\alpha}\right)^n h\left(\frac{s}{\alpha}\right)\right]^2 d\left(\frac{s}{\alpha}\right)} \leq \varepsilon^2 \int_0^\infty (\tau^n h(\tau))^2 d\tau \leq \varepsilon^2 L_n{}^2 < \infty ,$$

in which the upper bound L_n is independent of ε and α. For the second part, taking advantage of the property (10.26), we get

$$\frac{1}{\alpha^{2n+1}} \int\limits_{\sigma(\varepsilon)}^{\infty} |\mathbf{R}(s)|^2 h^2(\tfrac{s}{\alpha}) ds = \frac{1}{\alpha^{2(n-r)+1}} \int\limits_{\sigma(\varepsilon)}^{\infty} |\mathbf{R}(s)|^2 h^2(s) \left\{ \frac{h(\frac{s}{\alpha})}{\alpha^r h(s)} \right\}^2 ds$$

$$\leq \alpha^{2(r-n)-1} M^2_{\sigma(\varepsilon)} \left\|\mathbf{R}(s)\right\|_h .$$

(iv) In all, the integral in (10.28) meets the inequality

$$\frac{1}{\alpha^{2n+1}} \int\limits_{0}^{\infty} |\mathbf{R}(s)|^2 h^2(\tfrac{s}{\alpha}) ds \leq \varepsilon^2 L_n^{\ 2} + \alpha^w M^2_{\sigma(\varepsilon)} \left\|\mathbf{R}(s)\right\|_h , \tag{10.35}$$

with $w = 2(r - n) - 1 > 0$.[7] The right-hand side of this inequality can be kept as small as is desired, by choosing a sufficiently small ε and then stipulating an appropriate value of the retardation factor α. ■

This so-called *retardation theorem* states that a retarded process history can be approximated by its *Taylor* expansion, provided an influence function of a sufficiently high order exists. The approximation is *asymptotic*, that means the error is arbitrarily small for a sufficiently strong retardation (small α). In order to comprehend the physical meaning of this theorem, let it be mentioned that the *Taylor* expansion normally leads to errors of any size in an infinite interval: the error term in (10.24) does not, however, have the order of magnitude $\mathbf{o}(s^n)$ but $\mathbf{o}(\alpha^n)$. In this connection not only the memory norm but also the slowing down of the process is crucial. The error $\mathbf{o}(s^n)$ in (10.33) is shifted far enough back into the past to assume the magnitude of $\mathbf{o}(\alpha^n)$.

The retardation theorem can be applied to approximate the temporal course of strain processes.

For the difference history $\mathbf{E}^t_d(s) = \mathbf{E}(t - s) - \mathbf{E}(t)$ the derivatives of order k at $s = 0$ coincide with the material time derivatives of the current strain,

$$\mathbf{E}^{t(k)}_d(0) = (-1)^k \mathbf{E}^{(k)}(t) . \tag{10.36}$$

In this case the retardation theorem reads

$$\mathbf{E}^t_d(\alpha s) = \sum_{k=1}^{n} (-1)^k \frac{(\alpha s)^k}{k!} \mathbf{E}^{(k)}(t) + \mathbf{o}(\alpha^n) . \tag{10.37}$$

---

[7] COLEMAN & NOLL [1960], eq. (3.20).

During the *Taylor* expansion of the relative strain history

$$\mathbf{G}^{t}(s) = 2\mathbf{F}^{T-1}(t)\mathbf{E}_{d}^{t}(s)\mathbf{F}^{-1}(t) = \mathbf{F}^{T-1}(t)\mathbf{C}(t-s)\mathbf{F}^{-1}(t) - \mathbf{1} \tag{10.38}$$

the spatial strain rate tensors emerge as

$$\mathbf{A}_{k} = \mathbf{F}^{T-1}(t)\mathbf{C}^{(k)}(t)\mathbf{F}^{-1}(t) \qquad (k = 1, \cdots, n)\,, \tag{10.39}$$

which are known as *Rivlin-Ericksen* tensors.[8] Accordingly, the *Taylor* approximation of $\mathbf{G}^{t}(s)$ reads as

$$\mathbf{G}^{t}(s) = \sum_{k=1}^{n} (-1)^{k} \frac{(\alpha s)^{k}}{k!}\mathbf{A}_{k} + o(\alpha^{n})\,. \tag{10.40}$$

The *Rivlin-Ericksen* tensors are motivated by the retardation theorem. They play an important role when it comes to defining strain rates of a higher order, operating on the current configuration. At this point we shall just mention the most important properties, those which can be read straight off definition (10.39).

Theorem 10. 3

1) The *Rivlin-Ericksen* tensors can be attributed to spatial gradients of the velocity and acceleration fields and to higher-order acceleration gradients:

$$\mathbf{A}_{1} = \mathbf{L} + \mathbf{L}^{T} = 2\mathbf{D}\,, \qquad \mathbf{L} = \dot{\mathbf{F}}\mathbf{F}^{-1} = \mathrm{grad}\boldsymbol{v}(\boldsymbol{x}, t)\,, \tag{10.41}$$

$$\mathbf{A}_{2} = \mathbf{L}_{2} + \mathbf{L}_{2}^{T} + 2\mathbf{L}^{T}\mathbf{L}\,, \qquad \mathbf{L}_{2} = \ddot{\mathbf{F}}\mathbf{F}^{-1} = \mathrm{grad}\dot{\boldsymbol{v}}(\boldsymbol{x}, t)\,, \tag{10.42}$$

$$\vdots$$

$$\mathbf{A}_{n} = \mathbf{L}_{n} + \mathbf{L}_{n}^{T} + \sum_{i=1}^{n-1} \binom{n}{i}\mathbf{L}_{i}^{T}\mathbf{L}_{n-i}\,, \quad \mathbf{L}_{n} = \overset{(n)}{\mathbf{F}}\mathbf{F}^{-1} = \mathrm{grad}\overset{(n-1)}{\boldsymbol{v}}(\boldsymbol{x}, t)\,. \tag{10.43}$$

2) The *Rivlin-Ericksen* tensors are objective: under a change of frame we have

$$\mathbf{A}_{k}^{*} = \mathbf{Q}\mathbf{A}_{k}\mathbf{Q}^{T} \qquad (k = 1, \cdots, n)\,, \tag{10.44}$$

with $\mathbf{Q}\mathbf{Q}^{T} = \mathbf{1}$.

---

[8] See TRUESDELL & NOLL [1965], Sects. 24, 25.

3) The *Rivlin-Ericksen* tensors represent strain rates and higher-order time derivatives of strains:

$$d\boldsymbol{x}\cdot \mathbf{A}_k d\boldsymbol{y} = d\boldsymbol{X}\cdot \mathbf{C}^{(k)}(t)d\boldsymbol{Y} = \frac{d^k}{dt^k}\left[d\boldsymbol{x}(t)\cdot d\boldsymbol{y}(t)\right] \qquad (k = 1, \cdots, n). \tag{10.45}$$

□

Proof

The formulae (10.41), (10.42) and (10.43) are obtained from the definition (10.39):

$$\mathbf{A}_1 = \mathbf{F}^{T-1}(\mathbf{F}^T\mathbf{F})^{\cdot}\mathbf{F}^{-1} = \dot{\mathbf{F}}\mathbf{F}^{-1} + \mathbf{F}^{T-1}\dot{\mathbf{F}}^T ,$$

$$\mathbf{A}_2 = \mathbf{F}^{T-1}(\mathbf{F}^T\mathbf{F})^{\cdot\cdot}\mathbf{F}^{-1} = \ddot{\mathbf{F}}\mathbf{F}^{-1} + \mathbf{F}^{T-1}\ddot{\mathbf{F}}^T + 2\mathbf{F}^{T-1}\dot{\mathbf{F}}^T\,\dot{\mathbf{F}}\,\mathbf{F}^{-1} , \text{etc.}$$

The transformation behaviour (10.44) is the outcome of $\mathbf{F}^* = \mathbf{QF}$ and $\mathbf{C}^* = \mathbf{C}$. We finally obtain (10.45) by differenting $d\boldsymbol{x}\cdot d\boldsymbol{y} = d\boldsymbol{X}\cdot \mathbf{C}d\boldsymbol{Y}$ and inserting $d\boldsymbol{X} = \mathbf{F}^{-1}d\boldsymbol{x}$ and $d\boldsymbol{Y} = \mathbf{F}^{-1}d\boldsymbol{y}$. ■

The observations made so far allow an all-round definition of the material property of *viscoelasticity.*

Definition 10. 5

A *viscoelastic material* is generally represented by means of the reduced forms (10.1), (10.2), (10.3):

$$\tilde{\mathbf{T}}(t) = \mathbf{g}(\mathbf{E}) + \underset{s \geqq 0}{\mathfrak{G}}\left[\mathbf{E}_d^t(s); \mathbf{E}(t)\right] \qquad \text{(solid)},$$

$$\mathbf{T}(t) = \mathbf{f}(\mathbf{B}) + \underset{s \geqq 0}{\mathfrak{F}}\left[\mathbf{G}^t(s); \mathbf{B}(t)\right] \qquad \text{(isotropic solid)},$$

$$\mathbf{T}(t) = -\,p(\rho)\mathbf{1} + \underset{s \geqq 0}{\mathfrak{Q}}\left[\mathbf{G}^t(s); \rho(t)\right] \qquad \text{(fluid)}.$$

The *elastic* parts $\mathbf{g}(\mathbf{E})$, $\mathbf{f}(\mathbf{B})$ and $-p(\rho)\mathbf{1}$ serve to describe the *equilibrium stress* (*equilibrium relation*); the *memory* parts

$$\begin{aligned} \mathfrak{G} : \mathscr{H}_h \times \mathscr{S}ym(\mathbb{V}^3) &\longrightarrow \mathscr{S}ym(\mathbb{V}^3) \\ (\mathbf{E}_d^t(s), \mathbf{E}) &\longmapsto \underset{s \geqq 0}{\mathfrak{G}}\left[\mathbf{E}_d^t(s); \mathbf{E}\right] , \end{aligned}$$

$$\begin{aligned} \mathfrak{F} : \mathscr{H}_h \times \mathscr{S}ym^+(\mathbb{V}^3) &\longrightarrow \mathscr{S}ym(\mathbb{V}^3) \\ (\mathbf{G}^t(s), \mathbf{B}) &\longmapsto \underset{s \geqq 0}{\mathfrak{F}}\left[\mathbf{G}^t(s); \mathbf{B}(t)\right] , \end{aligned}$$

$$\begin{aligned} \mathfrak{Q}: \mathscr{H}_h \times \mathbb{R}^+ &\longrightarrow \mathscr{S}ym(\mathbb{V}^3) \\ (\mathbf{G}^t(s), \rho) &\longmapsto \underset{s \geqq 0}{\mathfrak{Q}}\left[\mathbf{G}^t(s); \rho\right] \end{aligned}$$

are rate-dependent functionals, vanishing for static processes. These functionals have the following characteristic:
An influence function $h(s)$ of the order $\mathrm{r} > \frac{1}{2}$ exists, so that the functionals $\mathfrak{G}$, $\mathfrak{F}$ and $\mathfrak{Q}$ possess continuity or differentiability properties in relation to the fading memory norm (10.12). ◊

This general definition leaves two degrees of freedom open, namely the order r of the influence function and the degree of continuity (differentiability) of the functional $\mathfrak{G}$, $\mathfrak{F}$ or $\mathfrak{Q}$. Special assumptions open up a range of possibilities for approximating the general reduced form of a constitutive equation by means of concrete representations, with a view to systematically deriving specific material models of viscoelasticity.

10. 1. 2 Continuity Properties and Approximations

A constitutive model of viscoelasticity can be represented by means of a memory functional possessing a property of continuity with respect to the norm defined by (10.12). Various assumptions regarding the degree of continuity are possible. Different physical meanings can then be derived from the mathematical definitions, as the following examples will demonstrate. As a representative example, let us consider the functional $\mathfrak{G}$, which permits any properties of material symmetry. The following argumentations also apply to the functional $\mathfrak{F}$, valid for isotropic solids, or to the functional $\mathfrak{Q}$, which provides a general framework for the viscoelastic behaviour of fluids.[9]

Theorem 10. 4
If the memory functional $\mathfrak{G}$ is continuous with respect to the fading memory norm (10.12), then $\mathfrak{G}$ has the *relaxation property*. If $\mathbf{E}_{\mathrm{d}}^{t+\delta}(s)$ is the static continuation of a given deformation history $\mathbf{E}_{\mathrm{d}}^{t}(s)$ and

$$\tilde{\mathbf{T}}^{\delta}(t) = \mathbf{g}(\mathbf{E}) + \underset{s \geqq 0}{\mathfrak{G}}\left[\mathbf{E}_{\mathrm{d}}^{t+\delta}(s); \mathbf{E}\right] \tag{10.46}$$

[9] Cf. TRUESDELL & NOLL [1965], Sect. 40.

the corresponding stress response, then it holds that

$$\lim_{\delta \to \infty} \underset{s \geqq 0}{\mathfrak{G}} \left[\mathbf{E}_{\mathrm{d}}^{t+\delta}(s); \mathbf{E}\right] = \mathbf{0} \tag{10.47}$$

and consequently

$$\lim_{\delta \to \infty} \widetilde{\mathbf{T}}^{\delta}(t) = \mathbf{g}(\mathbf{E}) \,. \tag{10.48}$$

□

Proof

In view of (10.20) the static continuation converges for $\delta \to \infty$ towards the zero function, $\mathbf{E}_{\mathrm{d}}^{t+\delta}(s) \to \mathbf{0}(s)$. The continuity of $\mathfrak{G}$ leads to

$$\mathfrak{G}\big[\lim_{\delta \to \infty} \mathbf{E}_{\mathrm{d}}^{t+\delta}(s); \mathbf{E}\big] = \lim_{\delta \to \infty} \mathfrak{G}\big[\mathbf{E}_{\mathrm{d}}^{t+\delta}(s); \mathbf{E}\big] = \mathbf{0} \,,$$

that is the relaxation property (10.47) or (10.48). ■

The following theorem expounds the consequences of a stronger assumption of continuity.

Theorem 10. 5

If the memory functional $\mathfrak{G}$ is continuously differentiable in relation to the norm (10.12) (r > 1/2), then it can be approximated by means of a linear functional of the integral type. In this case the general constitutive equation (10.1) allows the approximation

$$\widetilde{\mathbf{T}}(t) = \mathbf{g}(\mathbf{E}(t)) + \int_0^{\infty} \mathbb{L}(\mathbf{E}(t), s)[\mathbf{E}_{\mathrm{d}}^{t}(s)]\mathrm{d}s + o\Big(\big\|\mathbf{E}_{\mathrm{d}}^{t}(s)\big\|_h\Big) \,. \tag{10.49}$$

The kernel function $\mathbb{L}(\mathbf{E}, s)$ of the integral is a fourth order tensor, depending on the present strain $\mathbf{E}$ and the integration variable s). The tensor function $\mathbb{L}$: $(\mathbf{E}, s) \longmapsto \mathbb{L}(\mathbf{E}, s)$ satisfies the condition

$$\int_0^{\infty} \frac{\big|\mathbb{L}(\mathbf{E}, s)\big|^2}{h^2(s)} \mathrm{d}s < \infty \,, \tag{10.50}$$

with the Euclidean norm of $\mathbb{L}$,

$$\big|\,\mathbb{L}\,\big| = (L_{ijkl} L^{ijkl})^{1/2} \,. \tag{10.51}$$

□

Proof

The approximation of $\mathfrak{G}$ by way of a linear functional is the outcome of the assumption that $\mathfrak{G}$ is continuously differentiable: the *Taylor* expansion

$$\mathfrak{G}[\mathbf{E}_{\mathrm{d}}^{\mathrm{t}}(s);\mathbf{E}] = \mathfrak{G}[0(s);\mathbf{E}] + \mathrm{d}\mathfrak{G}[0(s);\mathbf{E}\,|\,\mathbf{E}_{\mathrm{d}}^{\mathrm{t}}(s)] + o\Big(\big\|\mathbf{E}_{\mathrm{d}}^{\mathrm{t}}(s)\big\|_h\Big)$$

leads to

$$\underset{s \geqq 0}{\mathfrak{G}}\,[\mathbf{E}_{\mathrm{d}}^{\mathrm{t}}(s);\mathbf{E}] = \underset{s \geqq 0}{\mathfrak{L}}\,[\mathbf{E}_{\mathrm{d}}^{\mathrm{t}}(s);\mathbf{E}] + o\Big(\big\|\mathbf{E}_{\mathrm{d}}^{\mathrm{t}}(s)\big\|_h\Big)\,, \tag{10.52}$$

where the functional $\mathfrak{L}$ is linear in $\mathbf{E}_{\mathrm{d}}^{\mathrm{t}}(s)$. Each component of $\mathfrak{L}$ is a linear functional of $\mathbf{E}_{\mathrm{d}}^{\mathrm{t}}(\cdot)$. Since these functionals are continuous, they are bounded and can therefore be represented in terms of an inner product along the lines of a representation theorem from functional analysis.[10] This can be summarised in the compact form of

$$\underset{s \geqq 0}{\mathfrak{L}}\,[\mathbf{E}_{\mathrm{d}}^{\mathrm{t}}(s);\mathbf{E}] = \Big\langle \widetilde{\mathbb{L}}(\mathbf{E},s)\,,\,\mathbf{E}_{\mathrm{d}}^{\mathrm{t}}(s)\Big\rangle_h\,. \tag{10.53}$$

Following the definition of the inner product (10.11), this leads to

$$\underset{s \geqq 0}{\mathfrak{L}}\,[\mathbf{E}_{\mathrm{d}}^{\mathrm{t}}(s);\mathbf{E}] = \int_0^\infty \widetilde{\mathbb{L}}(\mathbf{E},s)[\mathbf{E}_{\mathrm{d}}^{\mathrm{t}}(s)]h^2(s)\mathrm{d}s\,. \tag{10.54}$$

The tensor function $(\mathbf{E}, s) \longmapsto \mathbb{L}(\mathbf{E}, s)$ depends on the time variable $s > 0$ and on the present strain $\mathbf{E}(t)$. The linear functional $\mathfrak{L}$ is continuous by assumption, which implies that it is bounded. This implies a restrictive condition for the material function $\mathbb{L}(\mathbf{E}, s)$ in (10.49), if the *Cauchy-Schwarz* inequality - again in a compact notation - is applied:[11]

$$\Big|\Big\langle \widetilde{\mathbb{L}}(\mathbf{E},s)\,,\,\mathbf{E}_{\mathrm{d}}^{\mathrm{t}}(s)\Big\rangle_h\Big|^2 \leq \big\|\widetilde{\mathbb{L}}(\mathbf{E},s)\big\|_h^2\,\big\|\mathbf{E}_{\mathrm{d}}^{\mathrm{t}}(s)\big\|_h^2\,. \tag{10.55}$$

As a condition for the boundedness of $\mathfrak{L}$ we obtain the requirements

$$\big\|\widetilde{\mathbb{L}}(\mathbf{E},s)\big\|_h^2 = \int_0^\infty \big|\widetilde{\mathbb{L}}(\mathbf{E},s)\big|^2 h^2(s)\mathrm{d}s < \infty\,. \tag{10.56}$$

If we define a material function by

[10] See LJUSTERNIK & SOBOLEW [1979], § 2.

[11] BRONSTEIN et al. [1997], p. 27.

$$\mathbb{L}(\mathbf{E}, s) = \widetilde{\mathbb{L}}(\mathbf{E}, s)h^2(s) \,, \tag{10.57}$$

we obtain $\left|\widetilde{\mathbb{L}}(\mathbf{E}, s)\right|^2 h^2(s) = \left|\mathbb{L}(\mathbf{E}, s)\right|^2 h^{-2}(s)$, and (10.56) implies (10.50). ■

Definition 10. 6

A constitutive equation of the form

$$\widetilde{\mathbf{T}}(t) = \mathbf{g}(\mathbf{E}) + \int_0^\infty \mathbb{L}(\mathbf{E}, s)[\mathbf{E}_\mathrm{d}^\mathrm{t}(s)]\mathrm{d}s \tag{10.58}$$

or

$$\mathbf{T}(t) = \mathbf{f}(\mathbf{B}) + \int_0^\infty \mathbb{K}(\mathbf{B}, s)[\mathbf{G}^\mathrm{t}(s)]\mathrm{d}s \tag{10.59}$$

defines the theory of *finite linear viscoelasticity*.[12] ◊

In finite linear viscoelasticity the memory properties are described by means of the tensor-valued material function $\mathbb{L}(\mathbf{E}, s)$ or $\mathbb{K}(\mathbf{B}, s)$. This description can be substantiated and possibly simplified if representation theorems are known for these tensor functions.

For isotropic solids (10.59) is valid; the integrand in (10.59) is an isotropic function of the three variables $\mathbf{B}$, s, $\mathbf{G}^\mathrm{t}(\cdot)$, depending linearly on the third variable. The representation of an isotropic tensor function of this kind reads as follows:[13]

$$\mathbb{K}(\mathbf{B})[\mathbf{G}] = (\mathbf{k}_1\cdot\mathbf{G})\mathbf{1} + (\mathbf{k}_2\cdot\mathbf{G})\mathbf{B} + (\mathbf{k}_3\cdot\mathbf{G})\mathbf{B}^2 + \mathbf{k}_4\mathbf{G} + \mathbf{G}\mathbf{k}_4 \,, \tag{10.60}$$

$\mathbf{k}_\mathrm{i}$ being isotropic functions of the left *Cauchy-Green* tensor $\mathbf{B}$,

$$\mathbf{k}_\mathrm{i}(\mathbf{B}) = \varphi_{0\mathrm{i}}\mathbf{1} + \varphi_{1\mathrm{i}}\mathbf{B} + \varphi_{2\mathrm{i}}\mathbf{B}^2 \,. \tag{10.61}$$

These contain a total of 12 scalar material functions, which depend on the 3 basic invariants of $\mathbf{B}$, as well as on s,

$$\varphi_{\mathrm{ki}} = \varphi_{\mathrm{ki}}(\mathrm{I}_\mathbf{B}, \mathrm{II}_\mathbf{B}, \mathrm{III}_\mathbf{B}, s) \qquad (\mathrm{k} = 1, 2, 3, 4, \mathrm{i} = 0, 1, 2) \tag{10.62}$$

We add the representation of the equilibrium stress according to (9.39),

---

[12] Coleman & Noll [1961], p. 245.

[13] Coleman & Noll [1961], eq. (5.11).

$$\mathbf{f(B)} = \varphi_0(\mathrm{I_B}, \mathrm{II_B}, \mathrm{III_B})\mathbf{B} + \varphi_1(\mathrm{I_B}, \mathrm{II_B}, \mathrm{III_B})\mathbf{B} + \varphi_2(\mathrm{I_B}, \mathrm{II_B}, \mathrm{III_B})\mathbf{B}^2\ , \qquad (10.63)$$

so that the complete representation of the material properties within the context of finite linear viscoelasticity covers 15 material functions in all. This is, of course, a theoretical statement: with a view to practical applications, the amount of information resulting from the representation theory of tensor functions has to be radically decreased by further approximations and physically founded assumptions.

Finite linear viscoelasticity provides a fairly simple situation where a fluid is concerned: in this case the integrand in (10.59) is a linear isotropic tensor function of $\mathbf{G}^t$, which only depends on the scalars s and ρ: finite linear viscoelasticity of a fluid is represented by a constitutive equation in the form of

$$\mathbf{T}(t) = -\,p(\rho)\mathbf{1} + \int_0^\infty \{k_1(\rho, s)\mathbf{G}^t_\alpha(s) + k_2(\rho, s)[\mathrm{tr}\mathbf{G}^t_\alpha(s)]\mathbf{1}\,\}\mathrm{d}s\ , \qquad (10.64)$$

containing just three material functions in all.[14]

The constitutive equations of nonlinear elasticity and finite linear viscoelasticity are approximations of the general constitutive equations (10.1), (10.2) and (10.3) for simple materials, which are valid under suitable conditions pertaining to the continuity properties of the constitutive functionals. The quality of these approximations depends appreciably on the rate of the processes: the approximations become arbitrarily close for motions progressing slowly enough in an asymptotic sense.

Theorem 10. 6

1) If the functional $\mathfrak{G}$ is continuous and $\mathrm{r} > \frac{1}{2}$, then we have for retarded deformation processes the following asymptotic approximation of the general constitutive equation (10.1):

$$\tilde{\mathbf{T}}_\alpha(t) = \mathbf{g}(\mathbf{E}) + \mathbf{o}(1)\ . \qquad (10.65)$$

Differentiability of $\mathfrak{G}$ and $\mathrm{r} > \frac{3}{2}$ leads to the asymptotic approximation

$$\tilde{\mathbf{T}}_\alpha(t) = \mathbf{g}(\mathbf{E}) + \mathrm{O}(\alpha)\ . \qquad (10.66)$$

2) If the functional $\mathfrak{G}$ is continuously differentiable and $\mathrm{r} > \frac{1}{2}$, the asymptotic approximation

[14] COLEMAN & NOLL [1961], eq. (5.18).

$$\widetilde{\mathbf{T}}_\alpha(t) = \mathbf{g}(\mathbf{E}) + \int_0^\infty \mathbb{L}(\mathbf{E}, s)[\mathbf{E}^t_{\mathrm{d}\alpha}(s)]\mathrm{d}s + \mathbf{o}(1) \tag{10.67}$$

is valid. Continuous differentiability and $\mathrm{r} > \frac{3}{2}$ lead to

$$\widetilde{\mathbf{T}}_\alpha(t) = \mathbf{g}(\mathbf{E}) + \int_0^\infty \mathbb{L}(\mathbf{E}, s)[\mathbf{E}^t_{\mathrm{d}\alpha}(s)]\mathrm{d}s + \mathbf{o}(\alpha)\,. \tag{10.68}$$

□

Proof

1) The retardation theorem in the formulation (10.37) provides for $\mathrm{n} = 0$ $(\mathrm{r} > \frac{1}{2})$

$$\lim_{\alpha \to 0} \left\|\mathbf{E}^t_{\mathrm{d}\alpha}(s)\right\|_h = 0, \quad \text{i.e.} \quad \left\|\mathbf{E}^t_{\mathrm{d}\alpha}(s)\right\|_h = \mathrm{o}(1)$$

or $\underset{s \geqq 0}{\mathfrak{G}} \left[\mathbf{E}^t_{\mathrm{d}\alpha}(s); \mathbf{E}\right] = \mathbf{o}(1)$; that proves (10.65).

For $\mathrm{n} = 1$ $(\mathrm{r} > \frac{3}{2})$ we obtain from (10.37) $\left\|\mathbf{E}^t_{\mathrm{d}\alpha}(s)\right\|_h = \mathrm{O}(\alpha)$.

Differentiability implies $\underset{s \geqq 0}{\mathfrak{G}} \left[\mathbf{E}^t_{\mathrm{d}\alpha}(s); \mathbf{E}\right] = \mathbf{O}(\alpha)$, i.e. (10.66).

2) Continuous differentiability of $\mathfrak{G}$, given $\mathrm{n} = 0$, leads to (10.67) and, given $\mathrm{n} = 1$, to (10.68). ■

The theorem states that the theory of viscoelasticity contains elasticity as a limit case for slow motions. Wherever slow deformations occur, elasticity is an approximation of zeroth order of the general constitutive equation (10.1) and finite linear viscoelasticity a first order approximation. In principle, higher-order approximations can be created by presuming properties of higher-order differentiability. All approximations should be interpreted in the asymptotic sense, as is expressed by quoting the magnitude order of the error terms.

More restrictive approximations present themselves in the consequent application of the retardation theorem. For instance, (10.37) $\mathrm{n} = 1$ $(\mathrm{r} > \frac{3}{2})$ leads to the approximation of the whole deformation history by means of its tangent at $s = 0$, i.e. by the current strain rate (cf. Fig. 10. 2):

$$\mathbf{E}^t_{\mathrm{d}\alpha}(s) = -\,\alpha s\dot{\mathbf{E}}(t) + \mathbf{o}(\alpha)\,. \tag{10.69}$$

This supplies constitutive equations of the differential type. Inserting (10.69) into (10.68), we obtain the theory of *finite linear viscosity* as an approximation of finite linear viscoelasticity:

$$\widetilde{\mathbf{T}}_\alpha(t) = \mathbf{g}(\mathbf{E}) + \alpha\mathbb{H}(\mathbf{E})[\dot{\mathbf{E}}] + \mathbf{o}(\alpha)\,. \tag{10.70}$$

In this approximation

$$\mathbb{H}(\mathbf{E}) = -\int_0^\infty s\mathbb{L}(\mathbf{E}, s)\mathrm{d}s \tag{10.71}$$

is a *viscosity tensor*, which can still be dependent on the current strain $\mathbf{E}(t)$. The approximation of finite linear viscoelasticity for fluids using constitutive equations of the differential type is of particular interest.

Theorem 10. 7

1) If the functional $\boldsymbol{\Omega}$ is continuous and $\mathrm{r} > \frac{1}{2}$, we have the following asymptotic approximation of the general constitutive equation (10.3), valid for retarded deformation processes:

$$\mathbf{T}_\alpha(t) = -\,p(\rho)\mathbf{1} + \mathbf{o}(1)\,. \tag{10.72}$$

2) Continuous differentiability and $\mathrm{r} > \frac{3}{2}$ lead to

$$\mathbf{T}_\alpha(t) = -\,p(\rho)\mathbf{1} + \int_0^\infty \{k_1(\rho, s)\mathbf{G}_\alpha^t(s) + k_2(\rho, s)[\mathrm{tr}\mathbf{G}_\alpha^t(s)]\mathbf{1}\,\}\mathrm{d}s + \mathbf{o}(\alpha)\,, \tag{10.73}$$

and to[15]

$$\mathbf{T}_\alpha(t) = -\,p(\rho)\mathbf{1} + \alpha[2\eta(\rho)\mathbf{D} + 2\lambda(\rho)(\mathrm{tr}\mathbf{D})\mathbf{1}] + \mathbf{o}(\alpha)\,, \tag{10.74}$$

with the viscosities

$$\eta(\rho) = -\int_0^\infty sk_1(\rho, s)\mathrm{d}s\,, \qquad \lambda(\rho) = -\int_0^\infty sk_2(\rho, s)\mathrm{d}s\,, \tag{10.75}$$

depending on the mass density. □

---

[15] TRUESDELL & NOLL [1965], eq. (41.5).

Proof

1) The approximations (10.72) and (10.73) correspond to (10.65) and (10.68).

2) For n = 1 the retardation theorem in the formulation of (10.40) supplies the approximation of the deformation history by means of the spatial strain rate tensor:

$$\mathbf{G}^t(s) = -\,2\alpha\mathbf{D} + \mathrm{o}(\alpha)\,. \tag{10.76}$$

Substitution in (10.73) leads to (10.74) and (10.75). ■

This theorem's physically important claim lies in the established fact that the classical constitutive equation (5.18) of the *Navier-Stokes* fluid represents an asymptotic approximation of the general constitutive equation of a simple fluid with fading memory.

Generalisations can be drawn up systematically for finite linear visco-elasticity and viscosity by postulating higher-order differentiability of the constitutive functionals. Complete exploitation of the retardation theorem in this connection leads to asymptotic approximations of differential type, in which the state of stress is a function of the strain rates of first and higher order, that is a function of the *Rivlin-Ericksen* tensors.

Irrespective of the concrete representation of constitutive functionals, kinematic conditions of constraint may be introduced. In particular, the internal constraint *incompressibility* can considerably simplify the material description. This already became evident in the example of isotropic elasticity and is also valid for finite linear viscoelasticity.

Theorem 10. 8

The general constitutive equation

$$\mathbf{T}(t) = -\,p(x,t)\mathbf{1} + \mathbf{f}(\mathbf{B}) + \underset{s\,\geqq\,0}{\mathfrak{F}}\left[\mathbf{G}^t(s);\mathbf{B}(t)\right] \tag{10.77}$$

for incompressible isotropic solids permits the following asymptotic approximation for deformations of restricted size $\left(\delta = \|\mathbf{H}\| \ll 1\right)$ and slow motions:

$$\mathbf{T}_\alpha + p\mathbf{1} = \mu\Big[\Big(\beta + \tfrac{1}{2}\Big)\mathbf{B} + \Big(\beta - \tfrac{1}{2}\Big)\mathbf{B}^{-1}\Big]$$
$$+ \int_0^\infty \Big[\Phi(s)\mathbf{G}^t_\alpha(s) + \Psi(s)\big(\mathbf{B}\mathbf{G}^t_\alpha(s) + \mathbf{G}^t_\alpha(s)\mathbf{B}\big)\Big]\mathrm{d}s + \mathrm{O}(\delta^3) + \mathrm{o}(\alpha)\,. \tag{10.78}$$

□

Proof

Linear approximation of $\mathfrak{F}$ (with $r = \frac{3}{2}$) leads to

$$\mathbf{T}_\alpha + p\mathbf{1} = \mathbf{f}(\mathbf{B}) + \int_0^\infty \mathbb{K}(\mathbf{B}, s)[\mathbf{G}_\alpha^t(s)]ds + o(\alpha) . \tag{10.79}$$

In the representation (10.60) of the integrand $\mathbb{K}(\mathbf{B}, s)[\mathbf{G}_\alpha^t(s)]$ the first term $(\mathbf{k}\cdot\mathbf{G})\mathbf{1}$ is omitted, since the hydrostatic pressure p remains undetermined in the constitutive equation. In the remaining material functions the invariant $\mathrm{III}_\mathbf{B}$ drops out as an argument. Thereupon (10.79) merges into

$$\begin{aligned}\mathbf{T} + p\mathbf{1} &= \varphi_1(\mathrm{I}_\mathbf{B}, \mathrm{II}_\mathbf{B})\, \mathbf{B} + \varphi_{-1}(\mathrm{I}_\mathbf{B}, \mathrm{II}_\mathbf{B})\, \mathbf{B}^{-1} \\ &+ \int_0^\infty \Big\{\Big[\Big(\varphi_{02}\mathbf{1} + \varphi_{12}\mathbf{B} + \varphi_{22}\mathbf{B}^2\Big)\cdot\mathbf{G}^t\Big]\mathbf{B} \\ &+ \Big[\Big(\varphi_{03}\mathbf{1} + \varphi_{13}\mathbf{B} + \varphi_{23}\mathbf{B}^2\Big)\cdot\mathbf{G}^t\Big]\mathbf{B}^2 + \Big(\varphi_{04}\mathbf{1} + \varphi_{14}\mathbf{B} + \varphi_{24}\mathbf{B}^2\Big)\mathbf{G}^t \\ &+ \mathbf{G}^t\Big(\varphi_{04}\mathbf{1} + \varphi_{14}\mathbf{B} + \varphi_{24}\mathbf{B}^2\Big)\Big\}ds ,\end{aligned} \tag{10.80}$$

the index α now being omitted. With the norm of the displacement gradient,

$$\delta = \|\mathbf{H}\| = \sqrt{\mathbf{H}\cdot\mathbf{H}} ,$$

the asymptotic relations $\mathbf{B} = \mathbf{1} + O(\delta)$, $\mathbf{G}^t = O(\delta)$, and, in view of the incompressibility, also $\mathrm{I}_\mathbf{B} = 3 + O(\delta^2)$, $\mathrm{II}_\mathbf{B} = 3 + O(\delta^2)$ and $\mathrm{tr}\mathbf{G}^t = O(\delta^2)$ are valid (cf. (9.86), (9.87)).

Renewed consideration of the fact that pressure p is undetermined in the constitutive equation leads to the realisation that, as far as a second order approximation in δ is concerned, all material functions under the integral are irrelevant, with the exception of φ_{04} and φ_{14}. In these functions, however, the dependence on $\mathrm{I}_\mathbf{B}$ and $\mathrm{II}_\mathbf{B}$ ceases to apply. This implies the approximation (10.78), if the corresponding approximation (9.83) is borne in mind for the elastic part. ■

The constitutive equation (10.78) is an extension of the *Mooney-Rivlin* model to include *finite linear viscoelasticity.*

10. 2 Representation by Means of Internal Variables

10. 2. 1 General Concept

Rate-dependent functionals representing fading memory properties of the material in terms of viscoelasticity can also be represented implicitly within the framework of the theory of internal variables by means of ordinary differential equations, known as evolution equations. The evolution equations have to fulfil certain conditions, in order to ensure the relaxation property. The general formulation of conditions of this nature follows Coleman and Gurtin.[16]

Definition 10. 7

A viscoelastic material is defined by a stress-strain relation

$$\widetilde{\mathbf{T}} = \mathbf{f}(\mathbf{E}, q_1, \dots, q_N) \tag{10.81}$$

a set of evolution equations

$$\dot{q}_k(t) = f_k(\mathbf{E}(t), q_1(t), \dots, q_N(t)) \, , \quad k = 1, \dots, N \tag{10.82}$$

with the following characteristics:

1) For each state of strain $\mathbf{E}$ *equilibrium solutions* exist, i.e. solutions $\bar{q}_k$ of the system of equations

$$f_k(\mathbf{E}, \bar{q}_1, \dots, \bar{q}_N) = 0 \, , \quad k = 1, \dots, N \, . \tag{10.83}$$

The equilibrium solutions are functions of $\mathbf{E}$:

$$\mathbf{E} \longmapsto \bar{q}_k = \bar{q}_k(\mathbf{E}) \, . \tag{10.84}$$

2) These equilibrium solutions possess *global asymptotic stability*.
For every constant strain tensor $\mathbf{E}_0$ and arbitrary initial conditions $q_k(t_0) = q_{k0}$ the solution of the system of differential equations

$$\dot{q}_k(t) = f_k(\mathbf{E}_0, q_1(t), \dots, q_N(t)) \tag{10.85}$$

---

[16] Coleman & Gurtin [1967]. See also Lubliner [1969].

tends towards the equilibrium solution:

$$\lim_{t \to \infty} q_k(t) = \bar{q}_k(\mathbf{E}_0) \,. \tag{10.86}$$

◊

By inserting the general solution of the evolution equations (10.82) into the stress-strain relation (10.81), a stress functional with a relaxation or fading memory property arises. The set of equilibrium solutions constitutes the equilibrium relation in the sense of Definition 6. 2. The equilibrium stress is a function of the current strain $\mathbf{E}$:

$$\mathbf{E} \longmapsto \tilde{\mathbf{T}}_{eq} = \mathbf{f}_{eq}(\mathbf{E}) = \mathbf{f}(\mathbf{E}, \bar{q}_1(\mathbf{E}), \dots, \bar{q}_N(\mathbf{E})) \,. \tag{10.87}$$

Due to the properties of the evolution equations in viscoelasticity, slow processes take place close to states of equilibrium, the actual distance depending in each case on the rate of the process. It is possible to gain more insight into this fact by linearising the evolution equations within a neighbourhood of the equilibrium solution.

A *Taylor* expansion of the stress-strain relation can be based on a representation of the form

$$\tilde{\mathbf{T}} = \tilde{\mathbf{T}}_{eq} + \tilde{\mathbf{T}}_{ov} \,, \tag{10.88}$$

$$\tilde{\mathbf{T}} = \mathbf{f}_{eq}(\mathbf{E}) + \mathbf{f}_{ov}(\mathbf{E}, q_1, \dots, q_N) \tag{10.89}$$

in which the function $\mathbf{E} \longmapsto \tilde{\mathbf{T}}_{eq} = \mathbf{f}_{eq}(\mathbf{E})$ is the *equilibrium stress* and

$$\mathbf{E} \longmapsto \tilde{\mathbf{T}}_{ov} = \mathbf{f}_{ov}(\mathbf{E}, q_1, \dots, q_N) \,,$$

$$\mathbf{f}_{ov}(\mathbf{E}, q_1, \dots, q_N) = \mathbf{f}(\mathbf{E}, q_1, \dots, q_N) - \mathbf{f}(\mathbf{E}, \bar{q}_1, \dots, \bar{q}_N)$$

the so-called *overstress*, which vanishes by definition for the equilibrium solution $\bar{q}_k$,

$$\mathbf{f}_{ov}(\mathbf{E}, \bar{q}_1, \dots, \bar{q}_N) = \mathbf{0} \,. \tag{10.90}$$

With respect to the internal variables, the *Taylor* expansions

$$\tilde{\mathbf{T}} = \mathbf{f}_{eq}(\mathbf{E}) + \sum_{k=1}^{N} \frac{\partial \mathbf{f}_{ov}}{\partial q_k}(\mathbf{E}, \bar{q}_1, \dots, \bar{q}_N)[q_k - \bar{q}_k] + \mathrm{O}(\kappa^2) \tag{10.91}$$

and

$$\dot{q}_k(t) = \sum_{l=1}^{N} \frac{\partial f_k}{\partial q_l}(\mathbf{E}(t), \bar{q}_1(t), \dots, \bar{q}_N(t))[q_l(t) - \bar{q}_l(t)] + O(\kappa^2) \tag{10.92}$$

then create for a small distance

$$\kappa = \left\{ \sum_{k=1}^{N} [q_k(t) - \bar{q}_k(\mathbf{E})]^2 \right\}^{1/2} \tag{10.93}$$

from the equilibrium state $\bar{q}_k = \bar{q}_k(\mathbf{E})$ $(k = 1, \dots, N)$ an asymptotic approximation of the general constitutive model (10.81), (10.82).

The evolution equations (10.92) are linear in the internal variables $q_k(t)$ but depend nonlinearly on the state of strain. Since $\mathbf{E}$ is time-dependent, not only the equilibrium values of the internal variable are functions of time,

$$\bar{q}_k(t) = \bar{q}_k(\mathbf{E}(t)) , \tag{10.94}$$

but also the coefficients $\partial f_k / \partial q_j$ of the linear differential equations (10.92). From a physical point of view, these coefficients can be interpreted as reciprocal relaxation times (relaxation frequencies), depending on the deformation.

In the following investigation their dependence on the current state of strain shall be disregarded, i.e.

$$\frac{\partial f_k}{\partial q_l}(\mathbf{E}, \bar{q}_1(\mathbf{E}), \dots, \bar{q}_N(\mathbf{E})) \approx \frac{\partial f_k}{\partial q_l}(\mathbf{0}, \bar{q}_1(\mathbf{0}), \dots, \bar{q}_N(\mathbf{0})) .$$

Moreover, it is assumed that the functional matrix $\partial f_k / \partial q_j$ is negative definite and symmetric. This symmetry is justified from the point of view of thermodynamics (see Theorem 13. 11), whereas definiteness complies with the assumption of stability regarding the evolution equations (Definition 10. 7).

On these assumptions, the linearised evolution equations can be integrated explicitly.

Theorem 10. 9

Let the $(N \times N)$-matrix $\underset{\approx}{G} = (G_{kl})$,

$$G_{kl} = -\frac{\partial f_k}{\partial q_l}(\mathbf{0}, \bar{q}_1(\mathbf{0}), \dots, \bar{q}_N(\mathbf{0})) \tag{10.95}$$

of the evolution equations at $\mathbf{E} = \mathbf{0}$ be symmetric and positive definite. The solution of the linearised evolution equations (10.92) written in matrix notation,

$$\dot{\underset{\sim}{q}}(t) + \underset{\approx}{G}\,\underset{\sim}{q}(t) = \underset{\approx}{G}\,\bar{\underset{\sim}{q}}(\mathbf{E}(t))\,, \tag{10.96}$$

then reads

$$\underset{\sim}{q}(t) - \bar{\underset{\sim}{q}}(\mathbf{E}(t)) = \int_0^\infty \underset{\approx}{G}\mathrm{e}^{-\underset{\approx}{G}s}\Big\{\bar{\underset{\sim}{q}}(\mathbf{E}(t-s)) - \bar{\underset{\sim}{q}}(\mathbf{E}(t))\Big\}\mathrm{d}s\,. \tag{10.97}$$

$$\underset{\sim}{q} = \{q_1, \dots, q_N\}^{\mathrm{T}} \tag{10.98}$$

is the column vector of the internal variables and

$$\bar{\underset{\sim}{q}}(t) = \bar{\underset{\sim}{q}}(\mathbf{E}(t)) = \{\bar{q}_1(\mathbf{E}(t)), \dots, \bar{q}_N(\mathbf{E}(t))\}^{\mathrm{T}} \tag{10.99}$$

the vector of the accompanying state of equilibrium. The matrix

$$\mathrm{e}^{-\underset{\approx}{G}s} = \sum_{k=0}^{\infty} \frac{(-1)^k}{k!}\left(\underset{\approx}{G}s\right)^k \tag{10.100}$$

is likewise symmetric and positive definite. □

Proof

If $\underset{\sim}{a} = \{a_1, \dots, a_N\}^{\mathrm{T}}$ and $\underset{\sim}{b} = \{b_1, \dots, b_N\}^{\mathrm{T}}$ are two N-vectors (column vectors), the inner product in matrix notation is the scalar

$$\underset{\sim}{a}^{\mathrm{T}}\underset{\sim}{b} = \underset{\sim}{b}^{\mathrm{T}}\underset{\sim}{a} = \sum_{k=1}^{N} a_k b_k\,,$$

and the dyadic product is the $(N\times N)$-matrix $\underset{\sim}{a}\,\underset{\sim}{b}^{\mathrm{T}} = (a_i b_k)$.
The symmetric and positive definite $(N\times N)$-matrix $\underset{\approx}{G}$ defines N positive eigenvalues $\gamma_1, \dots, \gamma_N$ and a system $\{\underset{\sim}{v}_1, \dots, \underset{\sim}{v}_N\}$ of N orthonormal eigenvectors, $\underset{\sim}{v}_i^{\mathrm{T}}\underset{\sim}{v}_k = \delta_{ik}$. The matrix formed with the eigenvectors $\underset{\sim}{v}_k$,

$$\underset{\approx}{Q} = \sum_{k=1}^{N} \underset{\sim}{e}_k\,\underset{\sim}{v}_k^{\mathrm{T}}\,, \tag{10.101}$$

is orthogonal and decouples the differential equations (10.96), since it diagonalises the matrix $\underset{\approx}{G}$:

$$\underset{\approx}{Q}\underset{\approx}{G}\underset{\approx}{Q}^{\mathrm{T}} = \sum_{k=1}^{N} \lambda_k\,\underset{\sim}{e}_k\,\underset{\sim}{e}_k^{\mathrm{T}}\,. \tag{10.102}$$

Accordingly, (10.96) leads to

$$\underset{\approx}{Q}\dot{\underset{\sim}{q}}(t) + \left(\underset{\approx}{Q}\underset{\approx}{G}\underset{\approx}{Q}^{\mathrm{T}}\right)\underset{\approx}{Q}\underset{\sim}{q}(t) = \left(\underset{\approx}{Q}\underset{\approx}{G}\underset{\approx}{Q}^{\mathrm{T}}\right)\underset{\approx}{Q}\bar{\underset{\sim}{q}}(\mathbf{E}(t)) \,.$$

Given the definitions $\underset{\sim}{\beta} = \sum_{k=1}^{N} \beta_k \underset{\sim}{e}_k = \underset{\approx}{Q}\,\underset{\sim}{q} \qquad \left(\beta_k = \underset{\sim}{q}^{\mathrm{T}}\underset{\sim}{v}_k\right)$

and $\bar{\underset{\sim}{\beta}} = \sum_{k=1}^{N} \bar{\beta}_k \underset{\sim}{e}_k = \underset{\approx}{Q}\,\bar{\underset{\sim}{q}} \qquad \left(\bar{\beta}_k = \bar{\underset{\sim}{q}}^{\mathrm{T}}\underset{\sim}{v}_k\right)$,

we obtain the decoupled system

$$\dot{\beta}_k(t) + \gamma_k\beta_k(t) = \gamma_k\bar{\beta}_k(t) \,, \quad k = 1, 2, ..., N \,. \tag{10.103}$$

Integration leads via $\mathrm{e}^{-\gamma_k t}\left(\mathrm{e}^{\gamma_k t}\beta_k(t)\right)^{\boldsymbol{\cdot}} = \gamma_k\bar{\beta}_k(t)$ to

$$\beta_k(t) = \int_{-\infty}^{t} \gamma_k \mathrm{e}^{-\gamma_k(t-\tau)}\bar{\beta}_k(\tau)\mathrm{d}\tau \quad \text{and finally to}$$

$$\beta_k(t) = \int_0^{\infty} \gamma_k \mathrm{e}^{-\gamma_k s}\left[\bar{\beta}_k(t-s) - \bar{\beta}_k(t)\right]\mathrm{d}s + \bar{\beta}_k(t) \,.$$

This is the kth component of the matrix equation

$$\underbrace{\sum_k[\beta_k(t) - \bar{\beta}_k(t)]\underset{\sim}{e}_k}_{\underset{\sim}{\beta}(t) - \bar{\underset{\sim}{\beta}}(t)}$$

$$= \int_0^{\infty} \underbrace{\left(\sum_k \gamma_k \underset{\sim}{e}_k \underset{\sim}{e}_k^{\mathrm{T}}\right)}_{\underset{\approx}{Q}\underset{\approx}{G}\underset{\approx}{Q}^{\mathrm{T}}} \underbrace{\underbrace{\left(\sum_j \mathrm{e}^{-\gamma_j s} \underset{\sim}{e}_j \underset{\sim}{e}_j^{\mathrm{T}}\right)}_{\mathrm{e}^{-[\sum_j \gamma_j \underset{\sim}{e}_j \underset{\sim}{e}_j^{\mathrm{T}}]s}}}_{\mathrm{e}^{-\underset{\approx}{Q}\underset{\approx}{G}\underset{\approx}{Q}^{\mathrm{T}}s}} \underbrace{\left(\sum_i[\bar{\beta}_i(t-s) - \bar{\beta}_i(t)]\underset{\sim}{e}_i\right)}_{\bar{\underset{\sim}{\beta}}(t-s) - \bar{\underset{\sim}{\beta}}(t)}\mathrm{d}s \,, \text{ i.e.}$$

$$\underset{\sim}{\beta}(t) - \bar{\underset{\sim}{\beta}}(t) = \underset{\approx}{Q}\int_0^{\infty} \underset{\approx}{G}\left\{\underset{\approx}{Q}^{\mathrm{T}}\mathrm{e}^{-\underset{\approx}{Q}\underset{\approx}{G}\underset{\approx}{Q}^{\mathrm{T}}s}\right\}\left[\bar{\underset{\sim}{\beta}}(t-s) - \bar{\underset{\sim}{\beta}}(t)\right]\mathrm{d}s \,. \tag{10.104}$$

The definitions $\underset{\sim}{\beta} = \underset{\approx}{Q}\,\underset{\sim}{q}$ and $\bar{\underset{\sim}{\beta}} = \underset{\approx}{Q}\,\bar{\underset{\sim}{q}}$, combined with the isotropy of the matrix-valued exponential function,

$$\underset{\approx}{Q}^{\mathrm{T}} \mathrm{e}^{-\underset{\approx}{Q}\underset{\approx}{G}\underset{\approx}{Q}^{\mathrm{T}} s} \underset{\approx}{Q} = \mathrm{e}^{-\underset{\approx}{G} s} \tag{10.105}$$

lead to the predicted solution (10.97). ■

Given the solution (10.97) and the stress-strain relation (10.91), which is linear in the internal variables $q_k(t)$, one obtains the current state of stress as a functional of the equilibrium process:

$$\tilde{\mathbf{T}} = \mathbf{f}_{\mathrm{eq}}(\mathbf{E}) + \sum_{k=1}^{N} \frac{\partial \mathbf{f}_{\mathrm{ov}}}{\partial q_k}(\mathbf{E}, \bar{q}_1, \dots, \bar{q}_N) \int_0^{\infty} \underset{\approx}{G}\left\{\mathrm{e}^{-\underset{\approx}{Q}\underset{\approx}{G}\underset{\approx}{Q}^{\mathrm{T}} s}\right\}\left[\underset{\sim}{\bar{q}}(\mathbf{E}(t-s)) - \underset{\sim}{\bar{q}}(\mathbf{E}(t))\right] \mathrm{d}s \,. \tag{10.106}$$

A linear functional of the deformation history arises if the equilibrium values of the internal variables depend linearly on $\mathbf{E}$,

$$\bar{q}_k(\mathbf{E}) = \mathbf{L}_k \cdot \mathbf{E} \qquad (k = 1, \dots, N)\,, \tag{10.107}$$

where the $\mathbf{L}_k$ are tensor-valued constants. In this case we have

$$\tilde{\mathbf{T}} = \tilde{\mathbf{T}}_{\mathrm{eq}} + \tilde{\mathbf{T}}_{\mathrm{ov}} = \mathbf{f}_{\mathrm{eq}}(\mathbf{E}) + \int_0^{\infty} \mathbb{G}(\mathbf{E}, s)[\mathbf{E}_{\mathrm{d}}^{t}(s)] \mathrm{d}s\,, \tag{10.108}$$

with

$$\tilde{\mathbf{T}}_{\mathrm{ov}} = \int_0^{\infty} \mathbb{G}(\mathbf{E}, s)[\mathbf{E}_{\mathrm{d}}^{t}(s)] \mathrm{d}s \tag{10.109}$$

and

$$\mathbb{G}(\mathbf{E}, s)[\mathbf{E}_{\mathrm{d}}^{t}(s)] = \sum_{i,j,k=1}^{N} \frac{\partial \mathbf{f}_{\mathrm{ov}}}{\partial q_i} G_{ij}\left(\mathrm{e}^{-\underset{\approx}{G} s}\right)_{jk}\left(\mathbf{L}_k \cdot [\mathbf{E}(t-s) - \mathbf{E}(t)]\right). \tag{10.110}$$

This establishes the theory of *finite linear viscoelasticity* as an approximation of the general viscoelastic behaviour also in terms of the representation using internal variables.

Since the memory function $\mathbb{G}(\mathbf{E}, s)$ in the linear functional (10.108) is composed of decreasing exponential functions, the existence of an influence function of any order is guaranteed, and the retardation theorem (10.24) is applicable where slowed down deformations are concerned. In particular, for retarded processes ($0 < \alpha < 1$) a linear function of the strain rate emerges as an asymptotic approximation of the memory part (overstress):

$$\tilde{\mathbf{T}}_{\mathrm{ov}\alpha}(t) = \int_0^\infty \mathbb{G}(\mathbf{E}, s)[\mathbf{E}^t_{\mathrm{d}\alpha}(s)]\mathrm{d}s = \alpha\mathbb{H}(\mathbf{E})[\dot{\mathbf{E}}] + \mathrm{o}(\alpha) = \mathrm{O}(\alpha)\,, \tag{10.111}$$

$$\mathbb{H}(\mathbf{E}) = -\int_0^\infty s\mathbb{G}(\mathbf{E}, s)\mathrm{d}s\,. \tag{10.112}$$

$\mathbb{H}$ is the strain-dependent fourth-order viscosity tensor, already defined by (10.71). In this connection, however, we have to bear in mind that the approximations (10.108) and (10.111) are based on the simplification (10.95), i.e. a calculation omitting the time-dependence of the coefficients in the linearised evolution equations (10.92). These coefficients are interpreted as relaxation frequencies. Their implicit time-dependence is weakly pronounced in slow processes, so that there should be no difference in the asymptotic magnitude of the error term in (10.111).

As a conclusion, there is sufficient evidence that, in terms of the theory of viscoelasticity, slow processes take place in the close neighbourhood of the accompanying equilibrium process. According to solution (10.97) and in connection with the retardation theorem (10.2), the distance between them is governed by the asymptotic relation

$$q_{k\alpha}(t) - \bar{q}_k(\mathbf{E}(t)) = \mathrm{O}(\alpha)\,. \tag{10.113}$$

It should be noted that the restricting assumption (10.95) is not imperative for obtaining the asymptotic relation (10.113). To demonstrate that a nonlinear dependence of the relaxation frequencies G_{kl} and the equilibrium values $\bar{q}$ on the present strain can be incorporated, we confine ourselves to one single internal variable q, i.e. the evolution equation

$$\dot{q}(t) = f(\varepsilon(t), q(t))$$

and the first term of its *Taylor* expansion

$$\dot{q}(t) = \frac{\partial f}{\partial q}(\varepsilon(t), \bar{q}(\varepsilon(t)))[q(t) - \bar{q}(\varepsilon(t))]\,,$$

where $\bar{q}(\varepsilon)$ is the strain-dependent equilibrium value of q, defined by

$f(\varepsilon, \bar{q}) = 0\,.$

With the abbreviation $G(\varepsilon) = -\dfrac{\partial f}{\partial q}(\varepsilon, \bar{q}(\varepsilon))$ (analogous to (10.95))

we set about solving the differential equation

$$\dot{q}(t) + G(\varepsilon(t))q(t) = G(\varepsilon(t))\,\bar{q}(\varepsilon(t))\,.$$

A solution can be written as

$$q(t) = \int\limits_0^\infty \mathrm{e}^{-\int_0^s G(\varepsilon(t-\sigma))\mathrm{d}\sigma} G(\varepsilon(t-s))\bar{q}(\varepsilon(t-s))\mathrm{d}s$$ and with the identity

$$\int\limits_0^\infty \mathrm{e}^{-\int_0^s G(\varepsilon(t-\sigma))\mathrm{d}\sigma} G(\varepsilon(t-s))\mathrm{d}s = -\left[\mathrm{e}^{-\int_0^s G(\varepsilon(t-\sigma))\mathrm{d}\sigma}\right]_0^\infty = 1$$ as

$$q(t) - \bar{q}(\varepsilon(t)) = \int\limits_0^\infty \mathrm{e}^{-\int_0^s G(\varepsilon(t-\sigma))\mathrm{d}\sigma} G(\varepsilon(t-s))\left[\bar{q}(\varepsilon(t-s)) - \bar{q}(\varepsilon(t))\right]\mathrm{d}s\,.$$

If we consider slow processes $\big(\varepsilon(t-s) \longmapsto \varepsilon(t-\alpha s)\big)$ and apply the *retardation theorem* in the sense of (10.37) we arrive at the asymptotic relation (10.113): $q_\alpha(t) - \bar{q}(\varepsilon(t)) = 0(\alpha)$.

10. 2. 2 Internal Variables of the Strain Type

For the formulation of physically significant material models of viscoelasticity it is desirable to attribute concrete physical meanings to the internal variables q_k. To this end, however, it is not necessary to have immediate recourse to the microstructure of the material: the phenomenological modelling of the material behaviour can be suitably supported by conceiving the internal variables as stress or strain tensors, which refer to appropriately selected configurations.[17]

Even in the rheological models of classical linear viscoelasticity internal variables were employed to represent the states of displacement in damping elements. To illustrate this, the total strain in a *Maxwell* model is additively decomposed into one elastic and one viscous part,

[17] Cf. Reese & Govindjee [1998a].

$$\varepsilon = \varepsilon_{\mathrm{e}} + \varepsilon_{\mathrm{v}} \,, \tag{10.114}$$

the viscous strain ε_{v} corresponding to the displacement in the dashpot and the elastic strain ε_{e} to the deflection of the elastic spring. The decomposition, together with the elasticity relation $\varepsilon_{\mathrm{e}} = \sigma/\hat{E}$ and the flow rule $\dot{\varepsilon}_{\mathrm{v}} = \sigma/\eta$, leads to the one-dimensional model (5.104),

$$\frac{\dot{\sigma}}{\hat{E}} + \frac{\sigma}{\eta} = \dot{\varepsilon} \,. \tag{10.115}$$

The evolution equation

$$\eta\dot{\varepsilon}_{\mathrm{v}} = \hat{E}(\varepsilon - \varepsilon_{\mathrm{v}}) \tag{10.116}$$

for the *internal variable* ε_{v} then corresponds to the flow rule or to the equality of the forces in the spring and the dashpot.[18]

Transferring this very simple model structure to 3-dimensional finite deformations is made possible by introducing an intermediate configuration. On the one hand this is based on a multiplicative decomposition of the deformation gradient and, on the other hand, it implies an additive decomposition of the strain in conformity with the one-dimensional situation.[19] Within the concept of dual variables this intermediate configuration is defined by inserting $\mathbf{\Psi} = \mathbf{F}_{\mathrm{v}}$ and $\mathbf{\Phi} = \hat{\mathbf{F}}_{\mathrm{e}}$ into (8.76). This leads to the multiplicative decomposition

$$\mathbf{F} = \hat{\mathbf{F}}_{\mathrm{e}}\mathbf{F}_{\mathrm{v}} \tag{10.117}$$

and with respect to (8.84) to the strain tensor

$$\hat{\mathbf{\Gamma}} = \mathbf{F}_{\mathrm{v}}^{\mathrm{T}\,-1}\mathbf{E}\mathbf{F}_{\mathrm{v}}^{-1} \,,$$

which acts on the intermediate configuration defined by $\mathbf{F}_{\mathrm{v}}$ and decomposes according to (8.94):

$$\hat{\mathbf{\Gamma}} = \frac{1}{2}\mathbf{F}_{\mathrm{v}}^{\mathrm{T}\,-1}(\mathbf{F}^{\mathrm{T}}\mathbf{F} - \mathbf{1})\mathbf{F}_{\mathrm{v}}^{-1} = \frac{1}{2}\big[\hat{\mathbf{F}}_{\mathrm{e}}^{\mathrm{T}}\hat{\mathbf{F}}_{\mathrm{e}} - \mathbf{F}_{\mathrm{v}}^{\mathrm{T}\,-1}\mathbf{F}_{\mathrm{v}}^{-1}\big] \,. \tag{10.118}$$

---

[18] If nonlinear rate-dependence is to be modelled, it is expedient to assume a stress-dependent viscosity, $\eta = \eta(\sigma)$.

[19] A concept of this kind was applied with considerable success in LION [1997a], LION [1997b] and LION [1998].

This equation suggests the additive decomposition of the total strain,

$$\hat{\mathbf{\Gamma}} = \hat{\mathbf{\Gamma}}_e + \hat{\mathbf{\Gamma}}_v , \tag{10.119}$$

into a purely elastic part of the *Green* type,

$$\hat{\mathbf{\Gamma}}_e = \frac{1}{2}\left[\hat{\mathbf{F}}_e{}^T\hat{\mathbf{F}}_e - \mathbf{1}\right] , \tag{10.120}$$

and a purely viscous part of the *Almansi* type,

$$\hat{\mathbf{\Gamma}}_v = \frac{1}{2}\left[\mathbf{1} - \mathbf{F}_v{}^{T-1}\mathbf{F}_v{}^{-1}\right] . \tag{10.121}$$

The dual strain rate in the sense of (8.85) is expressed by

$$\overset{\triangle}{\hat{\mathbf{\Gamma}}} = \mathbf{F}_v{}^{T-1}\dot{\mathbf{E}}\mathbf{F}_v{}^{-1} = \mathbf{F}_v{}^{T-1}\left(\mathbf{F}_v{}^T\hat{\mathbf{\Gamma}}\mathbf{F}_v\right)^{\cdot}\mathbf{F}_v{}^{-1} ,$$

$$\overset{\triangle}{\hat{\mathbf{\Gamma}}} = \dot{\hat{\mathbf{\Gamma}}} + \hat{\mathbf{L}}_v{}^T\hat{\mathbf{\Gamma}} + \hat{\mathbf{\Gamma}}\hat{\mathbf{L}}_v , \tag{10.122}$$

where

$$\hat{\mathbf{L}}_v = \dot{\mathbf{F}}_v\mathbf{F}_v{}^{-1} \tag{10.123}$$

is the rate of the viscous deformation.

For the strain rate $\overset{\triangle}{\hat{\mathbf{\Gamma}}}$ we obtain from the table in Fig. 8. 1 the decomposition

$$\overset{\triangle}{\hat{\mathbf{\Gamma}}} = \overset{\triangle}{\hat{\mathbf{\Gamma}}}_e + \overset{\triangle}{\hat{\mathbf{\Gamma}}}_v = \hat{\mathbf{D}}_e + \hat{\mathbf{D}}_v . \tag{10.124}$$

The elastic part

$$\hat{\mathbf{D}}_e = \overset{\triangle}{\hat{\mathbf{\Gamma}}}_e = \dot{\hat{\mathbf{\Gamma}}}_e + \hat{\mathbf{L}}_v{}^T\hat{\mathbf{\Gamma}}_e + \hat{\mathbf{\Gamma}}_e\hat{\mathbf{L}}_v \tag{10.125}$$

inevitably contains the viscous deformation rate $\hat{\mathbf{L}}_v$, whereas the viscous part

$$\hat{\mathbf{D}}_v = \overset{\triangle}{\hat{\mathbf{\Gamma}}}_v = \dot{\hat{\mathbf{\Gamma}}}_v + \hat{\mathbf{L}}_v{}^T\hat{\mathbf{\Gamma}}_v + \hat{\mathbf{\Gamma}}_v\hat{\mathbf{L}}_v \tag{10.126}$$

merely consists of $\hat{\mathbf{L}}_\mathrm{v}$ in view of the identity

$$\overset{\triangle}{\hat{\mathbf{\Gamma}}}_\mathrm{v} = \dot{\hat{\mathbf{\Gamma}}}_\mathrm{v} + \hat{\mathbf{L}}_\mathrm{v}{}^\mathrm{T}\hat{\mathbf{\Gamma}}_\mathrm{v} + \hat{\mathbf{\Gamma}}_\mathrm{v}\hat{\mathbf{L}}_\mathrm{v} = \tfrac{1}{2}[\hat{\mathbf{L}}_\mathrm{v} + \hat{\mathbf{L}}_\mathrm{v}{}^\mathrm{T}] \,. \tag{10.127}$$

According to (8.86), the dual stress tensor $\hat{\mathbf{S}}_\mathrm{v}$ relating to $\hat{\mathbf{\Gamma}}$ is connected to the *Piola-Kirchhoff* stress tensor $\widetilde{\mathbf{T}}_\mathrm{v}$,

$$\hat{\mathbf{S}}_\mathrm{v} = \mathbf{F}_\mathrm{v}\widetilde{\mathbf{T}}_\mathrm{v}\mathbf{F}_\mathrm{v}{}^\mathrm{T} \,. \tag{10.128}$$

The stress and strain tensors are dual to one another in the sense of relation (8.108), i.e.

$$\widetilde{\mathbf{T}}_\mathrm{v}\cdot\mathbf{E} = \hat{\mathbf{S}}_\mathrm{v}\cdot\hat{\mathbf{\Gamma}} \,. \tag{10.129}$$

One natural method of generalising the one-dimensional evolution equation (10.116) is to assume proportionality between the viscous strain rate $\hat{\mathbf{D}}_\mathrm{v}$ and the stress $\hat{\mathbf{S}}_\mathrm{v}$, i.e.

$$\eta\hat{\mathbf{D}}_\mathrm{v} = \hat{\mathbf{S}}_\mathrm{v} \,. \tag{10.130}$$

The scalar proportionality factor η represents a viscosity. In turn, the viscosity can depend on process quantities, for instance on the stress, in order to incorporate a nonlinear rate-dependence, encountered in the experimental data given in Figs. 6. 5, 6. 10 and 6. 11:

$$\eta = \eta(\hat{\mathbf{S}}_\mathrm{v}, \cdots) \,. \tag{10.131}$$

To complete the generalised *Maxwell* model, the stress $\hat{\mathbf{S}}_\mathrm{v}$ has to be coupled with the elastic part

$$\hat{\mathbf{\Gamma}}_\mathrm{e} = \tfrac{1}{2}[\hat{\mathbf{F}}_\mathrm{e}{}^\mathrm{T}\hat{\mathbf{F}}_\mathrm{e} - \mathbf{1}] = \hat{\mathbf{\Gamma}} - \hat{\mathbf{\Gamma}}_\mathrm{v}$$

of the total strain by means of an elasticity relation:

$$\hat{\mathbf{S}}_\mathrm{v} = \hat{\mathbf{f}}_\mathrm{v}(\hat{\mathbf{\Gamma}}_\mathrm{e}) = \hat{\mathbf{f}}_\mathrm{v}(\hat{\mathbf{\Gamma}} - \hat{\mathbf{\Gamma}}_\mathrm{v}) \,. \tag{10.132}$$

This implies the evolution equation

$$\eta\overset{\triangle}{\hat{\Gamma}}_v = \hat{\mathbf{f}}_v(\hat{\Gamma} - \hat{\Gamma}_v)\,, \tag{10.133}$$

which describes the temporal development of the internal variable $\hat{\Gamma}_v$ in dependence on the total strain process $\hat{\Gamma}(t)$. This evolution equation can be regarded as a straight-forward generalisation of the well-motivated one-dimensional model (10.116).

As regards the integration of the evolution equation (10.133), it may be expedient to transform it to quantities related to the reference configuration (or to those of the current configuration). This is easy to accomplish using the kinematic relations available. (10.119) and (10.120) lead to

$$\hat{\Gamma} - \hat{\Gamma}_v = \tfrac{1}{2}\big[\hat{\mathbf{F}}_e^{\,T}\hat{\mathbf{F}}_e - \mathbf{1}\big] = \tfrac{1}{2}\big[\mathbf{F}_v^{\,T-1}\mathbf{F}^T\mathbf{F}\mathbf{F}_v^{\,-1} - \mathbf{1}\big]\,, \text{ i.e. to}$$

$$\hat{\Gamma} - \hat{\Gamma}_v = \tfrac{1}{2}\mathbf{F}_v^{\,T-1}(\mathbf{C} - \mathbf{C}_v)\mathbf{F}_v^{\,-1} = \mathbf{F}_v^{\,T-1}\mathbf{E}_e\mathbf{F}_v^{\,-1}\,. \tag{10.134}$$

$\mathbf{C}_v = \mathbf{F}_v^{\,T}\mathbf{F}_v$ is a right *Cauchy-Green* tensor and

$$\mathbf{E}_e = \tfrac{1}{2}(\mathbf{C} - \mathbf{C}_v) = (\mathbf{E} - \mathbf{E}_v)$$

a strain tensor defined as the difference between total and viscous strains. For the viscous strain rate, the relation (10.127) and the definition (10.123) of $\hat{\mathbf{L}}_v$ provide the transformation

$$\mathbf{F}_v^{\,T}\overset{\triangle}{\hat{\Gamma}}_v\mathbf{F}_v = \mathbf{F}_v^{\,T}\tfrac{1}{2}\big[\dot{\mathbf{F}}_v\mathbf{F}_v^{\,-1} + \mathbf{F}_v^{\,T-1}\dot{\mathbf{F}}_v^{\,T}\big]\mathbf{F}_v = \tfrac{1}{2}\big[\mathbf{F}_v^{\,T}\dot{\mathbf{F}}_v + \dot{\mathbf{F}}_v^{\,T}\mathbf{F}_v\big]\,,$$

with the result

$$\mathbf{F}_v^{\,T}\overset{\triangle}{\hat{\Gamma}}_v\mathbf{F}_v = \tfrac{1}{2}\dot{\mathbf{C}}_v = \dot{\mathbf{E}}_v\,. \tag{10.135}$$

The 2nd *Piola-Kirchhoff* stress based on the reference configuration according to (10.128) is finally given by

$$\tilde{\mathbf{T}}_v = \mathbf{F}_v^{\,-1}\hat{\mathbf{S}}_v\mathbf{F}_v^{\,T-1}\,.$$

As a result we obtain the elasticity relation

$$\tilde{\mathbf{T}}_{\mathrm{v}} = \mathbf{F}_{\mathrm{v}}^{-1}\hat{\mathbf{f}}_{\mathrm{v}}\Big(\tfrac{1}{2}\mathbf{F}_{\mathrm{v}}^{\mathrm{T}-1}[\mathbf{C} - \mathbf{C}_{\mathrm{v}}]\mathbf{F}_{\mathrm{v}}^{-1}\Big)\mathbf{F}_{\mathrm{v}}^{\mathrm{T}-1} \tag{10.136}$$

and the evolution equation (flow rule)

$$\eta\dot{\mathbf{C}}_{\mathrm{v}} = 2\,\mathbf{F}_{\mathrm{v}}^{\mathrm{T}}\hat{\mathbf{f}}_{\mathrm{v}}\Big(\tfrac{1}{2}\mathbf{F}_{\mathrm{v}}^{\mathrm{T}-1}[\mathbf{C} - \mathbf{C}_{\mathrm{v}}]\mathbf{F}_{\mathrm{v}}^{-1}\Big)\mathbf{F}_{\mathrm{v}}\,. \tag{10.137}$$

In connection with the polar decomposition

$$\mathbf{F}_{\mathrm{v}} = \mathbf{R}_{\mathrm{v}}\mathbf{U}_{\mathrm{v}} = \mathbf{R}_{\mathrm{v}}\sqrt{\mathbf{C}_{\mathrm{v}}}$$

it should be noted that the intermediate configuration contains a rotational part $\mathbf{R}_{\mathrm{v}}$. This rotational part drops out of the constitutive equations (10.136), (10.137), if the material function $\mathbf{f}_{\mathrm{v}}(\cdot)$ is an isotropic tensor function. Since the elasticity relation (10.132) is motivated as a nonlinear spring in a *Maxwell* model, the assumption of hyperelasticity appears to be natural:

$$\hat{\mathbf{S}}_{\mathrm{v}} = \rho_{\mathrm{R}}\frac{\mathrm{d}w_{\mathrm{v}}(\hat{\mathbf{\Gamma}}_{\mathrm{e}})}{\mathrm{d}\hat{\mathbf{\Gamma}}_{\mathrm{e}}}\,. \tag{10.138}$$

Converting this to the reference configuration along the lines of (10.134) we obtain

$$\hat{\mathbf{S}}_{\mathrm{v}} = \rho_{\mathrm{R}}\frac{\mathrm{d}w_{\mathrm{v}}(\mathbf{F}_{\mathrm{v}}^{\mathrm{T}-1}\mathbf{E}_{\mathrm{e}}\mathbf{F}_{\mathrm{v}}^{-1})}{\mathrm{d}(\mathbf{F}_{\mathrm{v}}^{\mathrm{T}-1}\mathbf{E}_{\mathrm{e}}\mathbf{F}_{\mathrm{v}}^{-1})} \tag{10.139}$$

as an intermediate result. This formula can be further simplified on the basis of the identity

$$\frac{\mathrm{d}f(\mathbf{ASA}^{\mathrm{T}})}{\mathrm{d}(\mathbf{ASA}^{\mathrm{T}})} = \mathbf{A}^{\mathrm{T}-1}\frac{\partial}{\partial\mathbf{S}}f(\mathbf{ASA}^{\mathrm{T}})\mathbf{A}^{-1}\,, \tag{10.140}$$

which, according to (9.60), is valid for scalar-valued tensor functions and invertible tensors $\mathbf{A}$. Applying this to the strain energy $w_{\mathrm{v}}(\hat{\mathbf{\Gamma}}_{\mathrm{e}})$ and observing

$$\hat{\mathbf{\Gamma}}_{\mathrm{e}} = \mathbf{F}_{\mathrm{v}}^{\mathrm{T}-1}\mathbf{E}_{\mathrm{e}}\mathbf{F}_{\mathrm{v}}^{-1}\,,\ \mathbf{E}_{\mathrm{e}} = \mathbf{E} - \mathbf{E}_{\mathrm{v}} = \tfrac{1}{2}(\mathbf{C} - \mathbf{C}_{\mathrm{v}})$$

yields

$$\hat{\mathbf{S}}_{\mathrm{v}} = \hat{\mathbf{f}}_{\mathrm{v}}(\hat{\mathbf{\Gamma}}_{\mathrm{e}}) = \rho_{\mathrm{R}}\mathbf{F}_{\mathrm{v}}\frac{\partial}{\partial \mathbf{E}_{\mathrm{e}}}w_{\mathrm{v}}(\mathbf{F}_{\mathrm{v}}{}^{\mathrm{T}-1}\mathbf{E}_{\mathrm{e}}\mathbf{F}_{\mathrm{v}}{}^{-1})\mathbf{F}_{\mathrm{v}}{}^{\mathrm{T}}. \tag{10.141}$$

In lieu of (10.136) and (10.137) we finally obtain the elasticity relation

$$\tilde{\mathbf{T}}_{\mathrm{v}} = \rho_{\mathrm{R}}\frac{\partial}{\partial \mathbf{E}_{\mathrm{e}}}w_{\mathrm{v}}(\mathbf{F}_{\mathrm{v}}{}^{\mathrm{T}-1}\mathbf{E}_{\mathrm{e}}\mathbf{F}_{\mathrm{v}}{}^{-1}) \tag{10.142}$$

and the evolution equation

$$\tfrac{1}{2}\eta\dot{\mathbf{C}}_{\mathrm{v}} = \rho_{\mathrm{R}}\mathbf{C}_{\mathrm{v}}\frac{\partial}{\partial \mathbf{E}_{\mathrm{e}}}w_{\mathrm{v}}(\mathbf{F}_{\mathrm{v}}{}^{\mathrm{T}-1}\mathbf{E}_{\mathrm{e}}\mathbf{F}_{\mathrm{v}}{}^{-1})\mathbf{C}_{\mathrm{v}} \tag{10.143}$$

for the internal variable $\mathbf{C}_{\mathrm{v}} = \mathbf{F}_{\mathrm{v}}{}^{\mathrm{T}}\mathbf{F}_{\mathrm{v}}$. Again, the rotational part $\mathbf{R}_{\mathrm{v}}$ drops out of these constitutive equations, if the strain energy $w_{\mathrm{v}}(\cdot)$ is an isotropic function.

The simplest example is a deviatoric elasticity relation like

$$\hat{\mathbf{S}}_{\mathrm{v}} = 2\mu_{\mathrm{v}}[\hat{\mathbf{\Gamma}}_{\mathrm{e}} - \tfrac{1}{3}(\mathrm{tr}\hat{\mathbf{\Gamma}}_{\mathrm{e}})\mathbf{1}], \tag{10.144}$$

μ_{v} being a shear modulus. The corresponding evolution equation

$$\eta\overset{\triangle}{\hat{\mathbf{\Gamma}}}_{\mathrm{v}} = 2\mu_{\mathrm{v}}[(\hat{\mathbf{\Gamma}} - \hat{\mathbf{\Gamma}}_{\mathrm{v}}) - \tfrac{1}{3}(\mathrm{tr}(\hat{\mathbf{\Gamma}} - \hat{\mathbf{\Gamma}}_{\mathrm{v}}))\mathbf{1}] \tag{10.145}$$

does not contribute to the volume strain as a result of $\mathrm{tr}\overset{\triangle}{\hat{\mathbf{\Gamma}}}_{\mathrm{v}} = 0$.
Transformation of (10.144) and (10.145) to the reference configuration then leads to the elasticity relation

$$\tilde{\mathbf{T}}_{\mathrm{v}} = \mu_{\mathrm{v}}\left\{\mathbf{C}_{\mathrm{v}}{}^{-1}\mathbf{C}\mathbf{C}_{\mathrm{v}}{}^{-1} - \mathbf{C}_{\mathrm{v}}{}^{-1} - \tfrac{1}{3}\left[\mathrm{tr}(\mathbf{C}\mathbf{C}_{\mathrm{v}}{}^{-1}) - 3\right]\mathbf{C}_{\mathrm{v}}{}^{-1}\right\},$$

$$\tilde{\mathbf{T}}_{\mathrm{v}} = \mu_{\mathrm{v}}\left\{\mathbf{C}_{\mathrm{v}}{}^{-1}\mathbf{C}\mathbf{C}_{\mathrm{v}}{}^{-1} - \tfrac{1}{3}\left[\mathrm{tr}(\mathbf{C}\mathbf{C}_{\mathrm{v}}{}^{-1})\right]\mathbf{C}_{\mathrm{v}}{}^{-1}\right\} \tag{10.146}$$

and the evolution equation

$$\eta\dot{\mathbf{C}}_{\mathrm{v}} = 2\mu_{\mathrm{v}}\left\{\mathbf{C} - \mathbf{C}_{\mathrm{v}} - \tfrac{1}{3}\left[\mathrm{tr}(\mathbf{C}\mathbf{C}_{\mathrm{v}}{}^{-1}) - 3\right]\mathbf{C}_{\mathrm{v}}\right\}, \text{ i.e.}$$

$$\eta\dot{\mathbf{C}}_{\mathrm{v}} = 2\mu_{\mathrm{v}}\left\{\mathbf{C} - \tfrac{1}{3}\left[\mathrm{tr}(\mathbf{C}\mathbf{C}_{\mathrm{v}}{}^{-1})\mathbf{C}_{\mathrm{v}}\right]\right\}. \tag{10.147}$$

10. 2. 3 A General Model of Finite Viscoelasticity

Following the preparatory work of the last section, we can now proceed to present a fairly general model of nonlinear viscoelasticity established on a sound physical basis within the phenomenological theory.[20] This model is a natural generalisation of the (2N + 1) parameter model of classical linear viscoelasticity: in its one-dimensional interpretation it consists of a nonlinear-elastic spring and a series of N nonlinear *Maxwell* elements in parallel (cf. Fig. 5. 7). The concept of dual variables provides a clear formalism for the formulation of the constitutive equations: a definition of N intermediate configurations is essential in view of the multiplicative decompositions

$$\mathbf{F} = \hat{\mathbf{F}}_{ek}\mathbf{F}_{vk}\,, \qquad k = 1, \dots, N\,. \tag{10.148}$$

In analogy to the procedure set out in the previous section, there are now N elasticity relations to be postulated, as well as N tensor-valued evolution equations, each based on its own individual intermediate configuration:

$$\hat{\mathbf{S}}_{\mathrm{ov}k} = \hat{\mathbf{f}}_{\mathrm{ov}k}(\hat{\mathbf{\Gamma}}_{ek}) = \rho_{\mathrm{R}}\frac{\mathrm{d}w_{\mathrm{v}k}(\hat{\mathbf{\Gamma}}_{ek})}{\mathrm{d}\hat{\mathbf{\Gamma}}_{ek}}\,, \tag{10.149}$$

$$\eta_k \overset{\triangle}{\hat{\mathbf{\Gamma}}}_{\mathrm{v}k} = \hat{\mathbf{S}}_{\mathrm{ov}k} = \rho_{\mathrm{R}}\frac{\mathrm{d}w_{\mathrm{v}k}(\hat{\mathbf{\Gamma}}_{ek})}{\mathrm{d}\hat{\mathbf{\Gamma}}_{ek}}\,, \qquad k = 1, \dots, N\,. \tag{10.150}$$

In these evolution equations the viscosities η_k may be material constants or functions of further quantities, characterising the mechanical process or its temporal history. The constitutive model of general viscoelasticity is completed by an assumption for the *equilibrium stress.* This corresponds to the parallel connected nonlinear spring, suggesting a function of the total strain:

$$\tilde{\mathbf{T}}_{\mathrm{eq}} = \mathbf{f}_{\mathrm{eq}}(\mathbf{E}) = \rho_{\mathrm{R}}\frac{\mathrm{d}}{\mathrm{d}\mathbf{E}}\,w_{\mathrm{eq}}(\mathbf{E})\,. \tag{10.151}$$

The sum of the stress tensors $\hat{\mathbf{S}}_{\mathrm{ov}k}$ forms the overstress, which is added to the equilibrium stress to give the total stress. However, when totalling these stresses, we have to bear in mind that the individual contributions have to refer to one and the same configuration in order to be added together. For this reason the stress tensors $\hat{\mathbf{S}}_{\mathrm{ov}k}$ have to be transformed to the reference configuration:

[20] See LION [1997a, b] and LION [1998]; cf. also the monograph LION [2000].

$$\tilde{\mathbf{T}} = \tilde{\mathbf{T}}_{\text{eq}} + \tilde{\mathbf{T}}_{\text{ov}} = \tilde{\mathbf{T}}_{\text{eq}} + \sum_{k=1}^{N} \mathbf{F}_{\text{v}k}^{-1}\hat{\mathbf{S}}_{\text{ov}k}\mathbf{F}_{\text{v}k}^{\text{T}-1}. \tag{10.152}$$

Considering (10.141) and (10.143) we finally obtain the following set of constitutive equations:

$$\tilde{\mathbf{T}} = \rho_{\text{R}} \frac{\text{d}}{\text{d}\mathbf{E}} w_{\text{eq}}(\mathbf{E}) + \rho_{\text{R}} \sum_{k=1}^{N} \frac{\partial}{\partial \mathbf{E}_{\text{e}k}} w_{\text{v}k}(\mathbf{F}_{\text{v}k}^{\text{T}-1}\mathbf{E}_{\text{e}k}\mathbf{F}_{\text{v}k}^{-1}), \tag{10.153}$$

$$\tfrac{1}{2}\eta_k \dot{\mathbf{C}}_{\text{v}k} = \rho_{\text{R}} \mathbf{C}_{\text{v}k} \frac{\partial}{\partial \mathbf{E}_{\text{e}k}} w_{\text{v}k}(\mathbf{F}_{\text{v}k}^{\text{T}-1}\mathbf{E}_{\text{e}k}\mathbf{F}_{\text{v}k}^{-1})\mathbf{C}_{\text{v}k}, \quad k = 1, \dots, N. \tag{10.154}$$

For $k = 1, \dots, N$ we have

$$\mathbf{E}_{\text{e}k} = \mathbf{E} - \mathbf{E}_{\text{v}k} = \tfrac{1}{2}(\mathbf{C} - \mathbf{C}_{\text{v}k}), \ \mathbf{C} = \mathbf{F}^{\text{T}}\mathbf{F} \tag{10.155}$$

and

$$\mathbf{C}_{\text{v}k} = \mathbf{F}_{\text{v}k}^{\text{T}}\mathbf{F}_{\text{v}k}. \tag{10.156}$$

Given the stress relation (10.153) combined with the evolution equations (10.154) we are in possession of a general constitutive model of nonlinear viscoelasticity, based on clear physical assumptions and principles. The material properties in these constitutive equations are represented in a concrete manner by means of the scalar-valued material functions $w_{\text{eq}}(\cdot)$ and $w_{\text{ov}k}(\cdot)$, and the viscosities η_k. The viscosities may be functions of appropriately selected process quantities (e.g. stresses and - possibly - further internal variables), in order to take nonlinearities in the rate-dependence into consideration. In principle, viscosities that depend on the process history are also feasible and may be physically meaningful.[21]

Contrary to the constitutive equations (10.81) and (10.82), which define the theoretical frame for representing viscoelasticity using internal variables, the internal variables of the constitutive modelling in this section are endowed with a concrete physical meaning, at least in the context of the phenomenological (macroscopic) theory.[22]

[21] A step in that direction is found in Haupt & Sedlan [2001].

[22] Consequences in view of the numerical solution of initial-boundary-value problems are developed in Hartmann [2001].

11 Plasticity

11.1 Rate-Independent Functionals

The theory of *plasticity*[1] models rate-independent material behaviour with hysteresis. Plasticity expands on the theory of elasticity by taking internal dissipation into consideration. However, internal friction is represented from a completely different viewpoint here, compared with viscoelasticity: whereas viscoelastic material behaviour is characterised by its fading memory properties, the theory of plasticity tends to express a perfect memory of a material body.

The marked memory capacity of a material, the modelling of which constitutes the theme of the plasticity theory, results from the rate-independence of the material response: slowing down the process history causes the deformation events to be shifted further back in time. Whereas a shifting of this kind in viscoelasticity weakens the effects of past deformation processes on the present stress, all the actions of the deformation history in rate-independent plasticity are retained in full. In particular the rate-independence of the material response excludes any relaxation property.

The theory of plasticity demands the representation of rate-independent functionals.

---

[1] See GREEN & NAGHDI [1965, 1971]. A comprehensive presentation of the theory of plasticity is given by LUBLINER [1990]. See also KHAN & HUANG [1995] and LEVITAS [1993]. The development of the plasticity theory is reviewed critically in NAGHDI [1990], where many references are to be found. BERTRAM [1999] develops a fairly general concept of elastoplasticity, discussing the extent to which the current state of finite plasticity fits into the scope of a constitutive theory within the tradition of rational mechanics.

Definition 11. 1

A functional is called *rate-independent* if the output quantity depends on the past history of the input process, yet is independent of the rate at which one and the same process history takes place. ◊

A general method for representing rate-independent functionals is the arclength description. In elementary mechanics the arclength description arises out of the possibility of replacing time by means of the arclength when portraying the motion of a point. The arclength is nothing other than the distance covered on the trajectory. An arclength can even be associated with a strain process.

Definition 11. 2

If $t \longmapsto \mathbf{E}(t)$ is a given strain process occurring at a material point $\mathscr{P} \in \mathscr{B}$, then the function

$$t \longmapsto s(t) = s(t_0) + \int_{t_0}^{t} \|\mathbf{E}'(\tau)\| \mathrm{d}\tau \quad \Longleftrightarrow \quad \dot{s}(t) = \|\dot{\mathbf{E}}(t)\| = \sqrt{\mathrm{tr}\big(\dot{\mathbf{E}}^2(t)\big)} \tag{11.1}$$

is termed *arclength* or *accumulated strain.* The function defined by the identity

$$t \longmapsto \mathbf{E}(t) = \bar{\mathbf{E}}\big(s(t)\big) \ , \tag{11.2}$$

namely

$$s \longmapsto \mathbf{E} = \bar{\mathbf{E}}(s) \tag{11.3}$$

is known as the *arclength representation* of the strain process $t \longmapsto \mathbf{E}(t)$. ◊

A strain process - like any curve in a finite dimensional space - clearly consists of an arclength representation of the strain tensor and the arclength itself, which is a function of time t:

$$t \longmapsto \mathbf{E}(t) \mathrel{\widehat{=}} \left\{\bar{\mathbf{E}}(s), s(t)\right\} . \tag{11.4}$$

For the geometry of the trajectory in the strain space, the curve parameter t is irrelevant, but not for the physical process $t \longmapsto \mathbf{E}(t)$; for this $\dot{s}(t)$ just stipulates the rate at which the trajectory is passed through. Accordingly, the reduced form (7.26) can also be regarded as a functional depending on two independent functions, namely on the trajectory $\sigma \longmapsto \bar{\mathbf{E}}(\sigma)$ $(0 \leq \sigma \leq s)$ and the "path history" $\tau \longmapsto s(\tau)$ $(0 \leq \tau \leq t)$:

$$\mathbf{E}(\tau)\Big|_{\tau \leqq t} \longmapsto \widetilde{\mathbf{T}}(t) = \underset{\substack{\sigma \leqq s(t) \\ \tau \leqq t}}{\mathfrak{R}} \left[\bar{\mathbf{E}}(\sigma(\tau)), s(\tau)\right]. \tag{11.5}$$

If a constitutive functional is *rate-independent*, then its dependence on the second argument, the function $\tau \longmapsto s(\tau)$, ceases to apply. In this case the current state of stress is obtained in the arclength representation:

$$t \longmapsto \widetilde{\mathbf{T}}(t) = \bar{\widetilde{\mathbf{T}}}(s(t)) = \underset{\sigma \leqq s(t)}{\mathfrak{F}} \left[\bar{\mathbf{E}}(\sigma)\right]. \tag{11.6}$$

Conversely, it becomes obvious that a constitutive equation in the form of (11.6) represents a rate-independent functional, since the arclength representation of the trajectory no longer contains information as to how fast the process was carried out. These arguments are summarised in the following theorem.

Theorem 11. 1

The current stress is a rate-independent functional of the strain history if and only if the constitutive equation can be represented in the general form of (11.6).[2] ■

The theorem states that in the theory of plasticity the arclength s replaces the time t as independent variable. More generally, every parameter which constitutes a monotonically increasing function of time t can be applied as an independent variable to replace the time. This fact opens up further possibilities for modelling material properties used in the *generalised arclength description* (see below).[3] The physical significance of Theorem 11. 1 is explained by the following

[2] The same statement was expounded along the same lines by PIPKIN & RIVLIN [1965]. See also OWEN & WILLIAMS [1968].

[3] The arclength representation was applied extensively by K.C. Valanis. In particular, he introduced a material-dependent arclength for the purpose of describing rate-independent hysteresis and nonlinear hardening behaviour under mechanical and thermomechanical points of view. See VALANIS [1971, 1974, 1975, 1977, 1980, 1981, 1995].

Correspondence Principle

For every rate-dependent functional in viscoelasticity there is a rate-independent functional, which is obtained when time is replaced by the (kinematic or generalised) arclength. Accordingly, there is a material model of *plasticity* to correspond to each material model of *viscoelasticity*. ♦

To begin with a very simple example, the application of this correspondence principle leads to a linear functional of the deformation history in the arclength representation. For isotropic material behaviour and small deformations a functional of this kind can be written in analogy to (5.51) as

$$\mathbf{T}(t) = \bar{\mathbf{T}}\big(s(t)\big) = \int_0^{s(t)} \Big\{2\mu(s - \bar{s})\mathbf{E}^{\mathrm{D}\prime}(\bar{s}) + K(s - \bar{s})\big[\mathrm{tr}\mathbf{E}'(\bar{s})\big]\mathbf{1}\Big\}\mathrm{d}\bar{s}\,. \tag{11.7}$$

It is plain to see that this functional is, as a whole, physically nonlinear, since the kinematic arclength $s(t)$ is a nonlinear functional of the strain history according to its definition (11.1).

In the following investigation the physical nonlinearity of the functional (11.7) is to be explained using one-dimensional deformation processes. For simplicity, we assume proportionality between the material functions $\gamma(s)$ and $\kappa(s)$ and equate

$$\kappa(s) = 2\gamma(s)\,\frac{(1+\nu)}{3(1-2\nu)} \tag{11.8}$$

($\nu = $ const.), which corresponds to

$$\bar{\mathbf{T}}\big(s(t)\big) = 2\int_0^{s(t)} \gamma(s - \bar{s})\Big[\mathbf{E}^{\mathrm{D}\prime}(\bar{s}) + \frac{1+\nu}{3(1-2\nu)}\big[\mathrm{tr}\mathbf{E}'(\bar{s})\big]\mathbf{1}\Big]\mathrm{d}\bar{s}\;[4]\,. \tag{11.9}$$

For uniaxial processes we have $\mathbf{E} = \varepsilon(t)\boldsymbol{e}_x \otimes \boldsymbol{e}_x - \varepsilon_{\mathrm{Q}}(t)\big(\boldsymbol{e}_y \otimes \boldsymbol{e}_y + \boldsymbol{e}_z \otimes \boldsymbol{e}_z\big)$ and

$\mathbf{T}(t) = \sigma(t)\boldsymbol{e}_x \otimes \boldsymbol{e}_x$. This leads to $\varepsilon_{\mathrm{Q}} = -\,\nu\varepsilon$ and the uniaxial stress relation

$$\sigma = \bar{\sigma}(s) = \int_0^{s} G(s - \bar{s})\bar{\varepsilon}\,'(\bar{s})\mathrm{d}\bar{s}\,, \tag{11.10}$$

[4] Cf. VALANIS [1995], eq. (63).

with

$$G(s) = 2(1 + \nu)\gamma(s) \tag{11.11}$$

and

$$s(t) = \sqrt{1 + 2\nu^2} \int_0^t |\varepsilon'(\tau)| \mathrm{d}\tau \quad \Longleftrightarrow \quad \dot{s}(t) = \sqrt{1 + 2\nu^2}\,|\dot{\varepsilon}(t)| \,. \tag{11.12}$$

The constitutive equations (11.7) and (11.10) have the same structure as the definition (5.51) of linear isotropic viscoelasticity and its one-dimensional implication (5.76). At this stage it is easy to recognise the correspondence principle mentioned above.

In order to illustrate the physical implication of this theory, let a decaying exponential function be chosen for the kernel function $R(s)$,

$$G(s) = E\,\mathrm{e}^{-\beta s} \tag{11.13}$$

(E and β are material constants), so the special functional

$$\sigma = \bar{\sigma}(s) = \int_0^s E\,\mathrm{e}^{-\beta(s - \bar{s})}\,\bar{\varepsilon}'(\bar{s})\mathrm{d}\bar{s} \,, \tag{11.14}$$

is discussed below. The following theorem demonstrates that relation (11.14) represents rate-independent hysteresis behaviour.

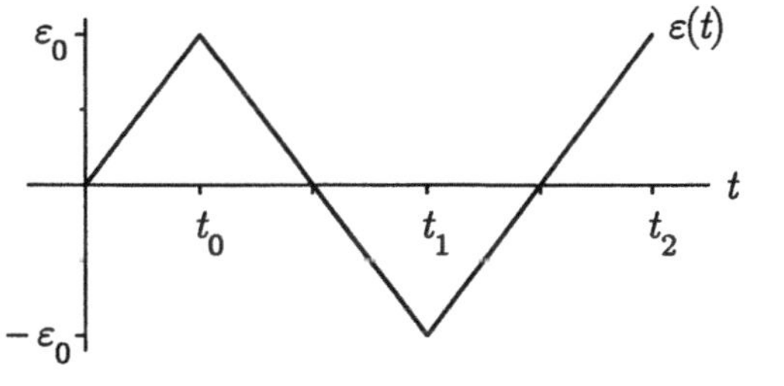

Figure 11. 1: Cyclic strain process

Theorem 11. 2

For a cyclic strain process with the strain amplitude ε_0 (see Figure 11. 1) the functional (11.14) produces the following stress response:

$\underline{0 \leq t \leq t_0}$

$$\sigma_0(\varepsilon) = \frac{E}{\beta}\left[1 - e^{-\beta\varepsilon}\right] \tag{11.15}$$

$\underline{t_0 \leq t \leq t_1}$

$$\sigma_1(\varepsilon) = \frac{E}{\beta}\left[-1 + 2e^{-\beta(\varepsilon_0 - \varepsilon)} - e^{-\beta(2\varepsilon_0 - \varepsilon)}\right] \tag{11.16}$$

$\underline{t_1 \leq t \leq t_2}$

$$\sigma_2(\varepsilon) = \frac{E}{\beta}\left[1 - 2e^{-\beta(\varepsilon_0 + \varepsilon)} + 2e^{-\beta(3\varepsilon_0 + \varepsilon)} - e^{-\beta(4\varepsilon_0 + \varepsilon)}\right] \tag{11.17}$$

Given unlimited continuation, a cyclic strain process with the strain amplitude ε_0 leads to a cyclic stress process with the amplitude

$$\sigma_\infty(\varepsilon_0) = \frac{E}{\beta}\tanh(\beta\varepsilon_0) \; . \tag{11.18}$$

□

Proof [5]

The functional (11.14) has to be evaluated at every time interval during which the strain process proceeds monotonically.

For $0 \leq t \leq t_0$ we initially have $\dot{\varepsilon}(t) > 0$ and (11.12) leads to

$$s(t) = \varepsilon(t) \; , \tag{11.19}$$

if we equate $\nu = 0$ for simplicity. Inserting this into (11.14) yields

$$\bar{\sigma}_0(s) = \int_0^s E\, e^{-\beta(s - \bar{s})}\, \bar{\varepsilon}'(\bar{s})\mathrm{d}\bar{s} \; , \tag{11.20}$$

$$\sigma_0(\varepsilon) = \int_0^\varepsilon E\, e^{-\beta(\varepsilon - \bar{\varepsilon})}\mathrm{d}\bar{\varepsilon} = \frac{E}{\beta}\left[1 - e^{-\beta\varepsilon}\right] ,$$

i.e. (11.15). In the next time interval, $t_0 \leq t \leq t_1$, we have $\dot{\varepsilon}(t) < 0$ and we obtain from (11.12) the arclength

[5] A similar analysis can be found in VALANIS [1971], pp. 542.

$$s(t) = s(t_0) + \int_{t_0}^{t} |\dot{\varepsilon}(\tau)| d\tau = \varepsilon_0 - \int_{t_0}^{t} \dot{\varepsilon}(\tau) d\tau \text{ , i.e.}$$

$$s(t) = 2\varepsilon_0 - \varepsilon(t) \ . \tag{11.21}$$

(11.14) yields the stress response

$$\bar{\sigma}_1(s) = \int_0^{s(t_0)} E\, e^{-\beta(s - \bar{s})}\, \bar{\varepsilon}'(\bar{s}) d\bar{s} - \int_{s(t_0)}^{s} E\, e^{-\beta(s - \bar{s})}\, \bar{\varepsilon}'(\bar{s}) d\bar{s} \ , \tag{11.22}$$

$$\sigma_1(\varepsilon) = \int_0^{\varepsilon_0} E\, e^{-\beta(2\varepsilon_0 - \varepsilon - \bar{\varepsilon})} d\bar{\varepsilon} - \int_{\varepsilon_0}^{2\varepsilon_0 - \varepsilon} E\, e^{-\beta(2\varepsilon_0 - \varepsilon - \bar{\varepsilon})} d\bar{\varepsilon} \ .$$

Evaluation of the integrals leads to (11.16).

In the third time interval, $t_1 \leq t \leq t_2$, we verify (11.17) in the same way from $\dot{\varepsilon}(t) > 0$,

$$s(t) = s(t_1) + \int_{t_1}^{t} \dot{\varepsilon}(\tau) d\tau = 4\varepsilon_0 - \varepsilon(t) \tag{11.23}$$

and

$$\sigma_2(\varepsilon) = \int_0^{\varepsilon_0} \Big\{E\, e^{-\beta(4\varepsilon_0 + \varepsilon - \bar{\varepsilon})}\Big\} d\bar{\varepsilon} - \int_{\varepsilon_0}^{3\varepsilon_0} \Big\{\cdots\Big\} d\bar{\varepsilon} + \int_{\varepsilon_0}^{4\varepsilon_0 + \varepsilon} \Big\{\cdots\Big\} d\bar{\varepsilon} \tag{11.24}$$

$$\sigma_2(\varepsilon) = \frac{E}{\beta}\Big\{\Big[e^{-\beta(4\varepsilon_0 - \varepsilon - \bar{\varepsilon})}\Big]\Big|_0^{\varepsilon_0} - \Big[e^{-\beta(4\varepsilon_0 - \varepsilon - \bar{\varepsilon})}\Big]\Big|_{\varepsilon_0}^{3\varepsilon_0}$$
$$+ \Big[e^{-\beta(4\varepsilon_0 - \varepsilon - \bar{\varepsilon})}\Big]\Big|_{3\varepsilon_0}^{4\varepsilon_0 + \varepsilon}\Big\} \ .$$

An unlimited continuation of the cyclic strain process leads to

$$\sigma_{2n-1}(\varepsilon) = \frac{E}{\beta}\Big\{ - 1 - 2e^{\beta\varepsilon}\Big[\sum_{k=1}^{2n-1} (-1)^k e^{-(2k-1)\beta\varepsilon_0} + \frac{1}{2} e^{-(4n-2)\beta\varepsilon_0}\Big]\Big\} \ ,$$

$$\sigma_{2n-1}(\varepsilon) = \frac{E}{\beta}\Big\{ - 1 + 2e^{\beta\varepsilon}\Big[\frac{e^{-\beta\varepsilon_0} + e^{-(4n-1)\beta\varepsilon_0}}{1 + e^{-2\beta\varepsilon_0}} - \frac{1}{2} e^{-(4n-2)\beta\varepsilon_0}\Big]\Big\} \tag{11.25}$$

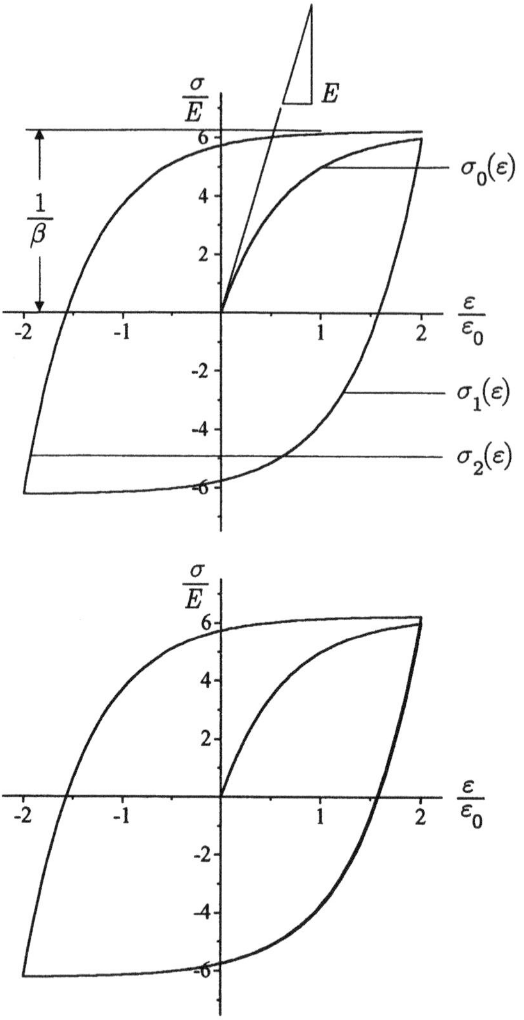

Figure 11. 2: Rate-independent hysteresis

and

$$\sigma_{2n}(\varepsilon) = \frac{E}{\beta}\left\{ 1 + 2e^{-\beta\varepsilon}\left[\sum_{k=1}^{2n} (-1)^k e^{-(2k-1)\beta\varepsilon_0} - \frac{1}{2}e^{-4n\beta\varepsilon_0}\right]\right\},$$

$$\sigma_{2n}(\varepsilon) = \frac{E}{\beta}\left\{ 1 - 2e^{-\beta\varepsilon}\left[\frac{e^{-\beta\varepsilon_0} - e^{-(4n+1)\beta\varepsilon_0}}{1 + e^{-2\beta\varepsilon_0}} + \frac{1}{2}e^{-4n\beta\varepsilon_0}\right]\right\}. \qquad (11.26)$$

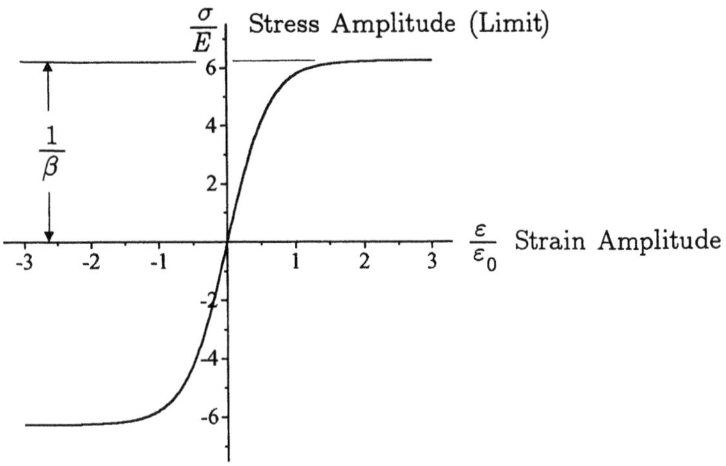

Figure 11. 3: Cyclic stress-strain curve

In passing to the limit $n \to \infty$ this implies

$$\lim_{n \to \infty} \sigma_{2n-1}(\varepsilon) = -\frac{E}{\beta}\,\frac{e^{\beta\varepsilon_0} + e^{-\beta\varepsilon_0} - 2e^{\beta\varepsilon}}{e^{\beta\varepsilon_0} + e^{-\beta\varepsilon_0}} \qquad (\dot{\varepsilon} < 0,\ \ \varepsilon_0 \geq \varepsilon \geq \varepsilon_0) \quad (11.27)$$

as well as

$$\lim_{n \to \infty} \sigma_{2n}(\varepsilon) = \frac{E}{\beta}\,\frac{e^{\beta\varepsilon_0} + e^{-\beta\varepsilon_0} - 2e^{-\beta\varepsilon}}{e^{\beta\varepsilon_0} + e^{-\beta\varepsilon_0}} \qquad (\dot{\varepsilon} > 0,\ \ \varepsilon_0 \geq \varepsilon \geq \varepsilon_0)\,. \quad (11.28)$$

This produces the stress amplitude (11.18), i.e.

$$\lim_{n \to \infty} \sigma_{2n}(\varepsilon_0) = \sigma_{\infty}(\varepsilon_0) = \frac{E}{\beta}\tanh(\beta\varepsilon_0)$$

$$\text{or} \quad \lim_{n \to \infty} \sigma_{2n-1}(-\varepsilon_0) = \sigma_{\infty}(-\varepsilon_0) = -\frac{E}{\beta}\tanh(\beta\varepsilon_0)\,. \qquad \blacksquare$$

The stress-strain curves σ_0, σ_1, σ_2 are shown in Fig. 11. 2. An open hysteresis loop is visible. Very soon this merges into a closed limit cycle, as demonstrated in (11.25), (11.26) and Fig. 11. 2 (lower part).
The cyclic curve $\sigma_{\infty}(\varepsilon_0)$ - the stress amplitude of the limit cycles depending on the strain amplitude - is illustrated in Fig. 11. 3.

In conjunction with the formulation of constitutive models of plasticity it may be expedient to begin with a *generalised arclength*, whose dependence on the process history is introduced as a tool for representing a material property. An example for the definition of a generalised arclength is the plastic arclength according to equation (5.188) or the plastic work according to equation (5.189).[6] A generalised arclength

$$t \longmapsto z(t) = \underset{\tau \leqq t}{\mathfrak{Z}} \left[\mathbf{E}(\tau)\right] \tag{11.29}$$

is distinguished by the following three characteristics:

$$1) \quad \dot{z}(t) > 0 \ \text{ for } \ \dot{\mathbf{E}}(t) \neq \mathbf{0}\,, \tag{11.30}$$

$$2) \quad \dot{z}(t) = 0 \ \text{ for } \ \dot{\mathbf{E}}(t) = \mathbf{0}\,, \tag{11.31}$$

$$3) \quad \dot{z} \longmapsto |\alpha|\dot{z} \quad \text{for} \quad \dot{\mathbf{E}} \longmapsto \alpha\dot{\mathbf{E}}\,. \tag{11.32}$$

An explicit representation of material properties in rate-independent plasticity then demands the specification of a generalised arclength

$$t \longmapsto z(t) = \underset{\tau \leqq t}{\mathfrak{Z}} \left[\mathbf{E}(\tau)\right]$$

and a functional relation

$$\tilde{\mathbf{T}}(t) = \bar{\tilde{\mathbf{T}}}(z(t)) = \underset{\zeta \leqq z(t)}{\mathfrak{G}} \left[\bar{\mathbf{E}}(\zeta)\right]. \tag{11.33}$$

11. 2 Representation by Means of Internal Variables

In order to represent a rate-independent functional within the framework of internal variables and evolution equations, one has to formulate these with respect to a generalised arclength $t \longmapsto z(t)$:

$$\tilde{\mathbf{T}} = \mathbf{f}(\mathbf{E}, q_1, \dots, q_N)\,, \tag{11.34}$$

---

[6] A representation by means of a generalised arclength - a constitutive function of the kinematic arclength - was proposed by VALANIS [1971] under the heading *Endochronic Theory of Plasticity*. See VALANIS [1971, 1975, 1977, 1980], RIVLIN [1981]; VALANIS [1981]; HAUPT [1977].

$$q_k'(z) = f_k(\mathbf{E}(z), q_1(z), \dots, q_N(z)) \qquad (k = 1, \dots, N), \tag{11.35}$$

$$z(t) = \underset{\tau \leqq t}{\mathfrak{Z}} [\mathbf{E}(\tau)]. \tag{11.36}$$

All solutions of evolution equations (11.35) are rate-independent functionals without any relaxation properties: in view of $\dot{q}_k(t) = q_k'(z)\dot{z}(t)$ the time-dependence of the internal variables is determined by the evolution equations

$$\dot{q}_k(t) = f_k(\mathbf{E}(t), q_1(t), \dots, q_N(t))\dot{z}(t) \qquad (k = 1, \dots, N), \tag{11.37}$$

which show that every deformation process is an equilibrium process. Due to the property (11.31) each static continuation $(\dot{\mathbf{E}}(t) = \mathbf{0})$ of any given process leads to temporarily constant internal variables q_k. The *correspondence principle* established in connection with Theorem 11. 1 is, of course, also valid within the theory of internal variables: every evolution equation in rate-dependent viscoelasticity can theoretically be adopted and used to represent a material model in rate-independent plasticity.

To illustrate this aspect by means of a simple example, let us take a closer look at the one-dimensional stress-strain relation

$$\sigma = E(\varepsilon - q) \tag{11.38}$$

in combination with the evolution equation

$$q'(z) = \beta(\varepsilon(z) - q(z)), \tag{11.39}$$

which is a transfer of the model (5.90), (5.91) of linear viscoelasticity on to the arclength representation (with $E = 0$ and $\hat{E} \overset{\text{Def}}{=} E$). The evolution equation (11.39) can be integrated: the equivalent formulation

$e^{-\beta z}\left(e^{\beta z} q(z)\right)' = \beta\varepsilon(z)$ produces

$$q(z) = \int_0^z \beta\, e^{-\beta(z-\zeta)}\varepsilon(\zeta)d\zeta = \varepsilon(z) - \int_0^z e^{-\beta(z-\zeta)}\varepsilon'(\zeta)d\zeta;$$

with (11.38) this implies

$$\sigma(z) = \int_0^z E\, e^{-\beta(z-\zeta)}\varepsilon'(\zeta)d\zeta. \tag{11.40}$$

The constitutive equations (11.38), (11.39) are equivalent to (11.14), if $z(t)$ is the kinematic arclength:

$$z(t) = \int_0^t |\dot{\varepsilon}(\tau)| \mathrm{d}\tau \quad \Longleftrightarrow \quad \dot{z}(t) = |\dot{\varepsilon}(\tau)| \, .$$

The model described by (11.38) and (11.39) (similar viscoelasticity) can be summarised in a single evolution equation for the stress:

$$\sigma'(z) = E\big(\varepsilon'(z) - q'(z)\big) = E\varepsilon'(z) - E\beta\big(\varepsilon(z) - q(z)\big) \text{ implies}$$

$$\sigma'(z) = E\varepsilon'(z) - \beta\sigma(z) \, . \tag{11.41}$$

In the time domain we have

$$\dot{\sigma}(t) = E\,\dot{\varepsilon}\,(t) - b\dot{z}(t)\sigma(t) \, . \tag{11.42}$$

Apart from the resemblance of its formulation, the differential equation (11.41) has nothing else in common with the physical predictions of the evolution equation (5.104) of the *Maxwell* model. It represents a rate-independent hysteresis for $\dot{z}(t) = \dot{s}(t)$. A detailed discussion on this property features in Theorem 11. 2.

It is typical for this hysteresis (depicted in Fig. 11. 2) that a stationary limit cycle is reached after very few periods, during the course of which the stress amplitude scarcely alters. That means it is presumably as good as impossible to represent phenomena of cyclic hardening by material models in the form of (11.10) or (11.38), (11.39) on the basis of the kinematic arclength, i.e. $z = s$.[7]

This situation can be rectified by introducing a generalised arclength, as mentioned above. An example is provided by the evolution equations

$$\dot{z}(t) = \frac{\dot{s}(t)}{1 + ap} \, , \tag{11.43}$$

$$\dot{p}(t) = \frac{\dot{s}(t)}{s_0} \Big(|\sigma(t)| - p(t) \Big) \, , \tag{11.44}$$

$$\dot{s}(t) = |\dot{\varepsilon}\,(t)| \, , \tag{11.45}$$

[7] Experimental evidence of cyclic hardening and appropriate constitutive modelling is presented in BRUHNS et al. [1992].

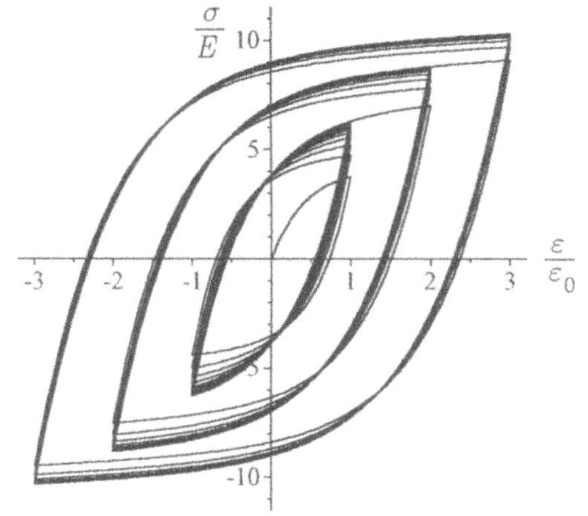

Figure 11. 4: Cyclic hardening with increasing strain amplitudes

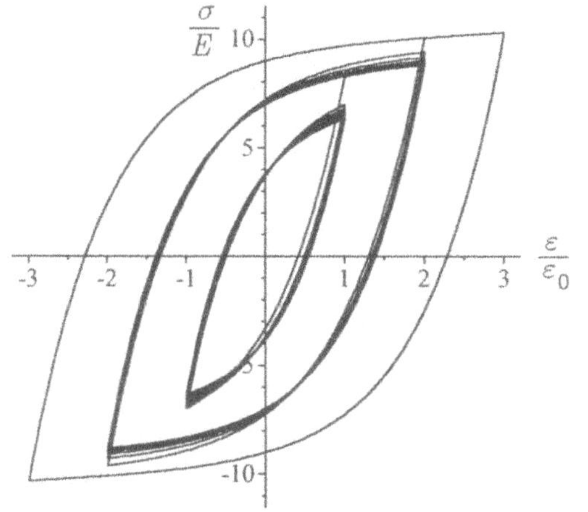

Figure 11. 5: Cyclic softening with decreasing strain amplitudes

in which the additional internal variable p depends on the process history by means of (11.44).[8]

[8]Evolution equations of this kind were developed by KAMLAH [1994]. See HAUPT et al. [1991, 1992a]; HAUPT & KAMLAH [1995]. The concept of a generalised arclength was successfully applied by LION [1994]; LÜHRS [1997], pp. 24; LÜHRS & HAUPT [1997] and HARTMANN et al. [1998].

As a result of these evolution equations, the generalised arclength $t \longmapsto z(t)$ is a rate-independent functional of the process history.

Now the one-dimensional material model consists of the nonlinearly coupled differential equations (11.42) - (11.45) and contains a total of four material parameters, namely E, b, a, and s_0. This system of equations cannot be solved in closed form. A numerical solution is illustrated in Fig. 11. 4. Various cyclic strain processes are given with different strain amplitudes. The strain amplitude is raised in steps, the mean strain being zero. Following each increase in the strain amplitude a new cyclic hardening begins: at each stage the stress response converges towards a new stationary hysteresis loop.

In Fig. 11. 5 the strain amplitude is decreased one step at a time, which leads to cyclic softening in the stress response at each stage. The curves in Figs. 11.4 and 11.5 illustrate the numerical solution of equations (11.42) - (11.45). As further numerical studies demonstrate, the behaviour of the model is similar in the case of nonzero mean strains. The limit values calculated for the stress amplitudes are independent of the mean strain. The mean stress always tends towards zero if the parameter b is off-zero. It is an essential characteristic of this model that every change in the strain amplitude is answered by a fresh process of cyclic hardening or softening.

A formulation of the equations (11.38), (11.39) for 3-dimensional finite deformations could look like this:

$$\mathbf{S} = 2\mu(\mathbf{y} - \mathbf{a})\,, \tag{11.46}$$

$$\overset{\triangledown}{\mathbf{y}} = \beta\dot{z}(t)(\mathbf{a} - \mathbf{y})\,. \tag{11.47}$$

Here, $\mathbf{a} = (1/2)(\mathbf{1} - \mathbf{F}\mathbf{F}^{\mathrm{T}})$ is the *Finger* tensor and $\mathbf{y}$ an internal variable of the same type of strain. The weighted *Cauchy* tensor $\mathbf{S}$ is the dual stress tensor to $\mathbf{a}$ within the scope of *family 2* (see (8.88) and (8.90), given $\mathbf{\Psi} = \mathbf{F}$). By eliminating $\mathbf{y}$ it is possible to obtain an evolution equation for $\mathbf{S}$:

Theorem 11. 3

The following evolution equation for the stress tensor is in accordance with the system of equations (11.46), (11.47):

$$\overset{\triangledown}{\mathbf{S}} = 2\mu\mathbf{D} - \beta\dot{z}(t)\mathbf{S}\,. \tag{11.48}$$

□

Proof

(11.46) leads to $\mathbf{y} = \frac{1}{2\mu}\,\mathbf{S} + \mathbf{a}$. By differentiation and the application of (1.131) we obtain

$$\overset{\triangledown}{\mathbf{y}} = \frac{1}{2\mu}\,\overset{\triangledown}{\mathbf{S}} + \overset{\triangledown}{\mathbf{a}} = \frac{1}{2\mu}\,\overset{\triangledown}{\mathbf{S}} - \mathbf{D}\,.$$

Insertion into (11.47) produces (11.48):

$$\frac{1}{2\mu}\,\overset{\triangledown}{\mathbf{S}} - \mathbf{D} = \beta\dot{z}(t)(\mathbf{a} - \mathbf{y}) = -\,\dot{z}(t)\beta\,\frac{1}{2\mu}\,\mathbf{S}\,. \qquad \blacksquare$$

It is also possible to formulate evolution equations with dual variables of the *family 1*.

Theorem 11. 4

The stress-strain relation

$$\mathbf{S} = 2\mu(\mathbf{A} - \mathbf{Y}) \tag{11.49}$$

and the evolution equation

$$\overset{\vartriangle}{\mathbf{Y}} = \beta\dot{z}(t)(\mathbf{A} - \mathbf{Y}), \tag{11.50}$$

give rise to the differential equation

$$\overset{\vartriangle}{\mathbf{S}} = 2\mu\mathbf{D} - \beta\dot{z}(t)\mathbf{S}\,. \tag{11.51}$$

□

Proof

(11.49) implies $\mathbf{Y} = \mathbf{A} - \frac{1}{2\mu}\,\mathbf{S}$ and $\overset{\vartriangle}{\mathbf{Y}} = \overset{\vartriangle}{\mathbf{A}} - \frac{1}{2\mu}\,\overset{\vartriangle}{\mathbf{S}} = \mathbf{D} - \frac{1}{2\mu}\,\overset{\vartriangle}{\mathbf{S}}$.

Accordingly, it holds that

$$\mathbf{D} - \frac{1}{2\mu}\,\overset{\vartriangle}{\mathbf{S}} = \beta\,\dot{z}(t)\,\frac{1}{2\mu}\,\overset{\vartriangle}{\mathbf{S}}\,. \qquad \blacksquare$$

A 3-dimensional generalisation of the evolution equations (11.43) to (11.45) for the generalised arclength reads as follows:

$$\dot{z}(t) = \frac{\dot{s}(t)}{1 + ap}\,, \tag{11.52}$$

$$\dot{p}(t) = \frac{\dot{s}(t)}{s_0}\Big(\|\mathbf{S}(t)\| - p(t)\Big)\,, \tag{11.53}$$

$$\dot{s}(t) = \|\mathbf{D}(t)\|\,. \tag{11.54}$$

The *correspondence principle*, i.e. the structural analogy to viscoelasticity, serves as a stimulation for further proposals regarding the formulation of material models in rate-independent plasticity: thus, a more general rate-independent functional can be set up by an analogous transfer of model (5.108), (5.109), or by transferring equations (5.110), (5.111). This corresponds to a superposition of a finite number of stresses to form a resultant total stress. A superposition of this kind enables the form of the stress-strain curve to be designed within wide limits so that the results of experiments can be optimally approximated. Even a continuous spectrum for the kernel function $G(s)$ or $G(z)$ in the integral (11.10) is feasible as a fairly general consequence of the correspondence principle (cf. (5.116)). Finally, the evolution equations introduced in Sects. 10. 2. 2 and 10. 2. 3 for strain-type internal variables are likewise transferable to rate-independent plasticity.

All material models in plasticity, which can be constructed in this way, have one common feature: they are basically unable to reflect a range of stress or strain, in which the material behaves in a purely elastic fashion: Every process of stress followed by unloading leads to a permanent deformation, in other words, every cyclic process is connected to a hysteresis. On the one hand, this property appears to be quite plausible; on the other hand we take note that the hysteresis effects observed below a critical amplitude are so small for numerous materials that it seems justified (especially in the case of metals) to ignore them. In these situations it makes sense to employ constitutive models which rely upon drawing a distinction between different cases, thus being capable of incorporating a so-called *elastic range*. In the case of rate-independent material behaviour this leads to the theory of *elastoplasticity*.

11. 3 Elastoplasticity

11. 3. 1 Preliminary Remarks

There are numerous instances in which the theory of plasticity with an *elastic range* approaches closely to reality. It is modelled by means of a *yield surface* in the stress space (or strain space) in conjunction with distinguishing between different cases. The classical theory of geometrically linear plasticity is an example of this kind of model. The basic idea is a yield function that delimits an elastic domain within the stress space by means of a yield surface. Yield and loading conditions, which provide criteria for the evolution of internal variables, are derived from the yield function. Moreover, assuming an associated flow rule, the evolution equation for the plastic strain can also be determined on the basis of the yield function.

The constitutive equations (5.178) - (5.186) of classical plasticity define a rate-independent functional within the scope of the generalised arclength representation. This becomes evident when one divides the flow rule (5.183) and the evolution equation (5.184) of kinematic hardening by the rate $\dot{s}_p(t)$ of the plastic arclength. In addition to the elasticity relation

$$\mathbf{T}^D = 2\mu(\mathbf{E}^D - \mathbf{E}_p) \qquad \tfrac{1}{3}(\mathrm{tr}\mathbf{T}) = \kappa(\mathrm{tr}\mathbf{E}) \tag{11.55}$$

we then obtain the evolution equations for the plastic strain $\mathbf{E}_p$ and the kinematic hardening variable (back stress) $\mathbf{X}$:

$$\mathbf{E}_p{}'(z) = \frac{1}{k}\frac{3}{2}\left[\mathbf{T}^D - \mathbf{X}^D\right], \tag{11.56}$$

$$\mathbf{X}'(z) = \frac{c}{k}\frac{3}{2}\left[\mathbf{T}^D - \mathbf{X}^D\right] - b\mathbf{X}. \tag{11.57}$$

In this case the generalised arclength is the accumulated plastic strain,

$$\dot{z}(t) = \sqrt{\tfrac{2}{3}}\,\|\dot{\mathbf{E}}_p(t)\|\,.$$

Its evolution equation completes the constitutive model:

$$\dot{z}(t) = \begin{cases} \dfrac{2\mu(\mathbf{T}^D - \mathbf{X}^D)\cdot\dot{\mathbf{E}}}{k(2\mu + c) - b(\mathbf{T}^D - \mathbf{X}^D)\cdot\mathbf{X} + \frac{2}{3}kk'(s_p)} & \text{for } f = 0 \text{ and } L > 0 \\ 0 & \text{otherwise} \end{cases} \tag{11.58}$$

The temporal evolution of the generalised arclength and internal variables depends on additional criteria, a typical feature of elastoplasticity. The criteria are based on the yield function

$$f(\mathbf{T}, \mathbf{X}, k) = \tfrac{1}{2}\mathrm{tr}\big((\mathbf{T}^D - \mathbf{X}^D)^2\big) - \tfrac{1}{3}[k(s_p)]^2 \tag{11.59}$$

and the loading function

$$L = (\mathbf{T}^{\mathrm{D}} - \mathbf{X}^{\mathrm{D}}) \cdot \dot{\mathbf{E}} \, . \tag{11.60}$$

Classical elastoplasticity is founded on the idea that a material behaves in an elastic manner under moderate loading, whereas loads above a critical level cause plastic flow, leaving permanent deformations.

An elastoplasticity model is frequently found to describe real material behaviour realistically. This applies particularly for metals which are characterised by their crystalline structure. Plastic flow processes in metals can be traced back to motions of *dislocations* in crystal lattices.[9] Complicated dynamic processes within the material itself are responsible for the creation and annihilation of dislocations, their propagation and interaction. These dynamic processes are ignored altogether in the phenomenological representation. All solutions of evolution equations define states of equilibrium. In reality these equilibrium states emerge from dynamic processes, whereas in the phenomenological description they form a continuous sequence and depend on the process history via the generalised arclength.

Accordingly, elastoplasticity can only reflect the most important macroscopic effects of the internal dynamics of dislocations. This manifestation is based on the quantitative definition of a critical load, below which no dislocation motions take place at all. The critical load is represented by a surface in the 6-dimensional space of stress tensors, the so-called *yield surface*. Equivalently, the critical loads may be represented by means of a yield surface in the space of strain tensors.[10]

The macroscopic effects of the dislocation motion, the plastic deformations, depend on the process history via the stress tensor $\mathbf{T}$ and the back stress $\mathbf{X}$. The motion of the dislocations is essentially driven by the stress $\mathbf{T}$; in the evolution equation (11.56), however, only the difference $\mathbf{T} - \mathbf{X}$ has any effect. The tensor $\mathbf{X}$ of *kinematic hardening* can be interpreted as a process-dependent obstacle impeding the motion of dislocations. *Isotropic hardening*, the function $k = k(s_{\mathrm{p}})$, is interpretable as a process-dependent yield stress in a uniaxial tension test. Isotropic hardening allows for the change in the critical load in dependence on the loading history. More detailed representation of isotropic hardening can be achieved by means of an evolution equation for the yield stress k, similar to the *Armstrong-Frederick* rule (5.184) or (11.57).[11]

---

9 See LUBLINER [1990], Chapter 2; HAVNER [1992].

10 Strain space formulations of finite elastoplasticity were developed and successfully applied by BESDO [1981, 1989].

11 See, for example, HAUPT & KAMLAH [1995], pp. 285.

A fundamental term in the phenomenological theory of elastoplasticity is the *yield function*, a scalar-valued function depending on stresses (or strains) and internal variables. The *yield function* determines the *yield surface* and the criteria for *loading* and *unloading*. It also determines the evolution equation for the plastic deformations, inasmuch as the so-called *associated flow rule* or *normality rule* (5.183) is assumed: following this assumption, the plastic strain rate tensor is proportional to the normal on the yield surface. The normality rule is a simplifying hypothesis that has been tried and tested in many, but not in all cases. For a physically consistent formulation of elastoplasticity the normality rule is not absolutely necessary and not adequate for all materials.[12]

Characteristic for elastoplasticity is the general postulate that every state of stress beyond the yield surface is excluded. Whereas the state of stress moves within the interior domain in the case of purely elastic processes, the stress tensor has to stay on the yield surface, whenever elastoplastic deformations take place. That means that the evolution equation for the plastic strain has to be constructed in such a way that the yield condition $f = 0$ is fulfilled identically. Therefore the flow rule (5.183) can only determine the plastic strain rate up to one scalar factor λ. The proportionality factor λ is consequently not a material parameter, but arises out of the requirement that the state of stress can never leave the yield surface during elastoplastic processes. This is the so-called *consistency condition*.[13]

In elastoplasticity the representation of inelastic material behaviour is inseparably linked with the geometry of the yield surface. The advantage of this approach is its clarity: the internal variables $\mathbf{X}$ and k have a geometrical meaning as the central point and the radius of the yield surface in the stress space; the term normality rule is already a geometrical interpretation. Hardening properties inducing a complicated dependence of the stress response on the process history can be geometrically interpreted as translation motion, expansion or even as reshaping of the yield surface. On the other hand, the drawback is that an experimental identification of the yield surface and its evolution in dependence on the process history is intricate, costly and apt to generate considerable room for uncertainty.

The representation theory of process-dependent yield surfaces, the detailed design of which is still subject to research, shall not be dealt with here.[14] Instead, the following sections concentrate only on a general representation of finite elastoplasticity within the scope of the concept of dual variables.

---

[12] See Ehlers [1993], pp. 389.

[13] The evolution equation (11.58) for the generalised arclength is constructed so as to satisfy the consistency condition.

[14] See, for example, Lubliner [1990], pp. 125; Ehlers [1993], pp. 366. More recent investigations are reported in Streilein [1997].

11.3.2 Stress-Free Intermediate Configuration

One fundamental idea which prevails throughout the general concept of elastoplasticity is the decomposition of the material model into two parts. The first and more simple part consists of an elasticity relation based on a stress-free intermediate configuration and responsible for determining the current stress as a function of the elastic strain. The second, by far more complicated part involves representing the evolution of this intermediate configuration in dependence on the process history.[15]

The kinematic foundation for realising both parts is the multiplicative decomposition of the deformation gradient into one elastic part $\hat{\mathbf{F}}_e$ and one plastic part $\mathbf{F}_p$:[16]

$$\mathbf{F} = \hat{\mathbf{F}}_e \mathbf{F}_p \, . \tag{11.61}$$

The elastic part $\hat{\mathbf{F}}_e$ serves to determine the current state of stress by means of an elasticity relation

$$\mathbf{T} = \mathbf{f}(\hat{\mathbf{F}}_e) \, . \tag{11.62}$$

The material function $\mathbf{f}(\cdot)$ meets the condition

$$\mathbf{f}(\mathbf{1}) = \mathbf{0} \, , \tag{11.63}$$

that means that the *intermediate configuration* defined by $(\boldsymbol{X}, t) \longmapsto \mathbf{F}_p(\boldsymbol{X}, t)$ is *locally stress-free.*

One can imagine that the stress-free intermediate configuration arises out of *local unloading.* At first sight, this supporting principle appears to be quite clear; upon closer examination, however, it inclines to be abstract.

At the end of a given deformation process, let the individual volume elements of a material body be separated and external forces be applied to the surfaces according to the free-body principle. We then imagine the stresses (surface forces) reduced to zero. This *local unloading* gives rise to a deformation, which is assumed to be related to the local state of stress according to an elasticity relation. The local deformation $\hat{\mathbf{F}}_e$ is the so-called *elastic part of the deformation gradient.* The residual deformation $\mathbf{F}_p$, relative

[15] The state of the art in modelling finite elastoplasticity is reviewed in NAGHDI [1990]. See also the papers SIMO [1988]; MIEHE & STEIN [1992]; MIEHE [1993] LE & STUMPF [1993].

[16] The idea for this decomposition originates from KRÖNER [1960]. See also LEE [1969].

to the undeformed reference configuration, is a permanent deformation, the so-called *plastic part of the deformation gradient.* This concept suggests that the individual volume elements can no longer be fitted together, since they changed their shape independently of each other during unloading. That means that the tensor field $\mathbf{F}_p$ does not satisfy the conditions of compatibility: $\mathbf{F}_p(X, t)$ is not generally representable as the gradient of a position vector; the same is true for $\hat{\mathbf{F}}_e$. The tensors $\hat{\mathbf{F}}_e$ and $\mathbf{F}_p$ are nonetheless frequently termed elastic and plastic deformation gradients respectively.[17]

In the polar decomposition $\mathbf{F}_p = \mathbf{R}_p\mathbf{U}_p$ the factor $\mathbf{U}_p$ is the plastic stretch tensor,

$$\mathbf{C}_p = \mathbf{F}_p{}^T\mathbf{F}_p = \mathbf{U}_p{}^2 \tag{11.64}$$

the plastic right *Cauchy-Green* tensor and

$$\mathbf{E}_p = \frac{1}{2}(\mathbf{F}_p{}^T\mathbf{F}_p - \mathbf{1}) \tag{11.65}$$

the corresponding plastic *Green* tensor. The *Green* tensor of the total deformation,

$$\mathbf{E} = \frac{1}{2}(\mathbf{F}^T\mathbf{F} - \mathbf{1}) = \frac{1}{2}(\mathbf{F}_p{}^T\hat{\mathbf{F}}_e{}^T\hat{\mathbf{F}}_e\mathbf{F}_p - \mathbf{1}),$$

induces the corresponding strain tensor $\hat{\mathbf{\Gamma}}$ operating on the intermediate configuration,

$$\hat{\mathbf{\Gamma}} = \mathbf{F}_p{}^{T-1}\mathbf{E}\mathbf{F}_p{}^{-1}; \tag{11.66}$$

this strain tensor decomposes additively into a purely elastic *Green* tensor,

$$\hat{\mathbf{\Gamma}}_e = \frac{1}{2}[\hat{\mathbf{F}}_e{}^T\hat{\mathbf{F}}_e - \mathbf{1}] \tag{11.67}$$

and a purely plastic *Almansi* tensor,

$$\hat{\mathbf{\Gamma}}_p = \frac{1}{2}[\mathbf{1} - \mathbf{F}_p{}^{T-1}\mathbf{F}_p{}^{-1}]: \tag{11.68}$$

$$\hat{\mathbf{\Gamma}} = \hat{\mathbf{\Gamma}}_e + \hat{\mathbf{\Gamma}}_p. \tag{11.69}$$

---

[17] The somewhat sophisticated interpretation of the multiplicative decomposition is replaced by the more abstract assumption of a *material isomorphism* in the work of BERTRAM [1999]. See also SVENDSEN [1998].

The strain tensor $\hat{\mathbf{\Gamma}}$ emerges from (8.84) in the light of the identification $\mathbf{\Psi} = \mathbf{F}_p$. In the same way the additive decomposition (11.69) emerges from (8.94) given $\mathbf{\Psi} = \mathbf{F}_p$ and $\mathbf{\Phi} = \hat{\mathbf{F}}_e$.
Since $\mathbf{F}_p$ is not a gradient field, nor do the tensors $\mathbf{C}_p$, $\mathbf{E}_p$ and $\hat{\mathbf{\Gamma}}_p$ fulfil any conditions of compatibility. The intermediate configuration is a non-Euclidean space, whose metric is the tensor field $\mathbf{C}_p(\boldsymbol{X}, t)$. The local unloading, which one can visualise in the physical interpretation of the intermediate configuration, gives rise to a relaxation of the continuous body, leading into a non-Euclidean space. Conversely, the elastic part of the deformation restores compatibility and thus puts the material continuum back into the Euclidean space. The elastic deformations required to this end invoke stresses, which are called *eigenstresses.*
The dual stress tensor associated to $\hat{\mathbf{\Gamma}}$ according to (8.86) is the 2nd *Piola-Kirchhoff* tensor with respect to the stress-free intermediate configuration:

$$\hat{\mathbf{S}} = \mathbf{F}_p\tilde{\mathbf{T}}\mathbf{F}_p^{\mathrm{T}} = \hat{\mathbf{F}}_e^{-1}\mathbf{S}\hat{\mathbf{F}}_e^{\mathrm{T}-1} \tag{11.70}$$

($\mathbf{S} = (\det\mathbf{F})\mathbf{T}$). The image of a material line element in the intermediate configuration, $\mathrm{d}\hat{\boldsymbol{x}} = \mathbf{F}_p\mathrm{d}\boldsymbol{X}$, varies with the rate $(\mathrm{d}\hat{\boldsymbol{x}})^{\cdot} = \hat{\mathbf{L}}_p\mathrm{d}\hat{\boldsymbol{x}}$.
The plastic deformation rate

$$\hat{\mathbf{L}}_p = \dot{\mathbf{F}}_p\mathbf{F}_p^{-1} \tag{11.71}$$

is known as the *plastic velocity gradient,* although $\hat{\mathbf{L}}_p$ is obviously not the gradient of a velocity field. The plastic deformation rate $\hat{\mathbf{L}}_p$ is connected to rates of rotation and stretch of the intermediate configuration as a result of

$$\hat{\mathbf{L}}_p = \dot{\mathbf{R}}_p\mathbf{R}_p^{\mathrm{T}} + \mathbf{R}_p(\dot{\mathbf{U}}_p\mathbf{U}_p^{-1})\mathbf{R}_p^{\mathrm{T}}\,. \tag{11.72}$$

Strain and stress rates, attributed to $\hat{\mathbf{\Gamma}} = \hat{\mathbf{\Gamma}}_e + \hat{\mathbf{\Gamma}}_p$ and $\hat{\mathbf{S}}$, are the outcome of the same transformations that led to these tensors (cf. (8.84) - (8.87)):

$$\overset{\triangle}{\hat{\mathbf{\Gamma}}} = \mathbf{F}_p^{\mathrm{T}-1}\dot{\mathbf{E}}\mathbf{F}_p^{-1} = \dot{\hat{\mathbf{\Gamma}}} + \hat{\mathbf{L}}_p^{\mathrm{T}}\hat{\mathbf{\Gamma}} + \hat{\mathbf{\Gamma}}\hat{\mathbf{L}}_p\,, \tag{11.73}$$

$$\overset{\triangledown}{\hat{\mathbf{S}}} = \mathbf{F}_p\dot{\tilde{\mathbf{T}}}\mathbf{F}_p^{\mathrm{T}} = \dot{\hat{\mathbf{S}}} - \hat{\mathbf{L}}_p\hat{\mathbf{S}} - \hat{\mathbf{S}}\hat{\mathbf{L}}_p^{\mathrm{T}}\,. \tag{11.74}$$

In analogy to the additive decomposition of the strain, $\hat{\mathbf{\Gamma}} = \hat{\mathbf{\Gamma}}_e + \hat{\mathbf{\Gamma}}_p$, a decomposition of the strain rate exists (cf. Fig. 8. 1):

$$\overset{\triangle}{\hat{\boldsymbol{\Gamma}}} = \overset{\triangle}{\hat{\boldsymbol{\Gamma}}}_{\mathrm{e}} + \overset{\triangle}{\hat{\boldsymbol{\Gamma}}}_{\mathrm{p}} = \hat{\mathbf{D}}_{\mathrm{e}} + \hat{\mathbf{D}}_{\mathrm{p}} . \tag{11.75}$$

In this decomposition

$$\hat{\mathbf{D}}_{\mathrm{e}} = \overset{\triangle}{\hat{\boldsymbol{\Gamma}}}_{\mathrm{e}} = \dot{\hat{\boldsymbol{\Gamma}}}_{\mathrm{e}} + \hat{\mathbf{L}}_{\mathrm{p}}{}^{\mathrm{T}} \hat{\boldsymbol{\Gamma}}_{\mathrm{e}} + \hat{\boldsymbol{\Gamma}}_{\mathrm{e}} \hat{\mathbf{L}}_{\mathrm{p}} \tag{11.76}$$

is the elastic part and

$$\hat{\mathbf{D}}_{\mathrm{p}} = \overset{\triangle}{\hat{\boldsymbol{\Gamma}}}_{\mathrm{p}} = \dot{\hat{\boldsymbol{\Gamma}}}_{\mathrm{p}} + \hat{\mathbf{L}}_{\mathrm{p}}{}^{\mathrm{T}} \hat{\boldsymbol{\Gamma}}_{\mathrm{p}} + \hat{\boldsymbol{\Gamma}}_{\mathrm{p}} \hat{\mathbf{L}}_{\mathrm{p}} \tag{11.77}$$

the plastic part of the strain rate. It is apparent that the elastic strain rate also depends on plastic deformations; by contrast, no elastic deformations occur in $\hat{\mathbf{D}}_{\mathrm{p}}$ at all: based on (11.68) and (11.71), the formula (11.77) leads to

$$\hat{\mathbf{D}}_{\mathrm{p}} = \tfrac{1}{2}\left[\hat{\mathbf{L}}_{\mathrm{p}} + \hat{\mathbf{L}}_{\mathrm{p}}{}^{\mathrm{T}}\right] . \tag{11.78}$$

The definitions given so far show that the scalar products of stress and strain tensors or their dual derivatives remain invariant when switching from the reference to the intermediate configuration (cf. (8.108) - (8.112)):

$$\tilde{\mathbf{T}} \cdot \mathbf{E} = \hat{\mathbf{S}} \cdot \hat{\boldsymbol{\Gamma}} , \tag{11.79}$$

$$\tilde{\mathbf{T}} \cdot \dot{\mathbf{E}} = \hat{\mathbf{S}} \cdot \overset{\triangle}{\hat{\boldsymbol{\Gamma}}} , \tag{11.80}$$

$$\dot{\tilde{\mathbf{T}}} \cdot \mathbf{E} = \overset{\nabla}{\hat{\mathbf{S}}} \cdot \hat{\boldsymbol{\Gamma}} , \tag{11.81}$$

$$\dot{\tilde{\mathbf{T}}} \cdot \dot{\mathbf{E}} = \overset{\nabla}{\hat{\mathbf{S}}} \cdot \overset{\triangle}{\hat{\boldsymbol{\Gamma}}} . \tag{11.82}$$

It is evident that the multiplicative decomposition (11.61) of the deformation gradient cannot be unique. Besides (11.61) we also have the decomposition

$$\mathbf{F} = \bar{\hat{\mathbf{F}}}_{\mathrm{e}} \bar{\mathbf{F}}_{\mathrm{p}} , \tag{11.83}$$

where

$$\bar{\hat{\mathbf{F}}}_{\mathrm{e}} = \hat{\mathbf{F}}_{\mathrm{e}} \bar{\mathbf{Q}}^{\mathrm{T}} \tag{11.84}$$

and

$$\bar{\mathbf{F}}_{\mathrm{p}} = \bar{\mathbf{Q}} \mathbf{F}_{\mathrm{p}} , \tag{11.85}$$

$\bar{\mathbf{Q}}$ being any orthogonal tensor. In view of the polar decomposition $\mathbf{F}_\mathrm{p} = \mathbf{R}_\mathrm{p}\mathbf{U}_\mathrm{p}$ and $\bar{\mathbf{F}}_\mathrm{p} = \bar{\mathbf{Q}}\mathbf{R}_\mathrm{p}\mathbf{F}_\mathrm{p}$ the orthogonal tensor $\bar{\mathbf{Q}}$ may be interpreted as an additional, optional rotation of the intermediate configuration.

The free choice of rotation $\bar{\mathbf{Q}}$ is to be borne in mind (in addition to the assumption of frame-indifference) when establishing the elasticity relation and when representing the temporal evolution of the intermediate configuration. All constitutive equations in elastoplasticity have to be invariant with respect to rotations $\bar{\mathbf{Q}}$ of the intermediate configuration as well as with respect to rotations $\mathbf{Q}$ of the reference system. By way of a simple conclusion from (4.40) and (11.84), (11.85), we obtain the transformation behaviour of the hitherto defined quantities under a change of frame and a rotation of the intermediate configuration.

Theorem 11. 5

During a change of frame, which is represented by the time-dependent rotation $\mathbf{Q}(t)$, we have the following transformation for the elastic part of the deformation gradient:

$$\hat{\mathbf{F}}_\mathrm{e} \longmapsto \mathbf{Q}\hat{\mathbf{F}}_\mathrm{e}\,. \tag{11.86}$$

The tensors $\mathbf{F}_\mathrm{p}$, $\hat{\mathbf{\Gamma}}_\mathrm{e}$, $\hat{\mathbf{\Gamma}}_\mathrm{p}$, $\hat{\mathbf{L}}_\mathrm{p}$, $\hat{\mathbf{D}}_\mathrm{e}$, $\hat{\mathbf{D}}_\mathrm{p}$, $\hat{\mathbf{S}}$, $\overset{\triangledown}{\hat{\mathbf{S}}}$ remain unchanged.

A local rotation of the intermediate configuration, represented by $\bar{\mathbf{Q}}(\boldsymbol{X}, t)$, induces the transformations

$$\hat{\mathbf{F}}_\mathrm{e} \longmapsto \hat{\mathbf{F}}_\mathrm{e}\bar{\mathbf{Q}}^\mathrm{T} \tag{11.87}$$

$$\mathbf{F}_\mathrm{p} \longmapsto \bar{\mathbf{Q}}\mathbf{F}_\mathrm{p} \qquad \hat{\mathbf{L}}_\mathrm{p} \longmapsto \dot{\bar{\mathbf{Q}}}\bar{\mathbf{Q}}^\mathrm{T} + \bar{\mathbf{Q}}\hat{\mathbf{L}}_\mathrm{p}\bar{\mathbf{Q}}^\mathrm{T}\,, \tag{11.88}$$

$$\hat{\mathbf{\Gamma}}_\mathrm{e} \longmapsto \bar{\mathbf{Q}}\hat{\mathbf{\Gamma}}_\mathrm{e}\bar{\mathbf{Q}}^\mathrm{T} \qquad \hat{\mathbf{\Gamma}}_\mathrm{p} \longmapsto \bar{\mathbf{Q}}\hat{\mathbf{\Gamma}}_\mathrm{p}\bar{\mathbf{Q}}^\mathrm{T}, \tag{11.89}$$

$$\hat{\mathbf{D}}_\mathrm{e} \longmapsto \bar{\mathbf{Q}}\hat{\mathbf{D}}_\mathrm{e}\bar{\mathbf{Q}}^\mathrm{T} \qquad \hat{\mathbf{D}}_\mathrm{p} \longmapsto \bar{\mathbf{Q}}\hat{\mathbf{D}}_\mathrm{p}\bar{\mathbf{Q}}^\mathrm{T}\,, \tag{11.90}$$

$$\hat{\mathbf{S}} \longmapsto \bar{\mathbf{Q}}\hat{\mathbf{S}}\bar{\mathbf{Q}}^\mathrm{T} \qquad \overset{\triangledown}{\hat{\mathbf{S}}} \longmapsto \bar{\mathbf{Q}}\overset{\triangledown}{\hat{\mathbf{S}}}\bar{\mathbf{Q}}^\mathrm{T}\,. \tag{11.91}$$

■

11. 3. 3 Isotropic Elasticity

In elastoplasticity the current value of the *Cauchy* stress is a function of the elastic part of the deformation:

$$\mathbf{S} = \mathbf{f}(\hat{\mathbf{F}}_e), \quad \mathbf{f}(\mathbf{1}) = \mathbf{0}. \tag{11.92}$$

The material function $\mathbf{f}$ is restricted by the assumption of frame-indifference:

Theorem 11. 6

The following representation of the elasticity relation is necessary and sufficient for fulfilling material frame-indifference:

$$\hat{\mathbf{S}} = \mathbf{g}(\hat{\mathbf{\Gamma}}_e). \tag{11.93}$$

□

Proof

According to the assumption of *frame-indifference*, $\mathbf{S} = \mathbf{f}(\hat{\mathbf{F}}_e)$ implies $\mathbf{S}^* = \mathbf{f}(\hat{\mathbf{F}}_e^*)$ with $\hat{\mathbf{F}}_e^* = \mathbf{Q}\hat{\mathbf{F}}_e$. For all orthogonal $\mathbf{Q}$ ($\det\mathbf{Q} = +1$) the material function $\mathbf{f}(\cdot)$ has to satisfy the restrictive condition

$$\mathbf{Q}\mathbf{f}(\hat{\mathbf{F}}_e)\mathbf{Q}^{\mathrm{T}} = \mathbf{f}(\mathbf{Q}\hat{\mathbf{F}}_e). \tag{11.94}$$

The choice of $\mathbf{Q} = \hat{\mathbf{R}}_e^{\mathrm{T}} = \hat{\mathbf{U}}_e\hat{\mathbf{F}}_e^{-1}$ $(\hat{\mathbf{F}}_e = \hat{\mathbf{R}}_e\hat{\mathbf{U}}_e)$ leads to

$$\hat{\mathbf{F}}_e^{-1}\mathbf{f}(\hat{\mathbf{F}}_e)\hat{\mathbf{F}}_e^{\mathrm{T}-1} = \hat{\mathbf{U}}_e^{-1}\mathbf{f}(\hat{\mathbf{U}}_e)\,\hat{\mathbf{U}}_e^{-1}.$$

Given $\hat{\mathbf{S}} = \hat{\mathbf{F}}_e^{-1}\mathbf{S}\,\hat{\mathbf{F}}_e^{\mathrm{T}-1}$ and $\hat{\mathbf{U}}_e^2 = \hat{\mathbf{C}}_e = \mathbf{1} + 2\hat{\mathbf{\Gamma}}_e$
we obtain $\hat{\mathbf{S}} = \mathbf{g}(\hat{\mathbf{\Gamma}}_e)$. ■

The arbitrary choice of $\bar{\mathbf{Q}}$ brings about a further restriction for the material function $\mathbf{f}(\cdot)$:

Theorem 11. 7

The elasticity relation $\hat{\mathbf{S}} = \mathbf{g}(\hat{\mathbf{\Gamma}}_e)$ is independent of any additional rotation $\bar{\mathbf{Q}}$ of the intermediate configuration, if and only if $\mathbf{g}(\cdot)$ is an isotropic tensor function. □

Proof

The transformation formulae $(11.89)_1$ and $(11.91)_1$ imply that, given the constitutive relation $\hat{\mathbf{S}} = \mathbf{g}(\hat{\Gamma}_e)$, the property $\bar{\mathbf{Q}}\hat{\mathbf{S}}\bar{\mathbf{Q}}^T = \mathbf{g}(\bar{\mathbf{Q}}\hat{\Gamma}_e\bar{\mathbf{Q}}^T)$ also holds, i.e. for all orthogonal $\bar{\mathbf{Q}}$ the identity

$$\mathbf{g}(\bar{\mathbf{Q}}\hat{\Gamma}_e\bar{\mathbf{Q}}^T) = \bar{\mathbf{Q}}\mathbf{g}(\hat{\Gamma}_e)\bar{\mathbf{Q}}^T . \tag{11.95}$$

■

Along the lines of this theorem, the elasticity relation can only be represented by an isotropic function if it is based on the initial formulation of $\mathbf{T} = \mathbf{f}(\hat{\mathbf{F}}_e)$. That means that in the assumptions made here elastic isotropy has to be present in the reference configuration.[18] According to the assumption of the stress-free intermediate configuration we have

$$\mathbf{g}(\mathbf{0}) = \mathbf{0} . \tag{11.96}$$

11. 3. 4 Yield Function and Evolution Equations

Motivated by the reduced form (11.93) for the elasticity relation, it seems expedient to formulate all constitutive equations of elastoplasticity in dual variables relating to the intermediate configuration. This facilitates the observation of both the objectivity postulate and the invariance with respect to the rotation of the intermediate configuration. The following equations present possibilities that embrace special cases and permit generalisations. The basis of finite elastoplasticity, apart from the elasticity relation (11.93), is a *yield function*

$$(\hat{\mathbf{S}}, \hat{\mathbf{X}}, k) \longmapsto F(\hat{\mathbf{S}}, \hat{\mathbf{X}}, k) .$$

Here $F(\cdot)$ is a scalar-valued isotropic function of the tensors $\hat{\mathbf{S}}$ and $\hat{\mathbf{X}}$. From a physical point of view, it makes sense to expect the yield function to depend on the difference between these two tensors:

$$(\hat{\mathbf{S}}, \hat{\mathbf{X}}, k) \longmapsto F(\hat{\mathbf{S}}, \hat{\mathbf{X}}, k) = f(\hat{\mathbf{S}} - \hat{\mathbf{X}}, k) . \tag{11.97}$$

---

[18] In this connection, anisotropic elasticity properties would have to be represented by isotropic tensor functions depending not only on the elastic strain but on other variables, known as structural tensors, besides. See Horz [1994], pp. 42; Horz et al. [1994]. Ways in which anisotropic elasticity may be taken into consideration without recourse to structural tensors are discussed in Section 13. 6.

In this case the internal variable $\hat{\mathbf{X}}$ can be interpreted as the centre of the yield surface in the stress space related to the intermediate configuration. A change in $\hat{\mathbf{X}}$ as the outcome of the process history corresponds to a translation of the yield surface and is known as *kinematic hardening*. Tensor $\hat{\mathbf{X}}$ is called the *hardening tensor* or *backstress tensor*. In concrete formulations the scalar k may be explained as the radius of the yield surface, whose variability is generally called *isotropic hardening*.

The plastic strain $\hat{\mathbf{\Gamma}}_{\mathrm{p}}$ is an internal variable, described by an evolution equation that determines the development of the intermediate configuration in dependence on the process history. This evolution equation is called the *flow rule*. According to the so-called *associated flow rule* (*normality rule*) the plastic strain rate is proportional to the gradient of the yield function:

$$\hat{\mathbf{D}}_{\mathrm{p}} = \overset{\triangle}{\hat{\mathbf{\Gamma}}}_{\mathrm{p}} = \begin{cases} \lambda \dfrac{\partial F}{\partial \hat{\mathbf{S}}} & \text{for } f = 0 \text{ and loading} \\ 0 & \text{otherwise} \end{cases} \tag{11.98}$$

Here the proportionality factor λ has to be calculated from the *consistency condition* $\dot{f}(t) = 0$. In place of the gradient of the yield function it is also possible to select another material function, i.e. a non-associated flow rule.[19] Even in the case of a non-associated flow rule the evaluation of the consistency condition is necessary for determining the proportionality factor λ.

Depending on the structure of the yield function $F(\cdot)$, the flow rule may include other internal variables as well, for instance $\hat{\mathbf{X}}$ or k. The tensor $\hat{\mathbf{X}}$ of *kinematic hardening* is the centre of the yield surface in view of (11.97). $\hat{\mathbf{X}}$ should be modelled as a rate-independent functional of the process history. It is crucial that $\hat{\mathbf{X}}$ remains constant during purely elastic processes in the same way as all other internal variables in elastoplasticity. It is therefore imperative to formulate the functional for $\hat{\mathbf{X}}$ with respect to the plastic arclength $s_{\mathrm{p}}(t)$ or with respect to a generalised arclength $z(t)$. One obvious method, which fulfils all invariance requirements, is an evolution equation along the lines of (11.48). Applying the same principles to quantities related to the intermediate configuration leads to the evolution equation

$$\overset{\triangledown}{\hat{\mathbf{X}}} = c\overset{\triangle}{\hat{\mathbf{\Gamma}}}_{\mathrm{p}} - b\,\dot{z}(t)\hat{\mathbf{X}} \, . \tag{11.99}$$

This evolution equation generalises the *Armstrong-Frederick* equation (5.184). It is evident that (11.99) defines a rate-independent functional. Moreover this

---

19 EHLERS [1993], pp. 389.

functional can be calculated explicitly by means of integration, as the following theorem shows.

Theorem 11. 8

By integration (11.99) leads to

$$\hat{\mathbf{X}}(z) = -\int_0^z \frac{c}{2}\, e^{-b(z-\zeta)}\, \mathbf{F}_p(z)[\mathbf{C}_p^{-1}(\zeta)]'\mathbf{F}_p^{T}(z)d\zeta\,. \tag{11.100}$$

□

Proof

Corresponding to the hardening variable $\hat{\mathbf{X}}$ (which is a stress tensor) we introduce a *Piola-Kirchhoff* tensor

$$\tilde{\mathbf{X}} = \mathbf{F}_p^{-1}\hat{\mathbf{X}}\mathbf{F}_p^{T-1}\,, \tag{11.101}$$

and the *Oldroyd* rate $\overset{\nabla}{\hat{\mathbf{X}}}$ transforms into the material time derivative

$$\dot{\tilde{\mathbf{X}}} = \mathbf{F}_p^{-1}\overset{\nabla}{\hat{\mathbf{X}}}\mathbf{F}_p^{T-1}\,. \tag{11.102}$$

It also holds that $\overset{\Delta}{\hat{\boldsymbol{\Gamma}}}_p = \frac{1}{2}(\dot{\mathbf{F}}_p\mathbf{F}_p^{-1} + \mathbf{F}_p^{T-1}\dot{\mathbf{F}}_p^{T}) = \frac{1}{2}\mathbf{F}_p^{T-1}\dot{\mathbf{C}}_p\mathbf{F}_p^{-1}$, i.e.

$$\mathbf{F}_p^{-1}\overset{\Delta}{\hat{\boldsymbol{\Gamma}}}_p\mathbf{F}_p^{T-1} = \frac{1}{2}\mathbf{C}_p^{-1}\dot{\mathbf{C}}_p\mathbf{C}_p^{-1} = -\frac{1}{2}(\mathbf{C}_p^{-1})^{\cdot}\,. \tag{11.103}$$

Accordingly, the evolution equation (11.99) is equivalent to

$$\dot{\tilde{\mathbf{X}}}(t) + b\dot{z}(t)\tilde{\mathbf{X}}(t) = -\frac{c}{2}(\mathbf{C}_p^{-1}(t))^{\cdot}\,. \tag{11.104}$$

Dividing by $\dot{z}(t)$ and introducing z, i.e.

$$\tilde{\mathbf{X}}(t) = \tilde{\mathbf{X}}(z(t))\,, \tag{11.105}$$

we obtain

$$\tilde{\mathbf{X}}'(z) + b\,\tilde{\mathbf{X}}(z) = -\frac{c}{2}(\mathbf{C}_p^{-1}(z))'\,. \tag{11.106}$$

Integration produces

$$\tilde{\mathbf{X}}(z) = \mathbf{F}_p^{-1}(z)\hat{\mathbf{X}}(z)\mathbf{F}_p^{T-1}(z) = -\int_0^z \frac{c}{2}\, e^{-b(z-\zeta)}\,[\mathbf{C}_p^{-1}(\zeta)]'d\zeta$$

or (11.100). ■

In view of a more detailed modelling of kinematic hardening properties it is possible to expand on the assumption (11.99) by writing $\hat{\mathbf{X}}$ as the sum of a finite number of partial backstresses,

$$\hat{\mathbf{X}} = \sum_{k=1}^{N} \hat{\mathbf{X}}_k \,, \tag{11.107}$$

and for every contribution $\hat{\mathbf{X}}_k$ postulating an evolution equation

$$\overset{\nabla}{\hat{\mathbf{X}}}_k = c_k \hat{\mathbf{D}}_p - b_k \dot{z}(t) \hat{\mathbf{X}}_k \,. \tag{11.108}$$

In compliance with this we then have the functional

$$\hat{\mathbf{X}}(z) = -\int_0^z \gamma(z - \zeta)\, \mathbf{F}_p(z)[\mathbf{C}_p^{-1}(\zeta)]' \mathbf{F}_p^{\mathrm{T}}(z) \mathrm{d}\zeta \tag{11.109}$$

with the kernel function

$$\gamma(z) = \sum_{k=1}^{N} \frac{1}{2} c_k \mathrm{e}^{-b_k z} \,. \tag{11.110}$$

11. 3. 5 Consistency Condition

The consistency condition is the requirement that every change in the intermediate configuration be compatible with the assumption that the state of stress remains on the yield surface:

$$\frac{\mathrm{d}}{\mathrm{d}t} F\big(\hat{\mathbf{S}}(t), \hat{\mathbf{X}}(t), k(t)\big) = 0 \,. \tag{11.111}$$

This requirement determines the proportionality factor λ in the flow rule as a function of the deformation rate and other variables. The precise result depends on the individual constitutive assumptions reached. An exemplary evaluation of the consistency condition may be based on the following assumptions.[20]

---

[20] Constitutive models of this structure have been applied in the context of analytical investigations by BONN [1992] and in connection with numerical solutions of boundary value problems by HARTMANN [1993].

Decomposition of the Deformation

$$\mathbf{F} = \hat{\mathbf{F}}_\mathrm{e}\mathbf{F}_\mathrm{p} \quad \Longleftrightarrow \quad \hat{\mathbf{\Gamma}} = \hat{\mathbf{\Gamma}}_\mathrm{e} + \hat{\mathbf{\Gamma}}_\mathrm{p} \tag{11.112}$$

Isotropic Elasticity Relation

$$\hat{\mathbf{S}} = \mathbf{g}(\hat{\mathbf{\Gamma}}_\mathrm{e}) = \rho_\mathrm{R}\frac{\mathrm{d}w(\hat{\mathbf{\Gamma}}_\mathrm{e})}{\mathrm{d}\hat{\mathbf{\Gamma}}_\mathrm{e}} \tag{11.113}$$

Yield Function

$$F(\hat{\mathbf{S}}, \hat{\mathbf{X}}, k) = f(\hat{\mathbf{S}} - \hat{\mathbf{X}}, k) \tag{11.114}$$

Flow Rule

$$\hat{\mathbf{D}}_\mathrm{p} = \begin{cases} \lambda\dfrac{\partial f}{\partial \hat{\mathbf{S}}} & \text{for } f = 0 \text{ and loading} \\ \mathbf{0} & \text{otherwise} \end{cases} \tag{11.115}$$

Kinematic Hardening

$$\overset{\triangledown}{\hat{\mathbf{X}}} = c\hat{\mathbf{D}}_\mathrm{p} - b\,\dot{s}_\mathrm{p}(t)\hat{\mathbf{X}} \tag{11.116}$$

Isotropic Hardening

$$k = k(s_\mathrm{p}) \tag{11.117}$$

Plastic Arclength

$$\dot{s}_\mathrm{p}(t) = \left\|\hat{\mathbf{D}}_\mathrm{p}\right\| \tag{11.118}$$

To evaluate the consistency condition in the case of a general isotropic elasticity relation, one has recourse to the following theorem.

Theorem 11. 9

If $\mathbf{E} \longmapsto \mathbf{g}(\mathbf{E})$ is an isotropic tensor function, the fourth-order tensor

$$\mathbb{C} = \mathbb{C}(\mathbf{E}) = \frac{\mathrm{d}}{\mathrm{d}\mathbf{E}}\mathbf{g}(\mathbf{E})$$

the derivative of $\mathbf{g}(\cdot)$ and $\mathbf{L}$ a second-order tensor, then the relation

$$\mathbf{L}\mathbf{g}(\mathbf{E}) + \mathbf{g}(\mathbf{E})\mathbf{L}^{\mathrm{T}} + \mathbb{C}\big[\mathbf{L}^{\mathrm{T}}\mathbf{E} + \mathbf{E}\mathbf{L}\big] = \mathbf{g}(\mathbf{E})(\mathbf{L} + \mathbf{L}^{\mathrm{T}}) + \mathbb{C}\big[(\mathbf{L} + \mathbf{L}^{\mathrm{T}})\mathbf{E}\big] \tag{11.119}$$

is valid. □

Proof

As an isotropic tensor function $\mathbf{g}(\cdot)$ is represented by (9.55),

$$\mathbf{g}(\mathbf{E}) = \psi_0\mathbf{1} + \psi_1\mathbf{E} + \psi_2\mathbf{E}^2 \,,$$

with $\psi_{\mathrm{i}} = \psi_{\mathrm{i}}(\mathrm{J}_1, \mathrm{J}_2, \mathrm{J}_3)$ $\quad (\mathrm{i} = 0, 1, 2)$ $\quad$ and $\quad \mathrm{J}_{\mathrm{k}} = \frac{1}{\mathrm{k}}\mathrm{tr}\mathbf{E}^{\mathrm{k}}$ $\quad (\mathrm{k} = 1, 2, 3)$.

From this representation the elasticity tensor $\mathbb{C}$ is calculated by way of a *Gateaux* derivative,

$$\mathbb{C}[\mathbf{H}] = \frac{\mathrm{d}}{\mathrm{d}s}\,\mathbf{g}(\mathbf{E} + s\mathbf{H})\bigg|_{s=0} \,,$$

$$\mathbb{C}[\mathbf{H}] = \sum_{\mathrm{k}=0}^{2}\bigg\{\Big(\frac{\partial\psi_{\mathrm{k}}}{\partial \mathrm{J}_1}\mathbf{1} + \frac{\partial\psi_{\mathrm{k}}}{\partial \mathrm{J}_2}\mathbf{E} + \frac{\partial\psi_{\mathrm{k}}}{\partial \mathrm{J}_3}\mathbf{E}^2\Big)\cdot\mathbf{H}\bigg\}\mathbf{E}^{\mathrm{k}} + \psi_1\mathbf{H}$$

$$+ \psi_2[\mathbf{E}\mathbf{H} + \mathbf{H}\mathbf{E}] \,. \tag{11.120}$$

For the left-hand side of (11.119) we obtain

$$\mathbf{L}\mathbf{g}(\mathbf{E}) + \mathbf{g}(\mathbf{E})\mathbf{L}^{\mathrm{T}} + \mathbb{C}\big[\mathbf{L}^{\mathrm{T}}\mathbf{E} + \mathbf{E}\mathbf{L}\big]$$

$$= \psi_0\,[\mathbf{L} + \mathbf{L}^{\mathrm{T}}] + \psi_1\,[\mathbf{L}\mathbf{E} + \mathbf{E}\mathbf{L}^{\mathrm{T}}] + \psi_2\,[\mathbf{L}\mathbf{E}^2 + \mathbf{E}^2\mathbf{L}^{\mathrm{T}}]$$

$$+ \sum_{\mathrm{k}=0}^{2}\bigg\{\Big[\Big(\frac{\partial\psi_{\mathrm{k}}}{\partial \mathrm{J}_1}\mathbf{1} + \frac{\partial\psi_{\mathrm{k}}}{\partial \mathrm{J}_2}\mathbf{E} + \frac{\partial\psi_{\mathrm{k}}}{\partial \mathrm{J}_3}\mathbf{E}^2\Big)\Big]\cdot\big[\mathbf{L}^{\mathrm{T}}\mathbf{E} + \mathbf{E}\mathbf{L}\big]\bigg\}\mathbf{E}^{\mathrm{k}}$$

$$+ \psi_1[\mathbf{L}^{\mathrm{T}}\mathbf{E} + \mathbf{E}\mathbf{L}\,] + \psi_2[\mathbf{E}(\mathbf{L}^{\mathrm{T}}\mathbf{E} + \mathbf{E}\mathbf{L}\,) + (\mathbf{L}^{\mathrm{T}}\mathbf{E} + \mathbf{E}\mathbf{L}\,)\mathbf{E}\,] \,,$$

and for the right-hand side

$$\mathbf{g}(\mathbf{E})(\mathbf{L} + \mathbf{L}^{\mathrm{T}}) + \mathbb{C}\big[(\mathbf{L} + \mathbf{L}^{\mathrm{T}})\mathbf{E}\big]$$

$$= \psi_0\,[\mathbf{L} + \mathbf{L}^{\mathrm{T}}] + \psi_1\,\mathbf{E}[\mathbf{L} + \mathbf{L}^{\mathrm{T}}] + \psi_2\,\mathbf{E}^2[\mathbf{L} + \mathbf{L}^{\mathrm{T}}]$$

$$+ \sum_{k=0}^{2} \left\{ \left[\left(\frac{\partial \psi_k}{\partial J_1} \mathbf{1} + \frac{\partial \psi_k}{\partial J_2} \mathbf{E} + \frac{\partial \psi_k}{\partial J_3} \mathbf{E}^2 \right) \right] \cdot \left[(\mathbf{L} + \mathbf{L}^T)\mathbf{E} \right] \right\} \mathbf{E}^k$$

$$+ \psi_1[(\mathbf{L} + \mathbf{L}^T)\mathbf{E}] + \psi_2[\mathbf{E}(\mathbf{L} + \mathbf{L}^T)\mathbf{E} + (\mathbf{L} + \mathbf{L}^T)\mathbf{E}^2] . \qquad \blacksquare$$

Based on this result, the consistency condition can now be evaluated and the proportionality factor λ calculated.

Theorem 11. 10

In connection with the assumptions (11.112) - (11.118) the consistency condition leads to the proportionality factor

$$\lambda = \frac{1}{N} \frac{\partial f}{\partial \hat{\mathbf{S}}} \cdot \mathbb{C}\left[\overset{\triangle}{\hat{\boldsymbol{\Gamma}}} \right], \tag{11.121}$$

with

$$N = \frac{\partial f}{\partial \hat{\mathbf{S}}} \cdot \mathbb{C}\left[\frac{\partial f}{\partial \hat{\mathbf{S}}} (\mathbf{1} + 2\hat{\boldsymbol{\Gamma}}_e) \right] + \left\{ c \left\| \frac{\partial f}{\partial \hat{\mathbf{S}}} \right\| - b \left(\frac{\partial f}{\partial \hat{\mathbf{S}}} \cdot \hat{\mathbf{X}} \right) \right\} \left\| \frac{\partial f}{\partial \hat{\mathbf{S}}} \right\| + 2 \left(\frac{\partial f}{\partial \hat{\mathbf{S}}} \right)^2 \cdot \hat{\mathbf{X}} - \frac{\partial f}{\partial k} k'(s_p) \left\| \frac{\partial f}{\partial \hat{\mathbf{S}}} \right\| . \tag{11.122}$$

The loading function

$$L = \frac{\partial f}{\partial \hat{\mathbf{S}}} \cdot \mathbb{C}\left[\overset{\triangle}{\hat{\boldsymbol{\Gamma}}} \right] \tag{11.123}$$

serves to formulate the *loading condition*:

$$L \begin{cases} > 0 \Leftrightarrow \text{loading} \\ = 0 \Leftrightarrow \text{neutral loading} \\ < 0 \Leftrightarrow \text{unloading} \end{cases} \tag{11.124}$$

$\square$

Proof

$\frac{\mathrm{d}}{\mathrm{d}t} f(\hat{\mathbf{S}}(t) - \hat{\mathbf{X}}(t), k(t)) = 0$, given the *Oldroyd* rate (11.74), leads to

$$0 = \frac{\partial f}{\partial \hat{\mathbf{S}}} \cdot (\dot{\hat{\mathbf{S}}} - \dot{\hat{\mathbf{X}}}) + \frac{\partial f}{\partial k} k'(s_p) \dot{s}_p(t) ,$$

$$0 = \frac{\partial f}{\partial \hat{\mathbf{S}}} \cdot \left\{ (\overset{\nabla}{\hat{\mathbf{S}}} - \overset{\nabla}{\hat{\mathbf{X}}}) + \hat{\mathbf{L}}_{\mathrm{p}}(\hat{\mathbf{S}} - \hat{\mathbf{X}}) + (\hat{\mathbf{S}} - \hat{\mathbf{X}})\hat{\mathbf{L}}_{\mathrm{p}}{}^{\mathrm{T}} \right\} + \frac{\partial f}{\partial k} k'(s_{\mathrm{p}}) \dot{s}_{\mathrm{p}}(t) .$$

In view of the isotropy of the yield function $f(\cdot)$ and $\hat{\mathbf{L}}_{\mathrm{p}} + \hat{\mathbf{L}}_{\mathrm{p}}{}^{\mathrm{T}} = 2\hat{\mathbf{D}}_{\mathrm{p}}$ this implies

$$\frac{\partial f}{\partial \hat{\mathbf{S}}} \cdot \overset{\nabla}{\hat{\mathbf{S}}} = \frac{\partial f}{\partial \hat{\mathbf{S}}} \cdot \overset{\nabla}{\hat{\mathbf{X}}} - 2\left[\frac{\partial f}{\partial \hat{\mathbf{S}}} (\hat{\mathbf{S}} - \hat{\mathbf{X}})\right] \cdot \hat{\mathbf{D}}_{\mathrm{p}} - \frac{\partial f}{\partial k} k'(s_{\mathrm{p}}) \| \hat{\mathbf{D}}_{\mathrm{p}} \| . \tag{11.125}$$

Moreover, based on the elasticity relation $\hat{\mathbf{S}} = \mathbf{g}(\hat{\boldsymbol{\Gamma}}_{\mathrm{e}})$

and $\overset{\triangle}{\hat{\boldsymbol{\Gamma}}}_{\mathrm{e}} = \dot{\hat{\boldsymbol{\Gamma}}}_{\mathrm{e}} + \hat{\mathbf{L}}_{\mathrm{p}}{}^{\mathrm{T}} \hat{\boldsymbol{\Gamma}}_{\mathrm{e}} + \hat{\boldsymbol{\Gamma}}_{\mathrm{e}} \hat{\mathbf{L}}_{\mathrm{p}}$ it is possible to calculate

$$\begin{aligned} \overset{\nabla}{\hat{\mathbf{S}}} &= \dot{\hat{\mathbf{S}}} - \hat{\mathbf{L}}_{\mathrm{p}} \hat{\mathbf{S}} - \hat{\mathbf{S}} \hat{\mathbf{L}}_{\mathrm{p}}{}^{\mathrm{T}} \\ &= \mathbb{C}\left[\overset{\triangle}{\hat{\boldsymbol{\Gamma}}}_{\mathrm{e}}\right] - \mathbb{C}[\hat{\mathbf{L}}_{\mathrm{p}}{}^{\mathrm{T}} \hat{\boldsymbol{\Gamma}}_{\mathrm{e}} + \hat{\boldsymbol{\Gamma}}_{\mathrm{e}} \hat{\mathbf{L}}_{\mathrm{p}}] - \hat{\mathbf{L}}_{\mathrm{p}} \mathbf{g}(\hat{\boldsymbol{\Gamma}}_{\mathrm{e}}) - \mathbf{g}(\hat{\boldsymbol{\Gamma}}_{\mathrm{e}}) \hat{\mathbf{L}}_{\mathrm{p}}{}^{\mathrm{T}} , \end{aligned}$$

$$\overset{\nabla}{\hat{\mathbf{S}}} = \mathbb{C}\left[\overset{\triangle}{\hat{\boldsymbol{\Gamma}}}_{\mathrm{e}}\right] - 2\mathbb{C}[\hat{\mathbf{D}}_{\mathrm{p}} \hat{\boldsymbol{\Gamma}}_{\mathrm{e}}] - 2\mathbf{g}(\hat{\boldsymbol{\Gamma}}_{\mathrm{e}}) \hat{\mathbf{D}}_{\mathrm{p}} , \tag{11.126}$$

employing the identity (11.119) in the last step. Using the decomposition $(11.112)_2$ or

$$\overset{\triangle}{\hat{\boldsymbol{\Gamma}}}_{\mathrm{e}} = \overset{\triangle}{\hat{\boldsymbol{\Gamma}}} - \overset{\triangle}{\hat{\boldsymbol{\Gamma}}}_{\mathrm{p}} = \overset{\triangle}{\hat{\boldsymbol{\Gamma}}} - \hat{\mathbf{D}}_{\mathrm{p}}$$

(11.125) and (11.126) lead to the conclusions

$$\begin{aligned} \frac{\partial f}{\partial \hat{\mathbf{S}}} \cdot \mathbb{C}\left[\overset{\triangle}{\hat{\boldsymbol{\Gamma}}}\right] = {} & \frac{\partial f}{\partial \hat{\mathbf{S}}} \cdot \mathbb{C}\left[\hat{\mathbf{D}}_{\mathrm{p}}\right] + 2\frac{\partial f}{\partial \hat{\mathbf{S}}} \cdot \mathbb{C}[\hat{\mathbf{D}}_{\mathrm{p}} \hat{\boldsymbol{\Gamma}}_{\mathrm{e}}] + 2\frac{\partial f}{\partial \hat{\mathbf{S}}} \cdot (\hat{\mathbf{S}} \hat{\mathbf{D}}_{\mathrm{p}}) \\ & + \frac{\partial f}{\partial \hat{\mathbf{S}}} \cdot \overset{\nabla}{\hat{\mathbf{X}}} - 2\left[\frac{\partial f}{\partial \hat{\mathbf{S}}} (\hat{\mathbf{S}} - \hat{\mathbf{X}})\right] \cdot \hat{\mathbf{D}}_{\mathrm{p}} - \frac{\partial f}{\partial k} k'(s_{\mathrm{p}}) \| \hat{\mathbf{D}}_{\mathrm{p}} \| , \end{aligned}$$

$$\begin{aligned} \frac{\partial f}{\partial \hat{\mathbf{S}}} \cdot \mathbb{C}\left[\overset{\triangle}{\hat{\boldsymbol{\Gamma}}}\right] = {} & \frac{\partial f}{\partial \hat{\mathbf{S}}} \cdot \mathbb{C}\left[\hat{\mathbf{D}}_{\mathrm{p}}\right] + 2\frac{\partial f}{\partial \hat{\mathbf{S}}} \cdot \mathbb{C}[\hat{\mathbf{D}}_{\mathrm{p}} \hat{\boldsymbol{\Gamma}}_{\mathrm{e}}] + 2\left(\hat{\mathbf{D}}_{\mathrm{p}} \frac{\partial f}{\partial \hat{\mathbf{S}}}\right) \cdot \hat{\mathbf{S}} \\ & + \frac{\partial f}{\partial \hat{\mathbf{S}}} \cdot \left[c\hat{\mathbf{D}}_{\mathrm{p}} - b \| \hat{\mathbf{D}}_{\mathrm{p}} \| \hat{\mathbf{X}}\right] - 2\left(\hat{\mathbf{D}}_{\mathrm{p}} \frac{\partial f}{\partial \hat{\mathbf{S}}}\right) \cdot (\hat{\mathbf{S}} - \hat{\mathbf{X}}) - \frac{\partial f}{\partial k} k'(s_{\mathrm{p}}) \| \hat{\mathbf{D}}_{\mathrm{p}} \| , \end{aligned}$$

$$\frac{\partial f}{\partial \hat{\mathbf{S}}} \cdot \mathbb{C}\left[\overset{\triangle}{\hat{\mathbf{\Gamma}}}\right] = \frac{\partial f}{\partial \hat{\mathbf{S}}} \cdot \mathbb{C}\left[\hat{\mathbf{D}}_{\mathrm{p}}(\mathbf{1} + 2\hat{\mathbf{\Gamma}}_{e})\right] + \frac{\partial f}{\partial \hat{\mathbf{S}}} \cdot \left[c\hat{\mathbf{D}}_{\mathrm{p}} - b\left\|\hat{\mathbf{D}}_{\mathrm{p}}\right\|\hat{\mathbf{X}}\right]$$

$$+ 2\left(\hat{\mathbf{D}}_{\mathrm{p}}\frac{\partial f}{\partial \hat{\mathbf{S}}}\right) \cdot \hat{\mathbf{X}} - \frac{\partial f}{\partial k}k'(s_{\mathrm{p}})\left\|\hat{\mathbf{D}}_{\mathrm{p}}\right\| .$$

Inserting the flow rule (11.115) yields the result

$$\frac{\partial f}{\partial \hat{\mathbf{S}}} \cdot \mathbb{C}\left[\overset{\triangle}{\hat{\mathbf{\Gamma}}}\right] = \lambda\left\{\frac{\partial f}{\partial \hat{\mathbf{S}}} \cdot \mathbb{C}\left[\frac{\partial f}{\partial \hat{\mathbf{S}}}(\mathbf{1} + 2\hat{\mathbf{\Gamma}}_{e})\right] + c\left\|\frac{\partial f}{\partial \hat{\mathbf{S}}}\right\|^{2} - b\left\|\frac{\partial f}{\partial \hat{\mathbf{S}}}\right\|\left(\frac{\partial f}{\partial \hat{\mathbf{S}}} \cdot \hat{\mathbf{X}}\right)\right.$$

$$\left. + 2\left(\frac{\partial f}{\partial \hat{\mathbf{S}}}\frac{\partial f}{\partial \hat{\mathbf{S}}}\right) \cdot \hat{\mathbf{X}} - \frac{\partial f}{\partial k}k'(s_{\mathrm{p}})\left\|\frac{\partial f}{\partial \hat{\mathbf{S}}}\right\|\right\} .$$

The numerator of the proportionality factor λ can be employed as a loading function. ■

Following the assumption of hyperelasticity the elasticity tensor $\mathbb{C}$ is symmetric, $\mathbb{C} = \mathbb{C}^{\mathrm{T}}$, and the result is

$$\lambda = \frac{1}{N}\,\mathbb{C}\left[\frac{\partial f}{\partial \hat{\mathbf{S}}}\right] \cdot \overset{\triangle}{\hat{\mathbf{\Gamma}}} \tag{11.127}$$

instead of (11.121).

One advantage of using dual variables and derivatives is the simplicity with which the whole set of constitutive equations is transformable, leaving just tensors which operate on the reference configuration or the current configuration.

Representation of the constitutive model by means of tensors operating on the reference configuration calls for the following relations:

$$\hat{\mathbf{S}} = \mathbf{F}_{\mathrm{p}}\tilde{\mathbf{T}}\mathbf{F}_{\mathrm{p}}^{\mathrm{T}} \qquad \overset{\nabla}{\hat{\mathbf{S}}} = \mathbf{F}_{\mathrm{p}}\dot{\tilde{\mathbf{T}}}\mathbf{F}_{\mathrm{p}}^{\mathrm{T}} \tag{11.128}$$

$$\hat{\mathbf{X}} = \mathbf{F}_{\mathrm{p}}\tilde{\mathbf{X}}\mathbf{F}_{\mathrm{p}}^{\mathrm{T}} \qquad \overset{\nabla}{\hat{\mathbf{X}}} = \mathbf{F}_{\mathrm{p}}\dot{\tilde{\mathbf{X}}}\mathbf{F}_{\mathrm{p}}^{\mathrm{T}} \tag{11.129}$$

$$\hat{\mathbf{\Gamma}} = \mathbf{F}_{\mathrm{p}}^{\mathrm{T}-1}\mathbf{E}\mathbf{F}_{\mathrm{p}}^{-1} \qquad \overset{\triangle}{\hat{\mathbf{\Gamma}}} = \mathbf{F}_{\mathrm{p}}^{\mathrm{T}-1}\dot{\mathbf{E}}\mathbf{F}_{\mathrm{p}}^{-1} \tag{11.130}$$

$$\hat{\mathbf{\Gamma}}_{\mathrm{p}} = \mathbf{F}_{\mathrm{p}}^{\mathrm{T}-1}\mathbf{E}_{\mathrm{p}}\mathbf{F}_{\mathrm{p}}^{-1} \qquad \overset{\triangle}{\hat{\mathbf{\Gamma}}}_{\mathrm{p}} = \mathbf{F}_{\mathrm{p}}^{\mathrm{T}-1}\dot{\mathbf{E}}_{\mathrm{p}}\mathbf{F}_{\mathrm{p}}^{-1} \tag{11.131}$$

$$\mathbf{E}_{\mathrm{p}} = \tfrac{1}{2}(\mathbf{F}_{\mathrm{p}}^{\mathrm{T}}\mathbf{F}_{\mathrm{p}} - \mathbf{1}) \tag{11.132}$$

$$\hat{\Gamma}_e = F_p^{T-1} E_e F_p^{-1} \tag{11.133}$$

$$E_e = E - E_p \tag{11.134}$$

The result is

Theorem 11. 11

The basic equations (11.112) - (11.118) of elastoplasticity can be written by way of the following equivalent formulation:

Decomposition of the Deformation

$$E_e = E - E_p \tag{11.135}$$

$$E_p = \frac{1}{2}(F_p^T F_p - 1) \tag{11.136}$$

Isotropic Elasticity Relation

$$\widetilde{T} = F_p^{-1} g(F_p^{T-1} E_e F_p^{-1}) F_p^{T-1} \tag{11.137}$$

$$\widetilde{T} = \rho_R \frac{\partial}{\partial E_e} w(F_p^{T-1} E_e F_p^{-1}) \tag{11.138}$$

Yield Function

$$F = f(F_p(\widetilde{T} - \widetilde{X}) F_p^T, k) \tag{11.139}$$

Flow Rule

$$\dot{E}_p = \lambda \frac{\partial}{\partial \widetilde{T}} f(F_p(\widetilde{T} - \widetilde{X}) F_p^T, k) \text{ for } f = 0 \text{ and loading} \tag{11.140}$$

Kinematic Hardening

$$\dot{\widetilde{X}} = c C_p^{-1} \dot{E}_p C_p^{-1} - b \dot{s}_p \widetilde{X} \tag{11.141}$$

Plastic Arclength

$$\dot{s}_p(t) = \| C_p^{-1} \dot{E}_p \| \tag{11.142}$$

Loading Function

$$L = \left(F_p^{T-1} \frac{\partial f}{\partial \widetilde{T}} F_p^{-1}\right) \cdot \mathbb{C}[F_p^{T-1} \dot{E} F_p^{-1}] \tag{11.143}$$

□

Proof

The plastic *Green* strain $\mathbf{E}_p$ is determined by the flow rule. $\mathbf{E}$ and $\mathbf{E}_p$ lead to $\mathbf{E}_e$ calculated according to (11.135). The elasticity relation (11.137) follows from (11.113), $(11.128)_1$ and (11.133). Hyperelasticity implies

$$\hat{\mathbf{S}} = \rho_R \frac{dw(\hat{\mathbf{\Gamma}}_e)}{d\hat{\mathbf{\Gamma}}_e} = \rho_R \frac{dw(\mathbf{F}_p^{T-1}\mathbf{E}_e\mathbf{F}_p^{-1})}{d(\mathbf{F}_p^{T-1}\mathbf{E}_e\mathbf{F}_p^{-1})},$$

$$\hat{\mathbf{S}} = \rho_R \mathbf{F}_p \frac{\partial}{\partial \mathbf{E}_e} w(\mathbf{F}_p^{T-1}\mathbf{E}_e\mathbf{F}_p^{-1})\mathbf{F}_p^{T}, \tag{11.144}$$

following the same line of thought in the last step as that leading up to (10.142). The same argument also yields

$$\frac{\partial f}{\partial \hat{\mathbf{S}}} = \frac{\partial f(\mathbf{F}_p(\tilde{\mathbf{T}} - \tilde{\mathbf{X}})\mathbf{F}_p^{T})}{\partial(\mathbf{F}_p\tilde{\mathbf{T}}\mathbf{F}_p^{T})} = \mathbf{F}_p^{T-1}\frac{\partial}{\partial \tilde{\mathbf{T}}} f(\mathbf{F}_p(\tilde{\mathbf{T}} - \tilde{\mathbf{X}})\mathbf{F}_p^{T})\mathbf{F}_p^{-1}, \tag{11.145}$$

enabling (11.115) together with $(11.131)_2$ to form the flow rule (11.140). The evolution equation (11.141) for kinematic hardening leads on from (11.116) given $(11.129)_2$ and $(11.131)_2$. The relations applying to the plastic arclength and the loading function are self-evident. ■

To transform the constitutive equations to the *current configuration* the following formulae are needed:

$$\hat{\mathbf{S}} = \hat{\mathbf{F}}_e^{-1}\mathbf{S}\hat{\mathbf{F}}_e^{T-1} \tag{11.146}$$

$$\overset{\triangledown}{\hat{\mathbf{S}}} = \hat{\mathbf{F}}_e^{-1}\overset{\triangledown}{\mathbf{S}}\hat{\mathbf{F}}_e^{T-1} = \hat{\mathbf{F}}_e^{-1}(\dot{\mathbf{S}} - \mathbf{LS} - \mathbf{SL}^T)\hat{\mathbf{F}}_e^{T-1} \tag{11.147}$$

$$\hat{\mathbf{X}} = \hat{\mathbf{F}}_e^{-1}\mathbf{X}\hat{\mathbf{F}}_e^{T-1} \tag{11.148}$$

$$\overset{\triangledown}{\hat{\mathbf{X}}} = \hat{\mathbf{F}}_e^{-1}\overset{\triangledown}{\mathbf{X}}\hat{\mathbf{F}}_e^{T-1} = \hat{\mathbf{F}}_e^{-1}(\dot{\mathbf{X}} - \mathbf{LS} - \mathbf{SL}^T)\hat{\mathbf{F}}_e^{T-1} \tag{11.149}$$

$$\hat{\mathbf{\Gamma}} = \hat{\mathbf{F}}_e^{T}\mathbf{A}\hat{\mathbf{F}}_e \tag{11.150}$$

$$\overset{\vartriangle}{\hat{\mathbf{\Gamma}}} = \hat{\mathbf{F}}_e^{T}\overset{\vartriangle}{\mathbf{A}}\hat{\mathbf{F}}_e = \hat{\mathbf{F}}_e^{T}(\dot{\mathbf{A}} + \mathbf{L}^T\mathbf{A} + \mathbf{AL})\hat{\mathbf{F}}_e = \hat{\mathbf{F}}_e^{T}\mathbf{D}\hat{\mathbf{F}}_e \tag{11.151}$$

$$\hat{\mathbf{\Gamma}}_e = \hat{\mathbf{F}}_e^{T}\mathbf{A}_e\hat{\mathbf{F}}_e = \hat{\mathbf{E}}_e \tag{11.152}$$

$$\mathbf{A}_{\mathrm{e}} = \frac{1}{2}\left(\mathbf{1} - \hat{\mathbf{F}}_{\mathrm{e}}^{\mathrm{T}-1}\hat{\mathbf{F}}_{\mathrm{e}}^{-1}\right) \tag{11.153}$$

$$\hat{\mathbf{E}}_{\mathrm{e}} = \frac{1}{2}\left(\hat{\mathbf{F}}_{\mathrm{e}}^{\mathrm{T}}\hat{\mathbf{F}}_{\mathrm{e}} - \mathbf{1}\right) \tag{11.154}$$

$$\hat{\mathbf{B}}_{\mathrm{e}} = \hat{\mathbf{F}}_{\mathrm{e}}\hat{\mathbf{F}}_{\mathrm{e}}^{\mathrm{T}} \tag{11.155}$$

$$\hat{\mathbf{\Gamma}}_{\mathrm{p}} = \hat{\mathbf{F}}_{\mathrm{e}}^{\mathrm{T}}\mathbf{A}_{\mathrm{p}}\hat{\mathbf{F}}_{\mathrm{e}} \tag{11.156}$$

$$\mathbf{A}_{\mathrm{p}} = \frac{1}{2}\left(\hat{\mathbf{F}}_{\mathrm{e}}^{\mathrm{T}-1}\hat{\mathbf{F}}_{\mathrm{e}}^{-1} - \mathbf{F}^{\mathrm{T}-1}\mathbf{F}^{-1}\right) \tag{11.157}$$

$$\overset{\triangle}{\hat{\mathbf{\Gamma}}}_{\mathrm{p}} = \hat{\mathbf{F}}_{\mathrm{e}}^{\mathrm{T}}\overset{\triangle}{\mathbf{A}}_{\mathrm{p}}\hat{\mathbf{F}}_{\mathrm{e}} = \hat{\mathbf{F}}_{\mathrm{e}}^{\mathrm{T}}(\dot{\mathbf{A}}_{\mathrm{p}} + \mathbf{L}^{\mathrm{T}}\mathbf{A}_{\mathrm{p}} \quad \mathbf{A}_{\mathrm{p}}\mathbf{L})\hat{\mathbf{F}}_{\mathrm{e}} \tag{11.158}$$

Equations (11.147), (11.149), (11.151) and (11.158) show that *Oldroyd* rates of tensors related to the current configuration are calculated on the basis of the velocity gradient $\mathbf{L}$ instead of the plastic deformation rate $\hat{\mathbf{L}}_{\mathrm{p}}$.[21] The outcome is

Theorem 11. 12

Employing variables operating on the current configuration the basic equations (11.112) to(11.118) of elastoplasticity take on the following form:

Decomposition of the Deformation

$$\mathbf{A}_{\mathrm{p}} = \mathbf{A} - \mathbf{A}_{\mathrm{e}} \tag{11.159}$$

Isotropic Elasticity Relation

$$\mathbf{S} = \hat{\mathbf{F}}_{\mathrm{e}}\mathbf{g}(\hat{\mathbf{F}}_{\mathrm{e}}^{\mathrm{T}}\mathbf{A}_{\mathrm{e}}\hat{\mathbf{F}}_{\mathrm{e}})\hat{\mathbf{F}}_{\mathrm{e}}^{\mathrm{T}} = \rho_{\mathrm{R}}\frac{\partial}{\partial\mathbf{A}_{\mathrm{e}}}w\left(\hat{\mathbf{F}}_{\mathrm{e}}^{\mathrm{T}}\mathbf{A}_{\mathrm{e}}\hat{\mathbf{F}}_{\mathrm{e}}\right) \tag{11.160}$$

Yield Function

$$F = f\left(\hat{\mathbf{F}}_{\mathrm{e}}^{-1}(\mathbf{S} - \mathbf{X})\hat{\mathbf{F}}_{\mathrm{e}}^{\mathrm{T}-1}, k\right) \tag{11.161}$$

Flow Rule

$$\overset{\triangle}{\mathbf{A}}_{\mathrm{p}} = \lambda\frac{\partial}{\partial\mathbf{S}}f\left(\hat{\mathbf{F}}_{\mathrm{e}}^{-1}(\mathbf{S} - \mathbf{X})\hat{\mathbf{F}}_{\mathrm{e}}^{\mathrm{T}-1}, k\right) \text{ for } f = 0 \text{ and loading} \tag{11.162}$$

[21] Cf. (11.73) - (11.77). According to Definition 8. 2 the operators $(\cdot)^{\nabla}$ and $(\cdot)^{\triangle}$ depend on the configuration to which they refer.

Kinematic Hardening

$$\overset{\triangledown}{\mathbf{X}} = c\,\hat{\mathbf{B}}_e\,\overset{\triangle}{\hat{\mathbf{A}}}_p\,\hat{\mathbf{B}}_e - b\dot{s}_p\mathbf{X} \tag{11.163}$$

Plastic Arclength

$$\dot{s}_p(t) = \left\| \hat{\mathbf{B}}_e\overset{\triangle}{\hat{\mathbf{A}}}_p \right\| \tag{11.164}$$

Plastic Strain Rate

$$\overset{\triangle}{\hat{\mathbf{A}}}_p = \frac{1}{2}\left(\hat{\mathbf{B}}_e^{-1}\right)^{\triangle} = -\frac{1}{2}\hat{\mathbf{B}}_e^{-1}\overset{\triangledown}{\hat{\mathbf{B}}}_e\hat{\mathbf{B}}_e^{-1} \tag{11.165}$$

Loading Function

$$L = \left(\hat{\mathbf{F}}_e^{\mathrm{T}}\frac{\partial f}{\partial \mathbf{S}}\hat{\mathbf{F}}_e\right)\cdot \mathbb{C}[\hat{\mathbf{F}}_e^{\mathrm{T}}\mathbf{D}\hat{\mathbf{F}}_e] \tag{11.166}$$

□

Proof

The elastic *Almansi* tensor $\mathbf{A}_e$ is determined by the flow rule (11.162) and (11.165) (see below). $\mathbf{A}$ and $\mathbf{A}_e$ lead to $\mathbf{A}_p$ calculated according to (11.157). The elasticity relation (11.160) leads on from (11.113) in combination with (11.152) and (11.146). The potential relation $(11.160)_2$ is obtained with the help of

$$\frac{dw(\hat{\mathbf{F}}_e^{\mathrm{T}}\mathbf{A}_e\hat{\mathbf{F}}_e)}{d(\hat{\mathbf{F}}_e^{\mathrm{T}}\mathbf{A}_e\hat{\mathbf{F}}_e)} = \hat{\mathbf{F}}_e^{-1}\frac{\partial w(\hat{\mathbf{F}}_e^{\mathrm{T}}\mathbf{A}_e\hat{\mathbf{F}}_e)}{\partial \mathbf{A}_e}\hat{\mathbf{F}}_e^{\mathrm{T}-1} .$$

Starting with the yield function

$$f = f(\hat{\mathbf{F}}_e^{-1}(\mathbf{S} - \mathbf{X})\hat{\mathbf{F}}_e^{\mathrm{T}-1}, k) ,$$

its gradient

$$\frac{\partial f}{\partial \hat{\mathbf{S}}} = \frac{\partial f(\hat{\mathbf{F}}_e^{-1}(\mathbf{S} - \mathbf{X})\hat{\mathbf{F}}_e^{\mathrm{T}-1}, k)}{\partial(\hat{\mathbf{F}}_e^{-1}\mathbf{S}\hat{\mathbf{F}}_e^{\mathrm{T}-1})} = \hat{\mathbf{F}}_e^{\mathrm{T}}\frac{\partial f}{\partial \mathbf{S}}\hat{\mathbf{F}}_e ,$$

leads to the flow rule (11.162). The evolution equations (11.163) and (11.164) are self-explanatory.

The flow rule is actually an evolution equation for the plastic strain. However, (11.159),

$$\mathbf{A}_p = \mathbf{A} - \mathbf{A}_e = \mathbf{A} - \tfrac{1}{2}(\mathbf{1} - \hat{\mathbf{B}}_e^{-1}) \,, \quad \text{implies}$$

$$\overset{\triangle}{\mathbf{A}}_p = \overset{\triangle}{\mathbf{A}} - \overset{\triangle}{\mathbf{A}}_e = \mathbf{D} + \tfrac{1}{2}(\hat{\mathbf{B}}_e^{-1})^{\cdot} - \tfrac{1}{2}\mathbf{L}^T(\mathbf{1} - \hat{\mathbf{B}}_e^{-1}) - \tfrac{1}{2}(\mathbf{1} - \hat{\mathbf{B}}_e^{-1})\mathbf{L} \quad \text{and}$$

$$\overset{\triangle}{\mathbf{A}}_p = \tfrac{1}{2}\Big\{(\hat{\mathbf{B}}_e^{-1})^{\cdot} + \tfrac{1}{2}\mathbf{L}^T\hat{\mathbf{B}}_e^{-1} + \hat{\mathbf{B}}_e^{-1}\mathbf{L}\Big\} \,, \text{ i.e.}$$

$$\overset{\triangle}{\mathbf{A}}_p = -\tfrac{1}{2}\hat{\mathbf{B}}_e^{-1}\Big\{\dot{\hat{\mathbf{B}}}_e - \mathbf{L}\hat{\mathbf{B}}_e - \hat{\mathbf{B}}_e\mathbf{L}^T\Big\}\hat{\mathbf{B}}_e^{-1} \,, \text{ that is (11.165).}$$

The loading function (11.166) leads on from (11.146) and (11.151). ■

Many materials, especially metals, are characterised by the property that the elastic strains are small, whereas the plastic deformations can be of any size. This considerably simplifies both kinematics and the material description. The simplification in kinematics is described in

Theorem 11. 13
In the case of small elastic strains the multiplicative decomposition (11.61) is approximately equal to the polar decomposition:

$$\mathbf{F} \approx \hat{\mathbf{R}}_e\mathbf{F}_p = \hat{\mathbf{R}}_e\mathbf{R}_p\mathbf{U}_p \,. \tag{11.167}$$

□

Proof
$\mathbf{F} = \hat{\mathbf{F}}_e\mathbf{F}_p = \hat{\mathbf{R}}_e\hat{\mathbf{U}}_e\mathbf{F}_p$ in combination with $\hat{\mathbf{U}}_e \approx \mathbf{1}$

leads to the approximation

$$\hat{\mathbf{F}}_e \approx \hat{\mathbf{R}}_e \,, \text{ or } \mathbf{F} \approx \hat{\mathbf{R}}_e\mathbf{F}_p = \hat{\mathbf{R}}_e\mathbf{R}_p\mathbf{U}_p \,.$$ ■

The corresponding simplification of the material description is illustrated by

Theorem 11. 14
For small elastic strains (and arbitrary rotations) the constitutive equations (11.112) - (11.118) and (11.159) - (11.166) can be represented in asymptotic approximation as follows:

Decomposition of the Deformation

$$\mathbf{A}_\mathrm{p} = \mathbf{A} - \mathbf{A}_\mathrm{e} \tag{11.168}$$

Isotropic Elasticity Relation

$$\mathbf{S} = \mathbf{g}(\mathbf{A}_\mathrm{e}) = \rho_\mathrm{R}\frac{\mathrm{d}w(\mathbf{A}_\mathrm{e})}{\mathrm{d}\mathbf{A}_\mathrm{e}} \tag{11.169}$$

Yield Function

$$f = f(\mathbf{S} - \mathbf{X}, k) \tag{11.170}$$

Flow Rule

$$\overset{\triangle}{\mathbf{A}}_\mathrm{p} = \lambda\frac{\partial}{\partial \mathbf{S}} f(\mathbf{S} - \mathbf{X}\,, k) \text{ for } f = 0 \text{ and loading} \tag{11.171}$$

Kinematic Hardening

$$\overset{\triangledown}{\mathbf{X}} = c\overset{\triangle}{\mathbf{A}}_\mathrm{p} - b\dot{s}_\mathrm{p}\mathbf{X} \tag{11.172}$$

Plastic Arclength

$$\dot{s}_\mathrm{p}(t) = \|\overset{\triangle}{\mathbf{A}}_\mathrm{p}\| \tag{11.173}$$

Plastic Strain Rate

$$\overset{\triangle}{\mathbf{A}}_\mathrm{p} = -\tfrac{1}{2}\overset{\triangledown}{\hat{\mathbf{B}}}_\mathrm{e} \tag{11.174}$$

Loading Function

$$L = \frac{\partial f}{\partial \mathbf{S}}\cdot \mathbb{C}[\mathbf{D}] \tag{11.175}$$

□

Proof

The elasticity relation (11.169) leads on from (11.160) with $\hat{\mathbf{F}}_\mathrm{e} = \hat{\mathbf{R}}_\mathrm{e}$ based on the isotropy of the tensor function $\mathbf{g}(\cdot)$ or $w(\cdot)$. The remaining statements are self-evident. ■

The above material model can be understood as the simplest generalisation of the constitutive equations (5.178) - (5.186) of classical elastoplasticity to large plastic strains.

12 Viscoplasticity

12.1 Preliminary Remarks

The theory of *viscoplasticity* depicts rate-dependent material behaviour with equilibrium hysteresis. Viscoplasticity is the most general form of material modelling. In principle, it incorporates all macroscopically observable phenomena.

A great number of suggestions pertaining to the representation of viscoplastic material behaviour have recently been put forward in the specialist literature published in this field. Since it is not possible to pay tribute to all these material models individually, this chapter is confined to just a few systematic points of view. Basically, there are three different approaches to choose from when modelling viscoplastic material properties:

- the demarcation of an elastic domain by means of a yield surface;
- a uniform formulation of evolution equations without yield surfaces and case distinction;
- the decomposition of the total stress into rate-independent equilibrium stress and rate-dependent overstress.

Introducing a yield function allows an elastic domain to be defined with the help of a *static yield surface*. States of stress outside the static yield surface lead to the evolution of inelastic deformations. This concept was devised by P. Perzyna[1] and applied or enlarged upon by numerous other authors, in

---

1 PERZYNA [1963]; PERZYNA [1966]; PERZYNA & WOJNO [1968]. Cf. PRAGER [1961], p. 126.

particular J. L. Chaboche.[2] One advantage of this theory is its clear structure within the general theory of internal variables. It can be understood as a generalisation of the classical theory of plasticity, which it includes as a special case (see below).

The second method of modelling viscoplasticity has no need for yield surfaces or the case distinctions connected with it. A system of nonlinear differential equations is set up to describe the material properties. Apart from the macroscopic quantities (stresses and strains), they also contain internal variables and describe process-dependent material behaviour in a uniform manner, i.e. making no distinctions between different cases. In the specific literature these constitutive models are called *unified viscoplasticity models*.[3] It is not intended to deal with these models in the text that follows.

A third way of constructing constitutive models of viscoplasticity is built on the a priori decomposition of the total stress into an equilibrium part (equilibrium stress) and a non-equilibrium part (overstress). In this concept, the relations between the equilibrium stress and the overstress become uncoupled to a large extent from each other, resulting in a modular structure of the constitutive model. The development of viscoplastity models based on an additive stress decomposition was a task carried out very successfully by E. Krempl.[4] The equilibrium stress in these constitutive models is dependent on the process history; the equilibrium relation in particular may also be influenced by the process rate.

Equilibrium stress represented by rate-independent functionals of the deformation history was first proposed by M. Korzeń[5] and further developed by A. Lion.[6] One advantage of this modelling structure besides the decoupling of the two parts of stress is the possibility of extending the constitutive modelling to thermomechanics (see Chap. 13).

In order to represent material models in viscoplasticity within the concept of internal variables, the stress tensor according to (7.121) is assumed to be a function of the strain tensor, also depending on a finite number of internal variables $q_1, \dots, q_N$:

$$\widetilde{\mathbf{T}} = \mathbf{f}(\mathbf{E}, q_1, \dots, q_N) .$$

[2] CHABOCHE [1977]; CHABOCHE & ROUSSELIER [1983]; CHABOCHE [1986, 1991, 1993].

[3] Unified viscoplasticity models are discussed in HART [1976]; MILLER [1976, 1987]; BODNER & PARTOM [1975]. A comparison is given in HARTMANN & KOLLMANN [1987].

[4] KREMPL [1979, 1987, 1998]; KREMPL et al. [1986]; KREMPL & LU [1984, 1989]; KREMPL & CHOI [1990]; KREMPL & BORDONARO [1998].

[5] KORZEŃ [1988]; HAUPT & KORZEŃ [1989].

[6] LION [1994]; HAUPT & LION [1995].

The temporal evolution of the internal variables $q_1, \dots , q_N$ is represented by means of a system of ordinary differential equations (see (7.122)):

$$\dot{q}_k(t) = f_k\big(\mathbf{E}(t), q_1(t), \dots , q_N(t)\big) \,, \qquad k = 1, \dots , N \,.$$

As suggested in (7.125), these *evolution equations* determine the current values of the internal variables q_k in dependence on the initial conditions and the strain process $\tau \longmapsto \mathbf{E}(\tau)$ $(\tau \leq t)$.

The physical significance of the evolution equations lies primarily in their *equilibrium solutions* and their *properties of stability*. In the constitutive theories discussed so far they have individual characteristics. In *viscoelasticity* the equilibrium solutions depend uniquely on the current state of strain and are asymptotically stable. In *plasticity* every solution of evolution equations is also an equilibrium solution. The equilibrium states in plasticity are functionals of the process history; this corresponds to a rate-independent hysteresis.

In *viscoplasticity* evolution equations also serve to represent relaxation properties leading to stable states of equilibrium. As opposed to viscoelasticity, however, there is no unique dependence of the equilibrium solutions on the current state of deformation: several states of equilibrium may be attributed to one and the same state of strain. This property of evolution equations corresponds to the *equilibrium hysteresis*, the modelling of which is made possible by viscoplasticity.

12. 2 Viscoplasticity with Elastic Domain

12. 2. 1 A General Constitutive Model

This section formulates a general constitutive model of viscoplasticity including the classical model (5.219) - (5.224) of geometrically linear viscoplasticity by way of a special case. A set of internal variables $(\mathbf{E}_{\mathrm{i}}, q_1, \dots , q_N)$ serves to represent the memory properties. Their physical meaning can be left open for the time being, with the exception of the inelastic strain

$$\mathbf{E}_{\mathrm{i}} = \tfrac{1}{2}\left(\mathbf{F}_{\mathrm{i}}^{\mathrm{T}}\mathbf{F}_{\mathrm{i}} - \mathbf{1}\right) . \tag{12.1}$$

This internal variable is a *Green* strain tensor, based on the multiplicative decomposition $\mathbf{F} = \hat{\mathbf{F}}_{\mathrm{e}}\mathbf{F}_{\mathrm{i}}$ of the deformation gradient in the sense of a stress-free intermediate configuration. The temporal evolution of the internal variables, in analogy to the classical model, is determined by a *yield function*, a scalar-valued function of the stress and the internal variables, $f = f(\widetilde{\mathbf{T}}, \mathbf{E}_\mathrm{i}, q_1, \ldots, q_N)$. The rate of change of all the internal variables is proportional to a power of the yield function, if it takes positive values; for negative values of the yield function the internal variables remain constant in time.

Definition 12. 1

The following set of constitutive equations defines a material model of *viscoplasticity with an elastic domain*:

Decomposition of the Deformation

$$\mathbf{E} = \mathbf{E}_\mathrm{e} + \mathbf{E}_\mathrm{i} \tag{12.2}$$

Elasticity Relation

$$\widetilde{\mathbf{T}} = \mathbf{f}(\mathbf{E}_\mathrm{e}, \mathbf{E}_\mathrm{i}, q_1, \ldots, q_N) \tag{12.3}$$

Yield Function[7]

$$f = f(\widetilde{\mathbf{T}}, \mathbf{E}_\mathrm{i}, q_1, \ldots, q_N) \tag{12.4}$$

Flow Rule

$$\dot{\mathbf{E}}_\mathrm{i}(t) = \frac{1}{\eta}\langle f\rangle^\mathrm{m}\mathbf{g}(\widetilde{\mathbf{T}}, \mathbf{E}_\mathrm{i}, q_1, \ldots, q_N) \tag{12.5}$$

Evolution Equations

$$\dot{q}_k(t) = \frac{1}{\eta}\langle f\rangle^\mathrm{m} f_k(\mathbf{E}_\mathrm{e}(t), \mathbf{E}_\mathrm{i}(t), q_1(t), \ldots, q_N(t)) \,, \quad k = 1, \ldots, N \tag{12.6}$$

For the *MacCauley* bracket $\langle\cdot\rangle$ definition (5.224) is valid:

$$\langle x\rangle = \begin{cases} x \text{ for } x \geq 0 \\ 0 \text{ for } x < 0 \end{cases}$$

◊

According to the physical meaning of the inelastic strain $\mathbf{E}_\mathrm{i}$, it is advisable from a physical point of view to introduce the difference $\mathbf{E}_\mathrm{e} = \mathbf{E} - \mathbf{E}_\mathrm{i}$ into the elasticity relation (12.3) as an independent variable in lieu of the total

[7] In these equations $f(\cdot)$ is dimensionless.

strain $\mathbf{E}$. Using the *yield function* $f(\cdot)$ according to (12.4) a surface, which is known as the *static yield surface*, is defined in the stress space by $f = 0$. The static yield surface demarcates an *elastic domain* in the *stress space*: for states of stress within this domain the material behaviour is purely elastic. The inelastic strains and all the other internal variables remain temporally constant during elastic processes. Instead of the stress tensor the elastic strain $\mathbf{E}_{\mathrm{e}}$ might be introduced into the yield function as an independent variable. In this case the elastic domain is defined in the *strain space.*

States of stress outside the elastic domain ($f > 0$) produce inelastic deformations in accordance with the *flow rule* (12.5). This flow rule determines an inelastic strain rate, whose absolute value increases with its distance from the static yield surface. If f is dimensionless in (12.5) and $\mathrm{g}(\cdot)$ a stress tensor, then the material parameter η is a *viscosity*. The exponent m is a material constant expressing a nonlinear rate-dependence (cf. (5.222) and (5.234)). The tensor-valued material function $\mathrm{g}(\cdot)$ completes the modelling of the inelastic strain rate. The special case of

$$\mathrm{g}(\widetilde{\mathbf{T}}, \mathbf{E}_{\mathrm{i}}, q_1, \dots, q_N) = k_0^{\,2}\frac{\partial f}{\partial \widetilde{\mathbf{T}}}\,, \tag{12.7}$$

k_0 being a constant with the dimension of a stress, defines an *associated flow rule.* In this case the function $\mathrm{g}(\cdot)$ is equal to the normal on the surface $f = c > 0$. A surface $f = c > 0$ is also termed *dynamic yield surface.* It is quite common practice to assume associated flow rules, though it may not always be expedient. No associated flow rule is required for the following general considerations.

The other internal variables also undergo a change together with the inelastic strain $\mathbf{E}_{\mathrm{i}}$. Their rate of change, according to the evolution equations (12.6), is proportional to the value of the yield function or zero. The appearance of the internal variables in the yield function allows the shape and location of the yield surface in the stress space to depend on the process history. The yield surface's history-dependence is known as *hardening.*

The set of constitutive equations (12.2) - (12.6) clearly fits into the general pattern of evolution equations (7.121), (7.122). It is also clear that model (12.2) - (12.6) does not belong to the viscoelasticity category: for one and the same state of strain $\mathbf{E}(t)$, the evolution equations (12.5), (12.6) admit more than one equilibrium value for the internal variables. The particular structure of this model of *viscoplasticity* is characterised by the prominent role of the yield function, which appears on the right-hand side of the evolution equations (12.5), (12.6) as their common factor. This circumstance lends the material model special properties.

12. 2. 2 Application of the Intermediate Configuration

The general constitutive equations (12.2) - (12.6) of viscoplasticity comply with the assumption of *material frame-indifference* and therefore provide a generally valid pattern, where specific material models of viscoplasticity can be constructed. To this end, the physical meaning of the internal variables ought to be clarified, at least within the framework of the macroscopic theory. For the formulation of the evolution equations the question as to whether the internal variables are scalar- or tensor-valued quantities, and which time derivatives should be chosen for tensor-valued internal variables is important. To clarify those questions, the notion of a stress-free intermediate configuration in connection with the concept of dual variables and derivatives has certain advantages. The methods used in viscoelasticity and plasticity are also applicable in viscoplasticity.

The physical justification for the decomposition (12.2) of the strain is based on the idea of the *local unloading* in exactly the same way as in plasticity.[8] The flow rule creates an inelastic state of strain $\mathbf{E}_i$, which is uniquely determined by the process history. The tensor field

$$\mathbf{E}_i(\boldsymbol{X}, t) = \tfrac{1}{2}(\mathbf{C}_i - \mathbf{1}) = \tfrac{1}{2}(\mathbf{F}_i{}^{\mathrm{T}}\mathbf{F}_i - \mathbf{1})$$

characterises the metric of the *intermediate configuration*. Corresponding to the inelastic strain tensor $\mathbf{E}_i$ of *Green* type there is a right stretch tensor $\mathbf{U}_i = \sqrt{2\mathbf{E}_i + \mathbf{1}}$. The uniqueness of this quantity prompts a formal definition of the then remaining elastic part

$$\bar{\hat{\mathbf{F}}}_e = \mathbf{F}\mathbf{U}_i^{-1}$$

of the deformation gradient $\mathbf{F}$. In this approach the multiplicative decomposition $\mathbf{F} = \bar{\hat{\mathbf{F}}}_e\mathbf{U}_i$ of the deformation gradient cannot be unique, since it is possible to introduce an orthogonal tensor $\bar{\mathbf{Q}}$ at any time, thus altering both factors without influencing the inelastic strain $\mathbf{E}_i$: $\mathbf{F} = \bar{\hat{\mathbf{F}}}_e\bar{\mathbf{Q}}^{\mathrm{T}}\bar{\mathbf{Q}}\mathbf{U}_i$. This leads to the kinematic decomposition

$$\mathbf{F} = \hat{\mathbf{F}}_e\mathbf{F}_i = \hat{\mathbf{F}}_e\mathbf{R}_i\mathbf{U}_i\,. \qquad (12.8)$$

In much the same way as in plasticity these arguments form the basis for decomposing the general material model into a functional relation, deter-

[8] Cf. Sect. 11. 3. 2.

mining the inelastic strain $\mathbf{E}_i$ (flow rule), and an elasticity relation. In this connection, no constitutive assumption is made for the rotation $\mathbf{R}_i$ occurring in the polar decomposition $\mathbf{F}_i = \mathbf{R}_i\mathbf{U}_i$. If one agrees to do without a constitutive equation for $\mathbf{R}_i$, then the entire material model ought to be invariant with respect to any choice of $\mathbf{R}_i$, i.e. invariant with respect to any rotation of the intermediate configuration.

It is easy to guarantee this property of invariance, provided isotropic tensor functions and *dual variables*, which refer to the inelastic intermediate configuration, are used to formulate the material model. In this sense, the *Green* strain tensor $\mathbf{E}$, in analogy to (11.66), is expressed in terms of the strain tensor

$$\hat{\mathbf{\Gamma}} = \mathbf{F}_i^{T-1}\mathbf{E}\mathbf{F}_i^{-1} \tag{12.9}$$

operating on the intermediate configuration. In connection with the multiplicative decomposition $\mathbf{F} = \hat{\mathbf{F}}_e\mathbf{F}_i$ definition (12.9) leads to the additive decomposition

$$\hat{\mathbf{\Gamma}} = \hat{\mathbf{\Gamma}}_e + \hat{\mathbf{\Gamma}}_i \tag{12.10}$$

into an elastic *Green* tensor and an inelastic *Almansi* tensor:

$$\hat{\mathbf{\Gamma}}_e = \frac{1}{2}\left(\hat{\mathbf{F}}_e^T\hat{\mathbf{F}}_e - \mathbf{1}\right), \qquad \hat{\mathbf{\Gamma}}_i = \frac{1}{2}\left(\mathbf{1} - \mathbf{F}_i^{T-1}\mathbf{F}_i^{-1}\right). \tag{12.11}$$

The associated strain rate is calculated from the *Green* strain rate using the same transformation as applied in (12.9),

$$\overset{\triangle}{\hat{\mathbf{\Gamma}}} = \mathbf{F}_i^{T-1}\dot{\mathbf{E}}\mathbf{F}_i^{-1}. \tag{12.12}$$

It is the covariant *Oldroyd* rate,

$$\overset{\triangle}{\hat{\mathbf{\Gamma}}} = \mathbf{F}_i^{T-1}\left(\mathbf{F}_i^T\hat{\mathbf{\Gamma}}\mathbf{F}_i\right)^{\cdot}\mathbf{F}_i^{-1} = \dot{\hat{\mathbf{\Gamma}}} + \hat{\mathbf{L}}_i^T\hat{\mathbf{\Gamma}} + \hat{\mathbf{\Gamma}}\hat{\mathbf{L}}_i, \tag{12.13}$$

calculated using the inelastic deformation rate

$$\hat{\mathbf{L}}_i = \dot{\mathbf{F}}_i\mathbf{F}_i^{-1}. \tag{12.14}$$

Corresponding to the additive decomposition of the total strain we then have the decomposition of the strain rate

$$\overset{\triangle}{\hat{\mathbf{\Gamma}}} = \overset{\triangle}{\hat{\mathbf{\Gamma}}}_{\mathrm{e}} + \overset{\triangle}{\hat{\mathbf{\Gamma}}}_{\mathrm{i}} \tag{12.15}$$

into an elastic part

$$\overset{\triangle}{\hat{\mathbf{\Gamma}}}_{\mathrm{e}} = \dot{\hat{\mathbf{\Gamma}}}_{\mathrm{e}} + \hat{\mathbf{L}}_{\mathrm{i}}^{\mathrm{T}}\hat{\mathbf{\Gamma}}_{\mathrm{e}} + \hat{\mathbf{\Gamma}}_{\mathrm{e}}\hat{\mathbf{L}}_{\mathrm{i}} \tag{12.16}$$

and an inelastic part

$$\overset{\triangle}{\hat{\mathbf{\Gamma}}}_{\mathrm{i}} = \dot{\hat{\mathbf{\Gamma}}}_{\mathrm{i}} + \hat{\mathbf{L}}_{\mathrm{i}}^{\mathrm{T}}\hat{\mathbf{\Gamma}}_{\mathrm{i}} + \hat{\mathbf{\Gamma}}_{\mathrm{i}}\hat{\mathbf{L}}_{\mathrm{i}} = \tfrac{1}{2}(\hat{\mathbf{L}}_{\mathrm{i}} + \hat{\mathbf{L}}_{\mathrm{i}}) \,. \tag{12.17}$$

The dual stress tensor to $\hat{\mathbf{\Gamma}}$ is, analogous to (11.70), the 2nd *Piola-Kirchhoff* tensor $\hat{\mathbf{S}}$ related to the intermediate configuration:

$$\hat{\mathbf{S}} = \mathbf{F}_{\mathrm{i}}\tilde{\mathbf{T}}\mathbf{F}_{\mathrm{i}}^{\mathrm{T}} = \hat{\mathbf{F}}_{\mathrm{e}}^{-1}\mathbf{S}\hat{\mathbf{F}}_{\mathrm{e}}^{\mathrm{T}-1} \,. \tag{12.18}$$

The attributed stress rate is calculated from the material derivative of $\tilde{\mathbf{T}}$ using the same transformation,

$$\overset{\triangledown}{\hat{\mathbf{S}}} = \mathbf{F}_{\mathrm{i}}\dot{\tilde{\mathbf{T}}}\mathbf{F}_{\mathrm{i}}^{\mathrm{T}} \,. \tag{12.19}$$

It is the contravariant *Oldroyd* rate

$$\overset{\triangledown}{\hat{\mathbf{S}}} = \mathbf{F}_{\mathrm{i}}(\mathbf{F}_{\mathrm{i}}^{-1}\hat{\mathbf{S}}\mathbf{F}_{\mathrm{i}}^{\mathrm{T}-1})^{\cdot}\,\mathbf{F}_{\mathrm{i}}^{\mathrm{T}} = \dot{\hat{\mathbf{S}}} - \hat{\mathbf{L}}_{\mathrm{i}}\hat{\mathbf{S}} - \hat{\mathbf{S}}\hat{\mathbf{L}}_{\mathrm{i}}^{\mathrm{T}} \,, \tag{12.20}$$

calculated using $\hat{\mathbf{L}}_{\mathrm{i}}$.

Given these strain and stress tensors and rates it is possible to formulate a model depicting finite viscoplasticity as follows:

Decomposition

$$\hat{\mathbf{\Gamma}} = \hat{\mathbf{\Gamma}}_{\mathrm{e}} + \hat{\mathbf{\Gamma}}_{\mathrm{i}} \qquad \overset{\triangle}{\hat{\mathbf{\Gamma}}} = \overset{\triangle}{\hat{\mathbf{\Gamma}}}_{\mathrm{e}} + \overset{\triangle}{\hat{\mathbf{\Gamma}}}_{\mathrm{i}} \tag{12.21}$$

Isotropic Elasticity

$$\hat{\mathbf{S}} = \hat{\mathbf{g}}(\hat{\mathbf{\Gamma}}_{\mathrm{e}}) \tag{12.22}$$

Yield Function

$$\hat{f} = \hat{f}(\hat{\mathbf{S}} - \hat{\mathbf{X}}, k) \tag{12.23}$$

Associated Flow Rule

$$\overset{\triangle}{\hat{\mathbf{\Gamma}}}_{\mathrm{i}} = \frac{1}{\eta}\langle \hat{f}(\hat{\mathbf{S}} - \hat{\mathbf{X}}, k)\rangle^{\mathrm{m}} \frac{\partial \hat{f}}{\partial \hat{\mathbf{S}}} \tag{12.24}$$

Kinematic Hardening

$$\overset{\triangledown}{\hat{\mathbf{X}}} = c\overset{\triangle}{\hat{\mathbf{\Gamma}}}_{\mathrm{i}} - b\sqrt{\tfrac{2}{3}}\,\big\|\overset{\triangle}{\hat{\mathbf{\Gamma}}}_{\mathrm{i}}\big\|\,\hat{\mathbf{X}} \tag{12.25}$$

The material functions $\hat{\mathbf{g}}(\cdot)$ and $\hat{f}(\cdot)$ are isotropic tensor functions.[9] All relations in this constitutive model are invariant under rotations of the intermediate configuration.

The material model (12.21) - (12.25) is a special case among the more general equations (12.2) - (12.6). To convince oneself of this fact, all that has to be done is to transform all the variables back to the reference configuration. This can be achieved via an analogous application of the transformation formulae (11.128) - (11.133), in which only the subscript $(\cdot)_{\mathrm{p}}$ is replaced by $(\cdot)_{\mathrm{i}}$. Instead of (12.21) - (12.25) one obtains the following set of equations:

$$\mathbf{E} = \mathbf{E}_{\mathrm{e}} + \mathbf{E}_{\mathrm{i}}\,, \qquad \dot{\mathbf{E}}(t) = \dot{\mathbf{E}}_{\mathrm{e}}(t) + \dot{\mathbf{E}}_{\mathrm{i}}(t)\,, \tag{12.26}$$

$$\tilde{\mathbf{T}} = \mathbf{F}_{\mathrm{i}}^{-1}\hat{\mathbf{g}}(\mathbf{F}_{\mathrm{i}}^{\mathrm{T}-1}\mathbf{E}_{\mathrm{e}}\mathbf{F}_{\mathrm{i}}^{-1})\mathbf{F}_{\mathrm{i}}^{\mathrm{T}-1}\,, \tag{12.27}$$

$$\hat{f} = \hat{f}(\mathbf{F}_{\mathrm{i}}(\tilde{\mathbf{T}} - \tilde{\mathbf{X}})\mathbf{F}_{\mathrm{i}}^{\mathrm{T}}, k)\,, \tag{12.28}$$

$$\dot{\mathbf{E}}_{\mathrm{i}}(t) = \frac{1}{\eta}\langle \hat{f}\rangle^{\mathrm{m}} \frac{\partial \hat{f}}{\partial \tilde{\mathbf{T}}}\,, \tag{12.29}$$

$$\dot{\tilde{\mathbf{X}}}(t) = \frac{1}{\eta}\langle \hat{f}\rangle^{\mathrm{m}} \left\{c\,\mathbf{C}_{\mathrm{i}}^{-1}\frac{\partial \hat{f}}{\partial \tilde{\mathbf{T}}}\mathbf{C}_{\mathrm{i}}^{-1} - b\sqrt{\tfrac{2}{3}}\,\Big\|\mathbf{F}_{\mathrm{i}}^{\mathrm{T}-1}\frac{\partial \hat{f}}{\partial \tilde{\mathbf{T}}}\mathbf{F}_{\mathrm{i}}^{-1}\Big\|\,\tilde{\mathbf{X}}\right\}, \tag{12.30}$$

$$\left(\mathbf{C}_{\mathrm{i}} = \mathbf{U}_{\mathrm{i}}^{2} = 2\mathbf{E}_{\mathrm{i}} + \mathbf{1}\right).$$

In view of the isotropy of the functions $\hat{\mathbf{g}}(\cdot)$ and $\hat{f}(\cdot)$ the orthogonal part $\mathbf{R}_{\mathrm{i}}$, occurring in the polar decomposition $\mathbf{F}_{\mathrm{i}} = \mathbf{R}_{\mathrm{i}}\mathbf{U}_{\mathrm{i}}$, is eliminated from the equations. The constitutive equations (12.26) - (12.30) have the same structure as equations (12.2) - (12.6) which were employed to define a fairly general model of viscoplasticity.

---

[9] This model was applied to the numerical simulation of metal-forming processes in LÜHRS et al. [1997].

12.3 Plasticity as a Limit Case of Viscoplasticity

The theory of viscoplasticity with elastic range has a special physical meaning. This is due to the fact that it contains the evolution equations of rate-independent plasticity as an asymptotic limit, namely for deformation processes evolving sufficiently slowly or, equivalently, for very small values of viscosity η. It is the aim of the following sections to verify this property.[10] This presupposes the special case of $\mathrm{m} = 1$.

12.3.1 The Differential Equation of the Yield Function

If we imagine a given strain process $t \longmapsto \mathbf{E}(t)$ in combination with initial conditions for the inelastic deformation $\mathbf{E}_\mathrm{i}$ and the internal variables q_k, the evolution equations (12.5), (12.6) together with the elasticity relation (12.3) determine the internal variables $q_1(t), \dots, q_N(t)$, the inelastic strain $\mathbf{E}_\mathrm{i}(t)$ and the stress response $\tilde{\mathbf{T}}(t)$ as functions of time. The solutions of equations (12.2) - (12.6) determine the temporal evolution of the yield function:

$$t \longmapsto f(t) = f(\tilde{\mathbf{T}}(t), \mathbf{E}_\mathrm{i}(t), q_1(t), \dots, q_N(t)) \,. \tag{12.31}$$

For every solution of the constitutive equations the function $f(t)$ has a certain characteristic, which is of vital importance in terms of the physical interpretation of the material model. In order to study this characteristic, we calculate the total time derivative of the yield function and apply the decomposition (12.2) and the elasticity relation (12.3):

$$\frac{\mathrm{d}}{\mathrm{d}t} f(\tilde{\mathbf{T}}(t), \mathbf{E}_\mathrm{i}(t), q_1(t), \dots, q_N(t)) = \frac{\partial f}{\partial \tilde{\mathbf{T}}} \cdot \dot{\tilde{\mathbf{T}}} + \frac{\partial f}{\partial \mathbf{E}_\mathrm{i}} \cdot \dot{\mathbf{E}}_\mathrm{i}(t) + \sum_k \frac{\partial f}{\partial q_k} \dot{q}_k(t) \,,$$

$$\dot{f}(t) = \frac{\partial f}{\partial \tilde{\mathbf{T}}} \cdot \left\{ \frac{\partial \mathbf{f}}{\partial \mathbf{E}_\mathrm{e}} [\dot{\mathbf{E}} - \dot{\mathbf{E}}_\mathrm{i}] + \frac{\partial \mathbf{f}}{\partial \mathbf{E}_\mathrm{i}} [\dot{\mathbf{E}}_\mathrm{i}(t)] + \sum_k \frac{\partial \mathbf{f}}{\partial q_k} \dot{q}_k(t) \right\}$$

$$+ \frac{\partial f}{\partial \mathbf{E}_\mathrm{i}} \cdot \dot{\mathbf{E}}_\mathrm{i}(t) + \sum_k \frac{\partial f}{\partial q_k} \dot{q}_k(t) \,.$$

If we take the evolution equations (12.6) and the flow rule (12.5) into account, given $\mathrm{m} = 1$, we obtain

[10] This verification follows an idea of KRATOCHVIL & DILLON [1969]. See HAUPT et al. [1992b]; TSAKMAKIS [1994], pp. 140.

$$\dot{f}(t) = \frac{\partial f}{\partial \tilde{\mathbf{T}}} \cdot \left\{ \frac{\partial \mathbf{f}}{\partial \mathbf{E}_{\mathrm{e}}} [\dot{\mathbf{E}}(t)] \right\} - \tfrac{1}{\eta} \langle f \rangle \left\{ \frac{\partial f}{\partial \tilde{\mathbf{T}}} \cdot \left[\left(\frac{\partial \mathbf{f}}{\partial \mathbf{E}_{\mathrm{e}}} - \frac{\partial \mathbf{f}}{\partial \mathbf{E}_{\mathrm{i}}} \right) [\mathbf{g}] \right] \right.$$

$$\left. - \frac{\partial f}{\partial \mathbf{E}_{\mathrm{i}}} \cdot \mathbf{g} - \sum_k \left(\frac{\partial f}{\partial \tilde{\mathbf{T}}} \cdot \frac{\partial \mathbf{f}}{\partial q_k} + \frac{\partial f}{\partial q_k} \right) f_k \right\} . \tag{12.32}$$

This leads to

Theorem 12. 1

For every mechanical process described by the constitutive equations (12.2) - (12.6), the yield function satisfies the differential equation[11]

$$\dot{f}(t) + \frac{K(t)}{\eta} \langle f(t) \rangle = R(t) , \tag{12.33}$$

with

$$R(t) = \frac{\partial f}{\partial \tilde{\mathbf{T}}} \cdot \left\{ \frac{\partial \mathbf{f}}{\partial \mathbf{E}_{\mathrm{e}}} [\dot{\mathbf{E}}(t)] \right\} = \left\{ \left(\frac{\partial \mathbf{f}}{\partial \mathbf{E}_{\mathrm{e}}} \right)^{\mathrm{T}} \left[\frac{\partial f}{\partial \tilde{\mathbf{T}}} \right] \right\} \cdot \dot{\mathbf{E}}(t) \tag{12.34}$$

and

$$K(t) = \frac{\partial f}{\partial \tilde{\mathbf{T}}} \cdot \left\{ \left(\frac{\partial \mathbf{f}}{\partial \mathbf{E}_{\mathrm{e}}} - \frac{\partial \mathbf{f}}{\partial \mathbf{E}_{\mathrm{i}}} \right) [\mathbf{g}] \right\} - \frac{\partial f}{\partial \mathbf{E}_{\mathrm{i}}} \cdot \mathbf{g} - \sum_k \left(\frac{\partial f}{\partial \tilde{\mathbf{T}}} \cdot \frac{\partial \mathbf{f}}{\partial q_k} + \frac{\partial f}{\partial q_k} \right) f_k . \tag{12.35}$$

■

The differential equation (12.33) is an identity that is valid for every solution of the constitutive equations. The considerations set out below are based on the assumption that the function $K(t)$ is always positive:

$$K(t) \geq K_0 > 0 . \tag{12.36}$$

This assumption is a restrictive condition for the material model as a whole. In the next section we show that this assumption refers to the stability of the equilibrium solutions of the constitutive equations. It leads to a relaxation property and is therefore called an *assumption of stability*.[12]

---

[11] Cf. KRATOCHVIL & DILLON [1969], pp. 3213.
[12] KRATOCHVIL & DILLON [1969], pp. 3213.

In the case of the constitutive model (12.21) - (12.25), formulated in relation to the intermediate configuration, the same steps have to be undertaken to calculate the functions $K(t)$ and $R(t)$. The outcome is outlined in

Theorem 12. 2

For every solution of the constitutive equations (12.21) - (12.25) the yield function fulfils the identity

$$\dot{\hat{f}}(t) + \frac{K(t)}{\eta}\langle \hat{f}(t)\rangle = R(t)\,, \tag{12.37}$$

with

$$R(t) = \mathbb{C}\Big[\frac{\partial \hat{f}}{\partial \hat{\mathbf{S}}}\Big]\cdot \overset{\vartriangle}{\hat{\mathbf{\Gamma}}} \tag{12.38}$$

and

$$\begin{aligned} K(t) = {} & \frac{\partial \hat{f}}{\partial \hat{\mathbf{S}}}\cdot \mathbb{C}\Big[\frac{\partial \hat{f}}{\partial \hat{\mathbf{S}}}\Big(\mathbf{1} + 2\hat{\mathbf{\Gamma}}_{\mathrm{e}}\Big)\Big] \\ & + \Big\{c\Big\|\frac{\partial \hat{f}}{\partial \hat{\mathbf{S}}}\Big\| - b\sqrt{\tfrac{2}{3}}\Big(\frac{\partial \hat{f}}{\partial \hat{\mathbf{S}}}\cdot \hat{\mathbf{X}}\Big)\Big\}\Big\|\frac{\partial \hat{f}}{\partial \hat{\mathbf{S}}}\Big\| + 2\Big(\frac{\partial \hat{f}}{\partial \hat{\mathbf{S}}}\hat{\mathbf{X}}\Big)\cdot \frac{\partial \hat{f}}{\partial \hat{\mathbf{S}}}\,. \end{aligned} \tag{12.39}$$

□

Proof

First of all we calculate the total time derivative of the yield function using (12.20) and (12.17),

$$\begin{aligned} \dot{\hat{f}}(\hat{\mathbf{S}}(t) - \hat{\mathbf{X}}(t), k) &= \frac{\partial f}{\partial \hat{\mathbf{S}}}\cdot\Big(\dot{\hat{\mathbf{S}}} - \dot{\hat{\mathbf{X}}}\Big) = \\ &= \frac{\partial \hat{f}}{\partial \hat{\mathbf{S}}}\cdot\Big(\overset{\triangledown}{\hat{\mathbf{S}}} - \overset{\triangledown}{\hat{\mathbf{X}}} + \hat{\mathbf{L}}_{\mathrm{i}}(\hat{\mathbf{S}} - \hat{\mathbf{X}}) + (\hat{\mathbf{S}} - \hat{\mathbf{X}})\hat{\mathbf{L}}_{\mathrm{i}}^{\mathrm{T}}\Big), \end{aligned}$$

which leads to

$$\dot{\hat{f}}(t) = \frac{\partial \hat{f}}{\partial \hat{\mathbf{S}}}\cdot\Big(\overset{\triangledown}{\hat{\mathbf{S}}} - \overset{\triangledown}{\hat{\mathbf{X}}}\Big) + 2\Big\{\frac{\partial \hat{f}}{\partial \hat{\mathbf{S}}}(\hat{\mathbf{S}} - \hat{\mathbf{X}})\Big\}\cdot \overset{\vartriangle}{\hat{\mathbf{\Gamma}}}_{\mathrm{i}}\,. \tag{12.40}$$

The material derivative of the elasticity relation (12.22) reads as

$$\dot{\hat{\mathbf{S}}} = \mathbb{C}\Big[\dot{\hat{\mathbf{\Gamma}}}_{\mathrm{e}}\Big],$$

the derivative

$$\mathbb{C} = \frac{\mathrm{d}}{\mathrm{d}\hat{\mathbf{\Gamma}}_{\mathrm{e}}} \hat{\mathrm{g}}(\hat{\mathbf{\Gamma}}_{\mathrm{e}}) \tag{12.41}$$

being the elasticity tensor. On the basis of (12.16) and (12.20) it follows that

$$\overset{\nabla}{\hat{\mathbf{S}}} = \mathbb{C}\Big[\overset{\Delta}{\hat{\mathbf{\Gamma}}}_{\mathrm{e}}\Big] - \hat{\mathbf{L}}_{\mathrm{i}}\hat{\mathbf{S}} - \hat{\mathbf{S}}\hat{\mathbf{L}}_{\mathrm{i}}^{\mathrm{T}} - \mathbb{C}\Big[\hat{\mathbf{L}}_{\mathrm{i}}^{\mathrm{T}}\hat{\mathbf{\Gamma}}_{\mathrm{e}} + \hat{\mathbf{\Gamma}}_{\mathrm{e}}\hat{\mathbf{L}}_{\mathrm{i}}\Big].$$

Given the identity (11.119),

$$\mathbf{L}\mathrm{g}(\mathbf{B}) + \mathrm{g}(\mathbf{B})\mathbf{L}^{\mathrm{T}} + \mathbb{C}\big[\mathbf{L}^{\mathrm{T}}\mathbf{B} + \mathbf{B}\mathbf{L}\big] = \mathrm{g}(\mathbf{B})(\mathbf{L} + \mathbf{L}^{\mathrm{T}}) + \mathbb{C}\big[(\mathbf{L} + \mathbf{L}^{\mathrm{T}})\mathbf{B}\big],$$

valid for every isotropic tensor function $\mathbf{B} \longmapsto \mathrm{g}(\mathbf{B})$, the incremental elasticity relation takes on the form analogous to (11.126),

$$\overset{\nabla}{\hat{\mathbf{S}}} = \mathbb{C}\Big[\overset{\Delta}{\hat{\mathbf{\Gamma}}}_{\mathrm{e}}\Big] - \hat{\mathbf{S}}(\hat{\mathbf{L}}_{\mathrm{i}} + \hat{\mathbf{L}}_{\mathrm{i}}^{\mathrm{T}}) - \mathbb{C}\Big[(\hat{\mathbf{L}}_{\mathrm{i}} + \hat{\mathbf{L}}_{\mathrm{i}}^{\mathrm{T}})\hat{\mathbf{\Gamma}}_{\mathrm{e}}\Big]$$

or

$$\overset{\nabla}{\hat{\mathbf{S}}} = \mathbb{C}\Big[\overset{\Delta}{\hat{\mathbf{\Gamma}}}\Big] - \mathbb{C}\Big[\overset{\Delta}{\hat{\mathbf{\Gamma}}}_{\mathrm{i}}\Big] - 2\hat{\mathbf{S}}\overset{\Delta}{\hat{\mathbf{\Gamma}}}_{\mathrm{i}} - 2\mathbb{C}\Big[\overset{\Delta}{\hat{\mathbf{\Gamma}}}_{\mathrm{i}}\hat{\mathbf{\Gamma}}_{\mathrm{e}}\Big]. \tag{12.42}$$

By introducing this result, the hardening law (12.25) and the flow rule (12.24) into the time derivative (12.40) the theorem's claim is verified. ■

The definition of the classical model of viscoplasticity is a special case of the general theory. It is based on the linearised *Green* strain tensor $\mathbf{E} = (1/2)\big[\mathrm{Grad}\boldsymbol{u} + (\mathrm{Grad}\boldsymbol{u})^{\mathrm{T}}\big]$ and comprises the constituents (5.219) - (5.223):

$$\mathbf{E} = \mathbf{E}_{\mathrm{e}} + \mathbf{E}_{\mathrm{i}}, \tag{5.219}$$

$$\mathbf{T} = 2\mu\Big[\mathbf{E}_{\mathrm{e}} + \frac{\nu}{1-2\nu}(\mathrm{tr}\mathbf{E}_{\mathrm{e}})\mathbf{1}\Big], \tag{5.220}$$

$$f(\mathbf{T} - \mathbf{X}, k) = \sqrt{\tfrac{3}{2}}\,\big\|\mathbf{T}^{\mathrm{D}} - \mathbf{X}^{\mathrm{D}}\big\| - k, \tag{5.221}$$

$$\dot{\mathbf{E}}_{\mathrm{i}}(t) = \frac{1}{\eta}\langle f\rangle\sqrt{\frac{3}{2}}\,\frac{\mathbf{T}^{\mathrm{D}} - \mathbf{X}^{\mathrm{D}}}{\big\|\mathbf{T}^{\mathrm{D}} - \mathbf{X}^{\mathrm{D}}\big\|}, \tag{5.222}$$

$$\dot{\mathbf{X}}(t) = c\dot{\mathbf{E}}_{\mathrm{i}}(t) - b\sqrt{\tfrac{2}{3}}\,\big\|\dot{\mathbf{E}}_{\mathrm{i}}(t)\big\|\mathbf{X}. \tag{5.223}$$

In this case the differential equation (12.33) for the yield function $f(t)$ contains the coefficient

$$K(t) = 3\mu + \frac{3}{2}c - b\sqrt{\frac{3}{2}}\,\frac{\mathbf{T}^{\mathrm{D}} - \mathbf{X}^{\mathrm{D}}}{\left\|\mathbf{T}^{\mathrm{D}} - \mathbf{X}^{\mathrm{D}}\right\|}\cdot\mathbf{X} \tag{12.43}$$

and the right-hand side

$$R(t) = \mu\sqrt{6}\,\frac{\mathbf{T}^{\mathrm{D}} - \mathbf{X}^{\mathrm{D}}}{\left\|\mathbf{T}^{\mathrm{D}} - \mathbf{X}^{\mathrm{D}}\right\|}\cdot\dot{\mathbf{E}}(t)\,. \tag{12.44}$$

It is obvious that the stability assumption (12.36) is confirmed. The evolution equation (5.223) for the backstress tensor $\mathbf{X}$ reads

$$\mathbf{X}'(s_{\mathrm{i}}) + b\mathbf{X}(s_{\mathrm{i}}) = c\mathbf{E}_{\mathrm{i}}{}'(s_{\mathrm{i}})\,,$$

when formulated with the arclength $s_{\mathrm{i}}(t)$ (accumulated inelastic strain), defined by

$$\dot{s}_{\mathrm{i}}(t) = \sqrt{\frac{2}{3}}\left\|\dot{\mathbf{E}}_{\mathrm{i}}(t)\right\|.$$

The solution $\mathbf{X}(s_{\mathrm{i}}) = \displaystyle\int_0^{s_{\mathrm{i}}} c e^{-b(s_{\mathrm{i}} - \sigma)}\mathbf{E}_{\mathrm{i}}{}'(\sigma)\mathrm{d}\sigma$

leads in view of $\left\|\mathbf{E}_{\mathrm{i}}{}'(s_{\mathrm{i}})\right\| = \sqrt{\frac{3}{2}}$ to the estimation

$$\left\|\mathbf{X}(s_{\mathrm{i}})\right\| \leq \sqrt{\frac{3}{2}}\int_0^{s_{\mathrm{i}}} c e^{-b(s_{\mathrm{i}} - \sigma)}\mathrm{d}\sigma\,,\ \text{that means to}$$

$$\left\|\mathbf{X}\right\| \leq \sqrt{\frac{3}{2}\frac{c}{b}}\,. \tag{12.45}$$

From

$$\left|\frac{\mathbf{T}^{\mathrm{D}} - \mathbf{X}^{\mathrm{D}}}{\left\|\mathbf{T}^{\mathrm{D}} - \mathbf{X}^{\mathrm{D}}\right\|}\cdot\mathbf{X}\right| \leq \left\|\mathbf{X}\right\| \leq \sqrt{\frac{3}{2}\frac{c}{b}} \tag{12.46}$$

one then obtains

$$K(t) \geq 3\mu > 0\,, \tag{12.47}$$

which suggests that the stability condition (12.36) is met by the classical model of viscoplasticity.

12. 3. 2 Relaxation Property

For an inelastic strain process, commencing at time t_0 with a state of stress on the yield surface and being otherwise open to choice, we have $f(t_0) = 0$ and $f(t) > 0$ for $t > t_0$, as well as the identity (12.33),

$$\dot{f}(t) + \frac{K(t)}{\eta} f(t) = R(t) .$$

A deformation process of this nature can be halted at any time $t > t_0$ and continued with a constant state of strain.

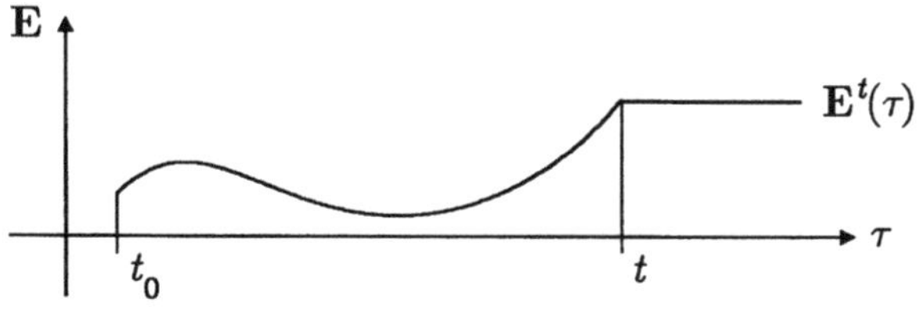

Figure 12. 1: Static continuation

Definition 12. 2
If $t \longmapsto \mathbf{E}(t)\ \left(t \geq t_0\right)$ is any given strain process, then the function

$$\tau \longmapsto \mathbf{E}^t(\tau) = \begin{cases} \mathbf{E}(\tau) \text{ for } t_0 \leq \tau \leq t \\ \mathbf{E}(t) \text{ for } \tau > t \end{cases} \tag{12.48}$$

is called the *static continuation* of $\mathbf{E}(\cdot)$, Fig. 12. 1 (cf. Definition 10. 3). ◊

Due to the constitutive equations (12.2) - (12.6) a static continuation leads to the response $\tau \longmapsto \left(\tilde{\mathbf{T}}^t(\tau),\ \mathbf{E}_i^{\ t}(\tau),\ q_1^{\ t}(\tau),\ \ldots\ ,\ q_N^{\ t}(\tau)\right)$ of the material model, implying the functions $f^t(\tau)$ and $K^t(\tau)$ according to (12.4) and (12.35) or (12.23) and (12.39) respectively. For $\tau > t$ these conform to the homogeneous differential equation

$$\frac{\mathrm{d}}{\mathrm{d}\tau} f^t(\tau) + \frac{K^t(\tau)}{\eta} f^t(\tau) = 0 \tag{12.49}$$

with the initial condition $f^t(t) = f(t)$. This initial-value problem determines the temporal evolution of the yield function during the static continuation,

$$f^t(\tau) = f(t)\,\mathrm{e}^{-\frac{1}{\eta}\int_t^\tau K^t(\sigma)\mathrm{d}\sigma}, \tag{12.50}$$

for which the estimation

$$f^t(\tau) \leq f(t)\,\mathrm{e}^{-\frac{K_0}{\eta}(\tau - t)} \tag{12.51}$$

is valid, based on the stability assumption (12.36). Accordingly, the yield function $f^t(\tau)$ tends towards zero as τ goes to infinity:

$$f_{\mathrm{eq}}(t) = \lim_{\tau \to \infty} f^t(\tau) = 0\,. \tag{12.52}$$

The vanishing limit value $f_{\mathrm{eq}}(t)$ of the yield function $f^t(\tau)$, which is brought about by every static continuation, characterises a set of equilibrium states lying on the static yield surface, which manifests itself during the course of the static continuation:

$$f_{\mathrm{eq}}(t) = 0 \iff \left\{\left(\tilde{\mathbf{T}}_{\mathrm{eq}}, \mathbf{E}_{\mathrm{ieq}}, q_{1\mathrm{eq}}, \ldots, q_{N\mathrm{eq}}\right)\right\}.$$

One of these equilibrium states is reached asymptotically by the material response:

$$\left(\tilde{\mathbf{T}}^t(\tau), \mathbf{E}_{\mathrm{i}}^t(\tau), q_1^t(\tau), \ldots, q_N^t(\tau)\right) \rightarrow \left(\tilde{\mathbf{T}}_{\mathrm{eq}}(t), \mathbf{E}_{\mathrm{ieq}}(t), q_{1\mathrm{eq}}(t), \ldots, q_{N\mathrm{eq}}(t)\right).$$

This is the *relaxation property*: during the static continuation of an inelastic process the material response relaxes to a stable state of equilibrium, which is attained asymptotically during the static continuation. Thus an *accompanying equilibrium process* is attributed to each inelastic deformation process:

$$t \longmapsto \left(\tilde{\mathbf{T}}_{\mathrm{eq}}(t), \mathbf{E}_{\mathrm{ieq}}(t), q_{1\mathrm{eq}}(t), \ldots, q_{N\mathrm{eq}}(t)\right). \tag{12.53}$$

It is in the nature of the equilibrium states in this accompanying process to depend on the whole deformation history.

12. 3. 3 Slow Deformation Processes

Considering the relaxation property one supposes that, in the case of motions taking place slowly enough, the solutions of equations (12.2) - (12.6) are to be found in the close vicinity of the accompanying equilibrium process. A mathematically sound proof of this supposition is presumably unknown to date. It is however possible to derive the *differential equations* for those processes that take place in the asymptotic limit of infinitely slow deformations. These differential equations are precisely those evolution equations found in rate-independent plasticity.

To verify this, we analyse another inelastic process beginning on the yield surface, but without static continuation. Now we have $f(t_0) = 0$ as well as $f(t) > 0$ and $\dot{\mathbf{E}}(t) \not\equiv \mathbf{0}$ for $t > t_0$. The differential equation for the yield function is now inhomogeneous, i.e.

$$\dot{f}(t) + \frac{K(t)}{\eta} f(t) = \left\{ \left(\frac{\partial \mathrm{f}}{\partial \mathbf{E}_{\mathrm{e}}} \right)^{\mathrm{T}} \left[\frac{\partial f}{\partial \tilde{\mathbf{T}}} \right] \right\} \cdot \dot{\mathbf{E}}(t) \, , \tag{12.54}$$

and has to be transformed into an integral equation for the next investigations:

$$f(t) = - \int_0^{t-t_0} \mathrm{e}^{-\frac{1}{\eta} \int_0^s K(t-\sigma) \mathrm{d}\sigma} \left\{ \left(\frac{\partial \mathrm{f}}{\partial \mathbf{E}_{\mathrm{e}}} \right)^{\mathrm{T}} \left[\frac{\partial f}{\partial \tilde{\mathbf{T}}}(t-s) \right] \right\} \cdot \frac{\mathrm{d}}{\mathrm{d}s} \mathbf{E}(t-s) \, \mathrm{d}s \, . \tag{12.55}$$

This integral relates the current value of the yield function to the strain history $s \longmapsto \mathbf{E}(t-s)$. Equation (12.55) obviously does not constitute a representation of the yield function as a linear functional of the strain history. Since the kernel of the integral is calculated from the outcome of the constitutive equations it consequently depends nonlinearly on the strain process. The integral equation (12.55) is merely an identity met by every inelastic process. The behaviour of the yield function resulting from a retardation of given deformation processes can be examined on the basis of this identity.

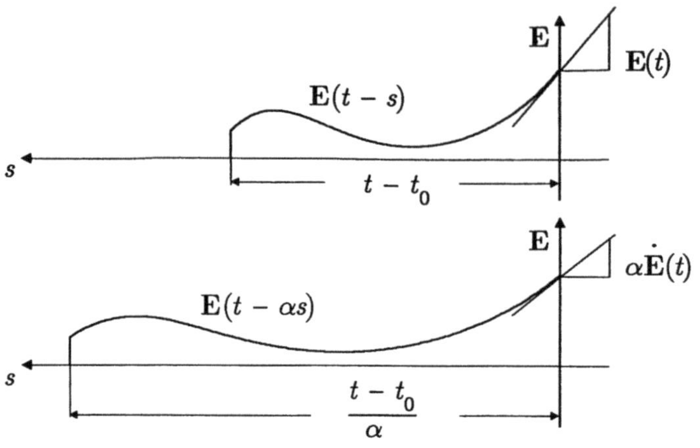

Figure 12. 2: Retarded strain history

Definition 12. 3
If t is a fixed time selected at will, $s \longmapsto \mathbf{E}(t - s)$ a given strain history and α ($0 < \alpha \leq 1$) a *retardation factor*, then the function

$$s \longmapsto \mathbf{E}(t - \alpha s)\Big|_{s \in [0, \frac{t - t_0}{\alpha}]} \tag{12.56}$$

is called the *strain history retarded by* α.[13] ◇

Whereas the original strain history is defined for $0 \leq s \leq t - t_0$, the retarded strain history $\mathbf{E}(t - \alpha s)$ extends for the time interval $0 \leq s \leq (t - t_0)/\alpha$ (see Fig. 12. 2).

By definition the retarded strain history assumes the property of the current strain rate being reduced by the factor α:

$$\frac{\mathrm{d}}{\mathrm{d}s}\mathbf{E}(t - \alpha s)\Big|_{s=0} = -\,\alpha\dot{\mathbf{E}}(t)\,. \tag{12.57}$$

The constitutive equations (12.2) - (12.7) determine the material response to the deformation process slowed down by α,

$$(t - s) \longmapsto \big(\widetilde{\mathbf{T}}_{\alpha}(t - s),\, \mathbf{E}_{i\alpha}(t - s),\, q_{1\alpha}(t - s),\, \dots,\, q_{N\alpha}(t - s)\big)\,,$$

[13] Cf. Definition 10. 4.

as well as the functions $f_\alpha(t)$ and $K_\alpha(t)$. We assume that $f_\alpha(t)$ is positive for all $\alpha > 0$: only those processes that remain inelastic for all retardations are subject to examination. The retarded yield function $f_\alpha(t)$ then fulfils equation (12.55), i.e.

$$f_\alpha(t) = -\int_0^{(t-t_0)/\alpha} e^{-\frac{1}{\eta}\int_0^s K_\alpha(t-\sigma)\mathrm{d}\sigma}\left\{\left(\frac{\partial \mathbf{f}_\alpha}{\partial \mathbf{E}_\mathrm{e}}\right)^\mathrm{T}\left[\frac{\partial f_\alpha}{\partial \tilde{\mathbf{T}}}(t-s)\right]\right\}\cdot\frac{\mathrm{d}}{\mathrm{d}s}\mathbf{E}(t-\alpha s)\,\mathrm{d}s\,. \tag{12.58}$$

Inserting the substitutions $s \longmapsto \bar{s} = \alpha s \quad (0 \le \bar{s} \le t - t_0)$ and

$\sigma \longmapsto \bar{\sigma} = \alpha\sigma \quad (0 \le \bar{\sigma} \le \bar{s})$, we obtain the equivalent formulation

$$f_\alpha(t) = \int_0^{t-t_0} e^{-\frac{1}{\alpha\eta}\int_0^{\bar{s}} K_\alpha(t-\frac{\bar{\sigma}}{\alpha})\mathrm{d}\bar{\sigma}}\, R_\alpha(t,\bar{s})\,\mathrm{d}\bar{s}\,, \tag{12.59}$$

where the abbreviation

$$R_\alpha(t,\bar{s}) = \left\{\left(\frac{\partial \mathbf{f}_\alpha}{\partial \mathbf{E}_\mathrm{e}}\right)^\mathrm{T}\left[\frac{\partial f_\alpha}{\partial \tilde{\mathbf{T}}}(t-\frac{\bar{s}}{\alpha})\right]\right\}\cdot\frac{\mathrm{d}}{\mathrm{d}t}\mathbf{E}(t-\bar{s}) \tag{12.60}$$

has been introduced. Equation (12.59) shows that only the product $\alpha\eta$ enters the expression for the yield function. Therefore, the passage to the limit $\alpha \to 0$, which is now of interest, is synonymous with $\eta \to 0$. Corresponding to this limit is a series of materials with an unlimited decrease in their viscosity. (12.59) implies that the yield function of the retarded process vanishes with α (or η), whenever the function $R_\alpha(t,\bar{s})$ is bounded. Provided that $R_\alpha(t,\bar{s}) \le M(t) < \infty$, the stability assumption (12.36) leads to the estimation

$$f_\alpha(t) \le M(t)\int_0^{t-t_0} e^{-\frac{K_0}{\alpha\eta}\bar{s}}\,\mathrm{d}\bar{s} = \frac{\alpha\eta}{K_0}M(t)\left[1 - e^{-\frac{K_0}{\alpha\eta}(t-t_0)}\right];$$

this indicates the asymptotic relation

$$f_\alpha(t) = O(\alpha\eta)\,. \tag{12.61}$$

Accordingly, based on the evolution equations (12.5), (12.6) the retarded inelastic strain rate and the time derivatives of the retarded stresses and internal variables vanish with $\alpha\eta$:

$$\left(\dot{\mathbf{E}}_{\mathrm{i}\alpha}(t)\,,\,\dot{\tilde{\mathbf{T}}}_{\alpha}(t)\,,\,\dot{q}_{1\alpha}(t),\,...\,,\,\dot{q}_{N\alpha}(t)\right) = \mathrm{O}(\alpha\eta)\,. \tag{12.62}$$

Thus the question arises as to whether finite limit values can be attributed to the *rescaled rates*

$$\frac{1}{\alpha}\dot{\mathbf{E}}_{\mathrm{i}\alpha}(t),\quad \frac{1}{\alpha}\dot{\tilde{\mathbf{T}}}_{\alpha}(t)\ \text{ and }\ \frac{1}{\alpha}\dot{q}_{k\alpha}(t)\,. \tag{12.63}$$

The rescaled rates constitute those velocities obtained when the slowing down of the deformation process is reversed. The rescaled strain rate is obviously equal to the original strain-rate $\dot{\mathbf{E}}(t)$, but all the other rescaled rates result from the outcome of the constitutive equations according to (12.63). In order to calculate the limit values it is necessary to examine the expression $\frac{1}{\alpha\eta}f_{\alpha}(t)$. The result is

Theorem 12. 1

The behaviour of the yield function in the face of a vanishing process rate or vanishing viscosity is characterised by the asymptotic relation

$$\frac{1}{\alpha\eta}f_{\alpha}(t) = \frac{R_{\alpha}(t,0)}{K_{\alpha}(t)} + \mathrm{o}(1)\,. \tag{12.64}$$

or by

$$\lim_{\alpha\eta\to 0}\frac{1}{\alpha\eta}f_{\alpha}(t) = \frac{R_0(t,0)}{K_0(t)} = \frac{1}{K_0(t)}\left\{\left(\frac{\partial\mathbf{f}_0}{\partial\mathbf{E}_{\mathrm{e}}}\right)^{\mathrm{T}}\left[\frac{\partial f_0}{\partial\tilde{\mathbf{T}}}(t)\right]\right\}\cdot\dot{\mathbf{E}}(t)\,. \tag{12.65}$$

□

Proof

Division of (12.59) by $\alpha\eta$ leads to

$$\frac{1}{\alpha\eta}f_{\alpha}(t) = \frac{1}{\alpha\eta}\int_0^{t-t_0} \mathrm{e}^{-\frac{1}{\alpha\eta}\int_0^{\bar{s}}K_{\alpha}(t-\frac{\bar{\sigma}}{\alpha})\mathrm{d}\bar{\sigma}}\,R_{\alpha}(t,\bar{s})\,\mathrm{d}\bar{s}\quad\text{or}$$

$$\frac{1}{\alpha\eta}f_{\alpha}(t) = -\int_0^{t-t_0}\frac{\mathrm{d}}{\mathrm{d}\bar{s}}\left\{\mathrm{e}^{-\frac{1}{\alpha\eta}\int_0^{\bar{s}}K_{\alpha}(t-\frac{\bar{\sigma}}{\alpha})\mathrm{d}\bar{\sigma}}\right\}\frac{1}{K_{\alpha}(t-\frac{\bar{s}}{\alpha})}\,R_{\alpha}(t,\bar{s})\,\mathrm{d}\bar{s}\,, \tag{12.66}$$

and integration by parts produces

$$\frac{1}{\alpha\eta}f_\alpha(t) = \frac{R_\alpha(t,0)}{K_\alpha(t)} - \left\{ e^{-\frac{1}{\alpha\eta}\int_0^{t-t_0} K_\alpha(t-\frac{\bar\sigma}{\alpha})d\bar\sigma} \right\} \left\{ \frac{R_\alpha(t, t-t_0)}{K_\alpha(t - \frac{t-t_0}{\alpha})} \right\}$$

$$+ \int_0^{t-t_0} \left\{ e^{-\frac{1}{\alpha\eta}\int_0^{\bar s} K_\alpha(t-\frac{\bar\sigma}{\alpha})d\bar\sigma} \right\} \frac{d}{d\bar s} \left\{ \frac{R_\alpha(t,\bar s)}{K_\alpha(t-\frac{\bar s}{\alpha})} \right\} d\bar s \, . \tag{12.67}$$

Provided that the functions $R_\alpha(t,\bar s)$ and $K_\alpha(t - \frac{\bar s}{\alpha})$ as well as their derivatives remain bounded, one subsequently arrives - once again under the stability assumption (12.36) - at the asymptotic relation

$$\frac{1}{\alpha\eta}f_\alpha(t) = \frac{R_\alpha(t,0)}{K_\alpha(t)} + o((\alpha\eta)^0) \, .$$

In the limit $\alpha\eta \to 0$ this leads to the statement (12.65). ■

The theorem shows that the internal variables' rescaled rates have finite limit values, if a constitutive model of viscoplasticity meets the restrictive condition (12.36). These limit values describe the evolution of the accompanying equilibrium process.

Definition 12. 4

The limit value of the rescaled inelastic strain rate is called the *plastic strain rate*:

$$\dot{\mathbf{E}}_p(t) = \lim_{\alpha\eta\to 0} \frac{1}{\alpha}\dot{\mathbf{E}}_{i\alpha}(t) \, . \tag{12.68}$$

◊

The plastic strain rate is precisely the inelastic strain rate of the accompanying equilibrium process. The evolution equation for the plastic strain $\mathbf{E}_p$ reads

$$\dot{\mathbf{E}}_p(t) = \left\{ \frac{1}{K_0(t)} \left(\frac{\partial \mathbf{f}_0}{\partial \mathbf{E}_e} \right)^T \left[\frac{\partial f_0}{\partial \tilde{\mathbf{T}}} \right] \cdot \dot{\mathbf{E}}(t) \right\} \mathbf{g}_0 \tag{12.69}$$

or

$$\dot{\mathbf{E}}_{\mathrm{p}}(t) = \frac{\left\{\left(\dfrac{\partial \mathbf{f}}{\partial \mathbf{E}_{\mathrm{e}}}\right)^{\mathrm{T}}\left[\dfrac{\partial f}{\partial \widetilde{\mathbf{T}}}\right]\right\}\cdot \dot{\mathbf{E}}(t)}{\dfrac{\partial f}{\partial \widetilde{\mathbf{T}}}\cdot\left\{\left(\dfrac{\partial \mathbf{f}}{\partial \mathbf{E}_{\mathrm{e}}} - \dfrac{\partial \mathbf{f}}{\partial \mathbf{E}_{\mathrm{p}}}\right)[\mathbf{g}]\right\} - \dfrac{\partial f}{\partial \mathbf{E}_{\mathrm{p}}}\cdot \mathbf{g} - \sum_k \left(\dfrac{\partial f}{\partial \widetilde{\mathbf{T}}}\cdot\dfrac{\partial \mathbf{f}}{\partial q_k} + \dfrac{\partial f}{\partial q_k}\right) f_k}\, \mathbf{g}\,. \tag{12.70}$$

The subscripts that have now become superfluous have been omitted in the last equation. In this equation the *flow rule of rate-independent plasticity* is recognisable.

Theorem 12. 2

The *consistency condition* of the theory of rate-independent plasticity,

$$\frac{\mathrm{d}}{\mathrm{d}t} f\big(\widetilde{\mathbf{T}}(t), \mathbf{E}_{\mathrm{p}}(t), q_1(t), \dots, q_N(t)\big) = 0\,, \tag{12.71}$$

combined with the following assumptions regarding

Decomposition of the Deformation

$$\mathbf{E} = \mathbf{E}_{\mathrm{e}} + \mathbf{E}_{\mathrm{p}}\,, \tag{12.72}$$

Elasticity Relation

$$\widetilde{\mathbf{T}} = \mathbf{f}(\mathbf{E}_{\mathrm{e}}, \mathbf{E}_{\mathrm{p}}, q_1, \dots, q_N)\,, \tag{12.73}$$

Yield Function

$$f = f(\widetilde{\mathbf{T}}, \mathbf{E}_{\mathrm{p}}, q_1, \dots, q_N)\,, \tag{12.74}$$

Flow Rule

$$\dot{\mathbf{E}}_{\mathrm{p}}(t) = \lambda\, \mathbf{g}(\widetilde{\mathbf{T}}, \mathbf{E}_{\mathrm{p}}, q_1, \dots, q_N)\,, \tag{12.75}$$

Evolution Equations

$$\dot{q}_k(t) = \lambda f_k\big(\widetilde{\mathbf{T}}, \mathbf{E}_{\mathrm{p}}, q_1, \dots, q_N\big)\,, \quad k = 1, \dots, N \tag{12.76}$$

leads to the proportionality factor

$$\lambda = \frac{R}{K}\,, \tag{12.77}$$

with R and K as in (12.34), (12.35), i.e.

$$\lambda = \frac{\left\{\left(\dfrac{\partial \mathbf{f}}{\partial \mathbf{E}_e}\right)^{\mathrm{T}}\left[\dfrac{\partial f}{\partial \widetilde{\mathbf{T}}}\right]\right\}\cdot \dot{\mathbf{E}}(t)}{\dfrac{\partial f}{\partial \widetilde{\mathbf{T}}}\cdot\left\{\left(\dfrac{\partial \mathbf{f}}{\partial \mathbf{E}_e} - \dfrac{\partial \mathbf{f}}{\partial \mathbf{E}_p}\right)[\mathbf{g}]\right\} - \dfrac{\partial f}{\partial \mathbf{E}_p}\cdot \mathbf{g} - \sum_k\left(\dfrac{\partial f}{\partial \mathbf{T}}\cdot\dfrac{\partial \mathbf{f}}{\partial q_k} + \dfrac{\partial f}{\partial q_k}\right)f_k}\,. \tag{12.78}$$

□

Proof

The time derivative of the yield function,

$$\dot{f}(t) = \frac{\mathrm{d}}{\mathrm{d}t} f(\widetilde{\mathbf{T}}(t), \mathbf{E}_p(t), q_1(t), \dots, q_N(t)) = \frac{\partial f}{\partial \widetilde{\mathbf{T}}}\cdot\dot{\widetilde{\mathbf{T}}} + \frac{\partial f}{\partial \mathbf{E}_p}\cdot\dot{\mathbf{E}}_p(t) + \sum_k \frac{\partial f}{\partial q_k}\dot{q}_k(t)$$

$$\dot{f}(t) = \frac{\partial f}{\partial \widetilde{\mathbf{T}}}\cdot\left\{\frac{\partial \mathbf{f}}{\partial \mathbf{E}_e}[\dot{\mathbf{E}}_e(t)] + \frac{\partial \mathbf{f}}{\partial \mathbf{E}_p}[\dot{\mathbf{E}}_p(t)] + \sum_k \frac{\partial \mathbf{f}}{\partial q_k}\dot{q}_k(t)\right\}$$

$$+ \frac{\partial f}{\partial \mathbf{E}_p}\cdot\dot{\mathbf{E}}_p(t) + \sum_k \frac{\partial f}{\partial q_k}\dot{q}_k(t)\,,$$

observing the decomposition (12.72), the evolution equations (12.76) and the flow rule (12.75), leads to the consistency condition in the form of

$$0 = \left\{\left(\frac{\partial \mathbf{f}}{\partial \mathbf{E}_e}\right)^{\mathrm{T}}\left[\frac{\partial f}{\partial \widetilde{\mathbf{T}}}\right]\right\}\cdot\dot{\mathbf{E}}(t)$$

$$+ \lambda\left\{\frac{\partial f}{\partial \widetilde{\mathbf{T}}}\cdot\left[\left(\frac{\partial \mathbf{f}}{\partial \mathbf{E}_p} - \frac{\partial \mathbf{f}}{\partial \mathbf{E}_e}\right)[\mathbf{g}]\right] + \frac{\partial f}{\partial \mathbf{E}_p}\cdot\mathbf{g} + \sum_k\left(\frac{\partial f}{\partial \widetilde{\mathbf{T}}}\cdot\frac{\partial \mathbf{f}}{\partial q_k} + \frac{\partial f}{\partial q_k}\right)f_k\right\}.$$

■

12. 3. 4 Elastoplasticity and Arclength Representation

It has already been ascertained that time can be replaced as an independent variable in all material models of rate-independent plasticity by the accumulated plastic strain or by another generalised arclength. Dividing all evolution equations by

$$\dot{z}(t) = \left\|\dot{\mathbf{E}}_{\mathrm{p}}(t)\right\| = \lambda\left\|\mathbf{g}\right\| = \frac{R}{K}\left\|\mathbf{g}\right\| \tag{12.79}$$

brings us to the arclength representation:[14]

$$\mathbf{E}_{\mathrm{p}}{}'(z) = \frac{\mathbf{g}(\widetilde{\mathbf{T}}, \mathbf{E}_{\mathrm{p}}, q_1, \dots, q_N)}{\left\|\mathbf{g}(\widetilde{\mathbf{T}}, \mathbf{E}_{\mathrm{p}}, q_1, \dots, q_N)\right\|}, \tag{12.80}$$

$$q_k{}'(z) = f_k(\mathbf{E}, \mathbf{E}_{\mathrm{p}}, q_1, \dots, q_N) \quad (k = 1, 2, \dots N), \tag{12.81}$$

$$\dot{z}(t) = \begin{cases} \frac{R}{K}\left\|\mathbf{g}\right\| & \text{for } f = 0 \text{ and } R > 0 \\ 0 & \text{otherwise} \end{cases} \tag{12.82}$$

The evolution equation (12.82) for the generalised arclength depends on yield and loading conditions. Due to the restrictive condition (12.36), a prerequisite for the material model from the start, the denominator on the right-hand side of (12.82) is always positive. To guarantee compatibility with the general requirements (11.30) - (11.32) imposed on the generalised arclength, the numerator for $f = 0$ and loading also has to be positive. Accordingly, with the loading function

$$R = \left\{\left(\frac{\partial \mathbf{f}}{\partial \mathbf{E}_{\mathrm{e}}}\right)^{\mathrm{T}}\left[\frac{\partial f}{\partial \widetilde{\mathbf{T}}}\right]\right\} \cdot \dot{\mathbf{E}}(t) \tag{12.83}$$

in (12.82) a *loading condition* is defined which can also be motivated by passage to the limit: relation (12.64) in conjunction with $K_\alpha(t) > 0$ causes the right-hand side of

$$K_\alpha(t)\,\frac{1}{\alpha\eta} f_\alpha(t) = \left\{\left(\frac{\partial \mathbf{f}_\alpha}{\partial \mathbf{E}_{\mathrm{e}}}\right)^{\mathrm{T}}\left[\frac{\partial f_\alpha}{\partial \widetilde{\mathbf{T}}}(t)\right]\right\} \cdot \dot{\mathbf{E}}(t) + 0(\alpha\eta) \tag{12.84}$$

to be positive for every finite value of $\alpha\eta > 0$, provided that the retarded process remains inelastic, which has also been taken for granted. Therefore, even in the limit case, it holds that

$$R = \left\{\left(\frac{\partial \mathbf{f}_0}{\partial \mathbf{E}_{\mathrm{e}}}\right)^{\mathrm{T}}\left[\frac{\partial f_0}{\partial \widetilde{\mathbf{T}}}(t)\right]\right\} \cdot \dot{\mathbf{E}}(t) \geq 0\,. \tag{12.85}$$

We can summarise by saying: the model of elastoplasticity, defined by (12.72) - (12.74) in connection with the evolution equations (12.80) - (12.83), is

[14] Cf. the formulation (11.55) - (11.58) of classical plasticity.

included in the viscoplasticity model (12.2) - (12.6) for sufficiently slow deformations or small viscosity values. In analogy to this, the constitutive model (12.21) - (12.25), that was formulated with variables of the intermediate configuration, brings us to the general model (11.112) - (11.118) of rate-independent elastoplasticity with the proportionality factor λ according to (11.121) and the loading function (11.123).[15] Corresponding statements apply to all special cases which can be derived from the general equations.

12. 4 A Concept for General Viscoplasticity

12. 4. 1 Motivation

The structure of the viscoplasticity model with an elastic domain, as discussed above, is not free from fundamental shortcomings. These may become apparent when complicated nonlinear rate-dependencies occur or when it is a question of presenting the temporal evolution of relaxation or creep processes over long periods of time in detail. In this respect the evolution equations do not imply sufficient degrees of freedom to warrant realistic modelling: in the flow rule (12.5) the strain rate $\dot{\mathbf{E}}_{\mathrm{i}}(t)$ is a function of the current inelastic strain itself. The evolution equation (12.5) is a first order differential equation for $\mathbf{E}_{\mathrm{i}}(t)$, which leads to quite special time developments of stress or strain during processes of relaxation or creep. With an evolution equation of this make-up many details of the process history, which might be important for a more detailed representation of the material response, are not taken into consideration. Many more possibilities are available for modelling the temporal evolution when differential equations of higher order are used for $\mathbf{E}_{\mathrm{i}}(t)$ or, equivalently, by introducing additional evolution equations for further internal variables which are directly connected with the development of $\mathbf{E}_{\mathrm{i}}$.

The methods of general viscoelasticity presented in Chap. 10 offer a wide range of possibilities in this field, even to the extent of incorporating continuous relaxation spectra.

In this context it seems expedient to model the equilibrium relation and the rate-dependent material behaviour separately. Experimental facts also suggest such a decoupling, since the accompanying equilibrium processes can be made available for direct observation as a result of relaxation tests (see Chap. 6).

---

[15] The algorithm of numerical integration, developed in HARTMANN et al. [1997] has this asymptotic property, too. See also LÜHRS [1997], p. 116; LÜHRS et al. [1997]; LÜHRS & HAUPT [1997].

12. 4. 2 Equilibrium Stress and Overstress

The basic assumption leading to a more general theory of viscoplasticity is an additive decomposition of the stress into an equilibrium stress and an overstress:

$$\tilde{\mathbf{T}} = \tilde{\mathbf{T}}_{\mathrm{eq}} + \tilde{\mathbf{T}}_{\mathrm{ov}} \,. \tag{12.86}$$

In the sense of a decoupled modelling of each of the stress parts, the *equilibrium stress* will be assumed to be a rate-independent functional of the process history. The assumption of $\tilde{\mathbf{T}}_{\mathrm{eq}}$ as a rate-independent functional is by no means absolutely imperative,[16] but plausible to a high degree. At least, no experimental facts are known that might directly contradict such an idea.

Within this conception, the representation methods of rate-independent *plasticity* are available for modelling the equilibrium stress. It is advantageous when applying the arclength description; the introduction of criteria in the form of yield and loading conditions is possible, but not necessary.

Whereas the equilibrium stress is represented as a rate-independent functional, the overstress has to be a rate-dependent functional of the process history. The overstress $\tilde{\mathbf{T}}_{\mathrm{ov}}$ has to disappear asymptotically for sufficiently slow processes or relax to zero during the course of any static continuation. To represent the rate-dependent functional of the overstress it is possible to apply the representation methods of viscoelasticity.

Separate modelling of equilibrium stress and overstress has already been suggested by the reduced form (7.28). In the sense of the concept under discussion, the tensor function $\mathbf{g}(\mathbf{E})$ in (7.28) should be replaced by a rate-independent functional with hysteresis property. A rate-dependent functional with a fading memory property has to be formulated for $\mathfrak{G}$.

The elements needed for constructing a general material model of viscoplasticity can be gleaned from Chaps. 10 and 11. One possibility is to represent the equilibrium stress within the framework of finite elastoplasticity, as given in the constitutive equations (11.135) - (11.143). Equations (10.153) and (10.154) are suitable for modelling the overstress, where $w_{\mathrm{eq}} = 0$ has to be substituted in the elasticity relation (10.153) in order to achieve relaxation to zero.

[16] The constitutive model in KREMPL et al. [1986], for example, does not have this characteristic.

12. 4. 3 An Example of General Viscoplasticity

This section is intended to illustrate and substantiate the concept of general viscoplasticity by means of an example. This example refers to the geometrically linear case of small deformations.

Definition 12. 5

The following set of constitutive equations models the behaviour of a *viscoplastic solid* within the frame of small deformations by decomposing the stress into an equilibrium stress and an overstress:[17]

$$\mathbf{T} = \mathbf{T}_{\mathrm{eq}} + \mathbf{T}_{\mathrm{ov}} \tag{12.87}$$

Equilibrium Stress

$$\mathbf{E} = \mathbf{E}_{\mathrm{e}} + \mathbf{E}_{\mathrm{p}} \tag{12.88}$$

$$\mathbf{T}_{\mathrm{eq}} = 2\mu_{\mathrm{eq}}\mathbf{E}_{\mathrm{e}}{}^{\mathrm{D}} + \kappa_{\mathrm{eq}}(\mathrm{tr}\mathbf{E}_{\mathrm{e}})\mathbf{1} \tag{12.89}$$

$$F = \tfrac{1}{2}\left\| \mathbf{T}_{\mathrm{eq}}^{\mathrm{D}} - \mathbf{X}^{\mathrm{D}} \right\|^2 - \tfrac{1}{3}g_0{}^2 \tag{12.90}$$

$$B = \left(\mathbf{T}_{\mathrm{eq}}^{\mathrm{D}} - \mathbf{X}^{\mathrm{D}}\right)\cdot\dot{\mathbf{T}}_{\mathrm{eq}}^{\mathrm{D}}(t) \tag{12.91}$$

$$\dot{\mathbf{E}}_{\mathrm{p}}(t) = \begin{cases} \lambda(\mathbf{T}_{\mathrm{eq}} - \mathbf{X})^{\mathrm{D}} & \text{for } F = 0 \text{ and } B > 0 \\ 0 & \text{otherwise} \end{cases} \tag{12.92}$$

$$\mathbf{X} = \sum_{k=1}^{\mathrm{M}} \mathbf{X}_k \tag{12.93}$$

$$\dot{\mathbf{X}}_k(t) = c_k\dot{\mathbf{E}}_{\mathrm{p}}(t) - b_k\dot{z}(t)\mathbf{X}_k \qquad (k = 1, ..., \mathrm{N}) \tag{12.94}$$

$$\dot{z}(t) = \frac{\dot{s}(t)}{1 + \alpha_p p} \tag{12.95}$$

$$\dot{s}(t) = \sqrt{\tfrac{2}{3}}\left\|\dot{\mathbf{E}}_{\mathrm{p}}(t)\right\| \tag{12.96}$$

---

[17] The constitutive model is taken from LION [1994]. See HAUPT & LION [1995]. A numerical treatment is described in HARTMANN [1998].

$$\dot{p}(t) = \frac{\dot{s}(t)}{s_p(1 + \alpha_\delta \delta)}(\|\mathbf{X}\| - p) \tag{12.97}$$

$$\dot{\delta}(t) = \frac{\dot{s}(t)}{s_\delta}(p - \delta) \tag{12.98}$$

Overstress

$$\mathbf{T}_{\mathrm{ov}} = \sum_{k=1}^{\mathrm{N}} \mathbf{T}_{\mathrm{ov}k} \tag{12.99}$$

$$\dot{\mathbf{T}}^{\mathrm{D}}_{\mathrm{ov}k}(t) = 2\mu_0 \lambda_k \dot{\mathbf{E}}^{\mathrm{D}}(t) - \frac{1}{z_{0k} M(\|\mathbf{T}^{\mathrm{D}}_{\mathrm{ov}}\|)} \mathbf{T}^{\mathrm{D}}_{\mathrm{ov}k} - \beta \lambda_k \dot{\mathbf{T}}^{\mathrm{D}}_{\mathrm{eq}}(t) \tag{12.100}$$

$$\mathrm{tr}\, \dot{\mathbf{T}}_{\mathrm{ov}k}(t) = 3(\kappa_0 - \kappa_{\mathrm{eq}}) \lambda_k \mathrm{tr}\, \dot{\mathbf{E}}(t) - \frac{1}{z_{0k} M(\|\mathbf{T}^{\mathrm{D}}_{\mathrm{ov}}\|)} \mathrm{tr}\, \mathbf{T}_{\mathrm{ov}k} \tag{12.101}$$

$$\sum_{k=1}^{\mathrm{N}} \lambda_k = 1 \tag{12.102}$$

$$M(\|\mathbf{T}^{\mathrm{D}}_{\mathrm{ov}}\|) = \exp\left(-\frac{\|\mathbf{T}^{\mathrm{D}}_{\mathrm{ov}}\|}{s_0}\right) \tag{12.103}$$

$$\beta = \left\langle \mathbf{T}_{\mathrm{ov}} \cdot \dot{\mathbf{T}}^{\mathrm{D}}_{\mathrm{eq}}(t) \right\rangle^0 = \begin{cases} 1 \text{ for } & \mathbf{T}_{\mathrm{ov}} \cdot \dot{\mathbf{T}}^{\mathrm{D}}_{\mathrm{eq}}(t) > 0 \\ 0 \text{ for } & \mathbf{T}_{\mathrm{ov}} \cdot \dot{\mathbf{T}}^{\mathrm{D}}_{\mathrm{eq}}(t) \leq 0 \end{cases} \tag{12.104}$$

◊

The representation of the rate-independent *equilibrium stress* in its basic features (12.88) - (12.92) follows the lines of conventional elastoplasticity. Beyond that, the stress tensor of kinematic hardening according to (12.93) is composed of a number of parts, for each of which an evolution equation of the *Armstrong-Frederick* type is postulated.

This makes a detailed modelling of the stress-strain curves possible. The evolution equations for the internal variables $\mathbf{X}_k$ are formulated with respect to the generalised arclength $t \longmapsto z(t)$. The introduction of the generalised arclength allows for the modelling of cyclic hardening and softening effects, occurring after a change in the strain amplitude. The evolution equation (12.95) for $z(t)$ is based on the plastic arclength $s(t)$ and includes the internal variable p. In the corresponding evolution equation

(12.97) another internal variable δ occurs. Its process dependence is expressed by evolution equation (12.98). The quantities μ_{eq}, κ_{eq}, g_0, c_k, b_k ($k = 1$, ..., N), α_p and s_δ are material parameters.

The rate-dependent *overstress* is the sum of several parts according to (12.99), which leads to a detailed representation of the relaxation and creep behaviour. The evolution equations (12.100), (12.101) may be motivated by *Maxwell* models; these are characterised by nonlinear viscosities depending on the overstress according to (12.103). This emphasises the strongly nonlinear rate-dependence of the overstress.[18] The corresponding material parameters are μ_0, κ_0, λ_k, z_{0k}, ($k = 1$, ..., N) and s_0, μ_0 and κ_0 being the spontaneous shear and bulk moduli. The presence of the term multiplied by β on the right-hand side of the evolution equation (12.100) causes the slope of the stress-strain curve to be determined by μ_0 in loading processes, starting with a finite rate from a state of equilibrium. The criterion (12.104) defining the factor β is included for reasons of thermomechanical consistency. A detailed explanation of this is given in Chap. 13.

For a numerical integration of the constitutive equations we used N = 2 in both the equilibrium stress and the overstress. The numerical values of the material parameters were identified on the basis of experimental data gleaned from torsion tests on thin-walled tubes made from XCrNi 18.9 steel. Some of this data is illustrated in Figs. 6. 5 - 6. 8. The material parameters were taken from the work of *Lion*[19] and set out in the following tabulation.

Equilibrium Stress

μ_{eq} MPa	κ_{eq} MPa	g_0 MPa	c_1 MPa	b_1 -	c_2 MPa	b_2 -	α_p MPa^{-1}	s_p -	α_δ MPa^{-1}	s_δ -
134000	664000	180	40370	650	2340	120	0.0078	0.065	0.004	2.4

Overstress

μ_0 MPa	κ_0 MPa	s_0 MPa	λ_1 -	z_{01} s	λ_2 -	z_{02} s
140000	683000	6.75	0.96	180	0.04	7900

---

[18] In particular, a sublinear rate-dependence follows from decreasing material function M(·).

[19] LION [1994], p. 117.

The following illustrations serve to exemplify the material responses predicted by the model for selected deformation processes. Monotonic and cyclic shear processes with hold times were given. We shall study the equilibrium stress first.

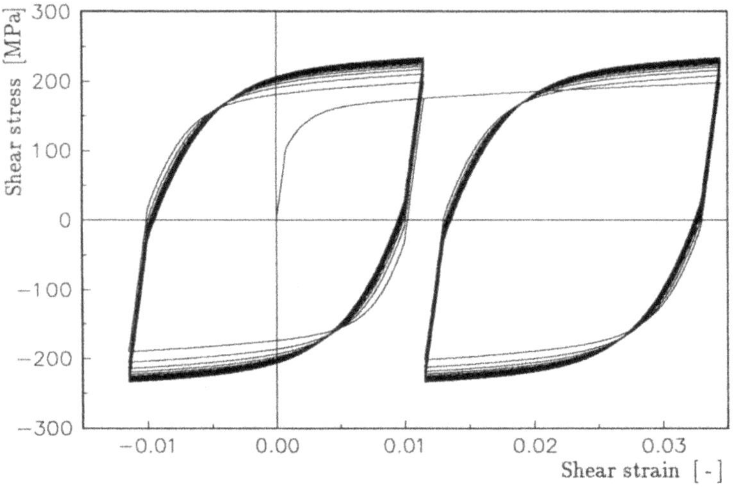

Figure 12. 3: Cyclic process with two different mean strain values; equilibrium stress

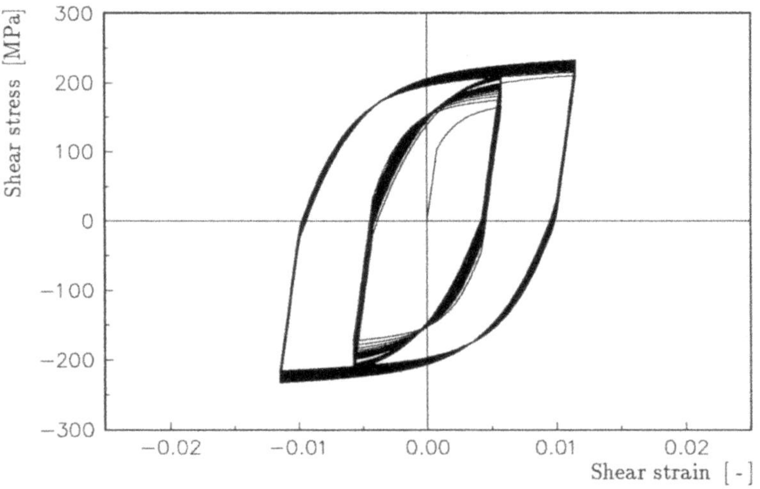

Figure 12. 4: Cyclic process with three different strain amplitudes (0.006 - 0.012 - 0.006); equilibrium stress

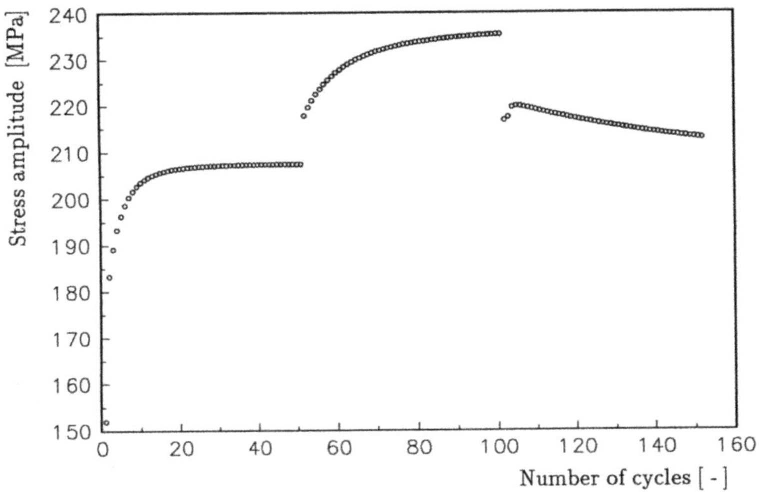

Figure 12. 5: Cyclic process with three different strain amplitudes (0.006 - 0.012 - 0.006); equilibrium stress, evolution of the stress amplitude

Figure 12. 3 shows the equilibrium stress response to a cyclic process with two different mean strains. The cyclic hardening process, the reproduction of which can be traced back to the introduction of the generalised arclength, is clearly recognisable. The generalised arclength $z(t)$, represented by the evolution equations (12.95) - (12.98), is a functional of the strain history.

Figure 12. 4 illustrates the equilibrium stress as a response to a cyclic shear process with three different amplitudes. The hardening effect is again recognisable, both at the beginning of the process and after the raising of the shear amplitude. Lowering the strain amplitude to the initial value leads to a decrease in the stress amplitude. This softening evolves more slowly than the hardening which preceded it, an effect which manifested itself in connection with the experimental investigation of the material in question. The non-symmetry in the evolution of the stress amplitude is demonstrated particularly well in Fig. 12. 5; its reproduction is primarily due to the evolution equation (12.98), which introduces an additional degree of freedom into the functional of the generalised arclength.

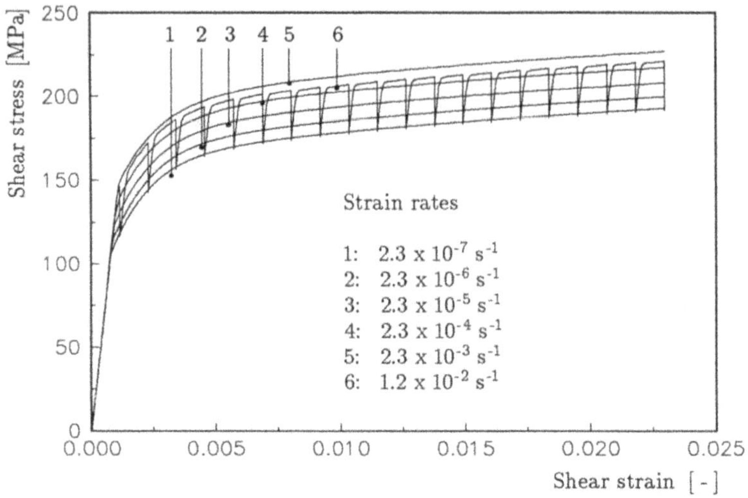

Figure 12. 6: Monotonic processes with different strain rates and inserted hold times; total stress

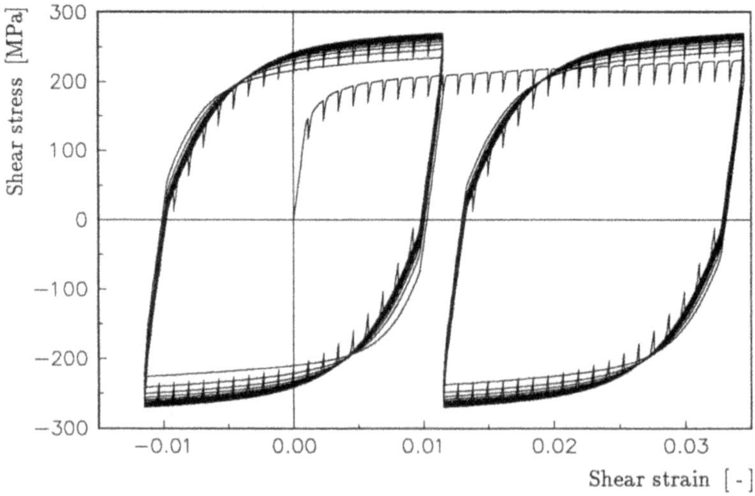

Figure 12. 7: Cyclic processes with two different mean strains (0.00 and 0.025) and inserted hold times; total stress

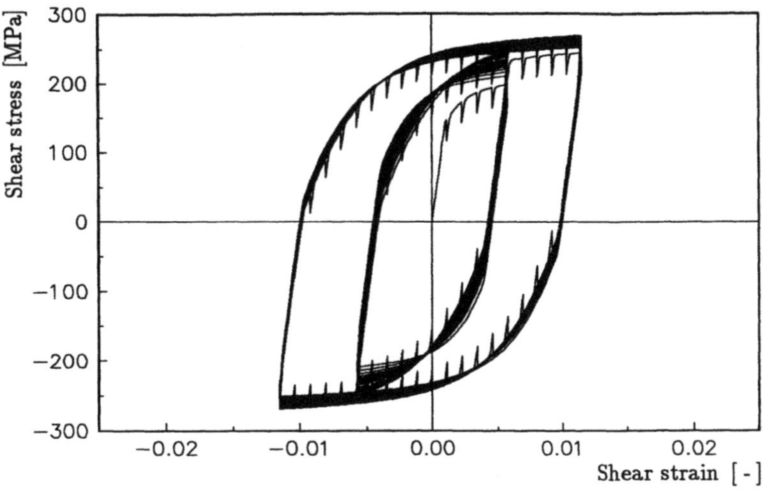

Figure 12. 8: Cyclic process with three different strain amplitudes (0.006 - 0.012 - 0.006) and inserted hold times; total stress

The response of the model as a whole to monotonic shear processes with different rates is reproduced in Fig. 12. 6. The nonlinearity of the rate-dependence is evident, as is the fact that the termination points of relaxation are located below the stress-strain curve belonging to the smallest shear velocity.

The total stress as the model's response to cyclic processes with different mean shears and amplitudes is illustrated in Figs. 12. 7 and 12. 8.

These figures show that the viscoplasticity model presented here is well suited for reproducing the main phenomena of inelastic material behaviour. This becomes apparent by comparing Fig. 12. 6 with Fig. 6. 5. Complete information and further details related to the compatibility of the numerical simulations with test data can be taken from the original paper.[20]

12. 4. 4 Conclusions Regarding the Modelling of Mechanical Material Behaviour

The concept of viscoplasticity described in the present section offers a number of different advantages. To begin with, one characteristic feature is the physical transparency. General viscoplasticity embraces the three more

[20] See Lion [1994], Figs. 2.4.1, 2.5.1, 2.5.4, 4.2.1.1, 4.2.1.4, 4.2.1.5 or the corresponding figures in Haupt & Lion [1995].

special categories of material description by construction: the theories of elasticity, viscoelasticity and plasticity are clearly recognisable as special cases of viscoplasticity; the theories of elasticity and plasticity emerge from the general model as asymptotic approximations for sufficiently slow deformation processes.

The concept of general viscoplasticity combines the individual methods involved in material modelling to form a whole. In this connection it also becomes clear that the classical theories of continuum mechanics are by no means arbitrary assumptions bearing no relation to one another. The constitutive equations of the classical theories can be regarded as asymptotic approximations of a fairly general material description. These approximations are applicable under certain restrictive conditions. The restrictions can relate to special properties of the constitutive functionals (mathematical structure, continuity), or to kinematic features of special classes of processes (slow motions, small deformations). The inverse procedure is the systematic generalisation of special assumptions concerning the material modelling. One recognises a hierarchy of constitutive theories and is accordingly well equipped for constructing material models that reflect the macroscopically evident material behaviour, with the desired precision on the one hand and reasonable expense on the other.

One particular aspect of general viscoplasticity lies in the potential of its thermodynamic interpretation. The accompanying equilibrium processes, depicted by the models of elasticity or plasticity, can be regarded as equilibrium processes from the point of view of thermodynamics. This presents the opportunity of opening up the delineation of the mechanical material properties towards a thermomechanical description, taking into account the temperature-dependence of the mechanical material behaviour and observing the basic principles of thermodynamics. This is to be discussed in the next chapter.

13 Constitutive Models in Thermomechanics

13.1 Thermomechanical Consistency

The macroscopic behaviour of material bodies is not determined by mechanical processes alone. Among the numerous non-mechanical effects thermomechanical energy transformations play a vital role. They change the temperature field and influence the mechanical behaviour of the material.

Elasticity parameters, viscosities, relaxation spectra, yield limits, rate-dependent and rate-independent hysteresis effects, every material property can depend to a greater or lesser degree on temperature. In order to take the temperature-dependence of the material behaviour into account it is not sufficient to simply insert the temperature into the material description as an additional parameter. The modelling of thermomechanical material behaviour has to observe the fundamental laws of mechanics. It also has to adhere to the natural laws of thermodynamics, i.e. the energy balance (first law) (3.18),

$$\dot{e}(\boldsymbol{X}, t) = -\frac{1}{\rho_{\mathrm{R}}}\mathrm{Div}\boldsymbol{q}_{\mathrm{R}} + r + \frac{1}{\rho_{\mathrm{R}}}\tilde{\mathbf{T}}\cdot\dot{\mathbf{E}} \tag{13.1}$$

and the principle of irreversibility (second law). The formulation of the principle of irreversibility usually employed in continuum thermomechanics has its roots in the entropy production (3.39),

$$\gamma = \dot{s}(\boldsymbol{X}, t) + \frac{1}{\rho_{\mathrm{R}}}\mathrm{Div}\Big(\frac{\boldsymbol{q}_{\mathrm{R}}}{\Theta}\Big) - \frac{r}{\Theta}, \tag{13.2}$$

and demands that this must be non-negative:

$$\gamma \geq 0 . \tag{13.3}$$

From the point of view of thermomechanics the *Clausius-Duhem* inequality (13.2) is a condition which every solution of the basic equations has to fulfil, if it is to represent a physically possible process at all.[1]

With respect to the use of the entropy inequality in the theory of material behaviour, the local heat supply is eliminated from the entropy production with the help of the energy balance. That brings us to the internal dissipation (3.46),

$$\Delta = \Theta\gamma , \tag{13.4}$$

$$\Delta = \frac{1}{\rho_R}\widetilde{\mathbf{T}}\cdot\dot{\mathbf{E}} - \dot{e} + \Theta\dot{s} - \frac{1}{\rho_R\Theta}\, q_R \cdot g_R \tag{13.5}$$

$\left(g_R = \mathrm{Grad}\Theta(X, t)\right)$.

In thermodynamics three alternative energy densities are employed besides the specific internal energy e:

Definition 13. 1

The energy density

$$\psi = e - \Theta s \tag{13.6}$$

is called specific *free energy*; the energy density

$$h = \frac{1}{\rho_R}\widetilde{\mathbf{T}}\cdot\mathbf{E} - e \tag{13.7}$$

is known as specific *enthalpy* (*complementary energy*) and the energy density

$$g = \frac{1}{\rho_R}\widetilde{\mathbf{T}}\cdot\mathbf{E} - \psi \tag{13.8}$$

specific *free enthalpy* (*complementary free energy*).

In thermodynamic theories of viscoelasticity, plasticity and viscoplasticity, where the inelastic strain tensor is an internal variable with a special status, the definition of the complementary energies is usually based on the elastic part of the strain, i.e.

[1] Cf. LUBLINER [1972]; LEHMANN [1984, 1985, 1988, 1991].

$$h = \frac{1}{\rho_R}\widetilde{\mathbf{T}}\cdot\mathbf{E}_{\mathrm{e}} - e \tag{13.9}$$

and

$$g = \frac{1}{\rho_R}\widetilde{\mathbf{T}}\cdot\mathbf{E}_{\mathrm{e}} - \psi\,. \tag{13.10}$$

◊

The term complementary energy for h and g is the outcome of the property

$$h + e = g + \psi = \frac{1}{\rho_R}\widetilde{\mathbf{T}}\cdot\mathbf{E}_{\mathrm{e}}\,,$$

$$h + e = g + \psi = \frac{1}{\rho_R}\widetilde{\mathbf{T}}\cdot\mathbf{E}\,, \tag{13.11}$$

that is to say, h and g complete the energy densities e and ψ to obtain the scalar product of stresses and strains. According to (8.108) this scalar product is invariant within the *family 1* of dual variables when undergoing a change of the configuration. It would be equally possible to define the complementary energies on the basis of the scalar product $\widetilde{\mathbf{t}}\cdot\mathbf{e}$, which is an invariant within *family 2* according to (8.113).

The different energy densities give rise to equivalent formulations for the *Clausius-Duhem* inequality or, equivalently, the *dissipation inequality*

$$\Delta \geq 0\,. \tag{13.12}$$

Theorem 13. 1

For the internal dissipation $\Delta = \Theta\gamma$, besides (13.5) we also have the representations

$$\Delta = \frac{1}{\rho_R}\widetilde{\mathbf{T}}\cdot\dot{\mathbf{E}} - \dot{\psi} - s\dot{\Theta} - \frac{1}{\rho_R\Theta}\,q_R\cdot g_R\,, \tag{13.13}$$

$$\Delta = -\frac{1}{\rho_R}\mathbf{E}\cdot\dot{\widetilde{\mathbf{T}}} + \dot{h} + \Theta\dot{s} - \frac{1}{\rho_R\Theta}\,q_R\cdot g_R\,, \tag{13.14}$$

$$\Delta = -\frac{1}{\rho_R}\mathbf{E}\cdot\dot{\widetilde{\mathbf{T}}} + \dot{g} - s\dot{\Theta} - \frac{1}{\rho_R\Theta}\,q_R\cdot g_R\,. \tag{13.15}$$

□

Proof

by inserting

$$-\dot{e} = -\dot{\psi} - (\Theta s)^{\cdot} = -\frac{1}{\rho_{\mathrm{R}}}(\widetilde{\mathbf{T}}\cdot\mathbf{E})^{\cdot} + \dot{h} = -\frac{1}{\rho_{\mathrm{R}}}(\widetilde{\mathbf{T}}\cdot\mathbf{E})^{\cdot} + \dot{g} - (s\Theta)^{\cdot}$$

into (13.5). ■

The balance relations of thermomechanics reveal that a total of 11 scalar constitutive equations has to be formulated to close the system of balance equations (momentum and energy balance), in the sense that the number of equations equals the number of unknown fields. When drawing up constitutive equations, one is at liberty as regards the choice of the independent variables. Once one has decided upon a suitable set of input variables, the *material response* determines all the other thermomechanical quantities.

Definition 13. 2

The temporal course of the independent variables, combined with the material response determined by means of the constitutive equations, is called a *thermomechanical process.* ◊

According to this definition, a thermomechanical process is still no solution of the initial-boundary-value problem, but merely a solution of the constitutive equations.

One way of systematically taking the dissipation inequality into consideration is to formulate all constitutive equations *a priori* in such a way that no thermomechanical process exists which might violate the dissipation inequality. This would also ensure that the initial-boundary-value problems resulting from the balance relations cannot define solutions that might contradict the principle of irreversibility.

We owe this procedure to B. D. Coleman and W. Noll.[2] It gives rise to a multitude of possibilities for establishing thermomechanical material models that describe macroscopically observed material behaviour in a realistic and physically comprehensible manner.[3]

---

[2] COLEMAN & NOLL [1963]. The same idea was developed further by COLEMAN [1964], COLEMAN & GURTIN [1967] and successfully applied by many other authors.

[3] An alternative procedure for evaluating the dissipation inequality is outlined in MÜLLER [1973], pp. 73.

Definition 13. 3

A set of constitutive equations, fulfilling the dissipation inequality (13.12) for all imaginable thermomechanical processes, is called *thermomechanically consistent.* ◊

From the four equivalent representations (13.5), (13.13) - (13.15) of the internal dissipation and from the demand for a thermomechanically consistent material description, it is possible to obtain indications as to which mechanical and thermodynamical quantities are to be regarded as independent variables, so that constitutive equations then have to be established for the remaining quantities. Here are the attributions one obtains:

$$(13.5) \quad \Longrightarrow \quad \left(\mathbf{E}, s, g_{\mathrm{R}}\right) \quad \rightarrow \quad \left(e, \tilde{\mathbf{T}}, \Theta, q_{\mathrm{R}}\right),$$

$$(13.13) \quad \Longrightarrow \quad \left(\mathbf{E}, \Theta, g_{\mathrm{R}}\right) \quad \rightarrow \quad \left(\psi, \tilde{\mathbf{T}}, s, q_{\mathrm{R}}\right),$$

$$(13.14) \quad \Longrightarrow \quad \left(\tilde{\mathbf{T}}, s, g_{\mathrm{R}}\right) \quad \rightarrow \quad \left(h, \mathbf{E}, \Theta, q_{\mathrm{R}}\right),$$

$$(13.15) \quad \Longrightarrow \quad \left(\tilde{\mathbf{T}}, \Theta, g_{\mathrm{R}}\right) \quad \rightarrow \quad \left(g, \mathbf{E}, s, q_{\mathrm{R}}\right).$$

In thermomechanics, just as in mechanics, the properties of *elasticity* are characterised by means of material functions. On the other hand, to describe the inelastic material behaviour (*viscoelasticity*, *plasticity*, *viscoplasticity*) it is necessary to incorporate the past history of the independent variables, since memory properties exist. The influence of the history of the input process on the material response is recorded in thermomechanics as well. This can be achieved either explicitly in the representation by means of functionals or implicitly in the representation using internal variables, whose functional dependence on the process history is represented by means of ordinary differential equations. In this chapter only the second method is to be described.[4]

---

[4] A general theory of thermoviscoelasticity, formulated by means of functionals, was first presented in COLEMAN [1964]. See COLEMAN & OWEN [1970] for a general theory.

13. 2 Thermoelasticity

13. 2. 1 General Theory

Thermoelasticity is the simplest constitutive theory that allows for a thermomechanically consistent representation of material behaviour.

Definition 13. 4

The following constitutive equations define a *thermoelastic material*:

Free Energy

$$\psi = \psi(\mathbf{E}, \Theta, g_{\mathrm{R}}) \tag{13.16}$$

Stress

$$\tilde{\mathbf{T}} = \tilde{\mathbf{T}}(\mathbf{E}, \Theta, g_{\mathrm{R}}) \tag{13.17}$$

Entropy

$$s = s(\mathbf{E}, \Theta, g_{\mathrm{R}}) \tag{13.18}$$

Heat Flux Vector

$$q_{\mathrm{R}} = q_{\mathrm{R}}(\mathbf{E}, \Theta, g_{\mathrm{R}}) \tag{13.19}$$

◊

This definiton leads to

Theorem 13. 2

The constitutive equations of thermoelasticity are thermomechanically consistent if and only if the following relations are valid:

$$\frac{\partial \psi}{\partial g_{\mathrm{R}}} = 0 \Longleftrightarrow \psi = \psi(\mathbf{E}, \Theta) \, , \tag{13.20}$$

$$\frac{1}{\rho_{\mathrm{R}}} \tilde{\mathbf{T}} = \frac{\partial}{\partial \mathbf{E}} \psi(\mathbf{E}, \Theta) \, , \tag{13.21}$$

$$s = - \frac{\partial}{\partial \Theta} \psi(\mathbf{E}, \Theta) \, , \tag{13.22}$$

$$- q_{\mathrm{R}} \cdot g_{\mathrm{R}} \geq 0 \, . \tag{13.23}$$

□

Proof

By inserting the total time derivative of the free energy,

$$\dot{\psi} = \frac{\mathrm{d}}{\mathrm{d}t}\,\psi(\mathbf{E}(t), \Theta(t), \boldsymbol{g}_\mathrm{R}(t)) = \frac{\partial\psi}{\partial\mathbf{E}}\cdot\dot{\mathbf{E}}(t) + \frac{\partial\psi}{\partial\Theta}\dot{\Theta}(t) + \frac{\partial\psi}{\partial\boldsymbol{g}_\mathrm{R}}\cdot\dot{\boldsymbol{g}}_\mathrm{R}(t)\,,$$

into the dissipation inequality (13.12), we obtain

$$\Delta = \Big(\frac{1}{\rho_\mathrm{R}}\tilde{\mathbf{T}} - \frac{\partial\psi}{\partial\mathbf{E}}\Big)\cdot\dot{\mathbf{E}}(t) - \Big(s + \frac{\partial\psi}{\partial\Theta}\Big)\dot{\Theta}(t) - \frac{\partial\psi}{\partial\boldsymbol{g}_\mathrm{R}}\cdot\dot{\boldsymbol{g}}_\mathrm{R}(t) - \frac{1}{\rho_\mathrm{R}\Theta}\,\boldsymbol{q}_\mathrm{R}\cdot\boldsymbol{g}_\mathrm{R} \geq 0\,.$$

The validity of this inequality for all conceivable thermomechanical processes can only be guaranteed if the factors belonging to the time derivatives $\dot{\mathbf{E}}(t)$, $\dot{\Theta}(t)$ and $\dot{\boldsymbol{g}}_\mathrm{R}(t)$ vanish. This proves the free energy's independence of the temperature gradient, the stress relation (13.21) and the entropy relation (13.22). What remains is the residual inequality (13.23), representing a restrictive condition for the constitutive equation of the heat flux vector.

■

According to the theorem's assertion just two material functions are required to describe a thermoelastic material in full, namely one scalar-valued function to represent the free energy, $\psi = \psi(\mathbf{E}, \Theta)$, and one vector-valued function for portraying the heat flux vector, $\boldsymbol{q}_\mathrm{R} = \boldsymbol{q}_\mathrm{R}(\mathbf{E}, \Theta, \boldsymbol{g}_\mathrm{R})$. Thermomechanical consistency then enforces the stress relation and the entropy relation, i.e. the free energy is the potential of stress and entropy.[5] Similar potential characteristics are also a feature of the other three energy functions, as the following theorem demonstrates.

Theorem 13. 3

For a thermomechanically consistent representation of thermoelastic material behaviour we can apply not only the free energy ψ but also the other energy functions e, h or g in conjunction with suitable potential relations:

1) *Internal Energy* $\quad e = e(\mathbf{E}, s)$

$$\frac{1}{\rho_\mathrm{R}}\tilde{\mathbf{T}} = \frac{\partial}{\partial\mathbf{E}}\,e(\mathbf{E}, s) \qquad \Theta = \frac{\partial}{\partial s}\,e(\mathbf{E}, s) \tag{13.24}$$

[5] The theorem is exemplary for the procedure adopted by the thermomechanical constitutive theory: the principal goal is to formulate the constitutive equations in such a way that thermodynamical consistency is identically satisfied. It should be noted that in many cases of physical interest the conditions available are only sufficient. This is, however, not a disadvantage from the point of view of engineering applications.

2) *Enthalpy* $h = h(\widetilde{\mathbf{T}}, s)$

$$\frac{1}{\rho_R}\mathbf{E} = \frac{\partial}{\partial \widetilde{\mathbf{T}}} h(\widetilde{\mathbf{T}}, s) \qquad \Theta = -\frac{\partial}{\partial s} h(\widetilde{\mathbf{T}}, s) \tag{13.25}$$

3) *Free Enthalpy* $g = g(\widetilde{\mathbf{T}}, \Theta)$

$$\frac{1}{\rho_R}\mathbf{E} = \frac{\partial}{\partial \widetilde{\mathbf{T}}} g(\widetilde{\mathbf{T}}, \Theta) \qquad s = \frac{\partial}{\partial \Theta} g(\widetilde{\mathbf{T}}, \Theta) \tag{13.26}$$

□

Proof

analogous to Theorem 13. 2 by means of differentiation and evaluation of the appropriate form of the dissipation inequality. ■

The potential properties of the energy functions e, ψ, h and g provide physical interpretations of their temporal changes. These emerge from the calculation of the total time derivatives observing the chain rule.

Theorem 13. 4

For the time rates of the energy functions the following *Gibbs* relations are valid:[6]

$$\dot{e}(\mathbf{E}, s) = \frac{1}{\rho_R}\widetilde{\mathbf{T}} \cdot \dot{\mathbf{E}} + \Theta \dot{s}\,, \tag{13.27}$$

$$\dot{h}(\widetilde{\mathbf{T}}, s) = \frac{1}{\rho_R}\mathbf{E} \cdot \dot{\widetilde{\mathbf{T}}} - \Theta \dot{s}\,, \tag{13.28}$$

$$\dot{\psi}(\mathbf{E}, \Theta) = \frac{1}{\rho_R}\widetilde{\mathbf{T}} \cdot \dot{\mathbf{E}} - s\dot{\Theta}\,, \tag{13.29}$$

$$\dot{g}(\widetilde{\mathbf{T}}, \Theta) = \frac{1}{\rho_R}\mathbf{E} \cdot \dot{\widetilde{\mathbf{T}}} + s\dot{\Theta}\,. \tag{13.30}$$

■

Hence, the rate of change of each specific energy (or thermodynamic potential) consists of one mechanical and one thermal part. The mechanical part is either the stress power or the complementary stress power. The thermal part comprises the temperature multiplied by the entropy rate or the entropy times the temperature rate.

---

6 See TRUESDELL & TOUPIN [1960], Sect. 251.

Potential	Representation	Property	*Gibbs* Relation
Internal energy e	$e(\mathbf{E}, s)$	$\frac{1}{\rho_R}\widetilde{\mathbf{T}} = \frac{\partial e}{\partial \mathbf{E}}$ $\Theta = \frac{\partial e}{\partial s}$	$\dot{e} = \frac{1}{\rho_R}\widetilde{\mathbf{T}} \cdot \dot{\mathbf{E}} + \Theta \dot{s}$
Free energy $\psi = e - \Theta s$	$\psi(\mathbf{E}, \Theta)$	$\frac{1}{\rho_R}\widetilde{\mathbf{T}} = \frac{\partial \psi}{\partial \mathbf{E}}$ $s = -\frac{\partial \psi}{\partial \Theta}$	$\dot{\psi} = \frac{1}{\rho_R}\widetilde{\mathbf{T}} \cdot \dot{\mathbf{E}} - s\dot{\Theta}$
Enthalpy $h = \frac{1}{\rho_R}\widetilde{\mathbf{T}} \cdot \mathbf{E} - e$	$h = h(\widetilde{\mathbf{T}}, s)$	$\frac{1}{\rho_R}\mathbf{E} = \frac{\partial h}{\partial \widetilde{\mathbf{T}}}$ $\Theta = -\frac{\partial h}{\partial s}$	$\dot{h} = \frac{1}{\rho_R}\mathbf{E} \cdot \dot{\widetilde{\mathbf{T}}} - \Theta \dot{s}$
Free enthalpy $g = \frac{1}{\rho_R}\widetilde{\mathbf{T}} \cdot \mathbf{E} - \psi$	$g(\widetilde{\mathbf{T}}, \Theta)$	$\frac{1}{\rho_R}\mathbf{E} = \frac{\partial g}{\partial \widetilde{\mathbf{T}}}$ $s = \frac{\partial g}{\partial \Theta}$	$\dot{g} = \frac{1}{\rho_R}\mathbf{E} \cdot \dot{\widetilde{\mathbf{T}}} + s\dot{\Theta}$

Table 13. 1: Thermodynamic potentials

As indicated by the *Gibbs* relations (13.27) - (13.30), the free energy ψ represents the strain energy for isothermal processes and the free enthalpy g the corresponding complementary energy. In the same way the internal energy or enthalpy represents the strain energy or complementary energy for isentropic (locally adiabatic) processes. The statements of the last three theorems are summarised in Table 13. 1.

The inequality (13.23) which is left after the evaluation of the dissipation inequality is called *heat conduction inequality.* The heat conduction inequality restricts the constitutive equation for the heat flux vector: it demands that the heat flux vector and the temperature gradient do not form an acute angle, that means: heat flux can only take place in the direction of decreasing temperature. One conclusion emerging from the heat conduction inequality is described by

Theorem 13. 5

The constitutive equation for the heat flux vector is subject to the restrictive condition

$$q_{\mathrm{R}}(\mathbf{E}, \Theta, \boldsymbol{0}) = \boldsymbol{0} \,. \tag{13.31}$$

Without a temperature gradient no heat flow can take place. □

Proof

According to the heat conduction inequality (13.23), the scalar-valued function

$$g_{\mathrm{R}} \longmapsto f(g_{\mathrm{R}}) = g_{\mathrm{R}} \cdot q_{\mathrm{R}}(\mathbf{E}, \Theta, g_{\mathrm{R}}) \leq 0 \tag{13.32}$$

has a relative extremum (maximum) at $g_{\mathrm{R}} = \boldsymbol{0}$. Accordingly, the derivative

$$\frac{\partial f}{\partial g_{\mathrm{R}}} = q_{\mathrm{R}}(\mathbf{E}, \Theta, g_{\mathrm{R}}) + \left(\frac{\partial q_{\mathrm{R}}}{\partial g_{\mathrm{R}}}\right)^{\mathrm{T}} g_{\mathrm{R}}$$

has to vanish for $g_{\mathrm{R}} = \boldsymbol{0}$. This leads to (13.31). ■

The potential relations of thermoelasticity bring about a simplification of the local energy balance (13.1): inserting the *Gibbs* relation (13.27) eliminates the stress power:

$$\Theta \dot{s} = -\frac{1}{\rho_{\mathrm{R}}} \mathrm{Div} q_{\mathrm{R}} + r \,. \tag{13.33}$$

Utilising the free energy converts this to an evolution equation for the temperature: given the total time derivative

$$\dot{s}(t) = -\frac{\mathrm{d}}{\mathrm{d}t} \frac{\partial \psi}{\partial \Theta}(\mathbf{E}, \Theta) = -\frac{\partial^2 \psi}{\partial \Theta^2} \dot{\Theta} - \frac{\partial^2 \psi}{\partial \Theta \partial \mathbf{E}} \cdot \dot{\mathbf{E}} \,,$$

(13.33) leads to the *equation of heat conduction:*

$$c_{\mathrm{d}} \dot{\Theta} = \Theta \frac{\partial^2 \psi}{\partial \Theta \partial \mathbf{E}} \cdot \dot{\mathbf{E}} - \frac{1}{\rho_{\mathrm{R}}} \mathrm{Div} q_{\mathrm{R}} + r \,. \tag{13.34}$$

In this partial differential equation

$$c_d = c_d(\mathbf{E}, \Theta) = -\,\Theta\,\frac{\partial^2\psi(\mathbf{E}, \Theta)}{\partial\Theta^2} \tag{13.35}$$

is the so-called *specific heat at constant deformation*. The term

$$\Theta\,\frac{\partial^2\psi}{\partial\Theta\partial\mathbf{E}}\cdot\dot{\mathbf{E}} = \frac{1}{\rho_R}\Theta\,\frac{\partial\widetilde{\mathbf{T}}(\mathbf{E}, \Theta)}{\partial\Theta}\cdot\dot{\mathbf{E}}\,, \tag{13.36}$$

which appears in the equation of heat conduction in addition to the heat flux represents the *thermoelastic coupling* effect. To complete the equation of heat conduction, a constitutive equation for the heat flux vector has to be inserted into (13.34). It is common practice to employ the *Fourier* model of isotropic heat conduction mentioned in (7.40),

$$\boldsymbol{q} = -\,\lambda\boldsymbol{g}\,, \tag{13.37}$$

where λ is the so-called *coefficient of heat conduction*, which can, in principle, depend on the temperature and the temperature gradient, i.e. $\lambda = \lambda(\Theta, \boldsymbol{g}\cdot\boldsymbol{g})$. The *Fourier* model of heat conduction meets the restrictive condition (13.31) and the heat conduction inequality (13.23), provided the coefficient of heat conduction is non-negative:

$$\lambda \geq 0\,. \tag{13.38}$$

Transformation of the *Fourier* model to the reference configuration, given $\boldsymbol{q}_R = (\det\mathbf{F})\mathbf{F}^{-1}\boldsymbol{q}$ and $\boldsymbol{g}_R = \mathbf{F}^T\boldsymbol{g}$, leads to a constitutive equation in the form of (13.19):

$$\boldsymbol{q}_R = -\,\lambda(\Theta, \boldsymbol{g}_R\cdot\mathbf{C}^{-1}\boldsymbol{g}_R)\sqrt{\det\mathbf{C}}\,\mathbf{C}^{-1}\boldsymbol{g}_R\,. \tag{13.39}$$

In view of (13.33), the *specific heat* or *heat capacity* can be defined more generally by means of an equation of the form

$$c = \Theta\,\frac{\partial s}{\partial\Theta} \tag{13.40}$$

(heat supply per temperature change), where either the strain or the stress is kept constant. For the specific heat at constant deformation we then have, equivalent to (13.35), the formula

$$c_{\mathrm{d}} = \Theta \, \frac{\partial s(\mathbf{E}, \Theta)}{\partial \Theta} \,, \tag{13.41}$$

for the application of which the entropy has to be written as a function of strain and temperature. In analogy thereto, (13.40) leads to the *specific heat at constant stress*,

$$c_{\mathrm{s}} = \Theta \, \frac{\partial s(\widetilde{\mathbf{T}}, \Theta)}{\partial \Theta} \,, \tag{13.42}$$

which requires the representation of the entropy as a function of stress and temperature.
Two special cases exist under the heading of general thermoelasticity, representing thermomechanical generalisations of classical theories. The special cases in question are the *thermoelastic fluid* and the *linear thermoelastic solid.*

13. 2. 2 Thermoelastic Fluid

The *elastic fluid* is defined according to (5.1) by a constitutive equation of the form

$$\mathbf{T} = -\, p\mathbf{1} \,, \tag{13.43}$$

where the hydrostatic pressure p depends on the specific volume $v = 1/\rho$ and the temperature Θ:

$$p = p(v, \Theta) \,. \tag{13.44}$$

According to (9.28) this gives rise to the stress power

$$\frac{1}{\rho_{\mathrm{R}}} \widetilde{\mathbf{T}} \cdot \dot{\mathbf{E}} = \frac{1}{\rho} \mathbf{T} \cdot \mathbf{D} = -\, p\dot{v} \,. \tag{13.45}$$

A natural definition of the complementary power is

$$-\, v\dot{p} = -\, (pv)^{\cdot} - (-\, p\dot{v}) \,,$$

although this definition does not quite fit into the general concept of thermoelasticity.

One calculates on the basis of the constitutive equation (13.43)

$$-\frac{1}{3\rho_R}\tilde{\mathbf{T}}\cdot\mathbf{C} = -\frac{1}{3\rho_R}(\det\mathbf{F})(\mathbf{F}^{-1}\mathbf{T}\mathbf{F}^{\mathrm{T}-1})\cdot(\mathbf{F}^{\mathrm{T}}\mathbf{F}) = \frac{p}{\rho} = pv\,, \tag{13.46}$$

i.e. $\frac{1}{\rho_R}\tilde{\mathbf{T}}\cdot\mathbf{E} \neq -pv$.

As in the case of solids, the definition of the free energy $\psi = e - \Theta s$ is just as common for thermoelastic fluids, whereas the enthalpy values differ from the definitions (13.7), (13.8) as a result of (13.46).

Definition 13. 5

The *enthalpy* of a thermoelastic fluid is defined by[7]

$$h = e + pv\,; \tag{13.47}$$

and the *free enthalpy* by [8]

$$g = h - \Theta s = e + pv - \Theta s\,. \tag{13.48}$$

◊

The definition provides the representations set out below for the internal dissipation:

$$\Delta = -p\dot{v} - \dot{e} + \Theta\dot{s} - \frac{1}{\rho\Theta}\boldsymbol{q}\cdot\boldsymbol{g}\,, \tag{13.49}$$

$$\Delta = -p\dot{v} - \dot{\psi} - s\dot{\Theta} - \frac{1}{\rho\Theta}\boldsymbol{q}\cdot\boldsymbol{g}\,, \tag{13.50}$$

$$\Delta = v\dot{p} - \dot{h} + \Theta\dot{s} - \frac{1}{\rho\Theta}\boldsymbol{q}\cdot\boldsymbol{g}\,, \tag{13.51}$$

$$\Delta = v\dot{p} - \dot{g} - s\dot{\Theta} - \frac{1}{\rho\Theta}\boldsymbol{q}\cdot\boldsymbol{g}\,. \tag{13.52}$$

The evaluation of the dissipation inequality $\Delta \geq 0$ thus leads to the following potential and *Gibbs* relations:

---

[7] BAEHR [1973], p. 74.

[8] BAEHR [1973], p. 167.

$$\underline{e = e(v, s):} \qquad p = -\frac{\partial e}{\partial v} \qquad \Theta = \frac{\partial e}{\partial s} \qquad \dot{e}(v, s) = -p\dot{v} + \Theta\dot{s} \tag{13.53}$$

$$\underline{\psi = \psi(v, \Theta):} \qquad p = -\frac{\partial \psi}{\partial v} \qquad s = -\frac{\partial \psi}{\partial \Theta} \qquad \dot{\psi}(v, \Theta) = -p\dot{v} - s\dot{\Theta} \tag{13.54}$$

$$\underline{h = h(p, s):} \qquad v = \frac{\partial h}{\partial p} \qquad \Theta = \frac{\partial h}{\partial s} \qquad \dot{h}(p, s) = v\dot{p} + \Theta\dot{s} \tag{13.55}$$

$$\underline{g = g(p, \Theta):} \qquad v = \frac{\partial g}{\partial p} \qquad s = -\frac{\partial g}{\partial \Theta} \qquad \dot{g}(p, \Theta) = v\dot{p} + s\dot{\Theta} \tag{13.56}$$

Under the assumption that constitutive functions are invertible, these equations inspire several equivalent representations for the elasticity relation of the fluid (13.43), e.g. $v = v(p, \Theta)$ or $p = p(v, s)$ etc. One also obtains constitutive equations such as $s = s(v, \Theta)$ or $s = s(p, \Theta)$ etc. In the thermodynamics of fluids a constitutive equation of the form

$$f_{\text{th}}(p, v, \Theta) = 0 \quad \text{and} \quad f_{\text{th}}(p, v, s) = 0$$

is called a *thermal equation of state*,[9] and a constitutive equation of the form

$$f_{\text{cal}}(s, \Theta, v) = 0 \quad \text{and} \quad f_{\text{cal}}(s, \Theta, p) = 0$$

a *caloric equation of state*.[10] In solid mechanics comparable definitions are applicable as well, although they are not so common.

Further material functions of physical significance can be derived by differentiation from the caloric or thermal equations of state, for example, specific heat coefficients according to (13.41) and (13.42) (see below). Moreover, we have

Definition 13. 6

If the specific volume is given as a function of temperature and pressure, the material function

$$\alpha = \frac{1}{v}\,\frac{\partial v(p, \Theta)}{\partial \Theta} \tag{13.57}$$

is called the *coefficient of thermal expansion*, and the material function

---

[9] BAEHR [1973], p. 24.

[10] BAEHR [1973], p. 66.

$$\kappa = -\frac{1}{v}\frac{\partial v(p,\Theta)}{\partial p} \tag{13.58}$$

the *coefficient of isothermal compressibility.* ◊

In the following it is assumed that all equations of state are invertible:

$$v = v(p, \Theta) \Longleftrightarrow p = p(v, \Theta) \ . \tag{13.59}$$

Definition 13. 6 then leads to

$$\frac{\partial p(v,\Theta)}{\partial v} = -\frac{1}{\kappa v} \tag{13.60}$$

and

$$\frac{\partial p(v,\Theta)}{\partial \Theta} = \frac{\alpha}{\kappa} \ . \tag{13.61}$$

The verification of equations of this kind rests on the differentiation of the state equations, observing the chain rule. In this particular case, we begin with the identity $p = p\big(v(p, \Theta), \Theta\big)$ and calculate

$$1 = \frac{\partial p(v, \Theta)}{\partial v}\frac{\partial v(p, \Theta)}{\partial p} = \frac{\partial p(v, \Theta)}{\partial v}(-\kappa v) \text{ and}$$
$$0 = \frac{\partial p(v, \Theta)}{\partial v}\frac{\partial v(p, \Theta)}{\partial \Theta} + \frac{\partial p(v, \Theta)}{\partial \Theta} = -\frac{1}{\kappa v}v\alpha + \frac{\partial p(v, \Theta)}{\partial \Theta} \ .$$

For thermoelastic fluids the local energy balance in the spatial representation reads

$$\dot{e} = -\frac{1}{\rho}\mathrm{div}\boldsymbol{q} + r - p\dot{v} \tag{13.62}$$

or, equivalently

$$\dot{h} = -\frac{1}{\rho}\mathrm{div}\boldsymbol{q} + r + v\dot{p} \ , \tag{13.63}$$

$$\Theta\dot{s} = -\frac{1}{\rho}\mathrm{div}\boldsymbol{q} + r \ . \tag{13.64}$$

Considering the state equations and the constitutive equation (13.37) for the heat flux vector, this leads to the equation of heat conduction.

Theorem 13. 6

For a thermoelastic fluid the *equation of heat conduction* is given by

$$c_v(v, \Theta)\dot{\Theta} + \Theta\frac{\alpha}{\kappa}v\,\mathrm{div}\boldsymbol{v} = \frac{1}{\rho}\,\mathrm{div}(\lambda\mathrm{grad}\Theta) + r \tag{13.65}$$

or, equivalently, by

$$c_p(p, \Theta)\dot{\Theta} - \Theta\alpha v\,\dot{p} = \frac{1}{\rho}\,\mathrm{div}(\lambda\mathrm{grad}\Theta) + r\,. \tag{13.66}$$

Here, the function

$$c_v(v, \Theta) = -\,\Theta\frac{\partial^2\psi}{\partial\Theta^2} = \Theta\frac{\partial s(v, \Theta)}{\partial\Theta} = \frac{\partial}{\partial\Theta}\,e(v, \Theta) \tag{13.67}$$

is the *specific heat at constant volume*, corresponding to a change in internal energy if the specific volume is held constant. The function

$$c_p(p, \Theta) = -\,\Theta\frac{\partial^2 g}{\partial\Theta^2} = \Theta\frac{\partial s(p, \Theta)}{\partial\Theta} = \frac{\partial}{\partial\Theta}\,h(p, \Theta) \tag{13.68}$$

corresponds to a change in enthalpy if the pressure is kept constant and is called the *specific heat at constant pressure*. The specific heat capacities are connected to each other by means of the relation

$$c_p - c_v = \Theta\frac{\alpha^2}{\kappa}v\,. \tag{13.69}$$

□

Proof

by calculating $\Theta\dot{s}$ and observing the potential relations as well as (13.61). First we calculate

$$\begin{aligned}\Theta\dot{s} &= -\,\Theta\frac{\mathrm{d}}{\mathrm{d}t}\Big(\frac{\partial\psi}{\partial\Theta}\Big) = -\,\Theta\frac{\partial^2\psi}{\partial\Theta^2}\dot{\Theta} - \Theta\frac{\partial^2\psi}{\partial\Theta\partial v}\,\dot{v} \\ &= -\,\Theta\frac{\partial^2\psi}{\partial\Theta^2}\dot{\Theta} + \Theta\frac{\partial p}{\partial\Theta}\,\dot{v} = -\,\Theta\frac{\partial^2\psi}{\partial\Theta^2}\dot{\Theta} + \Theta\frac{\alpha}{\kappa}\,\dot{v}\ ;\end{aligned}$$

in addition, the continuity equation (2.11) leads to

$$\dot{v} = -\,\frac{\dot{\rho}}{\rho^2} = \frac{\rho\,\mathrm{div}\,\boldsymbol{v}}{\rho^2} = v\,\mathrm{div}\boldsymbol{v}\,.$$

We finally arrive at

$$c_v(v, \Theta) = -\Theta \frac{\partial^2 \psi}{\partial \Theta^2} = \Theta \frac{\partial s(v, \Theta)}{\partial \Theta} = \frac{\partial}{\partial \Theta} e(v, \Theta) ,$$

making use of the potential relation $(13.54)_3$. $(13.67)_4$ leads on from $(13.53)_4$. In this case - as opposed to the natural representation $(13.53)_1$ - the internal energy is required as a function of v and Θ. In analogy to this procedure we have

$$\Theta \dot{s} = -\Theta \frac{\mathrm{d}}{\mathrm{d}t}\Big(\frac{\partial g}{\partial \Theta}\Big) = -\Theta \frac{\partial^2 g}{\partial \Theta^2}\dot{\Theta} - \Theta \frac{\partial^2 g}{\partial \Theta \partial p}\dot{p} = -\Theta \frac{\partial^2 g}{\partial \Theta^2}\dot{\Theta} - \Theta \frac{\partial v}{\partial \Theta}\dot{p}$$
$$= -\Theta \frac{\partial^2 g}{\partial \Theta^2}\dot{\Theta} - \Theta\, \alpha v\, \dot{p} .$$

Arising out of $(13.56)_3$ and $(13.55)_4$ it follows that

$$c_p(p, \Theta) = -\Theta \frac{\partial^2 g}{\partial \Theta^2} = \Theta \frac{\partial s(p, \Theta)}{\partial \Theta} = \frac{\partial}{\partial \Theta} h(p, \Theta) .$$

The enthalpy h has to be written as a function of the temperature here as well, in contrast to $(13.55)_1$. The identity $s = s(p, \Theta) = s\big(v(p, \Theta), \Theta\big)$ provides the relation between the heat capacities. Differentiation and taking $(13.54)_2$ and $(13.54)_3$ into account yields

$$\frac{\partial s(p, \Theta)}{\partial \Theta} = \frac{\partial s(v, \Theta)}{\partial v}\frac{\partial v(p, \Theta)}{\partial \Theta} + \frac{\partial s(v, \Theta)}{\partial \Theta}$$
$$= \frac{\partial p(v, \Theta)}{\partial \Theta}\frac{\partial v(p, \Theta)}{\partial \Theta} + \frac{\partial s(v, \Theta)}{\partial \Theta} = \frac{\alpha}{\kappa}(\alpha v) + \frac{\partial s(v, \Theta)}{\partial \Theta} . \qquad \blacksquare$$

The basic equations (5.5), (5.6) of a perfect fluid can now be generalised to thermomechanics. The complete system of differential equations for describing a thermoelastic fluid consists of the balance of linear momentum (5.5) and the balance of mass (2.11),

$$\frac{\partial \boldsymbol{v}}{\partial t} + (\mathrm{grad}\boldsymbol{v})\boldsymbol{v} = -v\, \mathrm{grad}\hat{p}(v, \Theta) + \boldsymbol{f} , \tag{13.70}$$

$$\dot{v} = v\, \mathrm{div}\boldsymbol{v} \tag{13.71}$$

as well as the equation of heat conduction in the formulation of (13.65) or (13.66). That makes 5 coupled differential equations altogether for determining the velocity field $\boldsymbol{v}(\boldsymbol{x}, t)$, the specific volume $(1/\rho) = v(\boldsymbol{x}, t)$, and the temperature field $\Theta(\boldsymbol{x}, t)$.

An important special case of a thermoelastic fluid is the so-called *perfect gas*, defined by the special thermal state equation (5.4), containing one material parameter $R > 0$,

$$p(v, \Theta) = \frac{R\Theta}{v} \Longleftrightarrow v(p, \Theta) = \frac{R\Theta}{p} . \tag{13.72}$$

Theorem 13. 7

A *perfect gas*, defined by the thermal state equation (13.72), is characterised by the following relations:

Thermal Expansion

$$\alpha = \frac{1}{v} \frac{\partial v(p,\Theta)}{\partial \Theta} = \frac{1}{\Theta} \tag{13.73}$$

Compressibility

$$\kappa = -\frac{1}{v} \frac{\partial v(p,\Theta)}{\partial p} = \frac{1}{p} \tag{13.74}$$

Heat Capacities

$$c_p - c_v = \Theta \frac{\alpha^2}{\kappa} v = R \tag{13.75}$$

$$c_v = c_v(\Theta) , \quad c_p = c_p(\Theta) \tag{13.76}$$

Internal Energy, Enthalpy

$$e(\Theta) = \int_{\Theta_0}^{\Theta} c_v(\Theta) \mathrm{d}\Theta , \quad h(\Theta) = \int_{\Theta_0}^{\Theta} c_p(\Theta) \mathrm{d}\Theta \tag{13.77}$$

Entropy

$$s(\Theta, v) = \int_{\Theta_0}^{\Theta} c_v(\Theta) \frac{\mathrm{d}\Theta}{\Theta} + R \ln \frac{v}{v_0} , \quad s(\Theta, p) = \int_{\Theta_0}^{\Theta} c_p(\Theta) \frac{\mathrm{d}\Theta}{\Theta} - R \ln \frac{p}{p_0} \tag{13.78}$$

□

Proof

The coefficients of expansion and compressibility according to (13.73) are the direct outcome of the definitions (13.57) and (13.58) applying the thermal state equation (13.72). By inserting these relations into (13.69) we obtain (13.75).

In order to convince oneself that the specific heats of a perfect gas can only depend on the temperature, one can take the relation

$$c_v(v, \Theta) = \frac{\partial}{\partial \Theta}\, e(v, \Theta)$$

and differentiate the identity $e(v, \Theta) = e(v, s(v, \Theta))$ with respect to v. Subject to the use of the potential relations and the thermal state equation (13.72), the outcome is

$$\frac{\partial e(v, \Theta)}{\partial v} = \frac{\partial e(v, s)}{\partial v} + \frac{\partial e(v, s)}{\partial s}\frac{\partial s(v, \Theta)}{\partial v} = -\,p + \Theta\frac{\partial p(v, \Theta)}{\partial \Theta} = 0\,.$$

In view of (13.75) this means that c_p depends only on the temperature as well. Irrespective of this argument, the same conclusion, reached on the basis of

$$c_p(p, \Theta) = \frac{\partial}{\partial \Theta}\, h(p, \Theta) \text{ and } h(p, \Theta) = h(p, s(p, \Theta))\,,$$

is $$\frac{\partial h(p,\Theta)}{\partial p} = \frac{\partial h(p,s)}{\partial p} + \frac{\partial h(p,s)}{\partial s}\frac{\partial s(p, \Theta)}{\partial p} = v + \Theta\Big(-\frac{\partial v(p,\Theta)}{\partial \Theta}\Big) = 0\,.$$

These results lead to $\dot{e} = c_v(\Theta)\dot{\Theta}$ and finally to $\dot{h} = c_p(\Theta)\dot{\Theta}$, i.e. (13.77). The representation formulae (13.78) for the entropy are obtainable by employing the results gained so far from

$$\Theta\dot{s} = \dot{e} + p\dot{v} = c_v(\Theta)\dot{\Theta} + p\dot{v} \Longrightarrow \dot{s} = c_v(\Theta)\,\frac{\dot{\Theta}}{\Theta} + R\frac{\dot{v}}{v}$$

or $$\Theta\dot{s} = \dot{h} - v\dot{p} = c_p(\Theta)\dot{\Theta} - v\dot{p} \Longrightarrow \dot{s} = c_p(\Theta)\,\frac{\dot{\Theta}}{\Theta} - R\frac{\dot{p}}{p}\,.$$ ■

According to the theorem, all four energy functions e, h, s and g can be calculated in the case of a perfect gas, provided one further material function is known, namely one of the specific heat capacities $c_v(\Theta)$ or $c_p(\Theta)$ as functions of the thermodynamic temperature.

13. 2. 3 Linear-Thermoelastic Solids

A thermomechanically consistent generalisation of classical linear elasticity is the theory of *linear thermoelasticity*.[11] This emanates from general thermoelasticity by geometrical and physical linearisation and the additional

[11] A detailed introduction to linear thermoelasticity is to be found in NOWACKI [1986]. See also the monograph CARLSON [1972].

assumption that small changes in temperature occur. The considerations set out below are accordingly based on the linearised strain tensor

$$\mathbf{E} = \frac{1}{2}\left(\mathrm{grad}\boldsymbol{u}(\boldsymbol{x}, t) + (\mathrm{grad}\boldsymbol{u}(\boldsymbol{x}, t))^{\mathrm{T}}\right) \tag{13.79}$$

and the difference in temperature

$$\vartheta(\boldsymbol{x}, t) = \Theta(\boldsymbol{x}, t) - \Theta_0 \,, \tag{13.80}$$

Θ_0 being a spatially and temporally constant reference temperature. Moreover, isotropic material properties are assumed in the following.
For isotropic material behaviour, the theory of linear thermoelasticity is defined by a constitutive equation for the free energy,

$$\rho\psi = \mu\left(\mathrm{tr}\mathbf{E}^{\mathrm{D}2}\right) + \frac{1}{2}K(\mathrm{tr}\mathbf{E})^2 - \rho b\vartheta(\mathrm{tr}\mathbf{E}) - \frac{1}{2}\rho c\vartheta^2 \,, \tag{13.81}$$

and the *Fourier model* for the heat flux vector,

$$\boldsymbol{q} = -\,\lambda\,\mathrm{grad}\vartheta(\boldsymbol{x}, t)\,. \tag{13.82}$$

By means of an appropriate application of the potential properties (13.21) and (13.22) this yields the linear elasticity relation

$$\mathbf{T} = 2\mu\mathbf{E}^{\mathrm{D}} + K(\mathrm{tr}\mathbf{E})\mathbf{1} - \rho b\vartheta\,\mathbf{1}$$

and the linear entropy relation

$$s = b(\mathrm{tr}\mathbf{E}) + c\vartheta\,.$$

It is evident that the identification

$$\rho b = 3K\alpha \tag{13.83}$$

applies, where α denotes the *coefficient of linear expansion* and

$K = \dfrac{2\mu(1 + \nu)}{3(1 - 2\nu)}$ the *isothermal bulk modulus* (cf. (13.57), (13.58)).

The material function

$$c_d = -\Theta \frac{\partial^2 \psi(\mathbf{E}, \Theta)}{\partial \Theta^2} = c\Theta \quad (\approx c\Theta_0 \text{ for } \vartheta \ll \Theta_0) \tag{13.84}$$

according to (13.35) is the specific heat at constant deformation. With these physically motivated parameters and quantities, the constitutive equations for the stress tensor and the entropy (the thermal and caloric equations of state) read as

$$\mathbf{T} = 2\mu\left[\mathbf{E} + \frac{\nu}{1-2\nu}(\mathrm{tr}\mathbf{E})\mathbf{1}\right] - 3K\alpha\vartheta\,\mathbf{1}\,, \tag{13.85}$$

$$s = c_d \frac{\vartheta}{\Theta_0 + \vartheta} + \frac{3K\alpha}{\rho}\,\mathrm{tr}\mathbf{E} \approx c_d \frac{\vartheta}{\Theta_0} + \frac{3K\alpha}{\rho}\,\mathrm{tr}\mathbf{E}\,, \tag{13.86}$$

and the general equation of heat conduction (13.34) converts to

$$c_d \dot{\vartheta} + \Theta_0 \frac{3K\alpha}{\rho}\,\mathrm{tr}\dot{\mathbf{E}} = \frac{\lambda}{\rho}\,\mathrm{div\,grad}\,\vartheta(\boldsymbol{x}, t) + r\,. \tag{13.87}$$

The balance of linear momentum

$$\rho \frac{\partial^2 \boldsymbol{u}}{\partial t^2} = \mu\left\{\Delta \boldsymbol{u} + \frac{1}{1-2\nu}\,\mathrm{grad\,div}\boldsymbol{u} - 2\frac{1+\nu}{1-2\nu}\,\alpha\,\mathrm{grad}\,\vartheta\right\} + \rho \boldsymbol{f} \tag{13.88}$$

combines with the equation of heat conduction (13.87) to form a coupled system of four differential equations for calculating the displacement field $\boldsymbol{u}(\boldsymbol{x}, t)$ and the temperature field $\vartheta(\boldsymbol{x}, t)$.
The specific heat at constant stress c_s can be calculated using

$$c_s(\mathbf{T}, \Theta) = \Theta \frac{\partial}{\partial \Theta}\, s(\mathbf{T}, \Theta) \approx \Theta_0 \frac{\partial}{\partial \vartheta}\, s(\mathbf{T}, \vartheta)\,. \tag{13.89}$$

With $\quad \mathrm{tr}\mathbf{T} = 3K\,\mathrm{tr}\mathbf{E} - 9K\alpha\vartheta \Longleftrightarrow \mathrm{tr}\mathbf{E} = \frac{1}{3K}\,\mathrm{tr}\mathbf{T} + 3\alpha\vartheta$

the entropy $\quad s(\mathbf{E}, \vartheta) = c_d \frac{\vartheta}{\Theta_0} + \frac{3K\alpha}{\rho}\,\mathrm{tr}\mathbf{E} \quad$ can be written in the form

$$s(\mathbf{T}, \vartheta) = c_d \frac{\vartheta}{\Theta_0} + \frac{9K\alpha^2}{\rho}\vartheta + \frac{\alpha}{\rho}\,\mathrm{tr}\mathbf{T}\,.$$

This leads to the relation

$$c_s - c_d = \Theta_0 \frac{9K\alpha^2}{\rho}\,, \tag{13.90}$$

which is comparable with (13.75) and to a formulation for the equation of heat conduction, equivalent to (13.87):

$$c_s \dot{\vartheta} + \Theta_0 \frac{\alpha}{\rho} \operatorname{tr} \dot{\mathbf{T}} = \frac{\lambda}{\rho} \operatorname{div} \operatorname{grad} \vartheta(\boldsymbol{x}, t) + r \, . \tag{13.91}$$

A constitutive model of thermoelasticity incorporating properties of *anisotropy* is not difficult to achieve. All one has to do is replace the special assumption (13.81) for the free energy by a general bilinear form

$$\rho\psi = \tfrac{1}{2}\mathbf{E}\cdot\mathbb{C}[\mathbf{E}] - \rho\mathbf{B}\cdot\mathbf{E}\vartheta - \tfrac{1}{2}\rho c\vartheta^2 \, . \tag{13.92}$$

$\mathbb{C}$ and $\mathbf{B}$ $(\mathbb{C} = \mathbb{C}^{\mathrm{T}}, \mathbf{B} = \mathbf{B}^{\mathrm{T}})$ are constant material tensors of fourth and second order, determinable on the basis of a given symmetry group, applying the approach practised in Sect. 9. 3. 2. The constitutive equation (13.82) for the heat flux vector has to be generalised by a formula along the lines of

$$q = -\,\boldsymbol{\Lambda} \operatorname{grad}\vartheta(\boldsymbol{x}, t) \, , \tag{13.93}$$

remembering that it is compulsory for the *heat conduction tensor* $\boldsymbol{\Lambda}$ to be positive definite in order to guarantee its compatibility with the heat conduction inequality (13.23).

13. 3 Thermoviscoelasticity

13. 3. 1 General Concept

Definition 10. 7 of the mechanical theory of viscoelasticity, utilising internal variables, can be applied to thermomechanics too. Representation (13.13) of the dissipation inspires the following

Definition 13. 7

A *thermoviscoelastic material* can be defined by the following constitutive equations:

Free Energy

$$\psi = \psi(\mathbf{E}, \Theta, g_{\mathrm{R}}, q_1, \dots, q_N) \tag{13.94}$$

Stress

$$\widetilde{\mathbf{T}} = \widetilde{\mathbf{T}}(\mathbf{E}, \Theta, g_{\mathrm{R}}, q_1, \dots, q_N) \tag{13.95}$$

Entropy

$$s = s(\mathbf{E}, \Theta, g_{\mathrm{R}}, q_1, \dots, q_N) \tag{13.96}$$

Heat Flux Vector

$$q_{\mathrm{R}} = q_{\mathrm{R}}(\mathbf{E}, \Theta, g_{\mathrm{R}}, q_1, \dots, q_N) \tag{13.97}$$

Evolution Equations

$$\dot{q}_k(t) = f_k(\mathbf{E}(t), \Theta(t), q_1(t), \dots, q_N(t)) \quad (k = 1, 2, \dots N) \tag{13.98}$$

The evolution equations are distinguished by the following properties:

1) For every state of strain and temperature $\Lambda = (\mathbf{E}, \Theta)$ *equilibrium solutions* exist, i.e. solutions $\bar{q}_j$ of the system

$$f_k(\mathbf{E}, \Theta, \bar{q}_1, \dots, \bar{q}_N) = 0\,, \quad k = 1, \dots, N \tag{13.99}$$

which uniquely depend on $\mathbf{E}$ and Θ:

$$\Lambda = (\mathbf{E}, \Theta) \longmapsto \bar{q}_k = \bar{q}_k(\mathbf{E}, \Theta)\,. \tag{13.100}$$

2) These equilibrium solutions are *asymptotically stable*:[12] for every pair $\Lambda_0 = (\mathbf{E}_0, \Theta_0)$ and any given initial conditions $q_k(t_0) = q_{k0}$ the solution of the system of differential equations

$$\dot{q}_k(t) = f_k(\mathbf{E}_0, \Theta_0, q_1(t), \dots, q_N(t)) \tag{13.101}$$

tends towards the equilibrium solution:

$$\lim_{t \to \infty} q_k(t) = \bar{q}_k(\mathbf{E}_0, \Theta_0) \quad (k = 1, 2, \dots N)\,. \tag{13.102}$$

◊

[12] Compare the stability assumption in Coleman & Gurtin [1967], pp. 601.

Since the current temperature gradient $\boldsymbol{g}_{\mathrm{R}}$ has to be contained in the constitutive equation for the heat flux vector, it is included in the general assumptions for ψ, $\widetilde{\mathbf{T}}$ and s as well as an independent variable. However, $\boldsymbol{g}_{\mathrm{R}}$ does not enter the list of variables in the evolution equations (13.98). A dependence of the evolution equations on $\boldsymbol{g}_{\mathrm{R}}$ would mean that the current values of the internal variables and, as a consequence, ψ, $\widetilde{\mathbf{T}}$ and s would also depend on the history of the temperature gradient. No assumption of this kind is intended.[13] The definition leads to

Theorem 13. 8

The dissipation inequality $\Delta \geq 0$ is fulfilled for all conceivable thermomechanical processes if and only if the following relations apply: [14]

Free Energy

$$\psi = \psi(\mathbf{E}, \Theta, \mathfrak{q}_1, \dots, \mathfrak{q}_N) \tag{13.103}$$

Stress Relation

$$\frac{1}{\rho_{\mathrm{R}}}\widetilde{\mathbf{T}} = \frac{\partial}{\partial \mathbf{E}}\, \psi(\mathbf{E}, \Theta, \mathfrak{q}_1, \dots, \mathfrak{q}_N) \tag{13.104}$$

Entropy Relation

$$s = -\frac{\partial}{\partial \Theta}\, \psi(\mathbf{E}, \Theta, \mathfrak{q}_1, \dots, \mathfrak{q}_N) \tag{13.105}$$

Dissipation Inequality

$$\delta - \frac{1}{\rho_{\mathrm{R}}\Theta}\, \boldsymbol{q}_{\mathrm{R}} \cdot \boldsymbol{g}_{\mathrm{R}} \geq 0 \tag{13.106}$$

$$\delta \geq 0 \tag{13.107}$$

The mechanical dissipation δ is given by

$$\delta = -\sum_{k=1}^{N} \frac{\partial \psi}{\partial \mathfrak{q}_k}\, \dot{\mathfrak{q}}_k \, . \tag{13.108}$$

□

[13] Cf. COLEMAN [1964], where the history of the temperature but only the present value of the temperature gradient enters the constitutive functionals.
[14] Cf. COLEMAN & GURTIN [1967], p. 601.

Proof

Calculating the total time derivative of the free energy provides the dissipation inequality

$$\Big(\frac{1}{\rho_{\mathrm{R}}}\tilde{\mathbf{T}} - \frac{\partial\psi}{\partial\mathbf{E}}\Big)\cdot\dot{\mathbf{E}}(t) - \Big(s + \frac{\partial\psi}{\partial\Theta}\Big)\dot{\Theta}(t) - \frac{\partial\psi}{\partial\boldsymbol{g}_{\mathrm{R}}}\cdot\dot{\boldsymbol{g}}_{\mathrm{R}}(t) +$$

$$- \sum_{k=1}^{N}\frac{\partial\psi}{\partial q_k}\dot{q}_k(t) - \frac{1}{\rho_{\mathrm{R}}\Theta}\,\boldsymbol{q}_{\mathrm{R}}\cdot\boldsymbol{g}_{\mathrm{R}} \geq 0\,. \qquad (13.109)$$

This inequality has to be met for all conceivable processes, that means for any choice of the current time derivatives of $\mathbf{E}$, Θ and $\boldsymbol{g}_{\mathrm{R}}$. Since neither the coefficients of $\dot{\mathbf{E}}$, $\dot{\Theta}$ and $\dot{\boldsymbol{g}}_{\mathrm{R}}$ nor the other terms depend on these time derivatives, the inequality might be violated for an appropriate choice of these rates; for this reason the coefficients have to vanish. This is accounted for in (13.103) - (13.105), and the residual inequality (13.106) is left over. Since the first term is independent of $\boldsymbol{g}_{\mathrm{R}}$, (13.107) also applies. It is evident that equations (13.103) - (13.107) (together with an appropriate constitutive equation for the heat flux vector) are sufficient for thermomechanical consistency. ■

A thermoviscoelastic material is described in full by specifying an *energy function* and a constitutive equation for the heat flux vector together with a set of evolution equations for internal variables. Thermomechanical consistency enforces the free energy's independence of the temperature gradient as well as the stress and entropy relations; these are generalisations of the potential relations of thermoelasticity. The residual inequality (13.107) is a restrictive condition for formulating evolution equations, and the more general heat conduction inequality (13.106) a restrictive condition for the heat flux vector. This condition is usually adopted in the stronger formulation

$$\boldsymbol{g}_{\mathrm{R}}\cdot\boldsymbol{q}_{\mathrm{R}}(\mathbf{E}, \Theta, \boldsymbol{g}_{\mathrm{R}}, q_1, \dots, q_N) \leq 0\,, \qquad (13.110)$$

which complies with the heat conduction inequality (13.23) of thermoelasticity. In complete analogy to thermoelasticity, the heat conduction inequality (13.110) leads to the condition

$$\boldsymbol{q}_{\mathrm{R}}(\mathbf{E}, \Theta, \boldsymbol{0}, q_1, \dots, q_N) = \boldsymbol{0} \qquad (13.111)$$

in this case as well: no heat flux occurs without a temperature gradient. A simple conclusion from Theorem 13. 8 is

Theorem 13. 9

For every thermomechanical process in viscoelasticity the generalised *Gibbs* relation

$$\dot{\psi}(\mathbf{E}, \Theta, q_1, \dots, q_N) = \frac{1}{\rho_R}\widetilde{\mathbf{T}}\cdot\dot{\mathbf{E}} - s\dot{\Theta} - \delta \tag{13.112}$$

and the inequality

$$\dot{\psi}(\mathbf{E}, \Theta, q_1, \dots, q_N) \leq \frac{1}{\rho_R}\widetilde{\mathbf{T}}\cdot\dot{\mathbf{E}} - s\dot{\Theta} \tag{13.113}$$

are valid. □

Proof

by calculating the temporal change of free energy, applying the stress and entropy relation and bearing in mind inequality (13.107):

$$\dot{\psi} = \frac{\mathrm{d}}{\mathrm{d}t}\,\psi(\mathbf{E}(t), \Theta(t), q_1(t), \dots, q_N(t)) = \frac{\partial\psi}{\partial\mathbf{E}}\cdot\dot{\mathbf{E}}(t) + \frac{\partial\psi}{\partial\Theta}\dot{\Theta}(t) + \sum_{k=1}^{N}\frac{\partial\psi}{\partial q_k}\dot{q}_k(t)\,,$$

$$\dot{\psi} = \frac{1}{\rho_R}\widetilde{\mathbf{T}}\cdot\dot{\mathbf{E}} - s\dot{\Theta} - \delta \leq \frac{1}{\rho_R}\widetilde{\mathbf{T}}\cdot\dot{\mathbf{E}} - s\dot{\Theta}\,. \quad ■$$

The rate of change of free energy is not larger than the stress power plus the thermal power $-s\dot{\Theta}$.

In viscoelasticity as well, the local energy balance passes into the heat conduction equation, representing an evolution equation for the temperature field.

Theorem 13. 10

For thermomechanical processes in viscoelasticity the local balance of energy turns into the equation of heat conduction,

$$\Theta\dot{s} = -\frac{1}{\rho_R}\mathrm{Div}q_R + r + \delta\,, \tag{13.114}$$

or

$$c_d\dot{\Theta} = \frac{1}{\rho_R}\Theta\frac{\partial\widetilde{\mathbf{T}}}{\partial\Theta}\cdot\dot{\mathbf{E}} - \frac{1}{\rho_R}\mathrm{Div}q_R + r - \sum_{k=1}^{N}\frac{\partial}{\partial q_k}(\psi + \Theta s)\dot{q}_k(t)\,. \tag{13.115}$$

The function

$$c_{\mathrm{d}} = -\,\Theta\frac{\partial^2}{\partial\Theta^2}\,\psi(\mathbf{E},\Theta,q_1,\dots,q_N) \tag{13.116}$$

is the *specific heat at constant deformation.* □

Proof

(13.114) follows on from (13.1) with $\dot{e} = (\psi + \Theta s)^{\cdot} = \dot{\psi} + s\dot{\Theta} + \Theta\dot{s}$ and (13.112). Differentiating the entropy leads to

$$\dot{s}(t) = -\,\frac{\mathrm{d}}{\mathrm{d}t}\frac{\partial\psi}{\partial\Theta}\big(\mathbf{E}(t),\Theta(t),q_1(t),\dots,q_N(t)\big) =$$
$$= -\,\frac{\partial^2\psi}{\partial\Theta^2}\dot{\Theta} - \frac{\partial^2\psi}{\partial\Theta\partial\mathbf{E}}\cdot\dot{\mathbf{E}} - \sum_{k=1}^{N}\frac{\partial^2\psi}{\partial\Theta\partial q_k}\dot{q}_k$$

and

$$s\dot{\Theta} = -\,\Theta\frac{\partial^2\psi}{\partial\Theta^2}\dot{\Theta} - \frac{1}{\rho_{\mathrm{R}}}\Theta\frac{\partial\widetilde{\mathbf{T}}}{\partial\Theta}\cdot\dot{\mathbf{E}} + \sum_{k=1}^{N}\Theta\frac{\partial s}{\partial q_k}\dot{q}_k(t)\,.$$ ■

Equation (13.116) shows that the specific heat c_{d} is not a constant. It generally depends on the thermodynamic process. As opposed to thermoelasticity, the equation of heat conduction (13.115) contains a further term that represents the dissipated energy. The mechanical dissipation δ causes a rise in temperature as a result of internal friction and, by so doing, represents a further contribution towards a thermomechanical coupling in addition to the thermoelastic effect. In the equation of heat conduction a constitutive assumption has to be inserted for the heat flux vector, such as (13.39).

To complete the thermomechanically consistent material description within the framework of viscoelasticity, the remaining inequality (13.107) also has to be satisfied, which amounts to restrictive demands on the evolution equations for the internal variables. For a general realisation of these restrictions, only sufficient conditions are known.

Theorem 13. 11

The following form of evolution equations is sufficient for the dissipation inequality $\delta \geq 0$:

$$\dot{q}_k(t) = -\,\sum_{k=1}^{N}\eta_{kl}\frac{\partial}{\partial q_l}\,\psi\big(\mathbf{E}(t),\Theta(t),q_1(t),\dots,q_N(t)\big)\quad(k = 1,2,\dots N)\,. \tag{13.117}$$

Here, the elements of the $N \times N$-matrix (η_{kl}) are material constants or functions that may depend on the state of stress, on temperature or on any other quantities characterising the thermomechanical process. Even a functional dependence on the process history is permitted, so long as this seems expedient from a physical point of view.[15] The only restriction is that the matrix (η_{kl}) has to be symmetric and positive definite to guarantee thermomechanical consistency:

$$\sum_{k,\,l=1}^{N} \eta_{kl}(\cdot) a_k a_l > 0 \tag{13.118}$$

for all N-vectors $(a_1, \dots, a_N)$. □

Proof

(13.108) and (13.117) lead to the mechanical dissipation

$$\delta = \sum_{k,\,l=1}^{N} \eta_{kl}(\cdot) \frac{\partial \psi}{\partial q_k} \frac{\partial \psi}{\partial q_l} , \tag{13.119}$$

which is strictly positive if the matrix (η_{kl}) is positive definite. ■

If the viscosity matrix (η_{kl}) does not depend on the internal variables q_l and if the $N \times N$-matrix

$$(G_{kl}) = \left(\sum_{l\,=1}^{N} \eta_{kj} \frac{\partial^2 \psi}{\partial q_j \partial q_l} \right) \tag{13.120}$$

is symmetric and positive as well, the theorem justifies the assumption concerning the symmetry and positiveness of the matrix

$$\underset{\approx}{G} = (G_{kl}) \, , \, G_{kl} = - \frac{\partial f_k}{\partial q_l} , \tag{13.121}$$

made in Theorem 10. 9 or (10.95).[16]

The concept described opens up a wide range of possibilities for consistent modelling of rate-dependent thermomechanical material behaviour without equilibrium hysteresis.

---

[15] KRAWIETZ [1986], p. 404; LION [1998].

[16] In the case of the (2N+1)-parameter model (5.93), (5.94) of linear viscoelasticity these conditions are trivially satisfied, since both matrices η_{kl} and $\partial^2\psi/\partial q_k \partial q_l$ are easily verified as being diagonal.

13. 3. 2 Thermoelasticity as a Limit Case of Thermoviscoelasticity

This section is devoted to the examination of the relationship between thermoviscoelasticity and thermoelasticity. In analogy to the mechanical theory of viscoelasticity it becomes apparent that thermoelasticity represents the equilibrium relation attributable to thermoviscoelasticity in terms of Definition 6. 2. One of the first arguments in favour of this is the thermodynamical interpretation of the relaxation property.

Definition 13. 8

If $t \longmapsto \Lambda(t) = \big(\mathbf{E}(t),\, \Theta(t)\big)$ is a strain and temperature process with a spatially constant temperature field $\big(g_{\mathrm{R}} = \mathit{0}\big)$, then the function $\tau \longmapsto \Lambda^t(\tau)$,

$$\Lambda^t(\tau) = \begin{cases} \Lambda(\tau) = \Big(\mathbf{E}(\tau)\,,\, \Theta(\tau)\Big) & \text{for } t_0 \leq \tau \leq t \\ \Lambda(t) \;= \Big(\mathbf{E}(t)\,,\, \Theta(t)\Big) & \text{for } \tau > t \end{cases} \tag{13.122}$$

is called the *isothermal static continuation* of the process history $\Lambda(\cdot)$. ◊

An isothermal static continuation, based on the constitutive equations (13.103) - (13.105) and (13.117), leads to the material response

$$\tau \longmapsto \big(\psi^t(\tau),\, \tilde{\mathbf{T}}^t(\tau),\, s^t(\tau),\, q_1{}^t(\tau), \dots,\, q_N{}^t(\tau)\big) \tag{13.123}$$

$\big(q_{\mathrm{R}}^t(\tau) = \mathit{0}\big)$, which describes a relaxation process. This process approaches a thermodynamic state of equilibrium, determined by the equilibrium solutions (13.100). As a result of the property (13.102) of the evolution equations, the dependent variables - free energy, stress and entropy - converge towards their equilibrium values:

$$\lim_{\tau \to \infty} \psi^t(\tau) = \psi_{\mathrm{eq}}\big(\mathbf{E}(t), \Theta(t)\big) = \psi\big(\mathbf{E}, \Theta, \bar{q}_1(\mathbf{E}, \Theta)\,, \dots,\, \bar{q}_N(\mathbf{E}, \Theta)\big)\,, \tag{13.124}$$

$$\lim_{\tau \to \infty} \tilde{\mathbf{T}}^t(\tau) = \tilde{\mathbf{T}}_{\mathrm{eq}}\big(\mathbf{E}(t), \Theta(t)\big) = \tilde{\mathbf{T}}\big(\mathbf{E}, \Theta, \bar{q}_1(\mathbf{E}, \Theta)\,, \dots,\, \bar{q}_N(\mathbf{E}, \Theta)\big)\,, \tag{13.125}$$

$$\lim_{\tau \to \infty} s^t(\tau) = s_{\mathrm{eq}}\big(\mathbf{E}(t), \Theta(t)\big) = s\big(\mathbf{E}, \Theta, \bar{q}_1(\mathbf{E}, \Theta)\,, \dots,\, \bar{q}_N(\mathbf{E}, \Theta)\big)\,. \tag{13.126}$$

This state of equilibrium is characterised according to

Theorem 13. 12

The *thermodynamic equilibrium*, reached asymptotically in a viscoelastic material during the course of an isothermic static continuation, is characterised by the following relations:

$$\psi^t(\tau) \to \psi_{\mathrm{eq}}\big(\mathbf{E}(t), \Theta(t)\big) = \text{minimum}\ , \tag{13.127}$$

$$\frac{1}{\rho_{\mathrm{R}}}\tilde{\mathbf{T}}_{\mathrm{eq}} = \frac{\partial}{\partial \mathbf{E}}\,\psi_{\mathrm{eq}}\big(\mathbf{E}, \Theta\big)\ , \tag{13.128}$$

$$s_{\mathrm{eq}} = -\frac{\partial}{\partial \Theta}\,\psi_{\mathrm{eq}}\big(\mathbf{E}, \Theta\big)\ . \tag{13.129}$$

□

Proof

The free energy cannot increase during an isothermal static continuation,

$\frac{\mathrm{d}}{\mathrm{d}\tau}\psi^t(\tau) \leq 0$, so the limit value $\psi_{\mathrm{eq}}\big(\mathbf{E}, \Theta\big) = \psi\big(\mathbf{E}, \Theta, \bar{q}_1(\mathbf{E}, \Theta), \dots , \bar{q}_N(\mathbf{E}, \Theta)\big)$

must be a relative minimum. In view of this *minimum property*, the partial derivatives vanish in the equilibrium state, i.e.

$$\frac{\partial \psi}{\partial q_k}(\mathbf{E}, \Theta, q_1, \dots , q_N)\bigg|_{q_k = \bar{q}_k(\mathbf{E}, \Theta)} = 0\ . \tag{13.130}$$

Since the potential relations

$$\frac{1}{\rho_{\mathrm{R}}}\tilde{\mathbf{T}}^t = \frac{\partial}{\partial \mathbf{E}}\,\psi\big(\mathbf{E}, \Theta, q_1{}^t(\tau), \dots , q_N{}^t(\tau)\big)\ , \tag{13.131}$$

$$s^t = -\frac{\partial}{\partial \Theta}\,\psi\big(\mathbf{E}, \Theta, q_1{}^t(\tau), \dots , q_N{}^t(\tau)\big) \tag{13.132}$$

are valid for all τ, the limit $\tau \to \infty$ yields

$$\frac{1}{\rho_{\mathrm{R}}}\tilde{\mathbf{T}}_{\mathrm{eq}} = \frac{\partial \psi}{\partial \mathbf{E}} + \left(\sum_{k=1}^{N} \frac{\partial \psi}{\partial q_k}\bigg|_{q_k = \bar{q}_k}\right)\frac{\partial \bar{q}_k}{\partial \mathbf{E}} = \frac{\partial}{\partial \mathbf{E}}\,\psi\big(\mathbf{E}, \Theta, q_1, \dots , q_N\big)\bigg|_{q_k = \bar{q}_k}$$

and

$$s_{\mathrm{eq}} = -\frac{\partial \psi}{\partial \Theta} - \left(\sum_{k=1}^{N} \frac{\partial \psi}{\partial q_k}\bigg|_{q_k = \bar{q}_k}\right)\frac{\partial \bar{q}_k}{\partial \Theta} = -\frac{\partial}{\partial \Theta}\,\psi\big(\mathbf{E}, \Theta, q_1, \dots , q_N\big)\bigg|_{q_k = \bar{q}_k}\ ,$$

since the partial derivatives with respect to q_k do not make any contribution according to (13.130). The derivatives with respect to $\mathbf{E}$ and Θ are total derivatives in the above relations:

$$\frac{1}{\rho_R}\tilde{\mathbf{T}}_{\text{eq}} = \frac{\partial}{\partial \mathbf{E}}\,\psi(\mathbf{E}, \Theta, \bar{q}_1(\mathbf{E}, \Theta), \dots, \bar{q}_N(\mathbf{E}, \Theta)) = \frac{\partial}{\partial \mathbf{E}}\,\psi_{\text{eq}}(\mathbf{E}, \Theta)\,, \tag{13.133}$$

$$s_{\text{eq}} = -\frac{\partial}{\partial \Theta}\,\psi(\mathbf{E}, \Theta, \bar{q}_1(\mathbf{E}, \Theta), \dots, \bar{q}_N(\mathbf{E}, \Theta)) = -\frac{\partial}{\partial \Theta}\,\psi_{\text{eq}}(\mathbf{E}, \Theta)\,. \tag{13.134}$$

■

The theorem describes the thermodynamic generalisation of the mechanical relaxation property. During the course of an isothermal static continuation, a state of thermodynamic equilibrium is reached. In this way an accompanying equilibrium process is associated to each thermomechanical process:

$$t \longmapsto \left(\mathbf{E}(t), \Theta(t), \psi_{\text{eq}}(\mathbf{E}, \Theta), \tilde{\mathbf{T}}_{\text{eq}} = \rho_R \frac{\partial \psi_{\text{eq}}}{\partial \mathbf{E}}, s_{\text{eq}} = -\frac{\partial \psi_{\text{eq}}}{\partial \Theta}\right). \tag{13.135}$$

It is important that the potential relations of thermoelasticity also apply in the equilibrium relation associated to thermoviscoelasticity.

A thermomechanical process of viscoelasticity takes place within the neighbourhood of the accompanying equilibrium process, which is all the smaller, the slower the given process unfolds. The following theorem is based on the assumption that the asymptotic relation (10.113) can be transferred to thermoviscoelasticity.

Theorem 13. 13

The *retardation* of a thermomechanical process, defined by

$$(\mathbf{E}(t), \Theta(t)) \longmapsto (\mathbf{E}_\alpha(t), \Theta_\alpha(t)) = (\mathbf{E}(\alpha t), \Theta(\alpha t)) \quad (0 < \alpha < 1) \tag{13.136}$$

and the corresponding material response

$$\left(\psi_\alpha, \tilde{\mathbf{T}}_\alpha, s_\alpha, q_{1\alpha}, \dots, q_{N\alpha}\right)(t)\,, \tag{13.137}$$

leads to the following asymptotic relations:

$$\psi_\alpha = \psi_{\text{eq}} + O(\alpha^2)\,, \tag{13.138}$$

$$\widetilde{\mathbf{T}}_\alpha = \widetilde{\mathbf{T}}_{\text{eq}} + \mathrm{O}(\alpha)\,, \tag{13.139}$$

$$s_\alpha = s_{\text{eq}} + \mathrm{O}(\alpha)\,, \tag{13.140}$$

$$\delta_\alpha = \mathrm{O}(\alpha^2)\,, \tag{13.141}$$

$$\dot{\psi}_\alpha = \frac{1}{\rho_{\mathrm{R}}}\widetilde{\mathbf{T}}_{\text{eq}} \cdot \dot{\mathbf{E}}_\alpha - s_{\text{eq}}\dot{\Theta}_\alpha + \mathrm{O}(\alpha^2)\,. \tag{13.142}$$

□

Proof

If the asymptotic relation (10.113), can be transferred to thermoviscoelasticity, we have

$$q_{k\alpha}(t) - \bar{q}_k(\mathbf{E}(t), \Theta(t)) = \mathrm{O}(\alpha)\,. \tag{13.143}$$

A *Taylor* expansion of free energy, bearing (13.130) in mind, reads as

$$\psi = \psi_{\text{eq}}(\mathbf{E}, \Theta) + \sum_{k,l=1}^{N} \frac{\partial^2 \psi}{\partial q_k \partial q_l}(q_k - \bar{q}_k)(q_l - \bar{q}_l) + \cdots\,. \tag{13.144}$$

Given (13.143) this gives rise to $\psi_\alpha = \psi_{\text{eq}}(\mathbf{E}, \Theta) + \mathrm{O}(\alpha^2)$.

Differentiation of the *Taylor* expansion with respect to $\mathbf{E}$ and Θ yields

$$\frac{1}{\rho_{\mathrm{R}}}\widetilde{\mathbf{T}} = \frac{\partial \psi_{\text{eq}}}{\partial \mathbf{E}} - 2\sum_{k,l=1}^{N} \frac{\partial^2 \psi}{\partial q_k \partial q_l}\frac{\partial \bar{q}_k}{\partial \mathbf{E}}(q_l - \bar{q}_l) + \mathrm{O}(\alpha^2) \quad \text{and} \tag{13.145}$$

$$s = -\frac{\partial \psi_{\text{eq}}}{\partial \Theta} + 2\sum_{k,l=1}^{N} \frac{\partial^2 \psi}{\partial q_k \partial q_l}\frac{\partial \bar{q}_k}{\partial \Theta}(q_l - \bar{q}_l) + \mathrm{O}(\alpha^2)\,, \tag{13.146}$$

which implies (13.139) and (13.140). The asymptotic relation (13.141) for δ_α follows from the definition (13.108),

$$\delta = -\sum_{k=1}^{N} \frac{\partial \psi}{\partial q_k} f_k \text{ together with } \frac{\partial \psi_\alpha}{\partial q_k} = \mathrm{O}(\alpha) \text{ and } f_k = \mathrm{O}(\alpha)\,. \tag{13.147}$$

The approximation (13.142) of the generalised *Gibbs* relation follows from the combination of these results. ■

If, in the case of strongly nonlinear evolution equations, it is not possible to start from the asymptotic relation (13.143), we should at least be able to assume

$$q_{k\alpha}(t) - \bar{q}_k(\mathbf{E}(t), \Theta(t)) = \mathrm{o}(1)\,, \tag{13.148}$$

which would lead to the weaker relations

$$\psi_\alpha = \psi_{\mathrm{eq}} + \mathrm{o}(1)\,, \quad \widetilde{\mathbf{T}}_\alpha = \widetilde{\mathbf{T}}_{\mathrm{eq}} + \mathrm{o}(1)\,, \quad s_\alpha = s_{\mathrm{eq}} + \mathrm{o}(1)\,, \quad \delta_\alpha = \mathrm{o}(1) \quad \text{and}$$

$$\dot{\psi}_\alpha = + \frac{1}{\rho_{\mathrm{R}}} \widetilde{\mathbf{T}}_{\mathrm{eq}} \cdot \dot{\mathbf{E}}_\alpha - s_{\mathrm{eq}} \dot{\Theta}_\alpha + \mathrm{o}(1)\,. \tag{13.149}$$

For the mechanical dissipation and *Gibbs* relation we would arrive at

$$\delta_\alpha = \mathrm{o}(\alpha) \quad \text{and} \quad \dot{\psi}_\alpha = + \frac{1}{\rho_{\mathrm{R}}} \widetilde{\mathbf{T}}_{\mathrm{eq}} \cdot \dot{\mathbf{E}}_\alpha - s_{\mathrm{eq}} \dot{\Theta}_\alpha + \mathrm{o}(\alpha)\,, \tag{13.150}$$

if at least $\dot{q}_{k\alpha}(t) = \mathrm{O}(\alpha)$ were valid.[17]

Summarising, we establish the following facts: thermoelasticity is a special case of thermoviscoelasticity. Moreover, thermoelasticity is the equilibrium relation that can be attributed to thermovisoelasticity in the sense of Definition 6. 2. It represents the accompanying equilibrium processes and is, in addition, an asymptotic approximation of thermoviscoelasticity, being arbitrarily close, as long as the thermomechanical processes develop slowly enough.

13. 3. 3 Internal Variables of Strain Type

The theory of internal variables forms a general framework, within which thermomechanically consistent material models can be formulated. To fill this frame with concrete assumptions it is advisable to give the internal variables a definite physical meaning. The essential preparation work has already been dealt with in Chap. 10.

The general model (10.149) - (10.152) of nonlinear viscoelasticity can be transferred to thermomechanics. A motivation for the thermomechanical generalisation is gained from the free energy of a one-dimensional linear

[17] Asymptotic relations of this kind are derived in COLEMAN [1964], eqs. (10.10), (10.16).

rheological model. In the case of isothermal processes, the free energy is nothing other than the potential energy of the elastic elements, which the model contains altogether.

It suffices to demonstrate the transfer for the special case of isothermal processes. For temporally and spatially constant temperature fields $\Theta(\boldsymbol{X}, t) = \Theta_0$ the dissipation inequality merges into the postulate that the rate of change of free energy cannot be greater than the stress power:

$$\tilde{\mathbf{T}} \cdot \dot{\mathbf{E}} - \rho_{\mathrm{R}} \dot{\psi} \geq 0 . \tag{13.151}$$

In fact, the evaluation of this more special dissipation inequality suggests that the problem of thermomechanical consistency is solved, once one has mastered the case of isothermal processes. Generalising the isothermal constitutive description by taking the temperature dependence of all material functions into consideration, leaves the structure of the stress relation and the evolution equations unchanged. The generalisation then gives rise to an entropy relation, which (together with the model of the heat flux vector) guarantees the thermomechanical consistency.[18] In keeping with this line of argumentation we have

Theorem 13. 14

The following constitutive model of isotropic thermoviscoelasticity, formulated for isothermal processes, is consistent from the thermomechanical point of view:

$$\psi = \psi_{\mathrm{eq}}(\mathbf{E}) + \sum_k \psi_{\mathrm{ov}k}(\hat{\boldsymbol{\Gamma}}_{\mathrm{e}k}) , \tag{13.152}$$

$$\tilde{\mathbf{T}}_{\mathrm{eq}} = \rho_{\mathrm{R}} \frac{\mathrm{d}\psi_{\mathrm{eq}}(\mathbf{E})}{\mathrm{d}\mathbf{E}} , \tag{13.153}$$

$$\hat{\mathbf{S}}_{\mathrm{ov}k} = \rho_{\mathrm{R}} \frac{\mathrm{d}\psi_{\mathrm{ov}k}}{\mathrm{d}\hat{\boldsymbol{\Gamma}}_{\mathrm{e}k}} , \tag{13.154}$$

$$\tilde{\mathbf{T}} = \rho_{\mathrm{R}} \frac{\mathrm{d}\psi_{\mathrm{eq}}(\mathbf{E})}{\mathrm{d}\mathbf{E}} + \rho_{\mathrm{R}} \sum_k \mathbf{F}_{\mathrm{v}k}{}^{-1} \frac{\mathrm{d}\psi_{\mathrm{ov}k}}{\mathrm{d}\hat{\boldsymbol{\Gamma}}_{\mathrm{e}k}} \mathbf{F}_{\mathrm{v}k}{}^{\mathrm{T}-1} , \tag{13.155}$$

[18] This simple argument does not apply if the dependence on the temperature history involves complex hysteresis effects, as is the case if phase transitions occur. Cf. HAUPT & HELM [1999].

$$\eta_k \overset{\triangle}{\hat{\Gamma}}_{\mathrm{v}k} = \rho_{\mathrm{R}} \hat{\mathbf{C}}_{\mathrm{e}k} \frac{\mathrm{d}\psi_{\mathrm{ov}k}}{\mathrm{d}\hat{\Gamma}_{\mathrm{e}k}} . \tag{13.156}$$

The energy densities $\psi_{\mathrm{eq}}(\mathbf{E})$ and $\psi_{\mathrm{ov}k}(\hat{\Gamma}_{\mathrm{e}k})$ are isotropic functions of their arguments.
For $k = 1, 2, 3, ...$ the elastic strains $\hat{\Gamma}_{\mathrm{e}k} = \frac{1}{2}(\hat{\mathbf{C}}_{\mathrm{e}k} - \mathbf{1})$ are based on the *Cauchy-Green* tensors

$$\hat{\mathbf{C}}_{\mathrm{e}k} = \hat{\mathbf{F}}_{\mathrm{e}k}{}^{\mathrm{T}} \hat{\mathbf{F}}_{\mathrm{e}k} , \text{ with } \hat{\mathbf{F}}_{\mathrm{e}k} \text{ from } \mathbf{F} = \hat{\mathbf{F}}_{\mathrm{e}k} \mathbf{F}_{\mathrm{v}k} .$$

The viscous strain rates

$$\overset{\triangle}{\hat{\Gamma}}_{\mathrm{v}k} = \dot{\hat{\Gamma}}_{\mathrm{v}k} + \hat{\mathbf{L}}_{\mathrm{v}k}{}^{\mathrm{T}} \hat{\Gamma}_{\mathrm{v}k} + \hat{\Gamma}_{\mathrm{v}k} \hat{\mathbf{L}}_{\mathrm{v}k} = \frac{1}{2}(\hat{\mathbf{L}}_{\mathrm{v}k} + \hat{\mathbf{L}}_{\mathrm{v}k}{}^{\mathrm{T}}) , \tag{13.157}$$

are derived from the inelastic *Almansi* strain

$$\hat{\Gamma}_{\mathrm{v}k} = \frac{1}{2}(\mathbf{1} - \mathbf{F}_{\mathrm{v}k}{}^{\mathrm{T}\,-1} \mathbf{F}^{\mathrm{T}\,-1}{}_{\mathrm{v}k}) \tag{13.158}$$

with $\hat{\mathbf{L}}_{\mathrm{v}k} = \dot{\mathbf{F}}_{\mathrm{v}k} \mathbf{F}_{\mathrm{v}k}{}^{-1}$.

The material functions $\psi_{\mathrm{eq}}(\cdot)$ and $\psi_{\mathrm{ov}k}(\cdot)$ are positive definite and the viscosities η_k positive.[19] □

Proof

First of all, we calculate the stress power

$$\widetilde{\mathbf{T}} \cdot \dot{\mathbf{E}} = (\widetilde{\mathbf{T}}_{\mathrm{eq}} + \widetilde{\mathbf{T}}_{\mathrm{ov}}) \cdot \dot{\mathbf{E}} = \widetilde{\mathbf{T}}_{\mathrm{eq}} \cdot \dot{\mathbf{E}} + \sum_k \hat{\mathbf{S}}_{\mathrm{ov}k} \cdot (\overset{\triangle}{\hat{\Gamma}}_{\mathrm{e}k} + \overset{\triangle}{\hat{\Gamma}}_{\mathrm{v}k}) ,$$

and insert the *Oldroyd* rates $\overset{\triangle}{\hat{\Gamma}}_{\mathrm{e}k} = \dot{\hat{\Gamma}}_{\mathrm{e}k} + \hat{\mathbf{L}}_{\mathrm{v}k}{}^{\mathrm{T}} \hat{\Gamma}_{\mathrm{e}k} + \hat{\Gamma}_{\mathrm{e}k} \hat{\mathbf{L}}_{\mathrm{v}k}$ to get

$$\widetilde{\mathbf{T}} \cdot \dot{\mathbf{E}} = \widetilde{\mathbf{T}}_{\mathrm{eq}} \cdot \dot{\mathbf{E}} + \sum_k \hat{\mathbf{S}}_{\mathrm{ov}k} \cdot \dot{\hat{\Gamma}}_{\mathrm{e}k} + \sum_k \hat{\mathbf{S}}_{\mathrm{ov}k} \cdot (2\hat{\Gamma}_{\mathrm{e}k} \hat{\mathbf{L}}_{\mathrm{v}k} + \overset{\triangle}{\hat{\Gamma}}_{\mathrm{v}k}) .$$

Combined with the change of free energy,

$$\dot{\psi}(t) = \frac{\mathrm{d}\psi_{\mathrm{eq}}(\mathbf{E})}{\mathrm{d}\mathbf{E}} \cdot \dot{\mathbf{E}} + \sum_k \frac{\mathrm{d}\psi_{\mathrm{ov}k}}{\mathrm{d}\hat{\Gamma}_{\mathrm{e}k}} \cdot \dot{\hat{\Gamma}}_{\mathrm{e}k} ,$$

[19] A constitutive theory of this kind is developed and applied in LION [1996, 1997a, 1997b]. Cf. REESE & GOVINDJEE [1998a, 1998b].

the dissipation inequality (13.151) then leads to the requirement

$$\Big(\widetilde{\mathbf{T}}_{\mathrm{eq}} - \rho_{\mathrm{R}}\frac{\mathrm{d}\psi_{\mathrm{eq}}(\mathbf{E})}{\mathrm{d}\mathbf{E}}\Big)\cdot\dot{\mathbf{E}} + \sum_k\Big(\hat{\mathbf{S}}_{\mathrm{ov}k} - \rho_{\mathrm{R}}\frac{\mathrm{d}\psi_{\mathrm{ov}k}}{\mathrm{d}\hat{\mathbf{\Gamma}}_{ek}}\Big)\cdot\dot{\hat{\mathbf{\Gamma}}}_{ek}$$

$$+ \sum_k \hat{\mathbf{S}}_{\mathrm{ov}k}\cdot\big(2\hat{\mathbf{\Gamma}}_{ek}\hat{\mathbf{L}}_{\mathrm{v}k} + \overset{\triangle}{\hat{\mathbf{\Gamma}}}_{\mathrm{v}k}\big) \geq 0\,. \tag{13.159}$$

The stress relations

$$\widetilde{\mathbf{T}}_{\mathrm{eq}} = \rho_{\mathrm{R}}\frac{\mathrm{d}\psi_{\mathrm{eq}}(\mathbf{E})}{\mathrm{d}\mathbf{E}}\,, \tag{13.160}$$

$$\hat{\mathbf{S}}_{\mathrm{ov}k} = \rho_{\mathrm{R}}\frac{\mathrm{d}\psi_{\mathrm{ov}k}}{\mathrm{d}\hat{\mathbf{\Gamma}}_{ek}} \tag{13.161}$$

are necessary and sufficient for its validity, together with the residual inequality

$$\sum_k\Big(\hat{\mathbf{C}}_{ek}\frac{\mathrm{d}\psi_{\mathrm{ov}k}}{\mathrm{d}\hat{\mathbf{\Gamma}}_{ek}}\Big)\cdot\overset{\triangle}{\hat{\mathbf{\Gamma}}}_{\mathrm{v}k} \geq 0 \tag{13.162}$$

($\hat{\mathbf{C}}_{ek} = 2\hat{\mathbf{\Gamma}}_{ek} + \mathbf{1}$). In this relation, the isotropy of the material functions $\psi_{\mathrm{ov}k}(\cdot)$ has been observed, leading to $\hat{\mathbf{S}}_{\mathrm{ov}k}\hat{\mathbf{\Gamma}}_{ek} = \hat{\mathbf{\Gamma}}_{ek}\hat{\mathbf{S}}_{\mathrm{ov}k}$ and finally to

$$\hat{\mathbf{S}}_{\mathrm{ov}k}\cdot\big(2\hat{\mathbf{\Gamma}}_{ek}\hat{\mathbf{L}}_{\mathrm{v}k} + \overset{\triangle}{\hat{\mathbf{\Gamma}}}_{\mathrm{v}k}\big) = \Big((2\hat{\mathbf{\Gamma}}_{ek} + \mathbf{1})\hat{\mathbf{S}}_{\mathrm{ov}k}\Big)\cdot\overset{\triangle}{\hat{\mathbf{\Gamma}}}_{\mathrm{v}k}\,. \tag{13.163}$$

The residual inequality (13.162) is identically satisfied with evolution equations of the form (13.156). The stress relation (13.155) is the thermomechanical generalisation of (10.152). The definiteness of the material functions $\psi_{\mathrm{ov}k}(\cdot)$ complies with the minimum property (13.127) of the free energy. ■

The free energy according to (13.152) is an isothermal strain energy, whose equilibrium part depends on the total deformation, whereas the viscoelastic parts $\psi_{\mathrm{ov}k}$ are positive functions of the elastic strain tensors $\hat{\mathbf{\Gamma}}_{ek}$ operating on the individual intermediate configurations, defined by $\mathbf{F}_{\mathrm{v}k}$ ($k = 1, 2, \dots$). The evolution equations (13.156) have clearly undergone a change in comparison with (10.150): thermomechanical consistency suggests the factor

$\hat{\mathbf{C}}_{ek}$, modifying the stress tensor $\hat{\mathbf{S}}_{ovk}$. This modification only ceases to apply for small elastic strains $\hat{\mathbf{\Gamma}}_{ek}$, which cannot as a general rule be anticipated. It should be noted, however, that the evolution equations (10.150) are thermodynamically consistent as well. Since the tensors $\hat{\mathbf{C}}_{ek}$ are symmetric and positive definite, based on (10.150) and (13.162) we have

$$\hat{\mathbf{C}}_{ek}\frac{\mathrm{d}\psi_{ovk}}{\mathrm{d}\hat{\mathbf{\Gamma}}_{ek}}\cdot\frac{\mathrm{d}\psi_{ovk}}{\mathrm{d}\hat{\mathbf{\Gamma}}_{ek}} = \left(\hat{\mathbf{U}}_{ek}\frac{\mathrm{d}\psi_{ovk}}{\mathrm{d}\hat{\mathbf{\Gamma}}_{ek}}\right)\cdot\left(\hat{\mathbf{U}}_{ek}\frac{\mathrm{d}\psi_{ovk}}{\mathrm{d}\hat{\mathbf{\Gamma}}_{ek}}\right) > 0 \quad \left(\hat{\mathbf{U}}_{ek}{}^2 = \hat{\mathbf{C}}_{ek}\right). \tag{13.164}$$

13. 3. 4 Incorporation of Anisotropic Elasticity Properties

The outcome of the previous section is such that anisotropic elasticity properties can be taken into consideration. In this context, the equilibrium stress, or ψ_{eq} in (13.152), can quite feasibly be an anisotropic function of the strain tensor. Attention is drawn to Chap. 9 for an appropriate representation of $\psi_{\mathrm{eq}}(\mathbf{E})$. Without being obliged to alter the structure of the evolution equations (13.156), it would, to a certain degree, be possible to incorporate anisotropic properties in the energy functions ψ_{ovk}, which would have to be isotropic functions of $\hat{\mathbf{\Gamma}}_{ek}$ and appropriate structural tensors. One example could be the strain energy according to (9.134) or (9.139), expressing a dependence of the elastic properties on a preferential direction or on several material directions. To apply these representations along the same lines, the strain tensor $\mathbf{E}$ has to be replaced by $\hat{\mathbf{\Gamma}}_{ek}$.

13. 3. 5 Incompressible Materials: An Extension of the Mooney-Rivlin Model to Thermoviscoelasticity

If a kinematic condition of internal constraint is assumed, the constitutive modeling modifies, because a part of the state of stress remains undetermined. In particular, the assumption of incompressibility implies that the constitutive equations determine the stress only up to a hydrostatic pressure.[20] This reduces the amount of information required for a description of material properties. For example, in the case of incompressible isotropic elasticity only two elasticity constants are

[20] Cf. Sect. 7. 4. 2.

necessary if a constitutive model is designed to represent all second-order effects.[21] This point of view led to the well-known *Mooney-Rivlin model*.[22]

In this section a generalisation of the Mooney-Rivlin model to thermo-viscoelasticity is outlined.[23] For simplicity, the discussion is restricted to isothermal processes, however, there is no difficulty to incorporate varying temperature fields within the common thermodynamic frame being applied throughout this book. Mathematically, the thermodynamic extension of the Mooney-Rivlin model rests on the statement of

Theorem 13. 15

If $\mathbf{A}$ is a symmetric positive definite second-order tensor with $\det\mathbf{A} = 1$, the first two basic invariants are bounded from below:[24]

$$\mathrm{I}_{\mathbf{A}} \geq 3 \tag{13.165}$$

$$\mathrm{II}_{\mathbf{A}} \geq 3 \tag{13.166}$$

□

Proof

Since $\mathbf{A}$ is positive symmetric and unimodular ($\det\mathbf{A} = 1$), its real and positive eigenvalues (λ_1, λ_2, λ_3) satisfy the condition $\lambda_1\lambda_2\lambda_3 = 1$, and we have for the first two invariants $\mathrm{I}_{\mathbf{A}}$ and $\mathrm{II}_{\mathbf{A}}$

$$\mathrm{I}_{\mathbf{A}} = \lambda_1 + \lambda_2 + \lambda_3 = \lambda_1 + \lambda_2 + \frac{1}{\lambda_1\lambda_2}$$

and

$$\mathrm{II}_{\mathbf{A}} = \lambda_1\lambda_2 + \lambda_2\lambda_3 + \lambda_3\lambda_1 = \lambda_1\lambda_2 + \frac{1}{\lambda_1} + \frac{1}{\lambda_2}\,.$$

In order to prove the theorem, it suffices to consider the function

$$(x, y) \longmapsto f(x, y) = x + y + \frac{1}{xy}$$

in the domain $(0, \infty) \times (0, \infty)$. The first-order derivatives

$$f_x = 1 - \frac{1}{x^2y} \quad \text{and} \quad f_y = 1 - \frac{1}{xy^2}$$

[21] Cf. Sect. 9. 2, Theorem 9. 12.

[22] Eq. (9.83).

[23] The presentation follows HAUPT & LION [2001a].

[24] Cf. (9.91).

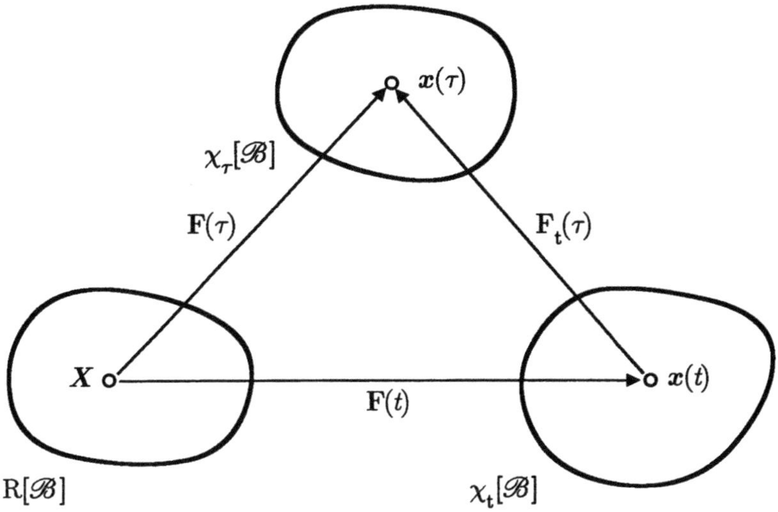

Figure 13. 1: Relative deformation gradient

indicate that there is an extremum at $(x, y) = (1, 1)$ with $f(1, 1) = 3$. This is a minimum because the matrix of the second-order derivatives is positive definite. ■

The following considerations start from the relative deformation gradient

$$\mathbf{F}_t(\tau) = \mathbf{F}(\tau)\mathbf{F}^{-1}(t) , \tag{13.167}$$

which operates on the current configuration (see Fig. 13. 1) and transforms the present material line element or surface element to the past configuration at time τ:

$$\mathrm{d}\boldsymbol{x}(\tau) = \mathbf{F}_t(\tau)\mathrm{d}\boldsymbol{x}(t) \tag{13.168}$$

$$\boldsymbol{n}(\tau) = \mathbf{F}_t^{\mathrm{T}-1}(\tau)\boldsymbol{n}(t) \tag{13.169}$$

The relative deformation gradient motivates the definition of the relative right *Cauchy-Green* tensor

$$\mathbf{C}_t(\tau) = \mathbf{F}_t^{\mathrm{T}}(\tau)\mathbf{F}_t(\tau) = \mathbf{F}^{\mathrm{T}-1}(t)\mathbf{C}(\tau)\mathbf{F}^{-1}(t) , \tag{13.170}$$

the relative *Green* strain[25]

$$\mathbf{E}_t(\tau) = \frac{1}{2}\big(\mathbf{C}_t(\tau) - \mathbf{1}\big) = \mathbf{F}^{\mathrm{T}-1}(t)\Big(\mathbf{E}(\tau) - \mathbf{E}(t)\Big)\mathbf{F}^{-1}(t) \tag{13.171}$$

and the relative *Piola* strain[26]

$$\mathbf{e}_t(\tau) = \frac{1}{2}\big(\mathbf{C}_t^{-1}(\tau) - \mathbf{1}\big) = \mathbf{F}(t)\Big(\mathbf{e}(\tau) - \mathbf{e}(t)\Big)\mathbf{F}^{\mathrm{T}}(t)\,. \tag{13.172}$$

These tensors have the property to represent the relative strain history in the sense of

$$\frac{1}{2}\Big(\mathrm{d}\boldsymbol{x}(\tau)\cdot\mathrm{d}\boldsymbol{y}(\tau) - \mathrm{d}\boldsymbol{x}(t)\cdot\mathrm{d}\boldsymbol{y}(t)\Big) = \mathrm{d}\boldsymbol{x}(t)\cdot\mathbf{E}_t(\tau)\mathrm{d}\boldsymbol{y}(t) \tag{13.173}$$

and

$$\frac{1}{2}\Big(\boldsymbol{n}(\tau)\cdot\boldsymbol{m}(\tau) - \boldsymbol{n}(t)\cdot\boldsymbol{m}(t)\Big) = \boldsymbol{n}(t)\cdot\mathbf{e}_t(\tau)\boldsymbol{m}(t)\,. \tag{13.174}$$

If a material body is assumed to be incompressible, the state of stress is

$$\mathbf{S} = -\,p\mathbf{1} + \mathbf{S}_{\mathrm{E}}\,, \tag{13.175}$$

where $p = p(\boldsymbol{x}, t)$ is not related to the process history by means of a constitutive equation; the constitutively determined stress $\mathbf{S}_{\mathrm{E}}$, the so-called extra-stress, decomposes into an equilibrium stress and an overstress:

$$\mathbf{S}_{\mathrm{E}} = \mathbf{S}_{\mathrm{eq}} + \mathbf{S}_{\mathrm{ov}}\,. \tag{13.176}$$

In terms of the second *Piola-Kirchhoff* stress $\widetilde{\mathbf{T}} = \mathbf{F}^{-1}\mathbf{S}\,\mathbf{F}^{\mathrm{T}-1}$ we have

$$\widetilde{\mathbf{T}} = -\,p\mathbf{C}^{-1} + \widetilde{\mathbf{T}}_{\mathrm{E}} = -\,p\mathbf{C}^{-1} + \widetilde{\mathbf{T}}_{\mathrm{eq}} + \widetilde{\mathbf{T}}_{\mathrm{ov}}\,, \tag{13.177}$$

and with the convective stress $\widetilde{\mathbf{t}} = \mathbf{F}^{\mathrm{T}}\mathbf{S}\,\mathbf{F}$ the general constitutive equation reads

$$\widetilde{\mathbf{t}} = -\,p\mathbf{C} + \widetilde{\mathbf{t}}_{\mathrm{E}} = -\,p\mathbf{C} + \widetilde{\mathbf{t}}_{\mathrm{eq}} + \widetilde{\mathbf{t}}_{\mathrm{ov}}\,. \tag{13.178}$$

---

[25] The relative *Green* strain has already been applied in the reduced form (7.80).
[26] See eq. (1.78) for the definition of the *Piola* strain tensor.

With these notations in mind, we state

Theorem 13. 16
The free energy functional

$$\rho_{\mathrm{R}}\psi(t) = -\int_{-\infty}^{t}\Big[\mu_{\mathrm{A}}(t-\tau)\frac{\mathrm{d}}{\mathrm{d}\tau}\operatorname{tr}\big(\mathbf{e}_{\mathrm{t}}(\tau)\big) + \mu_{\mathrm{B}}(t-\tau)\frac{\mathrm{d}}{\mathrm{d}\tau}\operatorname{tr}\big(\mathbf{E}_{\mathrm{t}}(\tau)\big)\Big]\mathrm{d}\tau \qquad (13.179)$$

or, equivalently,

$$\rho_{\mathrm{R}}\psi(t) = \mu_{\mathrm{A}}(\infty)\operatorname{tr}\mathbf{E}(t) + \mu_{\mathrm{B}}(\infty)\operatorname{tr}\mathbf{e}(t) - \int_{0}^{\infty}\Big[\mu_{\mathrm{A}}{}'(s)\operatorname{tr}\big(\mathbf{e}_{\mathrm{t}}(t-s)\big) + \mu_{\mathrm{B}}{}'(s)\operatorname{tr}\big(\mathbf{E}_{\mathrm{t}}(t-s)\big)\Big]\mathrm{d}s \qquad (13.180)$$

defines an isotropic incompressible viscoelastic material if the relaxation functions have the following properties:

$$\mu_{\mathrm{A,B}}(s) > 0\,, \qquad (13.181)$$

$$\frac{\mathrm{d}}{\mathrm{d}s}\mu_{\mathrm{A,B}}(s) \leq 0\,, \qquad (13.182)$$

$$\lim_{s\to\infty}\frac{\mathrm{d}}{\mathrm{d}s}\mu_{\mathrm{A,B}}(s) = 0\,, \qquad (13.183)$$

$$\frac{\mathrm{d}^2}{\mathrm{d}s^2}\mu_{\mathrm{A,B}}(s) \geq 0\,. \qquad (13.184)$$

The free energy (13.180) implies the stress relation

$$\mathbf{S}_{\mathrm{E}} = \mu_{\mathrm{A}}(\infty)\mathbf{B} - \mu_{\mathrm{B}}(\infty)\mathbf{B}^{-1} - \int_{0}^{\infty}2\Big[\mu_{\mathrm{A}}{}'(s)\,\mathbf{e}_{\mathrm{t}}(t-s) - \mu_{\mathrm{B}}{}'(s)\,\mathbf{E}_{\mathrm{t}}(t-s)\Big]\mathrm{d}s \qquad (13.185)$$

and the positive internal dissipation

$$\rho_{\mathrm{R}}\delta = \int_{0}^{\infty}\Big[\mu_{\mathrm{A}}{}''(s)\operatorname{tr}\big(\mathbf{e}_{\mathrm{t}}(t-s)\big) + \mu_{\mathrm{B}}{}''(s)\operatorname{tr}\big(\mathbf{E}_{\mathrm{t}}(t-s)\big)\Big]\mathrm{d}s\,. \qquad (13.186)$$

□

Proof

First, the equivalence of (13.180) and (13.179) will be verified. The relative strain tensors vanish at the present instant: $\mathbf{e}_t(t) = \mathbf{E}_t(t) = \mathbf{0}$. Assuming an undeformed state for $\tau \to \infty$, i.e., $\mathbf{F}(-\infty) = \mathbf{1}$, we conclude

$$\mathbf{e}_t(-\infty) = -\tfrac{1}{2}[\mathbf{1} - \mathbf{B}(t)] = -\mathbf{a}(t)\,, \tag{13.187}$$

$$\mathbf{E}_t(-\infty) = -\tfrac{1}{2}[\mathbf{1} - \mathbf{B}^{-1}(t)] = -\mathbf{A}(t) \tag{13.188}$$

and

$$\begin{aligned}\rho_R\psi(t) &= -\mu_A(t-\tau)\,\mathrm{tr}\,\mathbf{e}_t(\tau)\Big|_{-\infty}^{t} - \mu_B(t-\tau)\,\mathrm{tr}\,\mathbf{E}_t(\tau)\Big|_{-\infty}^{t} \\ &\quad + \int_{-\infty}^{t}\Big\{\frac{\partial}{\partial\tau}\mu_A(t-\tau)\,\mathrm{tr}\,\mathbf{e}_t(\tau) + \frac{\partial}{\partial\tau}\mu_B(t-\tau)\,\mathrm{tr}\,\mathbf{E}_t(\tau)\Big\}\mathrm{d}\tau \\ &= -\mu_A(\infty)\,\mathrm{tr}\,\mathbf{a}(t) - \mu_B(\infty)\,\mathrm{tr}\,\mathbf{A}(t) \\ &\quad - \int_0^{\infty}\Big[\mu_A{}'(s)\,\mathrm{tr}\big(\mathbf{e}_t(t-s)\big) + \mu_B{}'(s)\,\mathrm{tr}\big(\mathbf{E}_t(t-s)\big)\Big]\mathrm{d}s\,.\end{aligned}$$

This agrees with (13.180) if we observe $\mathrm{tr}\,\mathbf{a} = -\,\mathrm{tr}\,\mathbf{E}$ and $\mathrm{tr}\,\mathbf{A} = -\,\mathrm{tr}\,\mathbf{e}$. In order to apply the dissipation inequality (13.151) for isothermal processes, slightly modified for incompressibility,[27]

$$\widetilde{\mathbf{T}}_E \cdot \dot{\mathbf{E}} - \rho_R\dot{\psi} \geq 0\,, \tag{13.189}$$

the free energy (13.180) must be differentiated with respect to time. An intermediate calculation supplies

$$\frac{\mathrm{d}}{\mathrm{d}t}\,\mathrm{tr}\,\mathbf{e}_t(t-s) = 2[\mathbf{e}(t-s) - \mathbf{e}(t)]\cdot\dot{\mathbf{E}}(t) - \frac{\mathrm{d}}{\mathrm{d}s}\,\mathrm{tr}\,\mathbf{e}_t(t-s)\,, \tag{13.190}$$

$$\frac{\mathrm{d}}{\mathrm{d}t}\,\mathrm{tr}\,\mathbf{E}_t(t-s) = 2[\mathbf{E}(t-s) - \mathbf{E}(t)]\cdot\dot{\mathbf{e}}(t) - \frac{\mathrm{d}}{\mathrm{d}s}\,\mathrm{tr}\,\mathbf{E}_t(t-s) \tag{13.191}$$

and the relation

[27] The pressure p does not appear in the dissipation inequality, because it gives no contribution to the stress power.

$$\dot{\mathbf{e}}(t) = -\,\mathbf{C}^{-1}\dot{\mathbf{E}}(t)\mathbf{C}^{-1} \tag{13.192}$$

between the *Green* and *Piola* strain rates. This leads to

$$\rho_R\dot{\psi}(t) = \Big\{\mu_A(\infty)\mathbf{1} - \mu_B(\infty)\mathbf{C}^{-2} - \int_0^\infty 2\mu_A{}'(s)\left[\mathbf{e}(t-s) - \mathbf{e}(t)\right]\mathrm{d}s$$

$$+\,\mathbf{C}^{-1}(t)\int_0^\infty 2\mu_B{}'(s)\left[\mathbf{E}(t-s) - \mathbf{E}(t)\right]\mathrm{d}s\,\mathbf{C}^{-1}(t)\Big\}\cdot\dot{\mathbf{E}}(t)$$

$$+\int_0^\infty \Big\{\mu_A{}'(s)\frac{\mathrm{d}}{\mathrm{d}s}\,\mathrm{tr}\left(\mathbf{e}_t(t-s)\right) + \mu_B{}'(s)\frac{\mathrm{d}}{\mathrm{d}s}\,\mathrm{tr}\left(\mathbf{E}_t(t-s)\right)\Big\}\mathrm{d}s\,. \tag{13.193}$$

Inserting $\dot{\psi}$ into the inequality (13.189) yields according to the standard argumentation the stress relation

$$\widetilde{\mathbf{T}}_E = \mu_A(\infty)\mathbf{1} - \mu_B(\infty)\mathbf{C}^{-2} - \int_0^\infty \Big\{2\mu_A{}'(s)\left[\mathbf{e}(t-s) - \mathbf{e}(t)\right]$$

$$-\,2\mu_B{}'(s)\mathbf{C}^{-1}(t)[\mathbf{E}(t-s) - \mathbf{E}(t)]\mathbf{C}^{-1}(t)\Big\}\mathrm{d}s \tag{13.194}$$

and the internal dissipation

$$\rho_R\delta = -\int_0^\infty \Big\{\mu_A{}'(s)\frac{\mathrm{d}}{\mathrm{d}s}\,\mathrm{tr}\left(\mathbf{e}_t(t-s)\right) + \mu_B{}'(s)\frac{\mathrm{d}}{\mathrm{d}s}\,\mathrm{tr}\left(\mathbf{E}_t(t-s)\right)\Big\}\mathrm{d}s$$

$$= \int_0^\infty \Big\{\mu_A{}''(s)\,\mathrm{tr}\left(\mathbf{e}_t(t-s)\right) + \mu_B{}''(s)\,\mathrm{tr}\left(\mathbf{E}_t(t-s)\right)\Big\}\mathrm{d}s\,. \tag{13.195}$$

The stress relation (13.194) transforms with $\mathbf{S}_E = \mathbf{F}\,\widetilde{\mathbf{T}}_E\mathbf{F}^T$ to

$$\mathbf{S}_E = \mu_A(\infty)\mathbf{B} - \mu_B(\infty)\mathbf{B}^{-1} - \int_0^\infty \Big\{2\mu_A{}'(s)\mathbf{F}(t)[\mathbf{e}(t-s) - \mathbf{e}(t)]\mathbf{F}^T(t)$$

$$-\,2\mu_B{}'(s)\mathbf{F}^{T-1}(t)[\mathbf{E}(t-s) - \mathbf{E}(t)]\mathbf{F}^{-1}(t)\Big\}\mathrm{d}s\,, \tag{13.196}$$

which coincides with (13.185) if (13.171) and (13.172) are taken into account. Due to Theorem 13. 15, the equilibrium value of the free energy (13.180) is a relative minimum if the first derivatives of the relaxation functions are negative. Furthermore, the dissipation (13.186) is positive if the second-order derivatives of $\mu_{A,B}(\cdot)$ are positive. ■

According to the above theorem the constitutive model defined by means of the free energy (13.179) or (13.180) is thermodynamically consistent if the positive-valued relaxation functions have a negative slope and a positive curvature.

An advantage of this theory of viscoelasticity is the fact that no decompositions of the deformation are needed and that analytical methods, known from the classical theory of linear viscoelasticity, can be applied.[28]

Since the equilibrium part of the stress relation (13.185) corresponds to the *Mooney-Rivlin* model (9.83), the present theory can be understood to be one possible generalisation to viscoelasticity. If a more general behaviour of the equilibrium stress is to be modeled, the limit values $\mu_{A,B}(\infty)$ of the relaxation functions may be set to zero, and the equilibrium part of the free energy function, $\rho_R\psi_{eq} = \mu_A(\infty)\,\mathrm{tr}\mathbf{E}(t) + \mu_B(\infty)\,\mathrm{tr}\mathbf{e}(t)$, may be replaced by the general strain energy function $\psi_{eq}(\mathbf{B}) = \varphi(\mathrm{I}_{\mathbf{B}}, \mathrm{II}_{\mathbf{B}})$. The equilibrium stress is then the general stress relation of isotropic hyperelasticity,

$$\mathbf{S}_{eq} = 2\rho_R \mathbf{B}\frac{d\psi_{eq}(\mathbf{B})}{d\mathbf{B}}\,. \tag{13.197}$$

For completeness it should be mentioned that the above theory includes as special cases two versions of the classical *Maxwell* model[29] formulated in the context of finite strains. This provides

Theorem 13. 17

In view of a thermodynamically consistent formulation of the classical *Maxwell* model two possibilities exist, namely

$$\text{(Model A)}\quad \overset{\triangledown}{\mathbf{S}}_{ov} + \frac{1}{z_A}\mathbf{S}_{ov} = 2\mu_A\mathbf{D} \tag{13.198}$$

and

[28] Sect. 5. 5 and the literature cited there.

[29] See eq. (5.104) and Fig. 5. 5.

$$\text{(Model B)} \quad \overset{\triangle}{\mathbf{S}}_{\text{ov}} + \frac{1}{z_{\text{B}}}\mathbf{S}_{\text{ov}} = 2\mu_{\text{B}}\mathbf{D}\,. \tag{13.199}$$

In the case of (Model A) we have the free energy

$$\rho_{\text{R}}\psi_{\text{A}}(t) = \int_0^\infty \frac{\mu_{\text{A}}}{z_{\text{A}}}\,\mathrm{e}^{-\frac{s}{z_{\text{A}}}}\,\mathrm{tr}\left(\mathbf{e}_t(t-s)\right)\mathrm{d}s \tag{13.200}$$

and the stress relation

$$\mathbf{S}_{\text{ov}}(t) = \int_0^\infty 2\frac{\mu_{\text{A}}}{z_{\text{A}}}\,\mathrm{e}^{-\frac{s}{z_{\text{A}}}}\,\mathbf{e}_t(t-s)\mathrm{d}s\,. \tag{13.201}$$

(Model B) originates from the free energy

$$\rho_{\text{R}}\psi_{\text{B}}(t) = \int_0^\infty \frac{\mu_{\text{B}}}{z_{\text{B}}}\,\mathrm{e}^{-\frac{s}{z_{\text{B}}}}\,\mathrm{tr}\left(\mathbf{E}_t(t-s)\right)\mathrm{d}s\,, \tag{13.202}$$

implying the stress relation

$$\mathbf{S}_{\text{ov}}(t) = -\int_0^\infty 2\frac{\mu_{\text{B}}}{z_{\text{B}}}\,\mathrm{e}^{-\frac{s}{z_{\text{B}}}}\,\mathbf{E}_t(t-s)\mathrm{d}s\,. \tag{13.203}$$

The internal dissipation, related to models A) and B) is

$$\rho_{\text{R}}\delta_{\text{A}} = \int_0^\infty \frac{\mu_{\text{A}}}{z_{\text{A}}^2}\,\mathrm{e}^{-\frac{s}{z_{\text{A}}}}\,\mathrm{tr}\left(\mathbf{e}_t(t-s)\right)\mathrm{d}s\,, \tag{13.204}$$

$$\rho_{\text{R}}\delta_{\text{B}} = \int_0^\infty \frac{\mu_{\text{B}}}{z_{\text{B}}^2}\,\mathrm{e}^{-\frac{s}{z_{\text{B}}}}\,\mathrm{tr}\left(\mathbf{E}_t(t-s)\right)\mathrm{d}s\,. \tag{13.205}$$

□

Proof

The covariant *Oldroyd* rate of the *Cauchy* stress tensor has the property (8.70). Accordingly, (13.198) implies

$$\mathbf{F}^{-1}\overset{\triangledown}{\mathbf{S}}_{\text{ov}}\mathbf{F}^{\text{T}-1} + \frac{1}{z_{\text{A}}}\mathbf{F}^{-1}\mathbf{S}_{\text{ov}}\mathbf{F}^{\text{T}-1} = 2\mu_{\text{A}}\mathbf{F}^{-1}\mathbf{D}\mathbf{F}^{\text{T}-1}\,,$$

$$\dot{\tilde{\mathbf{T}}}_{\text{ov}} + \frac{1}{z_{\text{A}}}\tilde{\mathbf{T}}_{\text{ov}} = -\,2\mu_{\text{A}}\dot{\mathbf{e}}\,, \tag{13.206}$$

where (1.131) has been applied on the right-hand side. This differential equation is integrated to

$$\widetilde{\mathbf{T}}_{\text{ov}}(t) = -\int_{-\infty}^{t} 2\mu_{\text{A}}\, \mathrm{e}^{-\frac{t-\tau}{z_{\text{A}}}}\, \mathrm{e}'(\tau)\mathrm{d}\tau \tag{13.207}$$

$$= \int_{0}^{\infty} 2\frac{\mu_{\text{A}}}{z_{\text{A}}}\, \mathrm{e}^{-\frac{s}{z_{\text{A}}}} \left[\mathrm{e}(t-s) - \mathrm{e}(t)\right]\mathrm{d}s\,, \tag{13.208}$$

which corresponds to (13.201) in view of $\mathbf{S}_{\text{ov}} = \mathbf{F}\widetilde{\mathbf{T}}_{\text{ov}}\mathbf{F}^{\text{T}}$ and (13.172). (Model B) is treated in the same way: (13.199) implies

$$\mathbf{F}^{\text{T}}\overset{\triangle}{\mathbf{S}}_{\text{ov}}\mathbf{F} + \frac{1}{z_{\text{B}}}\mathbf{F}^{\text{T}}\mathbf{S}_{\text{ov}}\mathbf{F} = 2\mu_{\text{B}}\mathbf{F}^{\text{T}}\mathbf{D}\mathbf{F}\,,$$

$$\dot{\widetilde{\mathbf{t}}}_{\text{ov}} + \frac{1}{z_{\text{B}}}\widetilde{\mathbf{t}}_{\text{ov}} = 2\mu_{\text{B}}\dot{\mathbf{E}}\,, \tag{13.209}$$

where (8.72) and (1.130) are taken into account. Integration provides

$$\widetilde{\mathbf{t}}_{\text{ov}} = \int_{-\infty}^{t} 2\mu_{\text{B}}\, \mathrm{e}^{-\frac{t-\tau}{z_{\text{B}}}}\, \mathbf{E}'(\tau)\mathrm{d}\tau \tag{13.210}$$

$$= -\int_{0}^{\infty} 2\frac{\mu_{\text{B}}}{z_{\text{B}}}\, \mathrm{e}^{-\frac{s}{z_{\text{B}}}} \left[\mathbf{E}(t-s) - \mathbf{E}(t)\right]\mathrm{d}s\,, \tag{13.211}$$

which is equivalent to (13.203) in view of $\mathbf{S} = \mathbf{F}^{\text{T}-1}\widetilde{\mathbf{t}}_{\text{ov}}\mathbf{F}^{-1}$ and (13.171). The expressions (13.204) and (13.205) are evident. ■

It is possible to consider several *Maxwell* elements in parallel. According to the theorem this provides concrete representations for the relaxation functions in the sense of discrete or continuous relaxation spectra, i.e. the method, which was applied in the linear theory of viscoelasticity, applies in this theory as well.[30]

13. 4 Thermoviscoplasticity with Elastic Domain

13. 4. 1 Uniaxial Viscoplasticity

The term viscoplasticity denotes the theory of rate-dependent material behaviour with equilibrium hysteresis. For metallic materials this hysteresis

[30] See LION & HAUPT [2001].

is a consequence of inelastic deformations, which result from the motion of dislocations. A well-known micromechanical mechanism is the idea of a *Frank-Read* source:[31] During an elastic shear deformation, a dislocation line deforms under the influence of an external load. As the load increases, the stable state of equilibrium becomes unstable. The instability initiates a dynamic process: the dislocation propagates through the material until the next stable state of equilibrium is reached. This repeats periodically and leads to macroscopic inelastic deformations as a sequence of equilibrium states with changing stability properties.

This idea can be put into mathematical terms.[32] In order to do this, a process of isothermal simple shear is considered, $t \longmapsto \gamma(t)$, and a free energy depending on the shear strain γ and one internal variable κ,

$$\psi = \varphi(\gamma, \kappa)\,. \tag{13.212}$$

The dissipation inequality, $\tau\dot{\gamma}(t) - \rho\dot{\psi}(t) \geq 0$, i.e.,

$$\left[\tau - \rho\frac{\partial\varphi}{\partial\gamma}\right]\dot{\gamma}(t) - \rho\frac{\partial\varphi}{\partial\kappa}\,\dot{\kappa}(t) > 0\,, \tag{13.213}$$

leads to a stress relation

$$\tau = \rho\frac{\partial\varphi}{\partial\gamma}\,(\gamma, \kappa) \tag{13.214}$$

and an evolution equation for the internal variable:

$$\dot{\kappa}(t) = -\frac{\rho}{\eta}\frac{\partial\varphi}{\partial\kappa}(\gamma, \kappa)\,. \tag{13.215}$$

The derivative of the strain energy with respect to κ vanishes in the case of an equilibrium state,

$$\frac{\partial\varphi}{\partial\kappa}\,(\gamma, \bar{\kappa}) = 0\,. \tag{13.216}$$

An equilibrium state $(\gamma, \bar{\kappa})$ is stable if the second derivative is positive,

$$\frac{\partial^2\varphi}{\partial\kappa^2}\,(\gamma, \bar{\kappa}) > 0\,; \tag{13.217}$$

this corresponds to a local minimum of the free energy function. The qualitative behaviour of a *Frank-Read* source can be modelled on the basis of the particular free energy function

[31] Cf. PONTER, BATAILLE & KESTIN [1979].
[32] See GROLIG [1985].

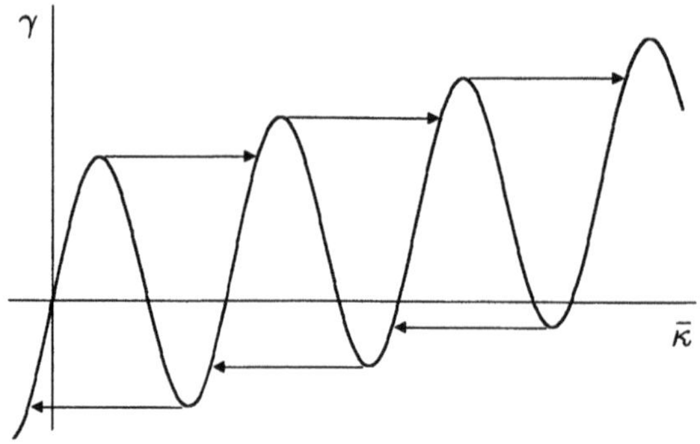

Figure 13. 2: Equilibrium states related to the evolution equation (13.220). Points, with positive slope of the curve are stable states of equilibrium, the others are unstable. The arrows symbolise dynamic processes, taking place whenever the stability of the equilibrium is lost.

$$\rho\psi = \frac{G}{2}(\gamma - \kappa)^2 + \frac{G_t}{2}\kappa^2 + \frac{GA}{k}(1 - \cos k\kappa)\,. \tag{13.218}$$

It contains a periodic (non-convex) term, in order to be qualified to respresent instability properties. The wave number k and the amplitude A may be related to the microstructure of the material. The other material constants are of macroscopic nature, G and G_t being a shear and tangent modulus. η is a viscosity. The free energy (13.218) implies the stress relation

$$\tau = G(\gamma - \kappa) \tag{13.219}$$

and the evolution equation

$$\dot{\kappa}(t) = \frac{G}{\eta}\left[\gamma - \left(1 + \frac{G_t}{G}\right)\kappa - A\sin k\kappa\right]. \tag{13.220}$$

Equilibrium states $(\gamma, \bar{\kappa})$ are equilibrium solutions of the evolution equation (13.220). The relation

$$\gamma = \left(1 + \frac{G_t}{G}\right)\bar{\kappa} + A\sin k\bar{\kappa} \tag{13.221}$$

represents the set of equilibrium states $(\gamma, \bar{\kappa})$. In this special case more than one value of $\bar{\kappa}$ is possible if a shear strain γ is prescribed, since (13.221) has more than one solution $\bar{\kappa}$. An equilibrium state $(\gamma, \bar{\kappa})$ is stable if

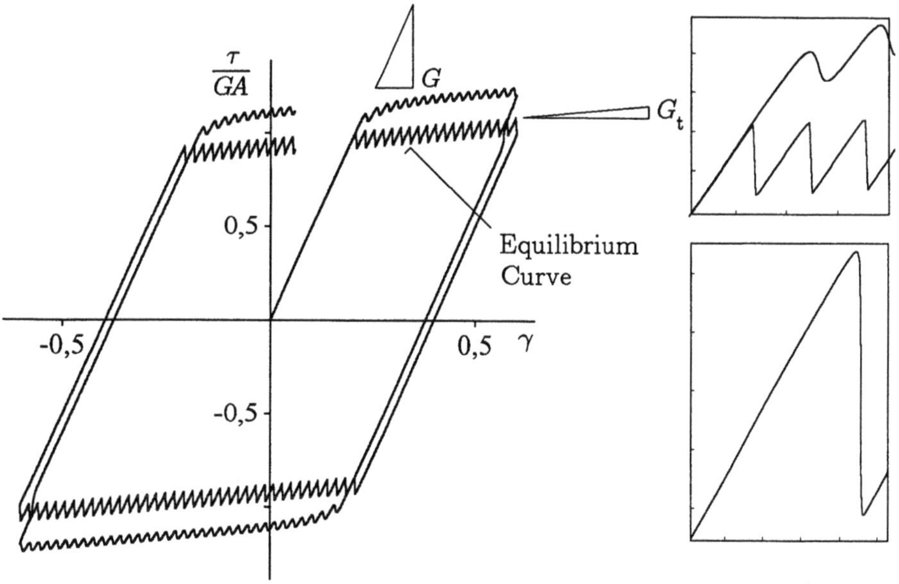

Figure 13. 3: Stress-strain curves according to (13.219), (13.220). Points with positive slope correspond to stable states of equilibrium.

$$\left(1 + \frac{G_t}{G}\right) + Ak \cos k\bar{\kappa} > 0\,, \tag{13.222}$$

and unstable otherwise. Fig. 13. 2 displays the set of equilibrium states in a $\bar{\kappa}\gamma$-plane. The points with positive slope of the curve are stable states of equilibrium, the others are unstable states.

If a shear process starting from zero is now considered, an unstable state is reached, followed by a dynamic process, whereby the internal variable κ approaches the next stable state of equilibrium. These events repeat during loading and similarly during unloading, thus leading to an equilibrium hysteresis. The hystereris in the state space affects the hysteresis of the stress-strain curve, which is obtained from a numerical integration of the evolution law (13.220). The response to a cyclic strain process is depicted in Fig. 13. 3. Numerically, the equilibrium curve belongs to a very small strain rate. Inserting finite strain rates, the rate-dependence of the model's response becomes apparent. Looking at the details of the equilibrium curve, only the points with positive slope represent states of equilibrium, whereas the other points belong to the dynamic process, leading from one stable state of equilibrium to the next.

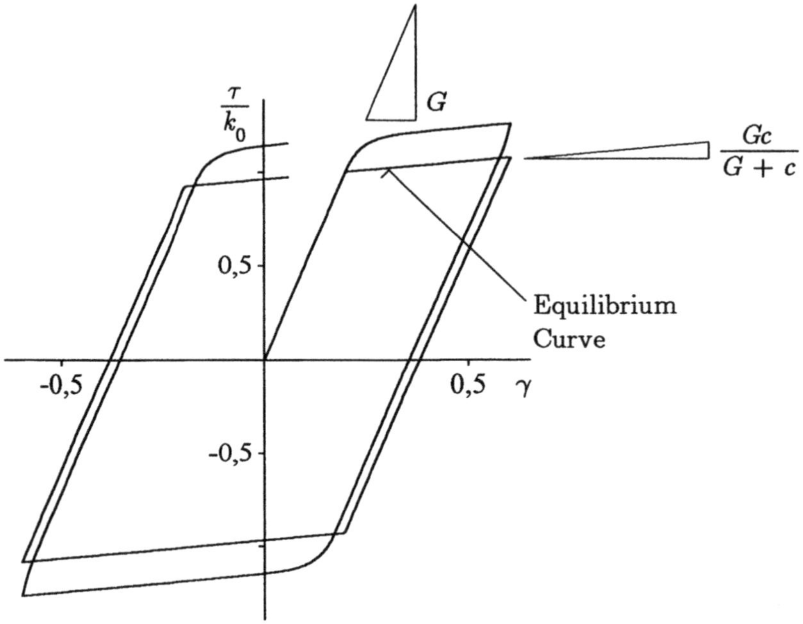

Figure 13. 4: Stress relation of classical viscoplasticity according to (13.225), (13.228) and (13.229) with b = 0

Usually, dynamic processes taking place on the microlevel, are completely ignored in phenomenological theories of material behaviour. Indeed, if the wave number k in (13.218) is increased, the amplitude of the oscillation becomes smaller and one is inclined to replace the oscillating curve by a smooth line.

This motivates the classical model of viscoplasticity,[33] which has already been discussed in Chapter 5. Its thermomechanical formulation is based on the definition of a free energy as a quadratic function of the elastic shear strain and one strain-like internal variable y for the incorporation of kinematic hardening:

$$\rho\psi = \frac{G}{2}(\gamma - \kappa)^2 + \frac{c}{2}(\kappa - y)^2 . \tag{13.223}$$

From the dissipation inequality $\tau\dot{\gamma}(t) - \rho\dot{\psi}(t) \geq 0$, i.e.

$$\Big[\tau - G(\gamma - \kappa)\Big]\dot{\gamma}(t) + \Big[G(\gamma - \kappa) - c(\kappa - y)\Big]\dot{\kappa}(t) + c(\kappa - y)\dot{y} > 0 \tag{13.224}$$

[33] Sect. 5. 8.

we infer a linear stress relation,

$$\tau = G(\gamma - \kappa)\,, \tag{13.225}$$

and the definition of a backstress:

$$\xi = c(\kappa - y) \quad \Longleftrightarrow \quad y = \kappa - \frac{1}{c}\,\xi\,. \tag{13.226}$$

The residual inequality

$$(\tau - \xi)\dot{\kappa}(t) + \xi\Big(\dot{\kappa} - \frac{1}{c}\,\dot{\xi}\Big) > 0 \tag{13.227}$$

suggests the formulation of evolution equations. Most famous and obvious are the proposals of Perzyna,

$$\dot{\kappa}(t) = \frac{c}{\eta}\left\langle\frac{1}{r_0}\left(|\tau - \xi| - k_0\right)\right\rangle^{\mathrm{m}}\frac{\tau - \xi}{|\tau - \xi|}\,, \tag{13.228}$$

and Armstrong-Frederick,

$$\dot{\xi}(t) = c\dot{\kappa} - b|\dot{\kappa}|\xi\,, \tag{13.229}$$

however, even other possibilities exist and may be convenient for practical use.

If the two evolution equations (13.228) and (13.229) are integrated, we obtain a stress-strain curve, shown in Fig. 13. 4.[34] Here, $b = 0$ has been set in order to compare the result with the foregoing model. As a consequence of $b = 0$, corresponding to linear kinematic hardening, the equilibrium curve consists of straight lines. The main difference to the previous stress strain relation is that the equilibrium curve represents a continuous sequence of stable equilibrium states.

These arguments may be summarised and interpreted as follows: on the microlevel there are dynamical processes leading to inelastic deformations and equilibrium hysteresis. On the macroscopic level the dynamic processes are ignored and replaced by their result, namely a continuous sequence of equilibrium states. These equilibrium states are locally stable and they depend on the history of the strain process.

[34] Cf. Theorem 5. 16 and Fig. 5. 19.

13. 4. 2 General Concept

The general constitutive model (12.2) - (12.6) of viscoplasticity can easily be adapted to thermomechanics. This leads to a special theory of thermo-viscoplasticity, which is a thermomechanically consistent generalisation of classical viscoplasticity (5.219) - (5.223). Moreover, the generalised theory suggests a thermodynamical interpretation of rate-independent plasticity: the equilibrium relation that is attributed to the constitutive model of thermoviscoplasticity in terms of Definition 6. 2, is the theory of rate-independent thermoplasticity. Thermoplasticity thus proves itself to be a kind of equilibrium thermodynamics, in which the dependent variables are rate-independent functionals of the process history.

Definition 13. 9

The following constitutive equations define a material model of *thermo-viscoplasticity with elastic domain*:

Decomposition of the Deformation

$$\mathbf{E} = \mathbf{E}_{\mathrm{e}} + \mathbf{E}_{\mathrm{i}} \tag{13.230}$$

Free Energy

$$\psi = \psi(\mathbf{E}_{\mathrm{e}}, \Theta, \mathbf{E}_{\mathrm{i}}, q_1, \dots, q_N) \tag{13.231}$$

Heat Flux Vector

$$q_{\mathrm{R}} = q_{\mathrm{R}}(\mathbf{E}_{\mathrm{e}}, \Theta, g_{\mathrm{R}}, \mathbf{E}_{\mathrm{i}}, q_1, \dots, q_N) \tag{13.232}$$

Yield Function

$$f = f(\widetilde{\mathbf{T}}, \Theta, \mathbf{E}_{\mathrm{i}}, q_1, \dots, q_N) \tag{13.233}$$

Flow Rule

$$\dot{\mathbf{E}}_{\mathrm{i}}(t) = \frac{1}{\eta}\langle f\rangle^{\mathrm{m}}\mathrm{g}(\widetilde{\mathbf{T}}, \Theta, \mathbf{E}_{\mathrm{i}}, q_1, \dots, q_N) \tag{13.234}$$

Evolution Equations (Hardening Models)

$$\dot{q}_k(t) = \frac{1}{\eta}\langle f\rangle^{\mathrm{m}} f_k(\mathbf{E}_{\mathrm{e}}(t), \Theta(t), \mathbf{E}_{\mathrm{i}}(t), q_1(t), \dots, q_N(t)) \,, \quad k = 1, \dots, N \tag{13.235}$$

McCauley Bracket: $\langle x \rangle = \begin{cases} x \text{ for } x \geq 0 \\ 0 \text{ for } x < 0 \end{cases}$

In the yield function (13.233) and the flow rule (13.234) the argument $\widetilde{\mathbf{T}}$ can also be replaced by $\mathbf{E}_e$, applying the stress relation (13.236). In the same way the elastic strain $\mathbf{E}_e$ in the evolution equations may be replaced by the stress $\widetilde{\mathbf{T}}$. ◊

The constitutive equations for stress and entropy then result from the requirement of thermomechanical consistency.

Theorem 13. 18

The following relations are necessary and sufficient for the validity of the dissipation inequality:

Stress Relation

$$\widetilde{\mathbf{T}} = \rho_R \frac{\partial \psi}{\partial \mathbf{E}_e} = \mathbf{f}(\mathbf{E}_e, \Theta, \mathbf{E}_i, q_1, \dots, q_N) \tag{13.236}$$

Entropy Relation

$$s = -\frac{\partial \psi}{\partial \Theta} = s(\mathbf{E}_e, \Theta, \mathbf{E}_i, q_1, \dots, q_N) \tag{13.237}$$

Dissipation Inequality

$$\Big(\widetilde{\mathbf{T}} - \rho_R \frac{\partial \psi}{\partial \mathbf{E}_i}\Big) \cdot \dot{\mathbf{E}}_i(t) - \Big(\sum_k \rho_R \frac{\partial \psi}{\partial q_k} \dot{q}_k(t)\Big) - \frac{1}{\Theta} q_R \cdot \mathrm{Grad}\Theta \geq 0 \tag{13.238}$$

This inequality leads to

$$\rho_R \delta = \Big(\widetilde{\mathbf{T}} - \rho_R \frac{\partial \psi}{\partial \mathbf{E}_i}\Big) \cdot \dot{\mathbf{E}}_i(t) - \Big(\sum_k \rho_R \frac{\partial \psi}{\partial q_k} \dot{q}_k(t)\Big) \geq 0\,. \tag{13.239}$$

□

Proof

Inserting the free energy into the dissipation,

$$\rho_R \Delta = \widetilde{\mathbf{T}} \cdot (\dot{\mathbf{E}}_e + \dot{\mathbf{E}}_i) - \rho_R \dot{\psi} - \rho_R s \dot{\Theta} - \frac{1}{\Theta} q_R \cdot g_R\,,$$

leads to

$$\Big(\widetilde{\mathbf{T}} - \rho_{\mathrm{R}}\frac{\partial\psi}{\partial\mathbf{E}_{\mathrm{e}}}\Big)\cdot\dot{\mathbf{E}}_{\mathrm{e}}(t) - \rho_{\mathrm{R}}\Big(s + \frac{\partial\psi}{\partial\Theta}\Big)\dot{\Theta}(t) +$$

$$+ \Big(\widetilde{\mathbf{T}} - \rho_{\mathrm{R}}\frac{\partial\psi}{\partial\mathbf{E}_{\mathrm{i}}}\Big)\cdot\dot{\mathbf{E}}_{\mathrm{i}}(t) - \Big(\sum_k \rho_{\mathrm{R}}\frac{\partial\psi}{\partial q_k}\dot{q}_k(t)\Big) - \frac{1}{\Theta}q_{\mathrm{R}}\cdot g_{\mathrm{R}} \geq 0\,.$$

Using the standard argumentation applied on several previous occasions, this gives rise to the theorem's statements. ■

The restrictive conditions for the evolution equations are easy to realise in analogy to (13.117).

Theorem 13. 19
The evolution equations

$$\dot{\mathbf{E}}_{\mathrm{i}}(t) = \frac{1}{\eta}\langle f\rangle^{\mathrm{m}}\Big(\widetilde{\mathbf{T}} - \rho_{\mathrm{R}}\frac{\partial\psi}{\partial\mathbf{E}_{\mathrm{i}}}\Big)\,, \tag{13.240}$$

$$\dot{q}_k(t) = -\frac{1}{\eta}\langle f\rangle^{\mathrm{m}}\frac{\partial\psi}{\partial q_k} \tag{13.241}$$

are sufficient for the validity of the dissipation inequality $\delta \geq 0$. □

Proof

$$\rho_{\mathrm{R}}\delta = \frac{1}{\eta}\langle f\rangle^{\mathrm{m}}\Big\{\mathrm{tr}\Big(\widetilde{\mathbf{T}} - \rho_{\mathrm{R}}\frac{\partial\psi}{\partial\mathbf{E}_{\mathrm{i}}}\Big)^2 + \rho_{\mathrm{R}}\sum_k\Big(\frac{\partial\psi}{\partial q_k}\Big)^2\Big\}\,. \tag{13.242}$$

■

In analogy to Theorem 13. 10 and equation (13.115) the local energy balance gives rise to the equation of heat conduction.

Theorem 13. 20
For the constitutive model (13.231) - (13.237) of thermoviscoplasticity the equation of heat conduction reads

$$\begin{aligned} c_{\mathrm{d}}\dot{\Theta} &= \frac{1}{\rho_{\mathrm{R}}}\,\Theta\frac{\partial\widetilde{\mathbf{T}}}{\partial\mathbf{E}_{\mathrm{e}}}\cdot\dot{\mathbf{E}}_{\mathrm{e}} - \frac{1}{\rho_{\mathrm{R}}}\mathrm{Div}q_{\mathrm{R}} + r + \\ &+ \Big(\frac{1}{\rho_{\mathrm{R}}}\widetilde{\mathbf{T}} - \frac{\partial(\psi + \Theta s)}{\partial\mathbf{E}_{\mathrm{i}}}\Big)\cdot\dot{\mathbf{E}}_{\mathrm{i}} - \sum_k\frac{\partial(\psi + \Theta s)}{\partial q_k}\dot{q}_k\,. \end{aligned} \tag{13.243}$$

The material function

$$c_{\mathrm{d}} = -\,\Theta \frac{\partial^2}{\partial \Theta^2}\psi(\mathbf{E}_{\mathrm{e}}, \Theta, \mathbf{E}_{\mathrm{i}}, q_1, \dots, q_N) \tag{13.244}$$

is the *specific heat at constant deformation.* □

Proof

The energy balance (13.1) is written in terms of the free energy and the additive decomposition of the total strain $\mathbf{E}$:

$$(\psi + \Theta s)^{\cdot} = -\frac{1}{\rho_{\mathrm{R}}}\mathrm{Div}\boldsymbol{q}_{\mathrm{R}} + r + \frac{1}{\rho_{\mathrm{R}}}\widetilde{\mathbf{T}}\cdot(\dot{\mathbf{E}}_{\mathrm{e}} + \dot{\mathbf{E}}_{\mathrm{i}})\,.$$

With the potential relations (13.236), (13.237) this leads to

$$\Theta\dot{s} = -\frac{1}{\rho_{\mathrm{R}}}\mathrm{Div}\boldsymbol{q}_{\mathrm{R}} + r + \Big(\frac{1}{\rho_{\mathrm{R}}}\widetilde{\mathbf{T}} - \frac{\partial\psi}{\partial\mathbf{E}_{\mathrm{i}}}\Big)\cdot\dot{\mathbf{E}}_{\mathrm{i}} - \sum_k \frac{\partial\psi}{\partial q_k}\dot{q}_k\,.$$

For the left-hand side we calculate

$$\dot{s}(t) = -\frac{\mathrm{d}}{\mathrm{d}t}\frac{\partial\psi}{\partial\Theta} = -\frac{\partial^2\psi}{\partial\Theta^2}\dot{\Theta} - \frac{\partial^2\psi}{\partial\Theta\partial\mathbf{E}_{\mathrm{e}}}\cdot\dot{\mathbf{E}}_{\mathrm{e}} - \frac{\partial^2\psi}{\partial\Theta\partial\mathbf{E}_{\mathrm{i}}}\cdot\dot{\mathbf{E}}_{\mathrm{i}} - \sum_k \frac{\partial^2\psi}{\partial\Theta\partial q_k}\dot{q}_k$$

and

$$\Theta\dot{s} = c_{\mathrm{d}}\dot{\Theta} - \frac{1}{\rho_{\mathrm{R}}}\Theta\frac{\partial\widetilde{\mathbf{T}}}{\partial\Theta}\cdot\dot{\mathbf{E}}_{\mathrm{e}} + \Theta\frac{\partial s}{\partial\mathbf{E}_{\mathrm{i}}}\cdot\dot{\mathbf{E}}_{\mathrm{i}} + \Theta\sum_k\frac{\partial s}{\partial q_k}\dot{q}_k\,.$$ ■

On the right-hand side of the heat conduction equation (13.243) several contributions, besides the thermoelastic coupling term and the local heat supply, can be distinguished. One of them is the *inelastic stress power*

$$p_{\mathrm{i}} = \frac{1}{\rho_{\mathrm{R}}}\widetilde{\mathbf{T}}\cdot\dot{\mathbf{E}}_{\mathrm{i}}\,. \tag{13.245}$$

The other terms can be related to the hardening properties: the internal variables $q_1, \dots, q_N$ together with the backstress tensor

$$\widetilde{\mathbf{X}} = \rho_{\mathrm{R}}\frac{\partial\psi}{\partial\mathbf{E}_{\mathrm{i}}} \tag{13.246}$$

constitute a macroscopic image of the dynamic processes taking place in the material and altering the microscopic structure. Accordingly, the power

$$p_{\mathrm{s}} = \frac{\partial(\psi + \Theta s)}{\partial\mathbf{E}_{\mathrm{i}}}\cdot\dot{\mathbf{E}}_{\mathrm{i}} + \sum_k\frac{\partial(\psi + \Theta s)}{\partial q_k}\dot{q}_k \tag{13.247}$$

is the rate of change of the internal energy due to structural changes within the material. Given the abbreviations (13.245) and (13.247), the equation of heat conduction takes the form

$$c_{\mathrm{d}}\dot{\Theta} = \frac{1}{\rho_{\mathrm{R}}}\Theta\frac{\partial\widetilde{\mathbf{T}}}{\partial\Theta}\cdot\dot{\mathbf{E}}_{\mathrm{e}} - \frac{1}{\rho_{\mathrm{R}}}\mathrm{Div}\,q_{\mathrm{R}} + r + p_{\mathrm{i}} - p_{\mathrm{s}}\,. \tag{13.248}$$

It states that only one part of the inelastic stress power causes a rise in temperature as a result of dissipation. The other part influences the microstructure of the material and is stored as internal energy.[35]

13. 4. 3 Application of the Intermediate Configuration

A physically transparent formulation of finite thermoviscoplasticity with an elastic domain can be achieved by introducing a stress-free intermediate configuration along the same lines as the procedure described in Sect. 12. 2. 2. This procedure can be transferred to thermomechanics and results in a concrete formulation of the general concept.

Theorem 13. 21

The following constitutive equations define a thermomechanically consistent constitutive model of *thermoviscoplasticity with thermoelastic domain*:[36]

Decomposition of the Deformation

$$\hat{\mathbf{\Gamma}} = \hat{\mathbf{\Gamma}}_{\mathrm{e}} + \hat{\mathbf{\Gamma}}_{\mathrm{i}} \qquad \overset{\triangle}{\hat{\mathbf{\Gamma}}} = \overset{\triangle}{\hat{\mathbf{\Gamma}}}_{\mathrm{e}} + \overset{\triangle}{\hat{\mathbf{\Gamma}}}_{\mathrm{i}} \tag{13.249}$$

Free Energy[37]

$$\rho_{\mathrm{R}}\psi = \rho_{\mathrm{R}}\varphi(\hat{\mathbf{\Gamma}}_{\mathrm{e}}, \Theta) + \frac{1}{2}\sum_{k} c_k \,\mathrm{tr}\left((\hat{\mathbf{\Gamma}}_{\mathrm{i}} - \hat{\mathbf{Y}}_k)^2\right) \tag{13.250}$$

Stress and Entropy Relation

$$\hat{\mathbf{S}} = \rho_{\mathrm{R}}\frac{\partial\varphi}{\partial\hat{\mathbf{\Gamma}}_{\mathrm{e}}} \qquad s = -\frac{\partial\psi}{\partial\Theta} \tag{13.251}$$

---

[35] Cf. LEHMANN [1984]; KAMLAH [1994], pp. 106; HAUPT et al. [1997]; HELM [1998]; KAMLAH & HAUPT [1998].

[36] Cf. TSAKMAKIS [1996]; KAMLAH & TSAKMAKIS [1999].

[37] The index $(\,)_k$ denotes the number of the internal variable, whereas the subscripts $(\,)_{\mathrm{e}}$, $(\,)_{\mathrm{i}}$ refer to elastic and inelastic strains.

Yield Function

$$f(\hat{\mathbf{P}}, \hat{\mathbf{P}}_{\hat{\mathbf{X}}}, k) = \sqrt{\tfrac{3}{2}}\,\|\hat{\mathbf{P}} - \hat{\mathbf{P}}_{\hat{\mathbf{X}}}\| - k(\Theta) \tag{13.252}$$

Flow Rule

$$\overset{\triangle}{\hat{\boldsymbol{\Gamma}}}_{\mathrm{i}} = \frac{1}{\eta}\langle f\rangle^{\mathrm{m}}\sqrt{\frac{3}{2}}\,\frac{\hat{\mathbf{P}} - \hat{\mathbf{P}}_{\hat{\mathbf{X}}}}{\|\hat{\mathbf{P}} - \hat{\mathbf{P}}_{\hat{\mathbf{X}}}\|} \tag{13.253}$$

Kinematic Hardening

$$\overset{\triangledown}{\hat{\mathbf{X}}}_k = c_k\overset{\triangle}{\hat{\boldsymbol{\Gamma}}}_{\mathrm{i}} - b_k\sqrt{\tfrac{2}{3}}\,\|\overset{\triangle}{\hat{\boldsymbol{\Gamma}}}_{\mathrm{i}}\|\hat{\mathbf{X}}_k \tag{13.254}$$

The material parameters c_k are assumed to be constants.[38] The parameters η, m and b_k may depend on the temperature and other process variables. The following definitions apply throughout:[39]

$$\hat{\mathbf{P}} = \left(\mathbf{1} + 2\hat{\boldsymbol{\Gamma}}_{\mathrm{e}}\right)\hat{\mathbf{S}} = \hat{\mathbf{C}}_{\mathrm{e}}\hat{\mathbf{S}}\,, \tag{13.255}$$

$$\hat{\mathbf{X}}_k = c_k(\hat{\boldsymbol{\Gamma}}_{\mathrm{i}} - \hat{\mathbf{Y}}_k)\,, \tag{13.256}$$

$$\hat{\mathbf{P}}_{\hat{\mathbf{X}}} = \sum_k \hat{\mathbf{X}}_k\left(\mathbf{1} + \tfrac{2}{c_k}\hat{\mathbf{X}}_k\right). \tag{13.257}$$

□

Proof

First, the stress power is calculated: given $\overset{\triangle}{\hat{\boldsymbol{\Gamma}}}_{\mathrm{e}} = \dot{\hat{\boldsymbol{\Gamma}}}_{\mathrm{e}} + \hat{\mathbf{L}}_{\mathrm{i}}{}^{\mathrm{T}}\hat{\boldsymbol{\Gamma}}_{\mathrm{e}} + \hat{\boldsymbol{\Gamma}}_{\mathrm{e}}\hat{\mathbf{L}}_{\mathrm{i}}$, we obtain

$$\tilde{\mathbf{T}}\cdot\dot{\mathbf{E}} = \hat{\mathbf{S}}\cdot\overset{\triangle}{\hat{\boldsymbol{\Gamma}}} = \hat{\mathbf{S}}\cdot(\overset{\triangle}{\hat{\boldsymbol{\Gamma}}}_{\mathrm{e}} + \overset{\triangle}{\hat{\boldsymbol{\Gamma}}}_{\mathrm{i}}) = \hat{\mathbf{S}}\cdot\dot{\hat{\boldsymbol{\Gamma}}}_{\mathrm{e}} + \hat{\mathbf{S}}\cdot(2\hat{\boldsymbol{\Gamma}}_{\mathrm{e}}\hat{\mathbf{L}}_{\mathrm{i}} + \overset{\triangle}{\hat{\boldsymbol{\Gamma}}}_{\mathrm{i}})\,,\ \text{i.e.}$$

$$\tilde{\mathbf{T}}\cdot\dot{\mathbf{E}} = \hat{\mathbf{S}}\cdot\dot{\hat{\boldsymbol{\Gamma}}}_{\mathrm{e}} + [(\mathbf{1} + 2\hat{\boldsymbol{\Gamma}}_{\mathrm{e}})\hat{\mathbf{S}}]\cdot\hat{\mathbf{L}}_{\mathrm{i}}\,. \tag{13.258}$$

In the case of isotropic elasticity we have $\hat{\boldsymbol{\Gamma}}_{\mathrm{e}}\hat{\mathbf{S}} = \hat{\mathbf{S}}\hat{\boldsymbol{\Gamma}}_{\mathrm{e}}$, and with

$\hat{\mathbf{C}}_{\mathrm{e}} = \mathbf{1} + 2\hat{\boldsymbol{\Gamma}}_{\mathrm{e}}$ and $\overset{\triangle}{\hat{\boldsymbol{\Gamma}}}_{\mathrm{i}} = \frac{1}{2}(\hat{\mathbf{L}}_{\mathrm{i}} + \hat{\mathbf{L}}_{\mathrm{i}}{}^{\mathrm{T}})$ the stress power merges into

[38] A temperature-dependence of the c_k would modify the entropy relation and the evolution equations (13.254).

[39] The tensor $\hat{\mathbf{P}}$ is called *Mandel* stress tensor, following MANDEL [1971], p. 127. Cf. TSAKMAKIS [1996], eqs. (5.12), (5.19).

$$\tilde{\mathbf{T}}\cdot\dot{\mathbf{E}} = \hat{\mathbf{S}}\cdot\dot{\hat{\mathbf{\Gamma}}}_{\mathrm{e}} + (\hat{\mathbf{C}}_{\mathrm{e}}\hat{\mathbf{S}})\cdot\overset{\triangle}{\hat{\mathbf{\Gamma}}}_{\mathrm{i}}\,. \tag{13.259}$$

Accordingly, the dissipation inequality reads as

$$\frac{1}{\rho_{\mathrm{R}}}\hat{\mathbf{S}}\cdot\dot{\hat{\mathbf{\Gamma}}}_{\mathrm{e}} + \frac{1}{\rho_{\mathrm{R}}}(\hat{\mathbf{C}}_{\mathrm{e}}\hat{\mathbf{S}})\cdot\overset{\triangle}{\hat{\mathbf{\Gamma}}}_{\mathrm{i}} - \dot{\psi} - s\dot{\Theta} - \frac{1}{\rho_{\mathrm{R}}\Theta}\,q_{\mathrm{R}}\cdot g_{\mathrm{R}} \geq 0\,. \tag{13.260}$$

In the time derivative of the free energy,

$$\rho_{\mathrm{R}}\dot{\psi} = \rho_{\mathrm{R}}\frac{\partial\varphi}{\partial\hat{\mathbf{\Gamma}}_{\mathrm{e}}}\cdot\dot{\hat{\mathbf{\Gamma}}}_{\mathrm{e}} + \rho_{\mathrm{R}}\frac{\partial\varphi}{\partial\Theta}\dot{\Theta} + \sum_k c_k(\hat{\mathbf{\Gamma}}_{\mathrm{i}} - \hat{\mathbf{Y}}_k)\cdot(\dot{\hat{\mathbf{\Gamma}}}_{\mathrm{i}} - \dot{\hat{\mathbf{Y}}}_k)\,,$$

the individual terms of the sum can be rearranged,

$$\begin{aligned}
&(\hat{\mathbf{\Gamma}}_{\mathrm{i}} - \hat{\mathbf{Y}}_k)\cdot(\dot{\hat{\mathbf{\Gamma}}}_{\mathrm{i}} - \dot{\hat{\mathbf{Y}}}_k)\\
&\quad = (\hat{\mathbf{\Gamma}}_{\mathrm{i}} - \hat{\mathbf{Y}}_k)\cdot\Big[(\overset{\triangledown}{\hat{\mathbf{\Gamma}}}_{\mathrm{i}} - \overset{\triangledown}{\hat{\mathbf{Y}}}_k) + \hat{\mathbf{L}}_{\mathrm{i}}(\hat{\mathbf{\Gamma}}_{\mathrm{i}} - \hat{\mathbf{Y}}_k) + (\hat{\mathbf{\Gamma}}_{\mathrm{i}} - \hat{\mathbf{Y}}_k)\hat{\mathbf{L}}_{\mathrm{i}}{}^{\mathrm{T}}\Big]\\
&\quad = (\hat{\mathbf{\Gamma}}_{\mathrm{i}} - \hat{\mathbf{Y}}_k)\cdot\Big[\overset{\triangle}{\hat{\mathbf{\Gamma}}}_{\mathrm{i}} - 2(\overset{\triangle}{\hat{\mathbf{\Gamma}}}_{\mathrm{i}}\hat{\mathbf{\Gamma}}_{\mathrm{i}} + \hat{\mathbf{\Gamma}}_{\mathrm{i}}\overset{\triangle}{\hat{\mathbf{\Gamma}}}_{\mathrm{i}}) - \overset{\triangledown}{\hat{\mathbf{Y}}}_k + 2(\hat{\mathbf{\Gamma}}_{\mathrm{i}} - \hat{\mathbf{Y}}_k)\overset{\triangle}{\hat{\mathbf{\Gamma}}}_{\mathrm{i}}\Big]\,,
\end{aligned}$$

where the relations $\overset{\triangle}{\hat{\mathbf{\Gamma}}}_{\mathrm{i}} = \dot{\hat{\mathbf{\Gamma}}}_{\mathrm{i}} + \hat{\mathbf{L}}_{\mathrm{i}}{}^{\mathrm{T}}\hat{\mathbf{\Gamma}}_{\mathrm{i}} + \hat{\mathbf{\Gamma}}_{\mathrm{i}}\hat{\mathbf{L}}_{\mathrm{i}}$ and $\overset{\triangledown}{\hat{\mathbf{\Gamma}}}_{\mathrm{i}} = \dot{\hat{\mathbf{\Gamma}}}_{\mathrm{i}} - \hat{\mathbf{L}}_{\mathrm{i}}\hat{\mathbf{\Gamma}}_{\mathrm{i}} - \hat{\mathbf{\Gamma}}_{\mathrm{i}}\hat{\mathbf{L}}_{\mathrm{i}}{}^{\mathrm{T}}$ have been applied. The result is

$$\begin{aligned}
(\hat{\mathbf{\Gamma}}_{\mathrm{i}} - \hat{\mathbf{Y}}_k)\cdot(\dot{\hat{\mathbf{\Gamma}}}_{\mathrm{i}} - \dot{\hat{\mathbf{Y}}}_k) &= \Big[(\hat{\mathbf{\Gamma}}_{\mathrm{i}} - \hat{\mathbf{Y}}_k)\big(\mathbf{1} + 2(\hat{\mathbf{\Gamma}}_{\mathrm{i}} - \hat{\mathbf{Y}}_k)\big)\Big]\cdot\overset{\triangle}{\hat{\mathbf{\Gamma}}}_{\mathrm{i}}\\
&\quad - (\hat{\mathbf{\Gamma}}_{\mathrm{i}} - \hat{\mathbf{Y}}_k)\cdot\Big[\overset{\triangledown}{\hat{\mathbf{Y}}}_k + 2(\overset{\triangle}{\hat{\mathbf{\Gamma}}}_{\mathrm{i}}\hat{\mathbf{\Gamma}}_{\mathrm{i}} + \hat{\mathbf{\Gamma}}_{\mathrm{i}}\overset{\triangle}{\hat{\mathbf{\Gamma}}}_{\mathrm{i}})\Big]\,.
\end{aligned} \tag{13.261}$$

Inserting this into the dissipation inequality and employing the definitions (13.255) - (13.257), we obtain the dissipation inequality

$$\begin{aligned}
&\Big(\frac{1}{\rho_{\mathrm{R}}}\hat{\mathbf{S}} - \frac{\partial\varphi}{\partial\hat{\mathbf{\Gamma}}_{\mathrm{e}}}\Big)\cdot\dot{\hat{\mathbf{\Gamma}}}_{\mathrm{e}} - \Big(s + \frac{\partial\psi}{\partial\Theta}\Big)\dot{\Theta} + \frac{1}{\rho_{\mathrm{R}}}\big(\hat{\mathbf{P}} - \hat{\mathbf{P}}_{\hat{\mathbf{X}}}\big)\cdot\overset{\triangle}{\hat{\mathbf{\Gamma}}}_{\mathrm{i}} +\\
&\quad + \frac{1}{\rho_{\mathrm{R}}}\sum_k \hat{\mathbf{X}}_k\cdot\Big[\overset{\triangledown}{\hat{\mathbf{Y}}}_k + 2(\overset{\triangle}{\hat{\mathbf{\Gamma}}}_{\mathrm{i}}\hat{\mathbf{\Gamma}}_{\mathrm{i}} + \hat{\mathbf{\Gamma}}_{\mathrm{i}}\overset{\triangle}{\hat{\mathbf{\Gamma}}}_{\mathrm{i}})\Big] \geq 0\,.
\end{aligned}$$

Using the usual line of argumentation, this implies the potential relations for stress and entropy. Eliminating the internal variables $\hat{\mathbf{Y}}_k$ by means of (13.256) and

$$\overset{\triangledown}{\hat{\mathbf{Y}}}_k = \overset{\triangledown}{\hat{\mathbf{\Gamma}}}_{\mathrm{i}} - \frac{1}{c_k}\overset{\triangledown}{\hat{\mathbf{X}}}_k = \overset{\triangle}{\hat{\mathbf{\Gamma}}}_{\mathrm{i}} - 2(\overset{\triangle}{\hat{\mathbf{\Gamma}}}_{\mathrm{i}}\hat{\mathbf{\Gamma}}_{\mathrm{i}} + \hat{\mathbf{\Gamma}}_{\mathrm{i}}\overset{\triangle}{\hat{\mathbf{\Gamma}}}_{\mathrm{i}}) - \frac{1}{c_k}\overset{\triangledown}{\hat{\mathbf{X}}}_k$$

finally leads to the inequality for the mechanical dissipation,

$$\left(\hat{\mathbf{P}} - \hat{\mathbf{P}}_{\hat{\mathbf{X}}}\right) \cdot \overset{\triangle}{\hat{\mathbf{\Gamma}}}_{\mathrm{i}} + \sum_k \hat{\mathbf{X}}_k \cdot \left(\overset{\triangle}{\hat{\mathbf{\Gamma}}}_{\mathrm{i}} - \frac{1}{c_k}\overset{\nabla}{\hat{\mathbf{X}}}_k\right) \geq 0 \, . \tag{13.262}$$

This is identically satisfied with the evolution equations (13.253), (13.254). ■

The internal variables $\hat{\mathbf{Y}}_k$ possess a clear physical meaning: the tensors $\hat{\mathbf{Y}}_k$, in the same way as $\hat{\mathbf{\Gamma}}_{\mathrm{i}}$, are strain tensors, which operate on the intermediate configuration and contribute to kinematic hardening. The stress tensors $\hat{\mathbf{X}}_k$ according to (13.256) are partial backstresses that form part of the resultant backstress tensor $\hat{\mathbf{P}}_{\hat{\mathbf{X}}}$, defined by (13.257).

No isotropic hardening was taken into account in the model. This can be done by introducing the argument of a scalar internal variable into the yield stress $k(\Theta)$ (analogous to (5.127), (5.131)). Furthermore, the viscosity η in the flow rule (13.253) may depend on temperature without affecting the thermodynamic consistency. A temperature-dependence of the coefficients c_k in the free energy (13.250) is also possible; this would influence the entropy relation $(13.251)_2$ and the evolution equations (13.254).

In order to derive the heat conduction equation, the material model is inserted into the energy balance. The outcome is

Theorem 13. 22

The equation of heat conduction relating to the material model (13.250) - (13.257) reads as

$$c_{\mathrm{d}}\dot{\Theta} = \frac{1}{\rho_{\mathrm{R}}}\Theta\frac{\partial \hat{\mathbf{S}}}{\partial \Theta} \cdot (\overset{\triangle}{\hat{\mathbf{\Gamma}}}_{\mathrm{e}} - 2\hat{\mathbf{\Gamma}}_{\mathrm{e}}\overset{\triangle}{\hat{\mathbf{\Gamma}}}_{\mathrm{i}}) - \frac{1}{\rho_{\mathrm{R}}}\mathrm{Div}\,q_{\mathrm{R}} + r$$

$$+ \frac{1}{\rho_{\mathrm{R}}}\left(\hat{\mathbf{P}} - \hat{\mathbf{P}}_{\hat{\mathbf{X}}}\right) \cdot \overset{\triangle}{\hat{\mathbf{\Gamma}}}_{\mathrm{i}} + \frac{1}{\rho_{\mathrm{R}}}\sqrt{\frac{2}{3}}\left\|\overset{\triangle}{\hat{\mathbf{\Gamma}}}_{\mathrm{i}}\right\| \sum_k \frac{b_k}{c_k}\mathrm{tr}(\hat{\mathbf{X}}_k^{\,2}) \, , \tag{13.263}$$

where the specific heat at constant deformation is given by

$$c_{\mathrm{d}} = -\,\Theta\frac{\partial^2}{\partial \Theta^2}\varphi(\hat{\mathbf{\Gamma}}_{\mathrm{e}}, \Theta) \, . \tag{13.264}$$

□

Proof

The definition (13.250) of the free energy in combination with the potential relations (13.251), the definitions (13.256), (13.257) and the rearrangement (13.261) give rise to the generalised *Gibbs* relation

$$\dot{\psi} = \frac{\partial \varphi}{\partial \hat{\mathbf{\Gamma}}_{\mathrm{e}}} \cdot \dot{\hat{\mathbf{\Gamma}}}_{\mathrm{e}} + \frac{\partial \varphi}{\partial \Theta} \dot{\Theta} + \frac{1}{\rho_{\mathrm{R}}} \hat{\mathbf{P}}_{\hat{\mathbf{X}}} \cdot \overset{\triangle}{\hat{\mathbf{\Gamma}}}_{\mathrm{i}} - \frac{1}{\rho_{\mathrm{R}}} \sum_k \hat{\mathbf{X}}_k \cdot \left(\overset{\triangle}{\hat{\mathbf{\Gamma}}}_{\mathrm{i}} - \frac{1}{c_k} \overset{\triangledown}{\hat{\mathbf{X}}}_k \right) .$$

With the evolution equations (13.254), this converts to

$$\dot{\psi} = \frac{1}{\rho_{\mathrm{R}}} \hat{\mathbf{S}} \cdot \dot{\hat{\mathbf{\Gamma}}}_{\mathrm{e}} - s\dot{\Theta} + \frac{1}{\rho_{\mathrm{R}}} \hat{\mathbf{P}}_{\hat{\mathbf{X}}} \cdot \overset{\triangle}{\hat{\mathbf{\Gamma}}}_{\mathrm{i}} - \frac{1}{\rho_{\mathrm{R}}} \sqrt{\frac{2}{3}} \|\overset{\triangle}{\hat{\mathbf{\Gamma}}}_{\mathrm{i}}\| \sum_k b_k(\Theta) \frac{1}{c_k} \mathrm{tr}(\hat{\mathbf{X}}_k^{\,2}) . \quad (13.265)$$

Insertion into the energy balance

$$\dot{\psi} + (\Theta s)^{\boldsymbol{\cdot}} = -\frac{1}{\rho_{\mathrm{R}}} \mathrm{Div} q_{\mathrm{R}} + r + \frac{1}{\rho_{\mathrm{R}}} \hat{\mathbf{S}} \cdot \dot{\hat{\mathbf{\Gamma}}}_{\mathrm{e}} + \frac{1}{\rho_{\mathrm{R}}} \hat{\mathbf{P}} \cdot \overset{\triangle}{\hat{\mathbf{\Gamma}}}_{\mathrm{i}}$$

leads to

$$\Theta \dot{s} = -\frac{1}{\rho_{\mathrm{R}}} \mathrm{Div} q_{\mathrm{R}} + r + \frac{1}{\rho_{\mathrm{R}}} \left(\hat{\mathbf{P}} - \hat{\mathbf{P}}_{\hat{\mathbf{X}}} \right) \cdot \overset{\triangle}{\hat{\mathbf{\Gamma}}}_{\mathrm{i}} + \frac{1}{\rho_{\mathrm{R}}} \sqrt{\frac{2}{3}} \|\overset{\triangle}{\hat{\mathbf{\Gamma}}}_{\mathrm{i}}\| \sum_k \frac{b_k}{c_k} \mathrm{tr}(\hat{\mathbf{X}}_k)^2 ,$$

and with

$$\Theta \dot{s} = -\Theta \frac{\partial^2 \varphi}{\partial \Theta^2} \dot{\Theta} - \Theta \frac{\partial^2 \varphi}{\partial \Theta \partial \hat{\mathbf{\Gamma}}_{\mathrm{e}}} \cdot \dot{\hat{\mathbf{\Gamma}}}_{\mathrm{e}} = c_{\mathrm{d}} \dot{\Theta} - \frac{1}{\rho_{\mathrm{R}}} \Theta \frac{\partial \hat{\mathbf{S}}}{\partial \Theta} \cdot (\overset{\triangle}{\hat{\mathbf{\Gamma}}}_{\mathrm{e}} - 2\hat{\mathbf{\Gamma}}_{\mathrm{e}} \overset{\triangle}{\hat{\mathbf{\Gamma}}}_{\mathrm{i}})$$

to (13.263). ■

In analogy to (13.243) and (13.248), the difference between the inelastic stress power and the absorption or release of energy, due to changes in the microstructure, becomes apparent in (13.263):[40]

$$p_{\mathrm{i}} - p_{\mathrm{s}} = \frac{1}{\rho_{\mathrm{R}}} \hat{\mathbf{P}} \cdot \overset{\triangle}{\hat{\mathbf{\Gamma}}}_{\mathrm{i}} - \frac{1}{\rho_{\mathrm{R}}} \left(\hat{\mathbf{P}}_{\hat{\mathbf{X}}} \cdot \overset{\triangle}{\hat{\mathbf{\Gamma}}}_{\mathrm{i}} - \sqrt{\frac{2}{3}} \|\overset{\triangle}{\hat{\mathbf{\Gamma}}}_{\mathrm{i}}\| \sum_k \frac{b_k}{c_k} \mathrm{tr}(\hat{\mathbf{X}}_k^{\,2}) \right) . \quad (13.266)$$

13. 4. 4 Thermoplasticity as a Limit Case of Thermoviscoplasticity

In Chap. 12 it was established that rate-independent plasticity can be interpreted as an equilibrium relation allocated to the theory of viscoplasticity with elastic domain. Moreover, plasticity evolves out of viscoplasticity as an asymptotic limit for sufficiently slow deformation processes or vanishing viscosity. Transferring this idea to thermomechanics leads to the perception that thermoplasticity is a kind of equilibrium thermodynamics. This equilibrium thermodynamics describes the accompanying equilibrium processes attributable to the thermomechanical processes of viscoplasticity.

[40] Cf. Lehmann [1984]; Kamlah [1994], pp. 106; Haupt et al. [1997]; Helm [1998]; Kamlah & Haupt [1998].

The following investigation refers to the thermomechanically consistent material model of thermoviscoplasticity, described by equations (13.231) - (13.237), (13.240) and (13.241). We equate m = 1 throughout.

$$\psi = \psi(\mathbf{E}_e, \Theta, \mathbf{E}_i, q_1, \dots, q_N) \,,$$

$$\tilde{\mathbf{T}} = \mathbf{f}(\mathbf{E}_e, \Theta, \mathbf{E}_i, q_1, \dots, q_N) \,, \qquad \mathbf{f} = \rho_R \frac{\partial \psi}{\partial \mathbf{E}_e} \,, \qquad s = -\frac{\partial \psi}{\partial \Theta} \,,$$

$$f = f(\tilde{\mathbf{T}}, \Theta, \mathbf{E}_i, q_1, \dots, q_N) \,,$$

$$\dot{\mathbf{E}}_i(t) = \frac{1}{\eta}\langle f\rangle \mathbf{g}(\tilde{\mathbf{T}}, \Theta, \mathbf{E}_i, q_1, \dots, q_N) \,, \qquad \mathbf{g} = \tilde{\mathbf{T}} - \rho_R \frac{\partial \psi}{\partial \mathbf{E}_i} \,,$$

$$\dot{q}_k(t) = \frac{1}{\eta}\langle f\rangle f_k(\mathbf{E}_e, \Theta, \mathbf{E}_i, q_1, \dots, q_N) \,, \quad f_k = -\frac{\partial \psi}{\partial q_k} \,, \qquad k = 1, \dots, N \,.$$

The technical details of the calculations and verifications largely correspond to those presented in full in Sect. 12. 3. Therefore, the following specifications can concentrate on the fresh aspects and results.
For every given deformation and temperature process $\Lambda(t) = \big(\mathbf{E}(t), \Theta(t)\big)$ the constitutive equations determine the free energy, the stress and all internal variables as well as the value of the yield function,

$$t \longmapsto \Lambda(t) \longmapsto f(t) = f\big(\tilde{\mathbf{T}}(t), \Theta(t), \mathbf{E}_i(t), q_1(t), \dots, q_N(t)\big) \,.$$

If one calculates the total time derivative,

$$\dot{f}(t) = \frac{\partial f}{\partial \tilde{\mathbf{T}}} \cdot \dot{\tilde{\mathbf{T}}} + \frac{\partial f}{\partial \Theta}\dot{\Theta}(t) + \frac{\partial f}{\partial \mathbf{E}_i} \cdot \dot{\mathbf{E}}_i(t) + \sum_k \frac{\partial f}{\partial q_k}\dot{q}_k(t) \,,$$

one obtains a series of terms, which are linear in the rates of the internal variables and therefore either proportional to the yield function or zero. In analogy to (12.33) the temporal evolution of the yield function is described by means of the differential equation

$$\dot{f}(t) + \frac{K(t)}{\eta}\langle f(t)\rangle = R(t) \,, \tag{13.267}$$

the right-hand side reading as

$$R(t) = \left\{\frac{\partial \mathbf{f}}{\partial \mathbf{E}_e}\left[\frac{\partial f}{\partial \tilde{\mathbf{T}}}\right]\right\} \cdot \dot{\mathbf{E}}(t) + \left\{\frac{\partial f}{\partial \tilde{\mathbf{T}}} \cdot \frac{\partial \mathbf{f}}{\partial \Theta} + \frac{\partial f}{\partial \Theta}\right\}\dot{\Theta}(t) \,. \tag{13.268}$$

The right-hand side of (13.267) is the scalar product of the time derivative of the strain-temperature vector $\boldsymbol{\Lambda}(t) = \big(\mathbf{E}(t),\, \Theta(t)\big)$ with the vector

$$\boldsymbol{\Sigma}(t) = \left(\frac{\partial \mathbf{f}}{\partial \mathbf{E}_{\mathrm{e}}}\left[\frac{\partial f}{\partial \tilde{\mathbf{T}}}\right],\ \frac{\partial f}{\partial \tilde{\mathbf{T}}} \cdot \frac{\partial \mathbf{f}}{\partial \Theta} + \frac{\partial f}{\partial \Theta} \right), \tag{13.269}$$

$$R(t) = \boldsymbol{\Sigma}(t) \cdot \dot{\boldsymbol{\Lambda}}(t)\ . \tag{13.270}$$

The coefficient

$$K(t) = \frac{\partial f}{\partial \tilde{\mathbf{T}}} \cdot \left\{ \left(\frac{\partial \mathbf{f}}{\partial \mathbf{E}_{\mathrm{e}}} - \frac{\partial \mathbf{f}}{\partial \mathbf{E}_{\mathrm{i}}} \right)[\mathbf{g}] \right\} - \frac{\partial f}{\partial \mathbf{E}_{\mathrm{i}}} \cdot \mathbf{g} - \sum_k \left(\frac{\partial f}{\partial \tilde{\mathbf{T}}} \cdot \frac{\partial \mathbf{f}}{\partial q_k} + \frac{\partial f}{\partial q_k} \right) \frac{\partial \psi}{\partial q_k}, \tag{13.271}$$

is determined by the solution of the constitutive equations as a function of time (cf. (12.35)). Once again we assume that the *stability condition*

$$K(t) \geq K_0 > 0 \tag{13.272}$$

holds, a restrictive assumption for the thermomechanical material model. The assumption of stability includes a relaxation property that leads to states of stable thermodynamic equilibrium. In order to verify this, we consider an inelastic process

$$t \longmapsto \boldsymbol{\Lambda}(t) = \big(\mathbf{E}(t)\,,\, \Theta(t)\big)\ , \quad f(t) > 0 \text{ for } t > t_0\,,$$

commencing at a time t_0. Its *isothermal static continuation*

$$\boldsymbol{\Lambda}^t(\tau) = \begin{cases} \boldsymbol{\Lambda}(\tau) = \big(\mathbf{E}(\tau)\,,\, \Theta(\tau)\big) & \text{for } t_0 \leq \tau < t \\ \boldsymbol{\Lambda}(t) = \big(\mathbf{E}(t)\,,\, \Theta(t)\big) & \text{for } \tau \geq t \end{cases} \tag{13.273}$$

beginning at time t evokes the response

$$\tau \longmapsto \big(\psi^t(\tau),\ \tilde{\mathbf{T}}^t(\tau),\ s^t(\tau),\ \mathbf{E}_{\mathrm{i}}{}^t(\tau),\ q_1{}^t(\tau), \ldots, q_N{}^t(\tau)\big) \longmapsto f^t(\tau)\,, K^t(\tau)$$

of the constitutive model and causes the yield function for $\tau \geq t$ to evolve as a solution of the homogeneous differential equation

$$\frac{\mathrm{d}}{\mathrm{d}\tau} f^t(\tau) + \frac{K^t(\tau)}{\eta} f^t(\tau) = 0 \tag{13.274}$$

with the initial condition $f^t(t) = f(t)$. According to the stability assumption (13.272), the solution tends towards zero as τ goes to infinity:

$$f^t(\tau) \to f_{\mathrm{eq}}(t) = 0 \quad \text{for} \quad \tau \to \infty \, .$$

The limit value $f_{\mathrm{eq}} = 0$ corresponds to a set of equilibrium states,

$$f_{\mathrm{eq}} = 0 \quad \Longleftrightarrow \quad \left\{ \left(\psi_{\mathrm{eq}}, \tilde{\mathbf{T}}_{\mathrm{eq}}, s_{\mathrm{eq}}, \mathbf{E}_{\mathrm{ieq}}, q_{1\mathrm{eq}}, \ldots, q_{N\mathrm{eq}} \right) \right\} ;$$

one of which is reached by the relaxation process:

$$\left(\psi^t, \tilde{\mathbf{T}}^t, s^t, \mathbf{E}_{\mathrm{i}}^t, q_1^t, \ldots, q_N^t \right)(\tau) \to \left(\psi_{\mathrm{eq}}, \tilde{\mathbf{T}}_{\mathrm{eq}}, s_{\mathrm{eq}}, \mathbf{E}_{\mathrm{ieq}}, q_{1\mathrm{eq}}, \ldots, q_{N\mathrm{eq}} \right)(t) \, . \tag{13.275}$$

In this sense every thermomechanical process,

$$t \longmapsto \boldsymbol{\Lambda}(t) = \left(\mathbf{E}(t) \, , \Theta(t) \right) \longmapsto \left(\psi(t), \tilde{\mathbf{T}}(t), s(t), \mathbf{E}_{\mathrm{i}}(t), q_1(t), \ldots, q_N(t) \right) ,$$

induces a sequence of thermodynamic equilibrium states, i.e. an *accompanying equilibrium process*

$$t \longmapsto \left(\psi_{\mathrm{eq}}(t), \tilde{\mathbf{T}}_{\mathrm{eq}}(t), s_{\mathrm{eq}}, \mathbf{E}_{\mathrm{ieq}}(t), q_{1\mathrm{eq}}(t), \ldots, q_{N\mathrm{eq}}(t) \right) .$$

In order to find the differential equations that determine these equilibrium states, an inelastic process is considered, just as in the mechanical theory, beginning at a time t_0 on the yield surface:

$$f(t_0) = 0 \, , \quad f(t) > 0 \, , \quad \dot{\boldsymbol{\Lambda}}(t) \not\equiv \mathbf{0} \text{ for } t > t_0 \, .$$

The differential equation for the yield function $f(t)$, which is now inhomogeneous,

$$\dot{f}(t) + \frac{K(t)}{\eta} f(t) = \boldsymbol{\Sigma}(t) \cdot \dot{\boldsymbol{\Lambda}}(t) \, , \tag{13.276}$$

implies the integral equation analogous to (12.55),

$$f(t) = - \int_0^{t - t_0} \mathrm{e}^{-\frac{1}{\eta} \int_0^s K(t - \sigma) \mathrm{d}\sigma} \boldsymbol{\Sigma}(t - s) \cdot \frac{\mathrm{d}}{\mathrm{d}s} \boldsymbol{\Lambda}(t - s) \mathrm{d}s \, . \tag{13.277}$$

This identity relates the current value of the yield function to the past history of the strain and temperature. In strict analogy to (12.56) it is possible to attribute a *retarded process history* with the retardation factor α $(0 < \alpha < 1)$ to every given strain-temperature history,

$$\Lambda(t-s)\Big|_{s\in[0,\,t-t_0]} \longmapsto \Lambda(t-\alpha s)\Big|_{s\in[0,\,\frac{t-t_0}{\alpha}]}\,, \tag{13.278}$$

which would evoke a different material response,

$$(t-s) \longmapsto \left(\psi_\alpha, \widetilde{\mathbf{T}}_\alpha, s_\alpha, \mathbf{E}_{\mathrm{i}\alpha}, q_{1\alpha}, \dots, q_{N\alpha}, f_\alpha, K_\alpha\right)(t-s)\,.$$

For all α, the potential relations

$$\widetilde{\mathbf{T}}_\alpha = \rho_{\mathrm{R}}\frac{\partial\psi_\alpha}{\partial\mathbf{E}_{\mathrm{e}}}\,, \quad s_\alpha = -\frac{\partial\psi_\alpha}{\partial\Theta} \tag{13.279}$$

still hold. For the following discussion, we assume that the given process remains inelastic for all retardations: $f_\alpha(t) > 0$ for all $\alpha \in (0, 1)$. The integral equation for the yield function of the retarded process,

$$f_\alpha(t) = \int\limits_0^{t-t_0} \mathrm{e}^{-\frac{1}{\alpha\eta}\int_0^{\bar{s}} K_\alpha(t-\frac{\bar{\sigma}}{\alpha})\mathrm{d}\bar{\sigma}}\, \boldsymbol{\Sigma}_\alpha\Big(t-\frac{\bar{s}}{\alpha}\Big)\cdot\frac{\mathrm{d}}{\mathrm{d}t}\Lambda(t-\bar{s})\mathrm{d}\bar{s}\,, \tag{13.280}$$

shows that the passages to the limit $\alpha \to 0$ and $\eta \to 0$ are equivalent. Instead of slowing down the thermomechanical process, it is also possible to imagine a given material replaced by a sequence of materials with decreasing viscosities. In analogy to Theorem 12. 1 or (12.64) we then obtain the asymptotic relation

$$\frac{1}{\alpha\eta}f_\alpha(t) = \frac{\boldsymbol{\Sigma}_\alpha(t)\cdot\dot{\Lambda}(t)}{K_\alpha(t)} + \mathrm{o}(1)\,. \tag{13.281}$$

Accordingly, for the rescaled rates of the internal variables we have finite limit values:

$$\lim_{\alpha\eta\to 0}\left(\frac{1}{\alpha}\dot{\mathbf{E}}_{\mathrm{i}\alpha}(t)\,,\ \frac{1}{\alpha}\dot{q}_{1\alpha}(t), \dots, \frac{1}{\alpha}\dot{q}_{N\alpha}(t)\right) < \infty\,. \tag{13.282}$$

We define the limit value of the rescaled inelastic strain rate as the *plastic strain rate*:

$$\dot{\mathbf{E}}_{\mathrm{p}}(t) = \lim_{\alpha\eta\to 0} \frac{1}{\alpha}\dot{\mathbf{E}}_{\mathrm{i}\alpha}(t) = \lim_{\alpha\eta\to 0} \frac{1}{\alpha\eta} f_\alpha\, \mathbf{g}(\widetilde{\mathbf{T}}_\alpha, \Theta_\alpha, \mathbf{E}_{\mathrm{i}\alpha}, q_{1\alpha}, \dots, q_{N\alpha}) . \tag{13.283}$$

What one obtains is

$$\dot{\mathbf{E}}_{\mathrm{p}}(t) = \lambda \mathbf{g}(\widetilde{\mathbf{T}}, \Theta, \mathbf{E}_{\mathrm{p}}, q_1, \dots, q_N) , \tag{13.284}$$

ascertaining that the proportionality factor

$$\lambda = \frac{\boldsymbol{\Sigma}\cdot\dot{\boldsymbol{\Lambda}}(t)}{K} , \tag{13.285}$$

with $\boldsymbol{\Sigma}(t)\cdot\dot{\boldsymbol{\Lambda}}(t)$ according to (13.270), (13.268) and K according to (13.271) is precisely the proportionality factor that one would obtain from the consistency condition

$$\frac{\mathrm{d}}{\mathrm{d}t} f(t) = f\big(\widetilde{\mathbf{T}}(t), \Theta(t), \mathbf{E}_{\mathrm{i}}(t), q_1(t), \dots, q_N(t)\big) = 0 \tag{13.286}$$

of rate-independent plasticity. This result gives rise to

Definition 13. 10

The following constitutive equations define a general material model of *thermoplasticity with thermoelastic domain*:

Decomposition of the Deformation

$$\mathbf{E} = \mathbf{E}_{\mathrm{e}} + \mathbf{E}_{\mathrm{p}} \tag{13.287}$$

Free Energy

$$\psi = \psi(\mathbf{E}_{\mathrm{e}}, \Theta, \mathbf{E}_{\mathrm{p}}, q_1, \dots, q_N) \tag{13.288}$$

Heat Flux Vector

$$q_{\mathrm{R}} = q_{\mathrm{R}}(\mathbf{E}_{\mathrm{e}}, \Theta, g_{\mathrm{R}}, \mathbf{E}_{\mathrm{p}}, q_1, \dots, q_N) \tag{13.289}$$

Stress Relation

$$\widetilde{\mathbf{T}} = f(\mathbf{E}_{\mathrm{e}}, \Theta, \mathbf{E}_{\mathrm{p}}, q_1, \dots, q_N) = \rho_{\mathrm{R}}\frac{\partial\psi}{\partial\mathbf{E}_{\mathrm{e}}} \tag{13.290}$$

Entropy Relation

$$s = s(\mathbf{E}_e, \Theta, \mathbf{E}_p, q_1, \dots, q_N) = -\frac{\partial \psi}{\partial \Theta} \tag{13.291}$$

Yield Function

$$f = f(\widetilde{\mathbf{T}}, \Theta, \mathbf{E}_p, q_1, \dots, q_N) \tag{13.292}$$

Flow Rule

$$\dot{\mathbf{E}}_p(t) = \begin{cases} \lambda \mathbf{g}(\widetilde{\mathbf{T}}, \Theta, \mathbf{E}_p, q_1, \dots, q_N) & \text{for } f = 0 \text{ and } \boldsymbol{\Sigma}\cdot\dot{\boldsymbol{\Lambda}} > 0 \\ 0 & \text{otherwise} \end{cases} \tag{13.293}$$

Evolution Equations (Hardening Model)

$$\dot{q}_k(t) = \lambda f_k(\mathbf{E}_e(t), \Theta(t), \mathbf{E}_p(t), q_1(t), \dots, q_N(t)) \, , \quad k = 1, \dots, N \tag{13.294}$$

The evolution equations (13.293), (13.294) employ the proportionality factor

$$\lambda = \frac{\boldsymbol{\Sigma}\cdot\dot{\boldsymbol{\Lambda}}(t)}{K} \, , \tag{13.295}$$

with

$$\boldsymbol{\Sigma}\cdot\dot{\boldsymbol{\Lambda}}(t) = \left\{\frac{\partial \mathbf{f}}{\partial \mathbf{E}_e}\left[\frac{\partial f}{\partial \widetilde{\mathbf{T}}}\right]\right\}\cdot\dot{\mathbf{E}}(t) + \left\{\frac{\partial f}{\partial \widetilde{\mathbf{T}}}\cdot\frac{\partial \mathbf{f}}{\partial \Theta} + \frac{\partial f}{\partial \Theta}\right\}\dot{\Theta}(t) \, ,$$

and

$$K = \frac{\partial f}{\partial \widetilde{\mathbf{T}}}\cdot\left\{\left(\frac{\partial \mathbf{f}}{\partial \mathbf{E}_e} - \frac{\partial \mathbf{f}}{\partial \mathbf{E}_p}\right)[\mathbf{g}]\right\} - \frac{\partial f}{\partial \mathbf{E}_p}\cdot\mathbf{g} - \sum_k \left(\frac{\partial f}{\partial \widetilde{\mathbf{T}}}\cdot\frac{\partial \mathbf{f}}{\partial q_k} + \frac{\partial f}{\partial q_k}\right)\frac{\partial \psi}{\partial q_k} \, .$$

◊

The ideas contained in this section form the basis of the following

Theorem 13. 23
The constitutive model (13.288) - (13.294) of thermoplasticity arises out of the thermoviscoplasticity model (13.231) - (13.237) as an asymptotic limit for thermomechanical processes evolving at a sufficiently slow rate or for vanishing viscosity. ■

It is obvious that the thermomechanical consistency of the thermoplasticity model is secured through the choice of

$$\mathbf{g}(\widetilde{\mathbf{T}}, \Theta, \mathbf{E}_{\mathrm{p}}, q_1, \dots, q_N) = \widetilde{\mathbf{T}} - \rho_{\mathrm{R}}\frac{\partial\psi}{\partial\mathbf{E}_{\mathrm{p}}} \tag{13.296}$$

and

$$f_k(\mathbf{E}_{\mathrm{e}}, \Theta, \mathbf{E}_{\mathrm{i}}, q_1, \dots, q_N) = -\frac{\partial\psi}{\partial q_k}, \qquad k = 1, \dots, N. \tag{13.297}$$

The physical meaning of Theorem 13. 23 lies in the ascertainment that the accompanying equilibrium processes of thermoviscoplasticity are represented by evolution equations coinciding with the evolution equations of thermoplasticity. Thus, rate-independent thermoplasticity is the equilibrium relation associated with the constitutive model of thermoviscoplasticity in the sense of Definition 6. 2.

In rate-independent thermoplasticity, in precisely the same way as in plasticity, it is possible to substitute time as an independent variable by means of a generalised arclength $t \longmapsto z(t)$, for example by means of the accumulated plastic strain, defined by

$$\dot{z}(t) = \left\|\dot{\mathbf{E}}_{\mathrm{p}}(t)\right\| = \lambda\left\|\mathbf{g}\right\| . \tag{13.298}$$

Dividing all evolution equations by $\dot{z}$ yields

$$\frac{1}{\dot{z}(t)}\dot{\mathbf{E}}_{\mathrm{p}}(t) = \mathbf{E}_{\mathrm{p}}{}'(z) \ , \quad \frac{1}{\dot{z}(t)}\dot{q}_k(t) = q_k{}'(z) \tag{13.299}$$

and gives rise to the arclength representation, in which the model of thermoplasticity assumes the following form:

$$\psi = \psi(\mathbf{E}_{\mathrm{e}}, \Theta, \mathbf{E}_{\mathrm{p}}, q_1, \dots, q_N) \, , \quad \widetilde{\mathbf{T}} = \rho_{\mathrm{R}}\frac{\partial\psi}{\partial\mathbf{E}_{\mathrm{e}}} \, , \quad s = -\frac{\partial\psi}{\partial\Theta} \, , \tag{13.300}$$

$$\mathbf{E}_{\mathrm{p}}{}'(z) = \frac{\mathbf{g}(\widetilde{\mathbf{T}}, \Theta, \mathbf{E}_{\mathrm{p}}, q_1, \dots, q_N)}{\left\|\mathbf{g}(\widetilde{\mathbf{T}}, \Theta, \mathbf{E}_{\mathrm{p}}, q_1, \dots, q_N)\right\|} \, , \tag{13.301}$$

$$q_k{}'(z) = f_k(\mathbf{E}, \Theta, \mathbf{E}_{\mathrm{p}}, q_1, \dots, q_N) \quad (k = 1, 2, \dots N) \, , \tag{13.302}$$

$$\dot{z}(t) = \begin{cases} \lambda\left\|\mathbf{g}\right\| & \text{for } f = 0 \text{ and } \boldsymbol{\Sigma}\cdot\dot{\boldsymbol{\Lambda}} > 0 \\ 0 & \text{otherwise} \end{cases} \tag{13.303}$$

The model of thermoplasticity in this formulation contains an evolution equation for the generalised arclength, which depends on distinctions between different cases. If we refrain from representing a thermoelastic domain, no case distinction is required.

It would also be possible to begin the investigation of thermoplasticity with the arclength representation and formulate the dissipation inequality based on the above equations:

$$\Big(\widetilde{\mathbf{T}} - \rho_{\mathrm{R}}\frac{\partial\psi}{\partial\mathbf{E}_{\mathrm{e}}}\Big)\cdot\dot{\mathbf{E}}_{\mathrm{e}}(t) - \rho_{\mathrm{R}}\Big(s + \frac{\partial\psi}{\partial\Theta}\Big)\dot{\Theta}(t) + \dot{z}(t)\Big(\widetilde{\mathbf{T}} - \rho_{\mathrm{R}}\frac{\partial\psi}{\partial\mathbf{E}_{\mathrm{i}}}\Big)\cdot\dot{\mathbf{E}}_{\mathrm{p}}(t)$$
$$- \dot{z}(t)\Big(\sum_k \rho_{\mathrm{R}}\frac{\partial\psi}{\partial q_k}\,q_k{}'(z)\Big) - \frac{1}{\Theta}\mathbf{q}_{\mathrm{R}}\cdot\mathrm{Grad}\Theta \geq 0\,. \qquad (13.304)$$

If we bear in mind that the rate of the generalised arclength $\dot{z}$ is a homogeneous function of the process rate,[41] it is not possible to derive the stress and entropy relation from this form of dissipation inequality by way of a necessary conclusion. It is only due to the asymptotic limit viscoplasticity → plasticity that the status of a necessary conclusion can be attributed to the stress and entropy relation.[42]

In view of the terminological structure of thermoplasticity it is evident that it cannot reflect relaxation behaviour: during an isothermal static continuation of any thermomechanical process all internal variables are constant in time. In this sense all solutions of evolution equations of thermoplasticity are *equilibrium solutions*, i.e. the theory of thermoplasticity is an *equilibrium thermodynamics*.

On the other hand, it is known fact that the evolution of plastic strains on the microlevel is connected with dynamic processes. These two contrasting aspects can nevertheless be reconciled, if we bear in mind that only equilibrium states are compared with each other in the arclength representation: the dynamical processes in between are disregarded in the phenomenological representation. All the while it is imperative to accept the abstract notion that the equilibrium states form a continuous sequence and depend on the process history via the arclength representation. After all, dissipation always takes place during these sequences of elastoplastic equilibrium states, a fact that digresses drastically from the conceptions of classical equilibrium thermodynamics.

The irreversibility of thermoelastoplastic processes is also expressed in the equation of heat conduction; this is similar to (13.248):

---

[41] Cf. (11.32).

[42] Cf. ACHARYA & SHAWKI [1996]; MIEHE [1995].

$$c_d\dot{\Theta} = \frac{1}{\rho_R}\Theta\frac{\partial\widetilde{\mathbf{T}}}{\partial\Theta}\cdot\dot{\mathbf{E}}_e - \frac{1}{\rho_R}\mathrm{Div}q_R + r + p_p - p_s\,. \tag{13.305}$$

On the right-hand side, besides the thermoelastic coupling term, we have the dissipative power as the difference between the plastic stress power p_p and the energy absorption (or release) resulting from structural changes in the material:

$$p_p - p_s = \dot{z}(t)\left(\frac{1}{\rho_R}\widetilde{\mathbf{T}}\cdot\mathbf{E}_p{}'(z) - \frac{\partial(\psi + \Theta s)}{\partial\mathbf{E}_p}\cdot\mathbf{E}_p{}'(z) - \sum_k\frac{\partial(\psi + \Theta s)}{\partial q_k}q_k{}'(z)\right). \tag{13.306}$$

In thermoplasticity, the energy dissipation is proportional to the rate of the generalised arclength, homogeneous in the process rate. Accordingly, the dissipated energy in cyclic processes depends only on the number of cycles and not on the velocity at which the cycles take place. This property is the characteristic of rate-independent hysteresis.

13. 5 General Thermoviscoplasticity

A separate representation of the material properties of equilibrium and non-equilibrium is expedient from a physical point of view; this also applies in the thermomechanical constitutive theory. In view of the result that rate-independent thermoplasticity is attributed to thermoviscoplasticity with elastic domain as its equilibrium relation, it is a natural step to formulate the material properties of thermodynamic equilibrium as an independent model and supplement it using rate-dependent parts, so as to represent the deviations from the thermodynamical equilibrium as well.[43]

The basic assumption is then a decomposition of the free energy into one equilibrium and one non equilibrium part:

$$\psi = \psi_{eq} + \psi_{ov}\,. \tag{13.307}$$

The equilibrium part ψ_{eq} is assumed to be a rate-independent functional of the process history. ψ_{eq} represents thermodynamic equilibrium; the arclength representation is suitable for modelling, either within the framework of the representation method using functionals or in the representation by

[43] The concept of general thermoviscoplasticity is explained in HAUPT [1995]. Successful applications are documented in various contributions: LION [1996]; LION [1997a]; LION [1997b]; LION [1997c]; LION [1998]; HAUPT & HELM [1999].

means of internal variables. The non-equilibrium part ψ_{ov} is a rate-dependent functional of the process history with relaxation to zero during static continuations. ψ_{ov} describes all deviations from thermodynamic equilibrium; for a concrete formulation the representation methods of thermo-viscoelasticity are applicable. The decomposition (13.307) of the free energy induces corresponding decompositions of stress and entropy:

$$\widetilde{\mathbf{T}} = \widetilde{\mathbf{T}}_{eq} + \widetilde{\mathbf{T}}_{ov}\,, \qquad s = s_{eq} + s_{ov}\,. \tag{13.308}$$

If we model the material behaviour in thermodynamic equilibrium and non-equilibrium separately, all methods of material theory discussed so far converge. The separate representation increases physical transparency and creates a multitude of possibilities for representing material properties, incorporating numerous facts that are observed in experiments. The problem of thermomechanical consistency can easily be solved in an uncoupled modelling process, since both plasticity and viscoelasticity offer a large supply of thermomechanically consistent models. Compared with other theories of viscoplasticity it is in most cases no problem to find an energy function that guarantees compatibility with the dissipation inequality in connection with suitable evolution equations for internal variables.

By construction, this general thermoviscoplasticity contains the constitutive theories of elasticity, viscoelasticity and plasticity clearly set out as special cases.

13. 5. 1 Small Deformations

Isothermal processes are investigated below to illustrate the method. The following example, valid for small deformations, generalises the mechanical model (12.87) - (12.104) to thermomechanics.

Theorem 13. 24

The constitutive equations (12.87) - (12.104)[44] combine with the free energy

$$\rho\psi = \mu_{eq}\mathrm{tr}\big(\mathbf{E}_e{}^{D2}\big) + \tfrac{1}{2}\kappa_{eq}\big(\mathrm{tr}\mathbf{E}_e\big)^2 + \sum_{k=1}^{M}\tfrac{1}{2}c_k\mathrm{tr}\Big(\big(\mathbf{E}_p{}^{D} - \mathbf{Y}_k{}^{D}\big)^2\Big)$$

[44] Cf. Haupt & Lion [1995].

$$+ \sum_{k=1}^{N} \mu_0 \lambda_k \operatorname{tr}\Big((\mathbf{E}^D - \mathbf{Z}_k{}^D)^2\Big) + \sum_{k=1}^{N} \tfrac{1}{2}(\kappa_0 - \kappa_{eq})\lambda_k \big[\operatorname{tr}(\mathbf{E} - \mathbf{Z}_k{}^D)\big]^2$$

$$\Big(\sum_{k=1}^{N} \lambda_k = 1\Big) \tag{13.309}$$

to form a thermomechanically consistent constitutive model of thermoviscoplasticity. □

Proof

by calculating the dissipation $\rho\delta = \mathbf{T}_{eq} \cdot (\dot{\mathbf{E}}_e + \dot{\mathbf{E}}_p) + \mathbf{T}_{ov} \cdot \dot{\mathbf{E}} - \rho_R \dot{\psi}$:

$$\rho\delta = \Big\{\mathbf{T}_{eq} - 2\mu_{eq}\mathbf{E}_e{}^D - \kappa_{eq}(\operatorname{tr}\mathbf{E}_e)\mathbf{1}\Big\} \cdot \dot{\mathbf{E}}_e$$

$$+ \Big\{\mathbf{T}_{ov} - \sum_{k=1}^{N} \lambda_k \Big[2\mu_0(\mathbf{E}^D - \mathbf{Z}_k{}^D) + (\kappa_0 - \kappa_{eq})[\operatorname{tr}(\mathbf{E} - \mathbf{Z}_k)]\mathbf{1}\Big]\Big\} \cdot \dot{\mathbf{E}}$$

$$+ \Big\{\mathbf{T}_{eq} - \sum_{k=1}^{M} c_k(\mathbf{E}_p{}^D - \mathbf{Y}_k{}^D)\Big\} \cdot \dot{\mathbf{E}}_p + \sum_{k=1}^{M} c_k(\mathbf{E}_p{}^D - \mathbf{Y}_k{}^D) \cdot \dot{\mathbf{Y}}_k$$

$$+ \Big\{\sum_{k=1}^{N} \lambda_k \Big[2\mu_0(\mathbf{E}^D - \mathbf{Z}_k{}^D) + (\kappa_0 - \kappa_{eq})[\operatorname{tr}(\mathbf{E} - \mathbf{Z}_k)]\mathbf{1}\Big]\Big\} \cdot \dot{\mathbf{Z}}_k \, .$$

The dissipation inequality $\delta \geq 0$ leads to the elasticity relations for the equilibrium stress (12.89),

$$\mathbf{T}_{eq} = 2\mu_{eq}\mathbf{E}_e{}^D + \kappa_{eq}(\operatorname{tr}\mathbf{E}_e)\mathbf{1} \, , \tag{13.310}$$

and the overstress (12.99),

$$\mathbf{T}_{ov} = \sum_{k=1}^{N} \mathbf{T}_{ovk} \, , \tag{13.311}$$

which is the sum of the partial overstresses

$$\mathbf{T}_{ovk} = \lambda_k \Big\{2\mu_0(\mathbf{E}^D - \mathbf{Z}_k{}^D) + (\kappa_0 - \kappa_{eq})[\operatorname{tr}(\mathbf{E} - \mathbf{Z}_k)]\mathbf{1}\Big\} \, . \tag{13.312}$$

With the *backstress* tensor

$$\mathbf{X} = \sum_{k=1}^{M} \mathbf{X}_k \,, \tag{13.313}$$

consisting of the *partial backstress* tensors

$$\mathbf{X}_k = c_k\left(\mathbf{E}_p{}^{D} - \mathbf{Y}_k{}^{D}\right) , \tag{13.314}$$

the mechanical dissipation reads as follows:

$$\begin{aligned}\rho\delta = {} & \left(\mathbf{T}_{eq} - \mathbf{X}^{D}\right)\cdot\dot{\mathbf{E}}_p + \sum_{k=1}^{M} \mathbf{X}_k{}^{D}\cdot\left(\dot{\mathbf{E}}_p - \frac{1}{c_k}\dot{\mathbf{X}}_k\right) \\ & + \sum_{k=1}^{N} \mathbf{T}_{ovk}\cdot\left(\dot{\mathbf{E}}^{D} - \frac{\dot{\mathbf{T}}^{D}_{ovk}}{2\mu_0\lambda_k}\right) + \sum_{k=1}^{N} \frac{1}{3}\left(\operatorname{tr}\mathbf{T}_{ovk}\right)\operatorname{tr}\left(\dot{\mathbf{E}} - \frac{\dot{\mathbf{T}}_{ovk}}{3(\kappa_0 - \kappa_{eq})\lambda_k}\right) .\end{aligned} \tag{13.315}$$

The evolution equations (12.92), (12.94) and (12.100), (12.101) finally give the dissipation the form

$$\begin{aligned}\rho\delta = {} & \lambda\operatorname{tr}\left(\mathbf{T}_{eq} - \mathbf{X}^{D}\right)^2 + \sum_{k=1}^{M} \frac{b_k}{c_k}\,\dot{z}(t)\,\operatorname{tr}\left(\mathbf{X}_k{}^{D2}\right) \\ & + \sum_{k=1}^{N} \frac{1}{2\mu_0\lambda_k z_{0k} M\left(\|\mathbf{T}^{D}_{ov}\|\right)}\operatorname{tr}\left(\mathbf{T}_{ovk}{}^{D2}\right) + \frac{\beta}{2\mu_0}\left(\mathbf{T}_{ov}\cdot\dot{\mathbf{T}}_{eq}\right) \\ & + \sum_{k=1}^{N} \frac{1}{9(\kappa_0 - \kappa_{eq})\lambda_k z_{0k} M\left(\|\mathbf{T}^{D}_{ov}\|\right)}\operatorname{tr}\left(\mathbf{T}_{ovk}{}^{2}\right) .\end{aligned} \tag{13.316}$$

■

The above argumentation shows that the way of modelling the generalised arclength $z(t)$ using the evolution equations (12.95) - (12.98) has no influence on the thermomechanical consistency. The dissipation (13.316) demonstrates that the last term on the right-hand side of the evolution equation (12.100) includes the factor β according to (12.104), for the purpose of achieving thermomechanical consistency. The term $-\beta\lambda_k\dot{\mathbf{T}}^{D}_{eq}(t)$ in (12.100) is necessary in order to create a slope of the stress-strain curve, determined by the spontaneous shear modulus μ_0, whenever loading from the thermodynamical equilibrium state takes place $\left(\mathbf{T}_{ov}\cdot\dot{\mathbf{T}}_{eq} > 0\right)$. A reproduction of stress-strain behaviour of this kind was motivated by experiments using steel XCrNi 18.9 (cf. Figs. 6. 5 and 12. 6). For $\mathbf{T}_{\text{ov}} \cdot \dot{\mathbf{T}}_{\text{eq}} < 0$ (unloading), however, the additional term can be dispensed with. Given the derivative of the equilibrium stress in (12.100) without factor β, it cannot be ruled out that the dissipation might assume negative values in the event of a stress reversal. The β - term does not apply, of course, in the case of materials where no such stress-strain curves are observed.

13. 5. 2 Finite Deformations

One way of representing the equilibrium behaviour is to use the constitutive model of thermoplasticity with yield surface. This emerges from thermoviscoplasticity with elastic domain according to (13.250) - (13.257) through $\eta \to 0$. For the sake of simplicity the temperature-dependence is completely ignored in this section. For isothermal processes the model of plasticity is given by the following constitutive equations:

$$\rho_{\text{R}}\psi_{\text{eq}} = \rho_{\text{R}}\varphi(\hat{\mathbf{\Gamma}}_{\text{e}}) + \frac{1}{2}\sum_k c_k \operatorname{tr}(\hat{\mathbf{\Gamma}}_{\text{p}} - \hat{\mathbf{Y}}_k)^2 , \tag{13.317}$$

$$\hat{\mathbf{S}}_{\text{eq}} = \rho_{\text{R}} \frac{\partial\varphi}{\partial\hat{\mathbf{\Gamma}}_{\text{e}}} , \tag{13.318}$$

$$\widetilde{\mathbf{T}}_{\text{eq}} = \mathbf{F}_{\text{p}}^{-1}\hat{\mathbf{S}}_{\text{eq}}\mathbf{F}_{\text{p}}^{\text{T}-1} , \tag{13.319}$$

$$f(\hat{\mathbf{P}}, \hat{\mathbf{P}}_{\hat{\mathbf{X}}}, k) = \sqrt{\tfrac{3}{2}}\|\hat{\mathbf{P}} - \hat{\mathbf{P}}_{\hat{\mathbf{X}}}\| - k , \tag{13.320}$$

$$\overset{\triangle}{\hat{\mathbf{\Gamma}}}_{\text{p}} = \begin{cases} \lambda\sqrt{\frac{3}{2}}\dfrac{\hat{\mathbf{P}} - \hat{\mathbf{P}}_{\hat{\mathbf{X}}}}{\|\hat{\mathbf{P}} - \hat{\mathbf{P}}_{\hat{\mathbf{X}}}\|} & \text{for } f = 0 \text{ and loading} \\ 0 & \text{otherwise} \end{cases} , \tag{13.321}$$

$$\overset{\triangledown}{\hat{\mathbf{X}}}_k = c_k\overset{\triangle}{\hat{\mathbf{\Gamma}}}_{\text{p}} - b_k\sqrt{\tfrac{2}{3}}\|\overset{\triangle}{\hat{\mathbf{\Gamma}}}_{\text{p}}\|\hat{\mathbf{X}}_k . \tag{13.322}$$

The following definitions are valid here:

$$\hat{\mathbf{P}} = \hat{\mathbf{C}}_{\text{e}}\hat{\mathbf{S}} , \quad \hat{\mathbf{X}}_k = c_k(\hat{\mathbf{\Gamma}}_{\text{p}} - \hat{\mathbf{Y}}_k) , \quad \hat{\mathbf{P}}_{\hat{\mathbf{X}}} = \sum_k \hat{\mathbf{X}}_k\Big(1 + \frac{2}{c_k}\hat{\mathbf{X}}_k\Big) . \tag{13.323}$$

Thermomechanical consistency is guaranteed in the sense of Theorems 13. 18 and 13. 20; it follows for the dissipation that

$$\rho_R \delta_{eq} = \lambda\sqrt{\frac{3}{2}}\left\|\hat{\mathbf{P}} - \hat{\mathbf{P}}_{\hat{\mathbf{X}}}\right\| + \lambda\sum_k \frac{b_k}{c_k}\,\mathrm{tr}(\hat{\mathbf{X}}_k)^2 . \tag{13.324}$$

To complete the model by means of a non-equilibrium part, equations (13.152) - (13.156), for which thermomechanical consistency is established according to Theorem 13. 14, can be employed. In doing so, we have to equate $\psi_{eq} = 0$ in (13.152) in order to guarantee relaxation to zero.

Decompositions of Deformation[45]

$$\mathbf{F} = \hat{\mathbf{F}}_{ek}\mathbf{F}_{vk}\,, \quad \hat{\mathbf{\Gamma}}_k = \mathbf{F}_{vk}^{T-1}\mathbf{E}\mathbf{F}_{vk}^{-1} = \hat{\mathbf{\Gamma}}_{ek} + \hat{\mathbf{\Gamma}}_{vk}\,, \tag{13.325}$$

$$\hat{\mathbf{\Gamma}}_{ek} = \frac{1}{2}[\hat{\mathbf{C}}_{ek} - \mathbf{1}]\,, \quad \hat{\mathbf{C}}_{ek} = \hat{\mathbf{F}}_{ek}^{T}\hat{\mathbf{F}}_{ek}\,, \quad \hat{\mathbf{\Gamma}}_{vk} = \frac{1}{2}[\mathbf{1} - \mathbf{F}_{vk}^{T-1}\mathbf{F}_{vk}^{-1}] \tag{13.326}$$

Non-Equilibrium Free Energy

$$\psi_{ov} = \sum_k \psi_{ovk}(\hat{\mathbf{\Gamma}}_{ek}) \tag{13.327}$$

Overstress

$$\hat{\mathbf{S}}_{ovk} = \rho_R \frac{d\psi_{ovk}}{d\hat{\mathbf{\Gamma}}_{ek}} \tag{13.328}$$

$$\tilde{\mathbf{T}}_{ov} = \rho_R \sum_k \mathbf{F}_{vk}^{-1}\frac{d\psi_{ovk}}{d\hat{\mathbf{\Gamma}}_{ek}}\mathbf{F}_{vk}^{T-1} \tag{13.329}$$

Flow Rules

$$\eta_k \overset{\triangle}{\hat{\mathbf{\Gamma}}}_{vk} = \rho_R \hat{\mathbf{C}}_{ek}\frac{d\psi_{ovk}}{d\hat{\mathbf{\Gamma}}_{ek}} \tag{13.330}$$

The dissipation associated with the non-equilibrium part then reads

$$\rho_R \delta_{ov} = \sum_k \eta_k\,\mathrm{tr}\left(\hat{\mathbf{C}}_{ek}\frac{d\psi_{ovk}}{d\hat{\mathbf{\Gamma}}_{ek}}\right)^2 . \tag{13.331}$$

When modelling the non-equilibrium part, it is also possible to do without the decompositions $\mathbf{F} = \hat{\mathbf{F}}_{ek}\mathbf{F}_{vk}$. The following alternative approach gives rise to evolution equations for *Cauchy* type overstresses.

[45] Cf. (10.117) - (10.127).

Theorem 13. 25

The *free energy*

$$\psi_{\mathrm{ov}} = \mu_0 \sum_k \lambda_k \mathrm{tr}(\mathbf{A} - \mathbf{Z}_k)^2 , \tag{13.332}$$

$$\sum_{k=1}^{\mathrm{N}} \lambda_k = 1 , \tag{13.333}$$

the *overstress*

$$\mathbf{S}_{\mathrm{ov}} = \sum_k \Big(1 + \frac{1}{\mu_0 \lambda_k} \mathbf{X}_{\mathrm{ov}k}\Big) \mathbf{X}_{\mathrm{ov}k} , \tag{13.334}$$

$$\tilde{\mathbf{T}}_{\mathrm{ov}} = \mathbf{F}^{-1} \mathbf{S}_{\mathrm{ov}} \mathbf{F}^{\mathrm{T}-1} , \tag{13.335}$$

and the *evolution equations*

$$\overset{\triangledown}{\mathbf{X}}_{\mathrm{ov}k} = 2\mu_0 \lambda_k \mathbf{D} - \frac{1}{z_{0k} M} \mathbf{X}_{\mathrm{ov}k} - \beta \lambda_k \overset{\triangledown}{\mathbf{S}}_{\mathrm{eq}} \tag{13.336}$$

with

$$\beta = \Big\langle \Big(\sum_k \mathbf{X}_{\mathrm{ov}k}\Big) \cdot \overset{\triangledown}{\mathbf{S}}_{\mathrm{eq}} \Big\rangle^0 \tag{13.337}$$

form a thermomechanically consistent constitutive model for representing material behaviour in non-equilibrium. Here, $\mathbf{A}$ is the *Almansi* strain tensor; the internal variables $\mathbf{Z}_k$ are tensors of the same type. □

Proof

by inserting the free energy ψ_{ov} into the dissipation inequality. This now reads

$$\mathbf{S}_{\mathrm{ov}} \cdot \mathbf{D} - \rho_{\mathrm{R}} \dot{\psi}_{\mathrm{ov}}(t) \geq 0 . \tag{13.338}$$

In analogy to (13.261) we take $\overset{\triangle}{\mathbf{A}} = \mathbf{D}$ into account and calculate

$$(\mathbf{A} - \mathbf{Z}_k) \cdot (\dot{\mathbf{A}} - \dot{\mathbf{Z}}_k)$$

$$= \Big[(\mathbf{A} - \mathbf{Z}_k)\Big(\mathbf{1} + 2(\mathbf{A} - \mathbf{Z}_k)\Big)\Big] \cdot \mathbf{D} - (\mathbf{A} - \mathbf{Z}_k) \cdot \Big[\overset{\triangledown}{\mathbf{Z}}_k + 2(\mathbf{DA} + \mathbf{AD})\Big] .$$

This prompts the definition

$$\mathbf{X}_{\mathrm{ov}k} = 2\mu_0\,\lambda_k(\mathbf{A} - \mathbf{Z}_k)\,, \tag{13.339}$$

and the result for the dissipation reads

$$\rho_{\mathrm{R}}\delta_{\mathrm{ov}} = \Big[\mathbf{S}_{\mathrm{ov}} - \sum_k \Big(\mathbf{1} + \frac{1}{\mu_0\lambda_k}\mathbf{X}_{\mathrm{ov}k}\Big)\mathbf{X}_{\mathrm{ov}k}\Big]\cdot\mathbf{D}$$

$$+ \sum_k \mathbf{X}_{\mathrm{ov}k}\cdot\Big[\overset{\triangledown}{\mathbf{Z}}_k + 2(\mathbf{DA} + \mathbf{AD})\Big]\,.$$

The definition (13.339) of the tensors $\mathbf{X}_{\mathrm{ov}k}$ now leads to

$$\overset{\triangledown}{\mathbf{Z}}_k + 2(\mathbf{DA} + \mathbf{AD}) = \overset{\triangledown}{\mathbf{A}} - \frac{1}{2\mu_0\lambda_k}\overset{\triangledown}{\mathbf{X}}_{\mathrm{ov}k} + 2(\mathbf{DA} + \mathbf{AD})$$

$$= \mathbf{D} - \frac{1}{2\mu_0\lambda_k}\overset{\triangledown}{\mathbf{X}}_{\mathrm{ov}k}\,,$$

whereby $\overset{\triangledown}{\mathbf{A}} = \dot{\mathbf{A}} - \mathbf{LA} - \mathbf{AL}^{\mathrm{T}}$ has been observed. The dissipation inequality provides the stress relation (13.334) and becomes

$$\rho_{\mathrm{R}}\delta_{\mathrm{ov}} = \sum_k \mathbf{X}_{\mathrm{ov}k}\cdot\Big(\mathbf{D} - \frac{1}{2\mu_0\lambda_k}\overset{\triangledown}{\mathbf{X}}_{\mathrm{ov}k}\Big) \geq 0\,. \tag{13.340}$$

With the evolution equations (13.336) for the partial overstresses the final result is the mechanical dissipation

$$\rho_{\mathrm{R}}\delta_{\mathrm{ov}} = \sum_k \frac{1}{2\mu_0\lambda_k z_{0k}M}\,\mathrm{tr}\,(\mathbf{X}_{\mathrm{ov}k})^2 + \frac{\beta}{2\mu_0}\Big(\sum_k \mathbf{X}_{\mathrm{ov}k}\Big)\cdot\overset{\triangledown}{\mathbf{S}}_{\mathrm{eq}}\,. \tag{13.341}$$

■

The theorem provides a thermomechanically consistent generalisation of the material model (12.99) - (12.104) to large deformations.

The purpose of the last term on the right-hand side of the evolution equations (13.336) is to let the change in the state of stress be determined by the spontaneous elasticity (μ_0), if a loading process evolves at a finite rate from an initial equilibrium state.[46] No provision was made for this in the constitutive model (13.317) - (13.330).

[46] Cf. the corresponding evolution equations (12.100) for small deformations.

13. 5. 3 Conclusion

The considerations so far have shown that there are various systematic methods for modelling thermomechanical material behaviour in a thermodynamically consistent manner, i.e. in compliance with the *dissipation inequality* expressing the *second law of thermodynamics.*

The general theory of viscoplasticity includes the more special constitutive theories (elasticity, viscoelasticity and plasticity) by construction. The material response consists of two parts as a general rule, each of which can be modelled separately: the *equilibrium response* is a rate-independent functional of the thermomechanical process history representing an accompanying process that comprises a sequence of thermodynamic equilibrium states. All deviations from thermodynamic equilibrium are represented by rate-dependent functionals with fading memory properties, which asymptotically tend towards zero for static continuations or slow processes. All the constitutive equations are formulated in such a way that the *Clausius-Duhem* inequality is satisfied for all thermodynamic processes which are conceivable in the context of a given material model. In principle, it is possible for a temperature-dependence of the material behaviour stipulated by experimental data to be incorporated within this framework. This requires a considerable amount of experiments in each individual case, however, to allow a realistic representation of the material parameters' dependence on temperature. Of course, it is possible that further effects of thermomechanical coupling exist, which necessitate specific extensions of the presented theories.[47]

A further aspect of the thermomechanically consistent material description concerns the reproduction of the thermomechanical conversion of energy, i.e. how the consequences drawn from constitutive modelling affect the formulation of the heat conduction equation. These questions still contain numerous problems connected with the provision of test data and basic modelling queries that remain to be solved.[48]

[47] One interesting example may be found in HAUPT & HELM [1999].

[48] CHRYSOCHOOS & BELMAJOUB [1992]; CHRYSOCHOOS et al. [1989]; GHONEIM [1990]; OLIFERUK et al. [1993]; KAMLAH & HAUPT [1998].

13. 6 Anisotropic Material Properties

13. 6. 1 Motivation

In connection with the concept of an intermediate configuration, isotropic functions have always been used so far. This applies both to the elasto-plasticity in Sect. 11. 3, to the viscoplasticity in Sect. 12. 2 and the thermomechanical generalisations of these theories in this chapter. The consequence resulting from using isotropic material functions is the fact that the evolution equations for the intermediate configurations are, without exception, evolution equations for stretch tensors, *Cauchy-Green* tensors or *Green* strain tensors. That means that only the metric of an intermediate configuration is physically significant; the orientation of the intermediate configuration or its rotational part ($\mathbf{R}_i$ in $\mathbf{F} = \mathbf{R}_i\mathbf{U}_i$) is dropped from all constitutive equations. As a general consequence, the constitutive modelling is restricted to materials with elasticity properties which are initially isotropic.

There are definitely possibilities for modelling certain anisotropies in material behaviour with the help of isotropic functions. One example of this is the modelling of kinematic hardening, which expresses the dependence of the plastic flow on direction due to the circumstance that the yield function is an isotropic function of two tensorial variables, namely the stress and the backstress. Anisotropies in elasticity properties can also be represented by means of isotropic functions depending not only on the elastic strain but also on structural tensors. However, the actual scope for representation is very limited so far. Elasticity relations with isotropic functions containing only the elastic strain tensor as an independent variable are in any case restricted to modelling just isotropic elasticity properties. In this respect, none of the material models discussed so far, in which intermediate configurations are employed, is in a position to take the elastic anisotropy behaviour of real materials into account.

The aim of the following explanations is to point out a way of overcoming these shortcomings, at least in part. This approach is based on the established representation possibilities for anisotropic elasticity, as set out in Chap. 9 for hyperelastic materials. Since the thermomechanically consistent material modelling takes priority anyway, the assumption of hyperelasticity does not constitute a restriction.

13. 6. 2 Axes of Elastic Anisotropy

An anisotropic hyperelastic solid is characterised by a symmetry group $\mathcal{g}_R \subset \mathcal{Orth}, \mathcal{g}_R \neq \mathcal{Orth}$: a reference configuration R exists, so that all rotations $\mathbf{Q} \in \mathcal{g}_R$ leave the strain energy or isothermal free energy invariant:

$$\mathbf{Q} \in \mathcal{g}_R \Longleftrightarrow \psi_e = \varphi_R(\mathbf{E}) = \varphi_R(\mathbf{QEQ}^T)\,.$$

The existence of various symmetry groups with physical meaning arises out of the crystalline structure of the material, i.e. out of the symmetry properties of the crystal lattices. Each type of lattice has its own symmetry group of rotations mapping the crystal onto itself. There is a complete system of scalars $J_k(\mathbf{E})$, which are invariant when $\mathbf{E}$ is subjected to the rotations of the symmetry group, and the free energy (or strain energy) can be written as a function of these invariants:[49]

$$\psi_e = \varphi_R(\mathbf{E}) = \varphi_R(J_1, J_2, \dots, J_N)\,. \tag{13.342}$$

If a symmetry group $\mathcal{g}_R \subset \mathcal{Orth}$ is known in connection with an undistorted reference configuration R, then it is possible to specify it when referring to another reference configuration $\hat{R}$. According to (7.62) the relation $\mathcal{g}_{\hat{R}} = \mathbf{P}\mathcal{g}_R\mathbf{P}^{-1}$ holds, $\mathbf{P} = \text{Grad}\,\hat{\boldsymbol{X}}(\boldsymbol{X})$ being the gradient of the mapping that leads from R to $\hat{R}$. It is evident that a general change of the reference configuration produces a group $\mathcal{g}_{\hat{R}}$ that does not, as a rule, consist merely of orthogonal tensors. The existence of a *natural reference configuration* R is required for the definition of a solid, i.e. for $\mathcal{g}_R \subset \mathcal{Orth}$.

Only in the special case of a change of reference configuration, consisting of a pure rotation itself, does the symmetrical group remain a subset of the group of orthogonal tensors. The transformation $\mathbf{P} = \mathbf{P}_o$ $(\mathbf{P}_o\mathbf{P}_o^T = \mathbf{1})$ leaves the symmetry group essentially unaltered,

$$\mathcal{g}_{\hat{R}} = \mathbf{P}_o\mathcal{g}_R\mathbf{P}_o^T\,, \tag{13.343}$$

since every element of $\mathcal{g}_R$ is subjected to one and the same rotation, $\hat{\mathbf{Q}} = \mathbf{P}_o\mathbf{Q}\mathbf{P}_o^T$. A physically significant consequence to be gained from these simple fact is

---

[49] Cf. Chap. 9.

Theorem 13. 26

Let $\mathcal{G}_R \subset \mathcal{O}rth$ be a symmetry group and $\mathbf{P}_o$ any orthogonal tensor. Then the following statements are valid:

1) If $\psi_e = \varphi_R(\mathbf{E})$ is a strain energy related to the reference configuration R with the symmetry property

$$\psi_e = \varphi_R(\mathbf{E}) = \varphi_R(\mathbf{QEQ}^T) \text{ for all } \mathbf{Q} \in \mathcal{G}_R ,$$

then $\psi_e = \varphi_{\hat{R}}(\hat{\mathbf{E}}) = \varphi_R(\mathbf{P}_o{}^T\hat{\mathbf{E}}\mathbf{P}_o)$ is the strain energy related to the reference configuration $\hat{R}$ with the symmetry property

$$\varphi_{\hat{R}}(\hat{\mathbf{E}}) = \varphi_{\hat{R}}(\hat{\mathbf{Q}}\hat{\mathbf{E}}\hat{\mathbf{Q}}^T) \text{ for all } \hat{\mathbf{Q}} \in \mathcal{G}_{\hat{R}} = \mathbf{P}_o\mathcal{G}_R\mathbf{P}_o{}^T .$$

2) If e represents the axis of rotation allocated to $\mathbf{Q} \in \mathcal{G}_R$, then $\hat{e} = \mathbf{P}_o e$ is the axis of rotation associated to $\hat{\mathbf{Q}} = \mathbf{P}_o\mathbf{Q}\mathbf{P}_o{}^T \in \mathcal{G}_{\hat{R}}$. □

Proof

1) $\hat{\mathbf{F}} = \mathbf{FP}^{-1} = \mathbf{FP}_o{}^T$ leads to $\hat{\mathbf{E}} = \mathbf{P}_o\mathbf{EP}_o{}^T$ and to

$$\varphi_R(\mathbf{E}) = \varphi_{\hat{R}}(\mathbf{P}_o\mathbf{EP}_o{}^T) \Longleftrightarrow \varphi_{\hat{R}}(\hat{\mathbf{E}}) = \varphi_R(\mathbf{P}_o{}^T\hat{\mathbf{E}}\mathbf{P}_o).$$

The symmetry property $\varphi_R(\mathbf{E}) = \varphi_R(\mathbf{QEQ}^T)$ for all $\mathbf{Q} \in \mathcal{G}_R$ implies

$$\varphi_R(\mathbf{P}_o{}^T\hat{\mathbf{E}}\mathbf{P}_o) = \varphi_R(\mathbf{QP}_o{}^T\hat{\mathbf{E}}\mathbf{P}_o\mathbf{Q}^T) ,$$

which is equivalent to

$$\varphi_{\hat{R}}(\hat{\mathbf{E}}) = \varphi_{\hat{R}}(\mathbf{P}_o\mathbf{QP}_o{}^T\hat{\mathbf{E}}\mathbf{P}_o\mathbf{Q}^T\mathbf{P}_o{}^T) \text{ for all } \hat{\mathbf{Q}} = \mathbf{P}_o\mathbf{QP}_o{}^T \in \mathcal{G}_{\hat{R}} = \mathbf{P}_o\mathcal{G}_R\mathbf{P}_o{}^T.$$

2) $\mathbf{Q}e = e$ leads to $\hat{\mathbf{Q}}\hat{e} = (\mathbf{P}_o\mathbf{QP}_o{}^T)(\mathbf{P}_o e) = \mathbf{P}_o e = \hat{e}$. ■

Definition 13. 11

In the representations $\psi_e = \varphi_R(\mathbf{E})$ and

$$\psi_e = \varphi_{\hat{R}}(\mathbf{P}_o\mathbf{EP}_o{}^T) \tag{13.344}$$

of the strain energy of a hyperelastic material, the orthogonal tensor $\mathbf{P}_o$ gives rise to a *rotation of the axes of elastic anisotropy.* ◊

The functions $\varphi_R(\cdot)$ and $\varphi_{\hat{R}}(\cdot)$ depend on the same system of invariants; however, in the case of $\varphi_R(\cdot)$ the invariants of the tensor $\mathbf{E}$ are formed, and in the case of the function $\varphi_{\hat{R}}(\cdot)$ the invariants of the tensor $\hat{\mathbf{E}} = \mathbf{P}_o\mathbf{E}\mathbf{P}_o^{\,T}$. Thus, the strain energy of an anisotropic elastic material can be written as

$$\psi_e = \hat{\varphi}(\hat{J}_1, \hat{J}_2, \dots, \hat{J}_N)\,, \tag{13.345}$$

where the arguments $\hat{J}_k = \hat{J}_k(\mathbf{P}_o\mathbf{E}\mathbf{P}_o^{\,T})$ $(k = 1, 2, \dots, N)$ are the complete set of scalars, which are invariant under all $\hat{\mathbf{Q}} \in \mathscr{g}_{\hat{R}} = \mathbf{P}_o\mathscr{g}_R\mathbf{P}_o^{\,T}$. The application of this representation in thermoelasticity provides

Theorem 13. 27

Let $\mathscr{g}_R \subset \mathcal{Orth}$ be a given symmetry group implying a basis of invariants $J_1(\mathbf{E}), J_2(\mathbf{E}), \dots, J_N(\mathbf{E})$ and $\boldsymbol{X} \longmapsto \mathbf{P}_o(\boldsymbol{X})$ a space-dependent orientation of the anisotropy axes (eigenvectors of the generators of the symmetry group). Then, the isothermal stress relation reads

$$\tilde{\mathbf{T}} = \rho_R \frac{\partial}{\partial \mathbf{E}}\, \varphi_{\hat{R}}(\mathbf{P}_o\mathbf{E}\mathbf{P}_o^{\,T})\,, \tag{13.346}$$

where the strain energy (the isothermal free energy) ψ_e is a function of the invariants $\hat{J}_k = \hat{J}_k(\mathbf{P}_o\mathbf{E}\mathbf{P}_o^{\,T})$:

$$\varphi_{\hat{R}}(\mathbf{P}_o\mathbf{E}\mathbf{P}_o^{\,T}) = \hat{\varphi}(\hat{J}_1, \hat{J}_2, \dots, \hat{J}_N)\,. \tag{13.347}$$

□

Proof

Inserting the strain energy into the dissipation inequality for isothermal conditions,

$$\tilde{\mathbf{T}}\cdot\dot{\mathbf{E}} - \rho_R\dot{\psi}(t) \geq 0\,, \tag{13.348}$$

leads to

$$\left\{\tilde{\mathbf{T}} - \rho_R\,\mathbf{P}_o^{\,T}\frac{d\hat{\varphi}(\mathbf{P}_o\mathbf{E}\mathbf{P}_o^{\,T})}{d(\mathbf{P}_o\mathbf{E}\mathbf{P}_o^{\,T})}\mathbf{P}_o\right\}\cdot\dot{\mathbf{E}} - \rho_R\dot{\psi}(t) \geq 0\,, \tag{13.349}$$

which implies

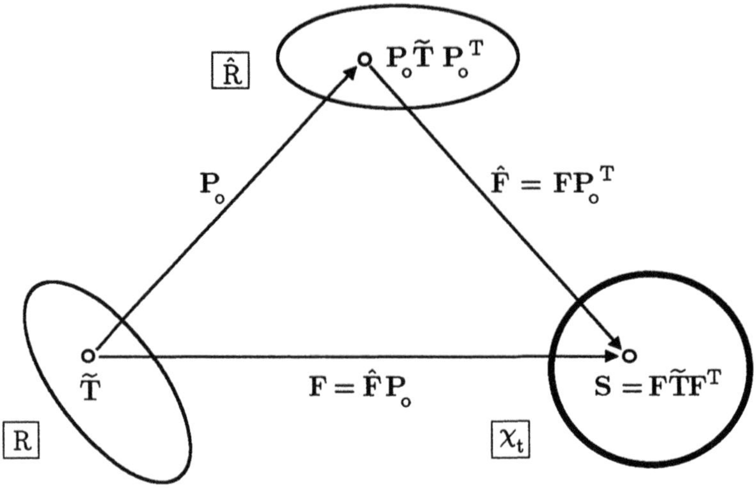

Figure 13. 5: Rotation of the reference configuration

$$\tilde{T} = \rho_R \, P_o^T \frac{d\hat{\varphi}(P_o E P_o^T)}{d(P_o E P_o^T)} P_o \tag{13.350}$$

or, equivalently,[50] (13.346). ■

The physical meaning of the theorem is explained as follows: the orientation $P_o(X)$ of the anisotropy axes defines a new reference configuration. The function $\psi_e = \hat{\varphi}(P_o E P_o^T)$ delivers the strain energy related to rotated reference configuration, and the derivative with respect to its argument the corresponding stress tensor. The stress relation (13.350) or (13.346) calculates the stress tensor operating on the original reference configuration R; this is sketched in Fig. 13. 5.

If the orientation P_o is constant in space, (13.346) supplies the stress relation of a single crystal; in the case of a spatial distribution $P_o = P_o(X)$ the elastic behaviour of a polycrystal is represented by (13.346). The axes of anisotropy are constant in time as far as elastic behaviour is concerned. However, the orientation may vary due to inelastic deformation processes. This will be accounted for in the next section.

[50] With application of eq. (9.60).

13. 7 Anisotropic Viscoplasticity

13. 7. 1 General Considerations

As it is common in plasticity and viscoplasticity, the following development starts from the multiplicative decomposition of the deformation gradient into elastic and inelastic parts:[51]

$$\mathbf{F} = \hat{\mathbf{F}}_e \mathbf{F}_i \tag{13.351}$$

This decomposition is not unique, because an arbitrary orthogonal transformation $\bar{\mathbf{Q}}$ can be inserted:

$$\mathbf{F} = \hat{\mathbf{F}}_e \bar{\mathbf{Q}}^T \bar{\mathbf{Q}} \mathbf{F}_i \tag{13.352}$$

The inelastic part $\mathbf{F}_i$ of the deformation gradient defines the intermediate configuration; in view of the polar decomposition, $\mathbf{F}_i = \mathbf{R}_i \mathbf{U}_i$, the inserted $\bar{\mathbf{Q}}$ changes the rotational part of the intermediate configuration. Since $\bar{\mathbf{Q}}$ is arbitrary, the rotation $\mathbf{R}_i$ of the intermediate state should not be the appropriate quantity to represent structural orientations within the material. Moreover, all constitutive equations set up in the sequel should be invariant with respect to the arbitrary choice of the rotation $\bar{\mathbf{Q}}$.[52]

The constitutive model of anisotropic viscoplasticity,[53] developed and discussed in this section, is based on a list of kinematic and dynamic quantities, which have already been discussed in earlier context:[54]

Decomposition of the Total Strain

$$\hat{\mathbf{\Gamma}} = \mathbf{F}_i^{T-1} \mathbf{E} \mathbf{F}_i^{-1} = \hat{\mathbf{\Gamma}}_e + \hat{\mathbf{\Gamma}}_i \tag{13.353}$$

---

[51] See Sect. 11. 3. 2 and Sect. 12. 2. 2.

[52] In Chapters 11 and 12 this invariance postulate was trivially satisfied, because all constitutive assumptions were formulated in the intermediate configuration using isotropic functions, see e.g. Theorem 11. 7. The consequence is, however, that only isotropic elasticity properties can be incorporated into the constitutive theory if structural variables are not introduced.

[53] A comprehensive survey of the extensive field of anisotropy in the context of inelastic deformations is given in Kocks et al. [1998].
A valuable introduction to anisotropic plasticity is found in Banabic et al. [2000]. This informative work covers fundamentals of constitutive modeling, basic facts from materials science and experimental identification as well as numerical simulation techniques. In particular, the relation to metal forming technology is thoroughly discussed and complemented with examples of application and numerous references.

[54] See Sects. 11. 3, 12. 2, 13. 4.

Elastic Strain

$$\hat{\boldsymbol{\Gamma}}_e = \frac{1}{2}\left(\hat{\mathbf{F}}_e^{\mathrm{T}}\hat{\mathbf{F}}_e - \mathbf{1}\right) \tag{13.354}$$

Inelastic Strain

$$\hat{\boldsymbol{\Gamma}}_i = \frac{1}{2}\left(\mathbf{1} - \mathbf{F}_i^{\mathrm{T}-1}\mathbf{F}_i^{-1}\right) \tag{13.355}$$

Inelastic Deformation Rate

$$\hat{\mathbf{L}}_i = \dot{\mathbf{F}}_i\mathbf{F}_i^{-1} \tag{13.356}$$

Total Strain Rate

$$\overset{\triangle}{\hat{\boldsymbol{\Gamma}}} = \mathbf{F}_i^{\mathrm{T}-1}\dot{\mathbf{E}}\mathbf{F}_i^{-1} = \dot{\hat{\boldsymbol{\Gamma}}} + \hat{\mathbf{L}}_i^{\mathrm{T}}\hat{\boldsymbol{\Gamma}} + \hat{\boldsymbol{\Gamma}}\hat{\mathbf{L}}_i = \overset{\triangle}{\hat{\boldsymbol{\Gamma}}}_e + \overset{\triangle}{\hat{\boldsymbol{\Gamma}}}_i \tag{13.357}$$

Elastic Strain Rate

$$\overset{\triangle}{\hat{\boldsymbol{\Gamma}}}_e = \dot{\hat{\boldsymbol{\Gamma}}}_e + \hat{\mathbf{L}}_i^{\mathrm{T}}\hat{\boldsymbol{\Gamma}}_e + \hat{\boldsymbol{\Gamma}}_e\hat{\mathbf{L}}_i \tag{13.358}$$

Inelastic Strain Rate

$$\overset{\triangle}{\hat{\boldsymbol{\Gamma}}}_i = \dot{\hat{\boldsymbol{\Gamma}}}_i + \hat{\mathbf{L}}_i^{\mathrm{T}}\hat{\boldsymbol{\Gamma}}_i + \hat{\boldsymbol{\Gamma}}_i\hat{\mathbf{L}}_i = \frac{1}{2}\left(\hat{\mathbf{L}}_i + \hat{\mathbf{L}}_i^{\mathrm{T}}\right) \tag{13.359}$$

Piola-Kirchhoff Stress

$$\hat{\mathbf{S}} = \mathbf{F}_i\tilde{\mathbf{T}}\mathbf{F}_i^{\mathrm{T}} = \hat{\mathbf{F}}_e^{-1}\mathbf{S}\,\hat{\mathbf{F}}_e^{\mathrm{T}-1}\,, \tag{13.360}$$

Mandel Stress

$$\hat{\mathbf{P}} = \hat{\mathbf{F}}_e^{\mathrm{T}}\hat{\mathbf{F}}_e\hat{\mathbf{S}} = \left(\mathbf{1} + 2\hat{\boldsymbol{\Gamma}}_e\right)\hat{\mathbf{S}} = \hat{\mathbf{F}}_e^{\mathrm{T}}\mathbf{S}\hat{\mathbf{F}}_e^{\mathrm{T}-1} \tag{13.361}$$

All these tensors operate on the intermediate configuration. Their transformation properties are laid down in

Theorem 13. 28

Let a change of frame be represented by an orthogonal tensor-valued function $t \longmapsto \mathbf{Q}(t)$ and a local rotation of the intermediate configuration by $(\boldsymbol{X}, t) \longmapsto \bar{\mathbf{Q}}(\boldsymbol{X}, t)$. Then, the following transformations apply for the various kinematic and dynamic quantities:

$$\mathbf{F} \quad \longmapsto \quad \hat{\mathbf{F}}^* = \mathbf{Q}\mathbf{F} \tag{13.362}$$

$$\hat{\mathbf{F}}_e \quad \longmapsto \quad \hat{\mathbf{F}}_e^{\,*} = \mathbf{Q}\hat{\mathbf{F}}_e\bar{\mathbf{Q}}^{\mathrm{T}} \tag{13.363}$$

$$\hat{\Gamma}_e \quad \longmapsto \quad \hat{\Gamma}_e^{\,*} = \bar{\mathbf{Q}}\hat{\Gamma}_e\bar{\mathbf{Q}}^{\mathrm{T}} \tag{13.364}$$

$$\mathbf{F}_i \quad \longmapsto \quad \mathbf{F}_i^{\,*} = \bar{\mathbf{Q}}\mathbf{F}_i \tag{13.365}$$

$$\hat{\Gamma}_i \quad \longmapsto \quad \hat{\Gamma}_i^{\,*} = \bar{\mathbf{Q}}\hat{\Gamma}_i\bar{\mathbf{Q}}^{\mathrm{T}} \tag{13.366}$$

$$\hat{\mathbf{L}}_i \quad \longmapsto \quad \hat{\mathbf{L}}_i^{\,*} = \dot{\bar{\mathbf{Q}}}\bar{\mathbf{Q}}^{\mathrm{T}} + \bar{\mathbf{Q}}\hat{\mathbf{L}}_i\bar{\mathbf{Q}}^{\mathrm{T}} \tag{13.367}$$

$$\hat{\mathbf{S}} \quad \longmapsto \quad \hat{\mathbf{S}}^* = \bar{\mathbf{Q}}\hat{\mathbf{S}}\bar{\mathbf{Q}}^{\mathrm{T}} \tag{13.368}$$

$$\hat{\mathbf{P}} \quad \longmapsto \quad \hat{\mathbf{P}}^* = \bar{\mathbf{Q}}\hat{\mathbf{P}}\bar{\mathbf{Q}}^{\mathrm{T}} \tag{13.369}$$

□

Proof

The transformation rules (13.362), (13.363), (13.365) are evident by definition, the other follow by simple calculation. ■

The specific stress power

$$\rho_{\mathrm{R}}\ell_{\mathrm{i}} = \widetilde{\mathbf{T}}\cdot\dot{\mathbf{E}} = \hat{\mathbf{S}}\cdot\overset{\triangle}{\hat{\Gamma}} \tag{13.370}$$

decomposes into elastic and inelastic parts according to (13.258) and the definition (13.361),

$$\rho_{\mathrm{R}}\ell_{\mathrm{i}} = \hat{\mathbf{S}}\cdot\dot{\hat{\Gamma}}_e + \hat{\mathbf{P}}\cdot\hat{\mathbf{L}}_i\,. \tag{13.371}$$

The total stress power is an objective scalar under a change of frame and a rotation of the intermediate configuration: $\ell_i = \ell_i^{\,*}$.

13. 7. 2 Free Energy Function

The basic assumption for the following thermomechanical model of viscoplasticity[55] is the definition of a free energy function. In the following only isothermal processes are considered; a generalisation to varying temperature fields is straight-forward if no phase transitions occur and the

[55] Cf. HAUPT & KERSTEN [2001].

conventional thermodynamic frame is retained. As usual, the free energy is assumed to be the sum of one elastic and one inelastic part:

$$\psi = \psi_e + \psi_i \tag{13.372}$$

Following the original idea of Ch. Tsakmakis,[56] the strain energy ψ_e should not only depend on the elastic part of the deformation gradient, but in addition on the orientation of the crystal. A possibility to consider this orientation was provided in the preceding section; accordingly, the orientation of the crystal lattice is incorporated by means of an orthogonal tensor-valued internal variable $\mathbf{\Phi}$, for which an appropriate evolution equation has to be formulated later:

$$\psi_e = \psi_e(\hat{\mathbf{F}}_e, \mathbf{\Phi}) \,. \tag{13.373}$$

Because the internal variable $\mathbf{\Phi}$ is introduced to represent the orientation of the crystal in the intermediate configuration, it must be transformed in the same manner as $\mathbf{F}_i(\boldsymbol{X}, t)$, namely like (13.365):[57]

$$\mathbf{\Phi} \longmapsto \mathbf{\Phi}^* = \bar{\mathbf{Q}}\mathbf{\Phi} \,. \tag{13.374}$$

This calls for a reduced form provided by

Theorem 13. 29

The elastic part (13.373) of the free energy satisfies the property of full invariance, i.e. it is invariant with respect to any given transformations $\bar{\mathbf{Q}}(\boldsymbol{X}, t)$ and $\mathbf{Q}(t)$ if and only if the representation

$$\psi_e = \varphi(\hat{\mathbf{F}}_e, \mathbf{\Phi}) = \hat{\psi}_e(\mathbf{\Phi}^{\mathrm{T}}\hat{\mathbf{\Gamma}}_e\mathbf{\Phi}) \tag{13.375}$$

is valid. □

Proof

Under the assumption that ψ_e fulfils the property of full invariance, the following identity holds for all $\mathbf{Q}$ and $\bar{\mathbf{Q}}$:

$$\varphi(\hat{\mathbf{F}}_e, \mathbf{\Phi}) = \varphi(\mathbf{Q}\hat{\mathbf{F}}_e\bar{\mathbf{Q}}^{\mathrm{T}}, \bar{\mathbf{Q}}\mathbf{\Phi}) \,.$$

---

[56] HÄUSLER [1999]; HÄUSLER, TSAKMAKIS & AIFANTIS [1999]; TSAKMAKIS [2001b, c].

[57] It will be explained below, that the transformation $\mathbf{\Phi}$ rotates the so-called isoclinic intermediate configuration, introduced in MANDEL [1973], into the intermediate configuration. See eq. (13.383).

Accordingly, the choice of $\mathbf{Q} = \bar{\mathbf{Q}}\hat{\mathbf{R}}_e^{\,T}$ with $\hat{\mathbf{R}}_e$ out of $\hat{\mathbf{F}}_e = \hat{\mathbf{R}}_e\hat{\mathbf{U}}_e$ leads to the identity

$$\psi_e(\hat{\mathbf{F}}_e, \boldsymbol{\Phi}) = \psi_e(\bar{\mathbf{Q}}\hat{\mathbf{U}}_e\bar{\mathbf{Q}}^T, \bar{\mathbf{Q}}\boldsymbol{\Phi}) ,$$

which has to apply for all $\bar{\mathbf{Q}}$. By choosing $\bar{\mathbf{Q}} = \boldsymbol{\Phi}^T$ one obtains

$$\psi_e(\hat{\mathbf{F}}_e, \boldsymbol{\Phi}) = \psi_e(\boldsymbol{\Phi}^T\hat{\mathbf{U}}_e\boldsymbol{\Phi}, \mathbf{1}) = \hat{\psi}_e(\boldsymbol{\Phi}^T\hat{\boldsymbol{\Gamma}}_e\boldsymbol{\Phi}) .$$

On the other hand, it is clear that the representation $(13.375)_2$ satisfies the requirement of full invariance:

$$\hat{\psi}_e(\boldsymbol{\Phi}^{*T}\hat{\boldsymbol{\Gamma}}_e^{\,*}\boldsymbol{\Phi}^*) = \hat{\psi}_e((\bar{\mathbf{Q}}\boldsymbol{\Phi})^T(\bar{\mathbf{Q}}\hat{\boldsymbol{\Gamma}}_e\bar{\mathbf{Q}}^T)(\bar{\mathbf{Q}}\boldsymbol{\Phi})) = \hat{\psi}_e(\boldsymbol{\Phi}^T\hat{\boldsymbol{\Gamma}}_e\boldsymbol{\Phi}) . \qquad \blacksquare$$

In view of the argumentation of Sect. 13. 6. 2 the theorem shows that the internal variable $\boldsymbol{\Phi}$ is given a clear physical meaning: $\boldsymbol{\Phi}$ describes a change in the orientation of the axes of elastic anisotropy in the sense of Definition 13. 11. In particular, the strain energy according to $(13.375)_2$ depends on the elastic strain tensor, rotated back from the intermediate configuration to that configuration, which was introduced by J. Mandel[58] as the *isoclinic intermediate configuration*. This is clear from Sect. 13. 6. 2 if one identifies the rotation $\mathbf{P}_o$ with $\boldsymbol{\Phi}^T$.

The orientation of the anisotropy axes may be space-dependent, if one imagines a *polycrystal*. Temporal changes of the anisotropy axes due to inelastic deformation processes are also possible, i.e. $\boldsymbol{\Phi} = \boldsymbol{\Phi}(X, t)$.

The result (13.375) constitutes a *reduced form* for the elastic part of the free energy, which gives rise to the elasticity relation by differentiation (see below). Both the material function $\hat{\psi}_e(\cdot)$ and the implied stress relation have properties of symmetry which are characterised through the type of the crystal lattice and the corresponding symmetry group $\mathscr{g}_{\hat{R}} = \boldsymbol{\Phi}^T\mathscr{g}_{R}\,\boldsymbol{\Phi}$. The results achieved in Chap. 9 are also applicable for a concrete representation of $\psi_e = \hat{\psi}_e(\boldsymbol{\Phi}^T\hat{\boldsymbol{\Gamma}}_e\boldsymbol{\Phi})$.

When evaluating the dissipation inequality, it should be noted that the *Mandel* stress tensor $\hat{\mathbf{P}} = \hat{\mathbf{C}}_e\hat{\mathbf{S}}$ is no longer symmetric,

$$\hat{\mathbf{P}} \neq \hat{\mathbf{P}}^T . \tag{13.376}$$

---

[58] MANDEL [1971], p. 44. See also MANDEL [1973], p. 728 and the work of Y. F. Dafalias: DAFALIAS [1983, 1985, 1987, 1990, 1998].

The isothermal dissipation inequality

$$\hat{\mathbf{S}}\cdot\dot{\hat{\mathbf{\Gamma}}}_{\mathrm{e}} + \hat{\mathbf{P}}\cdot\hat{\mathbf{L}}_{\mathrm{i}} - \rho_{\mathrm{R}}\Big(\frac{\mathrm{d}}{\mathrm{d}t}\,\hat{\psi}_{\mathrm{e}}(\mathbf{\Phi}^{\mathrm{T}}(t)\hat{\mathbf{\Gamma}}_{\mathrm{e}}(t)\mathbf{\Phi}(t)) + \dot{\psi}_{\mathrm{i}}(t)\Big) \geq 0 \tag{13.377}$$

leads – after some intermediate calculations – to

$$\Big\{\hat{\mathbf{S}} - \rho_{\mathrm{R}}\mathbf{\Phi}\,\frac{\mathrm{d}\hat{\psi}_{\mathrm{e}}(\mathbf{\Phi}^{\mathrm{T}}\hat{\mathbf{\Gamma}}_{\mathrm{e}}\mathbf{\Phi})}{\mathrm{d}(\mathbf{\Phi}^{\mathrm{T}}\hat{\mathbf{\Gamma}}_{\mathrm{e}}\mathbf{\Phi})}\,\mathbf{\Phi}^{\mathrm{T}}\Big\}\cdot\dot{\hat{\mathbf{\Gamma}}}_{\mathrm{e}}$$

$$+ \hat{\mathbf{P}}\cdot\hat{\mathbf{L}}_{\mathrm{i}} - 2\rho_{\mathrm{R}}\Big\{\hat{\mathbf{\Gamma}}_{\mathrm{e}}\mathbf{\Phi}\,\frac{\mathrm{d}\hat{\psi}_{\mathrm{e}}(\mathbf{\Phi}^{\mathrm{T}}\hat{\mathbf{\Gamma}}_{\mathrm{e}}\mathbf{\Phi})}{\mathrm{d}(\mathbf{\Phi}^{\mathrm{T}}\hat{\mathbf{\Gamma}}_{\mathrm{e}}\mathbf{\Phi})}\,\mathbf{\Phi}^{\mathrm{T}}\Big\}\cdot\Big(\dot{\mathbf{\Phi}}\mathbf{\Phi}^{\mathrm{T}}\Big) - \rho_{\mathrm{R}}\dot{\psi}_{\mathrm{i}}(t) \geq 0\,; \tag{13.378}$$

the standard argumentation together with (9.60) implies the stress relation

$$\hat{\mathbf{S}} = \rho_{\mathrm{R}}\mathbf{\Phi}\,\frac{\mathrm{d}\hat{\psi}_{\mathrm{e}}(\mathbf{\Phi}^{\mathrm{T}}\hat{\mathbf{\Gamma}}_{\mathrm{e}}\mathbf{\Phi})}{\mathrm{d}(\mathbf{\Phi}^{\mathrm{T}}\hat{\mathbf{\Gamma}}_{\mathrm{e}}\mathbf{\Phi})}\,\mathbf{\Phi}^{\mathrm{T}} = \rho_{\mathrm{R}}\,\frac{\partial\hat{\psi}_{\mathrm{e}}(\mathbf{\Phi}^{\mathrm{T}}\hat{\mathbf{\Gamma}}_{\mathrm{e}}\mathbf{\Phi})}{\partial\hat{\mathbf{\Gamma}}_{\mathrm{e}}}\,. \tag{13.379}$$

The interpretation of this stress relation is similar to (13.346), observing $\mathbf{P}_{\mathrm{o}} = \mathbf{\Phi}^{\mathrm{T}}$: the derivative of the strain energy with respect to its argument, which operates on the isoclinic intermediate configuration, yields the corresponding stress. After differentiation this stress tensor is rotated to the intermediate configuration. With (13.379) the inequality (13.378) leads to the remaining inequality

$$\hat{\mathbf{P}}\cdot\hat{\mathbf{L}}_{\mathrm{i}} - \hat{\mathbf{P}}\cdot\Big(\dot{\mathbf{\Phi}}\mathbf{\Phi}^{\mathrm{T}}\Big) - \rho_{\mathrm{R}}\dot{\psi}_{\mathrm{i}}(t) \geq 0\,. \tag{13.380}$$

13. 7. 3 Evolution Equations

For metallic materials with crystalline structure inelastic deformations originate from the motion of dislocations. The phenomenological effects are processes of simple shearing taking place along certain slip systems. The slip systems are uniquely determined from the geometry of the crystal. A slip system in the reference configuration consists of a plane with normal $\boldsymbol{N}_\alpha$ and a slip direction $\boldsymbol{S}_\alpha$. The index α denotes the number of the particular slip system. For instance, in the case of a face-centered cubic (fcc) crystal, sketched in Fig. 13. 6, we recognise 4 different octahedral planes as slip planes and for each plane 3 different slip directions, such that altogether 12 slip systems exist in this case.

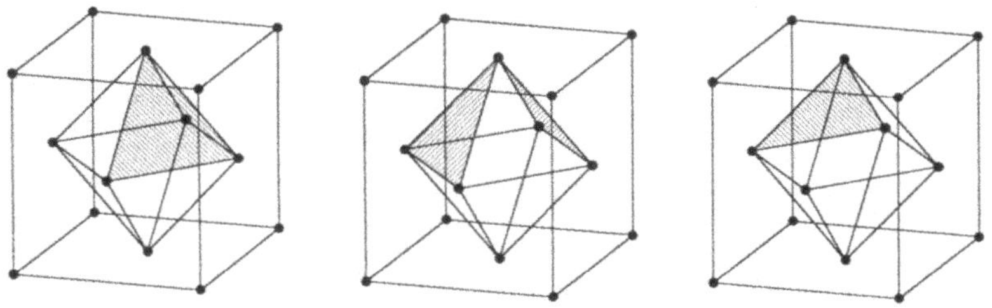

Figure 13. 6: Face-centered cubic crystal (fcc): 12 slip systems exist according to 4 different octahedral planes and 3 slip directions allocated to each plane

If $\dot{\kappa}_\alpha$ is the shear rate belonging to the slip system $\{\boldsymbol{N}_\alpha, \boldsymbol{S}_\alpha\}$, its tensor-valued contribution to the inelastic deformation rate is $\dot{\kappa}_\alpha \boldsymbol{S}_\alpha \otimes \boldsymbol{N}_\alpha$. All shear rates are superimposed to form an evolution equation for a tensor field $\widetilde{\mathbf{F}}_\mathrm{i}(\boldsymbol{X}, t)$:

$$\dot{\widetilde{\mathbf{F}}}_\mathrm{i} \widetilde{\mathbf{F}}_\mathrm{i}^{-1} = \sum_\alpha \dot{\kappa}_\alpha \boldsymbol{S}_\alpha \otimes \boldsymbol{N}_\alpha . \tag{13.381}$$

The shear deformations κ_α give rise to the inelastic part of the free energy: as a most simple constitutive assumption a quadratic form

$$\rho_\mathrm{R} \psi_\mathrm{i} = \sum_\alpha \frac{1}{2} c_\alpha (\kappa_\alpha - y_\alpha)^2 \tag{13.382}$$

is postulated, where the sum refers to all slip systems of the crystal. The scalars c_α are material constants and the y_α strain-like internal variables to account for hardening effects on the level of the slip systems.

Elementary arguments of geometry suggest that inelastic shear deformations taking place on slip systems influence their orientation in space. This is schematically illustrated in Fig. 13. 7 for the special case of only one slip system under simple tension or compression. As it is visualised in the figure, the shear direction rotates towards the direction of the loading force in the case of simple tension, whereas under compression the normal of the slip plane turns into the direction of the compressive load. Following Ch. Tsakmakis, the change of orientation is represented by the internal variable $\boldsymbol{\Phi}$ $(\boldsymbol{\Phi}\boldsymbol{\Phi}^\mathrm{T} = \mathbf{1})$.[59]

[59] This corresponds to the role of $\boldsymbol{\Phi}$ in the representation (13.375) of the strain energy.

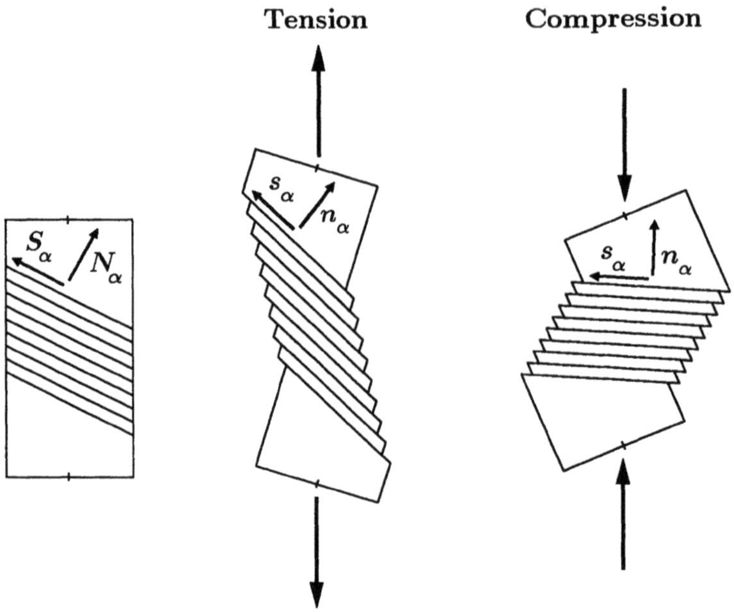

Figure 13. 7: Change in orientation under tension/compression due to single slip

Definition 13. 12

The field $(\boldsymbol{X},\, t) \longmapsto \widetilde{\mathbf{F}}_{\mathrm{i}}(\boldsymbol{X},\, t)$, defined by the evolution equation (13.381), is called the *isoclinic intermediate configuration.*[60] The internal variable $\boldsymbol{\Phi}$ represents the transformation of the isoclinic intermediate configuration $\widetilde{\mathbf{F}}_{\mathrm{i}}$ into the *intermediate configuration* $\mathbf{F}_{\mathrm{i}}(\boldsymbol{X},\, t)$:

$$\mathbf{F}_{\mathrm{i}} = \boldsymbol{\Phi}\widetilde{\mathbf{F}}_{\mathrm{i}} \tag{13.383}$$

◇

By definition, $\boldsymbol{\Phi}$ operates on the isoclinic intermediate configuration and rotates it into the intermediate configuration. Accordingly, $\boldsymbol{\Phi}$ transforms under a rotation $\bar{\mathbf{Q}}$ in the same way as $\mathbf{F}_{\mathrm{i}}$ does, namely according to (13.374),

$$\boldsymbol{\Phi} \longmapsto \boldsymbol{\Phi}^{*} = \bar{\mathbf{Q}}\boldsymbol{\Phi}\,,$$

[60] See MANDEL [1973], p. 728 (where $\boldsymbol{\Phi}$ is denoted as β) and corresponding remarks in DAFALIAS [1998], pp. 922, 923.

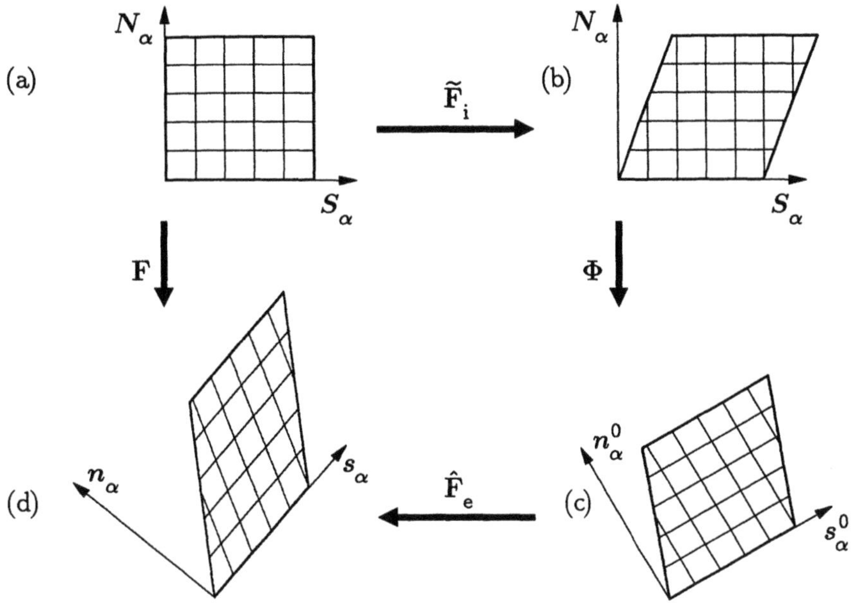

Figure 13. 8: Multiplicative decomposition of the deformation gradient
(a) reference configuration (b) isoclinic intermediate configuration
(c) intermediate configuration (d) current configuration

whereas the isoclinic intermediate configuration remains unchanged:

$$\widetilde{\mathbf{F}}_i^* = \widetilde{\mathbf{F}}_i \, . \tag{13.384}$$

Combining the decomposition (13.351) with (13.383), we obtain

$$\mathbf{F} = \hat{\mathbf{F}}_e \boldsymbol{\Phi} \widetilde{\mathbf{F}}_i \, . \tag{13.385}$$

Since $\widetilde{\mathbf{F}}_i$ leaves the slip system $\{\boldsymbol{N}_\alpha \, , \boldsymbol{S}_\alpha\}$ invariant, its image in the intermediate configuration is given by $\{\boldsymbol{n}^0_\alpha \, , \boldsymbol{s}^0_\alpha\}$ with

$$\boldsymbol{n}^0_\alpha = \boldsymbol{\Phi} \boldsymbol{N}_\alpha \, , \tag{13.386}$$

and

$$\boldsymbol{s}^0_\alpha = \boldsymbol{\Phi} \boldsymbol{S}_\alpha \, . \tag{13.387}$$

In the current configuration the image of the slip system α is $\{n_\alpha, s_\alpha\}$, with

$$n_\alpha = \hat{\mathbf{F}}_e^{\mathrm{T}-1} n_\alpha^0 = \hat{\mathbf{F}}_e^{\mathrm{T}-1} \boldsymbol{\Phi} N_\alpha \tag{13.388}$$

and

$$s_\alpha = \hat{\mathbf{F}}_e s_\alpha^0 = \hat{\mathbf{F}}_e \boldsymbol{\Phi} S_\alpha . \tag{13.389}$$

Fig. 13. 8 illustrates the relation between the various configurations. The configurations $\widetilde{\mathbf{F}}_i$ and $\mathbf{F}_i$ are related by (13.383). The corresponding relationship between their evolution equations is stated in

Theorem 13. 30
The evolution equation (13.381) for the isoclinic intermediate configuration,

$$\dot{\widetilde{\mathbf{F}}}_i \widetilde{\mathbf{F}}_i^{-1} = \sum_\alpha \dot{\kappa}_\alpha S_\alpha \otimes N_\alpha ,$$

is equivalent to the evolution equation

$$\dot{\mathbf{F}}_i \mathbf{F}_i^{-1} = \dot{\boldsymbol{\Phi}} \boldsymbol{\Phi}^{\mathrm{T}} + \left\{ \sum_\alpha \dot{\kappa}_\alpha s_\alpha^0 \otimes n_\alpha^0 \right\} \tag{13.390}$$

for the intermediate configuration; this equivalence holds for all functions $\boldsymbol{\Phi}(X, t)$. □

Proof
Differentiating (13.383) with respect to time yields

$$\dot{\mathbf{F}}_i \mathbf{F}_i^{-1} = \dot{\boldsymbol{\Phi}} \boldsymbol{\Phi}^{\mathrm{T}} + \boldsymbol{\Phi} \dot{\widetilde{\mathbf{F}}}_i \widetilde{\mathbf{F}}_i^{-1} \boldsymbol{\Phi}^{\mathrm{T}} .$$

This implies (13.390), if (13.381) is inserted, as well as (13.386), (13.387). The same steps in the reverse direction lead from (13.390) to (13.381). ■

While the evolution equation for the isoclinic intermediate configuration is not affected by any rotation $\bar{\mathbf{Q}}$ of the intermediate configuration, the evolution equation (13.390) is objective with respect to $\bar{\mathbf{Q}}$: from (13.374) and (13.367) we infer

$$\hat{\mathbf{L}}_i - \dot{\boldsymbol{\Phi}} \boldsymbol{\Phi}^{\mathrm{T}} \longmapsto \hat{\mathbf{L}}_i^* - \dot{\boldsymbol{\Phi}}^* \boldsymbol{\Phi}^{*\mathrm{T}} = \bar{\mathbf{Q}} \left(\hat{\mathbf{L}}_i - \dot{\boldsymbol{\Phi}} \boldsymbol{\Phi}^{\mathrm{T}} \right) \bar{\mathbf{Q}}^{\mathrm{T}} . \tag{13.391}$$

The objectivity of $\sum_\alpha \dot{\kappa}_\alpha \boldsymbol{s}_\alpha^0 \otimes \boldsymbol{n}_\alpha^0$ follows immediately from (13.386), (13.387) and (13.374).

The skew-symmetric part of the evolution equation (13.390) reads as

$$\hat{\mathbf{W}}_\mathrm{i} = \dot{\mathbf{\Phi}}\mathbf{\Phi}^\mathrm{T} + \mathbf{\Phi}\widetilde{\mathbf{W}}_\mathrm{i}\mathbf{\Phi}^\mathrm{T} \tag{13.392}$$

if we set

$$\hat{\mathbf{W}}_\mathrm{i} = \frac{1}{2}\Big[\dot{\mathbf{F}}_\mathrm{i}\mathbf{F}_\mathrm{i}^{-1} - (\dot{\mathbf{F}}_\mathrm{i}\mathbf{F}_\mathrm{i}^{-1})^\mathrm{T}\Big] \tag{13.393}$$

and

$$\widetilde{\mathbf{W}}_\mathrm{i} = \frac{1}{2}\Big[\dot{\widetilde{\mathbf{F}}}_\mathrm{i}\widetilde{\mathbf{F}}_\mathrm{i}^{-1} - (\dot{\widetilde{\mathbf{F}}}_\mathrm{i}\widetilde{\mathbf{F}}_\mathrm{i}^{-1})^\mathrm{T}\Big] = \frac{1}{2}\sum_\alpha \dot{\kappa}_\alpha\Big(\boldsymbol{S}_\alpha \otimes \boldsymbol{N}_\alpha - \boldsymbol{N}_\alpha \otimes \boldsymbol{S}_\alpha\Big). \tag{13.394}$$

This motivates

Definition 13. 13

In eq. (13.392) the tensor $\mathbf{\Phi}\widetilde{\mathbf{W}}_\mathrm{i}\mathbf{\Phi}^\mathrm{T}$ is called the *plastic spin* and $\dot{\mathbf{\Phi}}\mathbf{\Phi}^\mathrm{T}$ is called the *lattice spin*; the sum $\hat{\mathbf{W}}_\mathrm{i}$ of the plastic spin and lattice spin is called the *plastic material spin.* [61] ◊

Usually, co-rotation of $\hat{\mathbf{W}}_\mathrm{i}$ is tacitly assumed in crystal plasticity; we then have $\mathbf{\Phi} = \mathbf{1}$, so that the lattice spin vanishes and the plastic material spin equals the plastic spin. The assumption of lattice co-rotation will be dropped here and an evolution equation for the lattice spin will be formulated in order to allow for a difference between plastic material spin and plastic spin as a material property. This introduces an additional degree of freedom into the constitutive modelling of process-induced texture, which occurs in polycrystalline metals.

In combination with the definition (13.361) of the *Mandel* stress and the transformations (13.388), (13.389) the evolution equation (13.390) implies

$$\begin{aligned}\hat{\mathbf{P}}\cdot\Big(\hat{\mathbf{L}}_\mathrm{i} - \dot{\mathbf{\Phi}}\mathbf{\Phi}^\mathrm{T}\Big) &= \Big(\hat{\mathbf{F}}_\mathrm{e}^{\mathrm{T}}\mathbf{S}\hat{\mathbf{F}}_\mathrm{e}^{\mathrm{T}-1}\Big)\cdot\Big\{\sum_\alpha \dot{\kappa}_\alpha \boldsymbol{s}_\alpha^0 \otimes \boldsymbol{n}_\alpha^0\Big\} \\ &= \mathbf{S}\cdot\Big\{\sum_\alpha \dot{\kappa}_\alpha \boldsymbol{s}_\alpha \otimes \boldsymbol{n}_\alpha\Big\} = \sum_\alpha \boldsymbol{s}_\alpha\cdot(\mathbf{S}\boldsymbol{n}_\alpha)\dot{\kappa}_\alpha\,.\end{aligned}$$

[61] These definitions follow DAFALIAS [1998], p. 916.

The result is

$$\hat{\mathbf{P}}\cdot\left(\hat{\mathbf{L}}_{\mathrm{i}}-\dot{\boldsymbol{\Phi}}\boldsymbol{\Phi}^{\mathrm{T}}\right)=\sum_{\alpha}\tau_{\alpha}\dot{\kappa}_{\alpha}\,. \tag{13.395}$$

Definition 13. 14

The shear stresses

$$\tau_{\alpha}=\boldsymbol{s}_{\alpha}^{0}\cdot\left(\hat{\mathbf{P}}\boldsymbol{n}_{\alpha}^{0}\right)=\boldsymbol{s}_{\alpha}\cdot\left(\mathbf{S}\boldsymbol{n}_{\alpha}\right), \tag{13.396}$$

i.e. the projection of the *Mandel* stress tensor on the slip system in the intermediate configuration or the projection of the *Cauchy* stress tensor on the slip system in the current configuration are called resolved shear stresses or *Schmid* stresses.[62] ◊

Applying the definition and the result (13.395), the residual dissipation inequality (13.380) takes the simple form

$$\sum_{\alpha}\tau_{\alpha}\dot{\kappa}_{\alpha}-\rho_{\mathrm{R}}\dot{\psi}_{\mathrm{i}}(t)\geq 0\,, \tag{13.397}$$

which is most appropriate to motivate further evolution equations. Indeed, differentiating the inelastic free energy (13.382),

$$\rho_{\mathrm{R}}\dot{\psi}_{\mathrm{i}}(t)=\sum_{\alpha}c_{\alpha}(\kappa_{\alpha}-y_{\alpha})(\dot{\kappa}_{\alpha}-\dot{y}_{\alpha}) \tag{13.398}$$

and introducing the ***resolved back stresses***

$$x_{\alpha}=c_{\alpha}(\kappa_{\alpha}-y_{\alpha})\,, \tag{13.399}$$

the dissipation inequality (13.397) leads to the requirement

$$\sum_{\alpha}\left(\tau_{\alpha}-x_{\alpha}\right)\dot{\kappa}_{\alpha}+\sum_{\alpha}x_{\alpha}\left(\dot{\kappa}_{\alpha}-\frac{1}{c_{\alpha}}\dot{x}_{\alpha}\right)\geq 0\,, \tag{13.400}$$

which can be fulfilled in various ways. A simple possibility is a set of flow rules like[63]

$$\dot{\kappa}_{\alpha}=\frac{1}{\eta_{\alpha}}\left\langle\frac{|\tau_{\alpha}-x_{\alpha}|-k_{\alpha}}{r_{\alpha}}\right\rangle^{\mathrm{m}}\frac{\tau_{\alpha}-x_{\alpha}}{|\tau_{\alpha}-x_{\alpha}|} \tag{13.401}$$

---

[62] Cf. SCHMID & BOAS [1935].

[63] See e.g. PERZYNA [1966]; MÉRIC et al. [1991]; BERTRAM, BÖHLKE & KRASKA [1997]. An experimental investigation of material parameters related to flow rules of this kind can be found in the thesis DEBUSMANN [2000].

and *Armstrong-Frederick* equations for the back stresses:

$$\dot{x}_\alpha = c_\alpha \dot{\kappa}_\alpha - b_\alpha |\dot{\kappa}_\alpha| x_\alpha \,. \tag{13.402}$$

The quantities η_α, k_α, c_α, b_α, r_α and m are material parameters to be identified from appropriate experimental data.

According to (13.401) a resolved shear rate $\dot{\kappa}_\alpha$ occurs if the effective shear stress $|\tau_\alpha - x_\alpha|$ related to the particular slip system α exceeds the critical value k_α, the resolved yield stress. The resolved yield stresses may be material constants or depend on process quantities so to incorporate isotropic hardening on the slip system level.

The evolution equations (13.401) and (13.402) determine the shear rates $\dot{\kappa}_\alpha$, which complete the evolution equation (13.381) to calculate the isoclinic intermediate configuration. What remains to be set up is an evolution equation for the internal variable $\mathbf{\Phi}$, which rotates to the intermediate configuration and thus determines the orientation of the crystal lattice and the slip systems, respectively. From (13.397) it is recognised that the dissipation inequality does not impose any restriction on the choice of the lattice spin $\dot{\mathbf{\Phi}}\mathbf{\Phi}^{\mathrm{T}}$, however, the invariance properties of such a constitutive equation requires a careful investigation. This will be done in the next section.

13. 7. 4 Lattice Spin

Mathematically, a rotation $\bar{\mathbf{Q}}(\boldsymbol{X}, t)$ of the intermediate configuration can be compared with a local change of frame, while the origin remains fixed. Under such a change of frame in the sense of the active interpretation[64] a two-point tensor like $\mathbf{F}_{\mathrm{i}}$ or $\mathbf{\Phi}$ transforms according to (13.374),

$$\mathbf{\Phi}^* = \bar{\mathbf{Q}}\mathbf{\Phi} \,.$$

Now, three frames are considered: one so-called absolute frame, say frame (0), and two relative frames, frame (1) and frame (2).[65] The changes from the absolute to the relative frames are provided by the orthogonal transformations $\bar{\mathbf{Q}}_{10}(\boldsymbol{X}, t)$ and $\bar{\mathbf{Q}}_{20}(\boldsymbol{X}, t)$, and the change between frame (1) and frame (2) by

[64] Cf. Sect. 4. 4.

[65] The following argumentation is comparable to Sect. 4. 6, where the transformation properties of the absolute velocity and acceleration vectors are investigated.

$$\bar{Q}_{21} = \bar{Q}_{20}\bar{Q}_{10}{}^{T} . \tag{13.403}$$

The lattice rotation with respect to the frames (0), (1) and (3) may be denoted as $\boldsymbol{\Phi}_0$, $\boldsymbol{\Phi}_1$ and $\boldsymbol{\Phi}_2$, respectively.
Let $\boldsymbol{A}$ be a vector in the space of the natural reference configuration. ($\boldsymbol{A}$ can be a normal $\boldsymbol{N}_\alpha$ of a slip plane or a slip direction $\boldsymbol{S}_\alpha$.) We then have transformations like

$$\boldsymbol{a}_0 = \boldsymbol{\Phi}_0\boldsymbol{A} , \quad \boldsymbol{a}_1 = \boldsymbol{\Phi}_1\boldsymbol{A} , \quad \boldsymbol{a}_2 = \boldsymbol{\Phi}_2\boldsymbol{A} \tag{13.404}$$

$\left(\boldsymbol{a}_k = \boldsymbol{n}_\alpha^0, \boldsymbol{s}_\alpha^0\right)$. In view of the transformations

$$\boldsymbol{\Phi}_1 = \bar{Q}_{10}\boldsymbol{\Phi}_0 , \quad \boldsymbol{\Phi}_2 = \bar{Q}_{20}\boldsymbol{\Phi}_0 \tag{13.405}$$

the relations

$$\boldsymbol{a}_1 = \bar{Q}_{10}\boldsymbol{a}_0 , \quad \boldsymbol{a}_2 = \bar{Q}_{20}\boldsymbol{a}_0 \tag{13.406}$$

and

$$\boldsymbol{a}_2 = \bar{Q}_{21}\boldsymbol{a}_1 \tag{13.407}$$

are valid. Differentiation with respect to time yields

$$\dot{\boldsymbol{a}}_1 = \dot{\bar{Q}}_{10}\boldsymbol{a}_0 + \bar{Q}_{10}\,\dot{\boldsymbol{a}}_0 = \dot{\bar{Q}}_{10}\bar{Q}_{10}{}^{T}\boldsymbol{a}_1 + \bar{Q}_{10}\dot{\boldsymbol{a}}_0 \quad \text{or}$$

$$\dot{\boldsymbol{a}}_1 - \bar{\boldsymbol{\Omega}}_{10}\boldsymbol{a}_1 = \bar{Q}_{10}\dot{\boldsymbol{a}}_0 , \tag{13.408}$$

where the rate of rotation

$$\bar{\boldsymbol{\Omega}}_{10} = \dot{\bar{Q}}_{10}\bar{Q}_{10}{}^{T} \tag{13.409}$$

has been introduced. In its physical interpretation $-\bar{\boldsymbol{\Omega}}_{10}$ is the rotation rate of the frame (1) with respect to the absolute frame (0).[66] In complete analogy a second relation

$$\dot{\boldsymbol{a}}_2 - \bar{\boldsymbol{\Omega}}_{20}\boldsymbol{a}_2 = \bar{Q}_{20}\dot{\boldsymbol{a}}_0 \tag{13.410}$$

---

[66] Cf. Fig. 4. 4. Theorem 4. 5 states that the vectors of absolute velocity and acceleration are objective vectors if they are transformed between two relative frames. In particular, the tensor $-\boldsymbol{\Omega}$ occurring in eqs. (4.77) and (4.78) is the rate of rotation with respect to the inertial system.

is inferred with

$$\bar{\Omega}_{20} = \dot{\bar{Q}}_{20}\bar{Q}_{20}{}^{T} . \tag{13.411}$$

Combination of (13.410) and (13.408) together with (13.403) yields

$$\dot{a}_2 - \bar{\Omega}_{20}a_2 = \bar{Q}_{21}\left(\dot{a}_1 - \bar{\Omega}_{10}a_1\right) . \tag{13.412}$$

On the other hand, the rates $\dot{a}_1$ and $\dot{a}_2$ are calculated on the basis of (13.404) to

$$\dot{a}_1 = \dot{\Phi}_1\Phi_1{}^{T}a_1 , \quad \dot{a}_2 = \dot{\Phi}_2\Phi_2{}^{T}a_2 ; \tag{13.413}$$

inserting this into (13.412) and applying (13.403) leads to the result

$$\dot{\Phi}_2\Phi_2{}^{T} - \bar{\Omega}_{20} = \bar{Q}_{21}\left(\dot{\Phi}_1\Phi_1{}^{T} - \bar{\Omega}_{10}\right)\bar{Q}_{21}{}^{T} . \tag{13.414}$$

In terms of the notation

$$\Phi_1 = \Phi , \quad \bar{\Omega}_{10} = \bar{\Omega} , \quad \bar{Q}_{21} = \bar{Q} , \quad \Phi_2 = \Phi , \quad \bar{\Omega}_{20} = \bar{\Omega}^{*} \tag{13.415}$$

we can summarise the transformation rule (13.414) as the statement of

Theorem 13. 31
If $\bar{Q}(X, t)$ represents a change of frame, going from one relative frame to a second one, the rate of rotation with respect to the absolute frame is an objective tensor in the sense of

$$\dot{\Phi}^{*}\Phi^{*T} - \bar{\Omega}^{*} = \bar{Q}\left(\dot{\Phi}\Phi^{T} - \bar{\Omega}\right)\bar{Q}^{T} . \tag{13.416}$$

■

This can be applied in crystal viscoplasticity to set up an evolution equation for the orientation of the crystal lattice. As an assumption we postulate

$$\dot{\Phi}\Phi^{T} - \bar{\Omega} = \beta\,\Phi\widetilde{W}_{i}\Phi^{T} + \gamma\,\hat{P}_{A} , \tag{13.417}$$

with

$$\widetilde{W}_{i} = \frac{1}{2}\sum_{\alpha}\dot{\kappa}_{\alpha}\left(S_{\alpha} \otimes N_{\alpha} - N_{\alpha} \otimes S_{\alpha}\right) \tag{13.418}$$

and

$$\hat{\mathbf{P}}_{\mathrm{A}} = \tfrac{1}{2}\left(\hat{\mathbf{P}} - \hat{\mathbf{P}}^{\mathrm{T}}\right). \tag{13.419}$$

The quantities β and γ are material parameters. The left-hand side of (13.417) represents the absolute rate of rotation, consisting of the relative rate $\dot{\mathbf{\Phi}}\mathbf{\Phi}^{\mathrm{T}}$ plus the rotation rate of the relative frame, $-\bar{\mathbf{\Omega}} = -\dot{\bar{\mathbf{Q}}}_{10}\bar{\mathbf{Q}}_{10}{}^{\mathrm{T}}$. Since both sides of (13.417) are objective under a transformation from one relative frame to another, the postulate of full invariance is satisfied for the constitutive equation (13.417), as it is satisfied for all other equations of the present constitutive model. However, it should be noted that the evolution equation (13.417) does not meet the assumption of objectivity in the usual formulation, because it includes an information of the applied frame in the reference configuration, namely the rate of rotation $\bar{\mathbf{\Omega}}$. Instead, (13.417) is an observer-invariant relation in the sense of Definition 4. 10. For practical application we set $\bar{\mathbf{\Omega}} = \mathbf{0}$, i.e. $\bar{\mathbf{Q}}_{10} = \bar{\mathbf{Q}}_{10}(\boldsymbol{X})$ constant in time. Then, eq. (13.417) specialises to

$$\dot{\mathbf{\Phi}}\mathbf{\Phi}^{\mathrm{T}} = \beta\,\mathbf{\Phi}\widetilde{\mathbf{W}}_{\mathrm{i}}\mathbf{\Phi}^{\mathrm{T}} + \gamma\,\hat{\mathbf{P}}_{\mathrm{A}}\,. \tag{13.420}$$

13. 7. 5 Summary: A Constitutive Model of Anisotropic Viscoplasticity

The constitutive model of crystal viscoplasticity, developed in this section, is summarised as follows:

Decomposition of the Deformation

$$\mathbf{F} = \hat{\mathbf{F}}_{\mathrm{e}}\mathbf{F}_{\mathrm{i}} \tag{13.421}$$

$$\hat{\mathbf{\Gamma}} = \mathbf{F}_{\mathrm{i}}^{\mathrm{T}\,-1}\mathbf{E}\mathbf{F}_{\mathrm{i}}^{-1} = \hat{\mathbf{\Gamma}}_{\mathrm{e}} + \hat{\mathbf{\Gamma}}_{\mathrm{i}} \tag{13.422}$$

Isoclinic Intermediate Configuration

$$\dot{\widetilde{\mathbf{F}}}_{\mathrm{i}}\widetilde{\mathbf{F}}_{\mathrm{i}}^{-1} = \sum_{\alpha} \dot{\kappa}_{\alpha}\boldsymbol{S}_{\alpha} \otimes \boldsymbol{N}_{\alpha} \tag{13.423}$$

Intermediate Configuration

$$\mathbf{F}_{\mathrm{i}} = \mathbf{\Phi}\widetilde{\mathbf{F}}_{\mathrm{i}} \tag{13.424}$$

Free Energy

$$\psi = \psi_e + \psi_i \tag{13.425}$$

$$\psi_e = \hat{\psi}_e(\mathbf{\Phi}^T\hat{\mathbf{\Gamma}}_e\mathbf{\Phi}) = \hat{\varphi}(\hat{J}_1, \hat{J}_2, \dots, \hat{J}_N) \tag{13.426}$$

$$\hat{J}_k = \hat{J}_k(\mathbf{\Phi}^T\hat{\mathbf{\Gamma}}_e\mathbf{\Phi}) \tag{13.427}$$

$$\rho_R\psi_i = \sum_\alpha \frac{1}{2}c_\alpha(\kappa_\alpha - y_\alpha)^2 \tag{13.428}$$

Elasticity Relation

$$\hat{\mathbf{S}} = \rho_R \frac{\partial\hat{\psi}_e(\mathbf{\Phi}^T\hat{\mathbf{\Gamma}}_e\mathbf{\Phi})}{\partial\hat{\mathbf{\Gamma}}_e} = \rho_R \frac{\partial}{\partial\hat{\mathbf{\Gamma}}_e}\hat{\varphi}(\hat{J}_1, \hat{J}_2, \dots, \hat{J}_N) \tag{13.429}$$

Evolution Equations

$$\dot{\kappa}_\alpha = \frac{1}{\eta_\alpha}\left\langle \frac{|\tau_\alpha - x_\alpha| - k_\alpha}{r_\alpha} \right\rangle^m \frac{\tau_\alpha - x_\alpha}{|\tau_\alpha - x_\alpha|} \tag{13.430}$$

$$\dot{x}_\alpha = c_\alpha\dot{\kappa}_\alpha - b_\alpha|\dot{\kappa}_\alpha|x_\alpha \tag{13.431}$$

$$\dot{\mathbf{\Phi}}\mathbf{\Phi}^T = \beta\,\mathbf{\Phi}\left\{\frac{1}{2}\sum_\alpha \dot{\kappa}_\alpha\left(\boldsymbol{S}_\alpha \otimes \boldsymbol{N}_\alpha - \boldsymbol{N}_\alpha \otimes \boldsymbol{S}_\alpha\right)\right\}\mathbf{\Phi}^T + \gamma\frac{1}{2}\left(\hat{\mathbf{P}} - \hat{\mathbf{P}}^T\right) \tag{13.432}$$

The physical meaning of these constitutive equations may be summarised as follows: the material model applies to all kinds of crystal symmetry, where the geometry of the crystal determines the set of invariants occurring in the elasticity relation as well as the set of slip systems being decisive for the evolution of the isoclinic intermediate configuration.

While the above constitutive theory applies the basic concept introduced by Tsakmakis, it pursues a different way in the details of its formulation: plastic flow occurs on the level of the crystallographic slip systems; therefore the theory incorporates some direction towards micromechanics. In contrast to that Tsakmakis develops a phenomenological theory.[67]

---

[67] This is obvious from TSAKMAKIS [2001b], Sect. 2, where flow rules and hardening models are derived. These are applied in HÄUSLER & TSAKMAKIS [2001].

For $\beta = \gamma = 0$ the present constitutive model includes the classical theory of crystal plasticity in the sense of Asaro and other authors.[68] For vanishing β and γ the orientation $\mathbf{\Phi}$ of the crystal in the intermediate configuration is constant in time. However, even in this special case the orientation may be assumed to be space-dependent in order to describe the behaviour of a polycrystal. Non-vanishing parameters β and γ on the right-hand side of (13.432) provide the possibility to represent the effects of process-induced anisotropy (texture). In particular, the first term contributes to the lattice spin in proportionality to the plastic spin; the second term may be understood as a kind of torque, resulting from the antisymmetric part of the *Mandel* stress tensor. There is no restriction for the sign of the material parameters β and γ. Quite other assumptions for the lattice spin are also possible if suggested by experimental observations. On principle, a separate evolution equation for the lattice spin might be able to take into account the phenomenological effect of micromechanical interactions between different grains in the context of deforming polycrystals.

13. 7. 6 Numerical Simulations

On the basis of the above constitutive model numerical simulations are carried out for a face-centered cubic crystal, a symmetry property which is met in the context of various metals, such as Al, Cu Ni γ-Fe, Ag, Au and Pb. For cubic symmetry, 14 or 9 strain invariants form a basis to determine the strain energy according to the representation theory, cited in Chapter 9.[69] However, if small elastic strains are assumed and physical linearisation is applied in the elasticity relation, only invariants of first and second order have to be taken into account. Thus, no more than 3 invariants remain in this case; these are[70]

$$J_1 = E_{11}{}^2 + E_{22}{}^2 + E_{33}{}^2 , \tag{13.433}$$

$$J_2 = E_{11}E_{22} + E_{22}E_{33} + E_{33}E_{11} , \tag{13.434}$$

$$J_3 = E_{12}{}^2 + E_{23}{}^2 + E_{31}{}^2 . \tag{13.435}$$

---

[68] ASARO [1983a,b]; HAVNER [1992]; TEODOSIU [1997].

[69] See Sect. 9. 3. 2, in particular eqs. (9.120), (9.127) and (9.128).

[70] These invariants are equivalent to the first three of (9.127); cf. (9.153).

Parameter	Value	
C_1	170000	MPa
C_2	124000	MPa
C_3	76000	MPa
k_α	150	MPa
η_α	100000	s
m	2.5	–
c_α	0.0	MPa
b_α	0.0	–
β	0.0; 0.5; - 1.0	–
γ	0.0	–

Table 13. 2: Material parameters for numerical simulation

In view of (13.427) they should be expressed as

$$\hat{J}_1 = (\Phi^T \hat{\Gamma}_e \Phi)_{11}{}^2 + (\Phi^T \hat{\Gamma}_e \Phi)_{22}{}^2 + (\Phi^T \hat{\Gamma}_e \Phi)_{33}{}^2 , \tag{13.436}$$

$$\hat{J}_2 = (\Phi^T \hat{\Gamma}_e \Phi)_{11} (\Phi^T \hat{\Gamma}_e \Phi)_{22} + (\Phi^T \hat{\Gamma}_e \Phi)_{22} (\Phi^T \hat{\Gamma}_e \Phi)_{33} + (\Phi^T \hat{\Gamma}_e \Phi)_{33} (\Phi^T \hat{\Gamma}_e \Phi)_{11} , \tag{13.437}$$

$$\hat{J}_3 = (\Phi^T \hat{\Gamma}_e \Phi)_{12}{}^2 + (\Phi^T \hat{\Gamma}_e \Phi)_{23}{}^2 + (\Phi^T \hat{\Gamma}_e \Phi)_{31}{}^2 . \tag{13.438}$$

Thus, we have only three elasticity constants in this special case and the strain energy reads

$$\psi_e = \frac{1}{2} C_1 \hat{J}_1 + C_2 \hat{J}_2 + C_3 \hat{J}_3 . \tag{13.439}$$

The constitutive equations were integrated on the basis of a particular choice of the material parameters, listed in Table 13. 2.

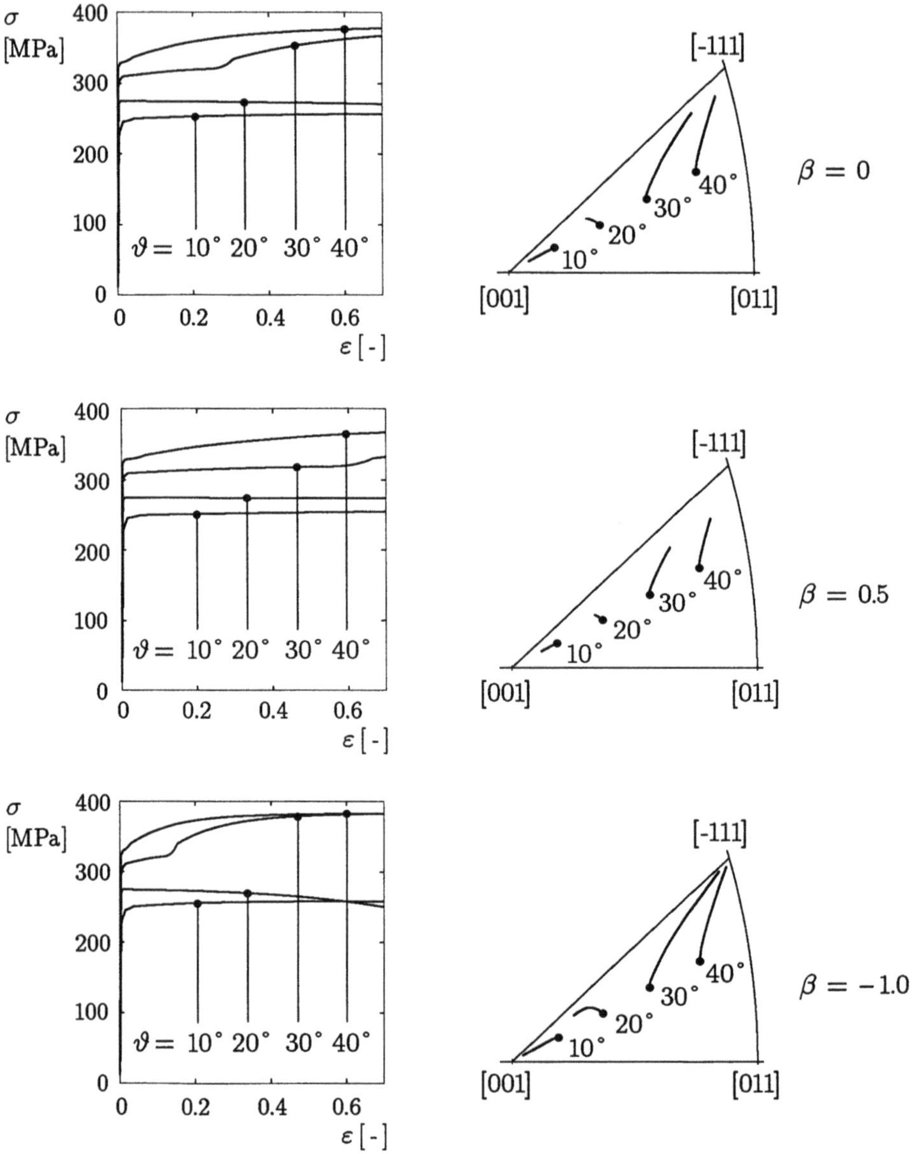

Figure 13. 9: Stress response of a single crystal to simple tension for 4 different initial orientations and 3 values of the parameter β.The evolution of the orientation is illustrated in the inverse polefigures. They show how the evolution is influenced by β.

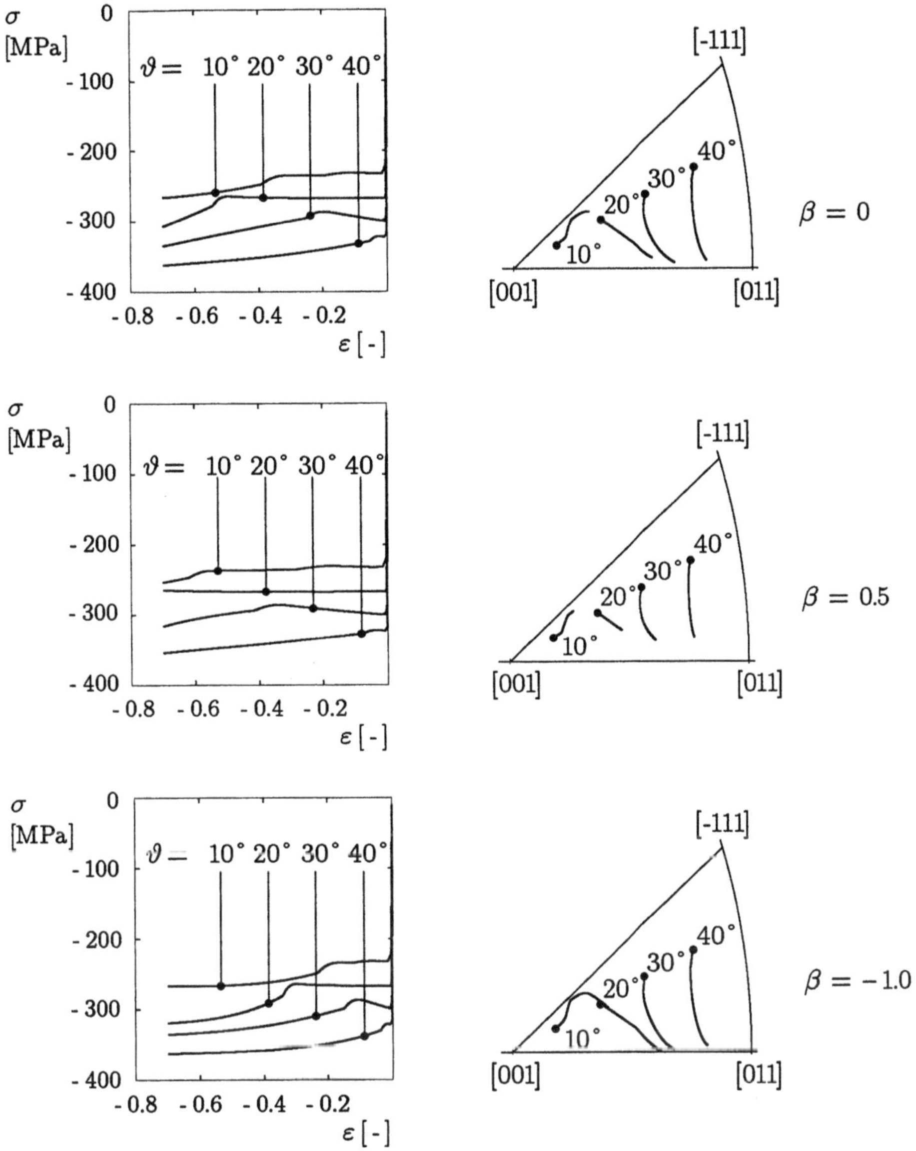

Figure 13. 10: Stress response of a single crystal to simple compression for 4 different initial orientations and 3 values of the parameter β

The numerical test calculations refer to strain-controlled homogeneous deformation processes. First, a uniaxial tension and compression test of a *single crystal* is analysed, corresponding to a total velocity gradient

$$\mathbf{L} \mathrel{\hat{=}} \begin{pmatrix} -0.5 & 0 & 0 \\ 0 & -0.5 & 0 \\ 0 & 0 & 1.0 \end{pmatrix} [10^{-3}\, \mathrm{s}^{-1}] \tag{13.440}$$

and

$$\mathbf{L} \mathrel{\hat{=}} \begin{pmatrix} 0.5 & 0 & 0 \\ 0 & 0.5 & 0 \\ 0 & 0 & -1.0 \end{pmatrix} [10^{-3}\, \mathrm{s}^{-1}] \,. \tag{13.441}$$

A familiar concept for a concrete representation of the internal variable $\mathbf{\Phi}$, the orientation of the crystal lattice, is the use of *Eulerian angles* $(\psi, \vartheta, \varphi)$.[71] Four different initial orientations were investigated, namely

$$\mathbf{\Phi}\Big|_{t=0} \mathrel{\hat{=}} \left\{\psi = 0^{\circ};\, \vartheta = 10^{\circ}; 20^{\circ}; 30^{\circ}; 40^{\circ};\, \varphi = -30^{\circ}\right\}. \tag{13.442}$$

Fig. 13. 9 illustrates the total stress in a single crystal of the different initial orientations as a function of the tensile strain up to 70 %. The first diagrams at the top of the figure display the stress response for $\beta = 0$, which corresponds to the classical theory of crystal viscoplasticity. The evolution of the orientation during the inelastic deformation process is depicted in the inverse polefigure. The diagrams in the middle and bottom of the figure present the stress response for the choice of $\beta = 0.5$ and $\beta = -1.0$. As the figure suggests, the rotation of the crystal is retarded for positive values and accelerated for negative values of β. In the case of compression we have similar results; these are sketched in Fig. 13. 10.

A further test calculation refers to a *polycrystal*. In the present context a polycrystal is defined by assuming the initial orientation to be space-dependent. In particular, the function $\mathbf{\Phi}(\boldsymbol{X}, t_0) = \mathbf{\Phi}_0(\boldsymbol{X})$ is piecewise constant, where each region of a constant value corresponds to a *grain*.

---

[71] SYNGE [1960], pp. 18, 19. Less familiar, but more convenient is the use of the *Eulerian parameters*, which motivates the application of the calculus of *quaternions* to represent the evolution of rigid-body rotations. Cf. SYNGE [1960], pp. 17, 20. In the present numerical simulations this concept was applied.

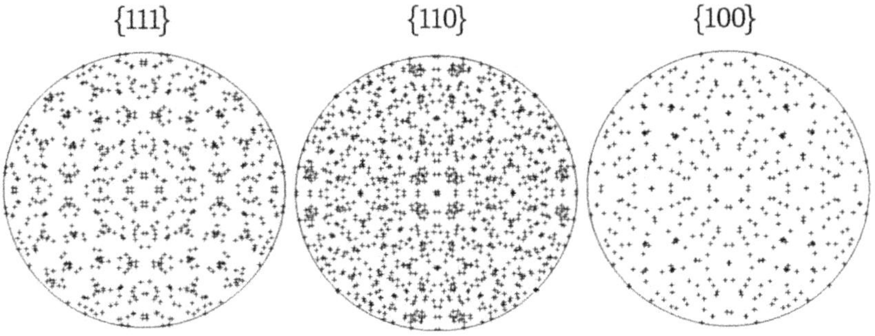

Figure 13. 11: Quasi-isotropic distribution of the initial orientation in a polycrystal (equal-area polefigures)

For the numerical simulation a total number of 192 grains is assumed and a quasi-isotropic distribution of the orientation.[72] The initial distribution is represented by means of equal-area polefigures, shown in Fig. 13. 11.

As a consequence of the space-dependent initial orientation the material is inhomogeneous, and a Finite-Element analysis should be applied to calculate the inhomogeneous strain and stress fields. Instead of this more consistent proceeding the simulation started from the homogeneous deformations (13.440) and (13.441); it was carried out on the level of the constitutive equations. A homogeneous deformation in combination with a finite number of grains and a random distribution of their orientations corresponds to the common Taylor-assumption or iso-strain assumption.[73]

On principle, the inhomogeneity of the elasticity properties implies an inhomogeneous stress field, which generally does not satisfy the equilibrium conditions. However, this inconsistency appears to be less important in view of the new distribution of orientation, which the material model predicts: preferred directions are recognised in the new orientation as a representation of the process-induced anisotropy. Fig. 13. 12 depicts the equal area polefigures for the planes {111}, {110} and {100}. They represent a final anisotropic distribution (texture), which corresponds to other results in the literature[74] and has been experimentally confirmed.[75]

[72] The distribution was taken from Müller [1993]. See also Wellerdick-Wojtasik [1997].

[73] Taylor [1938].

[74] See Müller [1993]; Wellerdick-Wojtasik [1997]; Anand et al. [1997]; Kraska [1997].

[75] Bronkhorst et al. [1992]; Kocks et al. [1998].

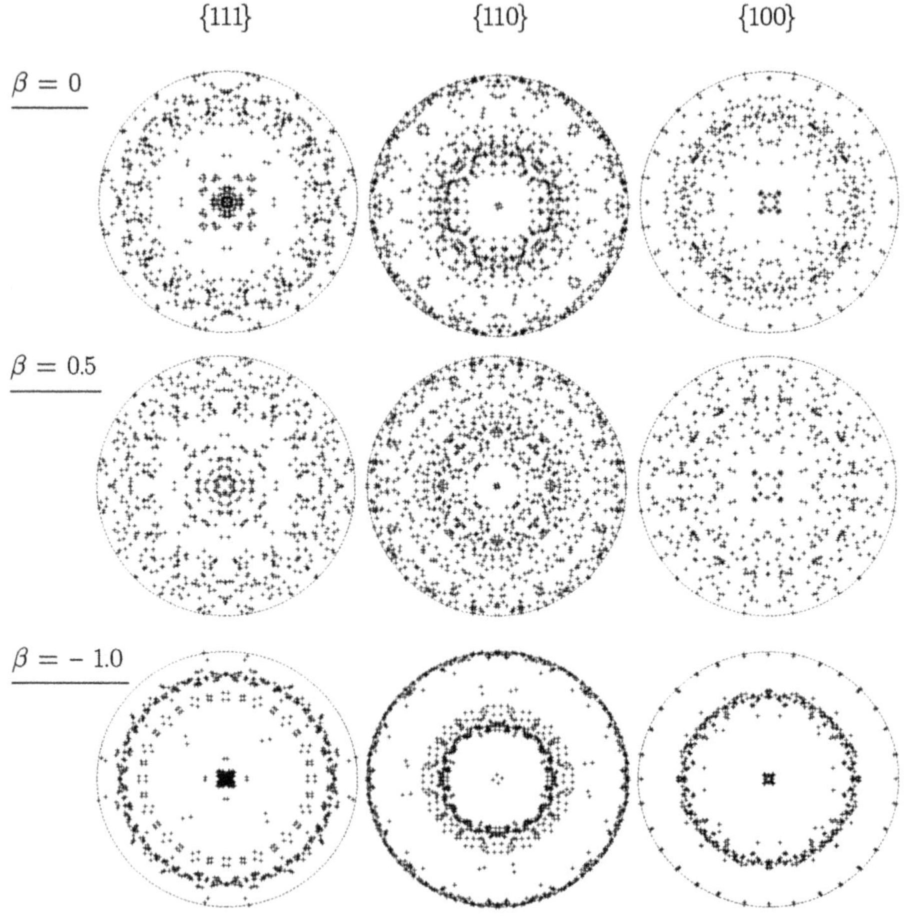

Figure 13. 12: Process-induced texture: stress response of a polycrystal to simple tension for 3 different values of the material parameter β. The process-induced texture is less pronounced for a positive value of b, whereas it becomes sharper if β is negative.

The diagrams in the middle and bottom of the figure demonstrate that the process-induced texture is less pronounced for a positive value of β, whereas it becomes sharper if β is negative. Similar results for compression are displayed in Fig. 13. 13.

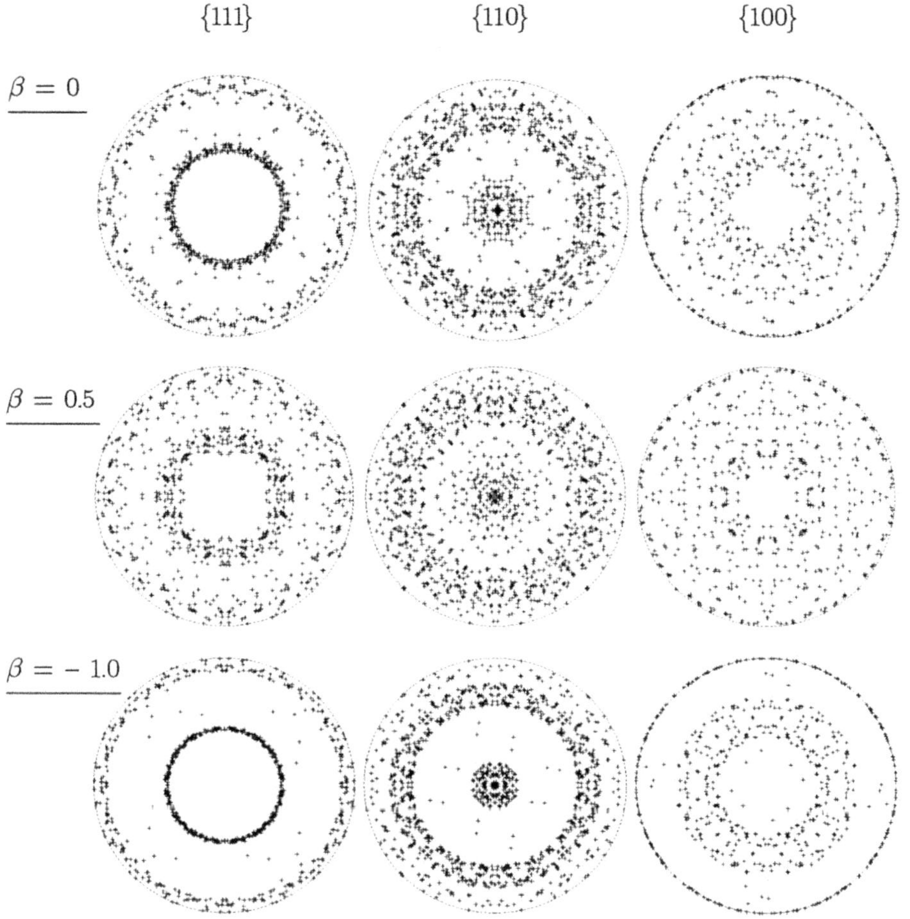

Figure 13. 13: Process-induced texture: stress response of a polycrystal to simple compression for 3 different values of the material parameter β. The process-induced texture is less pronounced for a positive value of b, whereas it becomes sharper if β is negative.

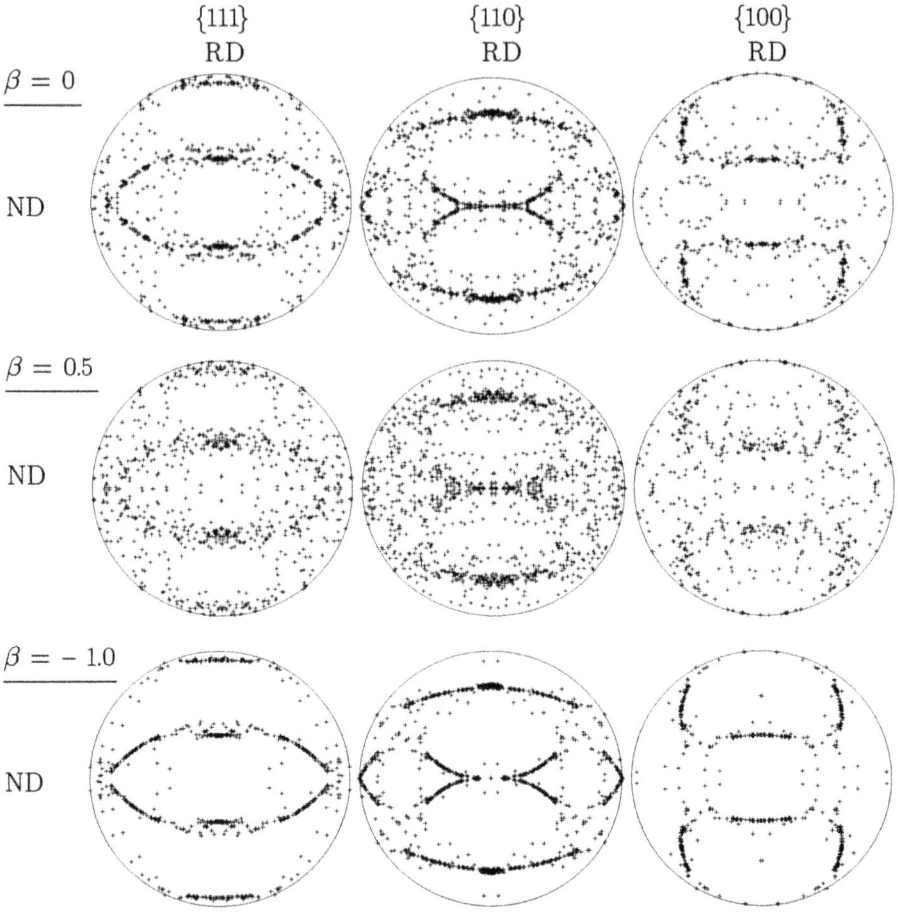

Figure 13. 14: Process-induced texture: stress response of a polycrystal to a rolling process with a thickness reduction of 70%
RD = Rolling Direction
ND = Normal Direction
The reduction of thickness in the rolling process is 70%.

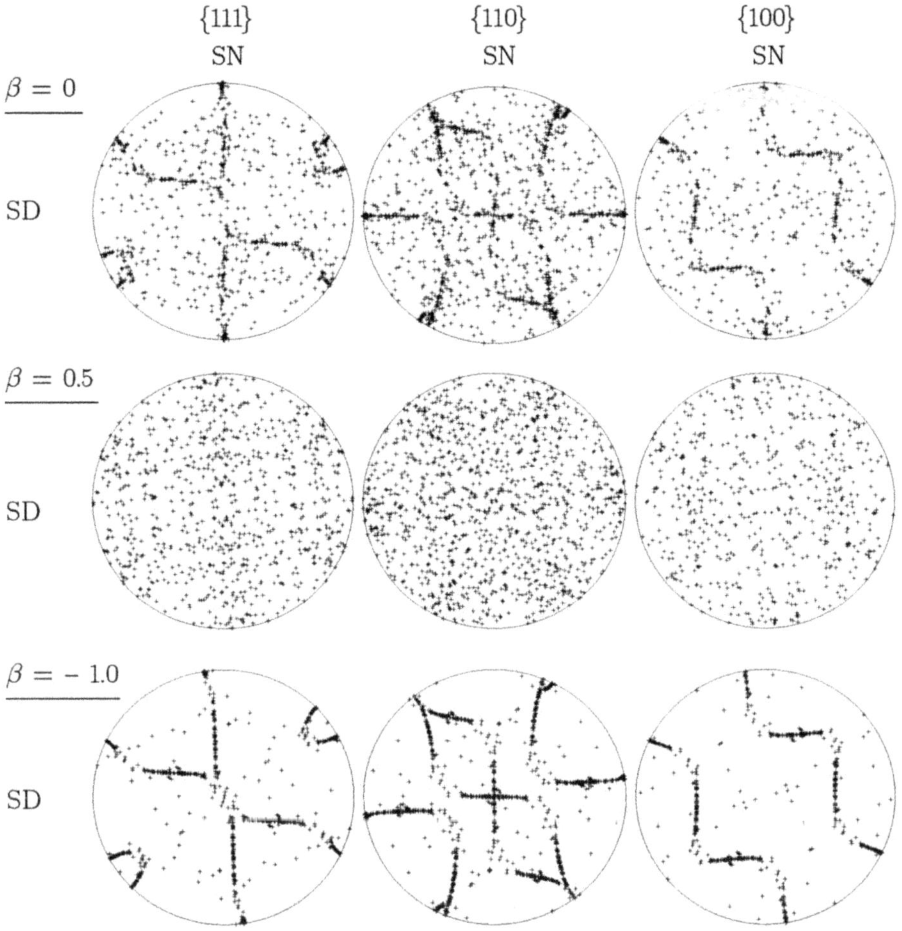

Figure 13. 15: Process-induced texture: stress response of a polycrystal to a process of simple shearing, ending at a shear deformation of 300%.
SN = Shear Normal
SD = Shear Direction
The final amount of shear deformation 300%.

Two further deformations were simulated on the basis of the Taylor-assumption, depicted in Figs. 13. 14 and 13. 15: a rolling process, corresponding to

$$\mathbf{L} \mathrel{\widehat{=}} \begin{pmatrix} \frac{1}{2}\sqrt{3} & 0 & 0 \\ 0 & 0 & 0 \\ 0 & 0 & -\frac{1}{2}\sqrt{3} \end{pmatrix} [\,10^{-3}\,\mathrm{s}^{-1}\,] \qquad (13.443)$$

and a process of simple shear,

$$\mathbf{L} \mathrel{\widehat{=}} \begin{pmatrix} 0 & \sqrt{3} & 0 \\ 0 & 0 & 0 \\ 0 & 0 & 0 \end{pmatrix} [\,10^{-3}\,\mathrm{s}^{-1}\,]\,. \qquad (13.444)$$

The influence of β on the intensity of the texture illustrates the physical meaning of this material parameter in the sense that texture is diminished for positive values and increased for negative values of β. The influence of the parameter γ remains to be investigated. Since there is no a priori restriction in view of the right-hand of (13.432), except its objectivity with respect to a rotation $\bar{\mathbf{Q}}$, alternative formulations of this evolution equation may be set up, if suggested by experimental facts.

13. 7. 7 Closing Remark

The above attempt at setting up a theory of viscoplasticity that incorporates anisotropic material behaviour can only be regarded as a rough outline, which is, by its very nature, incomplete. Despite the existence of extensive literature on the subject of crystal plasticity,[76] it is a fact that there are even a number of fundamental problems still pending a solution. The evolution of the axes of anisotropy should be further investigated. Hardening behaviour on the slip system level should be taken into account; the interaction of different slip systems in the context of hardening and cross hardening might be observed. A final goal is the formulation of evolution laws for appropriate quantities of inelastic deformation based on safe theoretical principles and experimental facts. It remains to be hoped that future research in the theory of materials may succeed in finding answers to those questions.

[76] See e.g. SCHMID & BOAS [1935]; HILL & HAVNER [1952]; KRÖNER [1958]; ASARO [1983a, 1983b]; HAVNER [1992]; HOSFORD [1993]; KHAN & HUANG [1995], pp. 343; BERTRAM & KRASKA [1995]; MIEHE [1996]; KALIDINDI et al. [1996]; ANAND et al. [1997], OLSCHEWSKI [1997]; BÖHLKE [2001].

References

ACHARYA, A.; SHAWKI, T.G.: The Clausius-Duhem Inequality and the Structure of Rate-Independent Plasticity. *Int. J. Plasticity* **12** [1996], 229-238

ALBER, H.-D.: *Materials with Memory. Initial-Boundary Value Problems for Constitutive Equations with Internal Variables.* Springer-Verlag [1998]

ALTENBACH, J.; ALTENBACH, H.: *Einführung in die Kontinuumsmechanik.* B.G. Teubner [1994]

ANAND, L.; BALASUBRAMANIAN, S.; KOTHARI, M.: Constitutive Modelling of Polycrystalline Metals at Large Strains. In: *Large Plastic Deformation of Crystalline Aggregates.* CISM Courses and Lectures No. 376. Springer-Verlag [1997], 109 - 172

ANTHONY, K.-H.: Die Reduktion von nichteuklidischen geometrischen Objekten in eine euklidische Form und physikalische Deutung der Reduktion durch Eigenspannungszustände in Kristallen. *Arch. Rat. Mech. Anal.* **37** [1970], 161-180

ARMSTRONG, P.J.; FREDERICK, C.O.: A Mathematical Representation of the Multiaxial Bauschinger Effect. General Electricity Generating Board, Report No. RD/B/N/731 [1966]

ASARO, R.J.: Micromechanics of Crystals and Polycrystals. In: Wu T.Y.; Hutchinson, J.W. (Eds.): *Advances in Applied Mechanics* **23** [1983a], 1-115

ASARO, R.J.: Crystal Plasticity. *Journal of Applied Mechanics* **50** [1983b], 921-934

BACKHAUS, E.: Progress Report on the Project DFG Ha 2075/1. University of Dresden, Institute of Mechanics [1997]

BAEHR, H.D.: *Thermodynamik.* Springer-Verlag [1973]

BALKE, H.: Zur Objektivität mikromechanisch motivierter Kriechgleichungen. *Technische Mechanik* **18** [1998], 203-208

BANABIC, D.; BUNGE, H.-J.; PÖHLAND, K.; TEKKAYA, A.E.: *Formability of Metallic Materials.* Springer-Verlag [2000]

BECKER, E.; BÜRGER, W.: *Kontinuumsmechanik.* B.G. Teubner [1975]

BELL, J.F.: The Experimental Foundation of Solid Mechanics. *Encyclopedia of Physics*, Vol. VIa/1. Springer-Verlag [1973]

BERTRAM, A.: An Alternative Approach to Finite Plasticity Based on Material Isomorhisms. *Int. J. Plasticity* **15** [1999], 353 - 374

BERTRAM, A.: *Axiomatische Einführung in die Kontinuummechanik.* B.I. Wissenschaftsverlag [1989]

BERTRAM, A.; HAUPT, P.: A Note of Andreussi-Guidugli's Theory of Thermomechanical Constraints in Simple Materials. *Bulletin de l'Académie Polonaise des Sciences* **6** [1976], 47-51

BERTRAM, A.; BÖHLKE, T.; KRASKA, M.: Numerical Simulation of Deformation Induced Anisotropy of Polycrystals. *Computational Materials Science* **9** [1997], 158 - 167

BERTRAM, A.; KRASKA, M.: Description of Finite Plastic Deformations in Single Crystals by Material Isomorphisms. In: Parker, D.F.; Enland, A.H. (Eds): *IUTAM Symposium on Anisotropy, Inhomogeneity and Nonlinearity in Solid Mechanics.* Kluwer Academic Publishers [1995], 77-99

BERTRAM, A.; SVENDSEN, B.: On Material Objectivity and Reduced Constitutive Equations. *Preprint* IFME 97/2, Institut für Mechanik der Universität Magdeburg [1997]

BESDO, D.: Zur Formulierung von Stoffgesetzen der Plastomechanik im Dehnungsraum nach Ilyushins Postulat. *Ingenieur-Archiv* **51** [1981], 1-8

BESDO, D.: Comparison of Strain- and Stress-Space Representations of Plasticity. In: F. Jinhong, S. Murakami (Eds.): *Advances in Constitutive Laws for Engineering Materials*, Vol. 1. Pergamon Press Oxford [1989], 173-175

BETTEN, J.: *Elastizität und Plastizitätstheorie.* Friedrich Vieweg & Sohn [1985]

BETTEN, J.: *Tensorrechnung für Ingenieure.* B.G. Teubner [1987]

BETTEN, J.: Anwendungen von Tensorfunktionen in der Kontinuumsmechanik anisotroper Materialien. *ZAMM Z. angew. Math. Mech.* **78** [1998], 1-15

BISHOP, R.L.; GOLDBERG, S.I.: *Tensor Analysis on Manifolds.* Dover Publications, Inc. [1968]

BODNER, S.R.; PARTOM, Y.: Constitutive Equations for Elastic-Viscoplastic Strain-Hardening Materials. *J. Appl. Mech.* **42** [1975], 385-389

BÖHLKE, T.: Crystallograhpic Texture Evolution and Elastic Anisotropy. Doctoral Thesis [2001], Otto-von-Guericke-Universität Magdeburg. Shaker Verlag, ISBN 3-2865-8758-8

BÖHLKE, T.; BRÜGGEMANN, C.: Graphical Representation of the Generalized Hooke's Law. *Technische Mechanik* **21** [2001], 145 - 158

BONN, R.: Rotationssymmetrische inhomogene Deformationen in der finiten Plastizitätstheorie unter Einbeziehung von Verfestigungsmodellen. Doctoral Thesis [1992], University of Kassel, Institute of Mechanics, Report 2/1992

BONN, R.; HAUPT, P.: Exact Solutions for Elastoplastic Deformations of a Thick-Walled Tube Under Internal Pressure, *Int. J. Plasticity* **11** [1995], 99-118

BOWEN, R.M.: *Introduction to Continuum Mechanics for Engineers.* Plenum Press [1989]

BOWEN, R.M.; WANG, C.C.: *Introduction to Vectors and Tensors. Vol. 1: Linear and Multilinear Algebra. Vol. 2: Vector and Tensor Analysis.* Plenum Press [1976]

BRAUER, R.: On the Relation between the Orthogonal Group and the Unimodular Group. *Arch. Rat. Mech. Anal.* **18** [1965], 97-99

BRÖCKER, TH.: JÄNICH, K.: *Einführung in die Differentialtopologie.* Springer-Verlag [1973]

BRONKHORST, C.A.; KALIDINDI, S.R.; ANAND, L.: Polycrystalline Plasticity and the Evolution of Crystallographic Texure in f.c.c. Metals. *Phil. Trans. R. Soc. London* **A 341** [1992], 443 - 477

BRONSTEIN, I.N.; SEMENDJAJEW, K.A.; MUSIOL, G.; MÜHLIG, H.: *Taschenbuch der Mathematik.* Verlag Harri Deutsch [1997]

BRUHNS, O.T.; LEHMANN, T.; PAPE, A.: On the Description of Transient Cyclic Hardening Behaviour of Mild Steel Ck 15. *Int. J. Plasticity* **8** [1992], 331-359

BURTH, K.; BROCKS, W.: *Plastizität, Grundlagen und Anwendungen für Ingenieure.* Vieweg Verlag [1992]

CAPUTO, M.; MAINARDI, F.: Linear Models of Dissipation in Anelastic Solids. *Riv. Il Nuovo Cimento (Ser. II), Vol. 1* [1971], 161 - 198

CARATHÉODORY, C.: Untersuchungen über die Grundlagen der Thermodynamik. Mathematische Annalen **67** [1909], 355-386

CARLSON, D.E.: Linear Thermoelasticity. *Encyclopedia of Physics,* Vol. VIa/2. Springer-Verlag [1972], 296-345

CARROLL, M.M.: Controllable Deformations of Incompressible Simple Materials. *Int. J. Engng. Sci.* **5** [1967], 515-525

CHABOCHE, J.-L.: Viscoplastic Constitutive Equations for the Description of Cyclic and Anisotropic Behaviour of Metals. *Bulletin de l'Académie Polonaise des Sciences Série Sc. et Tech.* **25** [1977], 33-42

CHABOCHE, J.-L.: Constitutive Equations for Cyclic Plasticity and Cyclic Viscoplasticity. *Int. J. Plasticity* **5** [1986], 247-302

CHABOCHE, J.-L.: On Some Modifications of Kinematic Hardening to Improve the Description of Kinematic Ratchetting Effects. *Int. J. Plasticity* **7** [1991], 661-678

CHABOCHE, J.-L.: Cyclic Viscoplastic Constitutive Equations, Part I: A Thermodynamically Consistent Formulation, Part II: Stored Energy - Comparison Between Models and Experiments. *J. Appl. Mech.* **60** [1993], 813-828

CHABOCHE, J.-L.; ROUSSELIER, G.: On the Plastic and Viscoplastic Constitutive Equations. *J. Pressure Vessel Tech.* **105** [1983], 153-164

CHADWICK, P.: *Continuum Mechanics.* John Wiley & Sons [1976],

CHADWICK, P.; VIANELLO, M.; COWIN, S.C.: A new Proof that the Number of Linear Elastic Symmetries is Eight. *J. Mech. Phys. Solids,* **49** [2001], 2471 - 2492

CHOI, S.H.; KREMPL, E.: Viscoplasticity Theory Based on Overstress: The Modeling of Biaxial Cyclic Hardening Using Irreversible Plastic Strain. In: McDowell, D.L.; Ellis, R. (Eds.): *Advances in Multiaxial Fatigue.* ASTM ATP 1191. Philadelphia [1993], 259-272

CHRISTENSEN, R.M.: *Theory of Viscoelasticity.* Academic Press [1982]

CHRYSOCHOOS, A.; BELMAHJOUB, F.: Thermographic Analysis of Thermomechanical Couplings. *Archives of Mechanics* **44** [1992], 55-68

CHRYSOCHOOS, A.; MAISONNEUVE, O.; MARTIN, G.; CAUMON, H.; CHEZEAUX, J.C.: Plastic and Dissipated Work and Stored Energy. *Nuclear Engineering and Design* **114** [1989], 323-333

COLEMAN, B.D.: Thermodynamics of Materials with Memory. *Arch. Rat. Mech. Anal.* **17** [1964], 1-46

COLEMAN, B.D.; GURTIN, M.E.: Thermodynamics with Internal State Variables. *J. Chem. Phys.* **47** [1967], 597-613

COLEMAN, B.D.; MIZEL, V.J.: Norms and Semi-Groups in the Theory of Fading Memory. *Arch. Rat. Mech. Anal.* **23** [1966], 87-274

COLEMAN, B.D.; NOLL, W.: An Approximation Theorem for Functionals, with Application in Continuum Mechanics. *Arch. Rat. Mech. Anal.* **6** [1960], 355-370

COLEMAN, B.D.; NOLL, W.: Foundations of Linear Viscoelasticity. *Rev. Mod. Phys.* **33** [1961], 239-249

COLEMAN, B.D.; NOLL, W.: The Thermodynamics of Elastic Materials with Heat Conduction and Viscosity. *Arch. Rat. Mech. Anal.* **13** [1963], 167-178

COLEMAN, B.D.; OWEN, D.R.: On the Thermodynamics of Materials with Memory. *Arch. Rat. Mech. Anal.* **36** [1970], 245-269

COWIN, S.: Properties of the Anisotropic Elasticity Tensor. *Q. J. Mech. Appl. Math.* **42** [1995], 210 - 224

DAFALIAS, Y.F.: A Missing Link in the Macroscopic Constitutive Formulation of Large Plastic Deformations. In: A. Sawczuk, G. Bianchi (eds.): *Plasticity Today. International Symposium on Recent Trends and Results in Plasticity.* Elsevier [1983], 135 - 151

DAFALIAS, Y.F.: The Plastic Spin. *J. Appl. Mech.* **52** [1985], 865 - 871

DAFALIAS, Y.F.: Issues on the Constitutive Formulation at Large Elastoplastic Deformations, Part I: Kinematics. *Acta Mechanica* **69** [1987], 119 - 138

DAFALIAS, Y.F.: On the Microscopic Origin of the Plastic Spin. *Acta Mechanica* **82** [1990], 31 - 48

DAFALIAS, Y.F.: Plastic Spin: Necessity or Redundancy? *Int. J. Plasticity* **14** [1998], 909 - 931

DAY, W.A.: *The Thermodynamics of Simple Materials with Memory.* Springer-Verlag [1972]

DE BOER, R: *Vektor- und Tensorrechnung für Ingenieure.* Springer-Verlag [1982]

DEBUSMANN, C.: Untersuchungen zur Parameteridentifikation für das Modell nach Nouailhas et al. und das kristallographische Modell nach Méric et al. am Beispiel der einkristallinen Nickel-Basis-Legierung CMSX-4. Doctoral Thesis [2000], TU Darmstadt. Shaker-Verlag Aachen

DROZDOV, A.: Fractional Differential Models in Finite Viscoelasticity. *Acta Mechanica* **124** [1997], 155 - 180

EHLERS, W.: Constitutive Equations for Granular Materials in Geomechanical Context. In: Hutter, K. (Ed.): *Continuum Mechanics in Environmental Sciences and Geophysics.* CISM Courses and Lectures No. 337. Springer-Verlag [1993], 313-402

ERDELYI; MAGNUS; OBERHELTINGER; TRICOMI: *Higher Transcendental Functions. Vol 3.* McGraw Hill Ltd. [1954]

ERINGEN, A.C.: *Nonlinear Theory of Continuous Media.* McGraw-Hill Book Company [1962]

ERINGEN, A.C.: *Mechanics of Continua.* John Wiley [1967]

FINDLEY, W.N.; LAI, J.S.; ONARAN, K.: *Creep and Relaxation of Nonlinear Viscoelastic Materials.* North-Holland Publishing Company [1976]

FORTE, S.; VIANELLO, M.: Symmetry Classes for Elasticity Tensors. *Journal of Elasticity* **43** [1996], 81 - 108

FOSDICK, R.L.: Dynamically Possible Motions of Incompressible Isotropic Simple Materials. *Arch. Rat. Mech. Anal.* **29** [1968], 272-288
FUNG, Y.C.: *A First Course in Continuum Mechanics.* Prentice-Hall [1977]
GEIRINGER, H.: Ideal Plasticity. *Encyclopedia of Physics,* Vol. VIa/3. Springer-Verlag [1973],403-533
GIESEKUS, H.: *Phänomenologische Rheologie.* Springer-Verlag [1994]
VAN DER GIESSEN, E.; KOLLMANN, F.G.: On Mathematical Aspects of Dual Variables in Continuum Mechanics. Part I: Mathematical Priciples. *ZAMM Z. angew. Math. Mech.* **76** [1996a], 447-462
VAN DER GIESSEN, E.; KOLLMANN, F.G.: On Mathematical Aspects of Dual Variables in Continuum Mechanics. Part II: Application in Nonlinear Solid Mechanics. *ZAMM Z. angew. Math. Mech.* **76** [1996b], 497-504
GHONEIM, H.: Analysis and Application of a Coupled Thermoviscoplasticity Theory. *Journal of Applied Mechanics* **57** [1990], 828-835
GREEN, A.E.; MCINNIS, M.C.: Generalized Hypo-Elasticity. *Proc. Roy. Soc. Edinburgh* **A 57** [1967], 220-230
GREEN, A.E.; NAGHDI, P.M.: A General Theory of an Elastic-Plastic Continuum. *Arch. Rat. Mech. Anal.* **18** [1965], 251-281
GREEN, A.E.; NAGHDI, P.M.: Some Remarks on Elastic-Plastic Deformation at Finite Strain. *Int. J. Eng. Sci.* **9** [1971], 1219-1229
GREEN, A.E.; NAGHDI, P.M; TRAPP, J.A.: Thermodynamics of a Continuum with Internal Constraints. *Int. J. Eng. Sci.* **8** [1970], 891-908
GREEN, A.E.; ZERNA, W.: *Theoretical Elasticity.* Oxford University Press [1968]
GROLIG, G.: Plastische Deformation als Ausdruck mikroskopischer Instabilität. Doctoral Thesis [1985], Technical University of Darmstadt
GROSS, B.: *Mathematical Structure of the Theories of Linear Viscoelasticity.* Hermann [1968]
GURTIN, M.E.: The Linear Theory of Elasticity. *Encyclopedia of Physics,* Vol. VIa/2. Springer-Verlag [1972], 1-295.
GURTIN, M.E.: *An Introduction to Continuum Mechanics.* Academic Press [1981]
HART, E.W.: Constitutive Relations for the Nonelastic Deformation of Metals. *Trans. ASME J. Engng. Mat. Technol.* **98** [1976], 193-202
HARTMANN, G.; KOLLMANN, F.G.: A Computational Comparison of the Inelastic Constitutive Models of Hart and Miller. *Acta Mech.* **69** [1987], 139-165
HARTMANN, S.: Lösung von Randwertaufgaben der Elastoplastizität. Ein Finite-Elemente-Konzept für nichtlineare kinematische Verfestigung bei kleinen und finiten Verzerrungen. Doctoral Thesis [1993], University of Kassel, Institute of Mechanics, Report 1/1993
HARTMANN, S.: Nichtlineare Finite-Elemente-Berechnung angewendet auf ein Viskoplastizitätsmodell mit Überspannungen. In: Hartmann, S.; Tsakmakis, Ch. (Eds): *Aspekte der Kontinuumsmechanik und Materialtheorie*, Berichte des Instituts für Mechanik (Bericht 1/1998), Gesamthochschul-Bibliothek Kassel [1998], 55-80
HARTMANN, S.: Computation in Finite Strain Viscoelasticity: Finite Elements Based on the Interpretation as Diffferential-Algebraic Equations. To appear in *Computer Methods in Applied Mechanics and Engineering* [2001]

HARTMANN, S.; LÜHRS, G.; HAUPT, P.: An Efficient Stress Algorithm with Applications in Viscoelasticity and Plasticity. *Int. Journal for Numerical Methods in Engineering* **40** [1997], 991-1013

HARTMANN, S.; KAMLAH, M.; KOCH, A.: Numerical Aspects of a non- proportional cyclic plasticity model under plane stress conditions. *Int. Journal for Numerical Methods in Engineering* **42** [1998], 1477-1498

HÄUSLER, O.: Anisotropes plastisches Fließen bei großen Deformationen. Doctoral Thesis [1999], Forschungszentrum Karlsruhe, FZKA 6351

HÄUSLER, O.; TSAKMAKIS, CH.: Description of Plastic Anisotropy Effects at Large Deformations. Part II: the Case of Transverse Isotropy. To appear in *Int. J. Plast.* [2001]

HÄUSLER, O.; TSAKMAKIS, CH.; AIFANTIS, E.C.: On Anisotropic Plastic Flow. In: O.T. Bruhns & E. Stein (Eds.): *Solid Mechanics and its Applications, IUTAM Symposium on Micro- and Macrostructural Aspects of Thermoplasticity.* Kluwer Academic Publishers [1999], 69 - 78

HAUPT, P: *Viskoelastizität und Plastizität.* Springer-Verlag [1977]

HAUPT, P: On the Mathematical Modelling of Material Behavior in Continuum Mechanics. *Acta Mech.* **100** [1993], 129-154

HAUPT, P: On the Thermodynamic Representation of Viscoplastic Material Behavior. *Proceedings of the ASME Materials Division* Vol. 1 (MD-Vol. 69-1), ISBN No. 0-7981-1759-8, The American Society of Mechanical Engineers [1995], 503-519

HAUPT, P.; HELM, D.: Thermomechanical Coupling in Viscoplasticity: Constitutive Modelling of Shape-Memory Behaviour. In: Maruszewski, B.T.; Muschik, W.; Radowicz, A. (eds.): *Trends in Continuum Physics TRECOP '98.* World Scientific Publishing Co. Pte. Ltd. [1999], 141-155

HAUPT, P.; HELM, D.; TSAKMAKIS, CH.: Stored Energy and Dissipation in Thermoviscoplasticity. *ZAMM Z. angew. Math. Mech.* **77** [1997], S119-S120

HAUPT, P.; KAMLAH, M.: Representation of Cyclic Hardening and Softening Properties Using Continuous Variables. *Int. J. Plasticity* **11** [1995], 267-291

HAUPT, P.; KAMLAH, M.; TSAKMAKIS, CH.: A New Model for a Representation of Hardening Properties in Cyclic Plasticity. J.-P. In: Boehler, A. S. Khan (eds.): *Anisotropy and Localization of Plastic Deformation. Proceedings of PLASTICITY '91: The Third International Symposium on Plasticity and its Current Applications.* Elsevier Applied Science [1991], 451-454

HAUPT, P.; KAMLAH, M.; TSAKMAKIS, CH: Continuous Representation of Hardening Properties in Cyclic Plasticity. *Int. J. Plasticity* **8** [1992a], 803-817

HAUPT, P.; KAMLAH, M.; TSAKMAKIS, CH: On the Thermodynamics of Rate-Independent Plasticity as an Asymptotic Limit of Viscoplasticity for Slow Processes. In: D. Besdo, E. Stein (eds.): *Finite Inelastic Deformations - Theory and Applications.* Springer-Verlag Berlin, Heidelberg [1992b], 107-116

HAUPT, P.; KERSTEN T.: On the Representation of Anisotropic Viscoplasticity. Submitted for publication [2001]

HAUPT, P.; KORZEŃ, M.: On the Mathematical Modelling of Material Behavior in Continuum Mechanics. In: F. Jinhong, S. Murakami (Eds.): *Advances in Constitutive Laws for Engineering Materials*, Vol. 1. Pergamon Press Oxford [1989], 456-459

HAUPT, P.; LION, A.: Experimental Identifikation and Mathematical Modelling of Viscoplastic Material Behavior. *Continuum Mechanics and Thermodynamics* 7 [1995], 73 - 96

HAUPT, P.; LION, A.: A generalisation of the Mooney-Rivlin Model to Finite Linear Viscoelasticity. In: D. Besdo, R.H. Schuster, J. Ihlemann (eds.): *Constitutive Models for Rubbber II.* Swets & Zeitlinger [2001a], 57 - 64

HAUPT, P.; LION, A.: On Finite Linear Viscoelasticity of Incompressible Isotropic Materials. Submitted for publication [2001b]

HAUPT, P.; LION, A.; BACKHAUS, E.: On the Dynamic Behaviour of Polymers During Finite Strains: Constitutive Modelling and Identification of Parameters. *Int. J. Sol. Struct.* **37** [1999], 3633 - 3646

HAUPT, P.; SEDLAN, K.: Viscoplasticity of Elastomeric Materials: Experimental Facts and Constitutive Modelling. *Archive of Applied Mechanics* **71** [2001], 89 - 109

HAUPT, P.; TSAKMAKIS, CH.: On the Principle of Virtual Work and Rate-Independent Plasticity. *Archive of Mechanics* **40** [1988], 403-414

HAUPT, P.; TSAKMAKIS, CH.: On the Application of Dual Variables in Continuum Mechanics. *Journal of Continuum Mechanics and Thermodynamics* **1** [1989], 165-196

HAUPT, P.; TSAKMAKIS, CH: Stress Tensors Associated with Deformation Tensors via Duality, *Archives of Mechanics* **48** [1996], 347-384

HAVNER, K.S.: *Finite Plastic Deformation of Crystalline Solids.* Cambridge University Press [1992]

HELM, D.: Experimentelle Untersuchungen und phänomenologische Modellierung thermomechanischer Kopplungseffekte in der Metallplastizität. In: HARTMANN, S.; TSAKMAKIS, CH. (Eds): *Aspekte der Kontinuumsmechanik und Materialtheorie*, Berichte des Instituts für Mechanik (Bericht 1/1998), Gesamthochschul-Bibliothek Kassel [1998], 81-105

HILL, R.: *The Mathematical Theory of Plasticity.* Oxford University Press [1950]

HILL, R.; HAVNER, K.S.: Perspectives in the Mechanics of Elastoplastic Crystals. *J. Mech. Phys. Solids*, **30** [1952], 349-354

HOHENEMSER, K.; PRAGER, W.: Über die Ansätze der Mechanik isotroper Kontinua. *ZAMM Z. angew. Math. Mech.* **12** [1932], 216-226

HOLZAPFEL, G, A.: *Nonlinear Solid Mechanics. A Continuum Approach for Engineers.* J. Wiley & Sons Ltd. [2000]

HORZ, M.: Linear-elastische Wellen in vordeformierten Medien. Doctoral Thesis [1994], University of Kassel, Institute of Mechanics, Report 2/1994

HORZ, M.; HAUPT, P.; HUTTER, K.: Zur Darstellung anisotroper Materialeigenschaften in der Elasto-Plastizität. *ZAMM Z. angew. Math. Mech.* **74** [1994], T243-T245

HOSFORD, W.F.: *The Mechanics of Crystals and Textured Polycrystals.* Oxford University Press [1993]

HUTTER, K.: The Foundations of Thermodynamics, its Basic Postulates and Implications. A Review Article. *Acta. Mech.* **27** [1977], 1-54

JAUNZEMIS, W.: *Continuum Mechanics.* The McMillan Company [1967]

KACHANOV, L.M.: *Foundations of the Theory of Plasticity.* North Holland Publishing Company [1971]

KALIDINDI, S.R.; BRONKHORST, C.A.; ANAND, L.: Crystallographic Texture Evolution in Bulk Deformation Processes of f.c.c. Metals. *J. Mech. Phys. Sol.* **40** [1992], 536 - 569

KAMLAH, M.: Zur Modellierung des Verfestigungsverhaltens von Materialien mit statischer Hysterese im Rahmen der phänomenologischen Thermomechanik. Doctoral Thesis [1994], University of Kassel, Institute of Mechanics, Report 3/1994

KAMLAH, M.; HAUPT, P.: On the Macroscopic Description of Stored Energy and Self-Heating during Plastic Deformation. *Int. J. Plasticity* **13** [1998], 893-911

KAMLAH, M.; TSAKMAKIS, CH.: Use of Isotropic Thermoelasticity Laws in Finite Deformation Viscoplasticity Models. *Cont. Mech. Thermodyn.* [1999], 73-88

KHAN, A.S.; HUANG, S.: *Continuum Theory of Plasticity.* John Wiley and Sons Inc., New York [1995]

KLINGBEIL, E.: *Tensorrechnung für Ingenieure.* B.I. Wissenschaftsverlag [1989]

KORZEŃ, M.: Beschreibung des inelastischen Materialverhaltens im Rahmen der Kontinuumsmechanik: Vorschlag einer Materialgleichung vom viskoelastisch-plastischen Typ. Doctoral Thesis [1988], Technical University Darmstadt

KOCKS, U.F.; THOMÉ, C.N.; WENK, H.R.: *Texture and Anisotropy. Preferred Orientations in Polycrystals and their Effect on Materials Properties.* Cambridge University Press [1998]

KRASKA, M.: Textursimulation bei großen inelastischen Verformungen mit der Technik des repräsentativen Volumenelements (RVE). Doctoral Thesis [1999], TU Berlin, Pro Business Verlag, ISBN 3-9805993-7-X

KRATOCHVIL, J.; DILLON, O.W.: Thermodynamics of Elasto-Plastic Materials as a Theory with Internal State-Variables. *J. Appl. Phys.* **40** [1969], 3207-3218

KRAWIETZ, A.: *Materialtheorie.* Springer-Verlag [1986]

KREISSIG, R.: *Einführung in die Plastizitätstheorie.* Fachbuchverlag [1992]

KREISSIG, R.: Auswertung inhomogener Verschiebungsfelder zur Identifikation der Parameter elastisch-plastischer Deformationsgesetze. Forschung im Ingenierwesen **64** [1998], 99-109

KREMPL, E.: An Experimental Study of Room-Temperature Rate-Sensitivity, Creep and Relaxation of AISI Type 304 Stainless Steel. *J. Mech. Phys. Solids* **27** [1979], 363-375

KREMPL, E.: Models of Viscoplasticity. Some Comments on Equilibrium (Back) Stress and Drag Stress. *Acta. Mech.* **69** [1987], 25-42

KREMPL, E.: Some General Properties of Solid Polymer Inelastic Deformation Behaviour and their Application to a Class of Clock Models. *J. Rheol.* **42** [1998], 713-725

KREMPL, E.; BORDONARO, C.M.: Non-Proportional Loading of Nylon 66 at Room Temperature. *Int. J. Plasticity* **14** [1998], 245-258

KREMPL, E.; CHOI, S.H.: Viscoplasticity Theory Based Overstress: The Modeling of Ratchetting and Cyclic Hardening of AISI Type 304 Stainless Steel. *Mechanics of Materials Laboratory, Rensselaer Polytechnical Institute*, Troy NY, Report MML 90-4. Troy [1990]

KREMPL, E.; LU, H.: The Hardening and Rate-Dependent Behavior of Fully Annealed AISI Type 304 Stainless Steel under Biaxial In-Phase and Out-of-Phase Strain Cycling at Room Temperature. *J. Eng. Mat. Techn.* **106** [1984], 376-382

KREMPL, E.; LU, H.: The Path and Amplitude Dependence of Cyclic Hardening of Type 304 Stainless Steel at Room Temperature. In: Brown, M.W.; Miller, K.J. (Eds.): *Biaxial and Multiaxial Fatigue*, EGF Publication 3. London [1989], 89-106

KREMPL, E.; MCMAHON, J.J.; YAO, D.: Viscoplasticity Based on Overstress with a Differential Growth Law for the Equilibrium Stress. *Mech. Mater.* **5** [1986], 35-48

KRÖNER, E.: Allgemeine Kontinuumstheorie der Versetzungen und Eigenspannungen. *Arch. Rat. Mech. Anal.* **4** [1960], 273-334

KRÖNER, E.: *Kontinuumstheorie der Versetzungen und Eigenspannungen.* Springer-Verlag [1958]

LAVENDA, B.H.: *Thermodynamics of Irreversible Processes.* THE MACMILLAN PRESS LTD [1998]

LE, K.CH.; STUMPF, H.: Constitutive Equations for Elastoplastic Bodies at Finite Strain: Thermodynamic Implementation. *Acta Mech.* **100** [1993], 155-170

LEE, E.H.: Elastic-Plastic Deformation at Finite Strains. *J. Appl. Mech.* **36** [1969], 1-6

LEHMANN, TH.: General Frame for the Definition of Constitutive Laws for Large Non-Isothermal Elastic-Plastic and Elastic-Visco-Plastic Deformations. In: Lehmann, Th. (Ed.): *The Constitutive Law in Thermoplasticity.* CISM Courses and Lectures No. 281. Springer-Verlag [1984], 379-463

LEHMANN, TH.: On a Generalized Constitutive Law in Thermoplasticity. In: Sawczuk, A.; Bianchi, G. (Eds.): *Plasticity Today.* London, New York [1985]

LEHMANN, TH.: On Thermodynamically Consistent Constitutive Laws in Plasticity and Viscoplasticity. *Archives of Mechanics* **40** [1988], 415-431

LEHMANN, TH.: Thermodynamical Foundations of Large Inelastic Deformations of Solid Bodies Including Damage. *Int. J. Plasticity* **7** [1991], 79-98

LEIGH, D.C.: *Nonlinear Continuum Mechanics.* McGraw-Hill [1968]

LEITMAN, M.J.; FISCHER, G.M.C.: The Linear Theory of Viscoelasticity. *Encyclopedia of Physics*, Vol. VIa/3. Springer-Verlag [1973], 1-123.

LEVITAS, V.I.: *Large Deformation of Materials with Complex Rheological Properties at Normal and High Pressure.* Nova Science Publishers, Inc., New York [1993]

LION, A.: Materialeigenschaften der Viskoplastizität: Experimente, Modellbildung und Parameteridentifikation. Doctoral Thesis [1994], University of Kassel, Institute of Mechanics, Report 1/1994

LION, A.: A Constitutive Model for Carbon-Black Filled Rubber: Experimental Investigations and Mathematical Representation. *Cont. Mech. Thermodyn.* **8** [1996], 153-169

LION, A.: A Physically Based Method to Represent the Thermomechanical Behaviour of Elastomers. *Acta Mech.* **123** [1997a], 1-25

LION, A.: On the Large Deformation Behaviour of Reinforced Rubber at Different Temperatures. *J. Mech. Phys. Solids* **45** [1997b], 1805-1834

LION, A.: On the Thermodynamics of Fractional Damping Elements. *Continuum Mechanics and Thermodynamics* **9** [1997c], 83 - 96

LION, A.: Thixotropic Behaviour of Rubber under Dynamic Loading Histories: Experiments and Theory. *J. Mech. Phys. Solids* **46** [1998], 895-930

LION, A.: Constitutive Modelling in Finite Thermoviscoplasticity: A Physical Approach Based on Rheological Models. *Int. J. Plasticity* [1999] (to appear)

LION, A.: Thermomechanik von Elastomeren. Experimente und Materialtheorie. Habilitation Thesis [2000], University of Kassel, Institute of Mechanics, Report 1/2000

LION, A.; HAUPT, P.: Finite Viscoelasticity: A Physical Approach based on Relaxation Spectra, Fractional Derivatives and Process-Dependent Viscosities. In: D. Besdo, R.H. Schuster, J. Ihlemann (eds.): *Constitutive Models for Rubber II*. Swets & Zeitlinger [2001], 65 - 72

LION, A.; SEDLAN K.: Eine Methode zur Formulierung thermodynamisch konsistenter Stoffgleichungen. In: HARTMANN, S.; TSAKMAKIS, CH. (Eds): *Aspekte der Kontinuumsmechanik und Materialtheorie*, Berichte des Instituts für Mechanik (Bericht 1/1998), Gesamthochschul-Bibliothek Kassel [1998], 153-174

LIPPMANN, H.: *Mechanik des plastischen Fließens*. Springer-Verlag [1981]

LIPPMANN, H.: *Angewandte Tensorrechnung*. Springer-Verlag [1993]

LIPPMANN, H.; MAHRENHOLTZ, O.: *Plastomechanik der Umformung metallischer Werkstoffe*. Springer-Verlag [1967]

LJUSTERNIK, L.A.; SOBOLEV, V.J.: *Elemente der Funktionalanalysis*. Verlag Harri Deutsch [1979]

LOCKETT, F.J.: *Nonlinear-Viscoelastic Solids*. Academic Press [1972]

LUBLINER, J.: On Fading Memory in Materials of Evolutionary Type. *Acta Mech.* **8** [1969], 75-81

LUBLINER, J.: On the Thermodynamic Foundations of Non-Linear Solid Mechanics. *Int. J. Non-Linear Mechanics* **7** [1972], 237-254

LUBLINER, J.: On the Structure of the Rate Equations of Materials with Internal Variables. *Acta Mech.* **17** [1973], 109-119

LUBLINER, J.: *Plasticity Theory*. Macmillan Publishing Company, New York, London [1990]

LÜHRS, G.: Randwertaufgaben der Viskoplastizität - Modellierung, Simulation und Vergleich mit experimentellen Daten aus zyklischen Prozessen und Umformvorgängen. Doctoral thesis [1997], University of Kassel, Institute of Mechanics, Report 1/1997

LÜHRS, G.; HARTMANN, S.; HAUPT, P.: On the Numerical Treatment of Finite Deformations in Elastoviscoplasticity. *Computer Methods in Applied Mechanics and Engineering* **144** [1997], 1-21

LÜHRS, G.; HAUPT, P.: On the Evolution of Inhomogeneous Deformations in Test Specimens under Cyclic Loading Conditions. In: Owen, D.R.J.; Oñate, E.; Hinton, E. (Eds.): *COMPUTATIONAL PLASTICITY (COMPLAS) Fundamentals and Applications*. CIMNE, Barcelona [1997], 945-950

LURIE, A.J.: *Nonlinear Theory of Elasticity*. North-Holland [1991]

MALVERN, L.E.: *Introduction to the Mechanics of a Continuous Medium*. Prentice-Hall [1969]

MANDEL, J.: Plasticité classique et viscoplasticité. CISM Courses and Lectures No. 97. Springer-Verlag [1971]

MANDEL, J.: Equations constitutives et directeurs dans les milieux plastiques et viscoplastiques. *Int. J. Sol. Struct.* **9** [1973], 725 - 740

MARSDEN, J.E.; HOFFMAN, M.J.: *Elementary Classical Analysis.* W.H Freeman & Company [1993]

MARSDEN, J.E.; HUGHES, T.J.R.: *Mathematical Foundations of Elasticity.* Prentice-Hall [1983]

MARSDEN, J.E.; TROMBA, A.J.: *Vektoranalysis.* Spektrum Akademischer Verlag Heidelberg [1995]. Original Title: Vector Calculus. W.H. Freeman and Company New York [1988]

MAUGIN, G.A.: Internal Variables and Dissipative Structures. *J. Non-Equilib. Thermodyn.* **15** [1990], 173-192

MAUGIN, G.; MUSCHIK, W.: Thermodynamics with Internal Variables, Part I: General Concepts Part II: Applications. *J. Non-Equilib. Thermodyn.* **19** [1994], 217-249

MÉRIC, L.; POUBANNE, P.; CHAILLETAUD, G.: Single Crystal Modeling for Structural Calculations: Part 1 - Model Presentation. *Journal of Engineering Materials and Technology* **113** [1991], 162 - 170

MIEHE, C.: Kanonische Modelle multiplikativer Elasto-Plastizität. Thermodynamische Formulierung und numerische Implementation. Universität Hannover, Institut für Baumechanik und numerische Mechanik Report F 93/1 [1993]

MIEHE, C.: A Theory of Large-Strain Isotropic Thermoplasticity Based on Metric Transformation Tensors. *Archive of Applied Mechanics* **66** [1995], 413-427

MIEHE, C.: Multisurface Thermoplasticity for Single Crystals at Large Strains in Terms of Eulerian Vector Updates. *Int J. Solids Structures* **33** [1996], 3103-3130

MIEHE, C.; KECK, J.: Superimposed Finite Elastic-Viscoelastic-Plastoelastic Stress Response with Damage in Filled Rubbery Polymers. Experiments, Modelling and Algorithmic Implementation. *Journal of the Mechanics and Physics of Solids* **48** [2000], 323 - 365

MIEHE, C.; STEIN, E.: A Canonical Model of Multiplicative Elasto-Plasticity. Formulation and Aspects of the Numerical Implementation. *European Journal of Applied Mechnics* **11** [1992], 25-43

MILLER, A.K.: An Inelastic Constitutive Model for Monotonic, Cyclic and Creep Deformation: Part I - Equations Development and Analytical Procedures; Part II - Application to type 304 stainless steel. *Trans. ASME J. Engng. Mat. Technol.* **98** [1976], 97-113

MILLER, A.K.: *Unified Constitutive Equations for Plastic Deformation and Creep of Engineering Alloys.* Elsevier [1987]

MIZEL, V.J.; WANG, C.-C.: A Fading Memory Hypothesis which Suffices for Chain Rules. *Arch. Rat. Mech. Anal.* **23** [1966], 124-134

MÜLLER, I.: On the Frame Dependence of Stress and Heat Flux. *Arch. Rat. Mech. Anal.* **45** [1972], 241-250

MÜLLER, I.: *Thermodynamik Die Grundlagen der Materialtheorie.* Bertelsmann Universitätsverlag [1973]

MÜLLER, I.: On the Frame Dependence of Electric Current and Heat Flux in a Metal. *Acta Mech.* **24** [1976], 117-128

MÜLLER, I.: *Thermodynamics.* Pitman [1985]

MÜLLER, M.: Plastische Anisotropie polykristalliner Materialien. VDI Reihe 18: Mechanik/Bruchmechanik, Bd. 137. Düsseldorf [1993]
MÜLLER, W.C.: Universal Solutions for Simple Thermodynamic Bodies. *Arch. Rat. Mech. Anal.* **35** [1969], 220-225
MUSCHIK, W.; EHRENTRAUT, H.: An Amendment to the Second Law. *J. Non-Equilib. Thermodyn.* **21** [1996], 175-192
MUSCHIK, W.: Objectivity and Frame Indifference, Revisited. *Archives of Mechanics* **50** [1998a], 541-547
MUSCHIK, W.: Irreversibility and Second Law. *J. Non-Equilib. Thermodyn.* **23** [1998b], 87-98
NAGHDI, P.M.: A Critical Review of the State of Finite Plasticity. *Journal of Applied Mathematics and Physics (ZAMP)* **41** [1990], 315-394
NOLL, W.: A Mathematical Theory of the Mechanical Behavior of Continuous Media. *Arch. Rat. Mech. Anal.* **2** [1958], 197-226
NOLL, W.: Proof of the Maximality of the Orthogonal Group in the Unimodular Group. *Arch. Rat. Mech. Anal.* **18** [1965], 100-102
NOLL, W.: A New Mathematical Theory of Simple Materials. *Arch. Rat. Mech. Anal.* **48** [1972], 1-50
NOWACKI, W.: *Theorie des Kriechens Lineare Viskoelastizität.* Franz Deuticke [1965]
NOWACKI, W.: *Thermoelasticity.* Pergamon Press [1986]
OGDEN, R.W.: *Nonlinear Elastic Deformations.* John Wiley and Sons [1984]
OLDHAM, K.B.; SPANIER, J.: *The Fractional Calculus: Theory and Applications of Differentiation and Integration to Arbitrary Order.* Academic Press London [1974]
OLDROYD, J.G.: On the Formulation of Rheological Equations of State. *Proc. Roy. Soc.* **A 200** [1950], 523-541
OLIFERUK, W.; SWIATNICKI, W.; GRABSKI, W.: Rate of Energy Storage and Microstructure Evolution During the Tensile Deformation of Austenitic Steel. *Mater. Sci. Eng.* **A161** [1993], 55-63
OLSCHEWSKI, J.: Der kubisch-flächenzentrierte Einkristall in Theorie und Experiment. BAM-V.31, *Report* 97/5 [1997], Unter den Eichen 87, 12205 Berlin
OWEN, D.R.: *A First Course in the Mathematical Foundations of Thermodynamics.* Springer-Verlag [1984]
OWEN, D.R.; WILLIAMS, W.O.: On the Concept of Rate-Independence. Quart. Appl. Math. **26** [1968], 321-329
PERZYNA, P.: The Constitutive Equations for Rate Sensitive Plastic Materials. *Quart. Appl. Math.* **20** [1963], 321-332
PERZYNA, P.: Fundamental Problems in Viscoplasticity. In: Kuerti, G. (Ed.): *Advances in Applied Mechanics*, Vol. 9. New York, London [1966], 243-377
PERZYNA, P.; WOJNO, W.: Thermodynamics of a Rate Sensitive Plastic Material. *Archives of Mechanics* **20** [1968], 499-511
PIPKIN, A.C.: *Lectures on Viscoelasticity Theory.* Springer-Verlag [1972]
PIPKIN, A.C.; RIVLIN, R.S.: Mechanics of rate-independent materials. *Zeitschrift für angewandte Mathematik und Physik (ZAMP)* **16** [1965], 313-326
PONTER, A.R.S.; BATAILLE, J.; KESTIN, J.: A Thermodynamic Model for the Time Dependent Deformation of Solids. *Journal de Méchanique* **18** [1979], 511 - 539
PRAGER, W.: *Einführung in die Kontinuumsmechanik.* Birkhäuser Verlag [1961]
PRAGER, W.; HODGE, P.G.: *Theorie ideal plastischer Körper.* Springer-Verlag [1954]

RASCHEWSKI, P.K.: *Riemannsche Geometrie und Tensoranalysis.* VEB Deutscher Verlag der Wissenschaften [1959]

RECKLING, K.A.: *Plastizitätstheorie und ihre Anwendung auf Festigkeitsprobleme.* Springer-Verlag [1967]

REESE, S.; GOVINDJEE, S.: A Theory of Finite Viscoelasticity and Numerical Aspects. *Int. J. Solids Structures* **35** [1998a], 3455-3482

REESE, S.; GOVINDJEE, S.: Theoretical and Numerical Aspects in the Thermo-Viscoelastic Material Behavior of Rubber-Like Polymers. *Mechanics of Time-Dependent Materials* **1** [1998b], 357-396

RIVLIN, R.S.: Some Comments on the Endochronic Theory of Plasticity. *Int. J. Solids Structures* **17** [1981], 231-248

RUBIN, M.: Plasticity Theory Formulated in Terms of Physically Based Micro-structural Variables - Part I. Theory - Part II. Examples. *Int. J. Solids Structures* **31** [1994], 2615-2634

SADIKI, A.; HUTTER, K.: On the Frame-Dependence and Form-Invariance of the Transport Equations for the Reynolds Stress Tensor and the Turbulent Heat Flux Vector: its Consequences on Closure Models in Turbulence Modelling. *Cont. Mech. Thermodyn.* **8** [1996], 341-349

SCHMID, E.; BOAS, W.: *Kristallplastizität mit besonderer Berücksichtigung der Metalle.* Springer, Berlin [1935]

SERRIN, J.: Mathematical Principles of Classical Fluid Mechanics. *Encylopedia of Physics*, Vol. VIII/1. Springer-Verlag [1959], 125-263

ŠILHAVÝ, M.: *The Mechanics and Thermodynamics of Continuous Media.* Springer-Verlag [1997]

SIMO, J.C.: A Framework of Finite Strain Elastoplasticity Based on Maximum Plastic Dissipation and Multiplicative Decomposition: Part I Continuum formulation. *Computer Methods in Applied Mechanics and Engineering* **66** [1988], 199-219

SIMO, J.C.; PISTER, K.S.: Remarks on Rate Constitutive Equations for Finite Deformation Problems: Computational Implications. *Computer Methods in Applied Mechanics and Engineering* **46** [1984], 201-215

SMITH, G.F.; RIVLIN, R.S.: The Strain Energy Function for Anisotropic Elastic Materials. *Trans. Amer. Math. Soc.* **88** [1958], 175-193

SPURK, J.: *Fluid Mechanics.* Springer-Verlag [1997]

STREILEIN, T.: Erfassung formativer Verfestigung in viskoplastischen Stoffmodellen. Report Nr. 97-83, Institut für Statik, TU Braunschweig, Beethovenstr. 51, D-38106 Braunschweig [1997]

SVENDSEN, B.: A Thermodynamic Formulation of Finite-Deformation Elasto-plasticity with Hardening Based on the Concept of Material Isomorphism. *Int. J. Plasticity* **14** [1998], 473-488

SVENDSEN, B.; BERTRAM, A.: On Frame-Indifference and Form-Invariance in Constitutive Theory. *Acta Mech.* **132** [1999], 195-207

SVENDSEN, B.; TSAKMAKIS, CH.: A Local Differential Geometric Formulation of Dual Stress-Strain Pairs and Time Derivatives. *Archives of Mechanics* **46** [1994], 49-91

SYNGE, J.L.: Classical Dynamics. In: Flügge, S. (Ed.): *Encyclopedia of Physics*, Vol. III/1. Springer-Verlag [1960], 1 - 225

SZABÓ, I.: *Geschichte der mechanischen Prinzipien.* Birkhäuser Verlag [1987]

TAYLOR, G.I.: Analysis of Plastic Strain in a Cubic Crystal. In: J.M. Lessels (Ed.): *Stephen Timoshenko 60th Anniversary Volume* [1938], 218 - 224

TEODOSIU, C.: Dislocation Modelling of Crystalline Plasticity. In: *Large Plastic Deformation of Crystalline Aggregates.* CISM Courses and Lectures No. 376. Springer-Verlag [1997], 21 - 80

THOMAS, T.Y.: Systems of Total Differential Equations Defined over Simply Connected Domains. *Annals Math.* **35** [1934], 730-734

TIMOSHENKO, S.P.; GOODIER, J.N.: *Theory of Elasticity.* McGraw Hill [1982]

TING, T.: *Anisotropic Elasticity: Theory and Applications.* Oxford University Press [1996]

TOBOLSKY, A. V.: *Mechanische Eigenschaften und Struktur von Polymeren.* Berliner Union [1967]

TRELOAR, L.R.G.: *The Physics of Rubber Elasticity.* Oxford University Press [1975]

TROSTEL, R.: *Mathematische Methoden der Festigkeitslehre.* Vorlesung an der TU Berlin [1966]

TROSTEL, R.: *Mechanik VII/1, Materialgleichungen spezieller Medien.* Technische Universität Berlin; Universiversitätsbibliothek, Abt. Publikationen [1990]

TROSTEL, R.: *Vektor- und Tensoralgebra.* Verlag Vieweg [1993]

TRUESDELL, C.: *The Elements of Continuum Mechanics.* Springer-Verlag [1985]

TRUESDELL, C.: *Rational Thermodynamics.* Springer-Verlag [1984a]

TRUESDELL, C.: *Thermodynamics for Beginners.* In: Truesdell, C. (Ed.): *Rational Thermodynamics.* Springer-Verlag [1984b], 82-106

TRUESDELL, C.; NOLL, W.: The Non-Linear Field Theories of Mechanics. In: Flügge, S. (Ed.): *Encyclopedia of Physics*, Vol. III/3. Springer-Verlag [1965]

TRUESDELL, C.; TOUPIN, R.A.: The Classical Field Theories. In: Flügge, S. (Ed.): *Encyclopedia of Physics*, Vol. III/1. Springer-Verlag [1960], 226 - 793

TSAKMAKIS, CH.: Über inkrementelle Materialgleichungen zur Beschreibung großer Deformationen. VDI Reihe 18: Mechanik/Bruchmechanik, Bd. 36. Dissertation Darmstadt [1987]

TSAKMAKIS, CH.: Kinematic Hardening Rules in Finite Plasticity - Part I: A Constitutive Approach - Part II: Some Examples. *Cont. Mech. Thermodyn.* **8** [1996], 215-246

TSAKMAKIS, CH.: General Strain and Stress Tensors, Associated by Duality. *Cont. Mech. Thermodyn.* **13** [2001a], 307 - 324

TSAKMAKIS, CH.: Description of Plastic Anisotropy Effects at Large Deformations. Part I: Restrictions imposed by the Second Law and the Postulate of Il'iushin. To appear in *Int. J. Plast.* [2001b]

TSAKMAKIS, CH.: Description of Plastic Anisotropy Effects at Large Deformations. Part II: The Case of Transverse Isotropy. To appear in *Int. J. Plast.* [2001c]

TSCHOEGL, N.W.: *The Phenomenological Theory of Linear Viscoelastic Material Behaviour.* Springer-Verlag [1989]

TSOU, J. C.; QUESNEL, D.J.: Internal Stress Measurements during the Saturation Fatigue of Polycrystalline Aluminum. *Materials Science and Engineering* **56** [1982], 289-299

VALANIS, K.C.: A Theory of Viscoplasticity without a Yield Surface. Part I - General Theory. Part II: Application to Mechanical Behavior of Metals. *Archives of Mechanics* **23** [1971], 517-233

VALANIS, K.C.: Effect of Prior Deformation on Cyclic Response of Metals. *J. Appl. Mech.* **41** [1974], 441-447

VALANIS, K.C.: On the Foundations of the Endochronic Theory of Viscoplasticity. *Archives of Mechanics* **27** [1975], 857-868

VALANIS, K.C.: Proper Tensorial Formulation of the Internal Variable Theory. The Endochronic Time Spectrum. *Archives of Mechanics* **29** [1977], 173-185

VALANIS, K.C.: Fundamental Consequences of a New Intrinsic Time Measure. Plasticity as a Limit of the Endochronic Theory. *Archives of Mechanics* **32** [1980], 171-191

VALANIS, K.C.: On the Substance of RIVLIN'S Comments on the Endochronic Theory. *Int. J. Solids Structures* **17** [1981], 249-265

VALANIS, K.C.: The Concept of Physical Metric in Thermodynamics. *Acta Mech.* **113** [1995], 169-184

WANG, C.-C.; TRUESDELL, C.: *Introduction to Rational Elasticity.* Noordhoff International Publishing [1973]

WEINITSCHKE, H.-J.: *Rationale Mechanik.* Vorlesung an der TU Berlin [1968]

WELLERDICK-WOJTASIK, N.: Theoretische und experimentelle Untersuchungen über die Fließflächenentwicklung bei großen Scherdeformationen. VDI Reihe 18: Mechanik/Bruchmechanik, Bd. 215. Düsseldorf [1997]

XIAO, H.: On Anisotropic Scalar Functions of a Single Symmetric Tensor. *Proceedings of the Royal Society of London* A **452** [1996a], 1545-1561

XIAO, H.: On Minimal Representations for Constitutive Equations of Anisotropic Elastic Materials. *Journal of Elasticity* **45** [1996b], 13-32

XIAO, H.: On Isotropic Extension of Anisotropic Tensor Functions. *ZAMM Z. angew. Math. Mech.* **76** [1996c], 205-214

XIAO, H.: On Isotropic Invariants of the Elasticity Tensor. *Journal of Elasticity* **46** [1997], 115-149

XIAO, H.; BRUHNS, O.T.; MEYERS, A.: Logarithmic Strain, Logarithmic Spin and Logarithmic Rate. *Acta Mech.* **124** [1997], 89-105

ZHANG, J.M.; RYCHLEWSKI, J.: Structural Tensors for Anisotropic Solids. *Archives of Mechanics* **42** [1990], 267-277

ZIEGLER, F.: *Mechanics of Solids and Fluids.* Springer-Verlag [1995]

Index

A
absolute acceleration 175
absolute frame 603
absolute frame of reference 172
absolute rate of rotation 606
absolute velocity 175
acceleration 22, 168
accompanying equilibrium process 490, 539, 541
accumulated plastic strain 233, 451, 575
accumulated strain 436
active interpretation 162, 164
additive decomposition 56, 427, 456
affine connection 27, 59
affine space 157
Almansi tensor 37, 40, 48, 50, 318, 331, 472, 481
analogy to viscoelasticity 450
anisotropic behaviour 378
anisotropic elasticity 545
anisotropic hyperelastic solid 388
anisotropic materials 394
anisotropic solid 304, 379, 587
anisotropic viscoplasticity 591, 606
anisotropy 530, 586
anisotropy axes 590, 595
approximation 366, 411, 414
arclength 233, 238, 326, 436
arclength representation 436, 498, 577
Armstrong-Frederick model 326, 461
arrow 158
associated flow rule 461, 479
assumption of stability 485
asymptotic approximation 366, 414, 425, 541
asymptotic limit 484, 574
asymptotic relations 539
asymptotic stability 419
axes of elastic anisotropy 588

B
backstress tensor 233, 461, 563
balance of entropy 127
balance of linear momentum 88, 95, 135, 529
balance of mass 79, 135
balance of mechanical energy 104, 345
balance of rotational momentum 88, 99
balance relations 3, 76, 177, 278
barotropic fluid 178
barycentric derivative 148
barycentric velocity 147
basic invariants 355, 379
Bauschinger effect 241
bilinear approximation 241
Boltzmann's principle 189
boundary conditions 4, 102, 130, 179, 183, 187
boundary-value problem 102, 110
boundedness 412
bulk modulus 186, 374, 528
bulk relaxation 191
bulk viscosity 182

C
caloric equation of state 522

cartesian coordinates 11, 60, 81, 160, 181, 184, 377
categories of material behaviour 269, 270
Cauchy elasticity 357
Cauchy heat flux vector 122
Cauchy stress 168, 280, 399
Cauchy stress principle 90
Cauchy stress vector 87, 104
Cauchy tensor 93, 96, 99, 318, 332
Cauchy-Green tensor 31, 40, 66, 285, 345, 373
Cayley-Hamilton equation 358, 364, 368
centre of mass 85, 88, 137, 169
centre of the yield surface 461
change of frame 160, 163, 164, 170, 458, 603
change of reference system 279
Christoffel symbol 59, 65
classical equilibrium thermodynamics 576
classical theories 3, 178, 249, 276
Clausius-Duhem inequality 125, 129, 511
closed systems 132
coaxial function 354
coefficient of compressibility 523
coefficient of expansion 522, 528
compatibility 64, 198, 456
complementary energy 510
complementary free energy 510
complementary stress power 520
complex modulus 217, 221
compressibility 372, 375
concentrations 145
concept of dual variables 318
condition of constraint 305
conditions of compatibility 41, 64, 67, 456
conditions of equilibrium 114
configuration 7
conjugate variables 330
conservation laws 137
conservation of energy 139, 147, 150
conservation of linear momentum 139, 146, 149
conservation of mass 83, 139, 146
conservation of mechanical energy 347
conservation of rotational momentum 139, 147
conservative forces 348
consistency condition 233, 235, 453, 461, 463, 496
constitutive equation 3, 275
constitutive functional 250, 281, 400, 437
constraint 305
continuity equation 79, 524
continuous relaxation spectrum 207, 211, 220
continuum mechanics 2
contravariant Oldroyd rate 48, 482
convective coordinates 38, 50, 64, 95, 108, 342
convective derivative 22
convective stress tensor 94, 332, 351
coordinate differential 25, 70
coordinate transformation 17
correspondence principle 438, 445, 450
Coulomb friction 272, 274
covariant derivative 61, 82, 98, 181
covariant Oldroyd rate 48, 481
creep 256, 257, 259, 263, 499
creep function 189, 191
crystal lattice 380, 452, 594, 595
crystal plasticity 608, 618
crystal viscoplasticity 606
cubic crystal 608
cubic system 383, 392
cumulative spectrum 207, 211, 225
curl 60
current configuration 9, 11, 343, 471
curvilinear coordinates 13, 82, 98
cyclic hardening 447
cyclic process 439, 504, 506
cyclic softening 447
cyclic stress-strain curve 443
cyclic test 262, 264, 268

D

dashpot 197, 272
decomposition 29, 44, 56, 335, 428, 454, 473, 475, 476, 480, 482, 577
decomposition of the deformation 464, 469, 471, 478, 560, 573, 591, 599
deformation gradient 23, 29, 39, 72, 283, 335, 427, 454, 458, 480
deformation history 284
derivative of fractional order 211
determinism 277, 279, 284, 305
differentiable manifold 9
differential 349
differential form 350
diffusion mass flux 148
diffusion stresses 149
diffusion velocities 147
discontinuity surface 140, 143
dislocation 452, 555
displacement field 529
displacement gradient 54

displacement vector 18, 54
dissipation 435, 511
dissipation inequality 511, 515, 532, 555, 561
dissipation modulus 216
divergence 60, 62, 99, 139
divergence theorem 97
dual derivatives 336, 337
dual strain rate 336, 428
dual strain tensor 336
dual stress rate 336
dual stress tensor 336
dual variables 331, 338, 427, 468, 481, 511
dynamic boundary condition 103, 131, 180, 183, 187
dynamic moduli 219
dynamic yield surface 479

E

efflux velocity 136
eigenstresses 456
eigenvalues 354
eigenvectors 354
elastic anisotropy 381, 586
elastic deformation gradient 454
elastic domain 475, 478, 479, 560
elastic fluid 178, 353, 520
elastic Green tensor 455
elastic material 346
elastic processes 453
elastic range 450
elastic region 228, 232
elastic-perfectly-plastic solid 227
elastic-plastic solid 231
elasticity 178, 198, 249, 271, 275, 311, 345
elasticity matrix 389
elasticity modulus 186
elasticity relation 429, 430, 433, 459, 478, 595
elasticity tensor 376, 468
elastomer 263
elastoplasticity 452, 469, 498
energy 347
energy balance 121, 124, 154, 509, 523
energy function 533
engineering mechanics 178
enthalpy 510, 516, 521
entropy 126, 168, 529, 531
entropy balance 139
entropy flux 152, 168
entropy inequality 125, 130, 154, 510
entropy production 125, 152, 509
entropy relation 515, 532, 561, 574
entropy supply 125, 126, 152
equation of heat conduction 518, 524, 529, 534, 562, 567
equilibrium 198, 260, 537, 556
equilibrium bulk modulus 192
equilibrium curve 261, 267
equilibrium hysteresis 263, 271, 475, 477
equilibrium part 577
equilibrium process 490
equilibrium relation 200, 260, 409, 541
equilibrium shear modulus 191
equilibrium solutions 419, 477, 531, 537, 576
equilibrium stress 193, 197, 200, 260, 409, 420, 433, 475, 476, 500, 501, 504
equilibrium thermodynamics 125, 560, 576
equivalence relation 157
equivalent classes 158
Euclidean metric 41
Euclidean norm 411
Euclidean space 11, 42, 58, 155
Euclidean vector space 12
Euler fluid 178
Eulerian angle 612
Eulerian representation 20
evolution equation 199, 233, 238, 431, 603
evolution equations 314, 419, 433, 445, 477, 531
experimental data 253
experimental identification 252

F

face-centered cubic crystal 597
fading memory 194, 197, 278
fading memory norm 410
fading memory property 398
family 1 336, 339, 340, 511
family 2 336, 341, 342, 448, 511
fcc 597
field equations 179, 240
filled rubber 263, 269
Finger tensor 37, 40, 50, 331, 448
finite deformations 394, 427
finite linear viscoelasticity 413, 418, 424
finite linear viscosity 416
finite viscoplasticity 482
first law 5, 119
flow rule 431, 461, 464, 469, 471, 478, 496, 560, 574, 602
fluid 299, 353, 397, 409
fluid mechanics 249
force 87, 168
form-invariant 292

Fourier model 519, 528
Fourier's Law 291
fractional derivative 211
fractional viscosity 212, 223
frame of reference 12, 13, 155, 159, 162
frame-indifference 277, 280, 284, 285, 286, 292, 301, 311, 372, 459, 480
Frank-Read source 555
Fréchet derivative 23
free energy 510, 530, 532, 542, 560, 573, 577, 594, 595
free energy functional 549
free enthalpy 510, 516, 521
free-body principle 2, 75, 87
full invariance 594, 606
functional 279
functional analysis 412
functionals 276, 312, 397, 435, 513

G

Galilei transformation 165
Gamma function 210
Gateaux derivative 22, 23, 350, 465
Gauß theorem 97, 99, 106, 112, 116, 124, 128, 133, 138, 140
general integral 314
general thermoviscoplasticity 578
general viscoplasticity 500, 501
generalised arclength 444, 446, 449, 451, 505, 575
generalised Gibbs relation 534, 567
generalised Maxwell model 212, 429
generating element 380, 381
geometric boundary condition 102, 131, 179, 183, 187
geometric linearisation 54, 57, 186
geometric vector 158
Gibbs relations 516, 521
grad 22
Grad 23
gradient 60
gradient operator 60
gradient vector 13
grain 612
gravitational mass 78
Green strain rate 47
Green tensor 32, 40, 50, 285, 331, 337, 345, 381, 477, 481
Green-McInnis rate 321, 343

H

hardening 479
hardening model 560, 574
hardening tensor 461
heat 168
heat capacity 519
heat conduction 291
heat conduction inequality 517, 530, 533
heat conduction tensor 530
heat flow 123, 518
heat flux vector 122, 515, 528, 531, 533, 560, 573
heat supply 120
Hilbert space 399, 405
hold times 257, 265, 506
Hooke solid 185
Hooke's law 186, 212, 273, 369
hydrodynamics 178
hydrostatic pressure 301, 308, 520
hyperelastic 346
hyperelasticity 351, 358, 378, 380, 468, 586
hysteresis 271

I

identification of material parameters 251
identification of material properties 252
incompatible configuration 69
incompressibility 178, 181, 308, 363, 366, 369, 417, 545
incompressible elastic material 363
incompressible fluid 185
incremental form 115
incremental stress power 117
inelastic Almansi strain 543
inelastic material behaviour 312
inelastic strain 477, 478, 481
inelastic stress power 563
inertia 292
inertial frame 155, 171
inertial mass 78
inextensibility 309
influence function 399, 424
initial conditions 4, 101, 130, 179, 183, 187, 195, 314, 323, 328, 419, 477, 484, 531
initial configuration 345
initial orientation 612, 613
initial-boundary-value problem 4, 101, 126, 177, 230, 249, 512
initial-value problem 490
inner product 399
integro-differential equation 195, 230
intermediate configuration 70, 72, 334, 336, 343, 427, 433, 454, 456, 477, 480, 483, 564, 567, 598

internal constraint 305
internal dissipation 130, 510, 549
internal energy 120, 168, 515, 564
internal variables 199, 313, 419, 426, 444, 476, 513, 533
invariance postulate 292
invariants 355, 358, 379, 381, 587
inverse polefigure 610
isentropic processes 517
iso-strain assumption 613
isochoric motion 82
isoclinic intermediate configuration 595, 598
isothermal processes 517, 542
isothermal static continuation 537, 570
isotropic 303
isotropic elastic body 352
isotropic elasticity 482
isotropic elasticity relation 464, 469, 471
isotropic function 290, 302, 353, 413, 431, 459, 460, 543
isotropic functional 302
isotropic hardening 452, 461, 464
isotropic hyperelasticity 360
isotropic solid 185, 186, 299, 302, 318, 397, 409, 413
isotropic thermoviscoelasticity 542
isotropy 388, 394

J

Jacobi matrix 24
Jaumann rate 320, 324, 326, 344

K

Kelvin model 203, 273
kinematic constraint 3, 276, 305
kinematic hardening 328, 452, 461, 464, 469, 472, 483, 567
kinematics 2
kinetic energy 105, 120, 347
Kirchhoff stress 332
Kirchhoff tensor 94

L

Lagrange stress vector 104
Lagrangian representation 21
Lamé equations 188
Laplace transformation 194, 195, 213
lattice rotation 604
lattice spin 601, 603
left Cauchy-Green tensor 31
left stretch tensor 31
line element 27
linear elasticity 185
linear kinematic hardening 241
linear momentum 84
linear thermoelasticity 527
linear viscoelasticity 188, 211, 283
linear viscosity 198, 212
linear-viscous fluid 182, 273
linear-elastic material 185, 366
linear-viscoelastic solid 189, 274
linearisation 365
linearised Green strain 487
linearised rotation tensor 56
linearised strain tensor 56
loading 263, 453
loading condition 228, 498
loading function 466, 469, 472, 498
local action 277, 281
local balance of momentum 97, 112
local balance of rotational momentum 112
local derivative 22
local unloading 454, 480
loss modulus 216, 218, 221

M

MacCauley bracket 245, 478
main open problem 278
Mandel stress tensor 592, 595, 602, 608
mass 78, 168
mass balance, local form 79
mass density 78
mass element 83
mass flow 136
material body 7, 75, 85
material coordinates 10, 15
material derivative 21, 83, 96, 349, 486
material element 8
material frame-indifference 156, 277, 286
material function 178, 189, 275, 314, 412, 413, 454, 483, 522, 543
material gradient 23
material line 26, 38
material line element 26, 33
material objectivity 277
material of grade n 282
material parameter 179, 182, 186, 228, 503, 565
material point 8
material properties 3, 252
material representation 20, 21, 81, 96, 103, 112, 124, 128, 153
material response 250, 435, 512, 537, 539, 572
material strain rate 47

material surface element 28, 35
material symmetry 3, 250, 276, 293
material velocity gradient 43
material volume element 28
mathematical modelling 272
Maxwell fluid 322, 344
Maxwell model 203, 212, 426, 552
Maxwell model of fractional order 212, 220
mean strain 266, 268, 504
mechanical dissipation 532, 536, 567
mechanical energy 105, 347
mechanical power 105
mechanics 1
memory part 409, 425
method of functionals 312
method of internal variables 312
metric coefficients 14, 16, 40, 59
metric tensor 40
Mittag-Leffler function 212, 214, 225
mixture 145
modified principle of determinism 305
moment of momentum 85
monoclinic system 381, 390
monotonic processes 506
Mooney-Rivlin material 368
Mooney-Rivlin model 418, 545, 552
motion 9, 15, 21, 23, 53, 65
Mullins effect 264
multi-component system 145
multiplicative decomposition 72, 335, 373, 427, 454, 473, 480
multiplicative decompositions 433

N

natural reference configuration 587
Navier equations 188
Navier-Poisson fluid 182
Navier-Stokes equations 184
Navier-Stokes fluid 290
Neo-Hooke material 369
Neo-Hooke model 375
new specimen 264
Newton fluid 182
non-equilibrium 583
non-equilibrium part 578
non-Euclidean space 63, 456
non-objectivity 175
nonlinear kinematic hardening 241
norm 399
normal strain 34
normality rule 232, 453, 461

O

objective 165, 168, 290
objective tensor 605
objective vector 175
objectivity 156, 167, 277, 606
observer-invariant 292, 311
observer-invariant relation 176, 289, 606
Oldroyd rate 48, 52, 321, 324, 327, 332, 337, 343, 344, 466, 471, 481, 553
open systems 132
orientation of the anisotropy axes 595
orientation of the crystal 594
orthogonal tensor 32
overstress 194, 197, 420, 425, 433, 475, 476, 500, 502

P

partial backstresses 567
partial differential equations 181, 184, 185, 188
particle 8, 12
passive interpretation 159, 164
passive system 260
path history 437
perfect fluid 178, 273, 525
perfect gas 179, 526
perfect memory 435
perfect plasticity 229
phenomenological representation 576
phenomenological theory 433
physical linearisation 365, 376, 378, 388, 608
physical observer 12
Piola tensor 35, 40, 50, 331, 337, 351
Piola-Kirchhoff heat flux 123
Piola-Kirchhoff stress tensor 369, 399
Piola-Kirchhoff stress vector 93
Piola-Kirchhoff tensor 93, 94, 100, 285, 332, 456
plastic Almansi tensor 455
plastic arclength 233, 464, 469, 472
plastic Cauchy-Green tensor 455
plastic deformation gradient 454
plastic deformation rate 456
plastic deformation 228
plastic Green tensor 455
plastic material spin 601
plastic spin 601
plastic strain 451
plastic strain rate 472, 495
plastic velocity gradient 456
plasticity 178, 227, 249, 271, 276, 311, 435
point space 12, 157, 158

point transformation 16, 41
Poisson number 186, 227, 366
polar decomposition 29, 431, 458, 473, 591
polefigure 613
polycrystal 595, 612
position vector 11, 14, 160, 172
postulate of full invariance 606
potential 346, 348
potential energy 348
potential relations 515, 521, 524, 539
power function 208
Prandtl-Reuß equations 227
principal stress 356
principle of d'Alembert 109, 110, 330, 349
principle of determinism 277
principle of equipresence 278
principle of frame-indifference 277
principle of intersection 75
principle of irreversibility 119, 147, 509
principle of local action 277
principle of material objectivity 156
principle of superposition 189
principle of virtual work 115
process-dependent yield surface 453
process-induced anisotropy 608
proper subgroup 304

Q
quadratic function 378

R
rate of free energy 542
rate of rotation 169, 175, 605
rate-dependence 201, 227, 258, 259, 263, 265, 429, 479, 503, 507
rate-dependent 191, 271, 397, 398, 475
rate-independence 243, 435
rate-independent 228, 271
rate-independent functional 435, 437, 445, 448, 451, 577
rate-independent hysteresis 439, 442, 446
rate-independent material behaviour 243
rate-independent plasticity 445, 496, 568
reaction stress 306, 364
reduced form 284, 286, 297, 397, 594, 595
reference configuration 9, 11, 294, 377, 430, 433, 483
reference system 155
relative acceleration 175
relative Cauchy-Green tensor 547
relative deformation gradient 547
relative derivative 52, 343
relative frame 172, 603
relative Green strain 548
relative motion 283
relative Piola strain 548
relative strain history 196, 397
relative velocity 175
relaxation 254, 255, 258, 263, 265, 499, 578
relaxation coefficient 206
relaxation frequency 207, 425
relaxation function 189, 191, 208, 210
relaxation property 410, 411, 419, 490, 539, 570
relaxation spectrum 206
relaxation time 206, 218
reloading 263
rescaled rates 494, 572
residual inequality 515, 544
resolved back stress 602
resolved shear stress 602
restrictive assumptions 250
restrictive condition 515
retardation 404, 539
retardation factor 403, 492, 572
retardation theorem 407, 424, 426
retarded process history 572
retarded strain history 403, 492
retarded yield function 493
reversible 347
Reynolds transport theorem 134
rheological model 197, 272
rhombic system 382, 390
Riemann-Christoffel tensor 41, 69
right Cauchy-Green tensor 31
right stretch tensor 31
rigid body 310
rigid body motion 32, 46, 292, 335
rigid-plastic solid 325
Rivlin-Ericksen tensor 408, 417
rocket 137
rolling process 616, 617
rotation 44, 46, 164, 380, 381, 458, 588
rotational momentum 85, 89

S
Schmid stress 602
second law 5, 119
second-order effects 367
shear modulus 186, 366, 367, 373
shear relaxation 191
shear rigidity 310
shear strain 34
shear viscosity 182

shearing 47
simple fluid 299
simple material 282, 286, 303
simple shear 617, 618
single crystal 610, 612
slip system 596
slow motions 250
slow processes 420, 491
small deformation rates 56
small deformations 53, 104, 187, 250, 501, 578
small elastic strains 473
solid 299, 409
solid mechanics 249
spatial coordinates 10
spatial divergence 96
spatial gradient 22
spatial representation 19, 21, 77, 81, 96, 123, 128, 153, 523
spatial strain rate 46, 182
spatial velocity gradient 43
specific entropy 126
specific heat 519, 524, 535, 563, 567
specific internal energy 120, 168
specific stress power 105
specific volume 353, 520, 525
spectral representation 354
spherical part 373
spin 46
spin tensor 45, 169
spontaneous elasticity 202, 584
spring 197, 272
stability 419, 485, 556
stability assumption 488
stability condition 570
standard solid 198, 199
static continuation 401, 410, 489, 537
static yield surface 475
stationary state 264
Stieltjes transformation 223
stiffness modulus 215
storage modulus 215, 218, 221
strain 32
strain control 252, 254, 265
strain energy 347, 349, 360, 366, 381, 517, 589
strain history 196, 287, 397
strain rate 45, 51, 108, 182, 339, 341, 428, 456, 481
strain space 479
strain tensors 331, 339, 341
strain-controlled process 260
stress 531
stress control 252, 256, 257
stress power 105, 107, 108, 330, 333, 346, 520, 542, 563, 593
stress rates 340, 342, 456, 482
stress relation 515, 532, 549, 561, 573, 589
stress response 201, 202, 411
stress space 232, 479
stress tensor 93, 340, 342
stress vector 87, 93
stress-free 377, 454, 477, 564
stress-strain curve 369, 370
stress-strain relation 196, 419
stretch 44, 46
stretching 46
stretching tensor 45
structural variables 387
subgroup 304
superposition 189
surface force 87
symmetry group 295, 353, 359, 360, 379, 380, 587, 595
symmetry properties 359, 377
system of forces 87

T

tangent modulus 238
tangent space 26, 40, 282
tangent vector 13, 25
Taylor expansion 358, 365, 367, 376, 404, 412, 420, 540
Taylor-assumption 613
temperature 125, 126, 168
temperature field 525, 529
temperature gradient 515, 518, 532
temperature-dependence 509, 567
tension test 253, 261, 264, 267
tensor function 353
test specimen 252
tetragonal system 382, 391
texture 608, 613, 618
theories of material behaviour 271, 272
theory of materials 3
theory of plasticity 227, 435
theory of viscoelasticity 397
thermal equation of state 522
thermodynamic equilibrium 538, 570
thermodynamic potentials 517
thermodynamic temperature 126
thermodynamics 5, 119
thermoelastic coupling 519
thermoelastic domain 564, 573

thermoelastic material 514
thermoelasticity 514, 537, 539, 541
thermomechanical consistency 515, 533, 536, 542, 544, 561, 574, 578, 580, 581, 582
thermomechanical process 512, 534
thermomechanically consistent 513, 514, 515, 527, 535, 541, 560, 564, 569, 578, 579, 583, 585, 586
thermomechanics 177, 476, 509, 513
thermoplasticity 560, 573, 576
thermoviscoelastic material 530
thermoviscoelasticity 537, 541
thermoviscoplasticity 554, 560, 564
time-dependent 191
torque 87
torsion test 253, 255
total differential 350
training process 264
transformation of coordinates 15
transformation of metric 41
transformation of points 15
translation 164
transport theorem 133
transport theorem, generalised 134
transverse isotropy 386, 387
triclinic system 381, 389

U

undistorted reference configuration 299
uniaxial process 438
uniaxial tension 195, 229, 612
unified viscoplasticity models 476
unimodular part 373
unimodular tensor 295, 299
unloading 228, 263, 453

V

vector space 12, 158
velocity 21, 42, 96, 168
velocity field 147, 525
velocity gradient 43, 51
virtual displacements 114
virtual strain rate 116
virtual stress work 113
virtual work 349
viscoelastic material 409
viscoelasticity 178, 188, 211, 249, 271, 276, 283, 311, 397, 413, 445, 499, 500, 530, 534
viscoplastic solid 244, 274
viscoplasticity 243, 246, 249, 271, 276, 311, 475, 478, 499, 554, 618
viscosity 198, 429, 479, 567
viscosity matrix 536
viscosity tensor 416, 425
viscous deformation rate 428
viscous fluid 273, 290
viscous solid 291
viscous strain rate 429, 543
Voigt matrix 378
volume force 87, 168
volume preserving motion 82
volume strain 372
vorticity tensor 45

W

weighted Cauchy tensor 94, 100

Y

yield condition 228, 453
yield function 453, 460, 464, 469, 471, 478, 482, 484, 560, 574
yield stress 229
yield surface 450, 452, 453, 475
Young's modulus 186

Z

Zaremba-Jaumann derivative 320
zero operator 63

www.ingramcontent.com/pod-product-compliance
Ingram Content Group UK Ltd.
Pitfield, Milton Keynes, MK11 3LW, UK
UKHW021304051126
16364UKWH00004B/12
* 9 7 8 3 6 6 2 0 4 7 7 6 7 *